18

Periodic Table

— New notation
— Previous IUPAC form
— CAS version

1 Group IA	2 IIA	3 IIIA IIIB	4 IVA IVB	5 VA VB	6 VIA VIB	7 VIIA VIIB	8 VIIA VIIB	9 VIIIA VIII	10	11 IB	12 IIB	13 IIIB IIIA	14 IVB IVA	15 VB VA	16 VIB VIA	17 VIIB VIIA	18 VIIIA
1 H 1.0079																	2 He 4.00260
3 Li 6.941	4 Be 9.01218											5 B 10.81	6 C 12.011	7 N 14.0067	8 O 15.9994	9 F 18.9984	10 Ne 20.179
11 Na 22.9898	12 Mg 24.305											13 Al 26.9815	14 Si 28.0855	15 P 30.9738	16 S 32.06	17 Cl 35.453	18 Ar 39.948
19 K 39.0983	20 Ca 40.08	21 Sc 44.9559	22 Ti 47.88	23 V 50.9415	24 Cr 51.996	25 Mn 54.9380	26 Fe 55.847	27 Co 58.9332	28 Ni 58.69	29 Cu 63.546	30 Zn 65.39	31 Ga 69.72	32 Ge 72.59	33 As 74.9216	34 Se 78.96	35 Br 79.904	36 Kr 83.80
37 Rb 85.4678	38 Sr 87.62	39 Y 88.9059	40 Zr 91.224	41 Nb 92.9064	42 Mo 95.94	43 Tc (98)	44 Ru 101.07	45 Rh 102.906	46 Pd 106.42	47 Ag 107.868	48 Cd 112.41	49 In 114.82	50 Sn 118.71	51 Sb 121.75	52 Te 127.60	53 I 126.905	54 Xe 131.29
55 Cs 132.905	56 Ba 137.33	57 La ★ 138.906	72 Hf 178.49	73 Ta 180.948	74 W 183.85	75 Re 186.207	76 Os 190.2	77 Ir 192.22	78 Pt 195.08	79 Au 196.967	80 Hg 200.59	81 Tl 204.383	82 Pb 207.2	83 Bi 208.980	84 Po (209)	85 At (210)	86 Rn (222)
87 Fr (223)	88 Ra 226.025	89 Ac ▲ 227.028	104 Unq (261)	105 Unp (262)	106 Unh (263)	107 Uns (262)											

★ Lanthanide series

58 Ce 140.12	59 Pr 140.908	60 Nd 144.24	61 Pm (145)	62 Sm 150.36	63 Eu 151.96	64 Gd 157.25	65 Tb 158.925	66 Dy 162.50	67 Ho 164.930	68 Er 167.26	69 Tm 168.934

▲ Actinide series

90 Th 232.038	91 Pa 231.036	92 U 238.029	93 Np 237.048	94 Pu (244)	95 Am (243)	96 Cm (247)	97 Bk (247)	98 Cf (251)	99 Es (252)	100 Fm (257)	101 Md (258)

F.V

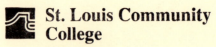

Inorganic Chemistry

James R. Bowser
SUNY College at Fredonia

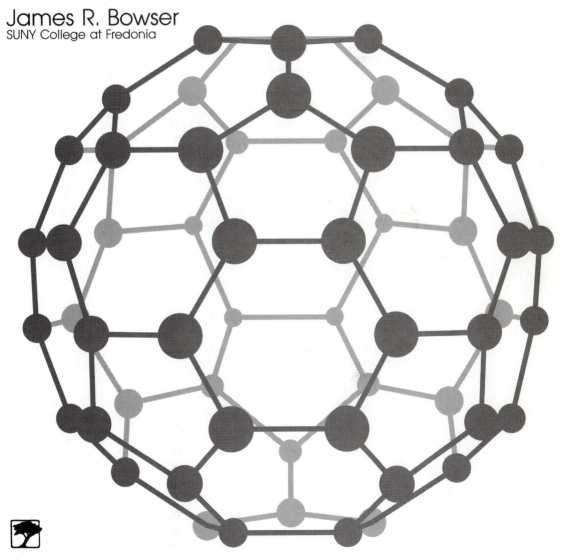

Brooks/Cole Publishing Company
Pacific Grove, California

To Nancy and Anne

Brooks/Cole Publishing Company
A Division of Wadsworth, Inc.

© 1993 by Wadsworth, Inc., Belmont, California 94002.

Printed in the United States of America

10 9 8 7 6 5 4 3 2 1

Library of Congress Cataloging-in-Publication Data

Bowser, James R.
    Inorganic chemistry/James R. Bowser.
        p. cm.
    Includes bibliographical references and index.
    ISBN 0-534-17532-5
    1. Chemistry, Inorganic. I. Title.
  QD151.5.B68 1993
  546—dc20

92-27732
CIP

Sponsoring Editor: Maureen A. Allaire, Harvey C. Pantzis
Editorial Assistant: Beth Wilbur
Production Service: Phyllis Niklas
Production Coordinator: Joan Marsh
Manuscript Editor: Phyllis Niklas
Permissions Editor: Carline Haga
Interior Design: Rogondino & Associates
Cover Design: Katherine Minerva
Interior Illustration: Rogondino & Associates, Katherine Minerva
Typesetting: Techset Composition Ltd.
Cover Printing: Phoenix Color Corporation
Printing and Binding: Arcata Graphics/Fairfield

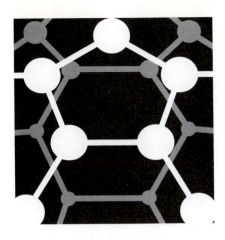

Preface

The general dissatisfaction with the presently available inorganic texts provided the main impetus for writing this book. This text will be useful in survey courses for advanced undergraduates and beginning graduate students, and also in courses dealing with certain selected topics.

An important goal of this book is to provide a sense of what it is like to be an inorganic chemist today. Its organization and content reflect three traits that characterize our discipline as it currently exists:

1. The domain of inorganic chemistry is incredibly broad.

This point is made to the student at the beginning of Chapter 1. Those of us who teach the subject appreciate the difficulties that arise, of course—how to decide what to cover and what to leave out? We believe we have succeeded in covering all the primary subjects. To keep the book to a reasonable length, some topics are given less attention than they deserve, while others are covered in detail. This allows for depth as well as breadth.

2. Contemporary inorganic chemistry is an intimate mix of theory and experiment.

Inorganic texts generally can be divided into two types: those that are fact-based and those that are theory(model)-based. This book falls into the second category, without apology. Any theory is fact-dependent, of course. In real life, experiment normally precedes and always supersedes theory. However, a student trying to learn the subject can easily be overwhelmed—and discouraged—by the vast amount of factual data within our discipline. Theoretical models serve to unify such data. This book therefore emphasizes theory (especially in the early chapters), with a generous dose of experimental methodology and results to illustrate and support the models described.

3. Inorganic chemistry is fun and exciting.

Inorganic chemists have made phenomenal gains in recent years, and it is important to convey this to our students. Twenty years ago, who would have believed that H_2 could act as a ligand? That so many biochemical processes are driven by enzymes whose activity requires a metal? That metal–carbon triple bonds and metal–metal quadruple bonds would become commonplace? That species such as Pb_5^{2-}, $Kr_2F_3^+Sb_2F_{11}^-$, $Fe_5C(CO)_{15}$, and $Mo_{12}PO_{40}^{3-}$ would be stable and isolable? Who would dare to predict what inorganic chemists *won't* accomplish in the next two decades?

I have tried to convey the excitement involved in the field of inorganic chemistry today, and I sincerely hope that this will motivate your students.

Organization

The topical coverage is divided into six parts:

Part I: Chapters 1 and 2 review important background information. You might choose to devote classroom time to all, most, or relatively little of this material, depending on your own situation. Experience suggests that, at the very least, it is useful for even well-prepared students to have this information at hand for consultation and review.

Part II: Chapters 3–6 discuss the bonding models of greatest importance to inorganic chemistry. Each of these chapters begins with material that most students will be familiar with, but then quickly moves to more advanced topics. In Chapter 7 these bonding models are used to explain the physical and chemical properties of the free elements, and Chapter 8 (Secondary Chemical Interactions) emphasizes intermolecular forces.

Part III: Chapters 9–11 discuss the primary types of chemical reactions, with examples presented from both aqueous and nonaqueous solutions.

Part IV: Chapters 12–14 survey the main group elements. They are organized in such a way as to demonstrate interrelationships throughout the periodic table, rather than only within the chemical families.

Part V: Chapters 15–18 deal with transition metals, with the primary focus on the structures, bonding, and reactions of both classical and organometallic complexes.

Part VI: Chapters 19–22 cover special topics. Their placement at the end of the book serves as an indirect review, since this material either illustrates or expands upon what came before.

This book is adequate for a 1 year course sequence in inorganic chemistry, which is becoming increasingly popular in American colleges and universities. However, an effort has been made to make each chapter stand alone in order to provide flexibility for those teaching a one-semester course. It should also be possible to change the order of coverage. For example, you may prefer to discuss the material on instrumental methods (in Chapter 21) earlier, either in its entirety, or on a piecemeal basis (topic-by-topic, as it complements other material).

Although many texts omit literature references, I have taken pains to include them. References enable us to give credit where credit is due, of course, but there is other value as well. Citations to the original literature provide the reader with historical perspective—a sense of how our knowledge evolved and of whether a given theory or experiment came before or after another. Citations to reviews provide avenues for curious students to explore a topic more deeply. For these reasons, references are sprinkled throughout the book. In addition, an up-to-date bibliography is provided for each chapter.

Each chapter ends with a set of questions and problems, which are designed to both test and extend the student's knowledge. The questions are plentiful (over 600 total) and vary in difficulty. Those marked with asterisks require a trip to the library to examine some recent literature. A solutions manual is available.

Acknowledgments

I am pleased to acknowledge those who have helped with this project, beginning with my home institution, SUNY College at Fredonia. Two colleagues in my department, Roy Keller and Tom Janik, provided helpful comments. I am also grateful to the University of Notre Dame for financial support and the use of its facilities during a sabbatical leave, and especially to Tom Fehlner, my host during that enjoyable time.

I thank the staff at Wadsworth and Brooks/Cole, including my editors, Jack Carey, Harvey Pantzis, and Maureen Allaire, and various others (Sue Belmessieri, Gerri Del Ré, Jennifer Kehr, and Nancy Miaoulis). Finally, I am grateful to my reviewers, whose comments helped avoid a myriad of embarrassments: Edwin H. Abbott, Montana State University; Michael Ashby, University of Oklahoma; Bruce Averill, University of Virginia; O. T. Beachley, SUNY at Buffalo; Romana Lashewycz-Rubycz, Hobart and William Smith Colleges; Dennis Lichtenberger, University of Arizona; Robert Lipschutz, Purdue University; Robert Parry, University of Utah;

Cortlandt Pierpont, University of Colorado; Paul Poskozim, Northeast Illinois University; Robert Stewart, Miami University of Ohio; Richard Treptow, Chicago State University; and Gary Wulfsberg, Middle Tennessee State University.

James R. Bowser
Fredonia, New York

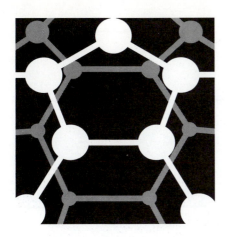

Brief Table of Contents

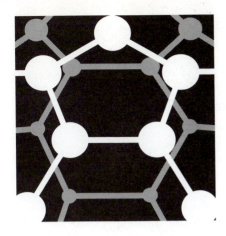

Contents

Part VI: Special Topics   629

19 Inorganic Cages and Clusters   630

Appendixes

PART

I

An Introduction to Quantum and Group Theories

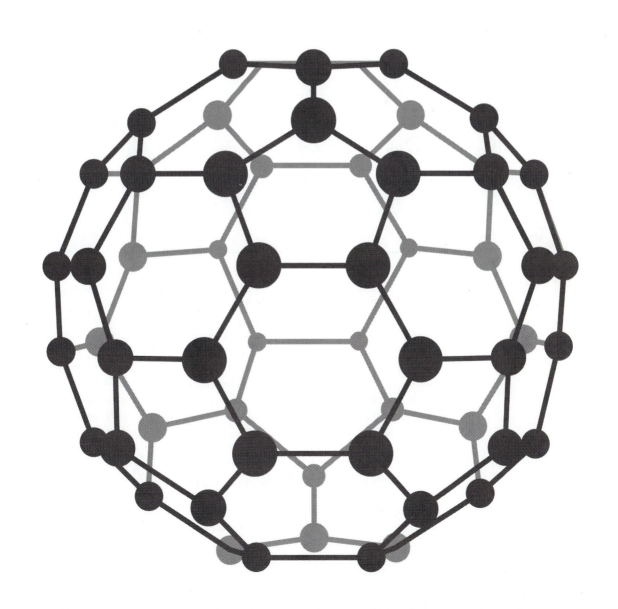

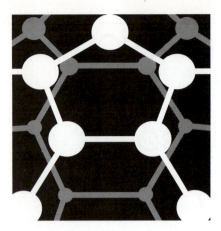

1

Atomic Theory and the Periodic Table

1.1   The Inorganic Domain

How should the term *inorganic chemistry* be defined? There seem to be about as many definitions as there are inorganic textbooks. Within these pages, inorganic chemistry is considered to be *the study of the structures, properties, reactivities, and interrelationships of the chemical elements and their compounds.* That covers a lot of ground! But the definition must be broad, because practicing inorganic chemists utilize a wide range of methods, and because the contemporary conception of the inorganic domain encompasses many different areas. Some inorganic chemists spend most of their working day with a computer. Others rarely stray far from their favorite spectrometer (or other type of instrument). Still others work at a laboratory bench, often with the aid of a vacuum manifold or dry box. Yet the theoretician, spectroscopist, and synthetic chemist may simply be using different approaches to study the same subject.

Inorganic chemistry was once much easier to catalog, but there has been a recent remarkable growth in what is included in the inorganic domain. This expansion has occurred in many different directions, and so there is now a great deal of overlap with the other traditional areas of chemistry (analytical, organic, and physical), as well as with biology, physics, and geology. This is illustrated in Figure 1.1.

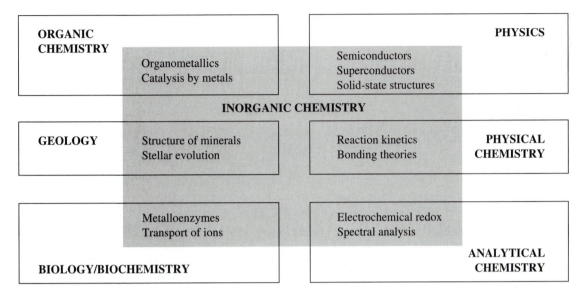

Figure 1.1 Interrelationships among inorganic chemistry and other scientific disciplines and subdisciplines.

What, then, do all inorganic chemists have in common? Perhaps the most important common denominator is the periodic table. The periodic table is valuable because it helps us organize the elements, of course, but it is also a powerful aid for understanding—and, what is more difficult, predicting—chemical structures, properties, and reactions. Hence, the periodic table is a focal point for this chapter. A summary of its theoretical basis begins in the next section. Later, some important periodic properties are discussed.

1.2 Electron Waves

The early experiments of J. Thomson (1897) and R. Millikan (1909) made possible the calculation of the rest mass of an electron, the accepted value being 9.11×10^{-31} kg. This value was used by Bohr in his theory of the atom, published in 1913.[1] In that model, the electron was treated as a charged particle in motion. Bohr's theory was quite successful for dealing with one-electron systems, particularly the hydrogen atom. Severe problems arose in applying it to other atoms, however, since neither the model nor

1. Bohr, N. *Philos. Mag.* **1913**, *26*, 476.

modifications were able to account adequately for electron–electron repulsions.

In 1924, Louis de Broglie proposed that matter has both wave and particulate properties.[2] The relationship

$$\lambda = \frac{h}{mv} \tag{1.1}$$

where h = Planck's constant (6.626×10^{-34} J·s), permits the calculation of the wavelength λ of an object having mass m and moving at velocity v. This wavelength is too small to be significant for objects of normal mass, so the wave properties go unnoticed. But the results are quite different for a particle of minuscule mass, such as the electron. Consider an electron moving at one-tenth the velocity of light. Substitution into de Broglie's equation gives $\lambda = 2.42 \times 10^{-11}$ m (or 24.2 pm), a wavelength in the X-ray region that is readily detectable. Thus, electrons exhibit a *duality*, behaving as both particles and waves. The next important advance in atomic theory resulted from the recognition that electrons exhibit wave properties.

1.3   The Schrödinger Equation

Erwin Schrödinger published his famous *wave equation* in 1926.[3] For a one-electron system such as a hydrogen atom, this equation can be written

$$\frac{\partial^2 \psi}{\partial x^2} + \frac{\partial^2 \psi}{\partial y^2} + \frac{\partial^2 \psi}{\partial z^2} + \frac{8\pi^2 m}{h^2}(E - V)\psi = 0 \tag{1.2}$$

Here, ψ represents the *wavefunction* for the electron; x, y, and z are Cartesian coordinates, which locate a position in space with reference to the nucleus at point $(0, 0, 0)$; E is the total energy of the system; and V is the potential energy. The usual application of this equation is to calculate either ψ (which has properties analogous to the amplitude of the electron wave) or ψ^2, the *probability density* of the wave, as a function of location with respect to the nucleus. Since there are an infinite number of points in space, there are an infinite number of solutions to the equation.

It is customary to place certain commonsense constraints, called *boundary conditions*, on this mathematical system. Four such conditions are:

1.   The hydrogen atom is finite in size; hence, ψ and ψ^2 must equal zero beyond some distance from the nucleus.

2.   The wavefunction is *normalized*. This means that, since the electron must be somewhere in space, its overall probability density must be unity.

2.   de Broglie, L. *Philos. Mag.* **1924,** *47,* 446.

3.   Schrödinger, E. *Ann. Physik.* **1926,** *81,* 109; *Phys. Rev.* **1926,** *28,* 1049.

3. The wavefunction ψ may have only one value at any point.

4. The wavefunction ψ must be a continuous function.

Solving equation (1.2) under these constraints is possible but mathematically difficult. The situation is improved somewhat by converting from Cartesian to polar coordinates, where a point in space is defined in terms of r, its distance from the origin (the nucleus), and two angles, θ and ϕ. This makes it possible to divide the overall wavefunction into three components:

$$\psi(r, \theta, \phi) = R(r) \cdot \Theta(\theta) \cdot \Phi(\phi) \tag{1.3}$$

The term $R(r)$ relates to the distance from the nucleus and is nondirectional; it is called the *radial* function. The last two terms are the *angular* functions and contain directionality information. It is useful to examine these terms individually.

Radial Wavefunctions

For the radial portion $R(r)$, a series of equations can be identified that meet the boundary conditions given above. Two *quantum numbers*, n and ℓ, arise such that there is a different n, ℓ combination for each equation. Each such combination represents an *orbital*. The equations for the first six orbitals of the hydrogen atom are given in Table 1.1.

Table 1.1 The radial wavefunction $R(r)$ for one-electron species

n	ℓ	$R(r)$
1	0	$2\left(\dfrac{Z}{a_0}\right)^{3/2} e^{-Zr/a_0}$
2	0	$\left(\dfrac{1}{2\sqrt{2}}\right)\left(\dfrac{Z}{a_0}\right)^{3/2}\left(1 - \dfrac{Zr}{2a_0}\right)e^{-Zr/2a_0}$
2	1	$\left(\dfrac{1}{2\sqrt{6}}\right)\left(\dfrac{Z}{a_0}\right)^{3/2} re^{-Zr/2a_0}$
3	0	$\left(\dfrac{2}{3\sqrt{3}}\right)\left(\dfrac{Z}{a_0}\right)^{3/2}\left(1 - \dfrac{2Zr}{3a_0} + \dfrac{2Z^2r^2}{27a_0^2}\right)e^{-Zr/3a_0}$
3	1	$\left(\dfrac{8}{27\sqrt{6}}\right)\left(\dfrac{Z}{a_0}\right)^{3/2}\left(\dfrac{Zr}{a_0} - \dfrac{Z^2r^2}{6a_0^2}\right)e^{-Zr/3a_0}$
3	2	$\left(\dfrac{4}{81\sqrt{30}}\right)\left(\dfrac{Z}{a_0}\right)^{7/2} r^2 e^{-Zr/3a_0}$

Note: For the hydrogen atom, $a_0 = 52.9$ pm (see p. 8).

The first, or *principal*, quantum number, n, is a positive integer ($n = 1, 2, 3, \ldots$). The potential energy of the electron is inversely related to the square of n [see equation (1.5), later in this section]. As a result, stability decreases as n increases:

$$n = 1 > n = 2 > n = 3 > \cdots$$

There is only one radial equation for the $n = 1$ case. There are two equations for $n = 2$ (the second *shell*); they lead to the 2s and 2p *subshells* (see below). There are three equations for $n = 3$ (corresponding to the 3s, 3p, and 3d subshells), and, in general, n subshells exist for a given principal quantum number.

The second, or *azimuthal*, quantum number, ℓ, provides information about orbital shape. The value of ℓ is dependent upon n; specifically, the allowed values of ℓ are the nonnegative integers up to a maximum of $n - 1$. The correspondences between this quantum number and the common orbital designations s, p, d, f, and g are

$\ell = 0 \rightarrow s$ orbital

$\ell = 1 \rightarrow p$ orbital

$\ell = 2 \rightarrow d$ orbital

$\ell = 3 \rightarrow f$ orbital

$\ell = 4 \rightarrow g$ orbital

It follows that the common orbital designations derive from the n and ℓ quantum numbers. For example, the orbital having $n = 1$ and $\ell = 0$ is the 1s orbital, and the septet of 4f orbitals have $n = 4$ and $\ell = 3$. Table 1.2 summarizes the relationships.

The expressions for $R(r)$ that were given in Table 1.1 can be used to construct plots of R versus r. The dependence, of R on the distance from the nucleus is graphed for the $n = 1$ case in Figure 1.2. It can be seen that this function is at a maximum at the nucleus and decreases rapidly with increasing r.

Table 1.2 Correspondence between the common orbital designations and the n and ℓ quantum numbers for $n = 1$–4

Designation	n	ℓ	Designation	n	ℓ
1s	1	0	4s	4	0
2s	2	0	4p	4	1
2p	2	1	4d	4	2
3s	3	0	4f	4	3
3p	3	1			
3d	3	2			

Figure 1.2
The radial
wavefunction for
hydrogen: plot of
R versus distance
from the nucleus for
the $1s$ orbital;
$a_0 = 52.9$ pm (see
text).

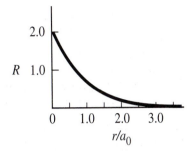

The function R has limited physical significance. More meaningful is R^2, which is directly related to the probability density ψ^2. The *surface probability* is the probability density at some given distance from the nucleus. It can be calculated for any distance r by multiplying R^2 by the number of points at that distance. Since the volume of a sphere is $V = \frac{4}{3}\pi r^3$, the number of points increases in proportion with r^2:

$$V = \frac{4}{3}\pi r^3 \qquad \frac{dV}{dr} = 4\pi r^2$$

It is therefore useful to graph $4\pi r^2 R^2$ (or just $r^2 R^2$) as a function of distance from the nucleus. Plots for the six cases having $n = 1, 2, 3$ are given in Figure 1.3.

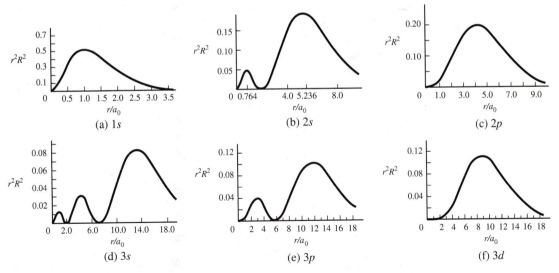

Figure 1.3 Plots of $r^2 R^2$ versus distance from the nucleus for hydrogen orbitals having $n = 1, 2, 3$. [Adapted with permission from Karplus, M.; Porter, R. N. *Atoms and Molecules*; Benjamin/Cummings: Menlo Park, CA, 1970.]

For the $1s$ orbital, the distance having the greatest surface density is symbolized by a_0, and is often taken to be the "size" of a hydrogen atom. Its calculated value is 52.9 pm.

Two probability density graphs are shown for $n = 2$, one for each of the $2s$ and $2p$ orbital types. In the $2s$ plot, r^2R^2 drops to zero at a distance of $2a_0$ (106 pm), but is greater than zero on either side. A surface having zero probability is a *node*; in this case the node is a spherical surface. For the $2p$ orbital, a node passes through the nucleus; this is an example of a *planar*, or *angular*, node. As it happens, any orbital has $n - 1$ total nodes, of which ℓ are of the angular type and the remainder $(n - \ell - 1)$ are radial.

A significant difference between the $2s$ and $2p$ orbitals (and, in fact, a general difference between s-type orbitals and all others) is that for s orbitals the maximum amplitudes are at the nucleus, while the other orbital types have angular nodes passing through the nucleus. This will later be seen to have important ramifications with respect to orbital stabilities and bonding properties.

Angular Wavefunctions

The two angular portions of the complete wavefunction for hydrogen might be considered separately; however, it is convenient to take them together.[4] The Θ and Φ functions provide information about the directionalities and boundaries of the various orbitals. The drawn boundaries are arbitrary, corresponding to surfaces at which the electron density drops below some chosen minimum value.

The angular wavefunctions for the most stable hydrogen-like orbitals are given in Tables 1.3 and 1.4. The ℓ quantum number recurs, and a third, m_ℓ, the *magnetic* quantum number, appears. This quantum number

Table 1.3 The angular function $\Theta(\theta)$ in terms of $\sin \theta$ and $\cos \theta$

ℓ	m_ℓ	$\Theta(\theta)$	ℓ	m_ℓ	$\Theta(\theta)$
0	0	$\dfrac{\sqrt{2}}{2}$	2	0	$\left(\dfrac{\sqrt{10}}{4}\right)(3\cos^2\theta - 1)$
1	0	$\left(\dfrac{\sqrt{6}}{2}\right)(\cos\theta)$	2	± 1	$\left(\dfrac{\sqrt{15}}{2}\right)(\sin\theta\cos\theta)$
1	± 1	$\left(\dfrac{\sqrt{3}}{2}\right)(\sin\theta)$	2	± 2	$\left(\dfrac{\sqrt{15}}{4}\right)(\sin^2\theta)$

4. Together, they belong to a class of functions known as *spherical harmonics* and are amenable to mathematical treatment.

Table 1.4 The angular function $\Phi(\phi)$ in terms of $\sin \phi$ and $\cos \phi$

m_ℓ	$\Phi(\phi)$	m_ℓ	$\Phi(\phi)$
0	$\dfrac{1}{\sqrt{2\pi}}$	+2	$\left(\dfrac{1}{\sqrt{\pi}}\right)(\cos 2\phi)$
+1	$\left(\dfrac{1}{\sqrt{\pi}}\right)(\cos \phi)$	−2	$\left(\dfrac{1}{\sqrt{\pi}}\right)(\sin 2\phi)$
−1	$\left(\dfrac{1}{\sqrt{\pi}}\right)(\sin \phi)$		

is dependent on ℓ according to the relationship $\ell \geq |m_\ell|$; that is, m_ℓ is an integer having an absolute value less than or equal to ℓ. If $\ell = 0$, then m_ℓ is necessarily zero. If $\ell = 1$, then m_ℓ can take on any of the three values $+1$, 0, or -1. This correlates with the fact that p orbitals are found in groups of three. If $\ell = 2$, then m_ℓ may equal $+2$, $+1$, 0, -1, or -2; thus, a d subshell contains five orbitals. In the same manner, f subshells contain seven orbitals, and g subshells have nine.

As can be seen from Tables 1.3 and 1.4, the product of the Θ and Φ functions has no angular (θ or ϕ) term for an s-type orbital (since $\ell = 0$ and therefore $m_\ell = 0$). Such orbitals are nondirectional (that is, spherical) in shape.

A p orbital has been described as shaped like a figure eight, but in three dimensions (see Figure 1.4, p. 10). Alternatively, consider a spherical balloon tied in the center; the result would approximate the shape of a $2p$ orbital. Note that a region of zero thickness—a node—would separate the two *lobes* of the balloon. In addition to the angular node passing through the nucleus, p-type orbitals having $n > 2$ also contain one or more internal nodes.

The p orbitals of a given shell are mutually perpendicular. Their orientations are taken to coincide with the x, y, and z axes of a Cartesian system, so the designations p_x, p_y, and p_z are commonly used. It follows that a p_x orbital has its maximum density along the x axis and an angular node in the yz plane.

Pictorial representations of hydrogen-like orbitals having $n \leq 3$ are given in Figure 1.4. In addition to these traditional representations, other graphical approaches are sometimes used in an effort to aid the visualization of orbital shapes and sizes. These include *contour plots* and *pseudo-three-dimensional* (or "bent-wire") diagrams. In a contour plot, lines are drawn to connect points having equal electron densities within an orbital. As is demonstrated for the $2p_x$ orbital in Figure 1.5, this can be done in either two or three dimensions. In pseudo-three-dimensional diagrams,

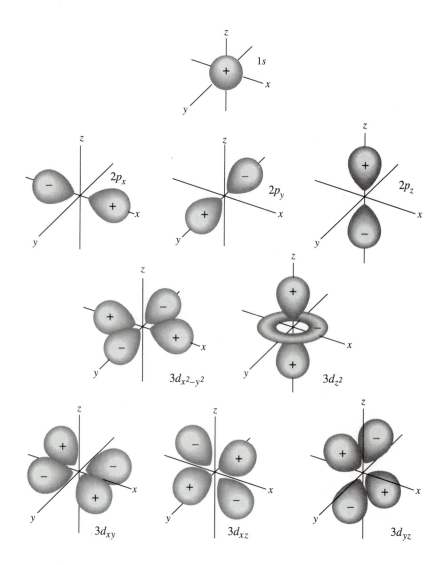

Figure 1.4
Representations of the shapes of hydrogen-like orbitals having $n \leq 3$.

regions having positive wave amplitudes are elevated and those with negative amplitudes are depressed on an otherwise two-dimensional surface (Figure 1.5e).

The d orbitals have two angular nodes (note that $\ell = 2$), and can roughly be described as three-dimensional four-leaf clovers. Theoretically, six such orbitals can be constructed. One (labeled $d_{x^2-y^2}$) has its maximum density in the xy plane and along the x and y axes. A second (d_{xy}) lies in the same plane, but between the two axes (see Figure 1.4). Four additional d orbitals can be visualized that are similar to these two, but with different orientations; their labels would be d_{xz}, $d_{z^2-x^2}$, d_{yz}, and $d_{z^2-y^2}$. However, the

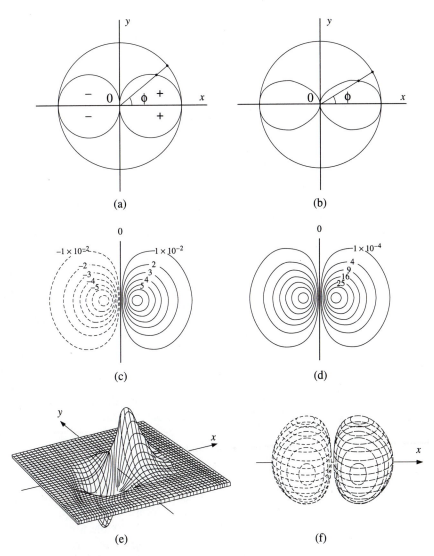

Figure 1.5 Graphical representations of the $2p_x$ orbital: (a) Polar plot of ψ_{2p_x} on the xy plane. (b) Polar plot of $(\psi_{2p_x})^2$ on the xy plane. (c) Two-dimensional contour plot of ψ_{2p_x} on the xy plane. (d) Two-dimensional contour plot of $(\psi_{2p_x})^2$ on the xy plane. (e) Pseudo-three-dimensional diagram of ψ_{2p_x}. (f) Three-dimensional contour diagram of ψ_{2p_x} along the x axis.
[Reproduced with permission from Kikuchi, Q.; Suzuki, K. *J. Chem. Educ.* **1985**, *62*, 206.]

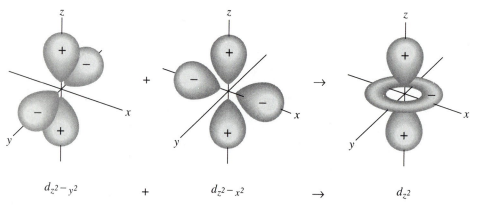

$$d_{z^2-y^2} \qquad + \qquad d_{z^2-x^2} \qquad \rightarrow \qquad d_{z^2}$$

Figure 1.6 The mixing of the $d_{z^2-x^2}$ and $d_{z^2-y^2}$ orbitals to produce the d_{z^2} linear combination.

wave equation allows for only five d orbitals in a given subshell. Hence, it is customary to construct a linear combination of $d_{z^2-x^2}$ and $d_{z^2-y^2}$, with the result being designated d_{z^2} and having a different shape from the other four d orbitals. It contains two large lobes along the z axis, both having the same sign of ψ, and a *torus* (or "doughnut") of opposite sign in the xy plane (Figure 1.6).

The seven f orbitals have, as would be expected, three angular nodes; their different shapes are again explained by the fact that some are linear combinations. (See the contour diagrams given in Figure 1.7.) No elements exist as yet for which g orbitals are occupied.

A set of the three quantum numbers n, ℓ, and m_ℓ is sufficient to define any orbital. However, a fourth quantum number is needed to uniquely identify an electron residing in that orbital due to the spin momentum of the electrons. This property results in two possible orientations in an external magnetic field (aligned with or in opposition to the field). The fourth quantum number is therefore the *spin* quantum number, m_s. Its allowed values are $+\frac{1}{2}$ and $-\frac{1}{2}$. It is possible for two electrons to occupy the same orbital, provided that their spin quantum numbers differ. This is one way to state the *Pauli exclusion principle*.[5] (An alternative statement is that no two electrons of an atom or ion may have the same four quantum numbers.)

Orbital Energies for One-Electron Systems

In calculating the energies of orbitals for one-electron systems, the wave equation yields results equivalent to those obtained by Bohr. The equation

5. Pauli, W. *Ann. Physik.* **1925**, *31*, 765.

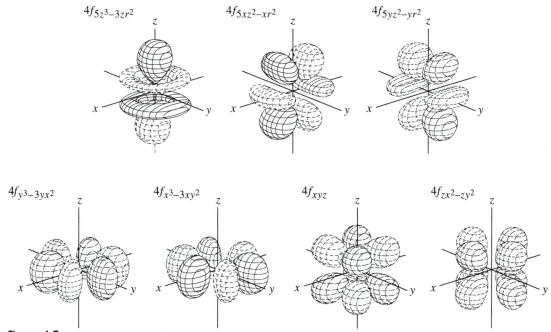

Figure 1.7
Three-dimensional contour surfaces for the seven 4f orbitals of hydrogen.
[Reproduced with permission from Kikuchi, Q.; Suzuki, K. *J. Chem. Educ.* **1985,**
62, 206.]

that is derived is

$$E = -\frac{2\pi^2 m_e q^4 Z^2}{n^2 h^2} \tag{1.4}$$

Here, m_e is the electron mass, q is the electron charge, Z is the nuclear charge, h is Planck's constant, and n is the principal quantum number. The various constants can be combined to give the simpler equation

$$E = -\frac{kZ^2}{n^2} \tag{1.5}$$

where k is about 13.6 eV. Note that orbital energies are always negative in sign. The energy of a free electron (not associated with any nucleus) is defined as zero, and its electrostatic attraction to a nucleus leads to decreased energy content (and increased stability). Since Z is constant for a given nucleus, the orbital energies of single-electron species depend solely on n. A listing of orbitals for such systems according to their relative stabilities is therefore

$$1s > 2s = 2p > 3s = 3p = 3d > 4s = 4p = 4d = 4f > \cdots$$

Equation (1.5) enables comparisons between different one-electron species. The cation He^+ contains two protons ($Z = 2$). Thus, the $1s$ orbital of He^+ lies at a calculated energy of $-54.4\,eV$, four times that of the hydrogen atom. It follows that removing the electron of He^+ requires four times the input of energy.

1.4   Many-Electron Systems—Penetration and Shielding

Up to this point the discussion has been limited to atoms and ions having only one electron. If two or more electrons are present, the Schrödinger approach must be modified to accommodate electron–electron repulsions. The form of the wave equation (using Cartesian coordinates) for the two-electron case is

$$\nabla_1^2 + \nabla_2^2 + \frac{8\pi^2 m}{h^2}\left(E + \frac{2q^2}{r_1} + \frac{2q^2}{r_2} - \frac{q^2}{r_{12}}\right)\psi = 0 \tag{1.6}$$

Here, ∇_1^2 and ∇_2^2 represent Laplacian operators [each equivalent to the first three terms of equation (1.2)] for electrons 1 and 2, respectively; r_1 and r_2 are the two electron–nucleus distances; and r_{12} is the distance between the electrons. Unfortunately, although this equation can be written, attempts to solve it with mathematical rigor have failed. This might seem to be a severe limitation; however, several approximate methods indicate that there are no major differences between orbitals in many-electron and one-electron systems. Furthermore, the assumption that hydrogen-like orbitals exist for all atoms and ions permits the prediction of many physical and chemical properties that are found to be consistent with experimental data. Thus, the wave equation (1.2) is generally applicable to all atoms and ions, and this is the primary reason for its popularity.

There is, however, one important difference that arises when two or more electrons are present. The s, p, d, and f orbitals of a given shell are not equivalent in energy for such systems. This is due to differences in *penetrating ability*—the ability of the electron(s) in a given orbital to penetrate the electron clouds of other orbitals, and thereby interact with the nucleus.

To understand the concept of penetration, it is first necessary to recall that all the orbitals of a given species have the same geometric center—the nucleus. It follows that the various orbitals must have considerable overlap in space (there is shared domain). Therefore, an electron in one orbital partially *shields*, or *screens*, electrons of other orbitals, decreasing the electrostatic attractions of the screened electrons for the nucleus. This causes the *effective nuclear charge*, symbolized as Z^*, to be less than the actual charge Z (the number of protons in the nucleus). For any given electron, the equation

$$Z^* = Z - \sum S \tag{1.7}$$

applies, where $\sum S$ corresponds to the sum of the shielding effects on the electron in question by the others.

As an example of how shielding affects stability, consider the $1s, 2s,$ and $2p$ orbitals. Calculations indicate that the $2s$ orbital shares more of the $1s$ domain nearest to the nucleus than do the $2p$ orbitals—that is, $2s$ penetrates $1s$ more effectively than does $2p$. (It also can be said that the $1s$ orbital screens $2p$ to a greater extent than it screens $2s$.) This leads to the prediction that for many-electron systems, the $2s$ orbital is more stable than the $2p$ orbitals—a prediction supported by various kinds of experimental evidence. Similarly, p orbitals penetrate more effectively than d orbitals, which in turn penetrate more than f orbitals.[6] Thus, the order of stability within a given shell is

$$s > p > d > f$$

and a general order of orbital stabilities for systems containing more than one electron is

$$1s > 2s > 2p > 3s > 3p > 4s \approx 3d > 4p > 5s \approx 4d > 5p > \cdots$$

1.5 Electron Configurations

The *aufbau* (building up) *principle* suggests that for many-electron species, orbitals are filled sequentially in the order given above. This provides an easy way to determine the most stable (ie, the *ground-state*) electron configuration for most elements. The experimental ground-state occupancies (as obtained from gas-phase spectroscopy) for the first 103 elements are given in Table 1.5 (pp. 16–17).

Several exceptions to the $n + \ell$ rule can be found in Table 1.5, the earliest of which appears in the first transition series. An interesting situation arises in comparing the $4s$ and $3d$ orbitals. The benefit of occupying an s-type rather than a d-type orbital almost exactly counterbalances the benefit of having a smaller principal quantum number. The $4s$ orbital is filled in preference to $3d$ for most neutral atoms. However, that is not the case for chromium and copper (see Table 1.5). Similar violations of the $n + \ell$ rule

---

6. Other screening effects must be considered as well. For example, what are the relative abilities of $2s$ and $2p$ orbitals to screen each other? If the $3s$ orbital is occupied, what is the influence of its electron(s)? The $s > p > d > f$ conclusion is found to be valid after all screening effects have been considered.

Table 1.5 Gas-phase electron configurations of the first 103 elements

At. No.	Element	Configuration	At. No.	Element	Configuration
1	H	$1s^1$	42	Mo	$[Kr]4d^55s^1$
2	He	$1s^2$	43	Tc	$[Kr]4d^55s^2$
3	Li	$[He]2s^1$	44	Ru	$[Kr]4d^75s^1$
4	Be	$[He]2s^2$	45	Rh	$[Kr]4d^85s^1$
5	B	$[He]2s^22p^1$	46	Pd	$[Kr]4d^{10}$
6	C	$[He]2s^22p^2$	47	Ag	$[Kr]4d^{10}5s^1$
7	N	$[He]2s^22p^3$	48	Cd	$[Kr]4d^{10}5s^2$
8	O	$[He]2s^22p^4$	49	In	$[Kr]4d^{10}5s^25p^1$
9	F	$[He]2s^22p^5$	50	Sn	$[Kr]4d^{10}5s^25p^2$
10	Ne	$[He]2s^22p^6$	51	Sb	$[Kr]4d^{10}5s^25p^3$
11	Na	$[Ne]3s^1$	52	Te	$[Kr]4d^{10}5s^25p^4$
12	Mg	$[Ne]3s^2$	53	I	$[Kr]4d^{10}5s^25p^5$
13	Al	$[Ne]3s^23p^1$	54	Xe	$[Kr]4d^{10}5s^25p^6$
14	Si	$[Ne]3s^23p^2$	55	Cs	$[Xe]6s^1$
15	P	$[Ne]3s^23p^3$	56	Ba	$[Xe]6s^2$
16	S	$[Ne]3s^23p^4$	57	La	$[Xe]5d^16s^2$
17	Cl	$[Ne]3s^23p^5$	58	Ce	$[Xe]4f^15d^16s^2$
18	Ar	$[Ne]3s^23p^6$	59	Pr	$[Xe]4f^36s^2$
19	K	$[Ar]4s^1$	60	Nd	$[Xe]4f^46s^2$
20	Ca	$[Ar]4s^2$	61	Pm	$[Xe]4f^56s^2$
21	Sc	$[Ar]3d^14s^2$	62	Sm	$[Xe]4f^66s^2$
22	Ti	$[Ar]3d^24s^2$	63	Eu	$[Xe]4f^76s^2$
23	V	$[Ar]3d^34s^2$	64	Gd	$[Xe]4f^75d^16s^2$
24	Cr	$[Ar]3d^54s^1$	65	Tb	$[Xe]4f^96s^2$
25	Mn	$[Ar]3d^54s^2$	66	Dy	$[Xe]4f^{10}6s^2$
26	Fe	$[Ar]3d^64s^2$	67	Ho	$[Xe]4f^{11}6s^2$
27	Co	$[Ar]3d^74s^2$	68	Er	$[Xe]4f^{12}6s^2$
28	Ni	$[Ar]3d^84s^2$	69	Tm	$[Xe]4f^{13}6s^2$
29	Cu	$[Ar]3d^{10}4s^1$	70	Yb	$[Xe]4f^{14}6s^2$
30	Zn	$[Ar]3d^{10}4s^2$	71	Lu	$[Xe]4f^{14}5d^16s^2$
31	Ga	$[Ar]3d^{10}4s^24p^1$	72	Hf	$[Xe]4f^{14}5d^26s^2$
32	Ge	$[Ar]3d^{10}4s^24p^2$	73	Ta	$[Xe]4f^{14}5d^36s^2$
33	As	$[Ar]3d^{10}4s^24p^3$	74	W	$[Xe]4f^{14}5d^46s^2$
34	Se	$[Ar]3d^{10}4s^24p^4$	75	Re	$[Xe]4f^{14}5d^56s^2$
35	Br	$[Ar]3d^{10}4s^24p^5$	76	Os	$[Xe]4f^{14}5d^66s^2$
36	Kr	$[Ar]3d^{10}4s^24p^6$	77	Ir	$[Xe]4f^{14}5d^76s^2$
37	Rb	$[Kr]5s^1$	78	Pt	$[Xe]4f^{14}5d^96s^1$
38	Sr	$[Kr]5s^2$	79	Au	$[Xe]4f^{14}5d^{10}6s^1$
39	Y	$[Kr]4d^15s^2$	80	Hg	$[Xe]4f^{14}5d^{10}6s^2$
40	Zr	$[Kr]4d^25s^2$	81	Tl	$[Xe]4f^{14}5d^{10}6s^26p^1$
41	Nb	$[Kr]4d^45s^1$	82	Pb	$[Xe]4f^{14}5d^{10}6s^26p^2$

(continued)

Table 1.5 (continued)

At. No.	Element	Configuration	At. No.	Element	Configuration
83	Bi	$[Xe]4f^{14}5d^{10}6s^26p^3$	94	Pu	$[Rn]5f^67s^2$
84	Po	$[Xe]4f^{14}5d^{10}6s^26p^4$	95	Am	$[Rn]5f^77s^2$
85	At	$[Xe]4f^{14}5d^{10}6s^26p^5$	96	Cm	$[Rn]5f^76d^17s^2$
86	Rn	$[Xe]4f^{14}5d^{10}6s^26p^6$	97	Bk	$[Rn]5f^97s^2$
87	Fr	$[Rn]7s^1$	98	Cf	$[Rn]5f^{10}7s^2$
88	Ra	$[Rn]7s^2$	99	Es	$[Rn]5f^{11}7s^2$
89	Ac	$[Rn]6d^17s^2$	100	Fm	$[Rn]5f^{12}7s^2$
90	Th	$[Rn]6d^27s^2$	101	Md	$[Rn]5f^{13}7s^2$
91	Pa	$[Rn]5f^26d^17s^2$	102	No	$[Rn]5f^{14}7s^2$
92	U	$[Rn]5f^36d^17s^2$	103	Lr	$[Rn]5f^{14}6d^17s^2$
93	Np	$[Rn]5f^46d^17s^2$			

Note: The configurations of many of the actinides ($Z \geq 89$) are uncertain.

arise for $5s$ and $4d$ (Nb, Mo, Ru, Rh, Pd, and Ag) and for $6s$, $5d$, and $4f$ (La, Ce, Gd, Pt, and Au).

When an electron is removed from an atom, the resulting cation has been changed in two ways. With one less electron, the penetration/shielding factor is reduced. Also, the size of the cation is smaller than that of its parent atom. Both changes make the cation more similar to a hydrogen atom than was the parent atom. Thus, the orbitals become more hydrogen-like, and the principal quantum number becomes the dominant factor in determining orbital stability. For most cations, then, the balance is shifted to favor the $3d$ over the $4s$, the $4d$ over the $5s$, and the $4f$ over the $5d$ and $6s$ orbitals. This is reflected in the ground-state configurations of such ions. For example, the lowest-energy configurations for Mn^0, Mn^+, and Mn^{2+} are

$$Mn^0: \quad [Ar]4s^23d^5$$
$$Mn^+: \quad [Ar]4s^13d^5$$
$$Mn^{2+}: \quad [Ar](4s^0)3d^5$$

In general, the correct ground-state configuration for a transition metal cation can be obtained by first writing the configuration for the parent atom and then removing the appropriate number of electrons from the orbital(s) having the highest value of n.

Hund's Rules

The ground-state electron configuration for carbon is $1s^22s^22p^2$, but this does not give the complete picture. Cases in which two or more *degenerate* orbitals (orbitals of equal energy) are partly filled give rise to different *states,* or subconfigurations. Such is the case for the 2*p* subshell of carbon. There are three viable possibilities:

1. The two 2*p* electrons might differ only in their m_ℓ quantum numbers (occupy different orbitals and have the same spin).

2. They might differ only in their m_s quantum numbers (occupy the same orbital but have opposite spins).

3. They might differ in both m_ℓ and m_s (different orbitals, opposite spins).

Hund's rules[7] can be used to determine which of these possibilities is actually the ground state. In general, the ground state is the one for which m_ℓ values differ and m_s values are the same. Hund's rules may be stated more formally as follows. For states arising from partly filled, degenerate orbitals:

1. The most stable state maximizes the number of unpaired electrons.

2. Among states of equal *multiplicity* (equal numbers of unpaired electrons), the most stable one maximizes the number of occupied orbitals.

The relative energies for carbon are:

$$\underline{\uparrow\downarrow} \quad \underline{} \quad \underline{} \quad \text{Excited state}$$

$$\underline{\uparrow} \quad \underline{\downarrow} \quad \underline{} \quad \text{Excited state} \qquad \uparrow \text{ Energy}$$

$$\underline{\uparrow} \quad \underline{\uparrow} \quad \underline{} \quad \text{Ground state}$$

The energy gap between the ground state and the first excited state is often called the *exchange energy*—the energy needed to cause an electron to "flip its spin." For the carbon atom, about 1.2 eV (roughly 120 kJ/mol) is required. The energy difference between the two excited states (sometimes called the *correlation energy*) derives from the fact that electrons in the same orbital repel one another more strongly than do those in different orbitals.

These rules are useful for predicting magnetic properties. For example, it is known that a neutral, ground-state atom of iron is *paramagnetic* (that is, there are one or more unpaired electrons). This is rationalized by writing the electron configuration ([Ar]$4s^2 3d^6$) and then applying Hund's rules:

$$\underline{\uparrow\downarrow} \quad \underline{\uparrow} \quad \underline{\uparrow} \quad \underline{\uparrow} \quad \underline{\uparrow}$$

Four unpaired electrons are indicated, which agrees exactly with the number observed in magnetic studies (see Chapter 21).

---

7. Hund, F. *Z. Physik.* **1925,** *33,* 345.

1.6 Term Symbols: Russell–Saunders Coupling

It should be clear that certain electron configurations comprise states of different energies, which arise from the different degrees and types of electron–electron interactions. A general term for such interactions is *coupling*. The term *orbital coupling* refers to the notion that the orbital (angular) momenta of the individual electrons in an atom or ion sum to an overall total momentum, symbolized as M_L, for the system:

$$M_L = \sum m_\ell \tag{1.8}$$

In the same way, *spin–spin coupling* refers to summing the spin momenta of the individual electrons to an overall spin momentum M_S:

$$M_S = \sum m_s \tag{1.9}$$

Analogous to the situation for atomic orbitals, the M_L and M_S values are quantized: $L \geq |M_L|$ and $S \geq |M_S|$. Thus, for $L = 2$ there are five possible values of M_L: $M_L = +2, +1, 0, -1,$ or -2.

Any state can then be defined by its L and S values.[8] To name a given state, Russell and Saunders[9] developed a shorthand notation consisting of a letter (derived from L) preceded by a numerical superscript (derived from S). The letter designations are the same as those for the atomic orbitals except that capital letters are used. Thus, $L = 0$ corresponds to an S state, $L = 1$ to a P state, etc. The numerical superscript is equal to $2S + 1$. As a simple example, a hydrogen atom in its electronic ground state ($1s^1$) has $L = 0$ and $S = \frac{1}{2}$, so the term symbol for that state is ^2S (read "doublet S"). Excited states for hydrogen include $^2S'$ ("doublet S prime"), ^2P, and ^2D.

The ground-state configuration for lithium is $1s^2 2s^1$; clearly, $L = 0$. The $1s$ electrons are spin-paired (this must be true for any filled subshell by the Pauli principle); therefore, $S = \frac{1}{2}$, and the ground-state term is again ^2S.

A vector model is sometimes useful for picturing states arising from p, d, or f subshells having two or more electrons. Consider the two vectors v_1 and v_2 (corresponding to, say, the two $2p$ electrons of carbon), each 1 unit in length. Depending on their relative orientations, these vectors might give any of three sums having unit lengths of 0, 1, and 2 (Figure 1.8). This leads to the prediction of three different types of atomic terms: S, P, and D. However, a more sophisticated approach is needed to determine their spin states and relative stabilities.

---

8. A third type of coupling, J or *spin–orbit* coupling, also occurs. For atoms having atomic numbers of 50 or below, however, spin–orbit coupling is less than 1% as strong as either orbital or spin–spin coupling; hence, it is ignored here. Interested readers may consult a reference such as Douglas, B. E.; Hollingsworth, C. A. *Symmetry in Bonding and Spectra*; Academic: Orlando, FL, 1985, for more information.

9. Russell, H. N.; Saunders, F. A. *Astrophys. J.* **1925**, *61*, 38.

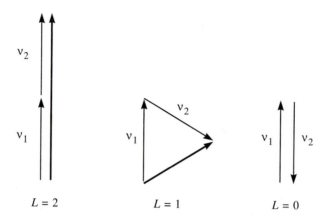

Figure 1.8
The vector model of Russell–Saunders coupling. The unit vectors v_1 and v_2 can couple to give any of three integral values of L.

The term symbol for the ground state of any species can be readily obtained using another useful model—a line and arrow diagram. Only partly filled subshells need to be included. Hund's rules are obeyed, and in placing electrons into orbitals one works from the most positive to the most negative m_ℓ. The diagram for a carbon atom is

$$m_\ell = \quad \underset{+1}{\uparrow} \quad \underset{0}{\uparrow} \quad \underset{-1}{}$$

Since this diagram shows $\sum m_\ell = M_L = 1$ and $\sum m_s = M_S = 1$, a ^3P ground-state term is indicated. Note that caution must be exercised in applying this model, since it is useful only for determining ground states. A more complex procedure must be followed to determine that the excited-state terms for carbon have the identities ^1D and ^1S.

The *degeneracy* of a given state is the number of equivalent combinations that it includes. Any state having $L = 1$ has three possible values of M_L ($+1, 0, -1$). For triplet ($S = 1$) terms, there are also three possible values (again, $+1, 0, -1$). The degeneracy of a ^3P state is therefore $3 \times 3 = 9$. The specific combinations, given in the form $[M_L, M_S]$, are

$$[+1, +1] \qquad [0, +1] \qquad [-1, +1]$$
$$[+1, 0] \qquad [0, 0] \qquad [-1, 0]$$
$$[+1, -1] \qquad [0, -1] \qquad [-1, -1]$$

You should be able to use this approach to show that the ground-state term symbol for a cobalt atom ($[\text{Ar}]4s^2 3d^7$) is ^4F, and that its degeneracy is $7 \times 4 = 28$.

$$m_\ell = \quad \underset{+2}{\uparrow\downarrow} \quad \underset{+1}{\uparrow\downarrow} \quad \underset{0}{\uparrow} \quad \underset{-1}{\uparrow} \quad \underset{-2}{\uparrow}$$

1.7   The Periodic Table

An examination of Table 1.5 shows similarities in the endings of certain electron configurations. For example, the configurations of Li, Na, K, Rb, and Cs all end with ns^1. Such end similarities are important, because it is those electrons that are most likely to be involved in chemical bonding.

The electrons at the level of the highest principal quantum number are called *valence* electrons; all others are *core* electrons.[10] The core electrons of any atom are highly stabilized and isolated from external attack by the electrons in outer orbitals. Thus, they are inert. In contrast, valence electrons are less stable and less isolated. The chemical properties of any species therefore depend on its valence electrons.

A set of elements having identical valence shell configurations is a *group* or *family*. The members of a given family occupy a vertical column of the periodic table. Groups are numbered in the recently revised IUPAC system from 1 to 18. (See the periodic table on the inside front cover.) Certain families have familiar names—the Group 1 elements are the alkali metals, Group 2 the alkaline earths, Group 17 the halogens, and Group 18 the noble gases. According to most authors, the transition metals comprise Groups 3–12. Elements 58–71 are the lanthanides, and elements 90–103 are the actinides.

What properties of atoms show periodic trends? Some of the most important are examined in the following subsections.

Orbital Energies for Many-Electron Species

Several useful methods have been developed for estimating the energies of orbitals. As early as 1930, Slater[11] described a simple set of rules to approximate shielding effects, and those shielding values could then be used to estimate orbital energies. More recently, Clementi and Raimondi extended this approach with the aid of computers.[12] Their general method, called the *self-consistent field (SCF) approach*, gives shielding constants that, as the name implies, are internally consistent. To illustrate, their equations for the first three orbitals are

$$S_{1s} = 0.300(1s - 1) + 0.0072(2s + 2p) + 0.0158(3s + 3p + 3d + 4s + 4p)$$
$$S_{2s} = 1.721 + 0.360(2s + 2p - 1) + 0.206(3s + 3p + 3d + 4s + 4p)$$
$$S_{2p} = 2.579 + 0.333(2p - 1) - 0.0773(3s) - 0.0161(3p + 4s)$$
$$- 0.0048(3d) + 0.0085(4p)$$

10.   For the transition metals, d electrons at the $n - 1$ level and/or f electrons at the $n - 2$ level are usually considered to be valence electrons as well.

11.   Slater, J. C. *Phys. Rev.* **1930**, *36*, 57.

12.   Clementi, E.; Raimondi, D. L. *J. Chem. Phys.* **1963**, *38*, 2868.

Here, S is the overall shielding factor for the electron in question. The remaining terms represent shielding contributions from the other electrons in the atom or ion, with the number of electrons in the various orbitals given in parentheses. Thus, for the boron atom, with the ground-state configuration $1s^2 2s^2 2p^1$,

$$S_{1s} = 0.300(2 - 1) + 0.0072(2 + 1) = 0.322$$
$$S_{2s} = 1.721 + 0.360(2 + 1 - 1) = 2.441$$
$$S_{2p} = 2.579 + 0.333(1 - 1) = 2.579$$

Next, the orbital energies can be estimated from the equation

$$E = \frac{-13.6\,\text{eV} \times (Z^*)^2}{(n^*)^2} \tag{1.10}$$

where Z^* is calculated using equation (1.7) and n^* is the *effective quantum number*. For the first three shells, $n^* = n$; hence, the orbital energies of boron are estimated to be

$$E_{1s} = \frac{-13.6(5.000 - 0.322)^2}{1^2} = -298 \text{ eV}$$

$$E_{2s} = \frac{-13.6(5.000 - 2.441)^2}{2^2} = -22.3 \text{ eV}$$

$$E_{2p} = \frac{-13.6(5.000 - 2.579)^2}{2^2} = -19.9 \text{ eV}$$

Table 1.6 gives the orbital energies for the first 19 elements as calculated by this method. Keep in mind that these numbers are produced from a model and should not be taken literally. Their value lies in the trends they establish.

Some observations that can be made from an examination of Table 1.6 are the following:

1. As atomic number increases, the energy of a given type of orbital becomes more negative (its stability increases). This is due to enhanced electrostatic attraction.

2. Going across a horizontal row of the periodic table, the shielding increase is incomplete—that is, the extra shielding by an added electron does not equal the effect of adding another proton to the nucleus. Thus, in comparing the $2p$ orbitals of B, C, and N, the value of Z^* increases (though by less than 1 unit). This explains the decrease in orbital energies.

3. Exceptions to the general order of orbital stability are sometimes encountered. For example, the $2p$ orbitals appear to become more stable than $2s$ for the elements beyond neon. The accuracy of the Clementi–Raimondi approach is questionable in this regard.

Table 1.6 Orbital energies (as estimated using the rules of Clementi and Raimondi) for the first 19 elements

Element	1s	2s	2p	3s	3p	4s
H	−14					
He	−39					
Li	−99	−6				
Be	−185	−12				
B	−298	−22	−20			
C	−437	−35	−32			
N	−604	−50	−48			
O	−797	−68	−66			
F	−1017	−89	−88			
Ne	−1264	−113	−113			
Na	−1536	−146	−159	−10		
Mg	−1833	−184	−213	−16		
Al	−2157	−225	−271	−24	−20	
Si	−2508	−271	−336	−34	−28	
P	−2884	−322	−408	−46	−36	
S	−3287	−376	−488	−59	−46	
Cl	−3717	−436	−574	−74	−57	
Ar	−4173	−499	−667	−91	−69	
K	−4655	−566	−767	−96	−90	−12

Note: In electron volts, rounded to the nearest integer.

4. The value for the 2s orbital of lithium, − 6 eV, is the smallest (least negative) in the table. The single electron in that orbital is especially vulnerable to removal by the input of a small amount of energy—that is, the first *ionization energy* of Li is small. The removal of a second electron must involve a much more stable (core) orbital, and requires considerably more energy—that is, the second ionization energy is large. Ionization energies will be discussed in more detail later in this section.

5. Other single, loosely held electrons are found at the valence levels of sodium (3s) and potassium (4s). This provides a basis for periodicity, suggesting that the three elements Li, Na, and K should be grouped into the same family of the periodic table (as, of course, they are).

6. The element having the most stable vacancy in an atomic orbital is fluorine (−88 eV for the one unoccupied position in the 2p subshell). This has a profound effect on the bonding characteristics of that element. We will return to this point when the concept of electronegativity is introduced in Section 1.9.

It is also worthwhile to compare the orbital energies of atoms to those of their cations and anions. For example, you should be able to calculate the following orbital energies for boron:

	B^+ ($1s^2 2s^2$)	B^0 ($1s^2 2s^2 2p^1$)	B^- ($1s^2 2s^2 2p^2$)
E_{1s}	-298 eV	-298 eV	-297 eV
E_{2s}	-29 eV	-22 eV	-16 eV
E_{2p}	—	-20 eV	-15 eV

Recall that these values are approximate. However, a clear trend can be seen. The removal of an electron reduces the shielding, resulting in increased stability for the valence orbitals. The reverse effect is observed when one or more electrons are added.

The Sizes of Atoms and Ions[13]

It is well-recognized that the volume occupied by any atom or ion depends on its environment, particularly with respect to its chemical bonds; hence, size effects will be discussed several times in later chapters. Nevertheless, it is possible to make some generalizations about size trends in the periodic table.

1. Size tends to decrease going across a period due to incomplete shielding (as was discussed above, the addition of each electron does not counter-balance the addition of a proton to the nucleus). Thus, a boron atom is larger than an atom of carbon, and carbon is larger than nitrogen. The radius of a fluorine atom (71 pm) is less than half that of lithium (157 pm), even though they belong to the same period.

2. Going down a family, size increases because of the added shell. For example, magnesium is larger than beryllium, and calcium is larger than magnesium.

3. For a given nuclear charge, size increases with the number of electrons. Said another way, for any element M, size increases as:

$$M^{2+} < M^+ < M^0 < M^- < M^{2-} < \cdots$$

This is not surprising, since increasing the number of electrons increases the shielding. Thus, F^0 is smaller than F^-, and Na^0 is larger than Na^+. The effect is particularly large if all the valence electrons are lost, because this changes the value of n for the outermost occupied orbital. For

---

13.   Mason, J. *J. Chem. Educ.* **1988**, *65*, 17.

example, Na^+ ($1s^2 2s^2 2p^6$) has only about half the radius of Na^0 ($1s^2 2s^2 2p^6 3s^1$).

4. For a given number of electrons, size decreases as the nuclear charge increases. For example, in the following series of 18-electron species, size decreases from left to right:

$$_{16}S^{2-} > {}_{17}Cl^- > {}_{18}Ar^0 > {}_{19}K^+ > {}_{20}Ca^{2+}$$

Here the difference in radius averages about 15% per proton.

5. The elements immediately following a transition series tend to be smaller than might otherwise be expected. For example, compare the pair Mg and Al ($[Ne]3s^2$ and $[Ne]3s^2 3p^1$) to Ca and Ga ($[Ar]4s^2$ and $[Ar]4s^2 3d^{10} 4p^1$). From the standpoint of the outermost shell, the difference is one electron in both cases. However, the difference in Z is only 1 for Mg/Al but 11 for Ca/Ga. As a result, aluminum has a slightly smaller radius than magnesium (about 11%), while gallium is considerably smaller than calcium (about 33%). A well-known effect of this type is the *lanthanide contraction*. A large decrease in size is observed in going from $_{57}La$ ($[Xe]6s^2 5d^1$) to $_{71}Lu$ ($[Xe]6s^2 4f^{14} 5d^1$). Here again the difference in configuration is in an inner subshell. The size differential leads to significant differences in chemical properties among the lanthanides (see Chapter 20).

The relative sizes of atoms and ions of the main group elements are pictured in Figure 1.9 (p. 26).

Ionization Energies

Ionization energy (IE) can be defined as *the energy required to remove an electron from a gaseous species*. This process is always endothermic. That is, $\Delta H > 0$ for the general equation

$$M^0(g) \rightarrow M^+(g) + 1e^- \qquad \Delta H = IE \tag{1.11}$$

Hence, ionization energies are always positive in sign.

It is possible, of course, to remove more than one electron from most atoms. Therefore, we may be interested in first, second, third, etc., ionization energies:

$$M^0(g) \rightarrow M^+(g) + 1e^- \qquad \Delta H = IE_1 \tag{1.12}$$

$$M^+(g) \rightarrow M^{2+}(g) + 1e^- \qquad \Delta H = IE_2 \tag{1.13}$$

$$M^{2+}(g) \rightarrow M^{3+}(g) + 1e^- \qquad \Delta H = IE_3 \tag{1.14}$$

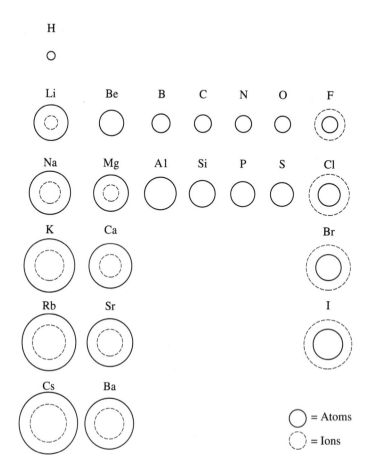

Figure 1.9
Relative atomic and ionic radii for some main group elements. (Ions are M^+ for Group 1, M^{2+} for Group 2, and X^- for Group 17.)

A given ionization energy might be expected to simply be the negative of some orbital energy,[14] and this is in fact the case for the hydrogen atom ($E_{1s} = -13.6\,\text{eV}$; $IE = +13.6\,\text{eV}$). This relationship does not hold for many-electron systems, however, since shielding factors are changed by the removal of an electron. Nevertheless, there is a strong correlation between the first ionization energy of an atom and the orbital energies of that atom and its monovalent cation.

Experimentally determined ionization energies are given in Table 1.7. Some periodic trends can be observed in Figure 1.10, where IE_1 for the first 20 elements is plotted as a function of atomic number. Ionization energies tend to increase going across a period and decrease going down a family, with certain exceptions. For example, IE_1 for boron is less than that for

---

14. This is sometimes referred to as *Koopmans' theorem* (see Chapter 21).

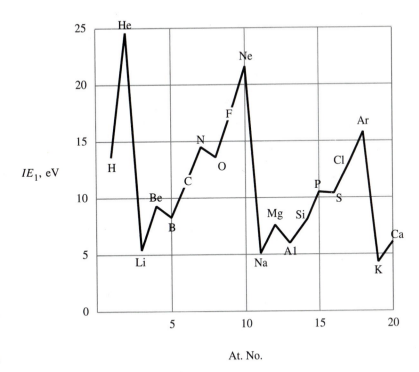

Figure 1.10

Plot of first ionization energy (IE_1) versus atomic number for the first 20 elements.

beryllium, because the removed electron is from a different subshell [Be (2s) versus the less stable B (2p)].

Another exception is found at the tops of Groups 15 and 16, where IE_1 for oxygen is seen to be less than that for nitrogen. Here, both electrons in question reside in 2p orbitals. Hund's rules predict these ground-state 2p subconfigurations for the free atoms:

$$\underline{\uparrow} \quad \underline{\uparrow} \quad \underline{\uparrow} \qquad \underline{\uparrow\downarrow} \quad \underline{\uparrow} \quad \underline{\uparrow}$$
$$\text{Nitrogen} \qquad\qquad \text{Oxygen}$$

Electron loss from oxygen is facilitated by the corresponding subtraction of correlation energy, since the ground state of O^+ does not contain that extra energy. Sulfur and phosphorus show a similar reversal.

Given earlier comments about shielding effects, it should not be surprising that for any atom,

$$IE_1 < IE_2 < IE_3 < \cdots$$

For example, the experimental values for boron are: $IE_1 = 8.30$ eV; $IE_2 = 25.15$ eV; $IE_3 = 37.93$ eV; $IE_4 = 259.4$ eV; and $IE_5 = 340.2$ eV. Why is there such a huge increase at IE_4? The answer is probably obvious to you: For the fourth and fifth ionizations, the electrons are removed from the $n = 1$

Table 1.7 First electron affinities and first, second, and third ionization energies (in electron volts) of the elements

At. No.	Element	EA_1	IE_1	IE_2	IE_3
1	H	0.75	13.598		
2	He	−0.22	24.587	54.416	
3	Li	0.62	5.392	75.638	122.451
4	Be	−2.5	9.322	18.211	153.893
5	B	0.28	8.298	25.154	37.930
6	C	1.27	11.260	24.383	47.887
7	N	0.07	14.534	29.601	47.448
8	O	1.46	13.618	35.116	54.934
9	F	3.40	17.422	34.970	62.707
10	Ne	−0.3	21.564	40.962	63.45
11	Na	0.55	5.139	47.286	71.64
12	Mg	−0.15	7.646	15.035	80.143
13	Al	0.46	5.986	18.828	28.447
14	Si	1.39	8.151	16.345	33.492
15	P	0.74	10.486	19.725	30.18
16	S	2.08	10.360	23.33	34.83
17	Cl	3.62	12.967	23.81	39.61
18	Ar	−0.36	15.759	27.629	40.74
19	K	0.50	4.341	31.625	45.72
20	Ca	−1.8	6.113	11.871	50.908
21	Sc	0.14	6.54	12.80	24.76
22	Ti	0.08	6.82	13.58	27.491
23	V	0.5	6.74	14.65	29.310
24	Cr	0.66	6.766	16.50	30.96
25	Mn	0	7.435	15.640	33.667
26	Fe	0.25	7.870	16.18	30.651
27	Co	0.7	7.86	17.06	33.50
28	Ni	1.15	7.635	18.168	35.17
29	Cu	1.23	7.726	20.292	36.83
30	Zn	−0.49	9.394	17.964	39.722
31	Ga	0.3	5.999	20.51	30.71
32	Ge	1.2	7.899	15.934	34.22

(continued)

At. No.	Element	EA_1	IE_1	IE_2	IE_3
33	As	0.8	9.81	18.633	28.351
34	Se	2.02	9.752	21.19	30.820
35	Br	3.36	11.814	21.8	36
36	Kr	−0.40	13.999	24.359	36.95
37	Rb	0.49	4.177	27.28	40
38	Sr	−1.7	5.695	11.030	43.6
39	Y	0	6.38	12.24	20.52
40	Zr	0.43	6.84	13.13	22.99
41	Nb	1.0	6.88	14.32	25.04
42	Mo	0.75	7.099	16.15	27.16
43	Tc		7.28	15.26	29.54
44	Ru	1.5	7.37	16.76	28.47
45	Rh	1.14	7.46	18.08	31.06
46	Pd	0.56	8.34	19.43	32.93
47	Ag	1.30	7.576	21.49	34.83
48	Cd	−0.33	8.993	16.908	37.48
49	In	0.3	5.786	18.869	28.03
50	Sn	1.25	7.344	14.632	30.502
51	Sb	1.05	8.641	16.53	25.3
52	Te	1.97	9.009	18.6	27.96
53	I	3.06	10.451	19.131	33
54	Xe	−0.42	12.130	21.21	32.1
55	Cs	0.47	3.894	25.1	
56	Ba	−0.5	5.212	10.004	
57	La	0.5	5.577	11.06	19.175
58	Ce		5.47	10.85	20.20
59	Pr		5.42	10.55	21.62
60	Nd		5.49	10.72	
61	Pm		5.55	10.90	
62	Sm		5.63	11.07	
63	Eu		5.67	11.25	
64	Gd		6.14	12.1	
65	Tb		5.85	11.52	
66	Dy		5.93	11.67	

(*continued*)

Table 1.7 (continued)

At. No.	Element	EA_1	IE_1	IE_2	IE_3
67	Ho		6.02	11.80	
68	Er		6.10	11.93	
69	Tm		6.18	12.05	23.71
70	Yb		6.254	12.17	25.2
71	Lu		5.426	13.9	
72	Hf	−0.8	7.0	14.9	23.3
73	Ta	0.32	7.89		
74	W	0.82	7.98		
75	Re	0.15	7.88		
76	Os	1.1	8.7		
77	Ir	1.6	9.1		
78	Pt	2.1	9.0	18.563	
79	Au	2.31	9.225	20.5	
80	Hg	−0.63	10.437	18.756	34.2
81	Tl	0.3	6.108	20.428	29.83
82	Pb	0.37	7.416	15.032	31.937
83	Bi	0.95	7.289	16.69	25.56
84	Po		8.48		
85	At				
86	Rn	−0.42	10.748		
87	Fr	0.46			
88	Ra		5.279	10.147	
89	Ac		6.9	12.1	
90	Th			11.5	20.0
91	Pa				
92	U				
93	Np				
94	Pu		5.8		
95	Am		6.0		

Sources: Ionization energies taken from Moore, C. E. *Ionization Potentials and Ionization Limits Derived from the Analyses of Optical Spectra*; NSRDS-NBS 34; National Bureau of Standards: Washington, DC, 1970. Electron affinities taken from compilations by Hotop, H.; Lineberger, W. C. *J. Phys. Chem. Ref. Data* **1985**, *14*, 731; and from Chen, E. C. M.; Wentworth, W. E. *J. Chem. Educ.* **1975**, *52*, 486.

(core) shell, rather than the $n = 2$ (valence) shell. All elements beyond helium show effects of this type.

Electron Affinities[15]

Electron affinity (EA) is often defined as *the energy released when an electron is added to a gaseous species.* In the form of an equation,

$$M(g) + 1e^- \rightarrow M^-(g) \qquad \Delta H = -EA \qquad \textbf{(1.15)}$$

Notice that because of the phrasing of the definition, the electron affinity is positive if the process is exothermic. Also, the reverse of equation (1.15) corresponds to an ionization energy, and the sign convention is such that electron affinity is sometimes called the "zeroth" ionization energy. This is an important point, because it emphasizes the relationship between the two concepts. Table 1.7 lists electron affinities and ionization energies.

Most elements have positive first electron affinities; that is, the addition of an electron to a neutral atom is usually exothermic. The elements of Groups 2 and 18 are exceptions. For these two families, electron gain requires starting a new subshell [np for Group 2 and $(n + 1)s$ for Group 18], which is unfavorable because of the low stabilities of those orbitals. Second electron affinities and beyond are always endothermic (usually strongly so) because of increased electron–electron repulsions. It is interesting to note that the oxide ion, O^{2-}, is thermodynamically unstable in the gas phase; the process

$$O(g) + 2e^- \rightarrow O^{2-}(g) \qquad \Delta H = -EA_1 - EA_2 \qquad \textbf{(1.16)}$$

is endothermic by 6.6 eV (over 600 kJ/mol). The fact that oxide ion is commonly found in solids will be explained in Chapter 6.

It is apparent from Table 1.7 that the general periodic trends—increasing going across a period and decreasing going down a family—are the same as for ionization energy. There are numerous exceptions, which for the most part are easily explained. For example, the first electron affinity of nitrogen is less than that of carbon. As can be deduced by writing the relevant subconfigurations, this is a correlation energy effect.

Somewhat surprisingly, the most negative (favorable) electron affinity of any element is that of chlorine. Why not fluorine? The best explanation appears to be that the small size of fluorine causes severe electron–electron repulsions in the anion. This effect is less significant for the larger chloride ion. A similar reversal occurs in Group 16, where the most exothermic EA_1 belongs to sulfur, the second member of the family.

15. Myers, R. T. *J. Chem. Educ.* **1990,** *67,* 307.

Valence State Ionization Energies

We mentioned earlier that although the orbital energies given in Table 1.6 are useful for certain trends they establish, their numerical values may be misleading. For many purposes a set of more accurate, experimentally determined values are needed, particularly for the valence electrons. Such data have been obtained from electronic spectroscopy and are called *valence state ionization energies (VSIE's)*.[16]

To illustrate the methodology, we will look again at carbon. As has already been shown, the $2p^2$ configuration comprises three distinct energy levels—the ^3P ground state and two excited states, ^1D and ^1S. It is possible to remove an electron from any of these levels. The ionization product C^+ has the ground-state configuration $2p^1$, and there is only one state, ^2P, for that configuration.

As can be seen in Figure 1.11, then, three different ionizations are possible from the electronic ground state of this system. The experimental ionization energies are 11.26 eV (electron removal from ^3P), 10.00 eV (^1D), and 8.58 eV (^1S). The relative populations of these states are the same as their degeneracies (9:5:1) at room temperature due to the Boltzmann

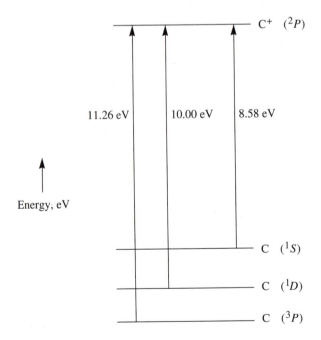

Figure 1.11
Determination of valence state ionization energies. Ionization from the ^3P, ^1D, and ^1S states of the $1s^2 2s^2 2p^2$ configuration of carbon.

Energy, eV

C$^+$ (^2P)

11.26 eV   10.00 eV   8.58 eV

C (^1S)

C (^1D)

C (^3P)

16. The terms *valence state ionization potential (VSIP)* and *valence orbital ionization potential (VOIP)* are also used.

distribution. Therefore, a weighted-average ionization energy can be calculated:

$$IE_{ave} = \frac{9(11.26) + 5(10.00) + 1(8.58)}{15} = 10.66 \text{ eV}$$

This is taken to be the VSIE for the $2p$ orbital of carbon. For the $2s$ orbital, the possible ionizations to the configuration $\ldots 2s^1 2p^2$ are measured and the weighted average calculated.

The values in Table 1.8 were determined in this manner. It can be seen that the same trends arise for VSIE's as for orbital energies, although the

Table 1.8   Valence state ionization energies (electron volts)

Element	$1s$	$2s$	$2p$	$3s$	$3p$	$4s$	$4p$
H	13.6	—	—	—	—	—	—
He	24.6	—	—	—	—	—	—
Li	—	5.4	—	—	—	—	—
Be	—	9.3	—	—	—	—	—
B	—	14.0	8.3	—	—	—	—
C	—	19.4	10.6	—	—	—	—
N	—	25.6	13.2	—	—	—	—
O	—	32.3	15.8	—	—	—	—
F	—	40.2	18.6	—	—	—	—
Ne	—	48.5	21.6	—	—	—	—
Na	—	—	—	5.1	—	—	—
Mg	—	—	—	7.6	—	—	—
Al	—	—	—	11.3	5.9	—	—
Si	—	—	—	14.9	7.7	—	—
P	—	—	—	18.8	10.1	—	—
S	—	—	—	20.7	11.6	—	—
Cl	—	—	—	25.3	13.7	—	—
Ar	—	—	—	29.2	15.8	—	—
K	—	—	—	—	—	4.3	—
Ca	—	—	—	—	—	6.1	—
Zn	—	—	—	—	—	9.4	—
Ga	—	—	—	—	—	12.6	6.0
Ge	—	—	—	—	—	15.6	7.6
As	—	—	—	—	—	17.6	9.1
Se	—	—	—	—	—	20.8	10.8
Br	—	—	—	—	—	24.1	12.5
Kr	—	—	—	—	—	27.5	14.3

Source: DeKock, R. L.; Gray, H. B. *Chemical Structure and Bonding*; Benjamin/Cummings: Menlo Park, CA, 1980; p. 227.

values themselves are considerably different; the VSIE's tend to be smaller in magnitude.

1.8   Charge–Energy Relationships

As might be deduced from earlier comments, a general relationship exists between the energy of a species and its charge. Ionization energies and electron affinities correspond to energy differences between integral charges which differ by 1 unit:

$$M^{2-} \xrightarrow[-e^-]{EA_2} M^- \xrightarrow[-e^-]{EA_1} M \xrightarrow[-e^-]{IE_1} M^+ \xrightarrow[-e^-]{IE_2} M^{2+} \xrightarrow[-e^-]{IE_3} \text{etc.} \tag{1.17}$$

It is therefore possible to construct a graph of energy versus charge for any element. Such a plot is shown for nitrogen in Figure 1.12.

This charge–energy relationship is approximated by the general equation

$$E = aq + bq^2 + cq^3 + dq^4 \tag{1.18}$$

Experimental ionization energies and electron affinities can be used to determined "best fit" values for the coefficients a, b, c, and d via the equations given in Table 1.9. The coefficients for nitrogen are: $a = 7.16$, $b = 6.21$, $c = 0.46$, and $d = -0.01$. You should take the time to verify that Table 1.9 and these values can be used to approximate the experimental data. For example, IE_2 for nitrogen is calculated to be 28.9 eV, compared to the actual value of 29.60 eV.

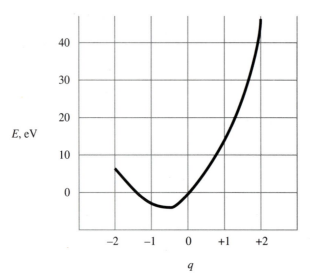

Figure 1.12
The total energy of nitrogen (in electron volts, referenced to $E = 0$ at $q = 0$) as a function of charge.

Table 1.9 Total potential energy expressions for an atom and its ions

q	Total Energy E, eV
-2	$-2a + 4b - 8c + 16d$
-1	$-a + b - c + d$
0	
$+1$	$+a + b + c + d$
$+2$	$+2a + 4b + 8c + 16d$
$+3$	$+3a + 9b + 27c + 81d$

Bracketing the rows: between -2 and -1: EA_2; between -1 and 0: EA_1; between 0 and $+1$: IE_1; between $+1$ and $+2$: IE_2; between $+2$ and $+3$: IE_3.

Source: Adapted from Iczkowski, R. P.; Margrave, J. L. *J. Am. Chem. Soc.* **1961**, *83*, 3547.

The optimum charge for nitrogen corresponds to the energy minimum of the curve of Figure 1.12. This can be calculated by finding the charge at which $dE/dq = 0$:

$$\frac{dE}{dq} = 0 = a + 2bq + 3cq^2 + 4dq^3 \qquad q = -0.62$$

The potential energy of nitrogen is therefore minimized at a charge of about -0.62 unit. Using equation (1.18), you should be able to show that this charge yields a stabilization of about 2.2 eV over the neutral atom. Hence, it is not surprising that nitrogen shows a tendency to increase its electron density in its reaction chemistry.

1.9 Electronegativity—An Introduction

Electronegativity was originally defined by Pauling as "the power of an atom in a molecule to attract electrons to itself."[17] Consider this definition carefully. What property of an atom would enable it to attract electron density? Atoms that possess *valence orbitals of high stability* should be successful in this regard. Thus, there is a strong correlation between the elements that have stable valence orbitals and those that exhibit high electronegativities. This can be seen by comparing the electronegativities listed in Table 1.10 to the VSIE's in Table 1.8.

What is the most electronegative element? The above reasoning suggests neon, since its valence orbitals (2*s* and 2*p*) are the most stable of any element.

---

17. Pauling, L. *J. Am. Chem. Soc.* **1932**, *54*, 3570; see also Pauling, L. *The Nature of the Chemical Bond*, 3rd ed.; Cornell University: Ithaca, NY, 1960; pp. 88*ff*.

Table 1.10 Electronegativities of the elements

At. No.	Element	Pauling	Allred–Rochow	At. No.	Element	Pauling	Allred–Rochow
1	H	2.20	2.20	35	Br	2.96	2.74
2	He			36	Kr		
3	Li	0.98	0.97	37	Rb	0.82	0.89
4	Be	1.57	1.47	38	Sr	0.95	0.99
5	B	2.04	2.01	39	Y(III)	1.22	1.11
6	C	2.55	2.50	40	Zr	1.33	
7	N	3.04	3.07	41	Nb	1.60	
8	O	3.44	3.50	42	Mo	2.16	
9	F	3.98	4.10	43	Tc	1.90	
10	Ne			44	Ru	2.20	
				45	Rh	2.28	
11	Na	0.93	1.01	46	Pd	2.20	
12	Mg	1.31	1.23	47	Ag(I)	1.93	1.42
				48	Cd	1.69	1.46
13	Al	1.61	1.47				
14	Si	1.90	1.74	49	In	1.78	1.49
15	P	2.19	2.06	50	Sn	1.96	1.72
16	S	2.58	2.44	51	Sb	2.05	1.82
17	Cl	3.16	2.83	52	Te	2.10	2.01
18	Ar			53	I	2.66	2.21
				54	Xe		
19	K	0.82	0.91				
20	Ca	1.00	1.04	55	Cs	0.79	0.86
				56	Ba	0.89	0.97
21	Sc(III)	1.36	1.20				
22	Ti	1.54	1.32	57	La	1.10	1.08
23	V	1.63	1.45	58	Ce	1.12	
24	Cr	1.66	1.56	59	Pr	1.13	
25	Mn	1.55	1.60	60	Nd	1.14	
26	Fe	1.83	1.64	61	Pm		
27	Co	1.88	1.70	62	Sm	1.17	
28	Ni	1.91	1.75	63	Eu		
29	Cu	2.00	1.75	64	Gd	1.20	
30	Zn	1.65	1.66	65	Tb		
				66	Dy	1.22	
31	Ga	1.81	1.82	67	Ho	1.23	
32	Ge	2.01	2.02	68	Er	1.24	
33	As	2.18	2.20	69	Tm	1.25	
34	Se	2.55	2.48	70	Yb		

(continued)

Table 1.10   (continued)

At. No.	Element	Pauling	Allred–Rochow	At. No.	Element	Pauling	Allred–Rochow
71	Lu	1.27		87	Fr	0.70	
				88	Ra	0.90	
72	Hf	1.30					
73	Ta	1.50		89	Ac	1.10	
74	W	2.36		90	Th	1.30	
75	Re	1.90		91	Pa	1.50	
76	Os	2.20		92	U	1.70	
77	Ir	2.20		93	Np	1.30	
78	Pt	2.28		94	Pu	1.30	
79	Au	2.54		95	Am	1.30	
80	Hg	2.00		96	Cm	1.30	
				97	Bk	1.30	
81	Tl	2.04		98	Cf	1.30	
82	Pb	2.33		99	Es	1.30	
83	Bi	2.02		100	Fm	1.30	
84	Po	2.00		101	Md	1.30	
85	At	2.20		102	No	1.30	
86	Rn						

Note: Values given for the transition metals correspond to the (II) state unless otherwise indicated.
Sources: Pauling values as revised by Allred, A. L. *J. Inorg. Nucl. Chem.* **1961**, *17*, 215; Allred–Rochow values from Allred, A. L.; Rochow, E. G. *J. Inorg. Nucl. Chem.* **1958**, *5*, 264.

However, no stable compounds of neon have as yet been reported. Therefore, Pauling's definition does not apply to that element (or to helium or argon). Among the remaining elements fluorine has the most stable valence orbitals, and by that criterion should (and does) have the greatest electronegativity. It is noteworthy that the periodic trends of electronegativity (increasing across a period and decreasing down a family) are the same as for ionization energy and electron affinity, and fluorine occupies an ideal location in the periodic table in that regard.

Two other trends of importance concern the variance of electronegativity with orbital type and with charge:

1. For the orbitals of a given atom or ion, electronegativity decreases as:

$$1s > 2s > 2p > 3s > \cdots$$

This derives from earlier discussions of orbital energies. [However, it is the electronegativity of the valence orbital(s) that is of greatest importance for considerations of chemical bonding.]

2. For a given element, electronegativity decreases with the number of electrons:

$$M^{2+} > M^+ > M > M^- > M^{2-} > \cdots$$

This also follows from orbital energies and makes intuitive sense as well—the more electron-poor a species is, the greater its electronegativity is expected to be.

The above comments paint a qualitative picture of electronegativity, but its most useful applications require quantitative comparisons. Several different electronegativity scales, calculated by various approaches, have been proposed. Pauling's original values were based on bond energies. His method will be described in Chapter 4. Allred and Rochow used effective nuclear charges to calculate the "electrostatic force" of a given atom's nucleus on its outermost electrons. The equation

$$\chi = \frac{3590Z^*}{r^2} + 0.744 \tag{1.19}$$

was used, where χ is the electronegativity and r is the covalent radius in picometers. The constants were chosen to conform with Pauling's scale, giving fluorine and cesium electronegativities of about 4.0 and 0.8, respectively. [Equation (1.19) cannot be used to obtain the values given in Table 1.10, however, since the effective nuclear charges were calculated by a method different from the one described earlier.]

In an early approach by Mulliken, three possible electron distributions (ie, three resonance structures) for a heteronuclear bond X–Z were considered:[18]

$$X^+Z^- \longleftrightarrow X\text{–}Z \longleftrightarrow X^-Z^+ \tag{1.20}$$

The importance of the first structure is logically related to the first ionization energy of X and the first electron affinity of Z. Similarly, the third structure must depend on IE_1 of Z and EA_1 of X. If X and Z have equal electronegativities, these two "ionic" structures are of equal importance. In that case,

$$IE(X) - EA(Z) = IE(Z) - EA(X) \tag{1.21}$$

which rearranges to

$$IE(X) + EA(X) = IE(Z) + EA(Z) \tag{1.22}$$

Consistent with this equality, Mulliken defined electronegativity as the

---

18. Mulliken, R. S. *J. Chem. Phys.* **1934**, *2*, 782. For a more recent discussion of Mulliken electronegativities, see Bratsch, S. G. *J. Chem. Educ.* **1988**, *65*, 34, 223.

average of the first ionization energy and the first electron affinity of an element:

$$\chi = \frac{IE_1 + EA_1}{2} \tag{1.23}$$

Since a single electron configuration can give rise to several ionization energies (recall the example of carbon), Mulliken argued that electronegativity is an orbital property as well as an atomic property.

The variance of electronegativity with charge was discussed earlier. If χ is defined as the change in energy with respect to charge, then from equation (1.18),

$$\chi = \frac{dE}{dq} = a + 2bq + 3cq^2 + 4dq^3 \tag{1.24}$$

Thus, the *nominal electronegativity* of an element (its electronegativity at $q = 0$) is equal to a in units of electron volts. The relationship between this method and Mulliken's definition can be seen by returning to Table 1.9. For atom M,

$$IE_1 = E(M^+) - E(M) = a + b + c + d \tag{1.25}$$

$$EA_1 = E(M) - E(M^-) = a - b + c - d \tag{1.26}$$

Therefore,

$$\chi \text{ (Mulliken)} = \frac{IE_1 + EA_1}{2} = \frac{2a + 2c}{2}$$

$$= a + c$$

In fact, Mulliken originally calculated an electronegativity of 7.62 eV for the $2p$ orbital of nitrogen; this exactly equals the sum of its a and c coefficients.

When the definition $\chi = dE/dq$ is applied to specific orbitals, the energy of the orbital is seen to be a function of its occupancy. If the last two terms of equation (1.24) are neglected, then

$$\chi = a + 2bq \tag{1.27}$$

This expression is convenient because the relationship between χ and q is linear. The coefficient a is then the nominal electronegativity of an orbital (according to Mulliken's definition), and b is a constant reflecting the ability of that orbital to accommodate additional electron density.

Pearson has redetermined Mulliken-type electronegativity values using a modified approach.[19] His results are given in Table 1.11.

---

19. Pearson, R. G. *Inorg. Chem.* **1988**, *27*, 734. Pearson actually used different symbolism than is employed here.

Table 1.11   Mulliken-type electronegativity parameters of atoms as redetermined by Pearson

At. No.	Element	a	b	At. No.	Element	a	b
1	H	7.18	6.43	34	Se	5.89	3.87
2	He			35	Br	7.59	4.22
				36	Kr		
3	Li	3.01	2.39				
4	Be	4.9	4.5	37	Rb	2.34	1.85
				38	Sr	2.0	3.7
5	B	4.29	4.01				
6	C	6.27	5.00	39	Y	3.19	3.19
7	N	7.30	7.23	40	Zr	3.64	3.21
8	O	7.54	6.08	41	Nb	4.0	3.0
9	F	10.41	7.01	42	Mo	3.9	3.1
10	Ne			43	Tc		
				44	Ru	4.5	3.0
11	Na	2.85	2.30	45	Rh	4.30	3.16
12	Mg	3.75	3.90	46	Pd	4.45	3.89
				47	Ag	4.44	3.14
13	Al	3.23	2.77	48	Cd	4.33	4.66
14	Si	4.77	3.38				
15	P	5.62	4.88	49	In	3.1	2.8
16	S	6.22	4.14	50	Sn	4.30	3.05
17	Cl	8.30	4.68	51	Sb	4.85	3.80
18	Ar			52	Te	5.49	3.52
				53	I	6.76	3.69
19	K	2.42	1.92	54	Xe		
20	Ca	2.2	4.0				
				55	Cs	2.18	1.71
21	Sc	3.34	3.20	56	Ba	2.4	2.9
22	Ti	3.45	3.37				
23	V	3.6	3.1	57	La	3.1	2.6
24	Cr	3.72	3.06	58	Ce		
25	Mn	3.72	3.72	59	Pr		
26	Fe	4.06	3.81	60	Nd		
27	Co	4.3	3.6	61	Pm		
28	Ni	4.40	3.25	62	Sm		
29	Cu	4.48	3.25	63	Eu		
30	Zn	4.45	4.94	64	Gd		
				65	Tb		
31	Ga	3.2	2.9	66	Dy		
32	Ge	4.6	3.4	67	Ho		
33	As	5.3	4.5	68	Er		

(continued)

Table 1.11 (continued)

At. No.	Element	a	b	At. No.	Element	a	b
69	Tm			79	Au	5.77	3.46
70	Yb			80	Hg	4.91	5.54
71	Lu						
				81	Tl	3.2	2.9
72	Hf	3.8	3.0	82	Pb	3.90	3.53
73	Ta	4.11	3.79	83	Bi	4.69	3.74
74	W	4.40	3.58				
75	Re	4.02	3.87				
76	Os	4.9	3.8				
77	Ir	5.4	3.8				
78	Pt	5.6	3.5				

Source: Pearson, R. G. *Inorg. Chem.* **1988**, *27*, 734.

The importance of the b parameter can be seen by comparing the values for fluorine, chlorine, and bromine given in Table 1.11. Fluorine has the largest nominal electronegativity among these elements, since its valence p orbitals are the most stable of the three. However, its orbitals are also comparatively small in volume. Destabilizing repulsions increase rapidly with increased electron density; this is reflected in its large b value relative to Cl or Br. It is interesting to use equation (1.27) to calculate χ at various negative charges (Table 1.12). The results suggest that both chlorine and bromine are more electronegative than fluorine at charges beyond about -0.5, since their more diffuse valence orbitals are less destabilized by the additional electron density.

The importance of the interrelationships among size, ionization energy, electron affinity, and electronegativity cannot be overemphasized. All depend on essentially the same factors, primary among which is valence orbital energy. For example, consider an atom having at least one highly stable

Table 1.12 The calculated effect of charge on the electronegativities of fluorine, chlorine, and bromine

q	$\chi(F)$	$\chi(Cl)$	$\chi(Br)$	q	$\chi(F)$	$\chi(Cl)$	$\chi(Br)$
0.0	10.41	8.30	7.59	-0.3	6.20	5.49	5.06
-0.1	9.01	7.36	6.75	-0.4	4.80	4.56	4.21
-0.2	7.61	6.43	5.90	-0.5	3.40	3.62	3.37

Note: Calculations based on equation (1.27), using a and b values from Table 1.11.

valence orbital. A considerable input of energy will be needed to remove an electron residing in that orbital—that is, the ionization energy will be large. If the orbital is unoccupied (or contains only one electron), it can serve as an attractive "home" for an extra electron; hence, a large electron affinity will be observed. If the orbital is involved in chemical bonding, its stability will be attractive to the shared electrons, and a high electronegativity will result.

Bibliography

Note: For convenience in literature searching, the end-of-chapter bibliographies are organized in chronological order, with the most recent entry first. When you want to obtain more information about a given topic, the most recent sources normally provide the best perspective, and usually represent good jumping-off points via their references to the chemical literature.

Emsley, J. *The Elements*; Oxford University: New York, 1989.

Gerloch, M. *Orbitals, Terms, and Spectra*; Wiley: New York, 1986.

Matthews, P. S. C. *Quantum Chemistry of Atoms and Molecules*; Cambridge University: New York, 1986.

Puddephatt, R. J.; Monaghan, P. K. *The Periodic Table of the Elements*, 2nd ed.; Clarendon: Oxford, 1986.

Atkins, P. W. *Molecular Quantum Mechanics*; Oxford University: New York, 1983.

DeKock, R. L.; Gray, H. B. *Chemical Structure and Bonding*; Benjamin/Cummings: Menlo Park, CA, 1980.

Lehmann, W. J. *Atomic and Molecular Structure: The Development of Our Concepts*; Wiley: New York, 1972.

Day, M. C.; Selbin, J. *Theoretical Inorganic Chemistry*, 2nd ed.; Van Nostrand Reinhold; New York, 1969.

Questions and Problems

1. Give a reasonable set of four quantum numbers for the valence electron(s) of:
   (a) Cs   (b) Al   (c) Y   (d) Nb^{2+}

2. Make a rough sketch that includes both the angular and radial nodes for each of the following orbital types:
   (a) $2s$ (b) $2p$ (c) $3d$ (d) $4p$

3. Write a set of m_ℓ and m_s quantum numbers for each of the ten electrons of a filled $3d$ subshell.

4. Give the ground-state electron configuration for:
   (a) Se   (b) Se^{2-} (c) V   (d) V^{2+} (e) Rh   (f) Rh^{3+}

5. Explain in your own words why the $3s$, $3p$, and $3d$ orbitals are equal in energy for hydrogen, but unequal for all other atoms.

6. Give the atomic number of the first element expected to have an occupied g orbital in its ground-state configuration.

7. A potassium atom and a scandium(II) ion are isoelectronic, but have different ground states. Use the concepts of shielding and penetration to rationalize.

8. Rationalize the observation that, in size:
   (a) $Mg > Mg^{2+}$ (b) $Ca > Zn$ (c) $H^- > He$
   (d) $As^{3-} > Se^{2-}$

9. Arrange the following species in order of increasing size: As, Sb, Se, and Se^+.

10. Calculate the third ionization energy of lithium.

11. Calculate the energy of an electron transition from the $n = 1$ to the $n = 2$ level of He^+.

12. Determine the number of unpaired electrons in the ground state of:
    (a) Ge   (b) Se   (c) Co   (d) Co^{2+} (e) Mo
    (f) Mo^{2+} (g) Mo^{3+}

13. Write the Russell–Saunders term symbol for the ground state of:
    (a) O   (b) Be   (c) Br^- (d) Mn^{2+} (e) Cu^{2+}
    (f) Zn^{2+} (g) Ir^{3+} (h) Pr

14. What is the degeneracy of a ^2D state? ^4P?

15. The d^2 configuration comprises five energy states, designated ^1S, ^1D, ^1G, ^3P, and ^3F in Russell–Saunders notation. Give the degeneracy of each state. Which is the ground state? Explain.

16. Arrange the $3d$ orbitals of Ni, Cu, and Cu^{2+} in order of increasing:
    (a) Size   (b) Stability   (c) Electronegativity

17. Decide which element has the greater second ionization energy, and defend your answer:
    (a) Li or Be   (b) B or C   (c) C or N   (d) Na or Mg

18. For each of the following pairs, choose the electron affinity that is the more thermodynamically favored; defend your answers.
    (a) EA_1 versus EA_2 of C   (b) EA_1 of Si versus EA_1 of Ge
    (c) EA_1 of Ge versus EA_1 of As

19. The periodic table is steadily being expanded by the synthesis of new elements.
    (a) Element 105 has been assigned the symbol Unp. What is its expected ground-state electron configuration?
    (b) What will be the atomic number of the next member of Group 15?
    (c) What will be the atomic number of the next alkali metal?

20. It is generally agreed that in electronegativity, $C > Si < Ge$ (see Table 1.10). Explain why this is true.

21. Use either of the electronegativity scales given in Table 1.10 to construct a plot of χ versus VSIE for the first 18 elements. What conclusions can be drawn from your plot?

22. Use the data given in Table 1.7 to graphically estimate the charge corresponding to minimum energy for oxygen.

23. The b values of boron and silicon are quite similar. Explain why.

24. Consider the compound ClF. Using the electronegativity values of a and b given in Table 1.11, at what charge distribution should $\chi(Cl) = \chi(F)$?

25. (a) At what charge does a fluorine atom have the same electronegativity as neutral iodine?
   (b) At what negative charge does iodine become the most electronegative halogen?
   (c) Is iodine likely to be found in compounds having the charge calculated in part (b)? Why not?

*26. L. C. Allen has proposed a three-dimensional periodic table, with the third dimension being electronegativity (*J. Am. Chem. Soc.* **1989**, *111*, 9003). After reading at least the first five pages of Allen's article, answer the following questions:
   (a) According to Allen, what is the most electronegative element? How does this relate to this chapter?
   (b) Boron and silicon have many chemical similarities. Since they belong to different families, this is not immediately obvious from a conventional periodic table. How about Allen's version?
   (c) Figure 2 in Allen's article identifies a "metalloid band." What do you notice about the elements located above that band compared to those located below?

*27. M. Laing has suggested a new arrangement of the periodic table (*J. Chem. Educ.* **1989**, *66*, 746). Answer the following questions after reading his arguments.
   (a) Laing's periodic table places carbon, silicon, titanium, and zirconium in the same column. How is this justified on the basis of electron configurations?
   (b) The formulas of the common binary chlorides of beryllium and magnesium are $BeCl_2$ and $MgCl_2$. Predict the formula for the binary chloride of cadmium.
   (c) What (if any) value do you see in aligning F and Cl with Mn, Tc, and Re?

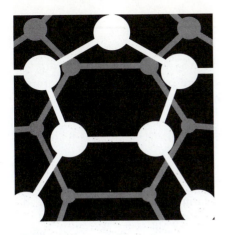

2

An Introduction to Symmetry and Group Theory

Symmetry is a common concept in everyday life, and all of us can differentiate between symmetric and nonsymmetric objects. But in chemistry it is useful to be more specific—that is, we want to know how a given species is symmetric. Certainly, the symmetry of the methane molecule is different from that of H_2O, although each is symmetric in its own way. The basic components—that is, the *elements* and *operations*—of symmetry are described in Section 2.1. Then, in the remainder of the chapter, some of the applications of this topic are explored.

2.1 Symmetry Operations and Elements

A *symmetry operation* carries an object from one position into another, equivalent position. A *symmetry element* is a geometric entity—a point, line, or plane—about which such an operation may be performed. As an example, consider a molecule of boron trifluoride. The boron atom lies at the center of an equilateral triangle of fluorines. If an imaginary rod (line) is inserted perpendicular to the plane of the triangle and passing through the boron, it can serve as a symmetry element, as shown:

$$F—B\begin{matrix} \cdot\cdot F \\ \\ F \end{matrix}$$

Equivalent positions are reached by rotating the rod by $120°$, $240°$, and $360°$; each of these acts is a symmetry operation.

The three most common types of symmetry elements are examined below.

Rotational Axes

A *rotational axis* (sometimes called a *proper axis of rotation*) is an imaginary line, rotation about which relates two or more equivalent positions of an object (as in the BF_3 molecule shown above). The symbol normally used for a rotational axis is C_n for rotation by $360°/n$. That is, rotation by $180°$ corresponds to a C_2 axis, rotation by $120°$ to a C_3 axis, rotation by $90°$ to a C_4 axis, etc. These are illustrated, using disks marked with circles in appropriate positions, in Figure 2.1. It can be seen that a rectangle has three

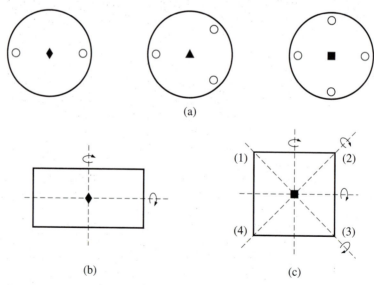

Figure 2.1 Illustrations of rotational axes. (a) Disks containing two- (◆), three- (▲), and fourfold (■) axes. (b) The three C_2 axes of a rectangle. The third axis is depicted by (◆) and is perpendicular to the plane of the paper. (c) One C_4 (■) and four C_2 axes of a square. The corners are numbered to show the results of the various rotations (see text).

C_2 axes, mutually perpendicular to one another. A square has one C_4 and four C_2 axes.

When an object has two or more rotational axes, one of them is normally perpendicular to all the others (though exceptions are found in certain highly symmetric bodies, as we will discuss below). In such cases, the axis of highest order is described as the *major axis*, and the others are C_2 axes perpendicular to C_{major}. It is useful to know that if an object has a major axis C_n and at least one perpendicular C_2, there are exactly n such C_2's.[1] (Note the square in Figure 2.1c, where a C_4 plus one C_2 axis correctly predicts a total of four C_2's.)

Any rotational axis of order three or greater actually implies more than one operation. For example, the presence of a C_4 axis requires that rotation by 90°, 180°, 270°, and 360° all must result in equivalent positions. By convention, rotations are always performed in the clockwise direction. Superscripts are used as needed to differentiate among the operations. Thus, for the C_4 series, rotation by 90° is symbolized by C_4^1 (rotation by $\frac{1}{4}$ of 360°); rotation by 180° is C_4^2 (equivalent to C_2); and rotation by 270° is C_4^3.

A special symbol is reserved for rotation by 360°: $C_4^4 = C_1 = E$. Obviously, every object arrives at an equivalent position when rotated by 360°. This is known as the *identity operation*, and while it may seem trivial it is necessary to include it for many purposes.

The corners of the square in Figure 2.1c are numbered to clarify the different effects of the C_4^n operations. The results for corner (1) are

$$C_4^1: \qquad (1) \rightarrow (2)$$
$$C_4^2 = C_2: \quad (1) \rightarrow (3)$$
$$C_4^3: \qquad (1) \rightarrow (4)$$
$$C_4^4 = E: \quad (1) \rightarrow (1)$$

Reflection (Mirror) Planes

A *mirror plane* (or *plane of reflection*) relates equivalent points on opposite sides of itself; this symmetry element is symbolized by σ. The act of reflection normally changes the sign of one of the three Cartesian coordinates of a given point. For example, reflection in the xy plane converts point (x, y, z) to $(x, y, -z)$.

---

1. For the logic behind this relationship, see Cotton, F. A. *Chemical Applications of Group Theory*, 3rd ed.; Wiley: New York, 1990; pp. 25–26.

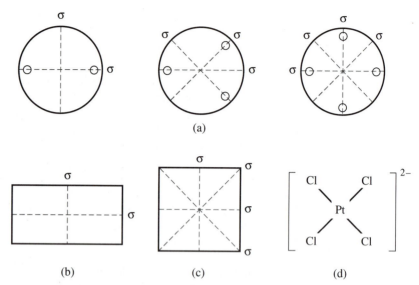

Figure 2.2   Illustrations of mirror planes. (a) Disks containing two, three, and four mirror planes. (b) The mirror planes of a rectangle. (c) The mirror planes of a square. (d) The square planar $[PtCl_4]^{2-}$ anion contains a C_4 rotational axis and four vertical mirror planes, two of which are dihedral (see text). Each figure also contains a mirror plane in the plane of the paper.

The disks of Figure 2.2a contain two, three, and four mirror planes, respectively. A rectangle has two perpendicular mirror planes, while a square has a total of four σ's.[2]

For bodies that contain both a rotational axis and one or more mirror planes, the plane is referenced to the axis. The object is assumed to be oriented so that the major axis is vertical. If there is a mirror plane perpendicular to that axis, it will lie in a horizontal position; such a plane is symbolized σ_h. Alternatively, the plane may contain the major axis, in which case it is a vertical plane, σ_v. Some thought should convince you that an object can have only one σ_h, but more than one σ_v.[3] In fact, bodies containing both a major axis C_n and at least one σ_v always have exactly n vertical planes. For example, the square planar complex $[PtCl_4]^{2-}$ shown in Figure 2.2d has a major C_4 axis contained by at least one vertical mirror plane; you should be able to verify that there are exactly four such planes. Two contain the Cl–Pt–Cl bond axes, while the other two bisect the dihedral

2.   Actually, any two-dimensional object has a mirror plane passing through all its points (in Figure 2.2, the plane of the paper). If this is included, then a rectangle has three and a square has five σ's.

3.   Except for the Platonic solids, which have more than one axis of high order; therefore, they may also have more than one σ_h (see Section 2.3).

angles formed by adjacent Pt–Cl bonds; the latter pair are sometimes denoted as σ_d's (for dihedral).

Centers of Inversion

An *inversion center* (symbolized by *i*) is a point—the geometric center of an object. If the translation of each point through the center of a body to an equal distance on the opposite side arrives at a symmetrically equivalent point, then that body has a center of inversion. The net effect is to convert any point (x, y, z) to $(-x, -y, -z)$. In Figure 2.3a, the first and third disks possess inversion centers, while the second does not (see figure caption).

A square and a rectangle each has a center of inversion, but a triangle does not. It should be obvious that an object can have no more than one inversion center.

Improper Rotational Axes

One other symmetry element must be mentioned in addition to those just described. An *improper axis of rotation* (S_n for rotation by $360°/n$) is, in a sense, a combined operation—rotation by C_n followed by reflection across a perpendicular plane:

$$C_n \times \perp\sigma \rightarrow S_n \tag{2.1}$$

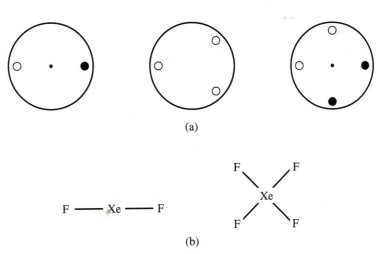

(a)

(b)

Figure 2.3 Illustrations of inversion centers. (a) The disks containing two and four circles have centers of inversion, but the one with three circles does not (○ signifies a circle on the top side and ● a circle on the bottom side of a disk). (b) Two binary fluorides of xenon, linear XeF_2 and square planar XeF_4, have centers of inversion.

Figure 2.4   Proper versus improper rotations of a rectangle. The proper C_2 rotation interconverts positions $(1)\leftrightarrow(4)$ and $(2)\leftrightarrow(3)$, while the improper S_2 axis interconverts $(1)\leftrightarrow(3)$ and $(2)\leftrightarrow(4)$. (Note that $S_2 = C_2 \times \sigma = i.$)

A tetrahedron is an example of an object having an improper axis. The sequence shown in equation (2.2) arrives at an equivalent position, while proper rotation (C_4) does not:

$$ \xrightarrow{\;C_4\;} \xrightarrow{\;\perp\sigma\;} \tag{2.2}$$

Therefore, the axis in question is of the S_4, rather than the C_4, type. Notice that the C_n and σ_h operations need not be present independently in order for the combined S_n operation to exist.

If you look closely at equation (2.2), you should be able to find a C_2 axis lying along the same line as S_4. This can be generalized: The presence of S_n (provided that n is an *even* integer) requires the existence of a coaxial $C_{n/2}$.

A rectangle with numbered corners provides another example for differentiating between proper and improper axes. As is shown in Figure 2.4, one of the C_2 axes interconverts corners $(1)\leftrightarrow(4)$ and $(2)\leftrightarrow(3)$, while an S_2 operation along the same axis interconverts $(1)\leftrightarrow(3)$ and $(2)\leftrightarrow(4)$. It also should be apparent that the inversion operation yields the same result, $(1)\leftrightarrow(3)$ and $(2)\leftrightarrow(4)$, as does the improper rotation. Thus, for the special case of the twofold improper axis, $S_2 = i$.

2.2   Point Groups

It is possible to organize objects into groups such that the members of any group have identical symmetry properties. Consider three objects: the letter E, a rectangular box having a bottom but no top, and a model of the water molecule. All of these objects possess the same four symmetry elements: E, C_2, and two σ_v's. Thus, they belong to the same *point group*.

In theory, there are an infinite number of point groups, but only about 20 are commonly encountered in chemistry. Each point group is identified by a capital letter, often accompanied by a number or number–letter subscript. For example, the three objects listed above all belong to the C_{2v} group.

It is important to be able to determine the point group of any molecule or other object. A systematic procedure for doing so is given below. The scheme begins by identifying all proper rotational axes of order greater than one. There must be one, more than one, or none at all. Thus, we begin with section A; if A does not apply, move to section B; etc. Within each section, we begin with number 1; if that does not apply, then move to number 2; and so forth.

A. *One and Only One C_n:*

1. Look for a mirror plane perpendicular to C_n (that is, a σ_h). If one is present, the point group is C_{nh}.

2. Look for a vertical mirror plane. If you find one, look next for an S_{2n} axis colinear with C_n (for C_2 look for S_4, etc.). If S_{2n} is present, then you have missed at least two perpendicular C_2's; the point group is D_{nd}.

3. If there are σ_v's but no S_{2n}, then the group is C_{nv}.

4. In the absence of any mirror planes or improper rotations the point group is C_n.

B. *More Than One C_n:*

1. For bodies having two or more axes of order three or greater:
    a. If there are six C_5 axes, the point group is I_h (for icosahedral) if an inversion center is present. It is I if there is no inversion center.
    b. If there are three C_4 axes, the point group is O_h (for octahedral) if i is present, and O if there is no inversion center.
    c. If there are four C_3 axes, the point group is T_d (for tetrahedral) if i is present, and T if there is no inversion center.

2. For bodies having one C_n ($n > 2$) plus n C_2 axes, or three perpendicular C_2's:
    a. The presence of a mirror plane perpendicular to C_{major} identifies the point group as D_{nh}.
    b. If there is a mirror plane that contains C_{major}, the point group is D_{nd}.
    c. The point group is D_n in the absence of any mirror planes.

C. *No Proper Rotational Axes Present:*

1. If there is an improper axis of order n ($n > 2$), the point group is S_n.

2. If a mirror plane is present, the point group is C_s.

3. If there is an inversion center, the point group is C_i.

4. In the absence of S_n, σ, and i, the group is C_1.

A flowchart for this procedure is given in Figure 2.5.

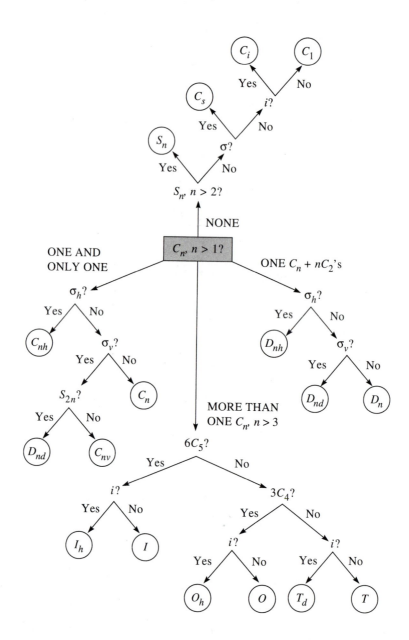

Figure 2.5
Flowchart for the assignment of point groups. The procedure begins by going in one of the four indicated directions leading away from the central rectangle.

Some examples are provided below to illustrate the use of this method.

1. **The Letter**: A

   There is one and only one C_2 axis. No perpendicular mirror plane is present, but there are two σ_v's; one bisects the letter and the other lies in the plane (ie, contains all the points) of the letter. There is no S_4 axis colinear with C_2. The point group is C_{2v}.

2. **A Light Bulb**:

   One rotational axis is present, and rotation by any amount about that axis yields an equivalent position—the axis is C_∞. There is no σ_h, but an infinite number of σ_v's can be found. The point group is $C_{\infty v}$. Can you think of any molecules that have these same symmetry properties (and hence also belong to the $C_{\infty v}$ group)?

3. **The Linear Dibromoaurate(I) Anion**: [Br–Au–Br]⁻

   There is one major axis (C_∞) and an infinite number of perpendicular C_2's. The mirror plane perpendicular to C_{major} identifies the point group as $D_{\infty h}$.

4. **Fluorogermane, GeH₃F**:

   One rotational axis (C_3) is present. There is no σ_h, but three σ_v's. Colinear S_6 is not present, so the point group is C_{3v}.

5. **Arsenic Pentabromide, AsBr₅ (a Trigonal Bipyramid)**:

   The rotational axes include one C_3 (the major axis) and three perpendicular C_2's. A horizontal mirror plane is present, so the point group is D_{3h}.

6. **The Trans Rotamer of Disilane, Si₂H₆**:

   The trans orientation yields a major axis (C_3) plus three perpendicular C_2's. There are three vertical,[4] but no horizontal mirror planes. The

---

4. More precisely, there are three dihedral planes; thus, the point group notation is D_{nd} rather than D_{nv}.

point group is therefore D_{3d}. The two other rotameric forms, gauche and eclipsed, belong to different point groups. (What are they?)

7. **The Crown-Shaped S$_8$ Molecule:**

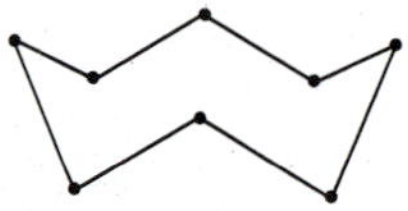

This species contains a major C_4 axis and four C_2's, each bisecting two of the S–S bonds. (There are also improper axes—can you identify them?) There is no σ_h, but four other mirror planes are present. The point group is D_{4d}.

8. **The Tetrachloroplatinate(II) Ion:**

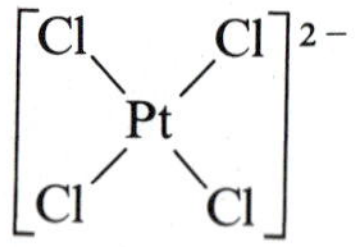

This planar anion contains the same symmetry elements as a square; they include C_4, four C_2's, σ_h, four σ_v's, and i. The point group is D_{4h}.

Other examples are presented systematically by point group in Table 2.1.

Table 2.1 The point groups of some inorganic species

Point Group	Example	Point Group	Example
C_1	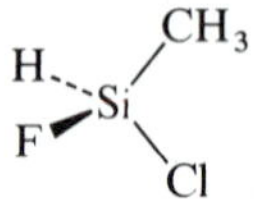	C_{3h}	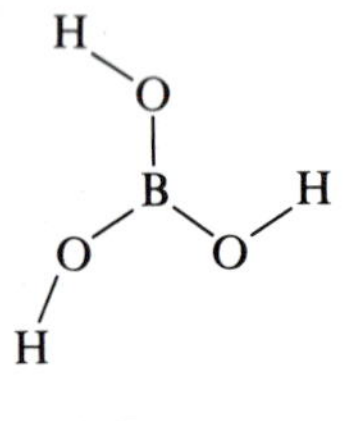
C_s	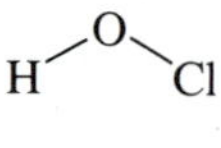		
		C_{3v}	
C_2	(Nonplanar conformation)		
		C_{4v}	
C_{2h}	(Trans rotamer)		
C_{2v}	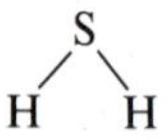	D_{2h}	

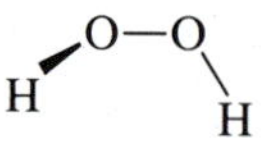

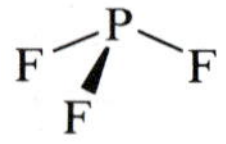

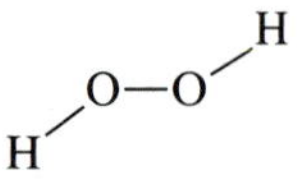

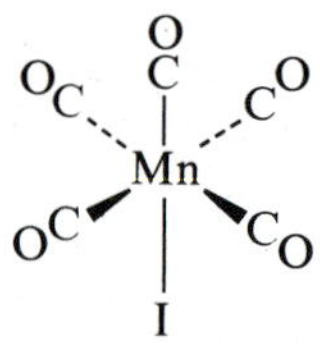

(continued)

Table 2.1 (coninued)

Point Group	Example	Point Group	Example
D_{2d}	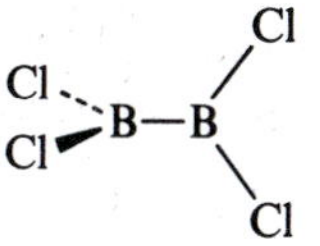 (Staggered conformation)	T_d	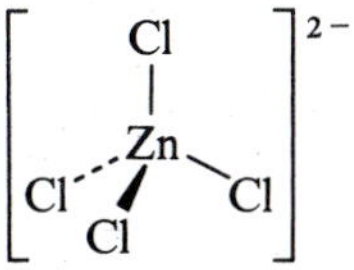
D_{3h}	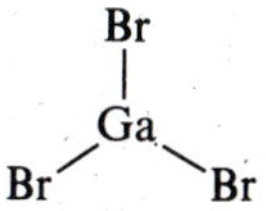	O_h	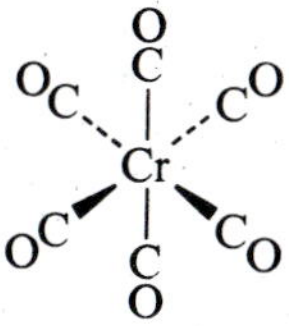
D_{3d}	$S_2O_6^{2-}$ 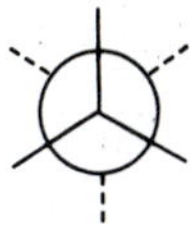 (Staggered conformation)	I_h	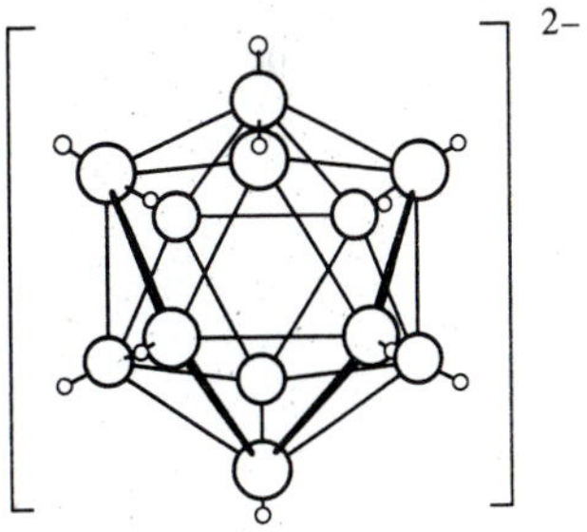 ($\bigcirc$ = BH) $B_{12}H_{12}^{2-}$ (Icosahedral)
D_{4h}			
D_{4d}		$C_{\infty v}$	$[N{-}C{-}O]^-$
	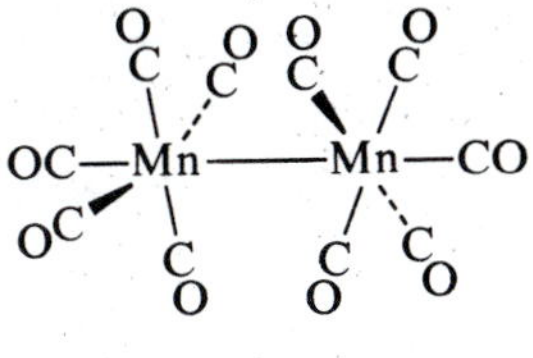	$D_{\infty h}$	$[I{-}Hg{-}I]^-$
D_{6h}	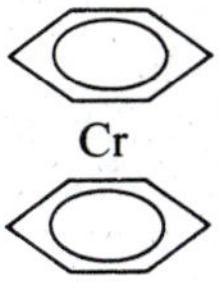 (Eclipsed conformation)		

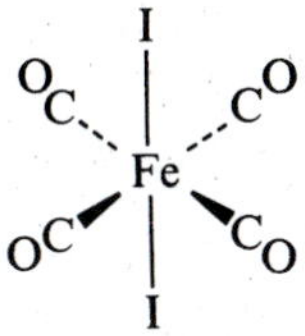

2.3 The Platonic Solids

As was implied earlier, certain highly symmetric objects have more than one axis of order three or greater. Plato recognized that there are five solid figures (polyhedra) for which all faces, edges, and vertices are symmetrically equivalent: the tetrahedron, cube, octahedron, pentagonal dodecahedron, and

| Tetrahedron | Cube | Octahedron | Pentagonal dodecahedron | Icosahedron |

Figure 2.6   The five Platonic solids.

Table 2.2   Number of faces, corners, and edges for the Platonic solids

Polyhedron	Faces	Corners	Edges
Tetrahedron	4 (triangles)	4	6
Cube	6 (squares)	8	12
Octahedron	8 (triangles)	6	12
Dodecahedron	12 (pentagons)	20	30
Icosahedron	20 (triangles)	12	30

icosahedron. These *Platonic solids* are pictured in Figure 2.6, and their structural characteristics are summarized in Table 2.2.

It is interesting to consider some of the interrelationships among these polyhedra. If the centers of the faces of an octahedron are connected, the result is a cube. The reverse is also true; that is, connecting the face centers of a cube creates an octahedron (Figure 2.7a). Thus, the octahedron and cube belong to the same point group, O_h. The pentagonal dodecahedron and icosahedron bear this same type of relationship, and both belong to the I_h group. Connecting the face centers of a tetrahedron merely gives a smaller tetrahedron. The relationship between a cube and a tetrahedron is shown in

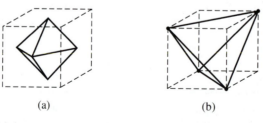

(a)   (b)

Figure 2.7   Interconversions of polyhedra. (a) Connecting the face centers of a cube forms an octahedron. (b) Connecting four corners of a cube produces a tetrahedron.

Figure 2.7b, where we can see that a tetrahedron is formed by connecting four of the eight corners of a cube.

Many important inorganic species have structures derived from the Platonic solids. For example, a series of divalent anions having the general formula $B_nH_n^{2-}$ is known. The $B_6H_6^{2-}$ anion is octahedral, while $B_{12}H_{12}^{2-}$ is icosahedral, as shown in Table 2.1.

2.4 The Combination of Operations: Multiplication Tables

It is often desirable to consider the overall result of two symmetry operations performed in sequence. This is the group theory equivalent to multiplication.

To illustrate, consider the rectangle in Figure 2.8, where the corners have been numbered to permit identification. If the operations labeled C_2 and C_2' are performed in sequence, the end result is to reverse positions $(1) \leftrightarrow (3)$ and $(2) \leftrightarrow (4)$. This same result could have been achieved from the single operation i. We therefore say that the *product* of C_2 followed by C_2' is i for this system:

$$C_2 \times C_2' = i \tag{2.3}$$

In the same manner, the product of C_2 and i is C_2', since both result in the interconversions $(1) \leftrightarrow (4)$ and $(2) \leftrightarrow (3)$:

$$C_2 \times i = C_2' \tag{2.4}$$

A *multiplication table* gives the products of all possible two-sequence combinations for a given point group. For example, consider the C_{2h} point group. Objects belonging to this point group possess four symmetry elements (ie, the group has an *order* of four): E, C_2, i, and σ_h. A molecule having C_{2h} symmetry is the trans rotamer of hydrogen peroxide. In Figure 2.9 the C_2 axis is vertical and in the plane of the paper; σ_h must therefore be perpendicular to that plane and pass through the center of each atom. The dots at the

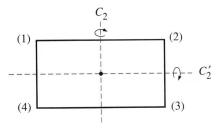

Figure 2.8 Examples of combined operations. The sequence $C_2 \times C_2'$ interconverts positions $(1) \leftrightarrow (3)$ and $(2) \leftrightarrow (4)$, while the sequence $C_2 \times i$ interconverts $(1) \leftrightarrow (4)$ and $(2) \leftrightarrow (3)$.

Figure 2.9
The orientations of the C_{2h} symmetry elements and the effects of the corresponding operations on the trans rotamer of H_2O_2.

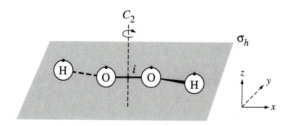

tops of the atoms in Figure 2.9 help clarify the different effects of C_2, i, and σ_h. The C_2 operation exchanges the positions of the two hydrogen and the two oxygen atoms, but leaves the dot at the top of each atom. Inversion also transposes the like atoms, but moves the dots to the bottoms. The horizontal reflection leaves all atoms in place, moving the dots from top to bottom. Hence, each operation is seen to have a unique effect.

The following products of the 16 two-sequence combinations of operations for the C_{2h} point group can be determined with the aid of Figure 2.9:

$$
\begin{array}{llll}
E \times E = E & C_2 \times E = C_2 & i \times E = i & \sigma_h \times E = \sigma_h \\
E \times C_2 = C_2 & C_2 \times C_2 = E & i \times C_2 = \sigma_h & \sigma_h \times C_2 = i \\
E \times i = i & C_2 \times i = \sigma_h & i \times i = E & \sigma_h \times i = C_2 \\
E \times \sigma_h = \sigma_h & C_2 \times \sigma_h = i & i \times \sigma_h = C_2 & \sigma_h \times \sigma_h = E
\end{array}
$$

All of this information can be condensed into the group multiplication table shown in Table 2.3. Each operation appears exactly once in each row and once in each column of the table; this is always the case for a true group.

Note that the C_{2h} group *commutes*; that is, the order in which two operations are carried out does not alter the end result. For any two operations A and B, $A \times B = B \times A$. This is not the case for all groups, so we follow the convention that the vertical column heading of a multiplication table identifies the operation to be performed first, and the horizontal row heading is the operation to be performed second.

Table 2.3   Multiplication tables for the C_{2h} and C_{3v} point groups

C_{2h}	E	C_2	i	σ_h
E	E	C_2	i	σ_h
C_2	C_2	E	σ_h	i
i	i	σ_h	E	C_2
σ_h	σ_h	i	C_2	E

C_{3v}	E	C_3^1	C_3^2	σ_v	σ_v'	σ_v''
E	E	C_3^1	C_3^2	σ_v	σ_v'	σ_v''
C_3^1	C_3^1	C_3^2	E	σ_v''	σ_v	σ_v'
C_3^2	C_3^2	E	C_3^1	σ_v'	σ_v''	σ_v
σ_v	σ_v	σ_v'	σ_v''	E	C_3^1	C_3^2
σ_v'	σ_v'	σ_v''	σ_v	C_3^2	E	C_3^1
σ_v''	σ_v''	σ_v	σ_v'	C_3^1	C_3^2	E

The C_{3v} point group has six elements: E, C_3^1, C_3^2, σ_v, σ_v', and σ_v''. An example of a molecule having C_{3v} symmetry is the pyramidal species NH_3. The group multiplication table is given in Table 2.3; it is readily apparent that it does not commute.

The six symmetry elements can be subdivided into three *classes*.[5] The identity element makes up one class, the two rotations a second class, and the mirror planes a third. By convention they are often listed in the shorthand notation E, $2C_3$, $3\sigma_v$.

2.5 Character Tables

A *character table* contains, in a highly symbolic form, information about how something of interest (an orbital, a bond, etc.) is affected by the operations of a given point group. Each point group has a unique character table, which is organized into a matrix. The column headings are the symmetry operations, grouped into classes. The horizontal rows are called the *irreducible representations* of the point group. The main body consists of characters (numbers), and a section on the right side of the table provides information about vectors and atomic orbitals. The character table for the C_{2h} group is given in Table 2.4.[6]

For our purposes it will not be necessary to derive the characters of any character table, although it is possible to do so.[7] We will want to be able to use these tables, however, so some explanation is in order. Character tables are organized according to certain conventions, the first of which has already been discussed: The major rotational axis defines the z direction. Among the numerical characters, a value of 1 indicates "symmetry" or "no change" with respect to the operation heading that column; -1 indicates

Table 2.4 The character table for the C_{2h} point group

C_{2h}	E	C_2	i	σ_h		
A_g	1	1	1	1	R_z	z^2, xy, x^2, y^2
B_g	1	-1	1	-1	R_x, R_y	xz, yz
A_u	1	1	-1	-1	z	
B_u	1	-1	-1	1	x, y	

5. For a summary of the distinguishing features of classes, see Cotton, F. A. *Chemical Applications of Group Theory*, 3rd ed.; Wiley: New York, 1990; pp. 13–16.

6. A set of character tables is given in Appendix IV.

7. For the method, see a source such as Cotton, F. A. *Chemical Applications of Group Theory*, 3rd ed.; Wiley: New York, 1990; Chapter 4; or Hatfield, W. E.; Parker, W. E. *Symmetry in Chemical Bonding and Structure*; Merrill: Columbus, OH, 1974; pp. 34*ff*.

Figure 2.10
The $+x$ to $-x$ vector in C_{2h} symmetry. The E and σ_h operations have no effect on this vector, while C_2 and i invert it.

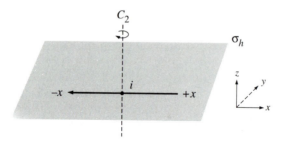

"asymmetry" or "reversal." To illustrate, consider a vector v_x running from points $+x$ to $-x$ of a Cartesian system (Figure 2.10). The C_{2h} point group symmetry is imposed on the vector such that the focus of each element is at the origin. That is, the C_2 axis passes through the origin, the horizontal mirror plane contains the origin, and the inversion center is at the origin. The σ_h axis therefore lies in the xy plane.

The identity operation does not change either the location or direction of v_x, of course, so the character 1 is assigned for that operation. The effect of C_2 is to invert the vector (reverse its direction); thus, the appropriate character for that column is -1. The i operation also gives an asymmetric result (-1), while σ_h has no effect (symmetric, character $= 1$). Reading across, then, the representation is:

	E	C_2	i	σ_h
v_x	1	-1	-1	1

This matches the representation labeled B_u in the C_{2h} character table. We therefore say that v_x "transforms into the B_u representation" or "has B_u symmetry" in the C_{2h} point group.

The headings for the horizontal rows are *Mulliken symbols*, and are assigned according to the following general rules:[8]

1. An A or B representation is singly *degenerate* (has a character of 1 in the identity column), an E representation is doubly degenerate (2 in the E column), and a T representation is triply degenerate. (The significance of degenerate representations will be discussed shortly.) The A symbol is used for symmetry with respect to C_{major} (ie, a character of 1 in the C_{major} column) and B for asymmetry toward C_{major} (-1 in that column).

8. Mulliken, R. S. *Phys. Rev.* **1933**, *43*, 279.

2. The subscripts $_g$ and $_u$ are abbreviations of the German words *gerade* (even) and *ungerade* (odd). They refer to symmetry and asymmetry with respect to inversion (1 and -1 in the i column, respectively).

3. Numerical subscripts relate to perpendicular C_2 axes in the D_n, D_{nh}, and D_{nd} point groups (1 for symmetry toward such operations and 2 for asymmetry). If the point group has vertical mirror planes but no secondary axes (C_{nv}), the same subscripts show symmetry or asymmetry with respect to σ_v.

4. Primes (′) and double primes (″) designate symmetry and asymmetry, respectively, toward horizontal mirror planes.

You should convince yourself that the Mulliken symbols given for the irreducible representations of the C_{2h} character table are consistent with these rules. (The primes and double primes are omitted in Table 2.4 because the representations have unique designations without them.)

We will now consider the symmetry properties of a set of three atomic p-type orbitals, again using the C_{2h} character table. If the geometric center of these orbitals (ie, the atomic nucleus) is placed at the origin, into what irreducible representation does each orbital transform? It should be apparent that p_x has the same symmetry properties as the $+x$ to $-x$ vector discussed earlier; that is, p_x belongs to the B_u representation. Here, the $1/(-1)$ convention can be applied to the sign of the wave amplitude (1 if unchanged and -1 if reversed by an operation). You should be able to determine that p_y also belongs to B_u, while p_z transforms as A_u. This provides a partial explanation for the far right section of the character table (Table 2.4), in which the symmetry properties of p- and d-type orbitals are indicated by the rows in which they are located.[9] Spherical s-type orbitals are totally symmetric, and so belong to the first irreducible representation of any point group (A_g in C_{2h}).

Next, we will examine a more complex system. The D_{3h} point group has an order of 12; the symmetry classes are E, $2C_3$, $3C_2$, σ_h, $2S_3$, and $3\sigma_v$. A character table for this group is given in Table 2.5 (p. 62).

Recalling that C_{major} (in this case, C_3) defines the z axis, it is easily determined that a p_z orbital transforms as A_2''. For p_x and p_y, however, a problem arises. For each of these, the C_3 operation neither leaves the orbital unchanged nor inverts it—thus, neither the character 1 nor -1 applies. The p_x and p_y orbitals are degenerate in C_{3v} symmetry. They must therefore be taken together; that is, the effects of the various operations

---

9. The symbols R_x, R_y, and R_z are also found in the right-hand section of Table 2.4. They refer to rotations about vectors oriented along the x, y, and z axes. This information is useful for certain purposes, such as in Raman spectroscopy.

Table 2.5 The character table for the D_{3h} point group

D_{3h}	E	$2C_3$	$3C_2$	σ_h	$2S_3$	$3\sigma_v$		
A_1'	1	1	1	1	1	1		$x^2 + y^2, z^2$
A_2'	1	1	-1	1	1	-1	R_z	
E'	2	-1	0	2	-1	0	(x, y)	$(x^2 - y^2, xy)$
A_1''	1	1	1	-1	-1	-1		
A_2''	1	1	-1	-1	-1	1	z	
E''	2	-1	0	-2	1	0	(R_x, R_y)	(xy, yz)

on them as a paired set can be determined. The resulting irreducible representation then has a character of 2 in the E column (see rule 1, above). The characters for $2C_3$ and $3C_2$ can be obtained with the aid of Figure 2.11, where we can see that rotation by 120° leaves a component of each orbital along the original axes. Specifically, if x' and y' are the new components, then:

$$\text{For } p_x\colon \quad x' = x(\cos 120°) \quad = -0.500$$
$$y' = x(-\sin 120°) = -0.866$$
$$\text{For } p_y\colon \quad x' = y(\sin 120°) \quad = 0.866$$
$$\underline{y' = y(\cos 120°) \quad = -0.500}$$
$$= -1.000$$

The character for $2C_3$ is therefore -1.

For a perpendicular C_2 rotation, the specifics depend on the orientation of the axis chosen. For example, if C_2 is placed along the x axis and C_2' along y, then:

$$\text{For } p_x\colon \quad x' = x(\cos 0°) \quad = +1.000$$
$$y' = x(-\sin 0°) = 0.000$$
$$\text{For } p_y\colon \quad x' = y(\sin 180°) = 0.000$$
$$y' = y(\cos 180°) = -1.000$$

The character is seen to be $+1.000 + (-1.000) = 0.$[10]

Similar reasoning may be applied to the remaining operations of the group. You should verify that the complete representation is given by:

	E	$2C_3$	$3C_2$	σ_h	$2S_3$	$3\sigma_v$	
$\Gamma(p_{x,y})$	2	-1	0	2	-1	0	$= E'$

---

10. It is somewhat more difficult to show (but equally true) that the character is also 0 if C_2 lies in any direction between these axes.

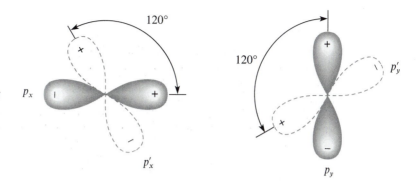

Figure 2.11
The effect of the C_3^1 operation on the degenerate $p_{x,y}$ pair in D_{3h} symmetry. Rotation by 120° converts p_x to the position labeled p'_x, and p_y to p'_y.

Because the p_x and p_y orbitals are mixed by symmetry, they transform together into a doubly degenerate irreducible representation. In this manner the existence and implications of such representations (sometimes referred to as representations having a dimensionality of two) can be understood. You should be able to show that there is triple degeneracy (T representation, with all three p orbitals mixed by symmetry) in the O_h point group.

2.6   Reducible Representations

Occasionally, a representation is encountered that has a dimensionality greater than one, yet is not one of the irreducible representations of the point group. It is often possible to resolve (as in resolving a vector into its components), or *reduce*, such a representation into two or more irreducible representations. As an example, consider the representation $\Gamma_{r(1)}$ in C_{2h} symmetry:

	E	C_2	i	σ_h
$\Gamma_{r(1)}$	2	0	2	0

It is apparent from inspection of Table 2.4 that $\Gamma_{r(1)}$ can be reduced to $A_g + B_g$. In the same way, a moment's reflection should convince you that for $\Gamma_{r(2)}$:

	E	C_2	i	σ_h	
$\Gamma_{r(2)}$	3	1	1	3	$= 2A_g + B_u$

For a more complex example, consider the planar BF_3 molecule (D_{3h} point group, Table 2.5). Imagine that we are seeking symmetry information about the three B–F bonds.[11] We can generate the three-dimensional reducible representation $\Gamma_{r(3)}$ by treating the bonds as vectors:

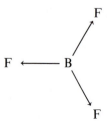

Then we determine how these vectors are affected by the various D_{3h} symmetry operations, according to the following code:

No effect $\rightarrow$ Character of 1

Inversion (reversal of vector direction) $\rightarrow$ -1

Translation (movement of vector) $\rightarrow$ 0

The character of the identity operation for this set of vectors is, of course, 3. For the C_3^1 and C_3^2 operations, all three vectors are moved; hence, the character of the $2C_3$ column is 0. Each C_2 axis lies along one of the vectors. For any of these rotations, the vector that coincides with the C_2 axis is unchanged while the other two are moved; the appropriate character is thus 1. By continuing this treatment, the complete representation is found to be:

	E	$2C_3$	$3C_2$	σ_h	$2S_3$	$3\sigma_v$
$\Gamma_{r(3)}$	3	0	1	3	0	1

With some effort (perhaps by trial and error) you should be able to determine that $\Gamma_{r(3)}$ reduces to $A_1' + E'$.

A more systematic method for reducing complex representations is available. It utilizes the equation

$$n = \frac{\chi(r) \cdot \chi(i) \cdot c}{g} \tag{2.5}$$

where n is the number of times a given irreducible representation appears, $\chi(r)$ is one of the characters of the reducible representation, $\chi(i)$ is the

11. Specific reasons for why such information might be useful, as well as some worked examples, will be provided in Chapter 3.

corresponding character of the irreducible representation, c is the order of the class (the coefficient of the column heading), and g is the order of the group. For $\Gamma_{r(3)}$, then,

$$n(A_1') = \frac{(3\cdot1\cdot1) + (0\cdot1\cdot2) + (1\cdot1\cdot3) + (3\cdot1\cdot1) + (0\cdot1\cdot2) + (1\cdot1\cdot3)}{12}$$

$$= \frac{12}{12} = 1$$

$$n(A_2') = \frac{(3\cdot1\cdot1) + (0\cdot1\cdot2) + (1\cdot(-1)\cdot3) + (3\cdot1\cdot1) + (0\cdot1\cdot2) + (1\cdot(-1)\cdot3)}{12}$$

$$= \frac{0}{12} = 0$$

$$n(E') = \frac{(3\cdot2\cdot1) + (0\cdot(-1)\cdot2) + (1\cdot0\cdot3) + (3\cdot2\cdot1) + (0\cdot(-1)\cdot2) + (1\cdot0\cdot3)}{12}$$

$$= \frac{12}{12} = 1$$

and so on. We again conclude that $\Gamma_{r(3)}$ reduces to $A_1' + E'$.

A nonnegative integer should result when calculating n using equation (2.5). If a fraction is obtained, then either an error has been made or the "reducible representation" is not valid.

Bibliography

Cotton, F. A. *Chemical Applications of Group Theory*, 3rd ed.; Wiley: New York, 1990.

Hargittai, I.; Hargittai, M. *Symmetry Through the Eyes of a Chemist*; V. C. H.: New York, 1986.

Wherrett, B. S. *Group Theory for Atoms, Molecules, and Solids*; Prentice-Hall: Englewood Cliffs, NJ, 1986.

Douglas, B. E.; Hollingsworth, C. A. *Symmetry in Bonding and Spectra*; Academic: Orlando, FL, 1985.

Kettle, S. F. A. *Symmetry and Structure*; Wiley: New York, 1985.

Murrell, J. N.; Kettle, S. F. A.; Tedder, J. M. *The Chemical Bond*, 2nd ed.; Wiley: New York, 1985.

Harris, D. C.; Bertolucci, M. D. *Symmetry and Spectroscopy*; Oxford University: New York, 1978.

Vincent, A. *Molecular Symmetry and Group Theory*; Wiley: New York, 1977.

Hatfield, W. E.; Parker, W. E. *Symmetry in Chemical Bonding and Structure*; Merrill: Columbus, OH, 1974.

Bernal, I.; Hamilton, W. C.; Ricci, J. S. *Symmetry: A Stereoscopic Guide for Chemists*; Freeman: San Francisco, 1972.

Orchin, M.; Jaffe, H. H. *Symmetry, Orbitals, and Spectra*; Wiley: New York, 1971.

Questions and Problems

1. Without looking at character tables, list the symmetry elements of the following point groups:
   (a) C_3 (order = 3)   (b) D_3 (order = 6)   (c) C_{4h} (order = 8)
   (d) D_{3h} (order = 12)

2. Explain in your own words the difference between a symmetry element and a symmetry operation.

3. We stated in this chapter that the symmetry operation $\sigma(xy)$ converts the point (x, y, z) to $(x, y, -z)$. Give the effect on point (x, y, z) of each operation:
   (a) $C_2(x)$ (b) $C_2(z)$ (c) $\sigma(yz)$

4. Assign a point group to each of the following:
   (a) A football (remember the laces!)

   (b) A football goal post:

   (c) The cheerleader shown here: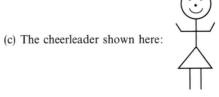

   (d) The megaphone shown here (without handle or insignia):

   (e) An equilateral triangle
   (f) An isosceles triangle (two equal sides)
   (g) A scalene triangle (all sides unequal)
   (h) A hexagon
   (i) An octagon

5. Assign a point group to each of the following molecules. The geometry about the central atom(s) is given in parentheses.

(a) Dichlorosilane, SiH_2Cl_2 (tetrahedral)
(b) Beryllium chloride, $BeCl_2$ (linear)
(c) Hydrogen cyanide, HCN (linear)
(d) Selenium difluoride, SeF_2 (angular)
(e) Stibine, SbH_3 (pyramidal)
(f) Fluorophosphine, PH_2F (pyramidal)
(g) Chlorostannane, H_3SnCl (tetrahedral)
(h) Chlorine trifluoride, ClF_3 (T-shaped)
(i) Chloroacetylene, $Cl—C\equiv C—H$ (linear)

(j) Allene,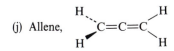

6. Determine the point group for a body containing four dots in the shape of a square, if:

(a) All dots are identical:

(b) Any one dot is shaded:

(c) Two opposite dots are shaded:

(d) Two adjacent dots are shaded:

7. Consider the octahedral species MX_6, where M is the central atom (point group O_h). What point group results if the symmetry is lowered by:
(a) Removing one X
(b) Replacing one X with a different group Z
(c) Removing two trans X's
(d) Removing two cis X's
(e) Removing three mutually adjacent X's

8. (a) Find six letters of the alphabet that belong to the C_{2v} point group.
(b) Find at least one letter that belongs to the D_{2h} group.
(c) Find at least one letter that belongs to the D_{4h} group.

9. The trans rotamer of H_2O_2 was cited as an example of a molecule having C_{2h} symmetry.
(a) What is the point group of the cis rotamer?

(b) What is the point group of the nonplanar equilibrium conformation?

10. Determine the point groups of two rotational isomers of B_2Cl_4:

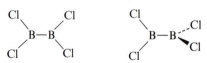

11. The correct point group is given for the most stable conformation of each of the following species. Sketch each molecule or ion.
    (a) NH_2F; C_s
    (b) I_2Br^-; $C_{\infty v}$
    (c) IF_4^+; C_{2v}
    (d) IF_4^-; D_{4h}
    (e) IF_5; C_{4v}
    (f) PF_4Cl; C_{2v}
    (g) $[Ni(CN)_4]^{2-}$; D_{4h}

12. The symmetry elements of the C_5 point group are E, C_5^1, C_5^2, C_5^3, and C_5^4. Give the product of:
    (a) $C_5^1 \times C_5^2$
    (b) $C_5^3 \times C_5^3$
    (c) $C_5^3 \times C_5^4$

13. Construct a multiplication table for the C_{2v} point group. The four symmetry elements are E, C_2, $\sigma_v(xz)$, and $\sigma_v'(yz)$. *Hint*: The water molecule belongs to this point group:

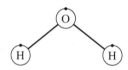

14. A partial character table for the D_{3d} group is given below:

	E	$2C_3$	$3C_2$	i	$2S_6$	$3\sigma_d$
Γ_1	1	1	1	1	1	1
Γ_2	1	1	-1	1	1	-1
Γ_3	2	-1	0	2	-1	0
Γ_4	1	1	1	-1	-1	-1
Γ_5	1	1	-1	-1	-1	1
Γ_6	2	-1	0	-2	1	0

    (a) What is the order of this group?
    (b) The six Mulliken symbols for this group are E_u, E_g, A_{1g}, A_{1u}, A_{2g}, and A_{2u}. Match these symbols with Γ_1–Γ_6.
    (c) Into what representation does a p_z orbital transform?
    (d) Demonstrate that the p_x and p_y orbitals are doubly degenerate and belong to Γ_6.

15. Assign the nine s, p, and d orbitals to their correct irreducible representations in the C_{4v} point group.

16. Reduce the following representations to their irreducible components in C_{3v} symmetry. (Use the character table in Appendix IV.)

	E	$2C_3$	$3\sigma_v$
(a)	3	0	-1
(b)	4	1	2
(c)	6	0	0
(d)	5	2	-1

17. Consider the four bonds of the tetrahedral molecule $TiCl_4$ to be vectors. Use these vectors to generate a reducible representation, and then reduce it.

18. Consider the five bonds of the trigonal bipyramidal anion GeF_5^- to be vectors. Use these vectors to generate a reducible representation, and then reduce it.

*19. There are many systematic ways to assign point groups. Flowcharts somewhat different from that given in this text have been suggested by M. Orchin and H. H. Jaffe (*J. Chem. Educ.* **1970**, *47*, 372) and by M. Zeldin (*J. Chem. Educ.* **1966**, *43*, 17). Examine their methods. Which one do you prefer? Why?

Models of Structure and Bonding

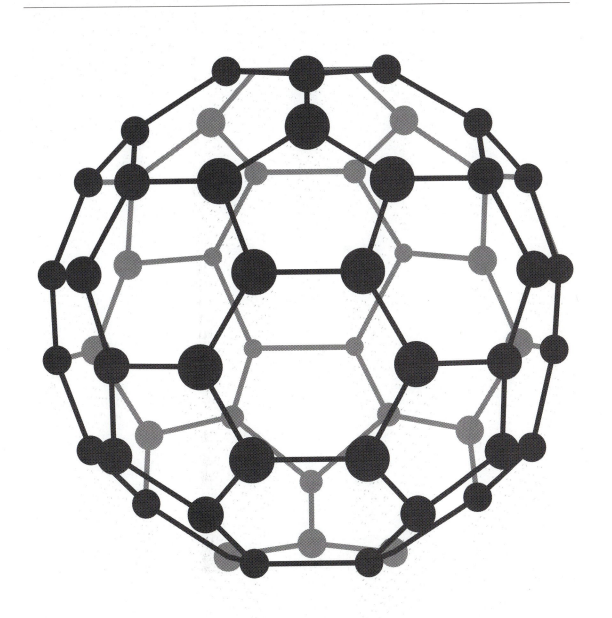

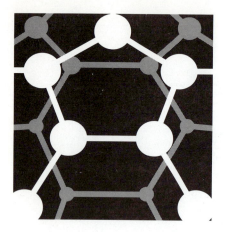

3

Molecular Orbital
Theory

As we saw in Chapter 1, the Schrödinger equation provides an accurate description of electron behavior for atomic hydrogen and other one-electron systems. When two or more electrons surround a single nucleus, approximations can be made that agree well with experimental data.

What about cases in which electrons are associated with two or more nuclei? This is a reasonable description for a chemical bond. It is again found that approximations can be made, based on the wave equation, that give results in agreement with what is known from experiment. One approach to describing chemical bonds will be examined in this chapter. The general concept is known as *molecular orbital (MO) theory*.[1] In its most basic form, MO theory presumes that atomic orbitals can be "mixed together" to create molecular orbitals; this process (for the case of stable aggregates) produces what we know as chemical bonds.

---

1. For historical perspectives of MO theory, see Mulliken, R. S. *Science* **1967**, *157*, 13; also Davidson, R. B. *J. Chem. Educ.* **1977**, *54*, 531.

3.1   The Linear Combination of Atomic Orbitals (LCAO) Approach

One aspect of MO theory is the *linear combination of atomic orbitals* (*LCAO*) method. As will be seen, this approach is especially useful for molecules and ions that have small numbers of atoms and/or high symmetries. The following statements summarize the primary tenets and assumptions of the LCAO approach.

1. Molecular orbitals are formed by the combination or interaction of "parent" atomic orbitals from two or more atoms.

2. Only valence electrons are involved in chemical bonding, and only valence orbitals are combined to produce molecular orbitals.

3. Orbitals are conserved during chemical bonding. The number of atomic orbitals mixed together always equals the number of molecular orbitals formed.

4. Molecular orbitals exhibit properties similar to those of atomic orbitals. For example, Hund's rules and the Pauli exclusion principle apply.

5. Only atomic orbitals having identical symmetry properties can interact with one another.

6. Orbital mixing is most significant when the parent orbitals have roughly the same energies. As the energy difference between parents increases, the effectiveness of *overlap* (see below) decreases.

Several of these ideas can be merged into one statement: The "daughter" molecular orbitals formed through chemical bonding retain most of the characteristics of their parent atomic orbitals. That is, the atomic orbitals are *perturbed*, but not totally changed. From this idea comes the equation

$$\psi = c_1\phi_1 \pm c_2\phi_2 \pm \cdots \tag{3.1}$$

Here ϕ represents a wavefunction for a parent atomic orbital, ψ is the resulting molecular orbital wavefunction, and each c is a constant called a *weighting coefficient*. The magnitude of c gives information about the relative importance of the corresponding atomic orbital in forming the molecular orbital. This equation may be considered a "recipe" for creating MO's, with the weighting coefficients giving the amount of each atomic orbital to be mixed together. The number of terms on the right side of the LCAO equation varies. In the most common case there are two terms, but it is not unusual (especially for π-type MO's) to have more.

The signs of the atomic orbital wavefunctions are important in applying equation (3.1). If the signs of two ϕ's match (both positive or both negative), then the two waves reinforce one another. This results in increased electron density between the nuclei, and therefore an increase in stability compared to the parent orbitals. In that case, the relationship between the parent and daughter orbitals can be symbolized as

$$\psi = c_1\phi_1 + c_2\phi_2 \tag{3.2}$$

In contrast, if the signs of the parent orbitals are mismatched (one positive and the other negative), the result is a decrease in electron density between the nuclei.

$$\psi = c_1\phi_1 - c_2\phi_2 \tag{3.3}$$

Such molecular orbitals are less stable than either of the parent orbitals. Because it is possible to combine two atomic orbitals in either way (sign match or mismatch), MO's often come in pairs—one *bonding* orbital, with increased electron density between the involved nuclei, for every *antibonding* orbital with decreased electron density (and a node) between the nuclei.

Molecular Orbital Designations

To distinguish between bonding and antibonding orbitals, an asterisk (*) is conventionally used as a superscript for all antibonding orbitals.

Recall that atomic orbitals are assigned the letter designations *s*, *p*, *d*, and *f* to differentiate among their shapes and/or number of angular nodes. In the same way, molecular orbitals are given the lowercase Greek letter designations σ (sigma), π (pi), and δ (delta). A σ MO can be formed from combinations of *s*-, *p*-, and/or *d*-type atomic orbitals; a π orbital can be formed by mixing *p* and/or *d* orbitals; and a δ orbital requires *d–d* interaction. A second distinction also can be made. A σ bond involves overlap of orbitals in one region of space; a π bond requires overlap in two regions, with a node between them; and a δ bond involves overlap in four regions (and two nodes). Figure 3.1 shows the most common types of orbital overlap.

As a very general rule, σ bonding tends to be stronger than π bonding, which in turn is stronger than δ bonding. It is also true that if two atoms share only one electron pair, the bond is best characterized as being of the σ type; if two bonds are present, one is a σ and one a π bond; and δ bonds are found only between atoms that are also engaged in σ and π interactions.

3.2 Overlap Using Only *s*-Type Orbitals

Consider the diatomic hydrogen molecule, H_2. The hydrogen atom has the ground-state configuration $1s^1$. Thus, only two orbitals (the $1s$ orbital of each atom) are available for bonding. These atomic orbitals mix to produce two molecular orbitals, one bonding and the other antibonding. The LCAO

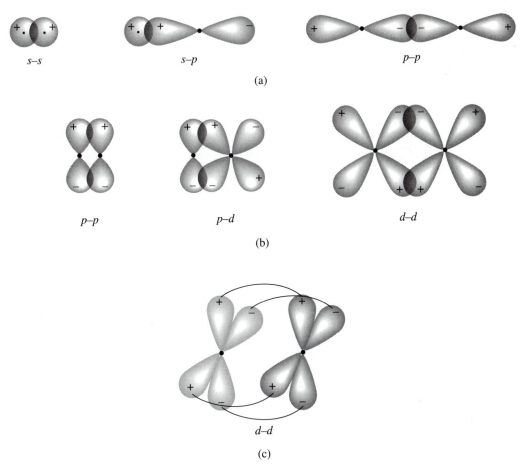

Figure 3.1 The most common types of orbital overlap: (a) σ-type; (b) π-type; (c) δ-type overlap.

equation for the bonding orbital is written as in equation (3.2):

$$\psi_b = c_1\phi_1 + c_2\phi_2$$

The functions ϕ_1 and ϕ_2 are known (see Tables 1.1, 1.3, and 1.4). What are c_1 and c_2? To answer this question, we must first recognize that we are really interested in the probability density ψ^2, rather than the amplitude ψ, of the wave. Squaring gives

$$\psi^2 = c_1^2\phi_1^2 + c_2^2\phi_2^2 + 2c_1c_2\phi_1\phi_2 \tag{3.4}$$

The last term of equation (3.4) is a *cross term*, or *overlap term*; its value is related to the degree of atomic orbital overlap, and therefore to the strength of the chemical bond being formed.

The total probability density can be determined by integrating over space. If $d\tau = dx\,dy\,dz$, then

$$\int \psi^2 \, d\tau = c_1^2 \int \phi_1^2 \, d\tau + c_2^2 \int \phi_2^2 \, d\tau + 2c_1 c_2 \int \phi_1 \phi_2 \, d\tau \qquad (3.5)$$

The next step is to *normalize* the atomic and molecular orbitals (ie, set them equal to 1). This simply means that the total probability of "finding" an electron in either an atomic or molecular orbital is unity. Then, since $\int \psi^2 \, d\tau = \int \phi_1^2 \, d\tau = \int \phi_2^2 \, d\tau = 1$,

$$1 = c_1^2 + c_2^2 + 2c_1 c_2 \int \phi_1 \phi_2 \, d\tau \qquad (3.6)$$

Next, we will let S_{12} equal the *overlap integral*, $\int \phi_1 \phi_2 \, d\tau$. This gives

$$1 = c_1^2 + c_2^2 + 2c_1 c_2 S_{12} \qquad (3.7)$$

However, it is often true that $2c_1 c_2 S_{12}$ is negligible with respect to the other two terms. If this simplifying assumption is made, then

$$1 = c_1^2 + c_2^2 \qquad (3.8)$$

Since the two parent atomic orbitals are identical (both being hydrogen $1s$ orbitals), it is reasonable that their contributions to the molecular orbital will be equal; that is, $c_1 = c_2$. In that case, $c_1 = c_2 = 1/\sqrt{2}$, and the final form for the normalized bonding orbital can be written as

$$\psi_b = \frac{1}{\sqrt{2}} (\phi_1 + \phi_2) \qquad (3.9)$$

What is the equation for the antibonding orbital? Following the same procedure, it is easy to show that

$$\psi^* = \frac{1}{\sqrt{2}} (\phi_1 - \phi_2) \qquad (3.10)$$

Pictorial representations for these two MO's are given in Figure 3.2. They are of the σ type, since overlap occurs in only one region of space. The labels σ_s and σ_s^* are assigned to the bonding and antibonding orbitals, respectively; the subscripts indicate that the parents were s-type orbitals.

Notice in Figure 3.2a that the bonding orbital has greater electron density between the nuclei than do its parent atomic orbitals. As stated earlier, this leads to enhanced stability. Conversely, σ_s^* has reduced electron density (and a node) between the nuclei; it is less stable than either of its parent orbitals. It can be shown that the increase in stabilization of the bonding orbital is approximately equal to the decrease in stabilization of the antibonding orbital, and in general that the sum of all the orbital

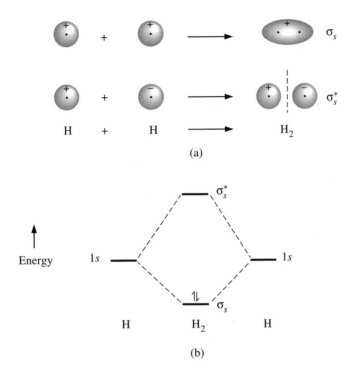

Figure 3.2
Molecular orbitals for H_2. (a) Sketches of the bonding (σ_s) and antibonding (σ_s^*) orbitals.
(b) The molecular orbital energy-level diagram. The destabilization of σ_s^* is slightly greater than the stabilization of σ_s because of electron–electron repulsion.

energies for a system does not change appreciably in going from atomic to molecular orbitals. (Actually, a more detailed treatment reveals that the antibonding orbital is destabilized slightly more than the bonding orbital is stabilized; this is reflected in Figure 3.2b.[2])

What can be said of the two electrons of H_2? The ground-state configuration logically places them in σ_s; they must therefore have opposite spins according to the Pauli principle. The antibonding orbital is vacant. The *molecular orbital energy-level diagram* that results is shown in Figure 3.2b. By convention, the parent atomic orbitals are placed on the two sides of the diagram, with the MO's in the middle.

Bond Length, Bond Energy, and Bond Order

The internuclear distance of H_2 is known from experiment to be 74.1 pm. This provides an interesting comparison to the atomic radius of the hydrogen atom, 52.9 pm. These values require that the relative relationship between the two original $1s$ orbitals after bond formation is as shown in Figure 3.3.

2. For a more detailed explanation, see Atkins, P. W. *Molecular Quantum Mechanics*; Oxford University: New York, 1983; p. 256.

Figure 3.3

The overlap of two $1s$ orbitals of atomic hydrogen in the formation of H_2. Notice the shared domain.

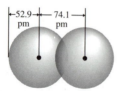

The term *overlap* is often used in bonding theory. It is apparent from Figure 3.3 that when a chemical bond is formed, one or more regions in space become shared domain—that is, orbitals of different atoms overlap. In general, the bond strength increases with the degree of overlap. From this an inverse relationship between bond length and bond strength can be inferred: All else being equal, as bonds are lengthened they are also weakened.

The theoretical prediction that the two electrons of H_2 are stabilized in going from the $1s$ to the σ_s orbital is consistent with the experimental fact that the reaction $H + H \rightarrow H_2$ is exothermic. The reverse process, involving the separation of the two nuclei, is endothermic; the energy required for separation is called the *bond energy* (more properly the *bond dissociation energy*, symbolized by D; see Chapter 4). The bond energy of H_2 has been experimentally measured as 432 kJ/mol.

It is often useful to discuss the *order* of a bond. Bond order can be calculated from the equation

$$\text{Bond order} = \frac{\text{Number of bonding } e^- - \text{Number of antibonding } e^-}{2}$$

$$(3.11)$$

Thus, the bond order of H_2 is $(2-0)/2 = 1.0$, which corresponds to a single bond in common terminology. As expected, there is a direct relationship between bond order and bond energy—species with high bond orders also tend to have high bond energies.

What is the effect of removing an electron from H_2 to produce H_2^+, the hydrogen molecule cation? The change in occupancy of σ_s reduces the bond order to $(1-0)/2 = 1/2$. We might expect the resulting bond energy to be exactly half of 432, or 216 kJ/mol, but in fact, it is 255 kJ/mol. This can be attributed to electron–electron repulsions in H_2, which, of course, are absent in H_2^+. The repulsion can be estimated as $255 - 216 = 39$ kJ/mol for the electronic ground state of H_2.

What happens to the bond length upon ionization of H_2? Since the bond order is decreased, the bond length should increase. This is indeed the case; the internuclear distance in H_2^+ is 106 pm.

Next, another system having only s-type electrons will be considered —the overlap of two helium atoms. Again, the two $1s$ atomic orbitals are mixed, giving the same general result as for H_2, one bonding (σ_s) and one

Table 3.1 Bond lengths (*d*) and bond dissociation energies (*D*) for the homonuclear diatomic molecules of the Group 1 elements

Compound	d, pm	D, kJ/mol	Compound	d, pm	D, kJ/mol
H_2	74	432	K_2	392	49
Li_2	267	110	Rb_2	432	47
Na_2	308	72	Cs_2	470	44

Note: All values are rounded to the nearest integer. The dissociation energies vary in accuracy, but uncertainties are generally ± 4 kJ/mol or less.
Sources: DeKock, R. L.; Gray, H. B. *Chemical Structure and Bonding*; Benjamin/Cummings: Menlo Park, CA, 1980; p. 229; Sanderson, R. T. *Chemical Bonds and Bond Energy*; Academic: New York, 1976; Chapter 3.

antibonding (σ_s^*) orbital. The total occupancy is four electrons (two from each He atom). Thus, both MO's are filled, and the electron configuration is $(\sigma_s)^2(\sigma_s^*)^2$. The bond order is $(2-2)/2 = 0$. This indicates that there is nothing gained from the overlap, and the bond energy is also 0.[3] It is therefore not surprising that this molecule has never been synthesized.

However, it is possible to study He_2^+, the monovalent cation of diatomic helium. The experimental bond length (108 pm) and dissociation energy (322 kJ/mol) are reasonable for the calculated bond order of 1/2.

The next simplest element is lithium, with the electron configuration $1s^2 2s^1$. Recall that the lithium $1s$ orbital is of the core type, lying low in energy and surrounded in space by the $2s$ domain; thus, both the $1s$ orbital and the $1s$ electrons can be ignored when considering the bonding in Li_2. There is one valence electron per atom. The overlap of two lithium $2s$ orbitals creates a bonding orbital and an antibonding orbital having the same general shapes as before. The electron configuration is $(\sigma_s)^2$, and the bond order is 1.0. The experimental values for the bond length and energy are 267 pm and 110 kJ/mol, respectively. This dissociation energy is considerably smaller than that of H_2, and reflects the less effective overlap of the larger lithium orbitals.

It should not be surprising that Be_2, with four valence electrons, has not been synthesized. Its ground-state configuration is $(\sigma_s)^2(\sigma_s^*)^2$ and, like He_2, its bond order is 0.

Generalizations can now be made for all the so-called *s*-block elements—that is, those of Groups 1 and 2 of the periodic table. For Group 1, the bond order of each of the homonuclear diatomic molecules is 1.0. Because size increases going down the family, the dissociation energies decrease and the bond lengths increase in going from Li_2 to Cs_2 (see Table 3.1). Each of

---

3. Actually, the process $He + He \rightarrow He_2$ is slightly endothermic; a net repulsive force arises from the interaction, since σ_s^* is destabilized more than σ_s is stabilized (see Atkins, P. W. *Molecular Quantum Mechanics*; Oxford University: New York, 1983; p. 256).

these elements is diatomic in the gas phase. However, as we will see in Chapter 7, the solid state permits a different type of bonding.

For the Group 2 elements, the bond orders of the diatomic molecules would be 0; hence, these elements are monatomic in the gas phase.

3.3 Overlap Using p-Type Orbitals

Boron has the electron configuration $1s^2 2s^2 2p^1$. There are three valence electrons, and this is the lightest element with an occupied p subshell. In constructing an MO diagram for B_2, then, it is necessary to consider a total of eight valence orbitals—the $2s$ and $2p$ subshells for each atom. The resulting diagram must therefore contain eight MO's.

Our general approach will be to overlap orbitals of the same type. (As will be seen shortly, this has its basis in group theory.) Thus, we begin by combining the two $2s$ orbitals, giving a now familiar result: σ_s (bonding) and σ_s^* (antibonding) MO's.

Next, consider the p orbitals. To be consistent with group theory convention, the z axis must define the bond axis. Then the relative orientations of the p_x, p_y, and p_z orbitals of the two atoms are as shown in Figure 3.4.

The two p_z orbitals can be combined. Their "end-to-end" interaction gives rise to overlap in one region of space (σ bonding) and forms MO's that can be labeled σ_s (bonding) and σ_s^* (antibonding). The latter has a node passing between the nuclei.

The two p_x and two p_y orbitals can be combined in a "side-to-side" manner. This is characteristic of π-type bonding, since it involves overlap in two regions of space. The two bonding MO's that result are degenerate (they differ in direction, but not in size, shape, or energy), and can together be assigned the label $\pi_{x,y}$. The antibonding pair is $\pi_{x,y}^*$. These orbitals are pictured in Figure 3.5.

The eight MO's have therefore been described. What are their relative energies? This is not a simple question to answer, and it will be seen that the correct order varies slightly from compound to compound. We can make some educated guesses, however. Since the $2s$ atomic orbital lies at lower energy than does $2p$, we may reasonably expect σ_s and σ_s^* to be the most

Figure 3.4

The relative orientations of the p_x, p_y, and p_z atomic orbitals of B_2.

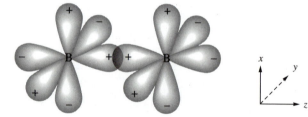

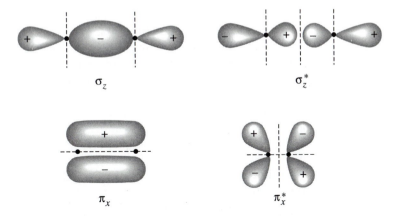

Figure 3.5
Pictorial diagrams of the MO's of B_2 having *p*-type parent orbitals. The π_y and π_y^* orbitals are identical in shape, size, and energy to π_x and π_x^*, but lie along the *y* axis.

stable of the MO's. You might also recall that σ bonding is generally stronger than π bonding. A tentative prediction is then that stability decreases from left to right:

$$\sigma_s > \sigma_s^* > \sigma_z > \pi_{x,y} > \pi_{x,y}^* > \sigma_z^* \quad (??)$$

However, spectroscopic and magnetic data indicate that the $\pi_{x,y}$ and σ_z levels are actually reversed; the correct energy-level diagram for B_2 is as given in Figure 3.6.

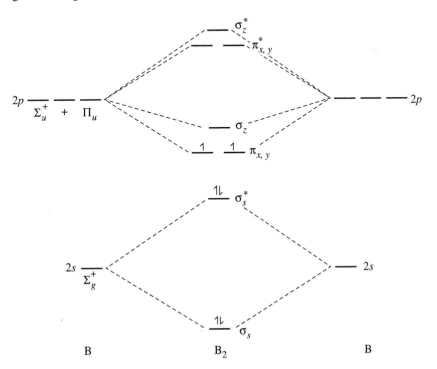

Figure 3.6
Molecular orbital diagram for B_2. This species is predicted to be paramagnetic, with a bond order of 1.0.

What conclusions can be drawn? The bond order is $(4 - 2)/2 = 1.0$. This indicates a stable bonding situation, best described as a single bond. The experimental dissociation energy of $B_2(g)$ is 274 kJ/mol. Is this consistent with the molecules previously examined? The bond energy of H_2 is over 50% greater, but hydrogen uses orbitals from the $n = 1$ quantum level; those smaller orbitals overlap more effectively than do boron's. In Li_2 ($D = 110$ kJ/mol) the overlap is between atomic orbitals having $n = 2$, but lithium is larger than boron. The consistency is seen to be quite good. It should also be noted that the bond length in B_2 (159 pm) is intermediate between those of H_2 and Li_2.

Another important conclusion can be drawn from the diagram. Hund's rules come into play, resulting in the prediction of two unpaired electrons per ground-state molecule. This has been shown experimentally to be correct; the B_2 molecule is paramagnetic. (Note that this results from the preferential filling of $\pi_{x,y}$ before σ_z.)

Recall that MO diagrams for the Group 1 and Group 2 diatomics have the same general form, but differ in the number of electrons. The same is true for the p-block elements. The homonuclear diatomics of Groups 13–18 all have MO's similar to those of B_2, but the number of valence electrons (and therefore the bond order) varies from group to group. You should verify that the bond orders deduced from Figure 3.6 are:

Group Number	13	14	15	16	17	18
Bond Order	1.0	2.0	3.0	2.0	1.0	0.0

The diatomics of Group 18, like those of Group 2, would have bond orders of 0. Consistent with this prediction, those elements are monatomic in the gas phase.

Experimental bond energies and lengths for the Group 13–17 diatomics are given in Table 3.2. Several trends are evident from the table. Some can be rationalized by MO theory, while others are better explained from the standpoint of localized bonding. Further discussion will therefore be delayed until Chapter 4.

The Role of Symmetry

The point group for any homonuclear diatomic species is $D_{\infty h}$. In groups having a C_{∞} axis the irreducible representations are usually symbolized by Greek letters instead of the more common designators $A, B, E,$ and T. In $D_{\infty h}$ symmetry, then, the s and p atomic orbitals transform as follows:

$$s \rightarrow \Sigma_g^+$$
$$p_{x,y} \rightarrow \Pi_u$$
$$p_z \rightarrow \Sigma_u^+$$

Table 3.2 Bond orders, lengths, and dissociation energies for the homonuclear diatomic compounds of Groups 13–17

Species	d, pm	D, kJ/mol
Group 13 (Bond order = 1.0)		
B_2	159	274
Group 14 (Bond order = 2.0)		
C_2	—	593
Si_2	225	314
Ge_2	—	272
Sn_2	—	192
Pb_2	—	96
Group 15 (Bond order = 3.0)		
N_2	110	942
P_2	189	483
As_2	229	380
Sb_2	221	295
Bi_2	—	195
Group 16 (Bond order = 2.0)		
O_2	121	494
S_2	189	425
Se_2	217	305
Te_2	256	223
Group 17 (Bond order = 1.0)		
F_2	142	154
Cl_2	199	240
Br_2	228	190
I_2	267	149

Note: All values are rounded to the nearest integer. The dissociation energies vary in accuracy, but uncertainties are generally ± 4 kJ/mol or less.

Sources: Dasent, W. E. *Inorganic Energetics*, 2nd ed.; Cambridge University: London, 1982; Chapter 4; DeKock, R. L.; Gray, H. B. *Chemical Structure and Bonding*; Benjamin/Cummings: Menlo Park, CA, 1980; p. 229.

As mentioned earlier, the LCAO approach requires that only orbitals having the same symmetry may be mixed. Given the above transformations it should be clear how the molecular orbital designations σ and π arise. (Recall that irreducible representations are conventionally given capital letter designations, while MO's are assigned lowercase symbols.) Also, the degeneracy of the π-type MO's is seen to be a consequence of the fact that p_x and p_y are degenerate (mixed by symmetry) in the $D_{\infty h}$ point group.

It should not be surprising that the labels of the various molecular orbitals are "correct" in the sense that the pictorial representation for each MO transforms into an irreducible representation having the expected symmetry type (σ or π). For example, for σ_s and σ_s^*:

$D_{\infty h}$	E	$2C_\infty$	$\infty\sigma_v$	i	$2S_\infty$	∞C_2	
σ_s	1	1	1	1	1	1	$= \Sigma_g^+$
σ_s^*	1	1	1	-1	-1	-1	$= \Sigma_u^+$

You should prove to yourself that $\pi_{x,y}$ transforms as Π_u, σ_z as Σ_g^+, $\pi_{x,y}^*$ as Π_g, and σ_z^* as Σ_u^+.

Configuration Interaction (Secondary s–p Mixing)

We mentioned earlier that the order of molecular orbital energies is not the same for all homonuclear diatomics. Specifically, in some species a reversal of orbital stabilities occurs so that σ_z becomes less stable than the $\pi_{x,y}$ pair. It is known from photoelectron spectroscopy that B_2, C_2, and N_2 have the order of orbital energies shown in Figure 3.6.[4] For O_2 and F_2, however, σ_z lies lower in energy than $\pi_{x,y}$. Why?

The LCAO approach involves the mixing of atomic orbitals from two or more different atoms. Since this is mathematically valid, it must be equally valid to mix orbitals from the same species, provided that the appropriate guidelines are followed (ie, only orbitals having identical symmetry properties can interact). For homonuclear diatomics both the σ_s and σ_z MO's belong to the Σ_g^+ irreducible representation; hence, *configuration interaction* (or *secondary mixing*) is possible between them. The result is to create two daughter orbitals, one of increased and the other of decreased stability (ie, the two orbital levels separate in energy).

An LCAO type of equation for s–p_z mixing is

$$\psi_{sp} = c_s\phi_s + c_z\phi_z \tag{3.12}$$

---

4. Photoelectron spectroscopy (PES) is discussed in Chapter 21. Another useful source is Fenske, R. F. *Prog. Inorg. Chem.* **1976**, *21*, 179.

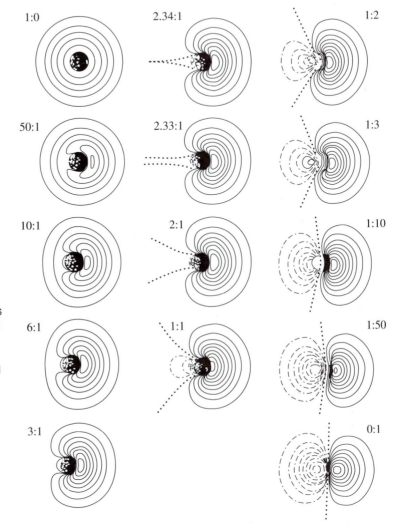

Figure 3.7
Contour plots of mixtures of *s* and p_z atomic orbitals in various proportions. Solid and dashed lines denote regions having positive and negative wave amplitudes, respectively; dotted lines are nodes. [Reproduced with permission from Verkade, J. G. *A Pictorial Approach to Molecular Bonding*; Springer-Verlag: New York, 1986; p. 37.]

The ratio of the weighting coefficients, c_s/c_z, can take on any value. This ratio influences the shape of the MO in question, as is evident from the contour plots shown in Figure 3.7.

Secondary mixing is also possible between σ_s^* and σ_z^*, since both have Σ_u^+ symmetry. The effect on the relative orbital energies is illustrated in Figure 3.8, where the σ_z level is shown to be initially more stable than $\pi_{x,y}$ on the assumption that σ bonding is inherently stronger than π bonding. Mixing destabilizes σ_z, however, and if such interaction is sufficiently strong, then the order of orbital stability is reversed.

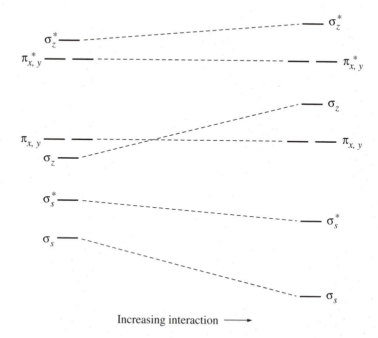

Figure 3.8
The effect of configuration interaction (secondary mixing) on the orbital energies of homonuclear diatomics. The σ_s and σ_z (with parents of Σ_g^+ symmetry) and σ_s^* and σ_z^* (Σ_u^+ symmetry) levels separate in energy.

Increasing interaction $\longrightarrow$

The question remains, however, of why the σ and π energies of O_2 and F_2 are not also reversed. Recall the importance of having an "energy match" between the two parent orbitals. For σ_s and σ_z this depends on how close their parents were in energy. That is, if the $2s$ and $2p_z$ atomic orbitals have approximately the same energy, then the same will be true for σ_s and σ_z. Conversely, if the atomic orbital parents lie at very different energies, then σ_s and σ_z will be widely separated.

The question therefore becomes one of determining the $2s$–$2p$ energy difference for the elements boron through fluorine. These values can be calculated from valence state ionization energies (Table 1.8), and are compiled in Table 3.3. It can be seen that a difference of somewhere between 12.4 and 16.5 eV results in a "crossover." The large differences for oxygen

Table 3.3 The $2s$–$2p$ energy gap for elements of the first short period

Element	$E\,(2s)$, eV	$E\,(2p)$, eV	ΔE, eV
B	−14.0	−8.3	5.7
C	−19.4	−10.6	8.8
N	−25.6	−13.2	12.4
O	−32.3	−15.8	16.5
F	−40.2	−18.6	21.6

and fluorine reduce σ_s–σ_z mixing in O_2 and F_2, causing these diatomics to have different ground-state configurations from the others.

Oxidation–Reduction of Diatomics: Frontier Orbitals

The correct MO diagram for the oxygen molecule is given in Figure 3.9. Note that the experimental observation that O_2 is paramagnetic is consistent with this diagram, and is explained by the lack of strong secondary mixing.

Let us examine Figure 3.9 from the standpoint of electron gain and loss. The least stable of the occupied orbitals is often referred to as the *highest occupied molecular orbital (HOMO)*; for O_2 it is the $\pi^*_{x,y}$ pair. Because it is antibonding, this level lies higher in energy than the parent oxygen atom $2p$ orbital. The first ionization energy of O_2 is therefore less than that of O (12.2 eV versus 13.6 eV for atomic oxygen). Also, removal of an electron from O_2 increases the bond order from 2.0 to 2.5. Thus, the bond energy of the oxygenyl cation, O_2^+, is greater (and the bond length shorter) than for O_2.

What if an electron is added to molecular oxygen to give O_2^-, the superoxide ion? The added electron enters the *lowest unoccupied molecular orbital (LUMO)*, once again the $\pi^*_{x,y}$ level. The new bond order is 1.5; reduction weakens and lengthens the O–O bond. Addition of a second electron produces the peroxide ion, O_2^{2-}, with a bond order of 1.0.

Table 3.4 gives experimental data for a variety of oxygen-containing diatomics. You should satisfy yourself that these values are readily rational-

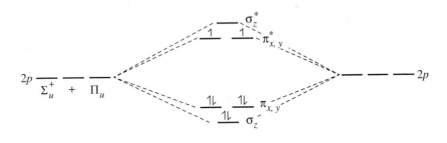

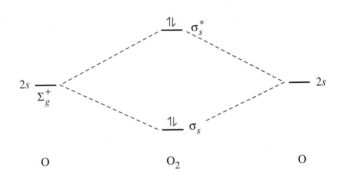

Figure 3.9
The molecular orbital diagram for O_2. The s–p_z mixing is limited by the large $2s$–$2p$ energy gap. The frontier orbitals are the $\pi^*_{x,y}$ pair, and the molecule is paramagnetic.

Table 3.4 Predicted bond orders and experimental bond lengths and dissociation energies for some diatomic oxygen-containing compounds and ions

Species	Bond Order	d, pm	D, kJ/mol
O_2^+	2.5	112	—
O_2	2.0	121	494
O_2^-	1.5	126	393
O_2^{2-}	1.0	149	—
OH	1.0	97	424
CO	3.0	113	1072
SiO	3.0	151	765
GeO	3.0	165	657
SnO	3.0	184	529
NO	2.5	115	628
NO^+	3.0	106	950
PO	2.5	147	592
AsO	2.5	162	473
SO	2.0	148	517
SeO	2.0	—	423

Note: All bond lengths and dissociation energies are rounded to the nearest integer. Energies vary in accuracy, but in most cases the uncertainties are ± 4 kJ/mol or less.
Sources: Dasent, W. E. *Inorganic Energetics*, 2nd ed.; Cambridge University: London, 1982; Chapter 4. DeKock, R. L.; Gray, H. B. *Chemical Structure and Bonding*; Benjamin/Cummings: Menlo Park, CA, 1980; Chapter 4.

ized by MO theory. The HOMO and LUMO are often grouped together as the *frontier orbitals*. As the site of either oxidation or reduction (and also the most likely orbitals to engage in further chemical bonding), the frontier orbitals are for most purposes the most important ones in any MO diagram.

3.4 Overlap Using *d*-Type Orbitals

The transition metals are characterized by the presence of partly filled d orbitals. Like the Group 1 elements, transition metals prefer three-dimensional arrays for their standard states, but may be diatomic in the gas phase. Can a molecular orbital picture incorporating d orbitals be generated for such species? Let us use chromium ($[Ar]4s^1 3d^5$) as a model case. The diatomic molecule Cr_2 is "stable" in the sense that it can be generated and studied spectroscopically under the appropriate conditions.

The valence orbitals of chromium can be taken to be the 4*s* and the five 3*d* orbitals, with 4*s* being slightly the most stable.[5] To construct an MO diagram for Cr_2, then, we begin by mixing the 4*s* orbitals to produce the well-known σ_s and σ_s^* MO's.

The *d* orbitals can be combined as follows:

1. **d_{z^2}–d_{z^2} (σ-Type) Interaction:** Taking the *z* axis to be the bond axis, the end-to-end overlap of the two d_{z^2} orbitals forms σ_z (bonding) and σ_z^* (antibonding) orbitals.

2. **d_{xz}–d_{xz} and d_{yz}–d_{yz} (π-Type) Interaction:** These orbitals engage in side-to-side overlap to produce two degenerate bonding and two degenerate antibonding MO's. The bonds are of the π type, since they involve overlap in two regions of space.

3. **d_{xy}–d_{xy} and $d_{x^2-y^2}$–$d_{x^2-y^2}$ (δ-Type) Interaction:** The four-lobe overlap of like orbitals gives two degenerate bonding (δ) and two degenerate antibonding (δ*) orbitals.

Sketches of these MO's are given in Figure 3.10.

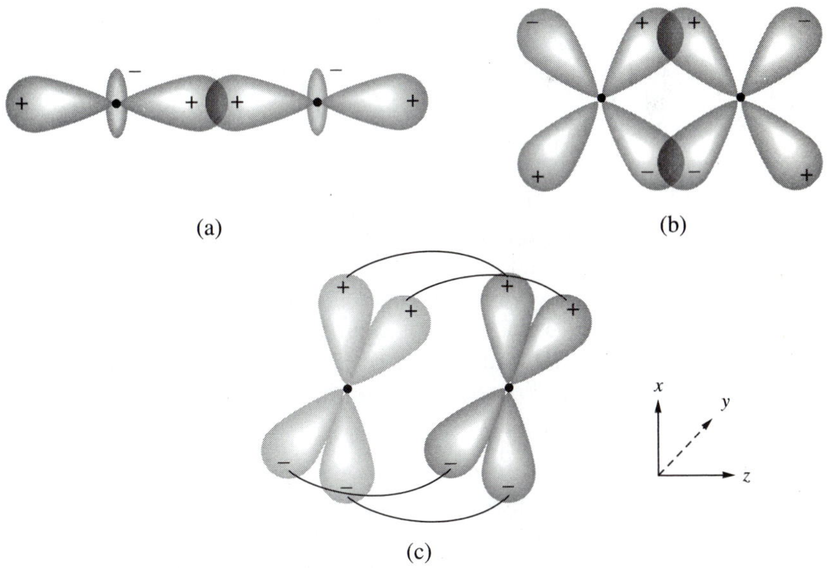

Figure 3.10
Types of *d* orbital overlap in homonuclear diatomics such as Cr_2. (a) σ bonding by overlap of two d_{z^2} orbitals. (b) π bonding by overlap of two d_{xz} (or two d_{yz}) orbitals. (c) δ bonding by overlap of two d_{xy} (or two $d_{x^2-y^2}$) orbitals.

(a)

(b)

(c)

5. This order is often reversed for later members of the first transition series and for metal cations. The 4*p* orbitals lie at higher energy and need not be considered in this example (but see Chapter 16).

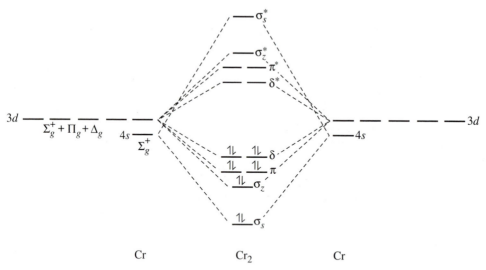

Figure 3.11 Molecular orbital diagram for Cr_2; only the $3d$ and $4s$ orbitals are considered. The indicated bond order is 6.0, and the molecule is predicted to be diamagnetic.

Following the general rule that σ bonding tends to be stronger than π bonding, which in turn is stronger than δ bonding, the relative energies of the ten molecular orbitals having d parent orbitals are predicted to be

$$\sigma < \pi < \delta < \delta^* < \pi^* < \sigma^*$$

Since there are 12 valence electrons, the six bonding orbitals are filled (see the MO diagram in Figure 3.11); hence, the calculated bond order is 6.0! However, the experimental bond energy is only 255 kJ/mol.[6] This small value (relative to the bond order) reflects the comparative weakness of the π- and δ-type overlap.

Molecular orbital diagrams for the homonuclear diatomics of all the d-block elements can be generated in this manner. A number of such species have bond orders of 4.0 or greater. The study of quadruple and higher-order bonding between transition metal centers is an exciting area of contemporary research.

3.5 Heteronuclear Diatomic Molecules

The mixing of orbitals from two different types of atoms is considered in this section. Nitric oxide, NO, is such a species. The valence orbitals (the $2s$

---

6. Busby, R.; Klotzbücher, W.; Ozin, G. A. *J. Am. Chem. Soc.* **1976**, *98*, 4013.

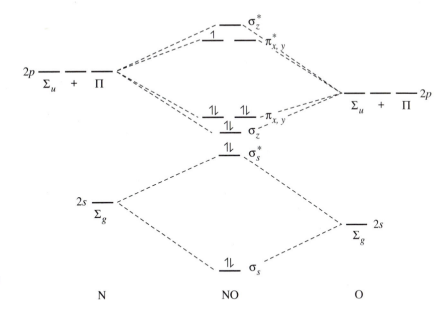

Figure 3.12
Qualitative molecular orbital diagram for NO. Note the skewing of the atomic orbitals.

and 2p subshells of each atom) interact to produce eight MO's, which are generally similar to those of O_2. Secondary mixing plays only a minor role because of the large 2s–2p energy gap of oxygen. The MO diagram (Figure 3.12) indicates a bond order of $(8 - 3)/2 = 2.5$. The experimental bond dissociation energy (628 kJ/mol) and bond length (115 pm) are consistent with a bond order between 2 and 3. (Compare to the O=O double bond of O_2 and the N≡N triple bond of N_2, as given in Table 3.2.)

Notice an important difference between the diagram in Figure 3.12 and those given earlier for homonuclear diatomics. The diagram for NO (and for any heteronuclear species) is lopsided, or skewed; that is, the atomic orbital energies are unequal on the two sides. Recall that the valence orbitals of oxygen are more stable than those of nitrogen (refer to Table 3.3). This is important because, since the energies of the parent atomic orbitals are not equal, their weighting coefficients also must be unequal; that is, for any given molecular orbital, $c_N \neq c_O$. For example, σ_s (and the other bonding orbitals) lies closer in energy to its oxygen than to its nitrogen parent; thus, $c_O > c_N$. For σ_s^* and the other antibonding orbitals, $c_O < c_N$.

This information is particularly significant for the frontier orbitals. Both the HOMO and LUMO lie at the $\pi_{x,y}^*$ level, which is primarily nitrogen in character. Thus, the unpaired electron resides mainly on the nitrogen atom, a prediction that is consistent with the reaction chemistry of nitric oxide; one-electron oxidation and reduction both typically involve attack at the nitrogen.

The combination of atoms having extremely different valence orbital energies may give rise to a different qualitative MO diagram. Consider carbon monoxide, CO. The relevant atomic orbital energies are:

Atom	E (2s), eV	E (2p), eV
C	−19.4	−10.6
O	−32.3	−15.8

As in O_2 and NO, there is limited secondary mixing; σ_z lies lower in energy than $\pi_{x,y}$. (However, some s–p interaction does occur; see below.) The large skewing of the atomic orbitals has a strong destabilizing effect on σ_s^*, resulting in the MO diagram shown in Figure 3.13. Notice that a reversal of orbital stabilities has occurred; σ_s^* is less stable than either σ_z or $\pi_{x,y}$. The bond order is 3.0 (ie, a triple bond is indicated), and all electrons are paired.

Figure 3.13 is in accord with experimental data in several ways. Carbon monoxide is diamagnetic, with a bond length (113 pm) and energy (1072 kJ/mol) consistent with triple bonding. Moreover, its chemical behavior is indicative of a HOMO that is predominantly carbon in parentage. This

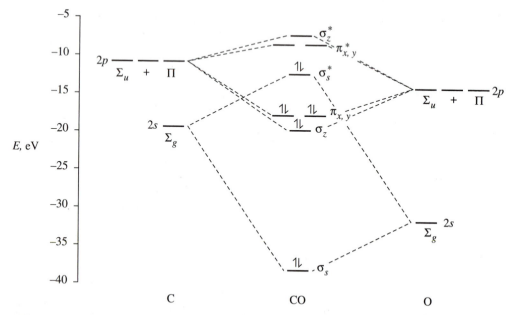

Figure 3.13 Molecular orbitals and their approximate energies for carbon monoxide, CO. The HOMO is the σ_s^* orbital, which is predominantly carbon in character.

compound is an excellent Lewis base (electron pair donor), particularly toward transition metals. Such interactions normally result in the linkage M–C–O, rather than M–O–C; that is, the donated electrons are from an orbital that is primarily carbon in character.

Although s–p mixing was ignored in the above discussion, it does play a role. Computer-generated contour diagrams of the MO's of CO (particularly σ_z and σ_s^*) exhibit shapes that are suggestive of s–p_z mixing, as shown in Figure 3.14.

Hydrogen Halides: Nonbonding Orbitals

The next molecule we will examine is hydrogen fluoride, HF. There are five valence atomic orbitals to consider, H (1s), F (2s), and the three F (2p)'s, and there are eight valence electrons.

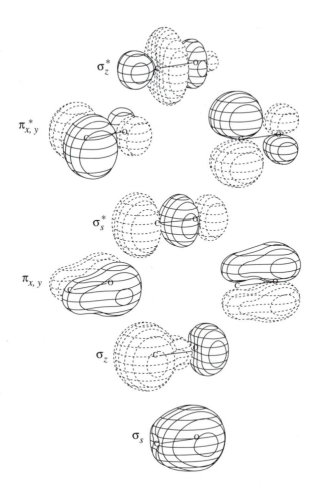

Figure 3.14
Contour diagrams for the molecular orbitals of CO. [Reproduced with permission from Jorgensen, W. L.; Salem, L. *The Organic Chemist's Book of Orbitals*; Academic: New York, 1973; p. 78.]

We must first determine which fluorine orbital(s) overlap with H (1s). The relevant symmetry transformations in the $C_{\infty v}$ point group are:

$$s \rightarrow \Sigma^+$$
$$p_x, p_y \rightarrow \Pi$$
$$p_z \rightarrow \Sigma^+$$

Thus, fluorine's 2s and $2p_z$ orbitals have the proper symmetry to overlap with H (1s), but F $(2p_x)$ and F $(2p_y)$ do not. This can be seen in Figure 3.15.

Bonding between H (1s) and either $2p_x$ or $2p_y$ is impossible because the region of positive $(+/+)$ overlap is exactly canceled by a region of negative $(+/-)$ overlap. The choice is therefore between F (2s) and F $(2p_z)$. The orbital energies are: H (1s), 13.6 eV; F (2s), 40.2 eV; and F (2p), 18.6 eV. Recalling that the effectiveness of overlap is greatest for orbitals having equal or nearly equal energies, the interaction is clearly stronger for $2p_z$. This logic leads to the MO diagram given in Figure 3.16.

Two electrons occupy the bonding orbital (σ) formed from H (1s)–F $(2p_z)$ overlap. The corresponding antibonding orbital is vacant, giving a bond order of 1.0. The six remaining electrons are formally *nonbonding*. Their environments (the fluorine 2s, $2p_x$, and $2p_y$ orbitals) are essentially unchanged in the conversion of the fluorine atom to the HF molecule, and are considered to be *lone pairs*. (In Chapter 4, we will see that an equivalent result is obtained from a localized bonding approach.) However, since 2s and $2p_z$ have the same symmetry, there is some secondary mixing. This destabilizes σ and stabilizes F (2s) so that the latter is not completely nonbonding.

The bonding electrons occupy an orbital that is primarily fluorine in character $(c_F > c_H)$. Fluorine therefore has a fraction over seven valence electrons, and so acquires a partial negative charge; the hydrogen is partially positive. The bond can be described as *polar covalent*.

As would be expected, the diagrams for HCl, HBr, and HI take the same form. However, as atomic size increases from F to I the H–X bonds lengthen and the bond energies decrease (see Table 3.5).

Figure 3.15

The orientations of fluorine's 2p orbitals relative to H (1s) in hydrogen fluoride. Nonzero overlap is possible only for p_z.

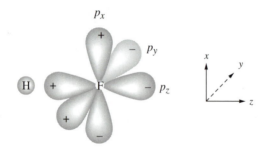

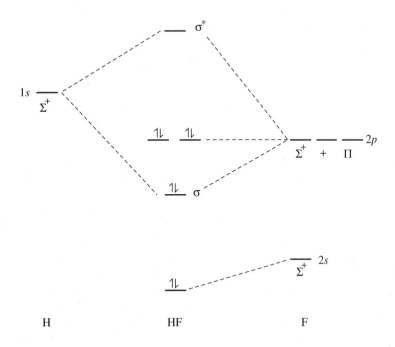

Figure 3.16
Qualitative molecular orbital diagram for hydrogen fluoride, HF. The primary bonding is between H ($1s$) and F ($2p_z$); there is some secondary interaction involving F ($2s$). The F ($2p_x$) and F ($2p_y$) orbitals are nonbonding.

Table 3.5 Bond lengths and dissociation energies for some hydrogen halides

Species	d, pm	D, kJ/mol	Species	d, pm	D, kJ/mol
HF	92	562	HBr	141	363
HCl	127	428	HBr$^+$	146	—
HCl$^+$	132	453	HI	161	295

Sources: Dissociation energies for neutral molecules are taken from Dasent, W. E. *Inorganic Energetics*, 2nd ed.; Cambridge University: London, 1982; p. 103. Remaining values from DeKock, R. L.; Gray, H. B. *Chemical Structure and Bonding*; Benjamin/Cummings: Menlo Park, CA, 1980; p. 259.

Only small changes in bond length and energy are observed upon the one-electron oxidation of any of the hydrogen halides. This is because the bond order is not altered by oxidation (a nonbonding electron is removed). There is generally a slight increase in dissociation energy because of reduced electron–electron repulsions. For example, D for HCl$^+$ is 453 kJ/mol, a 6% increase compared to HCl.

Lithium Fluoride: The Covalent–Ionic Transition

Lithium fluoride is valence isoelectronic with HF, and lithium's single valence electron is of the s type. To a first approximation, then, the MO diagram

Figure 3.17
Contour plot of the primary bonding orbital (σ) of LiF. The majority of the electron density is about the fluorine. [Reproduced with permission from Bader, R. F. W.; Bandrauk, A. D. *J. Chem. Phys.* **1968**, *49*, 1653.]

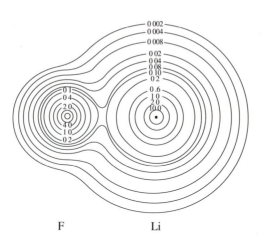

F   Li

for LiF contains the same orbitals as that of HF.[7] There is one major difference, however. Since the VSIE of the Li (2s) orbital is only 5.4 eV, the amount of skewing is great. As a result, the bonding orbital (σ) has much greater fluorine than lithium character ($c_F \gg c_{Li}$). (The reverse is true, of course, for σ^*.) It is exaggerating only slightly to argue that σ is virtually 100% fluorine. This suggests that all eight valence electrons "belong" to that atom; it has acquired a unit negative charge (and lithium a unit positive charge) through the bonding. Said another way, electron transfer from Li to F is implied. In this way the concept of ionic bonding is seen to be a natural part of MO theory.

Whenever a bond is formed between two different kinds of atoms, the resulting distribution of electron density is uneven; one nucleus obtains more than the other. The bond is described as covalent, polar covalent, or ionic depending on how uneven the distribution is. This in turn depends on the valence orbital energies of the bonded atoms.

Theoretical studies of lithium fluoride have been carried out in efforts to evaluate the electron distribution; these quantitative methods incorporate participation by the Li (2p) orbitals and other factors. A contour plot of the bonding orbital of LiF, reproduced from one such study, is given in Figure 3.17.

3.6 Polyatomic Molecules

We will next examine several compounds having the general formula MX_n, binary species having one central and two or more terminal atoms. The molecular orbitals of such compounds are *delocalized* (spread out over all

---

7. Here, "to a first approximation" means ignoring the 2p subshell of Li. This is an oversimplification, but is acceptable for qualitative purposes.

atoms in the system). The procedure for generating them is therefore slightly different from that for diatomics. One common approach has two stages. First, the terminal atoms are combined to give what are sometimes called *SALC's* (*symmetry-adapted linear combination orbitals*). These SALC's are then mixed with the appropriate atomic orbitals of the central atom.

Triatomic Molecules

Beryllium hydride, BeH_2, is a linear molecule (point group $D_{\infty h}$). The line defined by the three atoms contains C_∞, and so is taken to be the z axis. The terminal hydrogens contribute two valence orbitals. The signs of their wavefunctions may be either alike or different. This corresponds to two SALC's (ψ_1, from the like-sign possibility, and ψ_2, from the opposite-sign case), which are described by the equations

$$\psi_1 = \frac{1}{\sqrt{2}} (\phi_1 + \phi_2) \tag{3.13}$$

$$\psi_2 = \frac{1}{\sqrt{2}} (\phi_1 - \phi_2) \tag{3.14}$$

These SALC's are diagrammed in Figure 3.18a (p. 98). Using the $D_{\infty h}$ character table, their symmetries are found to be Σ_g^+ (ψ_1) and Σ_u^+ (ψ_2).

The central beryllium atom has four valence orbitals. The transformations of s- and p-type orbitals in $D_{\infty h}$ symmetry were given previously as Σ_g^+ (s), Π_u ($p_{x,y}$), and Σ_u^+ (p_z). Therefore, ψ_1 overlaps with Be ($2s$) to produce bonding and antibonding MO's (σ_g and σ_g^*, respectively), while ψ_2 has the proper symmetry to overlap with Be ($2p_z$) to form σ_u and σ_u^* orbitals. The $2p_x$ and $2p_y$ atomic orbitals, with π-type symmetry, have no hydrogen counterparts and so are nonbonding. Sketches of the bonding MO's and the completed molecular orbital diagram are given in Figures 3.18b and 3.18c.

The diagram contains six MO's: two bonding, two nonbonding, and two antibonding. All four electrons occupy σ-type bonding orbitals, so the situation corresponds to two single bonds in a Lewis-type formulation. The compound is diamagnetic. The LUMO's are the nonbonding $2p$ orbitals of Be. These relatively stable, empty orbitals are attractive sites for the addition of electron pairs from donor atoms. Therefore, BeH_2 is a potent Lewis acid, and much of its known chemistry derives from that property. (In fact, it polymerizes at room temperature; see Chapter 12.)

Next let us examine the water molecule.[8] The angular geometry of H_2O

---

8. For interesting discussions and other references, see Martin, R. B. *J. Chem. Educ.* **1988**, *65*, 668; and Laing, M. *J. Chem. Educ.* **1987**, *64*, 124.

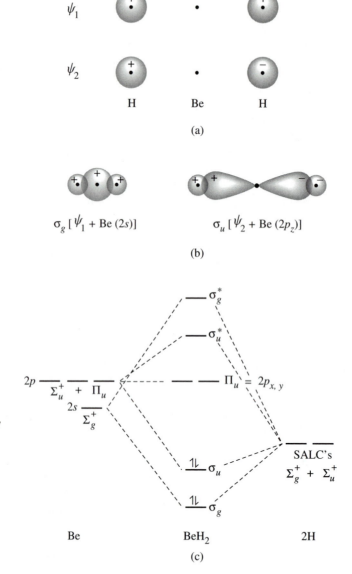

Figure 3.18
LCAO theory as applied to BeH$_2$. (a) Sketches of the two SALC's; ψ_1 has the proper symmetry to interact with Be (2s), while ψ_2 can interact with Be (2p$_z$). (b) The bonding molecular orbitals σ_g and σ_u. (c) The MO energy-level diagram.

(C_{2v} point group) leads to several differences in the MO diagram compared to BeH$_2$. You should be able to determine that the two SALC's belong to the A_1 (ψ_1, with matched signs for the atomic orbital wavefunctions) and B_2 (ψ_2, with opposite signs) representations. The s and p_z orbitals of the central atom each have A_1 symmetry, so both can interact with ψ_1. The primary overlap is with p_z because of its better energy match to H (1s), while

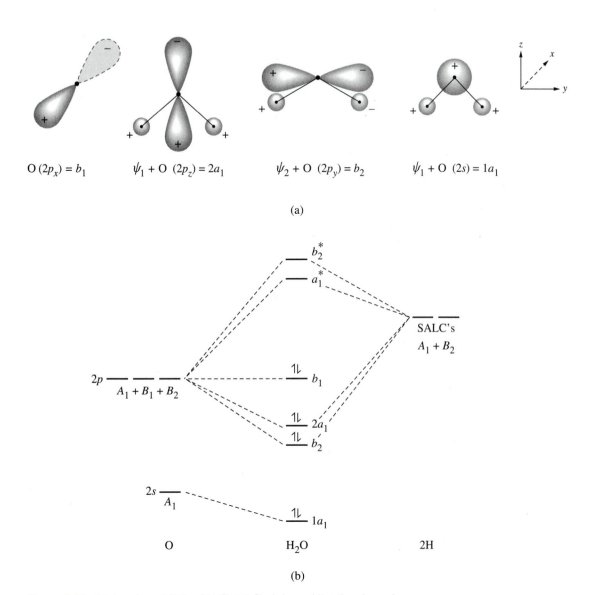

$O\ (2p_x) = b_1$ $\psi_1 + O\ (2p_z) = 2a_1$ $\psi_2 + O\ (2p_y) = b_2$ $\psi_1 + O\ (2s) = 1a_1$

(a)

(b)

Figure 3.19 Molecular orbitals of H_2O. (a) Sketches of the four lowest-energy orbitals. (b) Qualitative energy-level diagram.

secondary mixing occurs with $O\ (2s)$. The $O\ (2p_y)$ orbital has B_2 symmetry and interacts with ψ_2; $O\ (2p_x)$ belongs to B_1 and is nonbonding. Orbital sketches and the MO diagram are shown in Figure 3.19. The orbitals labeled $2a_1$ and b_2 are bonding, while $1a_1$ and b_1 are formally nonbonding (lone-pair) orbitals on the central oxygen ($1a_1$ has some bonding character, as noted above). The antibonding orbitals are vacant.

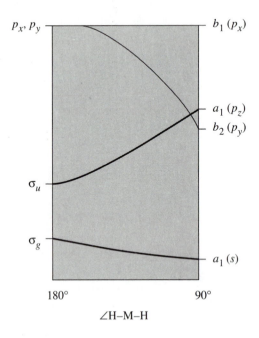

Figure 3.20
Walsh diagram for
MH_2 species.

Walsh Diagrams

The MO diagrams for BeH_2 and H_2O can be correlated using a *Walsh diagram*, where the energies of the MO's are plotted as a function of bond angle.[9] A Walsh diagram is given in Figure 3.20 for MH_2 molecules between angles of 180° and 90°. It can be seen that at $\angle H–M–H = 180°$ the orbital of p_y parentage ($b_2 \rightarrow \pi_y$) becomes nonbonding; the overlap has dropped to zero. It also should be clear that for an MH_2 system having four valence electrons (such as BeH_2) the optimal geometry is linear, because the two filled orbitals have their greatest overall stability at 180°. If eight electrons are present, however (as for H_2O), a nonlinear geometry is preferred.

Silane—A Five-Atom System

In Chapter 2 the three bonds of BF_3 were treated as vectors in order to obtain information about their symmetry properties. The same approach will be used here to determine the representations for the SALC's of the tetrahedral silane (SiH_4) molecule.

9. Walsh, A. D. *J. Chem. Soc.* **1953**, 2260, 2266, 2288, 2296, 2306.

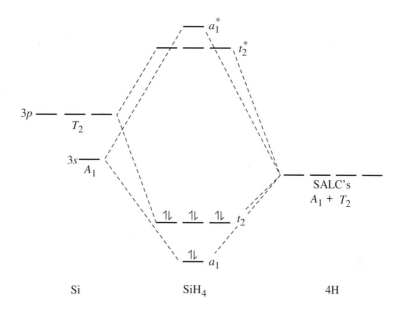

Figure 3.21
Molecular orbital
diagram for silane.

The four Si–H bond vectors give rise to the following reducible representation in T_d symmetry:

	E	$8C_3$	$3C_2$	$6S_4$	$6\sigma_d$
Γ_{SALC}	4	1	0	0	2

(Use the character table in Appendix IV to verify this; review Sections 2.5 and 2.6 if necessary.) We can reduce Γ_{SALC} to its irreducible components either by inspection or with the aid of equation (2.5):

$$\Gamma_{SALC} \to A_1 + T_2$$

Thus, the four SALC's divide into singly and triply degenerate subsets.

Next, we must determine which central atom orbitals have the proper symmetry to interact with these SALC's. Using the right side of the T_d character table, we find that silicon's valence orbitals transform as A_1 ($3s$) and T_2 ($3p_{x,y,z}$ as a degenerate trio).[10] Thus, there is a one-to-one correspondence between the symmetries of the central atom and the SALC's, a situation that leads to the maximum possible number of bonding orbitals. As can be seen in Figure 3.21, populating the resulting MO diagram with eight valence

---

10.   *d*-Type orbitals might be considered as well, and you might have noticed that the $d_{xy,xz,yz}$ subset also has T_2 symmetry. However, overlap by *np* orbitals is generally superior to that by *nd*; also, the energy match to the SALC's is much better with Si ($3p$) than with Si ($3d$).

electrons gives four occupied bonding and four vacant antibonding orbitals. This is consistent with the four single bonds commonly shown in Lewis formulations.

Octahedral Species

The vast majority of MX_6 systems have either octahedral or distorted octahedral shapes. Such geometries are particularly common for coordination compounds and complex ions (see Chapters 15 and 16). A discussion of the molecular orbital diagram of a generalized MX_6 species is therefore of value.

Treating the six M–X bonds as vectors in the O_h point group, the following reducible representation can be generated:

	E	$8C_3$	$6C_2$	$6C_4$	$3C_2$	i	$6S_4$	$8S_6$	$3\sigma_h$	$6\sigma_d$
Γ_{SALC}	6	0	0	2	2	0	0	0	4	2

(Here again, you should take the time to verify this. In doing so it will be useful to know that the second C_2 element is C_4^2, and that S_4 and S_6 are coaxial with the first C_2 and C_3, respectively.) In this case, Γ_{SALC} reduces to $A_{1g} + E_g + T_{1u}$.[11]

The atomic orbitals of the central atom transform as follows: $s \rightarrow A_{1g}$; $p_x, p_y, p_z \rightarrow T_{1u}$; $d_{z^2}, d_{x^2-y^2} \rightarrow E_g$; and $d_{xy}, d_{xz}, d_{yz} \rightarrow T_{2g}$. With nine atomic orbitals and six SALC's, there must be at least three nonbonding orbitals. Since no SALC's have T_{2g} symmetry, it is the $d_{xy,xz,yz}$ subset that is nonbonding. The symmetry matches allow for six bonding and six antibonding orbitals. The complete MO diagram is given in Figure 3.22. It can be seen that any number of electrons between 12 and 18 optimizes the bonding, since the 13th through 18th electrons enter nonbonding orbitals.

This analysis considers only σ bonding. Often, π overlap is important in octahedral complexes; however, that discussion will be delayed until Chapter 16.

Trihydrogen Ions: Three-Center Bonding[12]

Let us next apply LCAO–MO theory to a pair of trihydrogen ions, H_3^+ and H_3^-. We will assume a linear geometry.[13]

---

11. For pictorial diagrams of these SALC's, see Figure 16.14.

12. DeKock, R. L.; Bosma, W. B. *J. Chem. Educ.* **1988**, *65*, 194.

13. Actually, quantitative MO calculations indicate that a triangular geometry is preferred for H_3^+. See the reference in footnote 12 and citations therein.

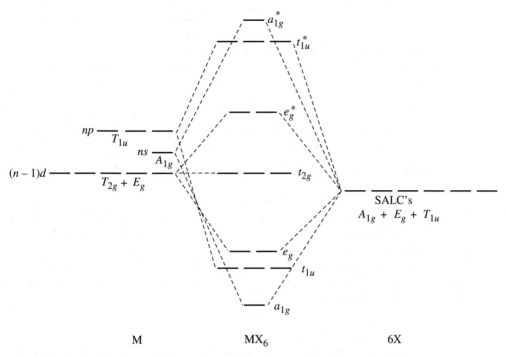

Figure 3.22 Idealized MO diagram (σ only) for the octahedral species MX_6. There are six bonding, three nonbonding, and six antibonding orbitals.

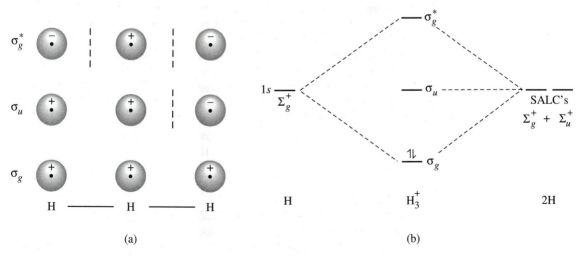

Figure 3.23 The LCAO–MO result for linear H_3^+. (a) Sketches of the three molecular orbitals. (b) The energy-level diagram. A three-center, two-electron bond is indicated.

It should be apparent that the two SALC's for H_3^+ are the same as for BeH_2. However, the central hydrogen has only one valence orbital, which transforms as Σ_g^+ in $D_{\infty h}$ symmetry. Thus, overlap with the SALC labeled ψ_1 in Figure 3.18a creates bonding and antibonding MO's, while ψ_2 is nonbonding. The resulting MO diagram is shown in Figure 3.23. Only the bonding orbital, σ_g, is occupied. All three hydrogens contribute to that orbital; in essence, this species contains one bond "spread out" over three atoms. Cases of this type are referred to as *three-center, two-electron* bonds.

The MO diagram for the H_3^- anion is similar, of course, except that the nonbonding orbital (σ_u) is also occupied. The H_3^- anion is said to contain a *three-center, four-electron* bond.

Bibliography

Verkade, J. G. *A Pictorial Approach to Molecular Bonding*; Springer-Verlag: New York, 1986.

Douglas, B. E.; Hollingsworth, C. A. *Symmetry in Bonding and Spectra*; Academic: Orlando, FL, 1985.

Kettle, S. F. A. *Symmetry and Structure*; Wiley: New York, 1985.

Murrell, J. N.; Kettle, S. F. A.; Tedder, J. M. *The Chemical Bond*, 2nd ed.; Wiley: New York, 1985.

Atkins, P. W. *Molecular Quantum Mechanics*; Oxford University: New York, 1983.

McQuarrie, D. A. *Quantum Chemistry*; University Science: Mill Valley, CA, 1983.

Ballhausen, C. J.; Gray, H. B. *Molecular Electronic Structures*; Benjamin/Cummings: Menlo Park, CA, 1980.

DeKock, R. L.; Gray, H. B. *Chemical Structure and Bonding*; Benjamin/Cummings: Menlo Park, CA, 1980.

Hatfield, W. E.: Parker, W. E. *Symmetry in Chemical Bonding and Structure*; Merrill: Columbus, OH, 1974.

Jorgensen, W. L.; Salem, L. *The Organic Chemist's Book of Orbitals*; Academic: New York, 1973.

Orchin, M.; Jaffe, H. H. *Symmetry, Orbitals, and Spectra*; Wiley: New York, 1971.

Questions and Problems

1. Briefly explain in your own words what is meant by the LCAO approach.

2. Differentiate among σ, π, and δ bonds via their characteristic numbers of lobes and nodes.

3. Use the $C_{\infty v}$ character table (Appendix IV) to answer the following questions about orbital overlap in that point group:
   (a) Which p-type orbital has the proper symmetry to form a σ bond with an s orbital from an adjacent atom?
   (b) Which d-type orbital has the proper symmetry to form a σ bond with an s orbital from an adjacent atom?
   (c) What two d orbitals can form a π bond with the $p_{x,y}$ pair?
   (d) What type of interaction is possible between two d_{xy} orbitals of adjacent atoms?

4. Which of the following species should be the most stable?

   BeH,   BeH$^+$,   BeH$^-$

   Defend your answer.

5. (a) Construct an MO diagram for lithium hydride, LiH. Clearly indicate the relative orbital energies, and calculate the bond order.
   (b) The experimental bond length of LiH is 160 pm, and the dissociation energy is 245 kJ/mol. Compare these values to those for H_2 and Li_2. Are they reasonable?
   (c) Which atom, Li or H, bears a partial negative charge in this compound? Explain.

6. For each of the following pairs, select the species having the greater bond order; defend your answers:
   (a) H_2 or H_2^+ (b) C_2 or C_2^+ (c) C_2 or C_2^{2-}
   (d) F_2 or F_2^+

7. Sodium and chlorine belong to the same period of the periodic table, and both Na_2 and Cl_2 have bond orders of 1.0. However, the dissociation energy of Cl_2 is more than three times greater than that of Na_2. Rationalize.

8. The bond energy of C_2 is more than twice that of B_2. (The ratio of bond orders is exactly 2:1.) The bond energy of N_2 is more than three times that of B_2. Explain.

9. The peroxide ion, O_2^{2-}, is isoelectronic with F_2. Which species do you believe contains the longer bond? Which bond is stronger? After answering, look up the experimental values (Tables 3.2 and 3.4).

10. Demonstrate that s–p mixing has no effect on calculated bond order for the homonuclear diatomics of Groups 13–17, but does influence the magnetic properties of those compounds.

11. The VSIE's of Al are 5.9 eV ($3p$) and 11.3 eV ($3s$). Do you expect $Al_2(g)$ to be diamagnetic or paramagnetic? Why?

12. (a) Explain why the first ionization energy of N_2 is greater than that of N.
    (b) Predict which is more favorable (ie, more positive): the first electron affinity of N_2 or that of N. Defend your answer.

13. Predict the bond order of:
    (a) Cu_2 (b) "Zn_2"

14. Speculate on the prospect of synthesizing the compound ScH. [*Hint*: Would the bond order be greater than 0?]

15. Early in the chapter we stated that δ bonds are found only between atoms that are also linked by both σ and π bonds. Carefully examine the MO diagram for Cr_2 (Figure 3.11). Do you see why this is true?

16. Construct an MO diagram for the BO^- ion, and use it to answer these questions:
    (a) What is the bond order of this anion?
    (b) Is this species diamagnetic or paramagnetic?
    (c) What are the HOMO and the LUMO?
    (d) Is the HOMO primarily boron or oxygen in character?

17. Predict which of the following reactions is the more favored and explain:

    $$NO + CN \longrightarrow NO^+ + CN^-$$

    $$NO + CN \longrightarrow NO^- + CN^+$$

18. Construct an MO diagram for the hydroxyl radical, $\cdot OH$, and use it to answer these questions:
    (a) What is the bond order?
    (b) How does this species compare in stability to HF?
    (c) Do you expect $\cdot OH$ to have a strong tendency to add one electron? Why?
    (d) Should this species have a strong tendency to add two electrons? Why not?

19. The two compounds BeO and LiF are isoelectronic. They differ in polarity, with BeO being the less polar of the two. Rationalize on the basis of VSIE's.

20. Use MO theory to explain why CO is a stronger Lewis base than is N_2. (The two are isoelectronic.)

21. (a) Construct an MO diagram for the singlet (diamagnetic) form of methylene, $:CH_2$. The H–C–H bond angle is about 100°.
    (b) Construct an MO diagram for the triplet (paramagnetic) form of methylene, $\cdot \overset{\cdot}{C}H_2$. This species is linear, with two unpaired electrons per molecule.

22. Do you expect the amide ion, NH_2^-, to be linear or bent? Defend your answer.

23. Use group theory to construct an MO diagram for the hypothetical square planar silane molecule (D_{4h} point group). Ignore d-orbital participation. Determine how many electrons occupy bonding orbitals, and compare to the tetrahedral geometry (Figure 3.21). Do you see why SiH_4 occurs in the tetrahedral form?

24. Construct an MO diagram for borane, BH_3, which belongs to the D_{3h} point group.
    (a) How many electrons occupy bonding orbitals?
    (b) Which of the p orbitals is nonbonding? Why?
    (c) Should BH_3 behave as a Lewis acid or a Lewis base? Explain why the reaction

    $$BH_3 + H^- \longrightarrow BH_4^-$$

    is exothermic.

25. Use group theory to construct an MO diagram for fluorosilane, SiH_3F. Begin by assigning the point group. Use the appropriate character table to generate a reducible representation for the four SALC's, and then reduce it. Finally, overlap these SALC's with appropriate atomic orbitals of silicon. Compare your diagram to that for SiH_4.

26. The MO diagram for the (hypothetical) linear species H_2Ne should be the same as that for BeH_2, except for the number of electrons.
    (a) How many bonding, nonbonding, and antibonding electrons are present?
    (b) Although there are more bonding than antibonding electrons, this species apparently is not stable. Speculate on the reason. [*Hint*: How do you expect H_2Ne to compare in stability to $H_2 + Ne$?]

27. Modify appropriate MO diagrams from the chapter to predict whether the following species are stable; explain your reasoning:
    (a) BH_2^+ (b) BH_2^- (c) Be_3

*28. Work through the paper by DeKock and Bosma concerning the trihydrogen system (footnote 12), with special emphasis on H_3^-.
    (a) Because the geometry is triangular (D_{3h}), there is no central atom. How is the approach used for the construction of MO's different from that for MX_n systems?
    (b) Explain in your own words why H_3^- favors triangular, rather than linear, geometry.

*29. Read the paper about "Proton Power..." (*J. Chem. Educ.* **1988**, *65*, 976). Answer these questions after examining the article:
    (a) In IE_1, H < H_2 < He. Why?
    (b) The imaginary movement of one proton from the neon nucleus to produce HF significantly stabilizes the p_z orbital but slightly destabilizes p_x and p_y. Why?
    (c) Can this approach be used to rationalize the existence of NaH? (With 12 protons, it can be derived from Mg.) What assumption must be made?

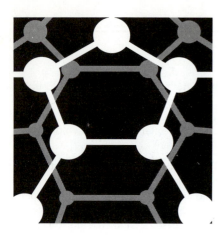

4

Covalent Bonds and Bond Energies

For small molecules and ions, molecular orbital theory generally gives the most accurate information we can obtain about covalent bonding. Unfortunately, MO theory becomes progressively more difficult to employ as molecules become more complex. For many-atom systems, chemists have long used an alternate approach that usually yields results similar to (or identical with) those of MO theory. In this method covalently bonded species are depicted using a system of elemental symbols, dots, and dashes. The pictorial representations that result are often called *Lewis structures*, since G. N. Lewis was a pioneer of this system.[1]

4.1 Guides for Writing Lewis Structures

Recall that only valence electrons participate in chemical bonding; hence, only those electrons are represented in Lewis structures. For example, since

---

1. Lewis, G. N. *J. Am. Chem. Soc.* **1916**, *38*, 762. For summaries of the contributions of Lewis and others to the development of covalent bonding theory, see Klein, D. J.; Trinajstić, N. *J. Chem. Educ.* **1990**, *67*, 633; Jensen, W. B. *J. Chem. Educ.* **1984**, *61*, 191; Pauling, L. *J. Chem. Educ.* **1984**, *61*, 201.

a phosphorus atom has five valence electrons, its Lewis structure is written

$:\overset{\cdot}{\underset{\cdot}{P}}\cdot$

For compounds and polyatomic ions, dashed lines are often used to designate bonded electron pairs and dots are used for nonbonded electrons. This is illustrated by the structure for hydrogen fluoride:

$H-\overset{\cdot\cdot}{\underset{\cdot\cdot}{F}}:$

Notice that this diagram indicates the presence of one bond, with three nonbonded (lone) electron pairs on fluorine. This is equivalent to the conclusions from the MO analysis (see Section 3.5).

The major tenet upon which Lewis structures are based is that atoms have a tendency to fill their valence orbitals. For example, a hydrogen atom has only one valence orbital ($1s$), which contains only one electron. Therefore, hydrogen has a tendency to form one covalent bond in order to acquire a share of a second electron and achieve a "stable duet." Most of the main group elements have four valence orbitals (one s- and three p-type) and thus bond to achieve *octets*. For nitrogen, with five valence electrons, this requires the formation of three bonds. Oxygen (six valence electrons) normally forms two bonds. Fluorine needs to form only one. Accordingly, the Lewis structure of OF_2 is

$:\overset{\cdot\cdot}{\underset{\cdot\cdot}{F}}-\overset{\cdot\cdot}{\underset{\cdot\cdot}{O}}-\overset{\cdot\cdot}{\underset{\cdot\cdot}{F}}:$

Some useful guidelines for writing Lewis structures are as follows:

1. For species having the general formula MX_n, M (the unique atom) is usually centrally located, and there are n M–X bonds. Thus, the nitrate ion, NO_3^-, has the framework

and not

or

2. Hydrogen and the halogens generally form one bond. The Group 16 elements most often form two bonds, Group 15 elements tend to form three bonds, and those of Group 14 typically form four bonds.

3. The Group 2 and Group 13 elements are often *electron-deficient*.[2] That is, they appear in covalent compounds surrounded by fewer than eight valence electrons. This can be rationalized using beryllium as an example. Since the Be atom has only two valence electrons, it can gain only two more by sharing. It is therefore limited to a maximum of four. In the same way, the members of Group 13 are normally limited to six valence electrons in their neutral compounds. (We will see later that such species often achieve octets through Lewis acid–base reactions.)

4. The elements beyond magnesium in the periodic table are *hypervalent* (expand their valence shells to contain more than eight electrons) in certain compounds and ions. This is accomplished through the use of *d* orbitals.

5. Because small internuclear distances are necessary for effective π-type overlap, carbon, nitrogen, oxygen, phosphorus, and sulfur are the main group elements that commonly form strong π bonds. Thus, double and triple bonds are likely only if some combination of two of those five elements are involved. (However, as we will see later in this chapter, under the appropriate conditions, other elements can form either "true" or partial π bonds as well.)

6. If the octet rule is obeyed, a useful equation is

$$\text{Number of bonds} = \frac{(\text{Total } e^- \text{ needed}) - (\text{Valence } e^- \text{ available})}{2} \qquad \textbf{(4.1)}$$

We demonstrate the use of this equation below.

The above statements are guides, not rules; each of the six can be violated. But they will help give you a feel for whether a given Lewis structure is reasonable. We now illustrate their use in several examples.

Diphosphine, P_2H_4: The framework must be

$$\begin{array}{ccc} \text{H} & & \text{H} \\ \diagdown & & \diagup \\ & \text{P---P} & \\ \diagup & & \diagdown \\ \text{H} & & \text{H} \end{array}$$

With reference to equation (4.1), the total e^- needed is the sum of the requirements for the individual atoms. Each hydrogen needs two electrons, while phosphorus requires an octet; thus, $4(2) + 2(8) = 24e^-$ are needed.

---

2. This statement applies primarily to Be in Group 2. The other members of that family form predominantly ionic bonds in their most common compounds.

Valence electrons available is the total number of valence electrons present, one from each hydrogen and five from each phosphorus, for a total of 14. The predicted number of bonds is $(24 - 14)/2 = 5$. Adding a lone pair to each phosphorus to complete its octet gives

$$\begin{array}{ccc} \text{H} & & \text{H} \\ \diagdown & & \diagup \\ & \ddot{\text{P}}-\ddot{\text{P}} & \\ \diagup & & \diagdown \\ \text{H} & & \text{H} \end{array}$$

Notice that the number of electrons represented in the diagram exactly equals the number available. This provides a useful check of the correctness of the structure.

The Bromite Ion, BrO$_2^-$:   Bromine is the central atom (guide 1). Since the final structure should show an octet about each atom, the number of electrons needed is 24. The electrons available must account for the negative charge, with the correct number being $7 + 2(6) + 1 = 20$.

(The logic of adding one electron should be clear before you continue. The bromite ion has a negative charge because it has one more electron than is calculated by simply summing those of the three neutral atoms. For a divalent anion, it is necessary to add two electrons in calculating the number of electrons available, and so on. In working with cations, we must subtract rather than add.)

The expected number of bonds is therefore $(24 - 20)/2 = 2$. Both Br–O linkages must be single bonds, giving

$$[:\ddot{\text{O}}\!-\!\ddot{\text{Br}}\!-\!\ddot{\text{O}}:]^-$$

Brackets are used to indicate that the species is an ion, and a superscript denotes the charge.

Hydrogen Cyanide, HCN: The carbon atom is centrally located in this molecule (H---C---N). Next, we use equation (4.1):

$$e^- \text{ needed} = 1(2) + 2(8) = 18$$
$$e^- \text{ available} = 1 + 4 + 5 = 10$$
$$\text{Number of bonds} = (18 - 10)/2 = 4$$

The list of elements forming multiple bonds (guide 5) includes carbon and nitrogen, but not hydrogen. There is only one possibility:

$$\text{H}-\text{C}\equiv\text{N}:$$

Sulfur Tetrafluoride, SF_4: The framework shows sulfur as the central atom and four S–F linkages. The electron count is

$$e^- \text{ needed} = 5(8) = 40$$
$$e^- \text{ available} = 6 + 4(7) = 34$$
$$\text{Number of bonds} = (40 - 34)/2 = 3(?)$$

There is a contradiction. At least four bonds are necessary to hold the molecule together, yet equation (4.1) indicates only three to be present.

The source of the problem can be found by rereading guide 6. The equation holds *only if the octet rule is obeyed.* The fact that we arrived at a contradiction is a strong indication that the central atom violates the octet rule. (Guide 4 indicates that sulfur is capable of doing so.) A different approach must be taken. When an octet rule violation occurs, the available electrons are allocated in this manner: First, form the minimum number of bonds necessary to hold the framework together, assuming that only single bonds are present. Next, complete the octets (or duets for hydrogens) for all terminal atoms. Remaining electrons (if any) are lone pairs on the central atom.

Of the 34 valence electrons of SF_4, 8 are needed to form four single bonds; 24 of the 26 remaining electrons are used to complete the octets of the fluorines. Two electrons remain, so there is a lone pair on sulfur. The correct structure is

$$
\begin{array}{c}
\ddot{\textrm{:F:}} \\
\diagdown \\
\ddot{\textrm{:F}}\!-\!\ddot{\textrm{S}}\!\cdot \\
\diagup \quad \diagdown \ddot{\textrm{F:}} \\
\ddot{\textrm{:F:}}
\end{array}
$$

Notice that sulfur does in fact have 10 valence electrons (four bonds plus one lone pair), violating the octet rule.

Resonance and Formal Charge

Sulfur Dioxide, SO_2: By this point you should be able to determine that the central atom is sulfur and that there are three bonds. One double and one single bond are therefore indicated. This can be accomplished in two ways:

$$:\!\ddot{\textrm{O}}\!-\!\ddot{\textrm{S}}\!=\!\ddot{\textrm{O}}\!: \quad \longleftrightarrow \quad :\!\ddot{\textrm{O}}\!=\!\ddot{\textrm{S}}\!-\!\ddot{\textrm{O}}\!:$$

These two structures are equivalent, but not identical. There are many molecules and ions for which more than one acceptable Lewis structure can be drawn; such species are said to exhibit *resonance.* It is important to

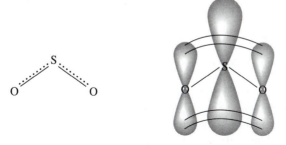

Figure 4.1
Delocalized π bonding in the SO_2 resonance hybrid. The π molecular orbital contains contributions from three atoms.

recognize that neither Lewis structure is correct. The SO_2 molecule is a *hybrid* (a combination of contributions from the individual structures). The S–O bonds are known from experiment to be identical in length, and are presumed equal in strength—between single and double. Said another way, the π bond is "spread out" over three atoms (Figure 4.1). As is shown above, a double-headed arrow is used to indicate resonance.

***The Thiocyanate Ion,* SCN⁻:** Three reasonable Lewis structures can be written for thiocyanate.

$$[:\ddot{S}=C=\ddot{N}:]^- \quad \longleftrightarrow \quad [:S\equiv C-\ddot{N}:]^- \quad \longleftrightarrow \quad [:\ddot{S}-C\equiv N:]^-$$

$$\text{(a)} \qquad\qquad\qquad\qquad \text{(b)} \qquad\qquad\qquad\qquad \text{(c)}$$

Like SO_2, this anion exhibits resonance. Notice that all the atoms of each structure have an octet of electrons, and each structure has exactly four bonds. There is an important difference between SCN⁻ and SO_2, however; in SCN⁻, the resonance structures are not equivalent. For this reason, they do not make equal contributions to the hybrid.

Which structure is the most important? We will answer this question through an analysis of *formal charges*. Formal charge may be defined as the charge assigned to a given atom assuming that the bonds it forms are completely (ie, nonpolar) covalent. One way to assign formal charges is to "break each bond in half"; that is, assign one electron of a given bond to each element. Comparison is then made of the number of valence electrons the atom possesses to its valence electrons in its neutral state.

As an example, breaking the bonds of structure (a) above gives

$$:\ddot{S}: \} :C: \{ :\ddot{N}:$$

Sulfur is seen to have six electrons after this treatment. Since a neutral sulfur atom also has six valence electrons, the electron count is unchanged; the formal charge of S is 0. Carbon is left with four electrons, so its formal charge is also 0. Nitrogen is allocated six electrons, and so is assigned a formal charge of -1.

The results for the other resonance structures are

Structure (b): $S = +1,$ $C = 0,$ $N = -2$

Structure (c): $S = -1,$ $C = 0,$ $N = 0$

The sum of the formal charges for each structure equals -1, the overall charge on the ion. This is always the case; that is, the total of all formal charges must equal the charge of the species. For neutral molecules, of course, this sum is 0.

How can formal charges be used to evaluate resonance structures? In general, the major contributors to a resonance hybrid (1) minimize nonzero formal charges; (2) avoid nonzero formal charges (especially those of like sign) on adjacent atoms; and (3) place negative formal charges on atoms of high electronegativity and positive formal charges on atoms of low electronegativity.

These rules are listed (more or less) in order of importance. Applying them to thiocyanate, structure (b) is clearly the least important of the three. The third criterion can be used to compare structures (a) and (c). Since nitrogen is more electronegative than sulfur, it can better accommodate the negative charge. The order of importance is therefore

(a) > (c) >> (b)

What are the implications? Since (a) and (c) are the most important structures, the C–N bond is expected to be strong—between a double and a triple bond. The C–S bond is weaker—between single and double. These predictions are supported by the reaction chemistry of thiocyanate; the C–S bond is typically cleaved more readily than the C–N linkage.

We can gain insight into the bonding in species that have an odd number of electrons through formal charge considerations. For example, the two best resonance structures that can be drawn for nitric oxide are

$$\ddot{\text{:}}\overset{\displaystyle\cdot}{\text{N}}=\ddot{\text{O}}\text{:} \quad \longleftrightarrow \quad \overset{\ominus}{\ddot{\text{:}}\text{N}}=\overset{\displaystyle\cdot\ \oplus}{\text{O}}\text{:}$$

One atom in each structure violates the octet rule. However, the first is seen to be superior according to formal charge criteria. (The nonzero formal charges are indicated by circled charges.) This suggests that the unpaired electron density is concentrated primarily on nitrogen, a prediction borne out by its reaction chemistry (see Chapter 14).[3]

Formal charges are also useful for the prediction and rationalization of the framework linkages (*topologies*) of molecules and ions. For example,

---

3. Recall that this same conclusion was reached through MO theory in Chapter 3.

consider the fulminate ion, CNO^-, where nitrogen is the central atom. This is a highly reactive species; it is used (in the form of salts such as lead fulminate) in blasting caps to provide the necessary energy to set off explosives. The source of energy is the conversion of CNO^- to NCO^-, the cyanate ion:

$$CNO^- \longrightarrow NCO^- \qquad \Delta H << 0 \qquad\qquad (4.2)$$

Why is this reaction exothermic? Like thiocyanate, three resonance structures can be drawn for fulminate (and also for the product, NCO^-). The major resonance structure for CNO^- is

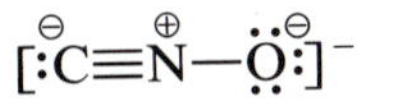

$$[:C \equiv \overset{\oplus}{N} - \overset{..}{\underset{..}{O}}:]^-$$

with formal charges judged unfavorable by all the criteria given above. For the product cyanate ion, the primary contributor to the hybrid is

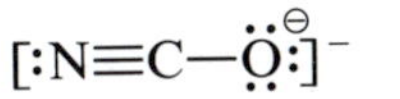

$$[:N \equiv C - \overset{..}{\underset{..}{O}}:]^-$$

which is clearly superior. Thus, N---C---O is the preferred framework, and this is reflected in the exothermicity of the fulminate-to-cyanate conversion.

Nitrous acid, HNO_2, is a member of an important class of compounds called *oxyacids* (ternary compounds containing hydrogen, oxygen, and one other element). Several topologies are possible for HNO_2:

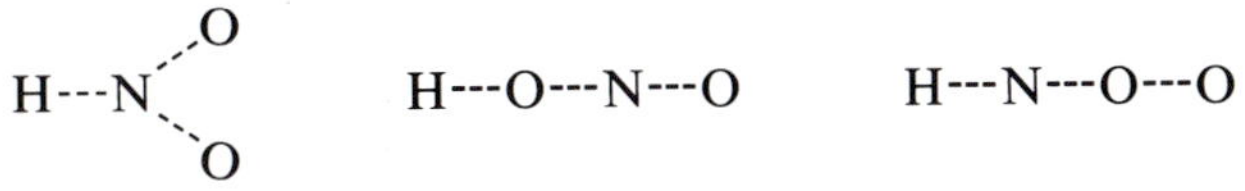

(a)　　　　　　(b)　　　　　(c)

It can be determined from equation (4.1) that four bonds are present. The best Lewis structures for these frameworks are

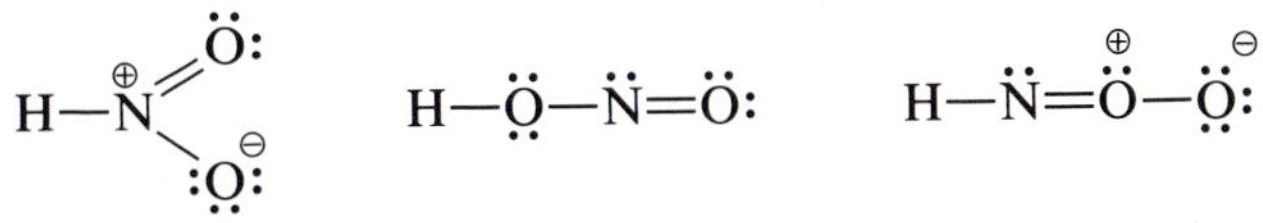

(a)　　　　　　(b)　　　　　(c)

Formal charge criteria suggest that structure (b) is the best option, and this is, in fact, the actual topology. Incidentally, a second, minor resonance structure can be written for (b) (what is it?), but the major contributor determines the topology.

Secondary Resonance

The Lewis diagram for boron trifluoride is

$$:\ddot{F}-B\overset{\displaystyle \ddot{F}:}{\underset{\displaystyle \ddot{F}:}{<}}$$

Here, each fluorine has an octet, and all formal charges are 0. Although the boron is electron-deficient, there is a way for it to achieve an octet. It involves shifting a fluorine lone pair into a bonding position:

$$:\ddot{F}-B\overset{\displaystyle \ddot{F}:}{\underset{\displaystyle \ddot{F}:}{<}} \longleftrightarrow \overset{\oplus}{:}F\overset{\ominus}{=}B\overset{\displaystyle \ddot{F}:}{\underset{\displaystyle \ddot{F}:}{<}}$$

This is a different kind of resonance structure, and, of course, there are two additional, equivalent structures having double bonds to the other fluorines. Note the nonzero formal charges that arise. These additional resonance structures violate all the formal charge criteria, and are therefore minor contributors to the hybrid. Their contributions are not insignificant, however. In fact, infrared and other types of spectroscopic data indicate that the B–F bonds in BF_3 are all equivalent, with bond strengths greater than expected for normal single bonds (see Chapters 13 and 21).

This is an example of *secondary resonance*. Secondary resonance structures are derived by converting lone pairs to bonding pairs. The extra bond (actually, a partial bond) is of the π type, and is sometimes called a *dative* bond, meaning one in which both electrons originated from the same atom.

Another example of secondary resonance occurs in the sulfate ion, SO_4^{2-}. Application of the Lewis structure guides gives

$$\left[\overset{\ominus}{:}\ddot{O}\overset{\displaystyle \overset{\ominus}{\ddot{O}}:}{\underset{\displaystyle \underset{\ominus}{\ddot{O}}:}{\underset{|}{-S\overset{(2+)}{-}\ddot{O}:}}}\overset{\ominus}{} \right]^{2-}$$

Here, each atom has an octet, but the formal charge distribution is unfavorable. These charges can be improved by secondary resonance:

$$\left[\overset{\ominus}{:}\ddot{O}\overset{\displaystyle \overset{\displaystyle :O:}{\|}}{\underset{\displaystyle \underset{\|}{:O:}}{-S-\overset{\ominus}{\ddot{O}}:}} \right]^{2-}$$

(Five additional, equivalent structures also can be drawn.) Since sulfur and oxygen both form strong π bonds, these secondary resonance structures are important contributors to the hybrid. The chemical behavior of SO_4^{2-} is consistent with a "bond-and-a-half" for each linkage. The S–O bonds are chemically and symmetrically equivalent, and each oxygen bears an average formal charge of $-\frac{1}{2}$.

How does sulfur accommodate the extra electrons? The most widely accepted explanation involves the participation of d orbitals in dative π bonding. Electron density from an oxygen lone pair residing in a formally nonbonding p-type orbital is donated into an empty sulfur d orbital having the proper symmetry for overlap. This is often described as a $(p \rightarrow d)\pi$ dative bond, and is illustrated in Figure 4.2.

When writing secondary resonance structures, caution must be exercised. For example, the tellurate ion, TeO_4^{2-}, is valence isoelectronic with sulfate. Secondary resonance is of lesser importance in this case, however, because tellurium's larger size makes it less efficient at π bonding. The elements most prone to hypervalency through secondary resonance (π-type Lewis acid behavior) are those of the second short period (especially Si, P, and S) and the transition metals.

Ionic Resonance

The Lewis diagram for LiF might be written as

$$Li-\ddot{\ddot{F}}:$$

where both atoms have 0 formal charges. A better description, however, is probably

$$[Li]^+[:\ddot{\ddot{F}}:]^-$$

in which the transfer of an electron from lithium to fluorine (resulting in an ionic bond) is indicated. This violates the formal charge rules, but is legitimate because of the large difference in electronegativity between Li and

Figure 4.2
Dative $(p \rightarrow d)\pi$ bonding in SO_4^{2-}. The darker orbital is an oxygen lone pair; the lighter orbital is a sulfur $3d$ orbital of the same symmetry.

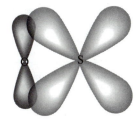

F. Recall that electron transfer between these elements is also consistent with MO theory (Section 3.5).

Hydrogen fluoride represents a somewhat different situation; it is best considered to be a resonance hybrid:

$$H-\ddot{\underset{..}{F}}: \quad \longleftrightarrow \quad [H]^+[:\ddot{\underset{..}{F}}:]^-$$

(a)   (b)

Which is the major contributor? The physical properties of pure HF (relatively low melting and boiling temperatures, low electrical conductivity, etc.) argue for the covalent structure. There is no question, however, that structure (b) plays a role. This is an example of *ionic resonance*. Ionic resonance structures are created by converting a bond pair to a lone pair on a highly electronegative element.

The extent of ionic resonance depends on the difference in electronegativity between the bonding atoms. For example, the resonance structure

$$[:\ddot{\underset{..}{X}}]^+[:\ddot{\underset{..}{F}}:]^-$$

is of greater importance for X = I than for X = F. We will return to this topic later in the chapter.

4.2   Homonuclear Bond Energies

Lewis structures provide a basis for understanding the overall bonding properties of any covalent compound. Next, we will examine individual bond linkages. Which is stronger, the F–F bond of F_2 or the Cl–Cl bond of Cl_2? Why are the bonds in SiF_4 stronger than those in $SiBr_4$? These types of questions are addressed in this section.

Bond Dissociation Energies

Let us begin with homonuclear diatomic molecules. The bond energies for such species are easily obtained; they are dissociation energies (D), experimentally determined by measuring the energy change for the reaction:

$$M_2(g) \quad \longrightarrow \quad 2M(g) \qquad \Delta H = D \tag{4.3}$$

A list of dissociation energies for homonuclear diatomics of the main group elements is given in periodic table format in Table 4.1.[4]

---

4. These values are corrected to a temperature of 0 K, and would be slightly larger (typically 3–5 kJ/mol) at 298 K. The so-called "zero-point energy" is excluded. For a more complete explanation, see Dasent, W. E. *Inorganic Energetics*, 2nd ed.; Cambridge University: London, 1982; Chapters 1 and 4.

Table 4.1 Bond dissociation energies for homonuclear diatomic molecules of the main group elements

Group 1	Group 13	Group 14	Group 15	Group 16	Group 17
H_2 432					
Li_2 110	B_2 274	C_2 593	N_2 942	O_2 494	F_2 154
Na_2 72		Si_2 314	P_2 483	S_2 425	Cl_2 240
K_2 49		Ge_2 272	As_2 380	Se_2 305	Br_2 190
Rb_2 47		Sn_2 192	Sb_2 295	Te_2 223	I_2 149
Cs_2 44		Pb_2 96	Bi_2 195		

Note: Values are in kilojoules per mole and are corrected to 0 K, in most cases accurate to ± 4 kJ/mol or less.
Sources: Dasent, W. E. *Inorganic Energetics*, 2nd ed.; Cambridge University: London, 1982; Chapter 4; DeKock, R. L.; Gray, H. B. *Chemical Structure and Bonding*; Benjamin/Cummings: Menlo Park, CA, 1980; p. 229.

Three observations become apparent from analysis of Table 4.1:

1. Bond order plays an important role. The Group 15 molecules (N_2, P_2, etc.) contain triple bonds. It is therefore not surprising that the Group 15 diatomic has the greatest dissociation energy for any period. The Group 14 and Group 16 compounds have bond orders of 2.0 and are intermediate in energy. The others contain only single bonds, and their dissociation energies are comparatively small.

2. There is a size effect. For example, Li_2 has a greater dissociation energy than does Na_2, which in turn is greater than that of K_2. The same trends are observed for the other groups, with the exception of the very low value for diatomic fluorine.

3. Lone-pair repulsions on adjacent atoms decrease bond energies. In particular, the bond is weakened if both bonded atoms have one or more lone pairs.[5] This effect is somewhat masked here (it will be seen more clearly later), but is apparent for the case of fluorine, the exception noted

---

5. It has been argued that bond weakening occurs even if only one of the involved atoms contains lone pairs. See Sanderson, R. T. *Polar Covalence*; Academic: New York, 1983; Chapter 3.

above. Since F_2 has three lone pairs on each atom (atoms that are quite small in size), severe repulsions are created between lone pairs on the adjacent fluorines, weakening the F–F bond. This bond is, in fact, weaker than the longer bonds in Cl_2 and Br_2.

Thermochemical Bond Energies

One problem involved in analyzing the values in Table 4.1 is that the σ and π contributions are not separable. It would be useful to be able to compare only σ bonds. For example, we might wish to compare the strength of the N–N bond of hydrazine, N_2H_4, to that of the O–O bond of hydrogen peroxide, H_2O_2. Unfortunately, such data are experimentally difficult to obtain for two reasons:

1. There is a substituent effect. For species having three or more atoms, the strength of a given type of bond varies with its environment because of steric, hybridization, and other factors (see below).

2. Because molecules containing more than one bond are being considered, the experimental measurement of bond dissociation energies is not simple. Even for cases in which stepwise dissociation can be observed, a different energy change is likely to accompany each step. For example, consider the stepwise O–H bond cleavage of water:

$$H\text{–}O\text{–}H \longrightarrow H + O\text{–}H \qquad \Delta H = 499 \text{ kJ/mol} \qquad \textbf{(4.4)}$$

$$H\text{–}O \longrightarrow H + O \qquad \Delta H = 428 \text{ kJ/mol} \qquad \textbf{(4.5)}$$

Table 4.2 Thermochemical single-bond energies for homonuclear bonding by elements of Groups 14–16

Group 14	Group 15	Group 16
C–C	N–N	O–O
356	159	143
Si–Si	P–P	S–S
226	200	266
Ge–Ge	As–As	Se–Se
188	177	193
Sn–Sn		
152		

Note: Values are in kilojoules per mole and are for bonding in the free elements except for N (N_2H_4) and O (H_2O_2).
Source: Dasent, W. E. *Inorganic Energetics*, 2nd ed.; Cambridge University: London, 1982; p. 108.

In such cases, it is customary to average the values; for H_2O the result is about 464 kJ/mol.

The term *thermochemical bond energy* (as opposed to dissociation energy) is used to indicate that a value was obtained indirectly. Table 4.2 lists thermochemical energies for the homonuclear single bonds of ten elements of Groups 14–16. In this table, the bond order factor has been eliminated. Size effects should cause bond energies to increase going across a period and to decrease going down a family. This expectation holds within Group 14, in which there are no lone pairs on adjacent atoms. But in Groups 15 and 16, it is the second member of each family (phosphorus and sulfur) that forms the strongest homonuclear single bond. This is another example of the adjacent atom lone-pair effect, which severely weakens the short N–N and O–O bonds.

Multiple Bonds: σ and π Contributions

Table 4.3 lists bond energies for some common homo- and heteronuclear multiple bonds. Again, these values can be rationalized by bond order, size, and adjacent atom lone-pair considerations.

All these bond energies contain simultaneous contributions from both σ and π bonds. Is there an accurate way to "partition" them into their σ

Table 4.3 The energies of some common homo- and heteronuclear multiple bonds

	C	N	O	P	S
C	807($\equiv$) C_2H_2 **593**(=) 586(=) C_2H_4				
N	881($\equiv$) HCN **750**($\doteq$)	942($\equiv$)			
O	1072($\equiv$) 805(=) CO_2	628($\doteq$) 469($\doteq$) NO_2	494(=) 300($\doteq$) O_3		
P	**511**($\doteq$)	**731**($\equiv$)	**592**($\doteq$)	**483**($\equiv$)	
S	**726**($\equiv$)	**481**($\doteq$)	**517**(=) 537($\doteq$) SO_2		**425**(=)

Note: Values are in kilojoules per mole. Dissociation energies are in boldface; others are thermochemical. Symbols in parentheses indicate the formal bond order: ($\doteq$) = 1.5, (=) = 2.0, ($\doteq$) = 2.5, and ($\equiv$) = 3.0.

Sources: Dasent, W. E. *Inorganic Energetics*, 2nd ed.; Cambridge University: London, 1982; Chapter 4; DeKock, R. L.; Gray, H. B. *Chemical Structure and Bonding*; Benjamin/ Cummings: Menlo Park, CA, 1980; Chapter 4; and Sanderson, R. T. *Chemical Bonds and Bond Energy*, 2nd ed.; Academic: New York, 1976; Chapters 7–9.

and π components? The general answer is no. We can make only rough estimates of the individual contributions, because of the synergism of σ and π bonding. As σ interaction becomes stronger, the bond distance decreases; this strengthens any π bonding between the two atoms. Similarly, strong π bonding enhances the σ interaction between the bonded atoms. Thus, for example, the σ contribution to the N–N bond energy of N_2 is surely greater than the thermochemical energy of the N–N single bond in N_2H_4.

Recognizing that the results will have only limited accuracy, rough estimates of σ and π contributions can be obtained by assuming that, for a given type of bond, the energy and length are inversely proportional.[6] Then the equation

$$E_\sigma = E_s\left(\frac{d_s}{d_m}\right) \tag{4.6}$$

can be used, where E_σ is the σ contribution to a multiple bond, E_s is the thermochemical single-bond energy, and d_s and d_m are the experimental single- and multiple-bond distances. For example, the experimental lengths of oxygen–oxygen single and double bonds are 148 and 121 pm, respectively. For the O=O bond of O_2, then,

$$E_\sigma = (143\ \text{kJ/mol})\left(\frac{148\ \text{pm}}{121\ \text{pm}}\right) = 175\ \text{kJ/mol}$$

Since the dissociation energy of O_2 is 494 kJ/mol, the π contribution is estimated to be about $494 - 175 = 319$ kJ/mol. The values listed in Table 4.4 were obtained in this manner.

The values in Table 4.4 should be used with considerable caution for the reasons given above. However, several trends are evident that are consistent with two generalizations:

1. For atoms without lone pairs (as in Group 14), homonuclear σ bonding is stronger than π bonding.

2. If lone pairs are present (Groups 15 and 16), the strength of σ bonding is at a maximum for the second element of the family; this appears to represent the best compromise between the size and adjacent atom lone-pair repulsion effects. As an illustration, the relative strengths of homonuclear σ and π bonds in Group 16 are plotted in Figure 4.3.

The free elements of Groups 15 and 16 clearly reflect these relative σ and π tendencies. The first member of each family exists as diatomic

---

6.   Other (logarithmic) relationships would probably be more accurate. However, they would also be more difficult to apply.

Table 4.4   Estimated σ and π contributions to the homonuclear double bonds of Groups 14 and 16

Compound	D, kJ/mol		
	Total	σ	π
Group 14			
C_2	593	410	183
Si_2	314	237	77
Group 16			
O_2	494	175	319
S_2	425	287	138
Se_2	305	208	97
Te_2	223	186	37

Note: The σ and π contributions were calculated using equation (4.6), data from Table 4.1, and homonuclear single-bond distances from Donohue, J. *The Structures of the Elements*; Wiley: New York, 1974; Chapters 7 and 9.

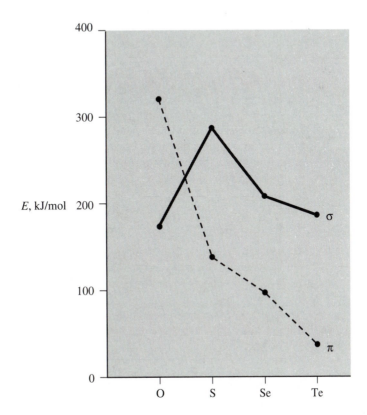

Figure 4.3
Plot of estimated σ and π contributions to the homonuclear double bonds of the Group 16 elements. Values are taken from Table 4.4.

molecules in its standard state (N_2 and O_2); these species are stabilized by strong π bonds. For the remaining members of each group, σ bonding dominates. For example, two of the three common allotropes of phosphorus have P_4 molecular units. The best Lewis structure that can be drawn for P_4 is

$$\text{:P} \overset{\ddot{\text{P}}}{\underset{\ddot{\text{P}}}{\boxed{}}} \text{P:}$$

where each phosphorus atom is σ bonded to the other three. This is superior to the diatomic $:P{\equiv}P:$ alternative, since three P–P σ bonds yield a greater total bond energy than one σ plus two π bonds.

In the same manner, elemental sulfur exists as S_8 units held together by σ bonding:

$$\text{:S:} \quad \begin{array}{c} \ddot{\text{S}}{-}\ddot{\text{S}}{-}\ddot{\text{S}} \\ \\ \ddot{\text{S}}{-}\ddot{\text{S}}{-}\ddot{\text{S}} \end{array} \quad \text{:S:}$$

Each sulfur atom forms two σ bonds. This is energetically superior to the oxygen-like alternative $:\ddot{S}{=}\ddot{S}:$, with one σ and one π bond per atom.

Recent Extensions to the List of π Bonding Elements[7]

In Section 4.1, we stated that only five main group elements commonly form π bonds: carbon, nitrogen, oxygen, phosphorus, and sulfur. Subsequent discussions of size effects help explain why this is the case. Over the past decade, however, inorganic chemists have synthesized and isolated compounds containing multiple bonds to a number of other main group elements. These species are generally stabilized by the presence of sterically bulky substituents such as t-butyl, mesityl ($2,4,6\text{-Me}_3\text{C}_6\text{H}_2-$), or bis(trimethylsilyl)methyl (($\text{Me}_3\text{Si})_2\text{CH}-$) groups, which inhibit decomposition reactions. For example, $\text{Me}_2\text{Si}{=}\text{SiMe}_2$ cannot be isolated anywhere near room temperature because of its facile dimerization:

$$2 \quad \begin{array}{c} \text{Me} \\ \\ \text{Me} \end{array}\!\!{\setminus}\!\!\diagup\!\!\text{Si}{=}\text{Si}\!\!\diagup\!\!{\setminus}\!\!\begin{array}{c} \text{Me} \\ \\ \text{Me} \end{array} \quad \longrightarrow \quad \begin{array}{c} \text{Me}_2\text{Si}{-}\text{SiMe}_2 \\ \quad | \quad\;\; | \\ \text{Me}_2\text{Si}{-}\text{SiMe}_2 \end{array} \tag{4.7}$$

However, the sterically hindered $(t\text{-Bu})_2\text{Si}{=}\text{Si}(t\text{-Bu})_2$ has been prepared and studied.

---

7. West, R. *Angew. Chem. Int. Ed. Engl.* **1987**, *26*, 1201; *Angew. Chem.* **1987**, *99*, 1231; Cowley, A. H.; Norman, N. C. *Prog. Inorg. Chem.* **1986**, *34*, 1; Cowley, A. H. *Polyhedron* **1984**, *3*, 389; *Acc. Chem. Res.* **1984**, *17*, 384; West, R.; Fink, M. J.; Michl, J. *Science* **1981**, *214*, 1343.

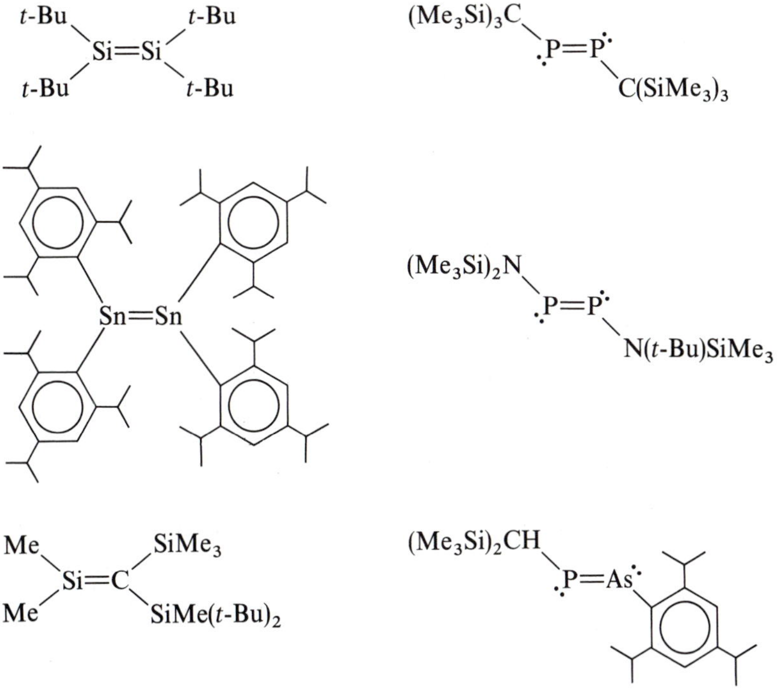

Figure 4.4
The structures of some isolable main group compounds containing unusual double bonds;

$\succ\!\!-$ = *i*-Pr.

Using this approach, compounds containing homonuclear Si=Si, Ge=Ge, Sn=Sn, P=P, and As=As double bonds (as well as heteronuclear linkages such as Si=C, Ge=C, Sn=C, P=As, P=Sb, and B=P) have been reported. Some specific examples are shown in Figure 4.4.

4.3 Heteronuclear Bond Energies

Electronegativity—An Alternate Approach

Electronegativity was first discussed in Chapter 1 in conjunction with orbital energies. At this point, a reexamination is in order.

The concept of electronegativity originated with Linus Pauling, and evolved from extensive examinations of bond energies. Pauling observed that the energy of a heteronuclear bond is always greater than either the geometric or arithmetic mean of the two homonuclear energies. That is, for any pair of elements X and Z,

$$E(X–Z) > [E(X–X) \cdot E(Z–Z)]^{1/2} \tag{4.8}$$

$$E(X–Z) > \frac{E(X–X) + E(Z–Z)}{2} \tag{4.9}$$

Why should this be true? The explanation lies with ionic resonance. For compound XZ, one of the ionic resonance structures (b) or (c) must be a significant contributor to the hybrid:

$$X-Z \longleftrightarrow X^+Z^- \longleftrightarrow X^-Z^+$$

(a)   (b)   (c)

If X is less electronegative than Z, then structure (b) is of importance. If X is the more electronegative element, then (c) is a significant contributor. In either case, the hybrid is more stable than is structure (a) alone.[8] This extra resonance energy is reflected in the observed bond energy. For a homonuclear molecule, the equal "pull" of the identical atoms on the bonded electrons allows for little resonance stabilization, and hence little additional bond energy.

The *ionic resonance energy*, symbolized by Δ, can be estimated using the equation

$$\Delta = E(X-Z) - [E(X-X) \cdot E(Z-Z)]^{1/2} \tag{4.10}$$

Pauling used bond energy data to construct a table similar to Table 4.5, in which energies are given in units of electron volts.

Pauling set the square root of Δ equal to the electronegativity difference between two given elements. If we make the arbitrary definition

$$\chi(H) = 2.2 \qquad \text{where } \chi = \text{Electronegativity} \tag{4.11}$$

then $\chi(F) = 4.0$, $\chi(O) = 3.7$, and $\chi(Cl) = 3.2$; these values approximate those of Table 1.10. This was the essence of Pauling's method and the source of the original electronegativity scale, although his data were considerably more refined.

Table 4.5 Ionic resonance and Pauling's approach to electronegativity

Bond X–Z	E_{obs}	$[E(X-X) \cdot E(Z-Z)]^{1/2}$	Δ	$\Delta^{1/2}$
H–F	5.82	2.67	3.15	1.77
H–Cl	4.44	3.34	1.10	1.05
H–O	4.81	2.58	2.23	1.49
H–C	4.31	4.06	0.25	0.50
C–F	5.08	2.43	2.65	1.63

Note: Bond energies are taken from Table 4.6 and converted to electron volts.

8. For the justification for this statement and an accompanying discussion, see Coulson, C. A. *Valence*, 2nd ed.; Oxford University: London, 1961.

Factors Influencing Heteronuclear Bond Energies

The thermochemical single-bond energies known to a reasonable degree of accuracy are given in Table 4.6, pp. 128–129. (Certain dissociation energies are also included for reference.)

The reader is encouraged to study Table 4.6 carefully while keeping the following observations in mind:

1. As was mentioned earlier, the energy of a given type of bond varies with its environment. More specifically, the average M–X bond energy for a series of compounds MX_n decreases with increasing n. For example, consider the Cl–F bond in three binary chlorine fluorides:

Compound	E(Cl–F), kJ/mol
ClF	251
ClF_3	174
ClF_5	152

This variation is probably caused by a combination of steric and hybridization factors. Evidence for a steric effect can be found in Figure 4.5, where the X–F bond energies for nine halogen fluorides are plotted.

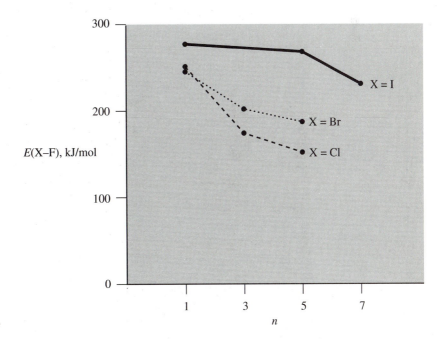

Figure 4.5
Plot of X–F bond energy versus n for binary halogen fluorides having the general formula XF_n.

Table 4.6 Single-bond energies for the main group elements

	H	B	C	Si	Ge	N
H	**432**	372 BH_3	416 CH_4	323 SiH_4	289 GeH_4	391 NH_3
Be						
B	372 BH_3		371 BMe_3 346 BEt_3			
Al			280 $AlMe_3$ 271 $AlEt_3$			
Ga			250 $GaMe_3$			
In			167 $InMe_3$			
C	416 CH_4	371 BMe_3 346 BEt_3	356 C(s) 331 C_2H_6	313 $Si\phi_4$	270 $GeMe_4$ 243 $GeEt_4$	302 Me_3N
Si	323 SiH_4		313 $Si\phi_4$	226 Si(s) 196 Si_2H_6		
Ge	289 GeH_4		270 $GeMe_4$ 243 $GeEt_4$		188 Ge(s) 163 Ge_2H_6	
Sn	253 SnH_4		223 $SnMe_4$ 194 $SnEt_4$			
Pb			160 $PbMe_4$ 129 $PbEt_4$			
N	391 NH_3		302 Me_3N			159 N_2H_4
P	321 PH_3		279 Me_3P 229 Et_3P			
As	297 AsH_3		239 Me_3As 190 Et_3As			
Sb	254 SbH_3		219 Me_3Sb 179 Et_3Sb			
O	464 H_2O		338 EtOH			190 NH_2OH
S	367 H_2S		300 EtSH			
Se	316 H_2Se		250 Et_2Se			
Te	266 H_2Te					
F	**562**	649 BF_3	490 CF_4	596 SiF_4	471 GeF_4	278 NF_3
Cl	**428**	447 BCl_3	325 CCl_4 349 $CHCl_3$	400 $SiCl_4$	339 $GeCl_4$	193 NCl_3
Br	**363**	372 BBr_3	279 CBr_4	325 $SiBr_4$	281 $GeBr_4$	
I	**295**	274 BI_3	217 CH_3I	248 SiI_4	216 GeI_4	
Xe						

Note: Values are in kilojoules per mole. Dissociation energies (0 K) are in boldface; others are thermochemical (298 K). Values vary in accuracy. Dissociation energies are generally within ± 4 kJ/mol; uncertainties for thermochemical energies are greater.

P	O	S	F	Cl	Br	I
321 PH_3	464 H_2O	367 H_2S	**562**	**428**	**363**	**295**
			635 BeF_2	466 $BeCl_2$	399 $BeBr_2$	319 BeI_2
			649 BF_3	447 BCl_3	372 BBr_3	274 BI_3
			587 AlF_3	425 $AlCl_3$	358 $AlBr_3$	284 AlI_3
				363 $GaCl_3$		
				309 $InCl_3$	269 $InBr_3$	210 InI_3
279 Me_3P	338 EtOH	300 EtSH	490 CF_4	325 CCl_4	279 CBr_4	217 CH_3I
229 Et_3P				349 $CHCl_3$		
			596 SiF_4	400 $SiCl_4$	325 $SiBr_4$	248 SiI_4
			471 GeF_4	339 $GeCl_4$	281 $GeBr_4$	216 GeI_4
				399 $SnCl_2$		
				315 $SnCl_4$	261 $SnBr_4$	
					260 $PbBr_2$	204 PbI_2
	190 NH_2OH		411 PbF_4	308 $PbCl_4$	207 $PbBr_4$	156 PbI_4
			278 NF_3	193 NCl_3		
200 P_4			503 PF_3	322 PCl_3	263 PBr_3	
197 P_2H_4			465 PF_5	260 PCl_5		
			486 AsF_3	309 $AsCl_3$	256 $AsBr_3$	
			387 AsF_5			
				314 $SbCl_3$	264 $SbBr_3$	
	143 H_2O_2		214 OF_2	207 OCl_2		
		266 S_8	342 SF_4	271 SCl_2		
			327 SF_6			
			303 SeF_6	241 $SeCl_2$		
			332 TeF_6	231 $TeCl_4$		
503 PF_3	214 OF_2	342 SF_4	**154**	**251**	**245**	**277**
465 PF_5	327 SF_6			174 ClF_3	202 BrF_3	
				152 ClF_5	187 BrF_5	268 IF_5
						231 IF_7
322 PCl_3	207 OCl_2	271 SCl_2	**251**	**240**	**215**	**208**
260 PCl_5			174 ClF_3			192 ICl_3
			152 ClF_5			
263 PBr_3			**245**	**215**	**190**	**175**
			202 BrF_3			
			187 BrF_5			
			277	**208**	**175**	**149**
			268 IF_5	192 ICl_3		
			231 IF_7			
	< 118 XeO_3					
			134 XeF_4			
			128 XeF_6			

Sources: Dasent, W. E. *Inorganic Energetics*, 2nd ed.; Cambridge University: London, 1982; Chapter 4; DeKock, R. L.; Gray, H. B. *Chemical Structure and Bonding*; Benjamin/Cummings: Menlo Park, CA, 1980; Chapter 4; Sanderson, R. T. *Chemical Bonds and Bond Energy*, 2nd ed.; Academic: New York, 1976; Chapters 7–9; Ashcroft, S. J.; Beech, G. *Inorganic Thermodynamics*; Van Nostrand Reinhold: New York, 1973; Chapters 3 and 5.

It is apparent that the decrease in bond energy with increasing n is greatest for the smallest central atom ($X = Cl$). In fact, the difference in $E(X–F)$ is 99 kJ/mol in going from ClF to ClF_5, but only 9 kJ/mol between IF and IF_5. (The influence of central atom hybridization on bond energies will be discussed in Chapter 5.)

2. Size is again an important factor. Examples such as

$$E(H_2O) > E(H_2S) > E(H_2Se) > E(H_2Te)$$

$$E(AlF_3) > E(AlCl_3) > E(AlBr_3) > E(AlI_3)$$

$$E(H_3C–CH_3) > E(H_3Si–SiH_3) > E(H_3Ge–GeH_3)$$

can be cited in this regard.

3. The destabilizing influence of adjacent atom lone-pair repulsions is again evident. Several cases can be identified in which this effect causes a reversal of the expected trend based on size; these include

$$E(PF_3) > E(NF_3)$$

$$E(CF_4) > E(NF_3) > E(OF_2)$$

$$E(H_3C–CH_3) > E(H_2H–NH_2) > E(HO–OH)$$

4. Upon quantitative study, the effect of ionic resonance can be discerned. Consider the dissociation energies for F–F, Cl–Cl, and Cl–F single bonds. The experimental values are:

Compound	D, kJ/mol
F_2	154
Cl_2	240
ClF	251

Following Pauling's methodology, the geometric mean of the F–F and Cl–Cl bond energies is $(154 \times 240)^{1/2} = 192$ kJ/mol. As expected, the heteronuclear bond energy listed above exceeds this mean (by over 30%). This indicates that ionic resonance plays a role in strengthening the Cl–F bond. Based on equation (4.10), it might be argued that about 192 kJ/mol of the bond energy is due to the covalent structure, with the difference (59 kJ/mol) being attributed to ionic resonance. Regardless of how accurate this estimate is, it should be clear that if ionic resonance is significant for Cl–F, in which the two atoms have only a small electronegativity difference, then it must contribute to the energies of all heteronuclear bonds.

4.4 Homonuclear Bond Lengths: Nonpolar Covalent Radii

As we mentioned in Chapter 3 (see Figure 3.3), the radius of a hydrogen atom is about 52.9 pm, while the internuclear distance in H_2 is about 74.1 pm. Without an advance knowledge of the latter value, we might have guessed that the two atomic spheres in H_2 just come in contact with—that is, just barely touch—one another. If this were the case, then the resulting internuclear distance would be $2(52.9) = 105.8$ pm, about 43% longer than the true value. This is a specific example of a general phenomenon. For homonuclear bonding between atoms of any element X,

$$d_{X-X} < 2r_X \tag{4.12}$$

where d is the internuclear distance and r is the radius of a single atom. Nevertheless, the notion that bonds between identical atoms can be "split in half" (half of the total distance being assigned to each atom) permits the generation of *nonpolar covalent radii*:

$$r_{cov, X} = \frac{d_{X-X}}{2} \tag{4.13}$$

A list of nonpolar covalent radii calculated using equation (4.13) is given in Table 4.7.

Unfortunately, this approach is limited to nonmetals and metalloids, because the interactions between the adjacent atoms of a bulk metal are not properly classed as covalent bonds. (They are better described as *metallic bonds*, and can be used to generate tables of metallic radii, as we will see in Chapter 7.) An alternate approach may be used to obtain covalent radii for

Table 4.7 Nonpolar covalent radii (r) as calculated from homonuclear bond length data for the nonmetals and metalloids

Element	r, pm	Compound	Element	r, pm	Compound
H	37	H_2	P	111 (—)	P_4
B	85	B(s)		95 ($\equiv$)	P_2
C	77 (—)	C(s)	O	74 (—)	H_2O_2
	67 ($=$)	C_2H_4		60 ($=$)	O_2
	60 ($\equiv$)	C_2H_2	S	102	S_8
Si	118	Si(s)	Se	117	Se_8
Ge	122	Ge(s)	F	71	F_2
N	73 (—)	N_2H_4	Cl	99	Cl_2
	61 ($=$)	N_2H_2	Br	114	Br_2
	55 ($\equiv$)	N_2	I	133	I_2

Source: Values were calculated from data given by Wells, A. F. *Structural Inorganic Chemistry*, 5th ed.; Clarendon: Oxford, 1984.

Table 4.8 Covalent radii for the metals and metalloids of Groups 2, 13, 14, and 15

Group 2	Group 13	Group 14	Group 15
Be	B		
102	82		
Mg	Al		
142	120		
	Ga	Ge	As
	121	118	119
	In	Sn	Sb
	146	140	135
	Tl	Pb	Bi
	148	147	149

Note: Values in picometers, as calculated from the lengths of their single bonds to tetrahedral carbon, using equation (4.14).

Source: Wade, K.; O'Neill, M. E. In *Comprehensive Organometallic Chemistry*; Wilkinson, G.; Stone, F. G. A.; Abel, E. W., Eds.; Pergamon: London, 1982; Volume 1, p. 10.

metals. In this method the lengths of metal–carbon bonds serve as the experimental data. The single-bond radius of carbon is taken to have a constant value of 77 pm (Table 4.7), and the equation

$$r_M = d_{M-C} - 77 \text{ pm} \qquad \textbf{(4.14)}$$

is used to estimate the "nonpolar" radius of the metal. Resulting values are compiled in Table 4.8.

4.5 Heteronuclear Bond Lengths

Merely adding two nonpolar covalent radii does not normally give accurate predictions for heteronuclear bond distances. For example, by adding the nonpolar radii (from Table 4.7) of hydrogen and oxygen together, the length of the H–O bond is predicted to be

$$d_{H-O} = r_H + r_O = 37 + 74 = 111 \text{ pm}$$

The actual value in the water molecule is 96 pm, a 16% difference.

Such a discrepancy is not unusual; rather, it is the norm. For virtually any heteronuclear bond X–Z,

$$d_{X-Z} < r_X + r_Z$$

The equation

$$d_{X-Z} = r_X + r_Z - k\Delta\chi \qquad \textbf{(4.15)}$$

can be used to more accurately estimate heteronuclear bond distances. Here, k is a constant equal to 8 pm for bonds involving any element of the first short period except carbon (for which $k = 0$), and $\Delta\chi$ is the electronegativity difference between elements X and Z on the Pauling scale. For the H–O bond, then, the predicted distance becomes

$$d_{\text{H–O}} = 37 + 74 - 8(1.24) = 101 \text{ pm}$$

reducing the error to about 5%.

4.6   The Prediction of Structures and Reactions from Bond Energies

Bond energies often can be used to predict or explain chemical structures and reactions. For example, two reasonable topologies may be suggested for the NOF_3 molecule. Their Lewis structures are

(a)   (b)

Structure (b) is favored on the basis of formal charge; however, structure (a) is the preferred form. Why?

If values from Table 4.6 are used to estimate the total bond energy for each structure, the following results are obtained:

Structure (a):   3(278 kJ/mol) + 190 kJ/mol = 1024 kJ/mol

            3 N–F   1 N–O

Structure (b):   2(278 kJ/mol) + 190 kJ/mol + 214 kJ/mol = 960 kJ/mol

            2 N–F   1 N–O   1 O–F

Thus, structure (a) is favored by about 64 kJ/mol.

Note that this is an unusual situation; the formal charge rules and bond energy estimates usually lead to the same predicted topology. The exceptional behavior of NOF_3 is best explained in terms of adjacent atom lone-pair repulsions. In structure (a), all four bonds involve the central nitrogen, which has no lone pairs. However, structure (b) has two central atoms. The oxygen lone pairs tend to weaken both the N–O and O–F bonds, destabilizing this structure.

A second example is provided by disulfur difluoride, S_2F_2, which has two known isomers with the following topologies:

$$
\begin{array}{cc}
 & S \\
 & | \\
F\text{---}S\text{---}S\text{---}F \qquad & F\text{---}S\text{---}F \\
(a) & (b)
\end{array}
$$

Structure (b) is the more stable of the two because of secondary resonance. The formation of the additional (partial) bond—an S–S π interaction—enhances the overall bond energy.

$$
\text{:}\overset{\ominus}{\underset{..}{S}}\text{:} \qquad\qquad \text{:}\overset{..}{S} \\
$$

Is the reaction below exothermic as written?

$$
H_2(g) + Cl_2(g) \longrightarrow 2\,HCl(g) \tag{4.16}
$$

Using Table 4.6, we obtain

Reactants: 432 kJ/mol + 240 kJ/mol = 672 kJ/mol

1 H–H bond   1 Cl–Cl bond

Product: 2(428 kJ/mol) = 856 kJ/mol

2 H–Cl bonds

So the reaction is exothermic. The estimated enthalpy change is −184 kJ/mol, compared to the experimental value of −185 kJ/mol (a much better agreement than can normally be expected!). This reaction is exothermic because ionic resonance stabilizes the heteronuclear H–Cl bond over the homonuclear bonds.

Observations of this type support the following generalization: There is a tendency for bonds to form between the elements having the highest and lowest electronegativities in any chemical system.[9] For system (4.16), this tendency favors the formation of hydrogen chloride from the free elements.

Finally, consider the possibility of preparing hydrogen iodide by the reaction

$$
2\,HBr + I_2 \longrightarrow 2\,HI + Br_2 \tag{4.17}
$$

Is this a viable synthetic route? An analysis of bond energies leads us to believe that it is not. You should be able to show that ΔH for the reaction

9. This generalization is sometimes superseded by another involving the relative "hardness" or "softness" of species; see Chapter 10.

is expected to be about $+95$ kJ/mol, and that the generalization stated above, which in this case favors the reactants, holds.

Bibliography

Webster, B. *Chemical Bonding Theory*; Blackwell: Oxford, 1990.

Verkade, J. G. *A Pictorial Approach to Molecular Bonding*; Springer-Verlag: New York, 1986.

Sanderson, R. T. *Polar Covalence*; Academic: New York, 1983.

Dasent, W. E. *Inorganic Energetics*, 2nd ed.; Cambridge University: London, 1982.

DeKock, R. L.; Gray, H. B. *Chemical Structure and Bonding*; Benjamin/Cummings: Menlo Park, CA, 1980.

Companion, A. *Chemical Bonding*, 2nd ed.; McGraw-Hill: New York, 1979.

Gimarc, B. M. *Molecular Structure and Bonding*; Academic: New York, 1979.

McWeeny, R. *Coulson's Valence*, 3rd ed.; Oxford University: London, 1979.

Sanderson, R. T. *Chemical Bonds and Bond Energy*; Academic: New York, 1976.

Dasent, W. E. *Nonexistent Compounds*; Marcel Dekker: New York, 1965.

Pauling, L. *The Nature of the Chemical Bond*, 3rd ed.; Cornell University: Ithaca, NY, 1960.

Questions and Problems

1. Write the single major Lewis structure for each of the following; indicate all nonzero formal charges:
   (a) HCN (b) CH_2O (c) IF_2^+ (d) IF_2^-
   (e) NHF_2 (f) $HOCl$ (g) $HOClO_2$ (h) $SiBr_4$
   (i) PCl_5 (j) S_2^{2-} (k) C_2H_2 (l) B_2Cl_4
   (m) C_3O_2 (n) $CH_3NH_3^+$ (o) XeF_2 (p) XeF_4

2. Explain in your own words what is meant by:
   (a) Hypervalency   (b) Resonance
   (c) Secondary resonance   (d) Ionic resonance

3. There is no discussion of transition metals in this chapter. Do you expect Lewis structures to be generally useful for compounds of those elements? Why not?

4. Write the three major resonance structures for the following, and evaluate their relative importance:
   (a) SO_3 (b) CO_2 (c) OCS

5. The fulminate and cyanate ions were discussed in this chapter. A third possible topology, with oxygen as the central atom, was ignored. What is the best

resonance structure that can be drawn for CON^- (including formal charges)? Was there justification for ignoring it?

6. Arrange the following species in order of increasing carbon–nitrogen bond energy, and explain your logic: CN^-, CNO^-, NCO^-.

7. Use formal charge criteria to predict the topology for nitrous oxide, N_2O.

8. In Chapter 3, MO theory was used to explain why carbon monoxide acts as a Lewis base (electron pair donor) through the carbon atom. The same observation can be explained using a formal charge argument. How?

9. The S–O bond strength in dimethylsulfoxide, Me_2SO, is between single and double. Explain.

10. (a) The known topology for dinitrogen pentoxide vapor is

$$O\diagdown \atop O\diagup N\text{---}O\text{---}N{\diagup O \atop \diagdown O}$$

Write all significant resonance structures, and identify the major contributor to the hybrid.
    (b) In the solid state, N_2O_5 is ionic. Use your knowledge of Lewis theory to predict the identities of the cation and anion.

11. The energy of a P–F bond is considerably greater than that of an N–F bond.
    (a) Explain, using a secondary resonance argument.
    (b) Explain, using an ionic resonance argument.
    (c) Explain, using an argument involving adjacent atom lone-pair repulsions.

12. Rationalize the observation that in bond energy, P–F > N–F but P–H < N–H.

13. How many equivalent resonance structures can be drawn for the oxalate ion, $C_2O_4^{2-}$? Sketch one.

14. The thiocyanate ion can be protonated to give HNCS. What is the most likely topology for this molecule? Argue your case.

15. Use bond energy data to estimate the enthalpy of formation of hydrogen peroxide, H_2O_2.

16. Two common allotropes of carbon are diamond, in which each atom forms four σ bonds, and graphite, in which each atom forms three σ bonds and one π bond. For elemental silicon and germanium, the standard states are diamond-like structures; the graphite-like lattice is unknown. Explain.

17. Examine Table 4.4 and suggest future candidates for possible homonuclear multiple bonding.

18. A $C{=}C$ double bond is roughly 13% shorter than a C–C single bond. An $N{=}N$ double bond is about 17% shorter than an N–N single bond. An $O{=}O$ double bond is about 19% shorter than an O–O single bond.
    (a) Rationalize these percentage differences on the basis of the relative strengths of the homonuclear σ and π bonds formed by these elements.

(b) A sulfur–sulfur double bond is only about 10% shorter than the corresponding single bond. Is this consistent with your answer to part (a)?

19. Estimate the σ and π contributions to the $P{\equiv}P$ triple bond of P_2. The bond distance in $P_2(g)$ is 189.4 pm; other necessary data can be found in Tables 4.3, 4.6, and 4.7.

20. Carbon dioxide consists of discrete molecules with carbon–oxygen multiple bonding. Silicon dioxide exhibits a polymeric structure held together solely by Si–O single bonds.
    (a) Write a Lewis structure for each species.
    (b) Rationalize the difference in behavior.

21. Use bond energies from Table 4.6 to estimate the Pauling electronegativity of sulfur. Compare your answer to the value given in Table 1.10.

22. Use appropriate bond energies between nitrogen and other elements of your choice to demonstrate the effect of each of the following factors on bond energy:
    (a) Bond order   (b) Size
    (c) Adjacent atom lone-pair repulsions

23. Nitrogen trifluoride, NF_3, is relatively stable toward decomposition to its elements. In contrast, NCl_3 can be dangerous to work with because of its tendency to decompose to $N_2 + Cl_2$. Use bond energies to rationalize.

24. For each of the following pairs, decide which sulfur–oxygen bond is longer and explain your logic:
    (a) SO_2 or SO_2^{2-} (b) SO_2 or SO_3 (c) H–O–S–H or H_2SO

25. (a) Use equation (4.15) to estimate the length of a nitrogen–fluorine single bond.
    (b) The experimental value in NF_3 is 136 pm. What is the percent error?

26. Use bond energies to predict the relative tendencies for the following reactions to occur:

$$2\,NH_3 \longrightarrow N_2H_4 + H_2$$

$$2\,PH_3 \longrightarrow P_2H_4 + H_2$$

What does your prediction indicate about the comparative tendencies of nitrogen and phosphorus to form homonuclear bonds?

27. Use bond energy data to estimate ΔH for the following reactions:

(a) $4\,H_2S_2 \longrightarrow S_8 + 4\,H_2$

(b) $2\,HCN + O_2 \longrightarrow 2\,HOCN$

(c) $HOCN \longrightarrow HNCO$

(d) $CH_4 + CF_4 \longrightarrow 2\,CH_2F_2$

*28. A theoretical study of the bonding in a variety of small phosphorus-containing compounds was reported by S. Mathieu, J. Navech, and J. C. Barthelat (*Inorg. Chem.* **1989**, *28*, 3099). Answer these questions after examining their article.
    (a) Write reasonable Lewis structures for

$$
\text{H—P—O} \qquad \text{H—P} \overset{\displaystyle O}{\underset{\displaystyle O}{\Big\langle}} \qquad \text{H—P} \overset{\displaystyle O}{\underset{\displaystyle O}{\Big\langle}}
$$

Include all nonzero formal charges.

(b) The authors predict a P–O bond distance of 163.8 pm for

$$
\text{H—P} \overset{\displaystyle O}{\underset{\displaystyle O}{\Big\langle}}
$$

but only 146.3 pm for H–P–O. Why the large difference?

(c) Which topology do you favor?

$$
\text{H—P} \overset{\displaystyle O}{\underset{\displaystyle O}{\Big\langle}} \qquad \text{or} \qquad \text{H—P—O—O}
$$

Why?

*29. An equation that interrelates bond order, bond length, and electronegativity has been advocated by L. Peter (*J. Chem. Educ.* **1986**, *63*, 123). Answer these questions after reading his arguments.

   (a) Use Peter's equation to estimate the length of the bond in ClF. The actual distance is 162.8 pm. What is the percent error?

   (b) The diatomic species SiO is valence isoelectronic with carbon monoxide; its internuclear distance is 166 pm. Is this consistent with a bond order of 3? (Answer quantitatively.)

   (c) Suggest another species for which Peter's equation might provide useful information about bond order, and carry your inquiry to some conclusion.

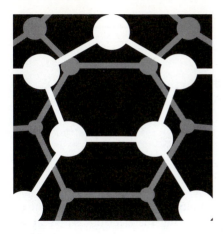

5

The Shapes and
Polarities of Molecules

Is there a way to explain the shapes and polarities of covalently bonded species? Consider three molecules having the general formula MH_2, where M is the central atom. For M = Be, the H–M–H bond angle is known to be exactly 180°. The angle is about 100° for M = C, and 92° for M = S. What causes these differences?

In this chapter, we will first introduce a model that rationalizes molecular structures. Then we will discuss some of the effects of structure on the physical and chemical properties of compounds.

5.1   Valence Shell Electron Pair Repulsion

One of the simplest methods for predicting molecular geometries is the *valence shell electron pair repulsion* (*VSEPR*) model. Its basic premise is that the shapes of covalently bonded species depend on the repulsions among the valence electron pairs of their central atom(s).[1]

1.   Sidgwick, N. V.; Powell, H. M. *Proc. Roy. Soc. A* **1957**, *176*, 153; Gillespie, R. J.; Nyholm, R. S. *J. Chem. Educ.* **1963**, *40*, 295. See also Gillespie, R. J.; Hargittai, I. *The VSEPR Model of Molecular Geometry*; Allyn & Bacon: Boston, 1991.

For binary compounds that conform to the general formula MX_n, the predictions of VSEPR are as follows:

1. ***Two Central Atom Electron Pairs (ie, MX_2):***   The Lewis structure for beryllium hydride is simply

    H–Be–H

    Being of like charge, the two valence electron pairs repel one another. This repulsion can be minimized by moving the electrons as far apart as possible. The bond angle that gives maximum separation is 180°, which produces a linear geometry. The experimental bond angle is in accord with this prediction.

2. ***Three Electron Pairs:***   The Lewis structure for boron triiodide is

    $$\ddot{\underset{\cdots}{I}} - B - \ddot{\underset{\cdots}{I}}$$

    Repulsions are minimized if the central atom's bonding orbitals (and thus the three iodines) form an equilateral triangle. Angles of 120° result, and the atom geometry can be described as trigonal planar.

    I—B $\rangle$ 120°

3. ***Four Electron Pairs:***   You are, of course, familiar with methane, CH_4. Its tetrahedral geometry, with bond angles of 109.5°, minimizes the repulsions among four electron pairs.

    109.5°

4. ***Five Electron Pairs:***   Phosphorus pentafluoride, PF_5, is an appropriate example for this category. Its shape is trigonal bipyramidal:

Unlike the structures described above, the bonds of a trigonal bipyramid are not symmetrically equivalent. There are two *axial* (or *apical*) bonds (to the circled fluorines) and three equatorial bonds (to the boxed fluorines). The angles are 120° between equatorial bonds and 90° between the axial and equatorial linkages.

5. **Six Electron Pairs**: A species having six central atom electron pairs is sulfur hexafluoride, SF_6. Octahedral geometry is exhibited by this molecule, with all bonds equivalent and positioned at 90° with respect to one another:

Sketches of the geometries described above are shown in Figure 5.1 (p. 142). Table 5.1 gives a summary of VSEPR predictions for central atoms having two through six electron pairs.

Table 5.1   Predictions of the VSEPR model for MX_n systems

Central Atom Electron Pairs				
Bond pairs	Lone pairs	Total	Atom Geometry	Bond Angle(s)
2	0	2	Linear	180°
3	0	3	Trigonal planar	120°
2	1	3	Angular	<120°
4	0	4	Tetrahedral	109.5°
3	1	4	Pyramidal	<109.5°
2	2	4	Angular	<109.5°
5	0	5	Trigonal bipyramidal	90°, 120°
4	1	5	Distorted tetrahedral	<90°, <120°
3	2	5	T-shaped	<90°
2	3	5	Linear	180°
6	0	6	Octahedral	90°
5	1	6	Square pyramidal	<90°
4	2	6	Square planar	90°
3	3	6	T-shaped	<90°
2	4	6	Linear	180°

6. **More Than Six Electron Pairs**: Species having more than six valence electron pairs surrounding a main group central atom are not common. This can be attributed mainly to steric crowding, since most of the stable

Number of
valence pairs

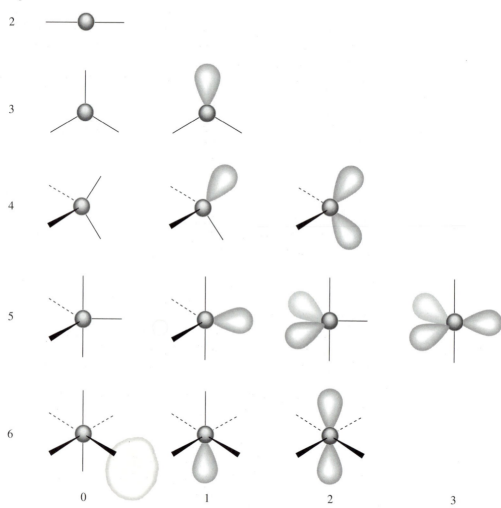

Number of lone pairs

Figure 5.1 Predicted geometries of MX_n molecules and ions having $n = 2$–6. [Adapted and used with permission from Gillespie, R. J. *J. Chem. Educ.* **1963**, *40*, 295.]

species of this type have large central and small terminal atoms. One known example is iodine heptafluoride, IF_7, with a pentagonal bipyramidal shape:

As in the trigonal bipyramid, this geometry contains two types of bonds—in this case, two axial and five equatorial. The angles are 72° (equatorial/equatorial) and 90° (axial/equatorial).

Geometries different from those given above are sometimes encountered; examples include square planar, square pyramidal, trigonal prismatic, and capped octahedral. Whenever such deviations from the VSEPR model occur there are usually good explanations for them. Some of these exceptions will be discussed as they arise in this and later chapters.

Most chemical species, of course, do not fit the simple formula MX_n. Cases in which the central atom contains one or more lone pairs, participates in π bonding, and/or is linked to more than one type of atom also must be considered. Some useful generalizations are given below.

Repulsions Involving Lone Pairs Exceed Repulsions Involving Only Bond Pairs. The ammonia molecule has, like methane, four valence electron pairs on the central atom. In NH_3, however, one of these is a lone pair.

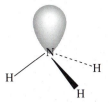

The H–N–H angles are 107.3°, indicative of slight distortion from a perfect tetrahedron. This can be rationalized by assuming that repulsions involving the lone pair push the bond pairs away from it, decreasing the H–N–H angles. The effect should be amplified when two lone pairs are present, as in the water molecule, and this is, in fact, observed. The bond angle in H_2O is only 104.5°.

These two examples represent the norm, and the lone pairs are said to be *stereochemically active*. There are cases, however, where a lone pair is "inactive"—that is, plays no apparent role in determining the geometry. This is especially common for species having the general formula $:MX_6$ (six bond

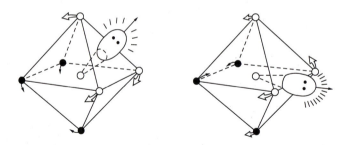

Figure 5.2   Two possible conformations for XeF_6. The indicated atom geometry is distorted octahedral, with the lone pair centering either a face or an edge. [Reproduced with permission from Gavin, R. M.; Bartell, L. S. *J. Chem. Phys.* **1968**, *48*, 2466.]

pairs and one lone pair). Anions such as $SbBr_6^{3-}$ and $TeCl_6^{2-}$ are known from X-ray diffraction studies to be perfect octahedra. However, the isoelectronic molecule XeF_6 has been a source of continuing debate. It is not octahedral, nor is it a pentagonal bipyramid with one vertex occupied by a lone pair. Several other possibilities have been suggested, including a distorted octahedron in which the lone pair centers either a face or an edge (see Figure 5.2). It is also possible that XeF_6 is *fluxional*, rapidly changing its shape between two or more conformations of approximately equal energy.

A lone pair may be stereochemically inactive because of participation in dative π bonding. For example, trisilylamine, $(H_3Si)_3N$, is planar about the central nitrogen atom, with Si–N–Si bond angles of precisely 120°. This geometry is consistent with three rather than four electron pairs on nitrogen, and presumably results from $(p \to d)$ π bonding (Figure 5.3), which delocalizes the lone-pair electron density into Si ($3d$) orbitals.

Multiple Bonds Do Not Change the Gross Geometries, but Repulsions Involving Double Bonds Are Greater Than Those of Single Bonds.   Phosgene has

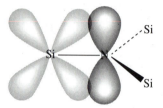

Figure 5.3   Dative $(p \to d)$ π bonding in $(H_3Si)_3N$. The darker orbital is a stereochemically inactive, "nonbonding" $2p$ orbital of nitrogen, which overlaps with three Si ($3d$) orbitals (only one of which is shown).

the formula $COCl_2$ and the Lewis structure

$$\ddot{:}\ddot{C}l-C-\ddot{C}l\ddot{:} \quad \text{with} \quad :\!O\!: \text{ double bonded to } C$$

This molecule is planar, with experimental bond angles of $124°$ ($\angle O–C–Cl$) and $112°$ ($\angle Cl–C–Cl$). Notice that these values average exactly to the trigonal planar angle expected for three electron pairs. Apparently, the π bond does not contribute to the overall geometry. This is typical. In the VSEPR model, the number of electron pairs is determined by counting each linkage to the central atom (regardless of whether it is a single, double, or triple bond) and each lone pair only once. (Said another way, only σ bonds are important in this regard.)

As noted above, however, the bond angles are distorted by $4°$ and $8°$ from the ideal value. This is easily explained: The $C{=}O$ bond region contains double the normal electron density, which naturally creates extra repulsion. The C–Cl bond pairs "swing away" from that region, widening $\angle O–C–Cl$ and narrowing $\angle Cl–C–Cl$.

Partial π bonding may create a similar effect. The I–O linkage in IOF_5 has partial double bond character:

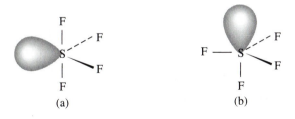

The geometry of IOF_5 is distorted octahedral. The four coplanar fluorines are displaced away from the oxygen, giving O–I–F bond angles of $98°$ compared to the $90°$ ideal.

Lone-Pair Repulsions Decrease as Bond Angles Increase. This generalization is most important for the trigonal bipyramidal framework. Sulfur tetrafluoride was shown in Chapter 4 to have five electron pairs about the central sulfur. The lone pair might conceivably be located in either an equatorial (a) or an axial (b) position. (The point group might be C_{2v} or C_{3v}.)

It is found experimentally that the lone pair preferentially occupies an equatorial position. That is, structure (a) (sometimes described as a *distorted tetrahedron*) shows the ground-state geometry, while structure (b) represents a conformation of slightly higher energy. Notice that structure (a) has two lone-pair/bond-pair repulsions at 90°, whereas (b) has three such repulsions. There are also interactions at 120° and 180° in both cases, but the repulsions at the most acute angle dominate.

Similarly, in ClF_3 (with three bond pairs and two lone pairs on chlorine), both lone pairs are equatorial; the result is a T-shaped structure. Xenon difluoride, with two bonding and three lone pairs on the central atom, is linear.

Table 5.2 lists bond length and bond angle data for several main group fluorides having five central atom electron pairs. The influence of lone-pair/bond-pair lone-pair/lone-pair repulsions in decreasing bond angles is evident from the table. It is also interesting to note the unequal bond lengths in these molecules, with the axial linkages invariably being longer than the equatorial bonds. This is a result of hybridization (see Section 5.2).

In an octahedral framework, two lone pairs can be oriented at either 90° (cis) or 180° (trans) with respect to one another. This situation arises for XeF_4. The trans conformation minimizes the lone-pair/lone-pair repulsion; hence, it is not surprising that the observed geometry is square planar:

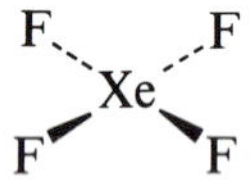

The Presence of Highly Electronegative Terminal Atoms Tends to Reduce Bond Angles. The bond angles in NF_3 are about 102°, roughly 5° smaller

Table 5.2   Molecular parameters for main group fluorides having five valence electron pairs on the central atom

| Compound | d_{M-F}, pm | | ∠ F–M–F, degrees | |
	Axial	Equatorial	Axial/Axial	Equatorial/Equatorial
PF_5	158	153	180	120
AsF_5	171	166	180	120
SF_4	164	154	179	103
SeF_4	177	168	169	101
ClF_3	170	160	175	—
BrF_3	181	172	172	—
KrF_2	188	—	180	—
XeF_2	198	—	180	—

Source: Compiled from data in Wells, A. F. *Structural Inorganic Chemistry*, 5th ed.; Clarendon: Oxford, 1984.

than in ammonia. This can be explained by recognizing that the high electronegativity of fluorine enables it to strip the central nitrogen of some of its valence electron density in the bonded pairs. The nitrogen lone pair then exerts a relatively stronger influence, causing the fluorines to swing away from it and reducing the F–N–F angles.

A similar effect is observed upon comparing the bond angle of OF_2 (103.2°) to that of water (104.5°). Here, the electronegativity difference between O and F is smaller than between N and F; consequently, there is a smaller difference in bond angle.

Sterically Large Substituents Increase Bond Angles.   This is self-evident for very large groups such as *t*-Bu and Me_3Si. The effect can manifest itself in less obvious cases as well, particularly if the terminal atoms are larger than the central atom. For example, the bond angle in OCl_2 is 110.8°, greater than in either H_2O or OF_2. (Electronegativity considerations predict an intermediate angle.) It appears that the short bonds formed by the small oxygen atom, coupled with the relatively large size of the terminal chlorines, cause steric repulsions that are reduced by the opening of the bond angle.

A similar trend is observed for the methyl halides:

CH_3X, X =	$\angle$ H–C–X, *degrees*
F	107.9
Cl	110.0
Br	111.2
I	111.4

The highly electronegative fluorine induces a small decrease in the H–C–X bond angle compared to the ideal tetrahedral value. An increasing steric effect for Cl, Br, and I then causes the angle to widen.

There are, of course, numerous compounds for which more than one of the above factors is operative. For example, consider nitrosyl fluoride:

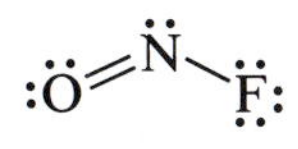

The ideal bond angle is 120°. The nitrogen lone pair and the high electronegativity of fluorine tend to decrease this angle, while the N=O double bond tends to increase it. The experimental value of 110° represents a compromise among these factors.

If fluorine is replaced by a larger halogen, still another consideration (the steric factor) is added. This, coupled with the reduced terminal atom electronegativity, results in an increase in bond angle.

O–N–X, X =	$\angle$, degrees
F	110
Cl	113
Br	117

Ozone is isoelectronic with ONF, but is symmetric. This molecule has two equivalent resonance structures, with a π bond spread out over two sites:

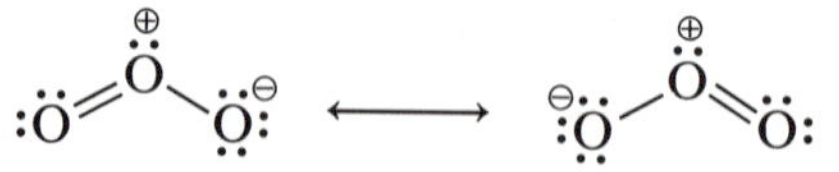

The ideal angle is again 120°. The π electron density tends to widen this angle, while the central oxygen's lone pair works in opposition. The experimental value is 117°.

Directed versus Undirected Covalent Bonding Models

The Lewis and VSEPR theories are sometimes grouped together as the *directed bonding model,* because they address the question of molecular geometries. This differentiates these methods from molecular orbital theory, which is an undirected model. The two approaches normally give similar or equivalent results; however, cases can be cited where they differ. One example is the water molecule. Both models identify a total of four occupied valence orbitals—two bonding and two nonbonding electron pairs. The directed model predicts that these four orbitals lie at only two different energy levels; that is, the bond pairs are chemically equivalent to one another, as are the lone pairs:

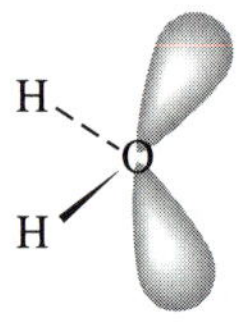

However, MO theory predicts four different energy levels (see Figure 3.19). Ultraviolet photoelectron spectroscopy (UV PES), an experimental tool capable of determining the energies of occupied orbitals in a covalent molecule (see Chapter 21), gives a photoelectron spectrum for water that contains four ionization bands. This supports the MO prediction.[2]

Numerous arguments have been made both in criticism and defense of theoretical aspects of the VSEPR model.[3] There is no question that it has considerable practical value, but its limitations must be kept in mind. It is possible for so many opposing factors to be involved for a given molecule that almost any geometry or bond angle might be rationalized, and at that point the model loses its usefulness. The model appears to work best when applied broadly, going no further than predicting the ideal angles of 180°, 120°, 109.5°, and 90°, with small variations due to lone pairs, π bonding, etc. Nevertheless, the basic ideas of VSEPR are easy to use, and it is applicable to a wide variety of chemical problems.

5.2   Hybridized Orbitals

We have now discussed the molecular shapes that minimize valence shell electron pair repulsions. An intriguing question arises at this point: How are these geometries achieved naturally while still maintaining the orbital overlap necessary for effective covalent bonding?

The answer is not at all obvious. Consider the case of BCl_3. The central boron has four valence orbitals. The $2s$ orbital is nondirectional, of course. The three $2p$ orbitals are mutually perpendicular, so their involvement in covalent bonding should result in bond angles of 90°. There are no valence atomic orbitals of boron (or any other atom) that lie at 120° angles with respect to one another, and 120° angles are required by VSEPR theory.

Such bond angles are achieved through *hybridization*, which may be defined as the mixing of two or more orbitals of the same atom.

Hybridization Involving Only *s*- and *p*-Type Orbitals

Equations similar to those used in the LCAO approach can be written for the mixing of one *s* with one *p* orbital of a given atom:

$$\psi_{hy} = c_s \phi_s \pm c_p \phi_p \tag{5.1}$$

Here, ψ_{hy} symbolizes the wavefunction for the hybrid orbital. In theory, any c_s/c_p ratio might be used. For the specific case of $c_s = c_p$, the wavefunctions

---

2.   Discussions of the bonding in H_2O are given by Martin, R. B. *J. Chem. Educ.* **1988**, *65*, 668; and Laing, M. *J. Chem. Educ.* **1987**, *64*, 124.

3.   Hall, M. B. *J. Am. Chem. Soc.* **1978**, *100*, 6333; Gillespie, R. J. *J. Chem. Educ.* **1974**, *51*, 367; Drago, R. S. *J. Chem. Educ.* **1973**, *50*, 244; and references cited therein.

can be normalized using a procedure similar to that of equations (3.4)–(3.10) to give

$$\psi_1 = \frac{1}{\sqrt{2}} (\phi_s + \phi_p) \tag{5.2}$$

$$\psi_2 = \frac{1}{\sqrt{2}} (\phi_s - \phi_p) \tag{5.3}$$

These two orbitals are designated as sp ($s^1 p^1$) hybrids to indicate that they derive from 1:1 combinations of s and p atomic orbitals.

Similarly, it is possible to combine one s with two p orbitals to yield three sp^2-type hybrids. If the two p orbitals are chosen to be p_x and p_y, it can be shown that after normalization the resulting hybrids are described by the equations[4]

$$\psi_1 = \frac{1}{\sqrt{3}} (\phi_s) + \frac{\sqrt{2}}{\sqrt{3}} (\phi_{p_x}) \tag{5.4}$$

$$\psi_2 = \frac{1}{\sqrt{3}} (\phi_s) - \frac{1}{\sqrt{6}} (\phi_{p_x}) + \frac{1}{\sqrt{2}} (\phi_{p_y}) \tag{5.5}$$

$$\psi_3 = \frac{1}{\sqrt{3}} (\phi_s) - \frac{1}{\sqrt{6}} (\phi_{p_x}) - \frac{1}{\sqrt{2}} (\phi_{p_y}) \tag{5.6}$$

These orbitals have identical shapes, sizes, and energies. They lie in the same plane (the xy plane) at angles of 120° with respect to one another. This gives rise to a perfect trigonal planar geometry, as is observed for species such as BCl_3. The p_z orbital is not involved in the bonding scheme; it is nonbonding and vacant. (Recall that the Lewis structure for BCl_3 shows only six electrons about the central atom, suggesting that one of boron's four valence orbitals is unused.)

The sp^2 hybridization is common for main group elements engaged in double bonding; examples such as $COCl_2$, ONF, O_3, SO_2, and SO_3 can be cited. In each case, the π bond is formed using a pure (unhybridized) p orbital from the central atom. Hybridized orbitals are normally limited to σ bonding only.

Using the same approach, the mixing of one s with three p orbitals produces a set of four sp^3-type hybrids, which again have identical shapes, sizes, and energies. They are oriented toward the corners of a tetrahedron, giving rise to bond angles of 109.5°.

---

4.   For the derivation see Day, M. C.; Selbin, J. *Theoretical Inorganic Chemistry*, 2nd ed.; Van Nostrand Reinhold: New York, 1969; pp. 192*ff.*

The mixing of *s*- and *p*-type orbitals in the appropriate proportions can be used to explain how the ideal bond angles for systems having two, three, and four central atom electron pairs are formed. But, as we will see later, these are not the only orbital combinations capable of producing these geometries (see Table 5.3, at the end of this section).

Unequal Hybridizations and Bent's Rule

While the *sp* ($c_s = 0.500$, or 50.0% *s* character), sp^2 ($c_s = 0.333$), and sp^3 ($c_s = 0.250$) hybridizations lead to regular, highly symmetric structures, they are only three of an infinite number of possible combinations. For example, consider the mixing of one *s* and three *p* orbitals to give approximately sp^3 hybrids using the following proportions:

For 3 orbitals: $c_s = 0.229$ and $c_p = 0.771$

For 1 orbital: $c_s = 0.313$ and $c_p = 0.687$

The total *s* character is $3(0.229) + 0.313 = 1.000$, and the total *p* character is $3(0.771) + 0.687 = 3.000$. Thus, the sp^3 label is appropriate as an average hybridization. However, we will demonstrate below that the three equivalent orbitals lie at angles of $107.3°$, rather than $109.5°$, with respect to one another. This is, in fact, the hybridization description for NH_3. The lone pair occupies the unique hybrid.

Recall that an *s* orbital is more electronegative than the *p* orbitals at the same quantum level. It follows that as the *s* character of a hybrid increases, the electronegativity of that orbital also increases. This reasoning leads to *Bent's rule*,[5] which may be paraphrased by two statements:

1. Substituents of high electronegativity tend to bond to hybrid orbitals of low *s* character.

2. Substituents of low electronegativity tend to bond to hybrid orbitals of high *s* character.

A highly electronegative terminal atom, by definition, has considerable ability to attract the electrons of a covalent bond. This is better achieved if its "competition"—the electronegativity of the central atom orbital—is low. From this it can be concluded that a substituent exerts an influence on the character of the hybrid with which it bonds. A filled orbital that does not overlap with a terminal atom orbital (ie, a lone pair) is normally high in *s* character (see the values for NH_3, above). The central atom "prefers" to place its own electrons in the most stable available orbital(s).

5. Bent, H. A. *Chem. Rev.* **1961**, *61*, 275; *J. Chem. Educ.* **1960**, *37*, 616; *J. Chem. Phys.* **1960**, *33*, 1258.

The relationship between the bond angle θ separating two or more equivalent s–p hybrids and their c_s and c_p values is

$$\cos \theta = \frac{-c_s}{c_p} \tag{5.7}$$

This equation can be checked for the three equivalent hybrids of NH_3 and the 107.3° bond angle given above.

As a second example, the observed bond angles in CH_2F_2 are 111.9° ($\angle H$–C–H) and 108.3° ($\angle F$–C–F). You should be able to demonstrate that this corresponds to $c_s = 0.272$ for the C–H hybrids and 0.239 for the C–F hybrids. Again, these results are consistent with Bent's rule.

A triangular plot showing the variance of bond angle with percent s character is given in Figure 5.4. Each vertex of that plot represents a "pure" orbital type; the three end lines correspond to s–p, s–d, and p–d mixtures; and all internal points represent hybrids having contributions from all three orbital types.

Bent's rule fails for systems containing partial or true π bonds. For example, the Br–P–Br bond angles in $POBr_3$ are 105.5°. Inserting this value into equation (5.7) gives $c_s = 0.211$ for the P–Br hybrids; therefore, by subtraction, $c_s = 0.367$ for the P–O orbital. In contradiction to Bent's rule, the more electronegative terminal atom (oxygen) is bonded to a phosphorus hybrid of relatively high electronegativity. Steric crowding cannot be the explanation, since the larger bromine atoms should tend to increase, rather than decrease, the Br–P–Br angles. The data are best accounted for by P–O π

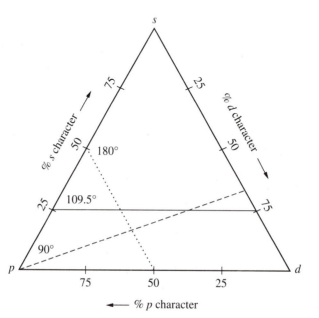

Figure 5.4
Triangular plot of selected bond angles versus hybridization for the mixing of s-, p-, and/or d-type atomic orbitals.

bonding. The additional electron density in that bond region increases the electron pair repulsion, and thereby widens the O–P–Br angles.

$$
\begin{array}{c}
\ddot{\text{:O:}} \\
\| \\
\text{P} \\
\ddot{\text{:Br}} \diagdown \ \diagdown \ \ddot{\text{Br:}} \\
\ddot{\text{:Br:}}
\end{array}
$$

Hybridization Schemes Incorporating *d* Orbitals

When a central atom has more than eight valence electrons, the hybridization scheme is assumed to include *d* orbitals. For example, a set of five hybrids forming a trigonal bipyramid can be constructed from the mixing of one *s*, three *p*, and one *d* (specifically, the d_{z^2}) orbital. Here, the weighting coefficients are usually taken to be:

Each of three equatorial hybrids: $\quad c_s = c_{p_x} = c_{p_y} = 0.333$

Each of two axial hybrids: $\quad\quad\quad c_{p_z} = c_{d_{z^2}} = 0.500$

Observe that the total orbital usage is one *s*, three *p*, and one *d* atomic orbital, as required for the sp^3d description. It is convenient for certain purposes to divide these orbitals into the sp^2 (equatorial) and *pd* (axial) subsets.

It should be clear that in electronegativity, $\chi_{eq} > \chi_{ax}$. Bent's rule therefore predicts that highly electronegative groups prefer the axial positions. This explains the observation that for the series $PF_{(5-x)}Br_x$ (where $x = 1$–3), the most stable isomers are

$$
\begin{array}{ccc}
\text{F} & \text{F} & \text{F} \\
\text{F}\diagdown\ | & \text{F}\diagdown\ | & \text{Br}\diagdown\ | \\
\quad\text{P}\!-\!\text{Br} & \quad\text{P}\!-\!\text{Br} & \quad\text{P}\!-\!\text{Br} \\
\text{F}\diagup\ | & \text{Br}\diagup\ | & \text{Br}\diagup\ | \\
\text{F} & \text{F} & \text{F}
\end{array}
$$

Since lone pairs tend to reside in orbitals of high *s* character, the equatorial positions are preferred for central atom lone pairs. Recall that the same result was predicted by VSEPR theory.

A related topic is the fluxionality of certain trigonal bipyramidal compounds and ions. The room-temperature ^{19}F NMR spectrum of PF_5 shows all fluorines to be equivalent, indicating that "scrambling" of the axial and equatorial positions must occur. The mechanism of this scrambling is believed to involve a square pyramidal intermediate, and is often referred to as the *Berry pseudorotation*.[6] One equatorial group remains in place while

6. Berry, S. *J. Chem. Phys.* **1960**, *32*, 923. See also Muetterties, E. L. *J. Am. Chem. Soc.* **1969**, *91*, 4115.

Figure 5.5
The Berry pseudorotation mechanism. By rotating about one equatorial bond, the two remaining equatorial and the two axial positions are interchanged.

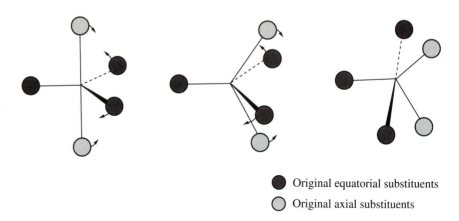

● Original equatorial substituents

○ Original axial substituents

the other two equatorial groups interchange with the axial substituents, as shown in Figure 5.5.

The efficiency of the Berry pseudorotation varies with the electronegativities of the substituents. For example, this process is much slower in dialkyltrifluorophosphoranes (R_2PF_3) than in PF_5, because it produces a less stable topology—at least one of the equatorial carbons must enter a disfavored axial position. Lone pairs can hinder the mechanism in the same way. For example, ClF_3 undergoes only slow interconversion at room temperature, since at least one lone pair must become axial upon pseudorotation.

The other common high-coordinate geometry, octahedral, can be achieved through sp^3d^2-type hybridization in which the d_{z^2} and $d_{x^2-y^2}$ orbitals are utilized. (Recall from Chapter 3 that these d orbitals have σ-type symmetry in the O_h point group.) For each of these six hybrids, $c_s = 0.167$ (1/6), $c_p = 0.500$ (1/2), and $c_d = 0.333$ (1/3). These weighting coefficients create hybrid orbitals of equal energy that lie at 90° angles with respect to one another, forming a regular octahedron.

The Role of Symmetry in Hybridization

We have seen that theoretical considerations allow for a variety of different orbital combinations, although some are unlikely because of the large energy differences between the atomic orbitals being hybridized. (Recall that the effectiveness of mixing decreases as the energy differential increases.) The possible combinations can be determined through group theory.

Bonds were treated as vectors in Chapters 2 and 3 for purposes of generating reducible representations. For example, it was found that the three bonds of BF_3 (D_{3h} symmetry) form a basis set for the reducible representation given at the top of the next page.

	E	$2C_3$	$3C_2$	σ_h	$2S_3$	$3\sigma_v$
Γ_r	3	0	1	3	0	1

This representation reduces to the irreducible representations $A'_1 + E'$. The significance of this procedure can be seen by examining the right side of the D_{3h} character table. An s-type orbital transforms as A'_1, while the degenerate $p_{x,y}$ pair belong to E'. This tells us that the mixing of central atom s, p_x, and p_y orbitals (ie, sp^2 hybridization) results in trigonal planar geometry (D_{3h} point group if the three substituents are identical).

Next, consider silane, SiH_4. The four bond vectors are oriented toward the corners of a tetrahedron, and the point group is, of course, T_d. The reducible representation is

	E	$8C_3$	$3C_2$	$6S_4$	$6\sigma_d$
Γ_r	4	1	0	0	2

which reduces to $A_1 + T_2$. Examining the right side of the T_d character table, we see that the s orbital transforms as A_1. There are two possibilities for T_2: the $p_{x,y,z}$ and the $d_{xy,xz,yz}$ subsets. Thus, the tetrahedral geometry can be achieved through either sp^3 or sd^3 hybridization, although when silicon is the central atom the better s–p energy match favors sp^3.

Table 5.3 lists possible orbital combinations for hybridizations leading to the most common molecular geometries.

Table 5.3 Symmetry-allowed hybrid combinations for the common molecular geometries

Coordination Number	Geometry	Point Group	Symmetry Combinations
2	Linear	$D_{\infty h}$	s, p_z p_z, d_{z^2}
3	Trigonal planar	D_{3h}	s, p_x, p_y s, d_{xy}, $d_{x^2-y^2}$
4	Tetrahedral	T_d	s, p_x, p_y, p_z s, d_{xy}, d_{xz}, d_{yz}
4	Square planar	D_{4h}	s, p_x, p_y, $d_{x^2-y^2}$ p_x, p_y, d_{z^2}, $d_{x^2-y^2}$
5	Trigonal bipyramidal	D_{3h}	s, p_x, p_y, p_z, d_{z^2} s, p_z, d_{xy}, $d_{x^2-y^2}$, d_{z^2}
5	Square pyramidal	C_{4v}	s, p_x, p_y, p_z, $d_{x^2-y^2}$ s, p_z, d_{xz}, d_{yz}, $d_{x^2-y^2}$ s, d_{z^2}, d_{xz}, d_{yz}, $d_{x^2-y^2}$
6	Octahedral	O_h	s, p_x, p_y, p_z, d_{z^2}, $d_{x^2-y^2}$

The Influence of Hybridization on Bond Lengths

How does hybridization play a role in determining bond lengths? Recall that a p orbital extends farther into space from the nucleus—that is, is more diffuse—than an s orbital.[7] It follows that orbitals high in s character are relatively compact and form short bonds. Increasing the p character of an s–p hybrid lengthens the bond. This can be generalized to include the even more diffuse d orbitals. Thus, for a particular central atom, bond length increases from left to right, as follows:

$$sp < sp^2 < sp^3 < sp^3d < sp^3d^2$$

This notion is supported by data for the carbon–hydrogen bonds of some simple organic compounds.

Compound	Hybridization	% s	d_{C-H}, pm
HC≡CH	sp	50	105.7
$H_2C{=}CH_2$	sp^2	33	107.9
$H_3C{-}CH_3$	sp^3	25	109.4

This bond shortening carries over to carbon–carbon single bonds; for example:

Compound	Hybridizations	d_{C-C}, pm
HC≡C–C≡CH	sp–sp	137
$H_2C{=}CH{-}CH{=}CH_2$	sp^2–sp^2	146
$H_3C{-}CH_2{-}CH_2{-}CH_3$	sp^3–sp^3	154

The above list requires some study. Convince yourself that only C–C single bonds are compared, so no bond order effect is operative.

The sp^2 equatorial "subhybrids" of a trigonal bipyramid are more compact than the axial (pd) subset, so the equatorial bonds are shorter than the axial bonds. (This was first encountered in Table 5.2.) Consider two binary phosphorus(V) halides:

PX$_5$, X =	d_{P-X}, Equatorial	d_{P-X}, Axial	% Difference
F	153.4	157.7	2.8
Cl	202	214	5.9

---

7. The s and p orbitals having the same principal quantum number have the same average distance from the nucleus. However, the compactness of spherical orbitals causes a significant portion of the electron density of an occupied p orbital to lie outside the corresponding s domain.

The greater percentage difference for PCl_5 suggests a steric effect (the bond-pair repulsions at 90° are reduced by extending the axial bonds).

5.3   Bond Polarity and Percent Ionic Character

In any heteronuclear bond, there is a drift of electron density from the less electronegative to the more electronegative element. The situation may be visualized as a tug-of-war in which one side grows weaker as it "wins." Recall from Chapter 1 that the χ of an atom decreases with increasing negative charge. It has therefore been reasoned that electron density "flows" until the electronegativities of the two bonding atoms are equalized; the bonded atoms thereby acquire equal and opposite partial charges. The greater these charges, the greater the *ionic character* of the bond, and hence the greater the bond energy because of ionic resonance.

Many methods have been proposed for estimating the ionic character of a heteronuclear bond. Two of the most important are discussed below.

The Equalization of Electronegativity[8]

The *equalization of electronegativity* approach is based on the notion that the electronegativity of an atom is a linear function of its charge. Recall equation (1.27), which we repeat here:

$$\chi = a + 2bq \tag{5.8}$$

a is the nominal electronegativity (the electronegativity of the neutral atom). Electronegativity equalization assumes that in a covalent bond the charge distribution is such that the bonded atoms have equal χ; that is, for compound X–Z, $\chi(X) = \chi(Z)$. Therefore,

$$a(X) + 2b(X) \cdot q(X) = a(Z) + 2b(Z) \cdot q(Z) \tag{5.9}$$

Recognizing that $q(X) = -q(Z)$, this gives

$$a(X) - 2b(X) \cdot q(Z) = a(Z) + 2b(Z) \cdot q(Z) \tag{5.10}$$

and therefore,

$$q(Z) = \frac{a(X) - a(Z)}{2[b(X) + b(Z)]} \tag{5.11}$$

---

8.   For a review of this topic, see Mortier, W. J. *Struct. Bonding* **1987**, *66*, 145.

As an example of the use of this method, let us estimate the ionic contribution to the K–Br bond of potassium bromide. The relevant electronegativity parameters from Table 1.11 are:

Atom	a	b
K	2.42	1.92
Br	7.59	4.22

The partial charge on potassium is then

$$q(K) = \frac{7.59 - 2.42}{2[4.22 + 1.92]} = +0.421$$

This estimate suggests that the charge distribution in KBr is $K^{+0.421}Br^{-0.421}$; that is, the bond is estimated to be about 42% ionic and 58% covalent.

Dipole Moments

A second method for predicting ionic contributions, again applicable mainly to heteronuclear diatomics, utilizes experimentally determined dipole moments. The equation

$$\mu = qd \tag{5.12}$$

where μ is the dipole, q is the charge separation, and d is the internuclear distance, is often employed.

Let us use hydrogen chloride as an example. The experimental dipole moment of HCl(g) is 1.1 D (debye) (or 1.1×10^{-8} esu·pm), and the internuclear distance is 127.4 pm. Therefore,

$$q = \frac{\mu}{d} = \frac{1.1 \times 10^{-8} \text{ esu·pm}}{127.4 \text{ pm}}$$

$$= 8.6 \times 10^{-11} \text{ esu}$$

Since the charge of an electron is 4.80×10^{-10} esu, this corresponds to a drift of 0.18 bonded electron, or an estimated ionic character of 18%.

Although generally limited to diatomics, it is sometimes possible to use the dipole method for other species. An example is the water molecule. The bond angle in H_2O is 104.5°. The O–H bond distance is 96 pm, and the experimental dipole moment is 1.87 D. By assuming that the molecular dipole is simply the resultant of the two individual bond moments, we can proceed according to Figure 5.6.

In the figure, x (the dipole for an individual bond) is the hypotenuse of a right triangle. One leg of that triangle is equal to $\mu/2$ and bisects the 104.5°

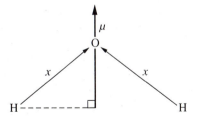

Figure 5.6 Dipole method for the estimation of the ionic character of the O–H bond of water. Here, μ is taken to be the vector sum of the two individual bond moments, x (see text).

bond angle. The opposite angle is $90.0° - (104.5°/2)$, or about $37.8°$. Then,

$$\sin 37.8° = 0.613 = \frac{\mu/2}{x}$$

Rearrangement gives $\mu = 1.23x$. Since $\mu = 1.87\,\text{D}$, then $x = 1.87/1.23$, or $1.52\,\text{D}$ per bond. Therefore,

$$q = \frac{x}{d}$$

$$= \frac{(1.52 \times 10^{-8}\,\text{esu·pm})(1e^-/4.80 \times 10^{-10}\,\text{esu})}{96\,\text{pm}}$$

$$= 0.33$$

From this we conclude that the O–H bond in water is about 33% ionic.[9]

This approach is impossible, of course, if the dipole contributions of individual bonds are canceled by molecular geometry. For example, BeF_2, BF_3, CF_4, PF_5, and SF_6 are all nonpolar molecules that contain polar bonds. The presence or absence of a molecular dipole has an important effect on the physical and chemical behavior of a given compound. Some specific properties related to dipole moments will be discussed in the next chapter and beyond.

Bibliography

Gillespie, R. J.; Hargittai, I. *The VSEPR Model of Molecular Geometry*; Allyn & Bacon: Boston, 1991.

9. This analysis is oversimplified; it ignores other factors of importance, such as lone-pair contributions and hybridization effects. For a more detailed discussion, see McWeeny, R. *Coulson's Valence*, 3rd ed.; Oxford University: London, 1979; pp. 153*ff*.

Sanderson, R. T. *Polar Covalence*; Academic: New York, 1983.

Coulson, C. A. *The Shape and Structure of Molecules*, 2nd ed.; Clarendon: Oxford, 1982 (revised by R. McWeeny).

Burdett, J. K. *Molecular Shapes*; Wiley: New York, 1980.

DeKock, R. L.; Gray, H. B. *Chemical Structure and Bonding*; Benjamin/Cummings: Menlo Park, CA, 1980.

·Fergusson, J. E. *Stereochemistry and Bonding in Inorganic Chemistry*; Prentice-Hall: Englewood Cliffs, NJ, 1974.

Gillespie, R. J. *Molecular Geometry*; Van Nostrand Reinhold: London, 1972.

Day, M. C.; Selbin, J. *Theoretical Inorganic Chemistry*, 2nd ed.; Van Nostrand Reinhold: New York, 1969.

Pauling, L. *The Nature of the Chemical Bond*, 3rd ed.; Cornell University: Ithaca, NY, 1960.

Questions and Problems

1. Describe the atom geometry and predict the bond angle(s) for each of the following species:
   (a) AlH_3
   (b) NCF
   (c) PF_2^-
   (d) NO_2^+
   (e) NO_2^-
   (f) SeF_4
   (g) I_3^-
   (h) SO_3^{2-}
   (i) XeO_3
   (j) XeF_5^+
   (k) ICl_4^-
   (l) B_2Cl_4
   (m) $H_2C{=}C{=}CH_2$
   (n) N_2F_2
   (o) SiF_6^{2-}
   (p) I_5^-

2. Give the point group of maximum symmetry for each of the idealized compounds MX_n, where $n = 2$–7.

3. The dipole moment of SbF_5 is 0. The dipole moment of IF_5 is 2.18 D. How are these experimental data consistent with VSEPR theory?

4. Use the VSEPR model to predict which of the following species have nonzero dipole moments: CS_2, H_2O_2, ClO_3^-, ClF_3, SeF_6.

5. In the solid state, PCl_5 consists of an ionic lattice with PCl_4^+ and PCl_6^- ions. Predict the structures of these ions.

6. For each of the following pairs, decide which has the greater bond angle and defend your answer:
   (a) NO_2^+ or NO_2^-
   (b) NO_2^- or NO_3^-
   (c) SO_3^{2-} or CO_3^{2-}
   (d) PF_3 or PCl_3

7. Carefully consider all relevant factors and then estimate the bond angles in the molecule NO_2Cl. Defend your reasoning.

8. The Si–O–Si bond angle of hexamethyldisiloxane, $(Me_3Si)_2O$, is about 150°. How does this compare to the predicted value? How might the discrepancy be explained?

9. Methyl isocyanate, CH_3NCO, is a bent molecule. Silyl isocyanate, H_3SiNCO, is linear. Suggest an explanation.

10. The carbanion $:C(CN)_3^-$ has a trigonal planar geometry in spite of the fact that there are four electron pairs about the central carbon. Draw a resonance structure (any one of three equivalent structures) that explains the stereo-chemical inactivity of the lone pair.

11. The CO_2^- ion can be prepared by the adsorption of CO_2 on a substrate, followed by ultraviolet irradiation. Microwave spectroscopy has been used to estimate a bond angle of $127°$ for this anion. Rationalize this observation.

12. Sketch the preferred structure for each of the following compounds, and briefly explain:
    (a) CH_3PF_4 (b) OSF_4 (c) XeO_2F_2

13. Predict the point group for the most stable topology of:
    (a) PF_4Cl (b) PF_3Cl_2 (c) SF_3Cl (d) IF_4^+

14. (a) After considering all relevant factors, predict the F–S–F and F–S–O bond angles in F_2SO. Explain your reasoning in detail.
    (b) The average of the three angles in this compound is about $102°$. Is this consistent with your predictions? If not, make appropriate adjustments and try again.

15. Give the hybridization for each central atom of:
(a) BCl_3	(b) CS_2	(c) AsF_5
(d) NO_2^+	(e) NO_2^-	(f) N_2H_4
(g) $(CH_3)_2O$	(h) SCl_2	(i) IF_2^+
(j) IF_2^-	(k) TeF_4^{2-}	(l) $HONO_2$
(m) $S_2O_3^{2-}$	(n) PCl_4^-	(o) $POCl_3$
(p) $H_2C{=}C{=}CH_2$		

16. The Cl–Si–Cl bond angle in dichlorosilane, SiH_2Cl_2, is $110.4°$.
    (a) Calculate c_s for the Si–Cl hybrid orbitals.
    (b) Use your answer to part (a) to calculate c_s for the Si–H hybrids.
    (c) Estimate the H–Si–H bond angle.
    (d) Are these values consistent with Bent's rule? Discuss.

17. The experimental bond angles for three binary compounds of arsenic are: AsH_3, $91.8°$; AsF_3, $96.2°$; and $AsCl_3$, $98.5°$. Rationalize and discuss.

18. Apply group theory to explore hybridization schemes for the two common five-coordinate geometries.
    (a) Show that the trigonal bipyramidal geometry (D_{3h} point group) can be achieved through sp^3d hybridization and that the specific d orbital involved is d_{z^2}.
    (b) Determine the d orbital involved in the hybridization scheme for the square pyramidal (C_{4v}) geometry.

19. The point group for GeH_3F is C_{3v}, not T_d. Use group theory to show that sp^3 hybridization is still viable.

20. Predict which is longer, and defend your answers:
    (a) The average P–F bond length in PF_3 or PF_5
    (b) The P–P bond in P_2H_4 or P_2F_4
    (c) The P–O bond in PO_2^- or PO_3^{3-}
    (d) The P–Cl bond in PCl_4^+ or PCl_6^-

21. Rationalize the following bond angles:
    (a) $\angle$ X–C–X of C_2X_4: X = H, 120°; X = F, 114°; X = Cl, 113.5°
    (b) $\angle$ X–C–X of COX_2: X = H, 115.8°; X = F, 108°; X = Cl, 111.3°

22. Apply the equalization of electronegativity approach to ClF using data from Table 1.11. What is the calculated charge on fluorine? What is the calculated percent ionic character of the bond?

23. Estimate the percent ionic character of an N–F bond by the electronegativity equalization method.

24. The dipole moment of NBr is about 0.78 D. Estimate the N–Br bond distance, and use that value to estimate the percent ionic character for this diatomic compound.

25. Given the dipole moment of H_2S (0.97 D), its bond angle (92°), and the H–S bond distance (135 pm), estimate the ionic character of the H–S bond.

26. The gas-phase dipole moment of the hydroxyl radical, ·OH, is 1.66 D, and the bond length for this species is 97.1 pm. Use this information to calculate the percent ionic character of the O–H bond, and compare your answer to that obtained in the text for the O–H bond of water.

27. Hydrogen azide, HN_3, is a compound of low thermal stability. It has the bond linkage H–N–N–N. Integrate your knowledge of covalent bond theory, resonance, and VSEPR theory to:
    (a) Predict the bond angles for this compound.
    (b) Predict which N–N bond is longer, and explain.
    (c) Predict how the bond angles and distances will change if HN_3 is deprotonated to the azide ion, N_3^-.

*28. The structure of the compound CH_3SF_3 has been determined by A. J. Downs and co-workers (*Inorg. Chem.* **1989**, *28*, 3286). After reading at least the first three pages of their article, answer these questions.
    (a) What is the point group of CH_3SF_3?
    (b) How would you describe the atom geometry? To what binary fluoride is it similar?
    (c) Is the sulfur lone pair stereochemically active or not?
    (d) Comment on the position of the methyl group (axial or equatorial).
    (e) The bond angle formed by the axial fluorines about the central atom is 175°, not 180°. Why?

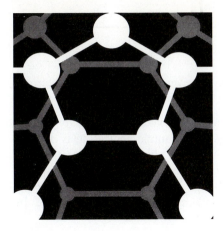

6

Ionic Bonding and the Solid State

Certain aspects of chemical bonding between elements having different electronegativities were discussed in previous chapters. Recall that an unbalanced distribution of electron density creates partial charges of opposite sign on the bonded atoms. For cases in which the electronegativity difference is so large that the electrostatic energy overwhelms the overlap (covalent) energy, the bond is customarily described as ionic.

How do the energetics of ionic bonding differ from those of covalency? What are the most important solid-state structures for ionic compounds, and why are those structures dominant? What physical and chemical properties result from such structures? Such questions are considered in this chapter.

6.1 The Formation of Free Ions

The generation of ions from neutral atoms obviously requires the gain or loss of one or more electrons. This can occur in at least two different ways:

1. Electron transfer may result from a collision between two atoms. This amounts to the well-known process of oxidation–reduction, of course.

As will shortly be demonstrated, however, electron transfer between isolated free atoms is energetically disfavored.

2. Electrons may be lost or gained under artificial conditions involving the input of a large amount of energy. For example, ionization is induced by various instrumental methods, including photoelectron spectroscopy and mass spectrometry. Such techniques are useful for studying species that are otherwise difficult to obtain. Two examples are the highly reactive cations H_3^+ and CH_5^+, both of which can be prepared by chemical ionization mass spectrometry.

$$H_2 + H^+ \longrightarrow H_3^+ \tag{6.1}$$

$$CH_4 + H^+ \longrightarrow CH_5^+ \tag{6.2}$$

The properties of these and other ions created under forcing conditions are discussed in other parts of this book. At present, we will turn our attention to the energetics of ion formation by electron transfer between atoms.

Is there any element for which the process described by equation (6.3) is thermodynamically favored?

$$2M(g) \longrightarrow M^+(g) + M^-(g) \tag{6.3}$$

This process can be separated into its oxidation and reduction components:

$$
\begin{array}{lll}
M(g) \longrightarrow M^+(g) + 1e^- & \Delta H = IE_1 \\
\underline{M(g) + 1e^- \longrightarrow M^-(g)} & \underline{\Delta H = -EA_1} \\
2M(g) \longrightarrow M^+(g) + M^-(g) & \Delta H = IE_1 - EA_1
\end{array}
$$

The value of ΔH will be negative for any element having a first electron affinity greater than its first ionization energy. However, an examination of Table 1.7 shows that there is no element in the periodic table for which this is the case. The most favorable electron affinity belongs to chlorine (3.62 eV), but its first ionization energy is over three times greater.

It can also be learned from Table 1.7 that the first ionization energy of every element exceeds the electron affinity of chlorine. There is therefore no combination of elements for which electron transfer is favored. The least endothermic possibility is electron transfer from cesium to chlorine:

$$Cs + Cl \longrightarrow Cs^+ + Cl^- \qquad \Delta H = IE_1\,(Cs) - EA_1\,(Cl)$$
$$= +3.894 - 3.62 = +0.27 \text{ eV}$$

This process is disfavored by 0.27 eV (or, since 1 eV = 96.487 kJ/mol, by 26 kJ/mol).[1]

It follows that electron transfer between two atoms to produce separated, gas-phase ions is never favored. How, then, are ionic compounds formed? Let us pursue the problem using a slightly different approach.

6.2 Electrostatic Energies

The force of attraction between two oppositely charged ions is given by Coulomb's law:

$$F = \frac{kq_+q_-}{d^2} \tag{6.4}$$

Here, d is the distance between the centers of gravity of the positive and negative charges. Since Energy = Force × Distance, the corresponding electrostatic energy is

$$E = \frac{kq_+q_-}{d} \tag{6.5}$$

In this equation, $k = 1/4\pi\epsilon$, where ϵ is the permittivity constant (for a vacuum, 8.854×10^{-12} C^2/m·J). By combining constants and using the appropriate conversion factors, equation (6.5) can be converted to the more convenient form

$$E = \frac{(1.389 \times 10^5 \text{ kJ·pm/mol})Z_+Z_-}{d} \tag{6.6}$$

where d is in picometers and Z_+ and Z_- represent the charges of the cation and anion, respectively. This equation enables us to calculate the electrostatic energy of attraction for any ion pair for which the interionic distance is known.

Returning to the Cs–Cl system, experimental studies indicate that d for CsCl(g) is 290.6 pm. Assuming the formulation Cs$^+$Cl$^-$, the electrostatic energy calculated using equation (6.6) is -478 kJ/mol, which is more than sufficient to overcome the ionization energy of cesium. Hence, the process described by equation (6.7) is expected to be exothermic.

$$\text{Cs(g)} + \text{Cl(g)} \longrightarrow \text{Cs}^+\text{Cl}^-\text{(g)} \tag{6.7}$$

1. Appendix I gives a variety of conversion factors, several of which will be used in this chapter.

However, neither cesium nor chlorine exist as free, gas-phase atoms at room temperature. Cesium is a metallic solid, while chlorine exists as $Cl_2(g)$. A more relevant equation is therefore

$$Cs(s) + \tfrac{1}{2}Cl_2(g) \longrightarrow Cs^+Cl^-(g) \qquad \Delta H = ?? \qquad (6.8)$$

We must consider the enthalpy of sublimation of cesium ($+76$ kJ/mol) and the bond dissociation energy of Cl_2 ($+240$ kJ/mol Cl_2, or $+120$ kJ/mol of chlorine atoms). The following sequence results, leading to a predicted ΔH of -255 kJ/mol for equation (6.8):[2]

$Cs(s) \longrightarrow Cs(g)$	$\Delta H_{sub} =$	$+76$ kJ/mol
$Cs(g) \longrightarrow Cs^+(g) + 1e^-$	$IE_1 =$	$+376$ kJ/mol
$\tfrac{1}{2}Cl_2(g) \longrightarrow Cl(g)$	$D(Cl\text{–}Cl)/2 =$	$+120$ kJ/mol
$Cl(g) + 1e^- \longrightarrow Cl^-$	$-EA_1 =$	-349 kJ/mol
$Cs(g) + Cl(g) \longrightarrow Cs^+Cl^-(g)$	$E_{calc} =$	-478 kJ/mol
$Cs(s) + \tfrac{1}{2}Cl_2(g) \longrightarrow Cs^+Cl^-(g)$	$\Delta H =$	-255 kJ/mol

Interactions in the Solid State: The Madelung Constant

There is an additional complication in the above system: Cesium chloride is a solid, not a gas, at room temperature (and for over 600°C beyond). The physical state is important because, unlike the gas phase, ions are packed closely together in the solid state. Thus, each ion experiences many electrostatic interactions simultaneously, and this greatly improves the energetics of ionic bonding. For solids, then, equation (6.6) should be modified to read

$$E = \frac{(1.389 \times 10^5 \text{ kJ·pm/mol})AZ_+Z_-}{d} \qquad (6.9)$$

where A is the *Madelung constant*.[3] The Madelung constant is an expression of all the electrostatic interactions (both attractive and repulsive) per mole of solid-state species.

How is A determined? As a simple example, consider a linear arrangement of ions in which the charges are alternated to maximize attractions. Such a situation is pictured in Figure 6.1, where the central cation is attracted to the two adjacent anions at distance d, repelled by two cations at distance

---

2. This treatment employs what is sometimes called *Hess' law*: The energy change for a process is independent of pathway. That is, the enthalpy changes of the individual steps can be summed to arrive at the overall value.

3. Madelung, E. *Physik. Z.* **1918**, *19*, 524. See also Quane, D. *J. Chem. Educ.* **1970**, *47*, 396.

Figure 6.1
Electrostatic interactions in the solid state: the determination of the Madelung constant for a linear arrangement of ions of alternating charge (see text).

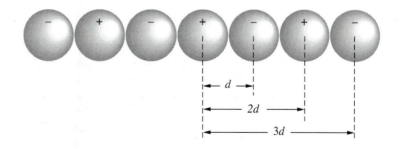

$2d$, attracted to two anions at $3d$, etc. The expression

$$A = 2 - \tfrac{2}{2} + \tfrac{2}{3} - \tfrac{2}{4} + \tfrac{2}{5} - \cdots$$

therefore applies. This infinite series equates to $A = 2(\ln 2) = 1.3863$. Thus, the stability of a group of ions arranged in a linear manner is 1.3863 times greater than one in which an equal number of ions are merely paired.

The solid-state structure of $CsCl(s)$ is a three-dimensional lattice in which each ion has eight nearest neighbors of opposite charge, six next-nearest neighbors of like charge, etc. (see Figure 6.2). The Madelung constant for such an arrangement is 1.76276. That is, the electrostatic energy per mole of $CsCl(s)$ is estimated to be about 1.8 times greater than that for $CsCl(g)$.[4] This is a significant gain in energy, and is the primary reason CsCl exhibits high melting temperatures (as do all ionic compounds).

Figure 6.2
A portion of the solid-state lattice of cesium chloride. Each ion has eight nearest neighbors of opposite charge, six next-nearest neighbors of the same charge, etc.

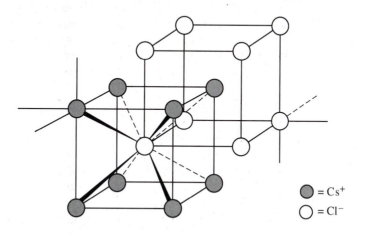

$\bigcirc = Cs^+$
$\bigcirc = Cl^-$

Madelung constants will be important to our discussion throughout this chapter. The values for five common ionic lattices are given in Table 6.1.

---

4. Well, almost. The internuclear bond distance in $CsCl(s)$ is 356.0 pm (roughly 20% greater than in the vapor state), which obviously affects this comparison.

Table 6.1 Values of the Madelung constant, A, for five common solid-state geometries

Lattice Type	+/− Coordination Numbers	A
ZnS	4/4	1.63805
NaCl	6/6	1.74756
CsCl	8/8	1.76276
TiO_2	6/3	2.4080
CaF_2	8/4	2.5194

Note: Various conventions are used in the literature that alter the numerical values of A. Those given here are appropriate for substitution into equations (6.9) and (6.11).

Corrections for Orbital Overlap

Still another refinement is necessary to adequately understand the energetic aspects of the ionic model. Equations (6.5), (6.6), and (6.9) indicate that electrostatic energy increases with decreasing internuclear distance. However, if the distance between an ion pair were to become too small, then the electron clouds of adjacent ions would begin to overlap; this would result in severe electron–electron repulsions. An ionic bond length is therefore an equilibrium distance; that is, it is the internuclear separation at which the opposing attractive and repulsive forces yield the minimum energy state. This is illustrated in Figure 6.3, where the attractive, repulsive, and net energies of the Cs^+Cl^- ion pair are plotted as a function of internuclear distance.

The overlap repulsion energy E_r obeys the relationship

$$E_r = \frac{k}{d^n} \tag{6.10}$$

where n (often referred to as the *Born exponent*) takes on values between 5 and 12, depending on the electron configuration of the ion (see Table 6.2). An average is taken for ion pairs having ions with different n values.

Table 6.2 The Born exponent, n, as a function of electron configuration

Configuration	n	Configuration	n
$1s^2$	5	$4s^24p^6$ or $5s^24d^{10}$	10
$2s^22p^6$	7	$5s^25p^6$ or $6s^25d^{10}$	12
$3s^23p^6$ or $4s^23d^{10}$	9		

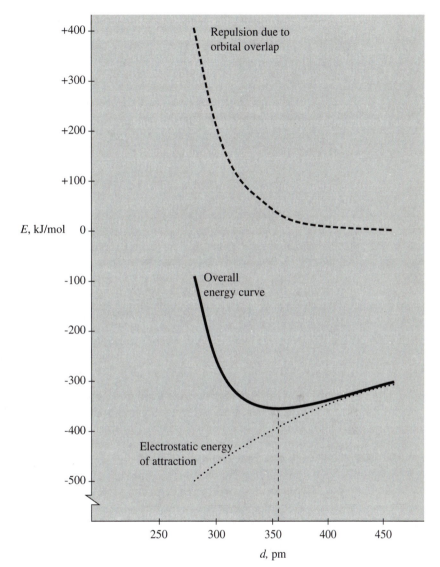

Figure 6.3
The energy of the Cs^+Cl^- ion pair as a function of internuclear distance. The minimum of the overall energy curve occurs at 356 pm, the equilibrium value of d for CsCl(s).

The combination of equations (6.6) and (6.10) permits the derivation of the equation

$$E = \frac{(1.389 \times 10^5 \ \text{kJ·pm/mol})AZ_+Z_-}{d}\left(1 - \frac{1}{n}\right) \tag{6.11}$$

which gives reasonable approximations for the net electrostatic energy. Alternatively, if the lattice type (and therefore the Madelung constant) is not known, then the *Kapustinskii equation*[5] can be used:

$$U = \frac{(1.202 \times 10^5 \text{ kJ·pm/mol})(\gamma)Z_+ Z_-}{d}\left(1 - \frac{34.5 \text{ pm}}{d}\right) \qquad (6.12)$$

Here, the numerical constant has been changed, and the symbol E replaced by U, where U is the *lattice energy*. This can be defined as the energy required to separate an ionic crystal into its component, gas-phase ions. Expressed as a chemical equation,

$$\text{M}_m\text{X}_x(\text{s}) \longrightarrow m\text{M}^{x+}(\text{g}) + x\text{X}^{m-}(\text{g}) \qquad \Delta H = U \qquad (6.13)$$

A consequence of this definition is that lattice energies are always positive.

The parameter γ in equation (6.12) represents the stoichiometric number of ions. For example, $\gamma = 2$ for CsCl and NH_4F, 3 for BaF_2 and Ca(CN)_2, and 4 for $\text{Ce(NO}_3)_3$. The Kapustinskii equation is useful because, even though it ignores the Madelung factor, a plot of U as calculated by equation (6.12) versus d yields a curve that closely resembles that predicted by theory (Figure 6.3).

Experimental Lattice Energies: Born–Haber Cycles

Can the lattice energy of an ionic compound be determined experimentally? There is no simple, direct method, but values can be obtained indirectly using a method developed by M. Born and F. Haber.[6] The only information not yet given that is needed to determine the experimental lattice energy of CsCl by this approach is ΔH_f°, its standard enthalpy of formation. Recall that ΔH_f° is defined as the enthalpy change for the formation of 1 mole of compound from its free elements when all species are in their standard states. Thus, the relevant equation for CsCl is

$$\text{Cs(s)} + \tfrac{1}{2}\text{Cl}_2(g) \longrightarrow \text{CsCl(s)} \qquad \Delta H = \Delta H_f^\circ = -407 \text{ kJ/mol} \qquad (6.14)$$

Let us now reorganize some of the data presented earlier in the chapter, add the equation for the standard enthalpy of formation, and then cancel those species that appear as both reactants and products.

---

5. Kapustinskii, A. F. *Z. Physik. Chem.* **1933**, *B22*, 257; *Quart. Rev.* **1956**, *10*, 283.

6. Born, M. *Verh. Dtsch. Physik. Ges.* **1919**, *21*, 679; Haber, F. *Verh. Dtsch. Physik. Ges.* **1919**, *21*, 750.

$$
\begin{array}{lll}
\text{Cs(s)} & \longrightarrow & \text{Cs(g)} & \Delta H_{sub} = +76 \text{ kJ/mol} \\
\text{Cs(g)} & \longrightarrow & \text{Cs}^+(g) + 1e^- & IE_1 = +376 \text{ kJ/mol} \\
\tfrac{1}{2}\text{Cl}_2(g) & \longrightarrow & \text{Cl(g)} & D(\text{Cl}_2)/2 = +120 \text{ kJ/mol} \\
\text{Cl(g)} + 1e^- & \longrightarrow & \text{Cl}^-(g) & -EA_1 = -349 \text{ kJ/mol} \\
\text{CsCl(s)} & \longrightarrow & \text{Cs(s)} + \tfrac{1}{2}\text{Cl}_2(g) & -\Delta H_f^\circ = +407 \text{ kJ/mol} \\
\hline
\text{CsCl(s)} & \longrightarrow & \text{Cs}^+(g) + \text{Cl}^-(g) & U = +630 \text{ kJ/mol}
\end{array}
$$

This is a *Born–Haber cycle*. By combining the energy changes for the component steps, the lattice energy of CsCl is found to be 630 kJ/mol. The graphical presentation of the above data shown in Figure 6.4 is helpful for visualizing the situation.

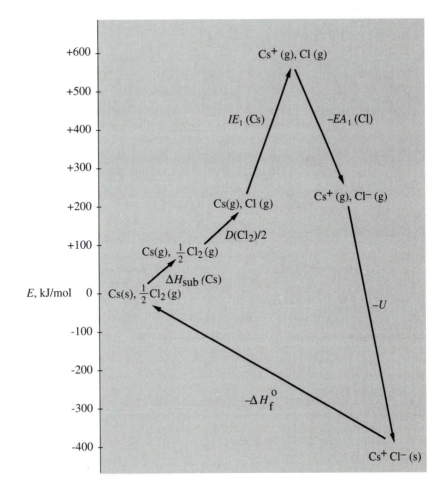

Figure 6.4
Enthalpy changes for the individual steps of the Born–Haber cycle for cesium chloride. By definition, $\Delta H = 0$ for the free elements in their standard states.

This can be generalized to the following relationship for the Group 1 fluorides and chlorides:

$$\Delta H_f^\circ = \Delta H_{sub}(M) + \frac{D(X_2)}{2} + IE_1(M) - EA_1(X) - U \qquad (6.15)$$

An additional term (ΔH_{sub} of X_2) is necessary for the bromides and iodides, since Br_2 and I_2 are not gases at room temperature; those values are $+31$ and $+62$ kJ/mol, respectively. For the Group 2 halides (with the empirical formula MX_2), $D(X_2)/2$ becomes $D(X_2)$, the $EA_1(X)$ term is doubled, and the second ionization energy of the metal must be included.

The solid-state bond distances and Born–Haber lattice energies are given for the Group 1 halides in Table 6.3; the lattice energies are plotted in Figure 6.5. It is evident that the energies are greatest for the fluorides and smallest for the iodides. This is due to a combination of factors; the X_2 bond energy, the electron affinity, and the size of the anion each play a role. The relatively small dissociation energy of F_2, the high electron affinity of fluorine, and the small size of F^- all promote high lattice energies for fluoride salts.

Lattice energies decrease going down Group 1. Here, there are three factors to consider: ΔH_{sub}, IE_1, and size. Examination of Figure 6.5 should convince you that the size effect dominates, since the periodic trends in sublimation and ionization energies are opposite those for lattice energy.

An interesting application of Born–Haber cycles was described by G. H. Purser, who evaluated the possibility of preparing ionic compounds

Table 6.3 Internuclear distances and Born–Haber lattice energies for the Group 1 halides

Compound	d, pm	U, kJ/mol	Compound	d, pm	U, kJ/mol
LiF	200.9	1034	LiBr	274.7	781
NaF	230.7	914	NaBr	298.1	728
KF	266.4	812	KBr	329.3	671
RbF	281.5	780	RbBr	343.4	654
CsF	300.5	744	CsBr	371.3	612
LiCl	256.6	840	LiI	302.5	718
NaCl	281.4	770	NaI	323.1	681
KCl	313.9	701	KI	352.6	632
RbCl	328.5	682	RbI	366.3	617
CsCl	356.0	630	CsI	395.0	584

Source: Solid-state internuclear distances taken from Wells, A. F. *Structural Inorganic Chemistry*, 5th ed.; Clarendon: Oxford, 1984; p. 444.

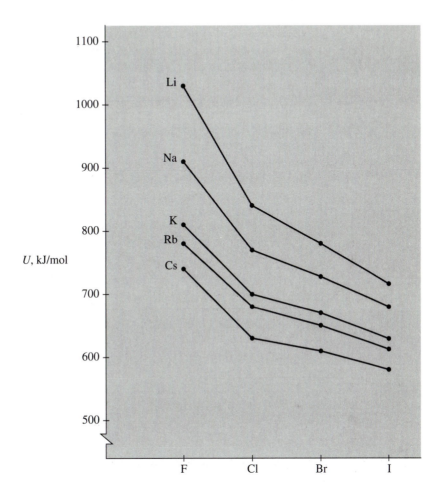

U, kJ/mol

Figure 6.5
Experimental
(Born–Haber) lattice
energies for the
Group 1 halides.

containing anions of the noble gases.[7] Based on equation (6.15), it is not unreasonable that species such as K^+Ne^- might exist. The two terms that involve the anion are $D(X_2/2)$, which is 0 for the noble gases, and $EA_1(X)$, which is unfavorable, but not strongly so. Thus, it is not surprising that of the twenty-five Group 1/Group 18 combinations considered, at least three (KNe, RbNe, and CsNe) were predicted to have $\Delta H_f^o < -60$ kJ/mol; hence, these three species may be isolable if appropriate syntheses can be developed.

The Stoichiometries of Ionic Compounds

Born–Haber cycles are useful for rationalizing the stoichiometries of ionic compounds. For example, we might wonder why cesium and chlorine atoms

7.   Purser, G. H. *J. Chem. Educ.* **1988**, *65*, 119.

combine in a 1:1 molar ratio rather than forming $CsCl_2$ (with a Cs^{2+} ion) or Cs_2Cl (with a divalent anion). Equations (6.11) and (6.12) suggest that these forms might be viable alternatives, since doubling either Z_+ or Z_- should double the lattice energy. Moreover, the Cs^{2+} ion is surely smaller than Cs^+. This should decrease the interionic distance and further enhance U. In fact, a reasonable estimate of the lattice energy for hypothetical $CsCl_2$ is roughly 1500 kJ/mol—a formidable amount of energy. Why, then, is that species unknown?

The loss of two electrons is required to form a Cs^{2+} ion. Equation (6.15) must be modified to

$$\Delta H_f^\circ = \Delta H_{sub}(Cs) + D(Cl_2) + IE_1(Cs) + IE_2(Cs) - 2EA_1(Cl) - U$$

The key to the problem is the second ionization energy of cesium, which is $+25.1$ eV, or $+2420$ kJ/mol. This huge value (over six times that of IE_1) is caused, of course, by the fact that the second electron is removed from a core, rather than a valence, orbital. The second ionization energy far exceeds the extra lattice energy that would be obtained for $CsCl_2$, and gives a rough estimate of about $+1200$ kJ/mol for ΔH_f°; thus, Cs_2Cl is extremely unstable.

In Chapter 1 we noted that, although the oxide ion should be thermodynamically unstable because of the unfavorable second electron affinity of oxygen, it is common in ionic lattices. This observation can now be explained. The high lattice energy of oxides more than offsets the energy requirement of EA_2 (about $+780$ kJ/mol). The presence in ionic-type lattices of certain other species of high charge, including N^{3-}, S^{2-}, and Al^{3+}, can be understood in the same manner.

You might wonder why entropy was not considered in the foregoing discussion. It is true that the highly ordered lattice of an ionic solid causes reactions such as

$$M(s) + \tfrac{1}{2}X_2(g) \longrightarrow M^+X^-(s) \qquad (6.16)$$

to be disfavored by entropy. However, ΔS° for equation (6.16) when $M^+X^- = NaCl$ is only $+27$ kJ/mol at 298 K—less than 4% of the lattice energy. The entropy effect is therefore minor, so it is customary to carry out Born–Haber and related calculations using enthalpies rather than free energies.

6.3 Covalency in Ionic Crystals

Inherent in the ionic bonding model is the assumption that the cationic and anionic charges are integral; that is, that electron transfer is complete. How reasonable is this notion? The best available evidence (see below) suggests that the binary compounds that are the most truly ionic are the Group 1 halides. Either equation (6.11) or (6.12) yields reasonable predictions for their lattice energies (often within 2% of the Born–Haber value), suggesting that these compounds have no appreciable covalent character in the solid state.

Equations (6.11) and (6.12) are less accurate for other binary salts, and it is instructive to examine some specific cases. Consider the hydrides of lithium, sodium, and potassium:

Compound	U_{calc}, kJ/mol [equation (6.12)]	U_{obs}, kJ/mol (Born–Haber)	Error
LiH	979	905	8.2%
NaH	846	811	4.3%
KH	737	714	3.2%

Electron transfer does not appear to be complete for these species. The small experimental lattice energies relative to the calculated values suggest that the actual charges are less than integral. Notice that the percentage error decreases as the electronegativity difference (and therefore the completeness of electron transfer) increases. It also can be concluded that covalent bonding does not compensate for the lost ionic energy.

A more extreme case is that of the alkaline earth oxides, which have presumed charges of 2+ for the cation and 2− for the oxide ion. Two representative examples are MgO and BeO:

Compound	U_{calc}, kJ/mol [equation (6.12)]	U_{obs}, kJ/mol (Born–Haber)	Error
MgO	3798	2585	46.9%
BaO	3058	1978	54.6%

Here, the ionic model fails miserably, with both compounds clearly retaining significant covalent character. If the actual lattice energies are inserted back into equation (6.12) and it is solved for Z_+ and Z_-, values of about 1.6 charge units are obtained (ie, $Z_+ = +1.6$ and $Z_- = -1.6$). Again, covalency does not appear to compensate.

The situation is different if the cation is of a transition or posttransition (Group 13 or 14) metal. Then the predicted lattice energy is often too low, as is the case for three silver salts:

Compound	U_{calc}, kJ/mol [equation (6.12)]	U_{obs}, kJ/mol (Born–Haber)	Error
AgF	840	954	−12.0%
AgCl	760	904	−15.9%
AgBr	733	895	−18.1%

The electron configuration of Ag^+ is $[Kr]5s^04d^{10}5p^0$. It therefore appears that any covalent interaction must involve (as for the Group 1 and 2 cations) s and/or p orbitals of a "new" quantum level. However, recall from Chapter 1 that shielding by d electrons is inefficient compared to that by the s and p orbital types. Therefore, the effective nuclear charges on the $5s$ and $5p$ orbitals of Ag^+ are considerably greater than those of the corresponding Group 1 and 2 ions (Rb^+ and Sr^{2+}). This results in contraction in the orbital sizes and an increase in their stabilities. Both changes lead to more efficient overlap with appropriate halogen orbitals, and thereby enhance the covalent contribution to the bond energy.

Interestingly, the covalent character of the silver halides seems to increase going from AgF to AgBr (note the error column above). This is as predicted by electronegativity arguments, and is also consistent with certain experimental data. For example, the melting point of AgCl is greater and the aqueous molar solubility is less than for AgF. These observations suggest that the AgCl lattice is the more covalent of the two.

Gold compounds are even more covalent than their silver analogues. Silver iodide was omitted from the above discussion because its lattice structure is different from the others, and so is not directly comparable. Nevertheless, its lattice is still predominantly ionic. In contrast, AuI is known to form zigzag chains in the solid state:

$$\diagdown Au \diagup I \diagdown Au \diagup I \diagdown Au \diagup I$$

The directional nature of these chains would be inefficient for ionic bonding, and provides convincing evidence that AuI(s) is best considered a covalent species.

The Covalent–Ionic Transition

For binary compounds conforming to the general formula MX_n (X = F, Cl, O, etc.), ionic character decreases as M varies from left to right across each period of the periodic table. This is as expected from electronegativities, since $\Delta\chi$ decreases in that direction.

Melting points provide a convenient (though imperfect) measure of ionic character, because the strong electrostatic attractions within an ionic lattice lead to high melting temperatures.[8] For this reason it is interesting

8. However, the same is true for three-dimensional networks held together by strong covalent bonds. For example, diamond (pure carbon) and silicon carbide (SiC, with a diamondlike lattice) both melt above 2900 K; see Chapter 7.

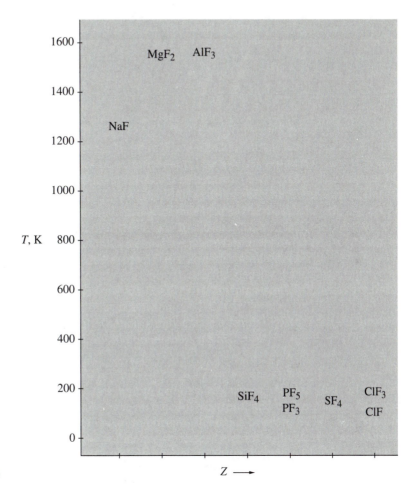

Figure 6.6
The melting
temperatures (K) of
some binary
fluorides of the
second short period.

to compare a series of compounds (eg, the fluorides) of a given period. The melting temperatures of some binary fluorides of the second period elements are plotted in Figure 6.6.

Only NaF, MgF_2, and AlF_3 can reasonably be classified as ionic. There is general agreement between melting point and electronegativity difference, but it is far from perfect; the transition across the series is smooth in electronegativity, but quite sudden in ionicity. This can be generalized: Electronegativity differences determine bond polarities, but have only an indirect effect on melting points. The large difference between the melting temperatures of AlF_3 and SiF_4 is indicative of a difference in the type of crystal lattice formed. Silicon tetrafluoride forms a nonionic lattice in which *London forces* play an important role (see Section 8.5).

6.4 Ionic Lattices Based on Cubic Geometries

Ionic lattices extend infinitely in three dimensions. Such lattices are best described by a *unit cell*, the smallest section which, by translation along its edges, generates the entire structure. The unit cells of most ionic lattices are either cubic or distorted from cubic in some way.[9] As a result, their specific characteristics normally can be understood through relationships to the cube and/or two related structures, the octahedron and the tetrahedron. The interrelationships among those geometries were described in Chapter 2 (see Figure 2.7 and Table 2.2).

Four types of locations, or *lattice positions*, are commonly occupied by ions: corners, edge centers, face centers, and internal (body) positions. These are illustrated in Figure 6.7.

In any cubic unit cell, each corner ion (or atom) is actually a part of eight different cells; therefore, such an ion counts only as $\frac{1}{8}$ ion for that cell. An edge ion is shared by four cells, and receives the multiplier $\frac{1}{4}$. In the same way, a facial ion counts as $\frac{1}{2}$. An internal body ion is counted as a whole ion.

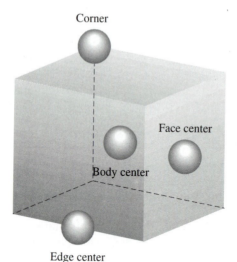

Figure 6.7
The four general types of atom/ion positions in a unit cell.

A Survey of Cubic Lattice Types

Only the corner positions are occupied in a *simple cubic* lattice. Since there are eight corners, each unit cell contains $8 \times \frac{1}{8} = 1$ atom. This type

---

9. The choice of a unit cell is arbitrary. Hence, an alternative statement might be, "The unit cells of most ionic lattices are conveniently chosen to be cubic...."

Figure 6.8
Three lattices
having cubic
unit cells:
(a) simple cubic;
(b) body-centered
cubic; (c) face-
centered cubic.

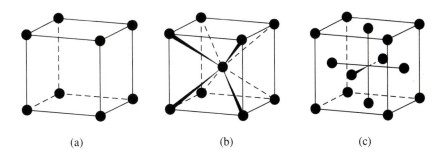

(a)   (b)   (c)

of lattice is quite rare in nature because of its relatively low density; it is inefficient in the sense that there is considerable empty space.[10] The coordination number (number of nearest neighbors) for a simple cubic lattice is 6, with the coordination geometry being octahedral.

In a *body-centered cubic* lattice, the corner atoms surround one body-centered atom. The population count is therefore $(8 \times \frac{1}{8}) + 1$, or 2 atoms per unit cell. The coordination number is 8.

A *face-centered cubic* lattice has all eight corners and the six face centers occupied, giving $(8 \times \frac{1}{8}) + (6 \times \frac{1}{2}) = 4$ atoms per cell. (This is also referred to as a *cubic close-packed* structure, as will be explained in Chapter 7.) The coordination number is 12. Diagrams of these fundamental lattice types are given in Figure 6.8.

In many ionic lattices the larger type of ion (usually the anion) forms a "sublattice" of one of these basic structures; the cations then occupy *interstices* (cavities, or holes) in that sublattice. For example, if the sublattice is of the simple cubic type, there is a cavity at the center of each cube. A cation filling that hole has eight anions as nearest neighbors (ie, it has cubic coordination); such is the case for CsCl(s) (Figure 6.2).

Cations usually occupy the interstices of a face-centered cubic sublattice in either of the following two ways:

1. They might occupy the body-centered and/or the twelve edge-centered positions. In this arrangement the interstices of the sublattice provide a total of $1 + (12 \times \frac{1}{4}) = 4$ vacancies. The coordination number of the cation is 6 for both the body- and edge-centered positions; hence, these are referred to as *octahedral holes*.

2. Alternatively, there are eight *tetrahedral holes* (coordination number 4) at internal positions. If a cube is divided into eight equal subcubes, these

---

10.   This generalization has been stated as follows: "The most probable structures will be those in which the most economical use is made of space." See Adams, D. M. *Inorganic Solids*; Wiley: New York, 1974; p. 41.

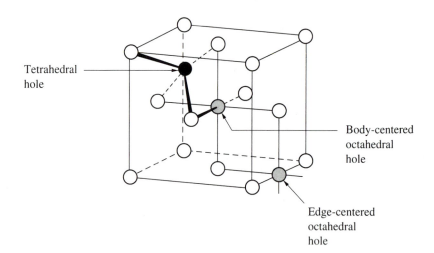

Figure 6.9

Tetrahedral and octahedral interstices in a face-centered cubic lattice. The unit cell contains a total of eight tetrahedral holes. The four octahedral holes are of two types: body-centered and edge-centered.

Tetrahedral hole

Body-centered octahedral hole

Edge-centered octahedral hole

tetrahedral holes center the subcubes. Eight cations can be accommodated per unit cell in this manner, so there are twice as many tetrahedral as octahedral holes. Figure 6.9 is helpful for visualizing the situation.

Next, we will examine the lattices of some common binary compounds having predominantly ionic character.

Some MX Systems—1:1 Stoichiometry

In crystalline cesium chloride, the chloride ions form a simple cubic sublattice and the Cs^+ ions fill the body-centered interstices (Figure 6.2). In so doing, the cations also comprise a simple cubic arrangement. The overall structure might therefore be described as two interpenetrating simple cubic sublattices, with the result resembling a body-centered cubic lattice.[11] For both types of ions, the coordination number is 8. The unit cell contains one cation and one anion, satisfying the 1:1 stoichiometry. Examples of other compounds that form this same type of lattice are CsBr, CsI, TlCl, TlBr, TlI, and more complex species such as $Ag[NbF_6]$ and $[Ni(H_2O)_6][SnCl_6]$.

The most stable lattice for the other Group 1 halides is called the *rock salt*, or sodium chloride, structure (pictured in Figure 6.10a). The anions form a face-centered cubic sublattice, with the sodiums filling the octahedral holes. (The alternative description of two interpenetrating face-centered cubic sublattices is also valid.) The overall result resembles a face-centered cubic lattice in which one type of ion occupies the corner and facial positions,

---

11. All atoms are identical in a true body-centered cubic lattice; thus, the CsCl structure is similar to, but strictly speaking is not, body-centered cubic.

Figure 6.10
Two common
lattices of 1:1
stoichiometry:
(a) sodium chloride
(rock salt);
(b) zinc blende
(two adjacent unit
cells shown).

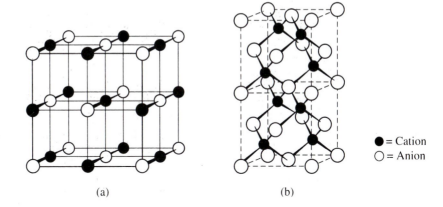

(a)   (b)

● = Cation
○ = Anion

giving $(8 \times \frac{1}{8}) + (6 \times \frac{1}{2}) = 4$ ions per unit cell. The counter ions center the twelve edges and occupy a body-centered position; this equates to $(12 \times \frac{1}{4}) + 1 = 4$ ions per cell, confirming the 1:1 stoichiometry. The coordination geometry is octahedral for both types of ions. The rock salt structure is extremely common, being preferred by all the Group 1 halides except CsCl, CsBr, and CsI. It is also adopted by the oxides of magnesium, calcium, strontium, and barium (among others) and by CaS, CeS, ScN, etc.

A third lattice common to 1:1 compounds is called the *zinc blende* structure after one type of naturally occurring zinc sulfide. The anion sublattice is again face-centered cubic. Zinc ions occupy half the tetrahedral interstices, resulting in coordination numbers of 4 for both types of ions (see Figure 6.10b). In addition to ZnS, some binary compounds that utilize the zinc blende lattice include BN, BP, AlP, CdS, HgSe, and CuI. This structure appears to be favored by species having appreciable covalency; the tetrahedral geometries are, of course, consistent with sp^3 hybridization.

It is interesting to compare the Madelung constants (Table 6.1) for these three lattices to their coordination numbers. Quite reasonably, high coordination is seen to correlate with large *A* values.

Some MX₂ and M₂X Systems—1:2 Stoichiometry

The *fluorite* lattice (named for the naturally occurring form of CaF₂) is common for compounds having the general formula MX₂. This structure is most easily described as a face-centered cubic sublattice of cations, with anions filling all the tetrahedral holes (see Figure 6.11a). The coordination numbers are 8 (in a cubic geometry) for the cations and 4 (tetrahedral) for the anions.

The anion and cation roles are reversed in certain species having the general formula M₂X; the resulting arrangement is the *antifluorite* lattice. A few of the binary systems that adopt fluorite or antifluorite lattices are SrF₂,

Figure 6.11
Two common ionic
lattices of 1:2
stoichiometry:
(a) fluorite;
(b) rutile.
[Reproduced with
permission from
Wells, A. F. *Structural
Inorganic Chemistry,*
5th ed.; Clarendon:
Oxford, 1984.]

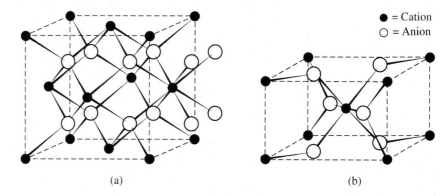

● = Cation
○ = Anion

(a)   (b)

PbF_2, MgH_2, UO_2, Mg_2Si, and all the Group 1 oxides and sulfides except Cs_2O and Cs_2S.

The *rutile* structure (named after one form of TiO_2) has metal ions in a body-centered cubic arrangement (two per unit cell). The anions lie in a lower-symmetry array, two internal and four on faces, creating an octahedral environment around each cation; see Figure 6.11b. The anions have co-ordination number 3, with a slightly distorted trigonal planar geometry. This is the first structure we have examined that does not have a truly cubic unit cell; it is *tetragonal* (ie, one dimension is different from the other two). The unit cell lengths in the parent TiO_2 crystal are 295.8 pm (two dimensions) and 459.4 pm. The rutile lattice is quite common for binary oxides and fluorides; examples include the dioxides of lead, tin, germanium, manganese, and chromium, and the difluorides of nickel, cobalt, zinc, manganese, and magnesium.

Summary descriptions of the structures discussed above are given in Table 6.4.

Table 6.4   Descriptions of six common lattices

Lattice Type	Sublattice Cation	Sublattice Anion	Coordination Number Cation	Coordination Number Anion	Geometry Cation	Geometry Anion
Cesium chloride	sc	sc	8	8	Cubic	Cubic
Rock salt	fcc	fcc	6	6	Octahedral	Octahedral
Zinc blende	fcc	fcc	4	4	Tetrahedral	Tetrahedral
Fluorite[a]	fcc	sc	8	4	Cubic	Tetrahedral
Antifluorite[a]	sc	fcc	4	8	Tetrahedral	Cubic
Rutile	bcc	—	6	3	Octahedral	Trigonal

Note: sc = simple cubic; fcc = face-centered cubic; bcc = body-centered cubic.
[a] Cation and anion sublattices have different dimensions.

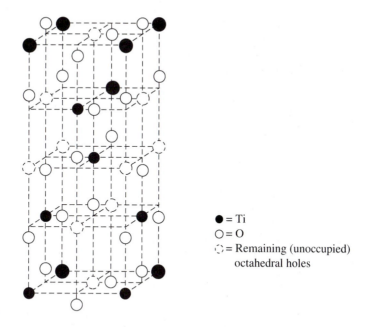

Figure 6.12
The anatase (TiO$_2$) lattice. The oxygens form a face-centered cubic sublattice, with cations filling half the octahedral interstices. [Reproduced with permission from Wells, A. F. *Structural Inorganic Chemistry*, 5th ed.; Clarendon: Oxford, 1984; p. 170.]

● = Ti
○ = O
◌ = Remaining (unoccupied) octahedral holes

Other Related Systems

Many compounds exhibit variations of the structures described above. A few are described here for illustration.

In *anatase* (a second crystalline form of TiO$_2$), the 1:2 stoichiometry is achieved by removing half the cations from the NaCl lattice. As shown in Figure 6.12, this results in the doubling of one dimension of the unit cell.

An example of a lattice accommodating 1:3 stoichiometry is that of CrCl$_3$, in which the anion sublattice is again face-centered cubic. The cations occupy one-third of the octahedral holes (ie, the rock salt lattice with two-thirds of the cations removed). Similarly, in UCl$_6$, the anion sublattice is the same, but cations occupy only one-sixth of the available sites. The coordination number is 6 for uranium but only 1 for chlorine—an unreasonable value unless we assume considerable covalency (note the high charge density of "U^{6+}").

The unit cell of ReO$_3$ is surprisingly simple; it is a cube in which the rheniums occupy the corner positions and the oxygens are edge-centered. This produces octahedral and linear coordination for the rheniums and oxygens, respectively (Figure 6.13, p. 184).

Many ternary salts having the general formula ABX$_3$ form the *perovskite* lattice. The unit cell is again cubic; one type of metal ion (A) occupies the corner positions, a second type (B) is body-centered, and the anions (X) are edge-centered (see Figure 6.14). Examples include CaTiO$_3$ (the mineral perovskite), SrSnO$_3$, BaCeO$_3$, KNbO$_3$, and KMnCl$_3$. Several recently

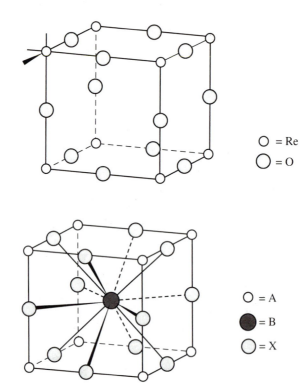

Figure 6.13
The rhenium trioxide lattice.

◯ = Re

◯ = O

Figure 6.14
The perovskite lattice.

◯ = A

⬤ = B

◯ = X

synthesized superconductors have perovskite or distorted perovskite lattices, and this is presently an area of intense study among solid-state chemists and physicists.[12]

Factors Influencing Lattice Choice

A logical question to ask at this point is, "What determines the lattice type for a given ion pair?" There appear to be at least three important factors.

1. *Stoichiometry.* It has been demonstrated that, although each lattice type is primarily associated with a specific cation/anion stoichiometry, variations sometimes result from incomplete occupancy of the interstices of a sublattice (compare NaCl, $CrCl_3$, and UCl_6). However, such structures may not yield the maximum possible electrostatic attractions, and so might be inferior to some other option. Hence, the cation/anion ratio plays a role in determining lattice choice.

12. For a readable and well-illustrated account, see Hazen, R. M. *Sci. Am.* **1988**, *258*(6), 74.

2. ***Coordination Number and Geometry.*** As will be seen in Chapters 15 and 16, Zn^{2+} has a strong preference for tetrahedral coordination. Other metal ions, such as Cr^{3+}, prefer octahedral geometry. Thus, it is not surprising that zinc normally utilizes a lattice in which its coordination number is 4 (either the zinc blende or closely related *wurtzite* lattice[13]), while ionic compounds containing Cr^{3+} typically occupy lattices in which the cation has octahedral coordination. (Note the example of $CrCl_3$ described above.) Coordination number is closely related to bond angle (90° for octahedral coordination, 109.5° for tetrahedral, etc.). The directionality requirements that arise from partial covalent bonding are often important in this regard, and as covalent character increases, the importance of coordination number and geometry tends to increase as well.

3. ***Relative Size.*** Every lattice type has a specific range of cation/anion size ratios for which it is best suited (ie, for which the electrostatic attractions are maximized with respect to the repulsions). For a series of compounds in which the stoichiometries are the same and the covalent character is minimal (such as the Group 1 halides), the size ratio becomes the major factor. Thus, it is prudent to next consider such size effects.

6.5 Ionic Radii

The distance between adjacent Cs^+ and Cl^- nuclei in crystalline cesium chloride is 356.0 pm. To a first approximation it can be assumed that the two oppositely charged ions are in contact with one another (to maximize the electrostatic attraction), but do not overlap. (Recall that the overlap of electron clouds leads to severe repulsion, and that covalency appears to be negligible in CsCl.) Thus, 356.0 pm is taken to be the sum of the *ionic radii* of Cs^+ and Cl^-, and, in general:

$$r_+ + r_- = d_{+/-} \qquad (6.17)$$

It is of considerable interest to determine individual values of r_+ and r_-, and many approaches to this problem have been attempted over the past 70 years. Unfortunately, different methods have often yielded contradictory results. This is partly because early methods assumed that ions are hard, rigid spheres whose sizes do not vary from crystal to crystal. As will shortly become evident, that does not appear to be true.

As early as 1920, Lande suggested that for an ion pair in which the anion is much larger than the cation (such as LiI), the anions must be in

---

13. In wurtzite, the anions are hexagonal close-packed rather than cubic close-packed. Hexagonal close packing will be described in Chapter 7.

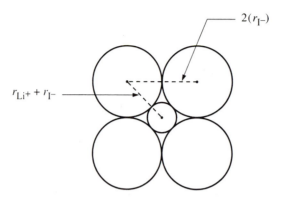

Figure 6.15
The Lande hypothesis for LiI. Ions are treated as hard spheres. Adjacent anions (larger circles) are assumed to be touching one another.

contact with one another.[14] In that case, r_- is exactly half the distance between adjacent anion nuclei (see Figure 6.15). For example, the I–I distance in LiI is 427.8 pm; thus, the radius of I^- is taken to be $427.8/2 = 213.9$ pm. The experimental Li–I distance is 302.5 pm. If we assume that the cations and anions are in contact, then r_{Li^+} must be $302.5 - 213.9 = 88.6$ pm.

Ionic radii for the other halide ions can then be obtained using that value and the known Li^+–X^- distances in the other lithium halide lattices. The results are compiled in Table 6.5. Unfortunately, they are not consistent with those from more recent (and, presumably, more accurate) methods.

A different procedure was developed by Pauling, who used data from isoelectronic cation/anion pairs such as Na^+ and F^-, and K^+ and Cl^- as a starting point.[15] The basic assumption of Pauling's method is that the radius of an ion is inversely related to the effective nuclear charge experienced

Table 6.5 Derivation of the ionic radii of halide ions from the solid-state internuclear distances of the lithium halides, using the method of Lande

Compound	d_{Li-X}, pm	$r_{X^-} = d_{Li-X} - 88.6$, pm
LiF	200.9	112.3
LiCl	256.6	168.0
LiBr	274.7	186.1
LiI	302.5	213.9

Note: The anionic radii are calculated on the assumption that $r_{Li^+} = 88.6$ pm (see text)

14. Lande, A. *Z. Physik* **1920**, *1*, 191.

15. Pauling, L. *J. Am. Chem. Soc.* **1927**, *49*, 765; *Proc. Roy. Soc. London* **1927**, *A114*, 181.

by its outermost electrons. That is, for isoelectronic ions,

$$Z^* \cdot r_{+/-} = k \tag{6.18}$$

Since Na^+ and F^- are isoelectronic, they have the same k value. Then the known Na^+–F^- distance in sodium fluoride (230.7 pm) and the effective nuclear charges of these ions (Pauling's methods yielded 6.48 for Na^+ and 4.48 for F^-) allow for the generation of two equations with two unknowns:

$$r_{Na^+} + r_{F^-} = 230.7 \text{ pm} \tag{6.19}$$

$$6.48 r_{Na^+} = 4.48 r_{F^-} \tag{6.20}$$

The radii calculated from these equations are 94.3 pm for Na^+ and 136.4 pm for F^-. Values for the isoelectronic pairs K^+ and Cl^-, Rb^+ and Br^-, and Cs^+ and I^- were determined in a similar manner.

These *univalent radii* have been widely published and used, but they are not completely satisfactory since experimental bond distances often vary from the predicted values by uncomfortable amounts. For example, the predicted internuclear distance in lithium iodide is $60 + 216 = 276$ pm. The actual distance is 302.5 pm, a difference of about 9%.

Pauling also recognized that this method gives unreasonable values for ions having charges greater than ± 1. This led to the development of his *crystal radii*,[16] which were calculated from ionic radii using the equation

$$\frac{r_c}{r_i} = Z^{-2/(n-1)} \tag{6.21}$$

Here, Z is the charge of the ion and n is the Born exponent (Table 6.2).

Ionic radii can sometimes be obtained experimentally through high-resolution X-ray diffraction. Electron density contour maps identify where the electron clouds of the ions "end" (their probability densities drop to a minimum), as well as the locations of their nuclei. As can be seen from Figure 6.16 (p. 188), the distance between these points is taken to be the ionic radius.

Among the first compounds to be examined in this manner were NaCl, LiF, CaF_2, and MgO. The results of these studies are summarized in Table 6.6 (p. 188).

What, then, is the size of a fluoride ion? Lande's approach gave a value of about 112 pm, Pauling estimated 136 pm, and the "measured" values are 109 and 110 pm. There is clearly no universal agreement on the size of F^-, or, for that matter, any ion. Ions in crystals may not be totally spherical, as illustrated by the electron density map of LiF in Figure 6.16. Ionic radii vary

---

16. Pauling, L. *The Nature of the Chemical Bond*, 3rd ed.; Cornell University: Ithaca, NY, 1960; pp. 511–519.

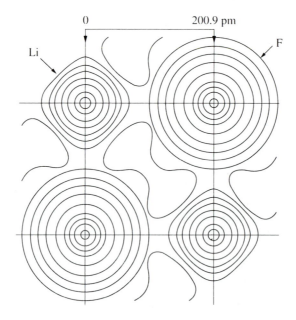

0   200.9 pm

Li   F

Figure 6.16
The distribution of electron density in LiF, as determined by high-resolution X-ray diffraction. [Reproduced with permission from Witte, H.; Wöefel, E. *Rev. Mod. Phys.* **1958**, *30*, 51.]

Table 6.6 Ionic radii, as determined by high-resolution X-ray diffraction

	Compound			
	NaCl	LiF	CaF$_2$	MgO
r_+, pm	118	92	126	102
r_-, pm	164	109	110	109

with their environments, just as covalent radii vary with bond order and electronegativity differences. Coordination number also plays a role. For example, the cesium and rubidium halides can be crystallized in either the rock salt or CsCl structure, depending on the conditions. The internuclear distances vary by about 3%, depending on the lattice type; an increase in coordination number (rock salt to CsCl) increases the distance.

The most generally useful values are probably those of Shannon,[17] which are based primarily on X-ray data. A compilation is given in Table 6.7, pp. 190–191.

17. Shannon, R. D. *Acta Crystallogr.* **1976**, *A32*, 751; Shannon, R. D.; Prewitt, C. T. *Acta Crystallogr.* **1969**, *B25*, 925. See also Shannon, R. D. In *Structure and Bonding in Crystals*; O'Keefe, M.; Navrotsky, A., Eds.; Academic: New York, 1981; Volume 2, Chapter 16.

The Relative Sizes of Ions: Radius Ratios

The relative sizes of a cation/anion pair can be expressed as a *radius ratio*:

$$\text{Radius ratio} = \text{R.R.} = \frac{r_+}{r_-} \qquad\qquad (6.22)$$

This ratio is normally less than unity, since cations are generally smaller than anions.

The influence of radius ratio on lattice stability can be seen using the rock salt structure as an example. One plane of the rock salt lattice (ie, a two-dimensional representation) is diagrammed in Figure 6.17, where the smaller and larger circles represent the cations and anions, respectively. For electrostatic attractions to be maximized, each cation must be in contact with four anions in the plane; this cannot occur if the cation is too small in relation to the anion, as in Figure 6.17a. On the other hand, if the radius ratio is sufficiently large, then the cation can be in contact with all anions and also separate them from one another. (That is, the cation is larger than the natural interstices in the anion sublattice; see Figure 6.17b.)

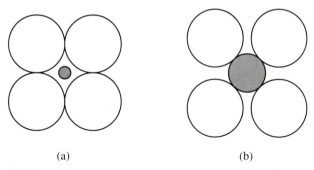

(a)   (b)

Figure 6.17   The effect of radius ratio on the stability of the rock salt lattice; shaded circles represent the cation. (a) $r_+ \ll r_-$. (b) $r_+ \approx r_-$. The anions cannot be in contact with one another in the latter case (see text).

The limiting case is therefore the ratio that just allows contact between all adjacent ions. This limiting ratio can be calculated with the aid of Figure 6.18 (p. 192). Since the triangle shown contains angles of 90°, 45°, and 45°,

$$\frac{2r_-}{2r_+ + 2r_-} = \sin 45° = 0.7071$$

$$2r_- = 0.7071(2r_+ + 2r_-) = 1.414(r_+) + 1.414(r_-)$$

$$\frac{r_+}{r_-} = \frac{0.586}{1.414} = 0.414 = \text{R.R.}$$

Table 6.7 The radii of some common monatomic ions

Ion	2	4	6	8	Ion	2	4	6	8
1+ Ions					Sr^{2+}			118	126
Ag^+	67	100	115		Ti^{2+}			86	
		102 SP			V^{2+}			79	
Au^+			137		Yb^{2+}			102	114
Cs^+			167	174	Zn^{2+}		60	74	
Cu^+	46	60	77		**3+ Ions**				
D^+	−10				Ac^{3+}			112	
H^+	−18				Al^{3+}		39	54	
Hg^+			119		As^{3+}			58	
K^+		137	138	151	Au^{3+}		68 SP	85	
Li^+		59	76	92	B^{3+}		11	27	
Na^+		99	102	118	Bi^{3+}			103	117
Rb^+			152	161	Ce^{3+}			101	114
Tl^+			150	159	Co^{3+}			54 LS	
2+ Ions								61 HS	
Ba^{2+}			135	142	Cr^{3+}			62	
Be^{2+}		27	45		Fe^{3+}		49 HS	55 LS	
Ca^{2+}			100	112				64 HS	
Cd^{2+}		78	95	110	Ga^{3+}		47	62	
Co^{2+}		58 HS	65 LS	90	In^{3+}		62	80	92
			74 HS		Ir^{3+}			68	
Cr^{2+}			73 LS		La^{3+}			103	116
			80 HS		Mn^{3+}			58 LS	
Cu^{2+}		57	73					64 HS	
		57 SP			Mo^{3+}			69	
Fe^{2+}		63 HS	61 LS	92 HS	Nb^{3+}			72	
		64 SP, HS	78 HS		P^{3+}			44	
Ge^{2+}			73		Rh^{3+}			66	
Hg^{2+}	69	96	102	114	Ru^{3+}			68	
Mg^{2+}		57	72	89	Sb^{3+}		76 PY	76	
Mn^{2+}		66 HS	67 LS	96	Sc^{3+}			74	87
			83 HS		Sm^{3+}			96	108
Ni^{2+}		55	69		Ta^{3+}			72	
		49 SP			Ti^{3+}			67	
Pb^{2+}		98 PY	119	129	Tl^{3+}		75	88	98
Pd^{2+}		64 SP	86		V^{3+}			64	
Pt^{2+}		60 SP	80		Y^{3+}			90	102
Sm^{2+}				127					

(continued)

Table 6.7 (*continued*)

Ion	2	4	6	8	Ion	2	4	6	8
	Coordination Number					Coordination Number			

Ion	2	4	6	8	Ion	2	4	6	8
4+ Ions					Ir^{5+}			57	
C^{4+}		15	16		Mn^{6+}		26		
Ce^{4+}			87	97	Mn^{7+}		25	46	
Ge^{4+}		39	53		Mo^{5+}		46	61	
Hf^{4+}		58	71	83	Mo^{6+}		41	59	
Mo^{4+}			65		Nb^{5+}		48	64	74
Pb^{4+}		65	78	94	Os^{6+}			55	
Po^{4+}			94	108	Os^{7+}			52	
Pt^{4+}			62		Os^{8+}		39		
Re^{4+}			63		Po^{6+}			67	
Rh^{4+}			60		Re^{6+}			55	
Ru^{4+}			62		Re^{7+}		38	53	
S^{4+}			37		Rh^{5+}			55	
Se^{4+}			50		Ru^{5+}			56	
Si^{4+}		26	40		Ru^{7+}		38		
Sn^{4+}		55	69	81	Ru^{8+}		36		
Ta^{4+}			68		S^{6+}		12	29	
Te^{4+}		66	97		Se^{6+}		28	42	
Th^{4+}			94	105	Ta^{5+}			64	74
Ti^{4+}		42	60	74	Tc^{7+}		37	56	
U^{4+}			89	100	Te^{6+}		43	56	
V^{4+}			58	72	U^{6+}	45	52	73	86
W^{4+}			66		V^{5+}		36	54	
Zr^{4+}		59	72	84	W^{6+}		42	60	
					Xe^{8+}		40	48	
"Ions" of High Positive Charge									
As^{5+}		34	46		**Anions**				
At^{7+}			62		Br^-			196	
Bi^{5+}			76		Cl^-			181	
Br^{7+}		25	39		F^-	128	131	133	
Cl^{7+}		8	27		I^-			220	
Cr^{6+}		26	44		O^{2-}	135	138	140	142
Fe^{6+}		25			S^{2-}			184	
I^{5+}			95		Se^{2-}			198	
I^{7+}		42	53		Te^{2-}			221	

Note: All radii in picometers, rounded to the nearest integer. Geometries are linear, tetrahedral, octahedral, and cubic, unless otherwise indicated; SP = square planar, PY = pyramidal, LS = low spin, and HS = high spin.

Source: Shannon, R. D. *Acta Crystallogr.* **1976**, *A32*, 751.

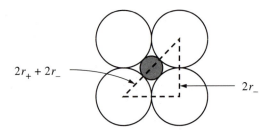

Figure 6.18   Calculation of the limiting radius ratio r_+/r_- for the rock salt lattice. The cation is shaded. All adjacent ions are in contact. One leg of the right isosceles triangle is $2r_-$, and the hypotenuse is $2r_+ + 2r_-$.

For any radius ratio less than 0.414, the stability of the rock salt lattice is diminished. All adjacent anions are in contact, but the cations do not "fill their holes" (they cannot touch all their nearest-neighbor anions). Hence, compounds are expected to utilize this lattice only if the radius ratio is about 0.414 or greater. (The same general relationship applies if the cation is larger than the anion, with the limiting ratio in that case being $1/0.414 = 2.41$.)

Limiting radius ratios for the other common lattice types can be determined using similar methods. They are compiled in Table 6.8.

As a test of the validity of this concept the radius ratios of a total of 227 compounds were calculated, and their predicted lattice types were compared to the actual structures.[18] The results are given for a sampling of

Table 6.8   Limiting radius ratios for five common ionic lattices

Lattice Type	R.R. $= r_+/r_-$	1/R.R.
1:1 Stoichiometry		
Zinc blende	0.225	4.44
Rock salt	0.414	2.41
Cesium chloride	0.732	1.37
1:2 Stoichiometry		
Rutile	0.414	2.41
Fluorite	0.732	1.37

18.   Nathan, L. C. *J. Chem. Educ.* **1985**, *62*, 215.

salts in Table 6.9. Examination of the table shows that the rules are broken by 9 of the 20 Group 1 halides—hardly an inspiring result! Overall, about 33% of the salts studied do not conform to the model. Thus, it is clear that, like most models in chemistry, the radius ratio is imperfect in its predictive ability. But this should not be a source of great surprise. For several compounds, the reason for its failure can be attributed to uncertainties in the ionic radii. For example, if Pauling's crystal radii had been used instead of Shannon's values, the prediction for ZnS would have been correct.

Of course, radius ratio is not the only factor to be considered. As was discussed earlier, other effects are important; this is especially true between the limiting ratios, where changeovers take place. For example, most ionic salts maintain some degree of covalent character. This covalent bonding involves p orbital overlap, and therefore is most effective when the X–M–X angles are 90°—exactly the angle formed by octahedral coordination in the NaCl structure. (The X–M–X angles in the cesium chloride lattice are about 70.5°.) Notice in Table 6.9 that all of the model's failures among the Group 1 halides involve lattices predicted to be of the CsCl type, but which exhibit the rock salt lattice instead.

Finally, under the appropriate conditions, many compounds (including most of the Group 1 halides) can be crystallized into more than one type of

Table 6.9   Calculated versus observed structures of salts, based on radius ratios

| Compound | R.R. | Lattice Type | | Compound | R.R. | Lattice Type | |
		Predicted	Observed			Predicted	Observed
LiF	0.76	CsCl	NaCl	CsF	(0.66)	NaCl	NaCl
LiCl	0.54	NaCl	NaCl	CsCl	(0.93)	CsCl	CsCl
LiBr	0.49	NaCl	NaCl	CsBr	0.99	CsCl	CsCl
LiI	0.44	NaCl	NaCl	CsI	0.88	CsCl	CsCl
NaF	0.97	CsCl	NaCl	BeO	0.47	NaCl	ZnS
NaCl	0.69	NaCl	NaCl	BeS	0.35	ZnS	ZnS
NaBr	0.64	NaCl	NaCl	MgO	0.68	NaCl	NaCl
NaI	0.56	NaCl	NaCl	MgS	0.51	NaCl	NaCl
KF	(0.78)	CsCl	NaCl	CaO	0.90	CsCl	NaCl
KCl	0.91	CsCl	NaCl	CaS	0.67	NaCl	NaCl
KBr	0.84	CsCl	NaCl	AlP	0.32	ZnS	ZnS
KI	0.74	CsCl	NaCl	TlCl	0.98	CsCl	CsCl
RbF	(0.72)	NaCl	NaCl	ZnS	0.52	NaCl	ZnS
RbCl	0.99	CsCl	NaCl	CdS	0.64	NaCl	ZnS
RbBr	0.91	CsCl	NaCl	HgS	0.68	NaCl	ZnS
RbI	0.81	CsCl	NaCl				

Note: Values in parentheses correspond to $1/\text{R.R.} = r_-/r_+$.
Source: Nathan, L. C. *J. Chem. Educ.* **1985**, *62*, 215.

lattice. (For example, the highest-density polymorphs—those of highest coordination numbers—are favored at high pressures.) This suggests that the energy difference between two lattice forms is very slight for many compounds, a situation that readily lends itself to exceptions.

6.6   Variations on the Themes: Some Ternary Systems

The solid-state structures of binary ionic compounds have been emphasized up to this point. Their lattices tend to be the most symmetric (and therefore most easily described and pictured) of any ionic materials. Next, we will describe a sampling of more complex systems. In many cases, we will find that the lattices are less symmetric variations of those discussed above.

Metal Cyanides

The cyanide ion is often grouped with NCO^-, CNO^-, SCN^-, and N_3^- as a *pseudohalogen* because of its chemical similarity to the halide ions. The lattices of several cyanide salts provide excellent illustrations of the interplay of stoichiometry, partial covalency, and radius ratio, resulting in structures that best suit the specific characteristics of those compounds.

It should not be surprising that the crystal structure of NaCN is of the rock salt type. Below 15°C, the cyanide ions are aligned parallel to one another (see Figure 6.19a). Free rotation of the anions occurs at higher temperature; this decreases the symmetry and increases the entropy of the crystal. The internuclear distance between adjacent sodiums is roughly

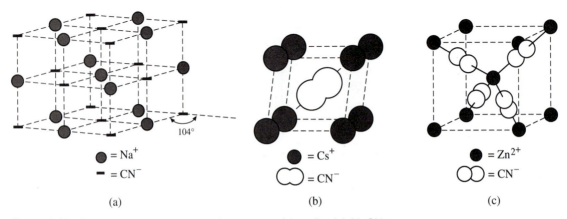

● = Na⁺	● = Cs⁺	● = Zn²⁺
▬ = CN⁻	⬭ = CN⁻	⬭ = CN⁻
(a)	(b)	(c)

Figure 6.19   The solid-state structures of some cyanide salts: (a) NaCN; (b) CsCN; (c) $Zn(CN)_2$. [Reproduced with permission from Wells, A. F. *Structural Inorganic Chemistry*, 5th ed.; Clarendon: Oxford, 1984; Chapter 22.]

415 pm. From this it has been estimated (how?) that the rotating cyanide ions behave as spheres having radii of about 190 pm, making CN^- roughly the same size as Br^-.

At low temperatures (below about $-55°C$), CsCN forms a lattice based on the cesium chloride structure in which the cubes are transformed into *rhombohedra* (ie, the angles are not all 90°); see Figure 6.19b. The cations still have coordination numbers of 8. The distortion is presumably caused by the nonspherical anions.

The lattice of $Zn(CN)_2$ must be amenable to the 1:2 stoichiometry, a small radius ratio (about 0.4), and zinc's preference for a tetrahedral environment. The solid-state structure of $Zn(CN)_2$ can be described in several ways. It is the zinc blende lattice with half the zinc ions removed. Or, it can be described as a face-centered cubic sublattice of anions with metals occupying one-quarter of the tetrahedral holes. As the unit cell is pictured in Figure 6.19c, the zincs form a body-centered cubic sublattice with cyanides occupying internal body positions. The low coordination numbers are consistent with the small radius ratio. Finally, the C–Zn–C bond angles are very close to the tetrahedral value.

As we saw earlier, the silver halides possess considerable covalent character, so the same might reasonably be assumed for AgCN. In fact, the solid-state structure of this species consists of infinite linear chains.

$$-Ag-C\equiv N-Ag-C\equiv N-$$

One- and two-dimensional structures are rarely, if ever, observed for ionic compounds—a three-dimensional option invariably exists that yields greater lattice energy. The structure of AgCN is therefore indicative of covalent bonding. The linearity is consistent with hybridizations of *sp* for the carbon and nitrogen atoms (as expected from the $C\equiv N$ triple bond) and either *sp* or *sd* for silver.

Some Oxygen-Containing Anions

The hydroxide ion is roughly the same size as F^- (about 133 pm for coordination number 6), and is very nearly spherical because of the great size discrepancy between hydrogen and oxygen. We might therefore expect metal hydroxides to parallel metal fluorides in their structural chemistry. However, the hydroxides introduce the additional complication of hydrogen bonding, so this parallel is not as strong as might be anticipated. (Hydrogen bonding is discussed in Chapter 8.) Thus, although KOH forms a rock salt lattice at temperatures above 250°C, its form at room temperature is more structurally complex and less symmetric. Sodium and lithium hydroxides also exhibit variations of the NaCl lattice, but are distorted in ways that promote hydrogen bonding by reducing intermolecular $O-H\cdots O$ distances.

Figure 6.20
The solid-state structure of LiOH. Shaded circles represent Li^+. [Reproduced with permission from Wells, A. F. *Structural Inorganic Chemistry*, 5th ed.; Clarendon: Oxford, 1984; p. 631.]

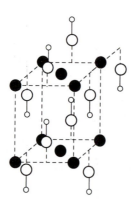

The LiOH lattice is pictured in Figure 6.20. Similarly, the structure of zinc hydroxide is a variant of the zinc blende lattice, with the distortions again allowing for enhanced hydrogen bonding.

The oxygen-containing anions SO_4^{2-}, SeO_4^{2-}, PO_4^{3-}, ClO_4^-, and IO_4^- are tetrahedral. The high symmetries of these ions generally lead to standard lattices. Of particular interest is the orthosilicate ion, SiO_4^{4-}, which is the simplest of a huge variety of silicate anions (see Chapter 13). It is commonly found in conjunction with calcium, magnesium, and/or iron in naturally occurring rocks and minerals such as the *olivines*, which can be formulated as $(Mg,Fe)_2SiO_4$. (The Mg^{2+}/Fe^{2+} ratio is variable.) The oxygens form an approximately face-centered cubic sublattice in these species. The silicons occupy tetrahedral holes (what fraction of them?), and the metals occupy octahedral vacancies.

Thermochemical Radii

A method for estimating the effective ionic radius for CN^- was suggested earlier in this section. A more general approach for determining the size of a polyatomic anion in a solid-state lattice utilizes the Kapustinskii equation. If the lattice energy for a salt is known, equation (6.12) can be solved (with some travail—try it!) for the interionic distance d. Then, by subtracting the counter ion's radius from d, a value for the polyatomic ion (called the *thermochemical radius*) is obtained.

A different approach was suggested by Solís-Correa and Gómez-Lara.[19] Their method takes covalency into account, and therefore gives somewhat different results from those obtained directly from the Kapustinskii equation. Values for selected oxyanions are given in Table 6.10.

19. Solís-Correa, H.; Gómez-Lara, J. *J. Chem. Educ.* **1987**, *64*, 942.

Table 6.10 Thermochemical radii for polyatomic oxyanions

Ion	Kapustinskii	Solís-Correa/Gómez-Lara	Ion	Kapustinskii	Solís-Correa/Gómez-Lara
CrO_4^{2-}	240	250	BiO_4^{3-}	268	264
MnO_4^-	240	270	SO_4^{2-}	230	214
CO_3^{2-}	185	198	ClO_3^-	200	195
SiO_4^{4-}	240	190	ClO_4^-	236	194
NO_2^-	155	178	BrO_3^-	191	214
NO_3^-	189	177	IO_3^-	182	237
PO_4^{3-}	238	219	IO_4^-	249	233

Note: All radii in picometers.
Source: Solís-Correa, H.; Gómez-Lara, J. *J. Chem. Educ.* **1987**, *64*, 942.

6.7 Defect and Nonstoichiometric Lattices

All crystals are not ideal, and it is useful to examine the most common kinds of variants that occur.

Random vacancies (ie, missing atoms or ions) can occur in crystals, provided that electroneutrality is maintained; this is possible if both anions and cations are absent. Such a situation is called a *Schottky defect*. On the other hand, an ion (usually a cation, because of its smaller size) may be out of place, appearing at a location not normally occupied and leaving a vacancy. This is a *Frenkel defect*. Both of these defects are illustrated schematically in Figure 6.21.

Schottky and Frenkel defects might be considered mistakes of nature; however, some benefits arise from them. A perfect crystalline solid represents

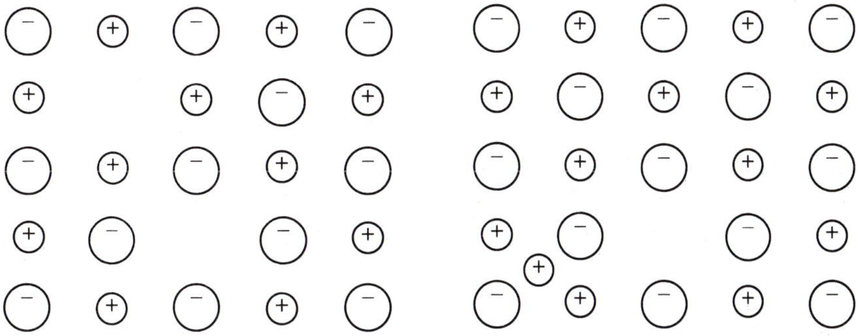

Figure 6.21
Illustrations of two common types of ionic lattice defects; (a) Schottky defect, which results from the absence of ions; (b) Frenkel defect, which results from the displacement of an ion.

(a)   (b)

the lowest possible entropy state. Lattice defects increase the randomness and therefore the entropy of the system. Because the importance of entropy increases with increasing temperature (recall that $\Delta G = \Delta H - T\Delta S$), lattice defects are most common when crystallization occurs at high temperatures.

A practical aspect of lattice defects concerns solid-state electrical conductivity. A solid material can act as a conductor only if it contains mobile, charged particles. The individual ions of a perfect ionic lattice are immobile, being locked in place by electrostatic attractions.[20] Lattice vacancies permit movement, however. For example, a cation can "hop" from its own interstice to an unoccupied site, thereby leaving its original interstice empty. (In essence, the vacancy appears to move.) It follows that the conductivity of such a material depends on the number of lattice defects per unit volume. By controlling the number of defects (eg, through crystallization at a specific temperature), it should be possible to obtain a desired conductivity. This is one method for the generation of semiconductors. However, *doping*, which we discuss below, is a more common approach. (We will have more to say about the chemical basis for semiconduction in Chapter 7.)

Lattice vacancies also can lead to variations in stoichiometry. One simple example involves the interaction of hydrogen with metals at high pressures. Titanium metal forms a face-centered cubic lattice. When Ti(s) is exposed to high pressures of $H_2(g)$, hydrogen atoms begin to occupy the tetrahedral lattice vacancies; this produces a structure similar to the fluorite lattice. The stoichiometry of TiH_x can range anywhere between $0 < x < 2$. A 1:1 stoichiometry corresponds to the filling of exactly half the available sites, although there is no particular stability to that situation.

A nonintegral stoichiometry often can be induced by heating a crystal to drive off a volatile product. For example, the following process occurs when zinc oxide is heated:

$$\text{ZnO} \xrightarrow{\;\Delta\;} \left(\frac{x}{2}\right)O_2 + \text{ZnO}_{(1-x)} \tag{6.23}$$

This is a redox reaction (O_2 is an oxidized product), so electrons must remain behind in the crystal. Those electrons occupy anionic sites in the lattice, and thereby create what are sometimes called *F centers* (from the German *farbe*, meaning "color"). Being extremely small, they migrate freely throughout the lattice. This has a major effect on the optical and electrical properties of the material.

Stoichiometry sometimes can be altered in a controlled way by *doping*—that is, the addition of a substituting ion. For example, crystalline

---

20. There are a few exceptions to this statement. For example, the silver ions in $RbAg_4I_5$ are mobile, enabling this compound to act as a solid-state conductor.

nickel oxide can be doped with lithium according to the equation

$$NiO + xLi_2O \longrightarrow Li_{2x}Ni^{II}_{(1-2x)}Ni^{III}_{2x}O_{(1+x)} \qquad (6.24)$$

Electron exchange between the Ni^{II} and Ni^{III} centers then enhances the ability of the crystal to carry an electrical charge. The electrical conductivity of NiO can be varied by a factor of more than 10^9 by incorporating lithium into up to 10% of the cationic sites. Other salts containing cations having two or more stable positive oxidation states show similar effects.

The complex binary iron–oxygen system can best be understood through an analysis of lattice vacancies. Radius ratio considerations lead to the prediction that FeO should exhibit the rock salt lattice, as in fact it does. When bonded to oxygen, however, iron has a preference for the Fe^{III} state. If some of the irons of FeO are trivalent, defect vacancies are necessary to maintain electrical neutrality. As a result, when prepared at atmospheric pressure, the stoichiometry of ferrous oxide is $Fe_{(1-x)}O$, where x ranges from about 0.05 to 0.16. Elevated temperatures and pressures are required to prepare stoichiometric FeO.

In one form of ferric oxide, γ-Fe_2O_3, the oxygen atoms remain in a face-centered cubic array. The Fe^{3+} ions occupy both octahedral and tetrahedral sites in the relative ratio of 2:1. *Magnetite*, Fe_3O_4, has the same oxygen sublattice; the cations occupy 50% of the octahedral sites (half as Fe^{2+} and half as Fe^{3+}) and one-eighth of the tetrahedral sites (as Fe^{3+}). You should be able to verify that this arrangement yields a 3:4 stoichiometry, as well as electroneutrality. The species γ-Fe_2O_3 and Fe_3O_4 are readily interconverted by simply adding or removing irons from appropriate lattice sites. A continuous interchange of electrons is believed to occur between the Fe^{II} and Fe^{III} centers in the octahedral positions of Fe_3O_4 at room temperature. This makes magnetite an excellent conductor of electricity, and is thought to be responsible for some of its remarkable magnetic properties as well.

Bibliography

West, A. R. *Basic Solid State Chemistry*; Wiley: New York, 1988.

Cox, P. A. *The Electronic Structure and Chemistry of Solids*; Oxford University: London, 1987.

Wells, A. F. *Structural Inorganic Chemistry*, 5th ed.; Clarendon: Oxford, 1984.

Dasent, W. E. *Inorganic Energetics*, 2nd ed.; Cambridge University: London, 1982.

O'Keefe, M.; Navrotsky, A., Eds. *Structure and Bonding in Crystals*; Academic: New York, 1981.

Ladd, M. F. C. *Structure and Bonding in Solid State Chemistry*; Halsted: New York, 1979.

Adams, D. M. *Inorganic Solids*; Wiley: New York, 1974.

Galasso, F. S. *Structure and Properties of Inorganic Solids*; Pergamon: Elmsford, NY, 1970.

Krebs, H. *Fundamentals of Inorganic Crystal Chemistry*; McGraw-Hill: New York, 1968.

Questions and Problems

1. Perform simple calculations to determine which pair of ions has the greatest energy of attraction: Na^+/F^-, Na^+/Te^{2-}, or Ba^{2+}/Te^{2-}. What do the results suggest about the relative importance of charge and size in determining lattice energies?

2. (a) Explain in your own words the significance of the Madelung constant, A.
   (b) Explain why A tends to be greatest for lattices that have high coordination numbers.

3. Use geometry and Figure 6.2 to demonstrate that, if the first term of the Madelung constant for the CsCl lattice is 8/1, then the second term is $-6/(\sqrt{3}/2)$.

4. For the rock salt lattice, the first three terms for the Madelung constant are

$$A = 6 - \frac{12}{\sqrt{2}} + \frac{8}{\sqrt{3}}$$

   (a) Carefully study the text diagram for the NaCl lattice (Figure 6.10a) and convince yourself that these terms are correct.
   (b) What is the value of A after three terms? How does this compare to the overall value of 1.7476?
   (c) Determine the fourth term. Find A after four terms and again compare to 1.7476.

5. The bond distance in NaCl(g) is 236.1 pm, while the internuclear distance in NaCl(s) is 281.4 pm. This is the norm. Gas-phase bond distances tend to be shorter than the corresponding values in solid-state lattices characteristic of ionic compounds. Explain.

6. Use the Kapustinskii equation to estimate the lattice energy of CaF_2 ($d = 248$ pm). Compare your answer to the Born–Haber value of 2611 kJ/mol.

7. Use data from Table 6.7 to estimate the lattice energy of Na_2O.

8. Copper(I) chloride utilizes the zinc blende lattice. Estimate the lattice energy of this species.

9. (a) Use a Born–Haber cycle to calculate the standard enthalpy of formation for the hypothetical ionic species MgF. The sublimation energy of Mg is 146 kJ/mol.
   (b) Presuming that your answer is negative, explain why this compound nevertheless does not exist.

10. Consider the possible existence of the ionic species $Na^+Na^-(s)$.
   (a) The sublimation enthalpy of sodium is about 107 kJ/mol. Use this and other values found in the text to estimate the minimum lattice energy necessary to give Na^+Na^- a negative standard enthalpy of formation.
   (b) Compare this U to values for the alkali metal halides given in Table 6.3. Might the required lattice energy actually be obtained?
   (c) Suggest other factors to be considered relative to the stability of Na^+Na^-, and briefly explain each.

11. Consider the hypothetical reaction

$$M(s) + Excess\ X(s) \longrightarrow MX_n(s)$$

   where the product is completely ionic. Assume the following values: $\Delta H_{sub}(M) = 160$ kJ/mol; $\Delta H_{sub}(X) = 130$ kJ/mol; $IE_1(M) = 450$ kJ/mol; $IE_2(M) = 1400$ kJ/mol; $EA_1(X) = +250$ kJ/mol; $EA_2(X)$ is prohibitively unfavorable; $r_{M^+} = 100$ pm; $r_{M^{2+}} = 80$ pm; $r_{X^-} = 120$ pm. Predict whether the observed value of n for the reaction product is 1 or 2.

12. In melting point, $NaF > NaCl$ and $CF_4 < CCl_4$. Why?

13. Rationalize the order of melting temperatures:
   (a) KF, 858°C; KCl, 770°C; KBr, 734°C; KI, 681°C
   (b) NaCl, 801°C; KCl, 770°C; RbCl, 718°C; CsCl, 645°C
   (c) MgO, 2852°C; MgF_2, 1261°C; NaF, 993°C

14. Predict which member of each pair has the higher melting point:
   (a) KF or CaF_2 (b) RbCl or RbBr (c) Li_2O or Li_2S
   (d) CuCl or $CuCl_2$ (e) PbF_2 or PbF_4

15. Calculate the length and volume of a unit cell of RbCl. (Table 6.3 and Figure 6.10a will be helpful.)

16. Consider a lattice patterned after that of NaCl, but having half the face-centered positions of the anion sublattice removed. What is the resulting stoichiometry?

17. A certain ionic lattice consists of a face-centered cubic sublattice of anions. Cations occupy one-fourth of the tetrahedral and one-sixth of the octahedral sites.
   (a) What is the empirical formula for this compound?
   (b) Assume that the anions are divalent. If all cations have the same charge, what is it?
   (c) If the anions are divalent and the cations in the tetrahedral sites have 2+ charges, what is the charge of the octahedral cations?

18. The yellow solid Ag_2HgI_4 has iodide ions in a cubic close-packed array, with Ag^+ and Hg^{2+} ions occupying tetrahedral holes. What percentage of the tetrahedral sites are filled?

19. Explain how the zinc blende and antifluorite lattices are interrelated.

20. Use information given in this chapter to calculate the density of RbF. [*Hint:* What are the mass and volume of one unit cell?]

21. Reexamine the perovskite lattice shown in Figure 6.14. Describe the coordination number and geometry about each type of ion.

22. Use Table 6.3 (along with some geometry) to calculate the distance between two nearest anion nuclei in KCl.

23. According to Shannon's compilation, the ionic radius of H^+ is *negative* (-18 pm). Speculate and discuss.

24. Calculate the radius ratio of ZnO and predict whether its lattice is of the ZnS, NaCl, or CsCl type.

25. Consider the following metal sulfides: CaS, BaS, CdS, and HgS. Which one has a radius ratio closest to the limiting value for the rock salt lattice?

26. The Sm^{2+} ion has a radius of about 127 pm. Predict the lattice type for SmF_2.

27. Ammonium chloride crystallizes in the CsCl lattice form. The cations and anions are in contact across the body diagonal of the unit cell, and the unit cell edge length is 386.5 pm.
    (a) What is the length of the body diagonal?
    (b) Given your answer to part (a), what is d_{N-Cl}?
    (c) What is the effective radius of the ammonium ion?
    (d) Estimate a radius ratio for NH_4Cl. What lattice is predicted for this ratio?
    (e) Estimate the lattice energy of NH_4Cl. Which of the Group 1 cations does the ammonium ion most resemble?

28. When "CrO" is crystallized under certain conditions, about 8% of the cationic sites are vacant (ie, the actual stoichiometry is $Cr_{0.92}O_{1.00}$). Nevertheless, the resulting crystals are electrically neutral. No impurities are present. Explain.

29. It is possible to dope a lattice of AgCl with Cd^{2+} ions because of the similar sizes of Ag^+ and Cd^{2+}. What type of defect results? Explain.

30. Reread the description of the magnetite lattice in Section 6.7 and satisfy yourself that the empirical formula is Fe_3O_4.

*31. A. A. Woolf has described a "graphical representation" of limiting radius ratios (*J. Chem. Educ.* **1989**, *66*, 509). Consider these questions after examining Woolf's article.
    (a) What is the general relationship between anion/cation coordination numbers and stoichiometry?
    (b) Considering only the three common lattices for 1:1 stoichiometry, which is best for covalence? Which is worst? Why? How does covalency sometimes cause incorrect predictions from the radius ratio model?

*32. McLafferty and co-workers used mass spectrometry to study oligomers of CsI in the gas phase (Amster, I. J.; McLafferty, F. W.; Castro, M. E.; Russell, D. H.; Cody, R. B.; Ghaderi, S. *Anal. Chem.* **1986**, *58*, 483). Consider these questions after reading their article.

(a) The highest abundances were found at *m/e* values of 393 and 387. To what ions do these masses correspond?

(b) The article also discusses "magic number arrangements" such as $Cs_nI_{n-1}^+$, where $n = 13, 22, 37$, and 62. What can you make of this? [*Hint*: Think cubic!]

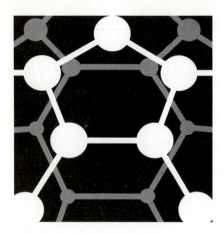

7

Occurrences, Abundances, and Structures of the Elements

Why are some isotopes stable and others radioactive? Why are some elements found in large quantities on Earth, while others occur in only minute amounts? Why are some elements found in nature as ions, others within covalent molecules, and still others as nonbonded atoms? What are the structures of the free elements in their condensed states? These basic questions about the chemical elements have fascinating answers, which will be discussed in this chapter.

7.1   Some Basic Nuclear Science—A Primer

Before the abundances of the elements—both on Earth and throughout the universe—can be discussed, it is first necessary to review some basic nuclear chemistry. The summary given here may be supplemented with information from a variety of other sources.[1]

Any atom is, of course, held together by the electrostatic attraction of its electron(s) for the positive nucleus. However, electrons are not required for a species to be stable. The deuterium cation ($^2H^+$) has no electrons, but is a stable entity. What is the nature of the force holding the neutron and proton of $^2H^+$ together? On a more general level, what forces bind any set of *nucleons* (protons plus neutrons) within an atomic nucleus?

The existence of many different subatomic particles has been documented. Seven of these are of primary importance to chemists and are listed in Table 7.1.

Free neutrons (those unassociated with any nucleus) exhibit several interesting properties. Though uncharged, they possess a nonzero magnetic moment. This suggests that they are composed of two or more smaller units, at least one of which must be charged.[2] In addition, the free neutron is unstable; it spontaneously decays to a proton, a beta particle (ie, an electron), and an antineutrino. The half-life of this process is 11–12 min.

$$\,^1_0n \longrightarrow \,^1_1p^+ + \,^0_{-1}e^- + \bar{v} \tag{7.1}$$

However, neutrons show no tendency to decompose in this manner when part of one of the 264 known stable nuclei. Why not? It is believed that neutrons and protons are bound within a nucleus by the continuous transfer of *pions* from one to the other. When a proton accepts a negative

Table 7.1 Fundamental subatomic particles

Particle	Symbol(s)	Rest Mass		Charge
		amu	g	
Proton	$^1_1p^+$, $^1H^+$	1.00728	1.673×10^{-24}	+1
Neutron	^1_0n	1.00867	1.675×10^{-24}	0
Electron	e^-, β^-	0.000549	9.110×10^{-28}	−1
Positron	β^+	0.000549	9.110×10^{-28}	+1
Pion	$\pi^{+/0/-}$	0.15	2.5×10^{-25}	+1, 0, −1
Neutrino	v	$<2 \times 10^{-8}$	0	0
Antineutrino	$\bar{v}$	$<2 \times 10^{-8}$	0	0

1. See, for example, Viola, V. E. *J. Chem. Educ.* **1990**, *67*, 723; Fergusson, J. E. *Inorganic Chemistry and the Earth*; Pergamon: Oxford, 1982; Chapter 1; Selbin, J. *J. Chem. Educ.* **1973**, *50*, 306, 380; Viola, V. E. *J. Chem. Educ.* **1973**, *50*, 311.

2. Actually, since the neutron is electrically neutral, at least two charged subunits (one positive and one negative) are required.

pion from a neutron, identities are exchanged; the former proton experiences an increase in mass and neutralization of its positive charge, thereby becoming a neutron:

$$_1^1p^+ + \pi^- \rightleftharpoons {}_0^1n^0 \qquad (7.2)$$

The nuclear force (also known as the *strong force*) depends on the effectiveness of the pion "glue" in holding the protons and neutrons together. At a distance of 10^{-3} pm (roughly the separation between adjacent protons and neutrons in a nucleus), the nuclear force is 30–40 times that of electrostatic repulsion. This force decreases very rapidly with increasing distance, however, and is effective only over distances smaller than the diameters of most nuclei. As a result, stabilization via pion transfer is significant only between nearest neighbors. Since all the protons of any nucleus undergo mutual repulsion, the size of a stable nucleus is limited; the addition of extra protons beyond a certain atomic number creates greater electrostatic repulsion than can be overcome by the pion forces.

It is also known that the ratio of neutrons to protons plays a role in determining nuclear stability. A plot of the number of protons versus the number of neutrons for the stable isotopes is given in Figure 7.1.

Stable isotopes of the lightest elements ($Z \leq 20$) often have equal numbers of protons and neutrons, presumably because that ratio yields the

Figure 7.1
Plot of number of protons versus number of neutrons for the stable nuclei. The solid line denotes equal numbers of protons and neutrons; each dot represents one of the 264 stable nuclei. [Reproduced with permission from Friedlander, G.; Kennedy, J. W.; Miller, J. M. *Nuclear and Radiochemistry*, 2nd ed.; Wiley: New York, 1964; p. 41.]

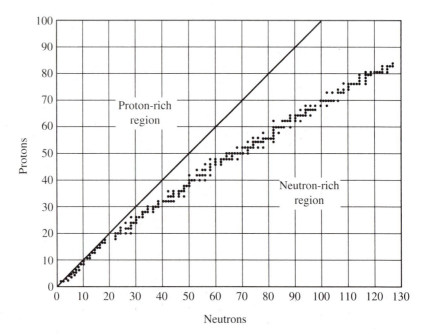

most effective pion exchange. For heavier isotopes, more neutrons than protons are required. The heaviest stable isotopes have n/p ratios of 1.50–1.56. This can be explained as a "dilution" effect, where proton–proton repulsions are minimized by interspersing the protons among greater numbers of neutrons.

Binding Energy

Although the exact nature of the nuclear force is not totally understood, it is possible to evaluate the *binding energy* for any nucleus. This energy derives from the *mass defect*, which is the difference between the mass an atom is expected to have (ie, the sum total of the masses of its individual protons, neutrons, and electrons) and its experimental value. The mass defect can be equated to the binding energy through Einstein's famous equation $\Delta E = \Delta mc^2$. A plot of binding energy per nucleon versus number of nucleons is given in Figure 7.2.

The maximum mass defect occurs for the isotope having 26 protons and 30 neutrons, ^{56}Fe. This represents the most stable known nucleus, and in that sense iron is the most stable element. Prior to that point in the curve, increasing the number of nucleons creates additional pion exchange, producing a greater stabilizing force. For heavier nuclei, the repulsions among large numbers of protons exceed the pion stabilization. It follows that *fusion*

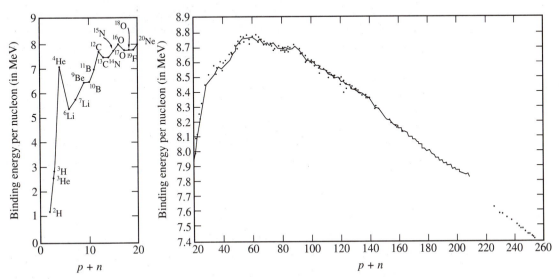

Figure 7.2   Plot of binding energy per nucleon versus the number of nucleons. Notable maxima occur at ^{4}He, ^{12}C, ^{16}O, and in the vicinity of ^{56}Fe.
[Reproduced with permission from Friedlander, G.; Kennedy, J. W.; Miller, J. M. *Nuclear and Radiochemistry*, 2nd ed.; Wiley: New York, 1964; pp. 28–29.]

reactions, in which two or more nuclei are merged into one, proceed with the release of energy for the light elements. Conversely, *fission* (the splitting of one nucleus into two smaller ones) releases energy for the heaviest elements.

Nuclear Energy Levels—Magic Numbers

We know that the electrons of a given atom or ion lie at different potential energy levels, leading to definite patterns of orbital occupancy ($1s$, $2s$, $2p$, etc.). The same is believed true for nucleons, and much effort has been expended toward devising models (analogous to the Bohr and Schrö-dinger theories for electrons) to describe and predict those energies. The situation for nucleons is complicated by several factors which will not be described here. Nevertheless, it is known that certain specific numbers of neutrons and/or protons lead to enhanced stability (just as certain numbers of electrons—those contained by the Group 18 atoms—result in enhanced stability). These *magic numbers*, as they are called, are 2, 8, 20, 28, 50, 82, and 126. Thus, maxima are observed in Figure 7.2 at ^{4}He and ^{16}O, both of which are doubly magic; that is, they have magic numbers of both protons and neutrons. (Some sources include 14 as a magic number as well, but it appears to be less strongly stabilizing than the others.)

There is also a clear preference for isotopes having even numbers of protons and/or neutrons, as is shown quantitatively in Table 7.2. This appears to be a pairing effect, and is again analogous to the situation for electrons. (The vast preponderance of chemically stable compounds have even numbers of electrons.)

Table 7.2 Even versus odd numbers of protons and neutrons for the 264 known stable nuclei

	Number of Nuclei	Percent
Both even	157	59
Even number of protons, odd number of neutrons	52	20
Odd number of protons, even number of neutrons	50	19
Both odd	5	2
	264	100

Elemental Abundances

Only 90 of the approximately 108 known elements occur naturally; they are the first 92, excluding technicium (Tc) and promethium (Pm). Of the naturally occurring elements, only the lightest 83 (through bismuth) have one or more stable isotopes. Said another way, no stable isotopes containing more than 83 protons have been identified.

There is great variation in the abundances of the naturally occurring elements, as can be seen in Table 7.3.

The preference for nuclei having magic numbers of protons and/or neutrons leads to greater abundances of such species, as shown in Figure 7.3 (p. 210). Maxima occur for doubly magic nuclei (most notably ^{4}He, ^{16}O, and ^{208}Pb). It is also apparent that nuclei with even, but not magic, numbers of neutrons and protons (eg, ^{12}C, ^{56}Fe, and ^{128}Te) often have high abundances.

Table 7.3 The estimated abundances of the elements in the universe

1 H	2.24×10^{10}	26 Fe	7.08×10^5	59 Pr	0.102
2 He	1.41×10^9	27 Co	1.78×10^3	60 Nd	0.380
3 Li	0.22	28 Ni	4.27×10^4	62 Sm	0.12
4 Be	0.032	29 Cu	2.57×10^2	63 Eu	0.1
5 B	<2.8	30 Zn	6.31×10^2	64 Gd	0.295
6 C	9.33×10^6	31 Ga	14	66 Dy	0.257
7 N	1.95×10^6	32 Ge	70.8	68 Er	0.13
8 O	1.55×10^7	37 Rb	8.91	69 Tm	0.041
9 F	8.12×10^2	38 Sr	17.8	70 Yb	0.2
10 Ne	8.32×10^5	39 Y	2.82	71 Lu	0.13
11 Na	4.27×10^4	40 Zr	12.6	72 Hf	0.14
12 Mg	8.91×10^5	41 Nb	1.8	74 W	1.1
13 Al	7.41×10^4	42 Mo	3.24	75 Re	<0.01
14 Si	1.0×10^6	44 Ru	1.51	76 Os	0.11
15 P	7.08×10^3	45 Rh	0.562	77 Ir	0.16
16 S	3.6×10^5	46 Pd	0.71	78 Pt	1.26
17 Cl	7.1×10^3	47 Ag	0.16	79 Au	<0.13
18 Ar	2.2×10^4	48 Cd	1.59	80 Hg	<2.8
19 K	3.24×10^3	49 In	1.00	81 Tl	0.18
20 Ca	5.01×10^4	50 Sn	2.2	82 Pb	1.91
21 Sc	24.5	51 Sb	0.22	83 Bi	<1.8
22 Ti	2.51×10^3	55 Cs	<1.8	90 Th	0.035
23 V	2.34×10^2	56 Ba	2.75	92 U	<0.09
24 Cr	1.15×10^4	57 La	0.302		
25 Mn	5.89×10^3	58 Ce	0.794		

Note: Values are normalized to Si $= 10^6$ atoms.
Source: Ross, J. E.; Aller, L. H. *Science* **1976**, *191*, 1223.

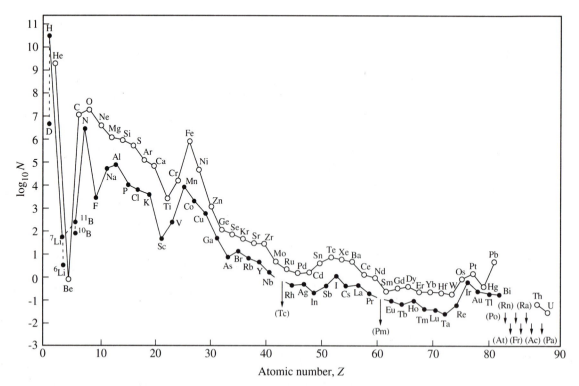

Figure 7.3 Plot of elemental abundances (logarithmic scale) versus atomic number. Light and dark circles represent elements having even and odd atomic numbers, respectively. [Reproduced with permission from Greenwood, N. N.; Earnshaw, A. *Chemistry of the Elements*; Pergamon: Oxford, 1984; p. 3. See also, Cameron, A. G. W. *Space Sci. Rev.* **1973**, *15*, 121.]

7.2 Nuclear Synthesis

Prevailing cosmological theory suggests that the various elements were created through several sequences of nuclear reactions. It is evident from Table 7.3 that hydrogen is the dominant element of the universe, and that natural abundance generally decreases with increasing nuclear size. Consistent with that knowledge, hydrogen is generally accepted as the parent of all other elements, which formed by nuclear fusion.

In addition to being the element of greatest abundance, hydrogen is thought to have been the first formed. Proponents of the *big bang theory* believe that hydrogen was produced from neutrons in a gigantic primordial reaction corresponding to equation (7.1). The big bang created temperatures in excess of 10,000,000 K, which permitted the onset of *hydrogen burning* reactions. Examples of some of these simplest fusion reactions are given by equations (7.3)–(7.5):

$$^1_1H + {}^1_1H \longrightarrow {}^2_1H + {}^{\ 0}_{+1}e + \nu \tag{7.3}$$

$$^1_1H + {}^2_1H \longrightarrow {}^3_2He + \gamma \tag{7.4}$$

$$^3_2He + {}^3_2He \longrightarrow {}^4_2He + 2^1_1H \tag{7.5}$$

Here, $^{\ 0}_{+1}e$ is a positron, ν is a neutrino, and γ is a gamma ray. The concentration of ^4_2He increases as these reactions occur. If the resulting temperature is sufficient ($> 10^8$ K), this eventually leads to *helium burning* reactions such as are described by equations (7.6)–(7.8):

$$^4_2He + {}^4_2He \longrightarrow {}^8_4Be + \gamma \tag{7.6}$$

$$^4_2He + {}^8_4Be \longrightarrow {}^{12}_{\ 6}C^* + \gamma \tag{7.7}$$

$$^4_2He + {}^{12}_{\ 6}C \longrightarrow {}^{16}_{\ 8}O + \gamma \tag{7.8}$$

Several aspects of equations (7.6)–(7.8) are noteworthy. First, ^8_4Be is quite unstable, with a half-life of about 10^{-16} s. Nevertheless, it is thought that a small, steady-state concentration of ^8_4Be exists under helium burning conditions. This explains the production of $^{12}_{\ 6}C^*$ by that process without invoking a statistically improbable three-body collision of ^4_2He atoms.

Also, note that $^{12}_{\ 6}C^*$ represents an excited state of the isotope. This species might subsequently relax to the isotopic ground state by the emission of energy. Alternatively, if the concentration and temperature ($> 5 \times 10^8$ K) are sufficiently high, *carbon burning* is possible. Examples of carbon burning reactions include

$$^{12}_{\ 6}C + {}^{12}_{\ 6}C \longrightarrow {}^{24}_{12}Mg + \nu \tag{7.9}$$

$$^{12}_{\ 6}C + {}^{12}_{\ 6}C \longrightarrow {}^{23}_{11}Na + {}^1_1p^+ \tag{7.10}$$

$$^{12}_{\ 6}C + {}^{12}_{\ 6}C \longrightarrow {}^{20}_{10}Ne + {}^4_2He \tag{7.11}$$

Still another possibility is initiation of the *carbon–nitrogen cycle*:

$$^{12}_{\ 6}C + {}^1_1H \longrightarrow {}^{13}_{\ 7}N + \gamma \tag{7.12}$$

$$^{13}_{\ 7}N \longrightarrow {}^{13}_{\ 6}C + {}^{\ 0}_{+1}e + \nu \tag{7.13}$$

$$^{13}_{\ 6}C + {}^1_1H \longrightarrow {}^{14}_{\ 7}N + \gamma \tag{7.14}$$

$$^{14}_{\ 7}N + {}^1_1H \longrightarrow {}^{15}_{\ 8}O + \gamma \tag{7.15}$$

$$^{15}_{\ 8}O \longrightarrow {}^{15}_{\ 7}N + {}^{\ 0}_{+1}e + \nu \tag{7.16}$$

$$^{15}_{\ 7}N + {}^1_1H \longrightarrow {}^{12}_{\ 6}C + {}^4_2He \tag{7.17}$$

A reaction competitive with equation (7.17) is

$$^{15}_{7}N + ^{1}_{1}H \longrightarrow ^{16}_{8}O \tag{7.18}$$

The carbon–nitrogen cycle generates helium, reinitiating the helium burning sequence.[3] In addition, several new elements are produced, including oxygen via equation (7.18). This permits the onset of *oxygen burning*, which generates certain moderately sized nuclei, including ^{28}Si and ^{31}S.

$$^{16}_{8}O + ^{16}_{8}O \longrightarrow ^{31}_{16}S + ^{1}_{0}n + \gamma \tag{7.19}$$

$$^{16}_{8}O + ^{16}_{8}O \longrightarrow ^{28}_{14}Si + ^{4}_{2}He + \gamma \tag{7.20}$$

The *α process* begins at still higher temperatures (above 10^9 K). This sequence takes its name from the fact that an α particle (ie, a $^{4}_{2}He$ nucleus) is involved as either a reactant or product in each step. The process begins with the energy-absorbing reaction

$$^{20}_{10}Ne \longrightarrow ^{16}_{8}O + ^{4}_{2}He \tag{7.21}$$

[The neon originates from equation (7.11).] The particles produced then undergo fusion with nuclei formed from carbon and oxygen burning. One possible sequence is given by equations (7.22)–(7.24):

$$^{24}_{12}Mg + ^{4}_{2}He \longrightarrow ^{28}_{14}Si + \gamma \tag{7.22}$$

$$^{28}_{14}Si + ^{4}_{2}He \longrightarrow ^{32}_{16}S + \gamma \tag{7.23}$$

$$^{32}_{16}S + ^{4}_{2}He \longrightarrow ^{36}_{18}Ar + \gamma \tag{7.24}$$

Evidence in support of the α process includes the fact that abundance steadily diminishes in the following series as the number of protons and neutrons increases by 2:

$$^{20}_{10}Ne > ^{24}_{12}Mg > ^{28}_{14}Si > ^{32}_{16}S > ^{36}_{18}Ar > ^{40}_{20}Ca$$

Among the ten lightest elements, only lithium, beryllium, and boron are not produced in any of the reactions described above. This is consistent with their low abundances (Table 7.3).

Isotopes heavier than ^{56}Fe must necessarily be formed by energetically disfavored processes. It is believed that a variety of capture (neutron and

---

3. Following equation (7.18), ^{16}O can undergo fission to form ^{12}C and ^{4}He. If that reaction is included as one in which helium is regenerated, the term *carbon–nitrogen–oxygen bi-cycle* applies.

proton) and emission (neutron, proton, or β particle) reactions are responsible.

The *e* (for *equilibrium*) *process* produces elements proximate to iron in the periodic table. Representative examples are given in equations (7.25)–(7.29):

$$^{56}_{26}\text{Fe} \longrightarrow {}^{55}_{25}\text{Mn} + {}^{1}_{1}p \tag{7.25}$$

$$^{56}_{26}\text{Fe} + {}^{1}_{0}n \longrightarrow {}^{57}_{26}\text{Fe} \tag{7.26}$$

$$^{57}_{26}\text{Fe} \longrightarrow {}^{57}_{27}\text{Co} + {}^{0}_{-1}e \tag{7.27}$$

$$^{57}_{27}\text{Co} + {}^{1}_{0}n \longrightarrow {}^{58}_{27}\text{Co} \tag{7.28}$$

$$^{58}_{27}\text{Co} \longrightarrow {}^{58}_{28}\text{Ni} + {}^{0}_{-1}e \tag{7.29}$$

The paired equations (7.26)–(7.27) and (7.28)–(7.29) represent what is sometimes referred to as the *s* (for *slow*) *process*, which consists of neutron capture followed by β emission. The designation arises because such cycles typically occur over a range of 10^2–10^5 years. Under certain circumstances (such as in areas of high neutron density within supernovas), an *r* (*rapid*) *process* can occur; in that case, many neutrons are captured simultaneously (or rather, within a matter of seconds and before β emission occurs). The heaviest naturally occurring elements (eg, uranium) are believed to have arisen through *r* processes. Because of the energy requirements of *s*- and *r*-type reactions, it is not surprising that the sum total of the abundances of all elements beyond iron in the periodic table (about three-fourths of the known elements) amounts to less than 0.2% of all matter.

7.3 The Elemental Composition of Earth

The Earth can be divided into subunits, as shown in Figure 7.4. At the center is the *core*, about 3500 km in thickness. It may be partly molten and partly solid, or it may contain two molten layers, but we know it consists primarily of iron and nickel. Our knowledge of the core is based on indirect evidence such as density, magnetic information, and waves produced by earthquakes. The core contains just over 30% of the total mass of the Earth.

Floating on the core is a *mantle* that is 2800 km thick and composed mainly of silicon, magnesium, iron, and other oxides. There is some direct evidence for the composition of the mantle from volcanic eruptions. The mantle makes up the majority (nearly 70%) of the Earth's mass.

The outermost region is the *crust*, which contains a large variety of elements. The crust shares the surface of the Earth with the *hydrosphere* (surface water), composed of hydrogen, oxygen, and 3–4% dissolved materials. The crust contains less than 1% of the Earth's mass; its approximate

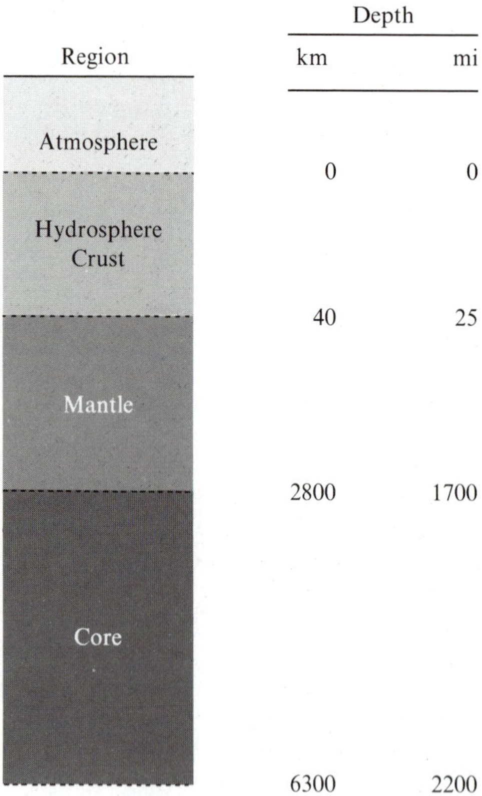

	Depth	
Region	km	mi
Atmosphere	0	0
Hydrosphere Crust	40	25
Mantle	2800	1700
Core	6300	2200

Figure 7.4
Schematic diagram showing the major subunits of the Earth (not to scale); approximate depths are given in both kilometers and miles.

composition is given in Table 7.4. The gaseous *atmosphere* completes the picture.

Comparison of Table 7.4 to Table 7.3 makes clear that, in relation to the overall universe, the Earth is deficient in hydrogen and helium. This is explained in part by the low molecular masses of both H_2 and He (and their resulting low densities in the vapor phase), which allow them to escape from the gravitational pull of Earth and into free space. The high abundance of iron on Earth is related, of course, to the great stability of the ^{56}Fe isotope. The fact that our solar system tends to have higher concentrations of the heavy elements than the universe as a whole suggests that our sun is not a first-generation star; that is, its matter has been part of several cycles of star formation and death.

It is interesting to compare Earth to the larger planets Saturn and Jupiter. The atmospheres of these so-called gaseous giant planets have much higher concentrations of H_2 and He than our own; this is consistent with the increased gravitational forces and lower temperatures on these planets. Also, certain nonmetals, such as carbon and nitrogen, are found in reduced

Table 7.4 The chemical composition of the Earth's crust

(a) By Element (% of Total Atoms)		(b) By Elemental Oxide (wt %, Continental Crust)	
Element	%	Oxide	%
O	46.6	SiO_2	59.3
Si	27.2	Al_2O_3	15.0
Al	8.13	CaO	7.2
Fe	5.00	FeO	5.6
Ca	3.63	MgO	4.9
Na	2.83	Na_2O	2.5
K	2.59	Fe_2O_3	2.4
Mg	2.09	K_2O	2.1
All others	1.9	All others	1.0
	100		100

Sources: Ronov, A. B.; Yaroshevskiy, A. A. *Geochem. Int.* **1976**, *13*, 89; Selbin, J. *J. Chem. Educ.* **1973**, *50*, 306.

states (CH_4 and NH_3). As we will see shortly, these elements typically occur in more oxidized forms on Earth.

7.4 The Chemical Forms of the Elements

Matter generally occurs as free atoms in interstellar space. However, numerous small molecules, including H_2, OH, CO, SO, SiO, CN, HCO, HCN, H_2O, H_2S, and NH_3, have been identified by spectroscopic analysis. Also, certain larger molecules, which may be considered precursors to carbon-based life, have been observed. These include HNCO, CH_3NH_2, $HC(O)NH_2$, CH_3OH, C_2H_5OH, $H_2C{=}CHCN$, and $CH_3C{\equiv}CCN$.

The chemistry of the Earth is more complex than that of outer space, because the greater proximity of atoms creates many more opportunities for chemical bonding. The naturally occurring forms of the elements can be explained by the bonding theories that were discussed in earlier chapters. A few specific examples are given below.

Hydrogen forms the strongest of all homonuclear single bonds. It is therefore not surprising that one of its natural states is as the diatomic molecule. Since this leads to very weak intermolecular forces (see Chapter 8), H_2 has a high volatility. As mentioned above, its low molecular mass causes it to escape the Earth's atmosphere; hence, the atmospheric concentration of H_2 is quite low. Hydrogen also forms strong heteronuclear covalent bonds with many other elements, most notably oxygen. This explains why these two elements compose the very substantial portion of the Earth's surface known as the hydrosphere.

The strong $N\equiv N$ triple bond causes nitrogen to be found in nature as N_2. Like H_2, weak intermolecular forces cause dinitrogen to be a gas at atmospheric temperatures. The vapor-phase density of N_2 is much greater than that of H_2, however, so it does not escape; it is, in fact, the dominant species in the atmosphere. Similarly, the strong homonuclear multiple bonding of oxygen causes it to occur in the atmosphere as $O_2(g)$. Of course, oxygen is also an important element in the hydrosphere and, like nitrogen, in the Earth's crust as well (see below).

It should be self-evident that the monatomic Group 18 elements are also found in the atmosphere.

The low ionization energies of the Group 1 metals and high electron affinities of the halogens cause these species to occur as ions. The high polarity of water draws these ions into the hydrosphere. Hence, these elements are normally found either as dissolved salts in rivers and seas or as crystalline solids in beds within the crust, deposited from previously existing bodies of water.

Geochemical Differentiation of Metals

The foregoing discussion indicates that the elements are highly *differentiated* on Earth; that is, they are segregated into areas. For example, most of the iron is found in the core, while sodium is localized in the crust. The metallic elements are divided into three groups based on their behavior in this regard, generally according to a scheme devised by Goldschmidt.[4]

In the early stages of the solar system's development, the high density of molten iron caused it to be drawn toward the center of the Earth (eventually forming the core). Metallic iron came in close contact with other metals during that process. This promoted certain oxidation–reduction reactions, converting easily reduced metals such as nickel, platinum, silver, and gold to the free state. These free metals either alloyed with the iron and were drawn with it into the core (nickel) or remained in the crust. (Gold, silver, and platinum occur as free metals even today.) Thus, metals whose cations are reduced by Fe(s) comprise one of the three subgroups (the *siderophiles*) mentioned above.

The less easily ionized metals remained in positive oxidation states in the crust, typically associated with oxygen or sulfur via ionic or polar covalent bonding. Small and electropositive metals such as magnesium and aluminum prefer oxygen; they are classed as *lithophiles*. Larger and more electronegative metals (copper, zinc, mercury, lead, etc.) favor sulfur; they

---

4. Goldschmidt, V. M. *J. Chem. Soc.* **1937**, 655. See also Goldschmidt, V. M. *Geochemistry*; Clarendon: Oxford, 1954; Chapter 2.

are *chalcophiles*. These different preferences are explained by hard–soft acid–base theory in Chapter 10.

Oxyanion Salts

We saw in Chapter 6 that polyatomic anions are often found in ionic lattices. Thus, nonmetals such as carbon (as CO_3^{2-}), nitrogen (as NO_3^-), phosphorus (as PO_4^{3-}), sulfur (as SO_4^{2-}), and especially silicon (as SiO_4^{4-} or in numerous other anions) are found in the crust as well. Their occurrence as oxyanions is explained by the strong covalent bonds formed between these elements and oxygen. Note that these anions suffer no adjacent atom lone-pair repulsions and are stabilized by either true or partial π bonding.

Many of the oxyanions form water-insoluble salts with Al^{3+}, Mg^{2+}, and/or $Fe^{2+/3+}$. This accounts for their presence in the Earth's crust, rather than in the hydrosphere (though small concentrations are found in seawater as well).

The primary chemical forms of the elements are presented in periodic table format in Figure 7.5.

The next several sections will focus on the forms of the elements in their free states—that is, the states of the elements when only homonuclear bonds are present.[5]

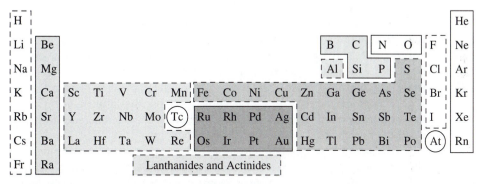

Figure 7.5 Generalized occurrences of the elements of the Earth's crust, hydrosphere, and atmosphere. □ = atmospheric gases; ⬚ = soluble ionic salts (hydrosphere); □ = crustal salts (carbonates, phosphates, silicates, sulfates, etc.); ■ = free crustal elements; ▦ = oxides (lithophiles); ▩ = sulfides (chalcophiles). Circled elements do not occur naturally on Earth. Most elements are actually found in more than one of the above categories. [Adapted from a figure by Fergusson, J. E. *Inorganic Chemistry and the Earth*; Pergamon: Oxford, 1982; p. 3.]

---

5. For a comprehensive review of the solid-state structures of the elements, see Donohue, J. *The Structures of the Elements*; Wiley: New York, 1974.

7.5 Structures and Properties of the Group 18 Elements

Recall that the noble gases exist as monatomic atoms in their standard states. The absence of homonuclear bonding is readily explained by either molecular orbital or Lewis theory. Upon sufficient cooling, of course, the noble gases can be crystallized. What are their solid-state structures?

The relevant question is this: What is the optimal way to organize spherical atoms into a condensed state? Recall from Chapter 6 that efficiency (the incorporation of the largest possible number of atoms into a given volume) is an important consideration. The most efficient schemes in this regard are the two types of close packing described below.

Cubic and Hexagonal Close-Packed Structures

To explain close packing we will begin in two dimensions. Imagine a large number of spherical objects of equal size (eg, tennis balls). The number of spheres that can be placed around a central sphere in the same plane is exactly 6 (see Figure 7.6a). Adding height to the model, a second layer might be placed on top of the first by fitting the new spheres into the interstices of the original layer. This increases the coordination number of the central sphere from 6 to 9. Another layer might be placed below the original in a like manner, adding another 3 coordinations, for an overall total of 12. The addition of more layers then creates an infinite lattice.[6]

It is found on closer examination that the first, third, fifth, etc., layers may or may not eclipse one another. If they do, the result is *hexagonal close packing* (hcp), giving an ABABAB... pattern. If the first and third layers are staggered, then it happens that the first and fourth must be eclipsed (ABCABCABC...); this is *cubic close packing* (ccp) (Figure 7.6b). Although it is not immediately apparent, the cubic close-packed lattice is identical to the structure described earlier as face-centered cubic; the difference is merely one of perspective.

It is possible to compare close-packed to other lattices on the basis of the amount of "empty space" in each structure. Recall that the ccp lattice (face-centered cubic) contains four atoms per unit cell, with adjacent atoms in contact along the face diagonals (see Figure 7.7).

A face diagonal is therefore equal to four atomic radii, and also to $\sqrt{2} \cdot l$, where l is the edge length. Hence, $l = 4r/\sqrt{2}$, and the total volume of the unit cell is

$$V_{\text{tot}} = l^3 = \left(\frac{4r}{\sqrt{2}}\right)^3 = 22.63r^3 \qquad (7.30)$$

---

6. The question of how to assemble identical spheres in the most dense manner possible has generated considerable interest (professional and recreational) among mathematicians. See Gardner, M. *Mathematical Circus*; Knopf: New York, 1979; pp. 35*ff*.

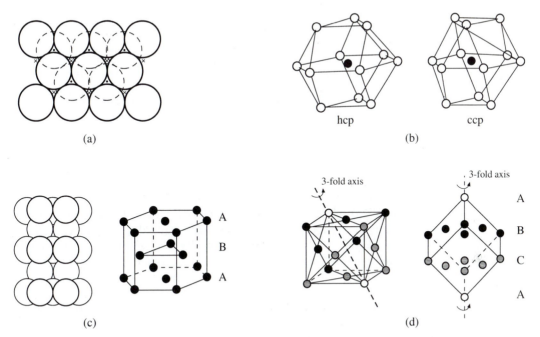

(a)

hcp   ccp

(b)

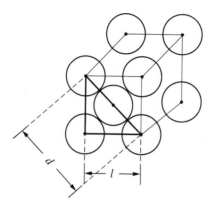

(c)   (d)

Figure 7.6 The two types of close packing. (a) Two superimposed layers; the atoms of the second layer occupy interstices of the first. (b) Arrangement of the twelve nearest neighbors about one atom (shaded) for the hcp and ccp lattices. (c) Unit cell for the hcp lattice. (d) Unit cell for the ccp lattice. [Reproduced with permission from Adams, D. M. *Inorganic Solids*; Wiley: New York, 1974; pp. 42–44.]

Figure 7.7
The geometry of the ccp (face-centered cubic) lattice. The relationship between the face diagonal d and the edge length l is
$$d = \sqrt{2} \cdot l.$$

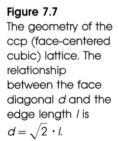

The volume occupied by the four spheres is

$$V_{occ} = 4(\tfrac{4}{3}\pi r^3) = 16.76 r^3 \tag{7.31}$$

Thus, the percentage of occupied volume is $(16.76 r^3 / 22.63 r^3) \times 100\% = 74\%$.

A similar treatment yields the same value for the hcp lattice. In contrast, analyses of the simple and body-centered cubic lattices yield occupied volumes of only 52% and 68%, respectively.

Returning to the Group 18 elements, low-temperature X-ray diffraction shows that solid helium, neon, argon, krypton, and xenon all form cubic close-packed structures. Since atomic size increases going down a family, the size of the unit cell increases in the same order. Experimental data are given in Table 7.5. It should be noted that the cell dimensions are temperature-dependent; as the temperature is increased, thermal vibrations cause each atom to require more space. For example, the unit cell edge length of xenon is 618 pm near 60 K and 627 pm near 120 K, a difference of about 1.5%.

These data make it possible to calculate the densities of the solid states of these elements. The basic approach is to find the mass and volume of the unit cell. For example, argon has an atomic mass of 39.95 amu. Since the ccp lattice has four atoms per unit cell, the cell mass is 4(39.95) = 159.8 amu. The density is then

$$\frac{159.8 \text{ amu}}{1.50 \times 10^8 \text{ pm}^3} \times \frac{1 \text{ g}}{6.022 \times 10^{23} \text{ amu}} \times \frac{10^{30} \text{ pm}^3}{1 \text{ cm}^3} = 1.77 \text{ g/cm}^3$$

van der Waals Radii

The cell dimensions of the Group 18 lattices also provide an experimental way to evaluate their atomic sizes. Recall that chemists employ several

Table 7.5   Unit cell dimensions for the ccp lattices and van der Waals radii of the Group 18 elements

Element	Cell Edge l, pm	Cell Volume V, pm^3	van der Waals Radius r, pm
He	424	7.62×10^7	150
Ne	446	8.87×10^7	158
Ar	531	1.50×10^8	188
Kr	564	1.79×10^8	199
Xe	613	2.30×10^8	217

Note: Radii are calculated from cell dimensions and rounded to the nearest integer.
Source: Unit cell edge lengths (4 K) taken from Donohue, J. *The Structures of the Elements*; Wiley: New York, 1974; Chapter 2.

different measures of this type, including covalent and ionic radii. The distance of nearest approach by nonbonded atoms relates to the *van der Waals radius* of a species.

The van der Waals radii of the Group 18 elements are given in Table 7.5 and are calculated from the dimensions of the unit cells. For example, the edge length for the He(s) cubic close-packed lattice is 424 pm. Since the face diagonal is equal to $4r = \sqrt{2} \cdot l$,

$$r = \frac{\sqrt{2}(424)}{4} = 150 \text{ pm}$$

It is possible to crystallize helium into a body-centered cubic (bcc) lattice having a unit cell edge length of 411 pm. The geometry of this structure is such that adjacent atoms are in contact along the body diagonal, which corresponds to $\sqrt{3} \cdot l$. For the equation

$$r = \frac{\sqrt{3}(411)}{4} \tag{7.32}$$

a value of 178 pm is calculated—about 19% larger than that obtained for cubic close packing.

Some sources give van der Waals radii for elements other than those of Group 18.[7] However, the lattices of other nonmetals contain covalent bonds and are generally distorted from ideal geometries (see Section 7.7). These factors complicate the determination of such radii.

7.6 The Free Metals

Lattice Structures

Like the noble gases, the solid states of metals contain spherical atoms of uniform size. It might therefore be expected that metals should crystallize into either of the two close-packed lattice types, and this is often the case. Body-centered cubic structures are also common.

Allotropes are different structural forms of the same element. Most metals exhibit allotropy, with the most stable lattice type depending on the temperature and pressure. Lanthanum is a good example. The hcp lattice is dominant below 150°C. A transition to cubic close packing is observed between 150 and 254°C. (The ccp lattice also can be obtained at room temperature and high pressure.) Another transition, this time to body-centered cubic, occurs at 864°C. In each case, of course, the cell dimensions

---

7. For example, see Biondi, A. *J. Chem. Phys.* **1964**, *68*, 441.

change. The unit cell edge lengths for hcp are 375 and 607 pm, the edge length for ccp is 531 pm, and the edge length for bcc is 426 pm.

So-called *metallic radii* can be generated for the free metals in precisely the same manner as van der Waals radii. (A distinction is made between the two because of the very different kinds of interatomic forces involved; see below.) Not surprisingly, metallic radii vary with lattice type. In Table 7.6 the preferred structures and calculated radii are given for elements that form metallic-type lattices. The periodic trends apparent in these data are as expected from earlier discussions (see Chapter 1).

Metallic Bonding

An important difference between metals and the Group 18 elements is that the interactions between metal atoms are much stronger. This explains why all metals are solids or liquids at and well above room temperature. A specific type of interaction, often called *metallic bonding*, differentiates the metals from the other elements. Metallic bonding can be explained either by analogy to ionic bonding or through molecular orbital theory.

Recall that an ionic lattice is composed of alternating cations and anions. For most metals, the first (and sometimes the second and third)

Table 7.6 Lattice types and metallic radii for metals

Li	Be												
b	h												
157	112												
Na	Mg											Al	
b	h											c	
191	160											143	
K	Ca	Sc	Ti	V	Cr	Mn	Fe	Co	Ni	Cu	Zn	Ga	
b	h	h	h	b	b	*	b	h	c	c	h	*	
235	197	164	147	135	129	137	126	125	125	128	137	153	
Rb	Sr	Y	Zr	Nb	Mo	Tc	Ru	Rh	Pd	Ag	Cd	In	
b	h	h	h	b	b	h	h	c	c	c	h	c	
250	215	182	160	147	140	135	134	134	137	144	152	167	
Cs	Ba	La	Hf	Ta	W	Re	Os	Ir	Pt	Au	Hg	Tl	Pb
b	b	h	h	b	b	h	h	c	c	c	*	h	c
272	224	188	159	147	141	137	135	136	139	144	155	171	175
		Ce					Th						
		c					c						
		182					180						

Note: Radii are given in picometers and are corrected for 12-coordination. b = body-centered cubic; h = hexagonal close-packed; c = cubic close-packed. Metals marked with * have lattices that are either very distorted or different from the categories described in the text.
Source: Data taken from Wells, A. F. *Structural Inorganic Chemistry*, 5th ed.; Clarendon: Oxford, 1984; Chapter 29.

ionization energies are relatively small. Thus, a metal atom can "eject" one or more valence electrons and thereby form a cation.[8] In the ionic model, then, a metallic lattice is viewed as consisting of metal cations, with electrons as the "anions."

Some thermodynamic aspects of this model have been explored by Rioux, who has estimated that the lattice energy of calcium (treated as $Ca^{2+}, 2e^-$) is 21.6 eV (about 2080 kJ/mol) for a fluorite-type lattice.[9] Another simple method utilizes the following Born–Haber cycle:

$$Ca(s) \longrightarrow Ca(g) \qquad \Delta H_{sub} = \quad 178 \text{ kJ/mol} \qquad (7.33)$$

$$Ca(g) \longrightarrow Ca^+(g) + 1e^- \qquad IE_1 = \quad 590 \text{ kJ/mol} \qquad (7.34)$$

$$Ca^+(g) \longrightarrow Ca^{2+}(g) + 1e^- \qquad IE_2 = 1145 \text{ kJ/mol} \qquad (7.35)$$

$$Ca^{2+}, 2e^- \longrightarrow Ca(s) \qquad \Delta H_f^\circ = \quad 0 \text{ kJ/mol} \qquad (7.36)$$

$$Ca^{2+}, 2e^- \longrightarrow Ca^{2+}(g) + 2e^- \qquad U = 1913 \text{ kJ/mol} \qquad (7.37)$$

The lattice energies calculated from these two approaches are in reasonable agreement, and the large lattice energy is consistent with the high melting temperature (850°C) of metallic calcium. In fact, as can be seen in Table 7.7, the general relationship

$$U_{met} = \Delta H_{sub} + IE_1 \qquad (7.38)$$

gives an excellent correlation between lattice energy and melting point for the Group 1 metals, and the analogous equation

$$U_{met} = \Delta H_{sub} + IE_1 + IE_2 \qquad (7.39)$$

Table 7.7 Melting temperatures and calculated lattice energies for the metals of Groups 1 and 2

Element	mp, K	U, kJ/mol	Element	mp, K	U, kJ/mol
Group 1			Group 2		
Li	453	679	Be	1556	2970
Na	371	603	Mg	923	2334
K	336	508	Ca	1123	1913
Rb	312	484	Sr	1043	1779
Cs	302	452	Ba	977	1650

Note: Lattice energies are calculated from equations (7.38) and (7.39).

8. Since all ionization energies are endothermic, this statement would not be appropriate for a gas-phase metal atom.

9. Rioux, F. *J. Chem. Educ.* **1985**, *62*, 383.

works reasonably well for Group 2. (The melting points of beryllium and magnesium are unexpectedly low; this is explained below.)

There is, however, at least one significant difference between metallic and ionic lattices. The size of an electron is minuscule compared to an actual anion. As a result, in a metallic lattice the cations are in contact with one another, but the electron "anions" are not locked in place. Instead, they move freely among the interstices and are delocalized (the descriptive phrase *sea of electrons* is often used). This explains the high thermal conductivities exhibited by free metals, since the rapidly moving, delocalized electrons are extremely efficient at dispersing thermal energy. The solid-state electrical conductivity of most metals also can be explained: It results from the presence of mobile, charged particles within a lattice. The conductivities of the Group 1 and Group 2 metals are given in Table 7.8, where it can be seen that electrical conductivities decrease from Na to Cs, but the value for lithium is anomalously low. The general trend can be explained via a size argument. Conductivity should depend on the concentration of delocalized electrons. Since the number of delocalized electrons per atom is constant in Group 1 (one valence electron per atom), increasing the atomic volume should decrease the conductivity. (The electrons must travel farther to carry a given charge.) Lithium has a comparatively low conductivity because of its high ionization energy relative to the others; that is, it "cheats" (on average, delocalizes less than 1 mole of electrons per mole of atoms), so its conductivity is lower than expected.

The same rationale can be applied to the Group 2 elements, where the conductivities of the first two elements, beryllium and magnesium, are low in comparison to the others. This suggests that their higher ionization energies limit delocalization to fewer than 2 moles of electrons per mole of atoms. This also explains the comparatively low melting temperatures of Be and Mg. If fewer than two electrons are delocalized, then the resulting lattice energy must be lowered accordingly.

The ionic model of metallic bonding is simple to use, but it is not without problems. The transition elements and the metals of Group 13 do not show

Table 7.8   Electrical conductivities for the metals of Groups 1 and 2

Element	Conductivity, $\Omega^{-1}\ cm^{-1}$	Element	Conductivity, $\Omega^{-1}\ cm^{-1}$
Group 1		Group 2	
Li	1.2×10^5	Be	Small
Na	2.4×10^5	Mg	2.2×10^5
K	1.6×10^5	Ca	2.6×10^5
Rb	8.0×10^4	Sr	4.3×10^4
Cs	5.3×10^4	Ba	—

Note: Conductivities were measured at 0°C.

the smooth melting point and conductivity trends that are predicted. For that and other reasons, a different approach—derived from molecular orbital theory—is probably superior.

Band Theory

Recall from Chapter 3 that the LCAO method requires the number of molecular orbitals in any system to equal the number of atomic orbitals used in creating them. The overlap of two atomic orbitals produces two MO's, one bonding and one antibonding. If three atomic orbitals are used, three molecular orbitals (generally of three different energies) result. It follows that the mixing of an infinite number of atomic orbitals creates an infinite number of MO's. The resulting MO diagram typically contains two *bands* (one bonding and one antibonding band), as shown in Figure 7.8.

This is the basic molecular orbital description for the lattice of a free metal. An approximately infinite number of atoms are present, each contributing one or more valence orbitals to form the MO's. The extent to which the bands are filled depends on the number of valence orbitals and valence electrons. For sodium (or any of the alkali metals), each atom can be assumed to contribute one s-type orbital. Since there is one valence electron per atom, the number of delocalized electrons equals the number of molecular orbitals; hence, exactly one-half of the MO's (the bonding band) are filled. This amounts to $\frac{1}{2}$ bond per atom, a stable bonding situation. In this sense the bonding is equivalent to that in $Na_2(g)$, one bond for every two atoms.

The electrical conductivities of the Group 1 metals also can be explained using band theory. For these elements the energy gap (sometimes called the

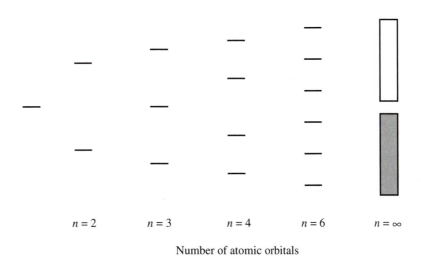

Figure 7.8
Molecular orbitals resulting from the overlap of *n* equivalent atomic orbitals. For an infinite number of atomic orbitals, bonding (■) and antibonding (□) bands are created.

$n = 2$ $n = 3$ $n = 4$ $n = 6$ $n = \infty$

Number of atomic orbitals

insulation band) between the bonding and antibonding bands is relatively small. Therefore, sufficient energy exists at room temperature for electron transitions to occur, populating a few of the antibonding orbitals according to the Boltzmann distribution. These antibonding electrons are free to move throughout the lattice via the antibonding orbitals, providing a mechanism by which charge is carried.

The same treatment leads to a different prediction for the Group 2 elements. We might have again assumed that the overlap of n atomic orbitals from n atoms creates $n/2$ bonding and $n/2$ antibonding orbitals. However, since there are $2n$ valence electrons, both the bonding and antibonding bands must be filled. The predicted bond energy would then be 0. But, not only does Mg(s) exist, it is very stable (mp = 650°C). What has gone wrong?

Recall that the valence level of magnesium contains both s- and p-type orbitals. The p orbitals of adjacent atoms can overlap to form bonding and antibonding bands (labeled ψ_p and ψ_p^* to differentiate them from the s-type bands). This complicates the MO diagram, as seen in Figure 7.9. The fact that a portion of the ψ_p band lies lower in energy than part of ψ_s^* causes electron density to flow from ψ_s^* to ψ_p; hence, both of those bands are partially populated. The result is an increase in bond strength (to a maximum of 1.0 bond per atom for complete ψ_p occupancy). The electrical conductivity of magnesium is also now explained.

It follows that for the Group 3 metals, three valence electrons and three valence orbitals (one ns and two np's) should be considered. Some thought should convince you that the maximum overlap for this group amounts to $\frac{3}{2}$ bonds per atom.

Enthalpies of Atomization

Probably the best way to evaluate the strength of homonuclear interaction is through the *atomization enthalpy*—that is, the energy required to convert 1 mole of atoms from the elemental standard state to monatomic, vapor-phase atoms. For diatomic gases, ΔH_{atm} is simply half the bond dissociation energy:

$$\tfrac{1}{2}X_2(g) \longrightarrow X(g) \qquad \Delta H = \Delta H_{atm} \tag{7.40}$$

For metals, ΔH_{atm} is the sublimation energy:

$$M(s) \longrightarrow M(g) \qquad \Delta H_{sub} = \Delta H_{atm} \tag{7.41}$$

Band theory predicts that ΔH_{atm} should increase in the order Na < Mg < Al, since that is the order of increasing bond energy, and this is indeed the case. In fact, ΔH_{atm} increases through at least the fifth member of any period. This is strongly suggestive of d orbital participation in the bonding,

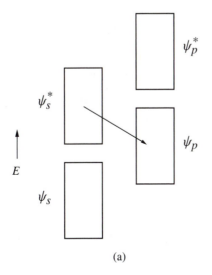

(a)

Figure 7.9
Band theory as applied to the Group 2 metals. (a) Bands produced from orbitals having s- and p-type parent orbitals. Electron density flows from ψ_s^* to ψ_p. (b) Qualitative plot of potential energy versus internuclear distance for the ns and np bands. The shaded regions represent occupied orbitals.

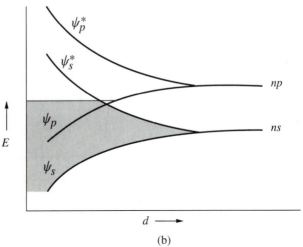

(b)

since if only s- and p-type orbitals were involved, the maximum would be expected in Group 4. (For Group 5, and beyond, electron pairing would occur, producing nonbonding lone pairs on adjacent atoms.)

However, the experimental atomization enthalpies of all metals cannot be explained by such a simple model. Values for the metals of the first long period (K through Zn) are plotted in Figure 7.10. While the trend for the first five of these elements is readily explained, the remaining values are not. In general, it is fair to say that our understanding of the bonding in the solid states of the transition elements is incomplete.

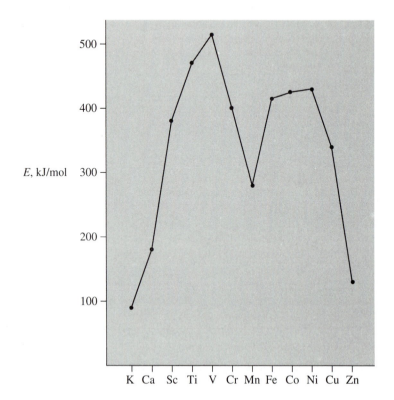

E, kJ/mol

Figure 7.10
The atomization
energies of
potassium through
zinc.

7.7 The Diatomic Elements

For reasons discussed earlier, seven of the nonmetals (hydrogen, nitrogen, oxygen, fluorine, chlorine, bromine, and iodine) exist as X_2 molecules in their standard states. These covalent units persist in the elemental lattices, which have approximate close packing; however, since the molecules are not spherical, the lattices exhibit distortions of one kind or another.

Both cubic and hexagonal close-packed forms of $H_2(s)$ have been observed at very low temperatures (<5 K). The edge length of the ccp unit cell is 533 pm, which corresponds to an effective van der Waals radius of 188 pm for H_2.

The halogen lattices can be described by the term *orthorhombic*, meaning a lattice having three unequal sides and all 90° angles (see Figure 7.11). The stacking of the rodlike molecules alternates in a regular fashion, with four molecules (eight atoms) per unit cell.

The interatomic distances are given in Table 7.9. The shortest distance represents the covalent bond length in each case and can be compared to the gas-phase values of 142, 199, 228, and 267 pm for F_2, Cl_2, Br_2, and I_2, respectively.

Figure 7.11
One plane of the orthorhombic lattices of the halogens. [Adapted and reproduced with permission from Donohue, J. *The Structures of the Elements*; Wiley: New York, 1974; p. 396.]

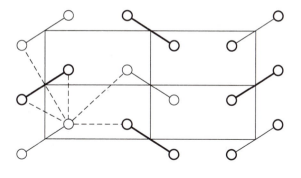

Table 7.9 Internuclear distances in the solid-state structures of the halogens

Element	a	b	c
F	149	282	320
Cl	182	332	374
Br	227	332	380
I	272	356	406

Note: Distances are in picometers. Distance a is the intramolecular (bonding) distance; b and c are nearest approaches by nonbonded atoms.
Source: Data compiled by Donohue, J. *The Structures of the Elements*; Wiley: New York, 1974; Chapter 10.

Both $N_2(s)$ and $O_2(s)$ have been crystallized in several allotropic forms, all of which are distortions of close-packed or body-centered cubic structures. For example, the allotrope α-O_2 has a *body-centered monoclinic* lattice (three different cell dimensions, with one angle not equal to 90°). The molecules are aligned parallel to one another, two per unit cell. The cell is $540 \times 509 \times 343$ pm, with an internal angle of 132°.

7.8 The Network Elements

When an element forms strong covalent bonds to itself that number more than two linkages per atom, its standard state is likely to be a *network solid*—that is, a three-dimensional lattice held together by covalent bonds. What elements are the best candidates for networking? Recall from Chapter 4 that small atoms that lack adjacent atom lone-pair repulsions form strong homonuclear bonds. Carbon, and to a lesser extent, silicon, germanium, and tin share these characteristics.

Figure 7.12
Solid-state structures
of two common
allotropes of
carbon:
(a) diamond;
(b) graphite.
[Reproduced with
permission from
Donohue, J. *The
Structures of the
Elements*; Wiley:
New York, 1974;
pp. 253, 257.]

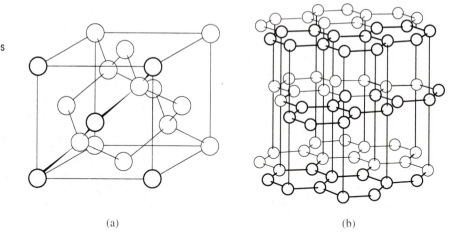

(a)   (b)

Carbon forms strong homonuclear bonds of both the σ and π types, and has a strong preference for forming four bonds per atom. Hence, its two best-known allotropes are both network polymers. In the first (diamond), each carbon forms four σ bonds, while in the second (graphite), each atom forms one π and three σ bonds.

The diamond lattice can be constructed from sp^3 hybridized carbon atoms, with all bond angles being 109.5°. This causes units of six bonded carbon atoms to close on themselves, forming puckered rings reminiscent of cyclohexane; the rings fuse together to form the three-dimensional network.[10] The unit cell is closely related to that of zinc blende; the replacement of both types of ions of ZnS by carbon yields the diamond lattice (Figure 7.12a).

Two of the best-known physical characteristics of diamond are its hardness and its high melting temperature. The hardness is explained by the difficulty of displacing any atom from its most stable location. The remarkable melting point (4030°C) also results from the high bond energy, since covalent bonds must be broken to destroy the lattice.

In graphite, each carbon has sp^2 hybridization. This leads to bond angles of 120° and planar six-membered rings; thus, the relationship between diamond and cyclohexane is the same as between graphite and benzene (see Figure 7.12b). Like benzene, graphite has alternating single and double bonds, giving rise to considerable electron delocalization. The fact that delocalization persists throughout each lattice layer, rather than only within individual rings as in C_6H_6, causes graphite to be an electrical conductor.

---

10.   Adamantane, $C_{10}H_{16}$, is an even better (but less well-known) analogy.

This makes graphite valuable for a variety of purposes; for example, it is an ideal electrode for certain electrochemical cells.

The flat, fused, six-membered rings create planar sheets that stack on top of one another. Both of the common types of layer stacking (ccp and hcp) are known. The interactions between sheets are clearly nonbonding, as the interplanar distance is 335 pm. (The C–C bond distance is only 142 pm.)

The lubricating properties of graphite were once explained as the sliding of sheets in relation to one another. However, the discovery that graphite fails to act as a lubricant at very low pressures led to an alternate explanation. The distance between planes is sufficient for small molecules such as N_2 and O_2 to become embedded in the lattice. Each layer "floats" on these molecules, allowing it to be easily displaced from the adjacent sheet. This idea is supported by the fact that graphite serves as an intercalation host for electropositive metals and a variety of small molecules (see Chapter 8).

Which is the more stable form, diamond or graphite? A simple approach to answering this question is to consider the relative energies of four σ bonds versus one π and three σ bonds. Evidence was presented in Chapter 4 to suggest that carbon's homonuclear σ bonds are stronger than its π bonds. However, the huge resonance stabilization of graphite tips the balance in its favor, and it is the more stable allotrope by just under 3 kJ/mol.

Silicon is positioned directly below carbon in the periodic table and shares its preference for tetravalence. Thus, it is not surprising that elemental silicon also forms a network polymer. Unlike carbon, however, there is only one common elemental form. The relative weakness of Si–Si π bonding precludes the graphite type of structure. As a result, silicon (and germanium as well) form diamondlike lattices.

Since bonds weaken with increasing size, hardness and melting point are expected to decrease in the order C > Si > Ge. This is, in fact, the case. The observed melting temperatures are 4030°C for carbon (diamond allotrope), 1412°C for silicon, and 958°C for germanium.

The standard-state lattices of the remaining Group 14 elements, tin and lead, are of the metallic rather than the network type, although one allotrope of tin (the α, or gray, form) has the diamond structure.

Buckminsterfullerene: A New Form of Carbon

A recent development that has received enormous attention involves a new allotrope of carbon composed of 60 atom aggregates. The geometry of the C_{60} molecule (named *buckminsterfullerene*, or "bucky ball") is properly described as truncated icosahedral; for practical purposes, it is roughly spherical (see Figure 7.13).

The bonding in C_{60} features an extensive network of fused five- and six-membered rings. (The structural surface consists of 12 pentagons and 20 hexagons.) As in graphite, each carbon has sp^2 hybridization, with three

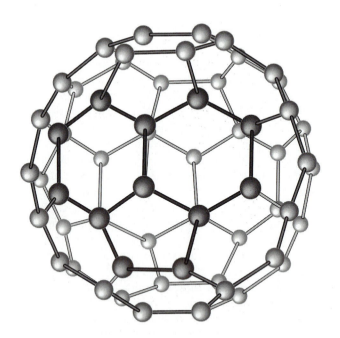

Figure 7.13
The structure of buckminster-fullerene, C_{60}.

σ bonds and one π bond; there is complete π delocalization. All the atoms are in symmetrically equivalent positions.

This form of carbon can be produced in several ways, including the laser-induced vaporization of graphite and the controlled burning of benzene. The synthetic methodology has improved to the point that macroscopic quantities of pure C_{60} can be produced. As a result, it is now possible to study its chemical and physical properties, and many exciting avenues of research are being explored. For example, C_{60} has been shown to act as a ligand and also as a three-dimensional host for metal ions such as K^+ and La^{3+}. Alkali metal salts become superconducting at low temperatures, while salts such as $\phi_4P^+C_{60}^-$ act as semiconductors.[11]

7.9   Semiconduction

The electrical properties of the Group 14 elements (especially diamondlike allotropes and related species) are of special interest, and are nicely explained by band theory. Four valence orbitals are utilized by each atom. A group of n atoms therefore generates $4n$ molecular orbitals ($2n$ bonding and $2n$

---

11.   For a review of C_{60} chemistry, see Dederich, F.; Whetton, R. L. *Angew. Chem. Int. Ed. Engl.* **1991**, *30*, 678.

antibonding). The $4n$ valence electrons just fill the bonding band, leaving the antibonding orbitals vacant.

The magnitude of the band gap is crucial to the electrical properties. If the covalent bonds are strong, the resulting gap is large. In that event, the electrons are "trapped" in the lower band and are thereby immobilized; thus, the material acts as an electrical resistor. Weaker bonding produces a smaller gap. Then there may be sufficient thermal energy at room temperature to allow some minority of electrons to enter the antibonding band. These excited electrons are mobile, and conductivity results. Intermediate cases lead to *semiconduction*, in which the conductivity varies markedly with temperature.

This model predicts that conductivity should increase going down Group 14, since homonuclear bond energies decrease in that direction. The experimental data given in Table 7.10 verify this. The first four values in the table are for diamond-type lattices (the conductivity of graphite is much higher, $7 \times 10^2 \, \Omega^{-1} \, cm^{-1}$). Lead is not strictly comparable, because as noted earlier, it forms a metallic (ccp) lattice; its conductivity is included for reference only.

Silicon and germanium are *intrinsic* semiconductors (semiconducting in the pure state), which makes them extremely valuable in materials science. A few of the myriad of applications include photovoltaic cells, transistors, and rectifiers.[12]

Table 7.10 Electrical conductivities of the Group 14 elements

Element	Conductivity, $\Omega^{-1} \, cm^{-1}$	Element	Conductivity, $\Omega^{-1} \, cm^{-1}$
C	1×10^{-6}	Sn	1×10^{0}
Si	2×10^{-5}	Pb	5×10^{4}
Ge	2×10^{-2}		

Note: Values are for diamond-type lattices except that of lead, which is of the metallic type. All conductivities were measured at 25°C.

Extrinsic Semiconduction and Doping

Recall from Chapter 6 that the conducting ability of certain solids can be altered by doping (the addition of impurities). Semiconduction induced in this manner is *extrinsic*. Perhaps the best-known examples involve crystalline silicon. The addition of small concentrations of atoms having three valence electrons (often boron, which is roughly the same size as silicon) creates vacancies in the bonding band. Such vacancies increase the mobility of the

12. See Bell, D. A. *Electronic Devices and Circuits*; Reston: Reston, VA, 1980; Horowitz, M. *Practical Design of Solid State Devices*; Reston: Reston, VA, 1979.

bonded electrons, and thereby increase the conductivity. This is a *p*-type (*positive*-type) semiconductor, since the dopant contains fewer valence electrons than the lattice atoms. The opposite approach also can be taken. By adding atoms having five valence electrons (eg, arsenic), population of the antibonding band is achieved. This is *n*-type (*negative*-type) semiconduction, and again increases the conductivity relative to the pure element.

The Influence of Covalency/Ionicity on Lattice Type

The structural tendencies of a large number of species conforming to the general formula $A^N B^{8-N}$ (N = number of valence electrons) have been cataloged. Such species range from highly ionic (eg, $Na^I Cl^{VII}$) through polar covalent ($Ga^{III} As^V$) to nonpolar covalent (diamond). As we discussed in Chapter 6, there is a relationship between the covalent/ionic character of a compound and its solid-state structure. In general, the preferred lattice changes from rock salt to either diamond or zinc blende as the ionic character decreases. The typical sequence is

Ionic $\longleftrightarrow$ Covalent

Rock salt $\longleftrightarrow$ Wurtzite $\longleftrightarrow$ Diamond or zinc blende

This is evident from Table 7.11. Tetrahedral coordination (the zinc blende, wurtzite, and diamond lattices) dominates for 1:1 species having less than about 78% ionic character, while octahedral geometry is favored beyond that point. As a general rule, the eight-electron compounds that are the best

Table 7.11 Structure preferences and estimates of ionic character for selected $A^N B^{8-N}$ systems

Crystal	Structure	Estimated % Ionic
C (diamond)	D	0
BN	Z	22
GaAs	Z	31
AlN	W	45
ZnO	W	62
ZnSe	Z, W	68
CuBr	Z, W	74
MgS	W, R	79
NaCl	R	94
RbI	R	95

Note: D = diamond; Z = zinc blende; W = wurtzite; and R = rock salt lattice.
Source: Estimates of ionic character from Phillips, J. C. *Rev. Mod. Phys.* **1970**, *42*, 317.

intrinsic semiconductors utilize the intermediate (zinc blende and wurtzite) lattices.

7.10 Elemental Boron

With only three valence electrons, we might expect B(s) to be limited to only three bonds per atom, making it electron-deficient. In fact, the bonding in elemental boron involves delocalized molecular orbitals in which each valence electron pair is shared by more than two nuclei.

Recall that the LCAO–MO approach permits more than two atomic orbitals to contribute to a bond, provided that symmetry constraints are met. For example, H_3^+ has a three-center, two-electron bond (see Figure 3.23). This same type of bonding is observed in elemental boron;[13] however, the electron delocalization occurs on a massive scale. The result is a three-dimensional network, with clusters of 12 atoms each (icosahedra) being the basic lattice units. The fundamental difference between this structure and the graphite lattice results from the different types of electron delocalization. Three-center bonding is most efficient for nonplanar arrangements of atoms, while alternating $p–p$ π bonds require planarity.

At least 16 allotropes of boron have been identified, and all are based in some way or another on icosahedra. One example, named R-105 (for *rhombohedral*) boron, consists of units having three icosahedra fused along common faces. The individual bond lengths vary between 172 and 190 pm (see Figure 7.14).

Figure 7.14
The R-105 allotrope of elemental boron. (a) An icosahedral unit. (b) The fusion of three icosahedra to form a trimeric group. [Reproduced with permission from Donohue, J. *The Structures of the Elements*; Wiley: New York, 1974; pp. 67–68.]

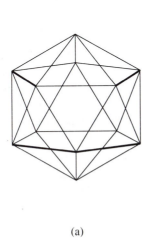

(a)

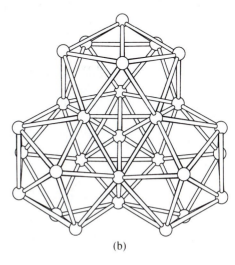

(b)

13. Multicenter bonding also stabilizes the boron hydrides and derivatives; see Chapter 19.

As the only nonmetal of Group 13, the structural chemistry of boron is unique. Aluminum and the remaining elements of the family crystallize in close-packed lattices held together by metallic bonds.

7.11   Other Elements Having Finite Molecular Units

The majority of the elements have now been accounted for, with only the nonmetals and metalloids of Groups 15 and 16 remaining. All these elements exhibit a variety of allotropes, most of which are composed of molecular units having between 4 and 12 atoms.

Group 15: Phosphorus, Arsenic, Antimony, and Bismuth

The structural chemistry of elemental phosphorus is complex, with at least four crystalline and several amorphous forms having been studied.[14] The liquid and gas phases are dominated by P_4 units in which each atom forms three single bonds. The equilibrium

$$P_4 \; \rightleftharpoons \; 2P_2 \tag{7.42}$$

is established above about 800°C.

Tetrahedral units survive in the crystalline form known as white phosphorus. This allotrope can be converted to a second form, red phosphorus, either by heating or by irradiation with X-rays. (This has hindered the structural analysis of white phosphorus.) There are several red-colored amorphous modifications, as well as a crystalline form; however, black phosphorus is the thermodynamically stable allotrope. It consists of puckered layers, with an average interlayer separation of 332 pm, roughly 49% greater than the intralayer (P–P bond) distances.

Arsenic, antimony, and bismuth are generally similar. Their standard-state lattices are isostructural with black phosphorus. Arsenic vapor exists as As_4 units, but Sb_4 and Bi_4 are unstable in the gas or liquid phases—a fact that is readily understood from the relatively weak homonuclear bonds formed by these elements.

Group 16: Sulfur, Selenium, and Tellurium

Sulfur probably has the most extensive allotropy of any element. This can be rationalized by two observations:

---

14. See Corbridge, D. E. C. *The Structural Chemistry of Phosphorus*; Elsevier: Amsterdam, 1974; and references cited therein.

1. The S–S single bond energy is very high, with only H–H and C–C homonuclear single bonds being stronger. Thus, extensive σ bonded networks can be constructed.

2. The valences of hydrogen (one bond per atom) and carbon (four bonds) lead to clearly defined structural preferences—diatomic units for hydrogen and either diamond or graphite networks for carbon. In contrast, each sulfur atom tends to form two bonds. There is no dominant structure for that situation; molecules having nearly any number of atoms can be accommodated.

Consider the application of equation (4.1) to a system having n sulfur atoms:

$$\text{Total } e^- \text{ needed} = 8n$$
$$\text{Valence } e^- \text{ available} = 6n$$
$$\text{Number of bonds} = \frac{8n - 6n}{2} = n$$

The equation works (leads to octets for all atoms) for any value of n. Thus, it is not surprising that $S_2, S_3, S_4, S_6, S_7, S_8, S_9, S_{10}, S_{12}$, and S_{20} molecular units have all been observed.

The best known is S_8. In orthorhombic sulfur, eight-membered rings occupy a crown conformation; the rings are then stacked in a complex manner (see Figure 7.15). The internal bond angles average 108°, close to the tetrahedral ideal.

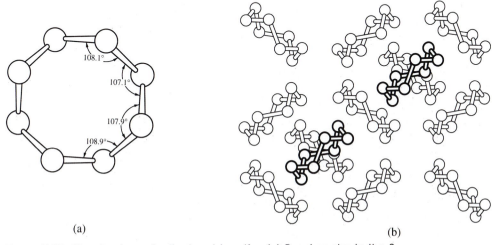

(a)   (b)

Figure 7.15 The structure of orthorhombic sulfur. (a) Bond angles in the S_8 ring. (b) Packing diagram. [Reproduced with permission from Donohue, J. *The Structures of the Elements*; Wiley: New York, 1974; pp. 332–333.]

Analogous to O_2 and ozone, S_2 and S_3 have been observed in the gas phase. The larger members of the series (S_4–S_{20}) all contain cyclic units. It is also possible to polymerize elemental sulfur to S_x chains containing over 200,000 atoms.

The heavier members of Group 16 form weaker homonuclear bonds, and their ionization energies are relatively small compared to sulfur. Hence, they are more metallic in character, and their structures reflect this. Three allotropes of selenium contain eight-membered rings that differ only in their packing. There is also a "metallic" gray form, which is actually intermediate between covalency and a cubic type of lattice more typical of metals. The standard-state structure of tellurium is similar.

Interestingly, the liquid- and gas-phase chemistry of selenium does not closely parallel that of sulfur. For example, although many ring sizes (eg, 5, 6, 7, and 8) are found in liquid selenium, it is the hexamer, rather than the octamer, that dominates.

Bibliography

Cox, P. A. *The Elements: Their Origin, Abundance, and Distribution*; Oxford University: London, 1989.

Wells, A. F. *Structural Inorganic Chemistry*, 5th ed.; Clarendon: Oxford, 1984.

West, A. R. *Solid State Chemistry and Its Applications*; Wiley: New York, 1984.

Fergusson, J. E. *Inorganic Chemistry and the Earth*; Pergamon: Oxford, 1982.

Henderson, P. *Inorganic Geochemistry*; Pergamon: Oxford, 1982.

Ladd, M. F. C. *Structure and Bonding in Solid State Chemistry*; Halsted: New York, 1979.

Smith, R. A. *Semiconduction*; Cambridge University: New York, 1978.

Parish, R. V. *The Metallic Elements*; Longman: New York, 1977.

Adams, D. M. *Inorganic Solids*; Wiley: New York, 1974.

Donohue, J. *The Structures of the Elements*; Wiley: New York, 1974.

Powell, P.; Timms, P. *Chemistry of the Non-Metals*; Chapman and Hall: London, 1974.

Taylor, R. J. *The Origin of the Chemical Elements*; Wykeham: London, 1972.

Questions and Problems

1. Explain in your own words the nature of the force that binds a group of protons and neutrons together into an atomic nucleus.

2. No stable nuclei are known that have two or more protons and no neutrons. Rationalize.

3. The following isotopes have unusually high stability: ^4_2He, $^{40}_{20}\text{Ca}$, $^{56}_{26}\text{Fe}$, $^{118}_{50}\text{Sn}$, and $^{208}_{82}\text{Pb}$. Calculate their neutron/proton ratios. What trend is observed? Discuss.

4. Each of the following nuclei is unstable. Determine their n/p ratios and use these to predict the mode of decomposition (beta or positron decay) for each:
   (a) ^7_4Be (b) $^{11}_6\text{C}$ (c) $^{14}_6\text{C}$

5. Calculate the binding energy of ^4He in both electron volts and kilojoules per mole, given the following masses: $^0_{-1}e$, 0.0005486 amu; ^1_1p, 1.00728 amu; ^1_0n, 1.00867 amu; ^4He, 4.00260 amu.

6. Consider the following isotopes: ^3H, ^{39}K, ^{52}Cr, ^{60}Ni, and ^{132}Xe.
   (a) Which contain(s) a magic number of protons?
   (b) Which contain(s) a magic number of neutrons?

7. Only one stable isotope of bismuth is known. Predict its mass number and justify your answer.

8. Write nuclear equations for the following:
   (a) The production of ^{32}S by oxygen burning
   (b) The production of ^{31}P by oxygen burning
   (c) The production of ^{65}Zn from ^{63}Cu by an s process

9. Explain these observations in your own words:
   (a) Most terrestrial chlorine is found in the hydrosphere.
   (b) Nickel is found in the Earth's core.
   (c) Silver is found in the crust as the free element, but calcium occurs as Ca^{2+} in various salts.
   (d) Most of the phosphorus found in the Earth's crust is in the form of the phosphate ion.

10. Classify each of the following metals as a siderophile, lithophile, or chalcophile:
    (a) Al   (b) Hg   (c) Sn   (d) Pt

11. Use geometry to prove that the percent occupancy of a simple cubic lattice is about 52%.

12. Calculate the percent occupancy of a body-centered cubic lattice, and compare your answer to the value given in this chapter (68%).

13. It is possible to crystallize H_2 in a cubic close-packed lattice having an edge length of 533.8 pm. Calculate the density of H_2 in this state.

14. One allotrope of fluorine consists of F_2 molecules in a slightly distorted ccp lattice. The unit cell edge length is 677 pm. Calculate the effective van der Waals radius of F_2.

15. Chromium forms a body-centered cubic lattice in which the length of the unit cell is 288.4 pm.
    (a) Calculate the density of Cr(s).
    (b) Calculate the metallic radius of chromium. Compare your answer to the value given in Table 7.6.

16. Use data from Table 1.7 to determine the sum of the first and second ionization

energies for beryllium, magnesium, and calcium. Relate the values found to the melting temperatures and electrical conductivities of these elements.

17. Rationalize the following melting point trends:
    (a) Ca > Sr > Rb   (b) Al > Mg > Na   (c) Zn > Cd > Hg

18. It can be argued that hybridization is implicit in the band theory approach for magnesium. Discuss.

19. The conductivities of many semiconductors increase in a regular manner with temperature. For example, a plot of $1/T$ (in Kelvin) versus conductivity is often linear. Use band theory to explain this qualitatively.

20. An icosahedral B_{12} unit contains 36 valence electrons, sufficient to form 18 localized two-center bonds. Determine the number of B–B interactions (nearest neighbors only) in an icosahedron of borons, and use that value to estimate the average B–B bond order for such a structure. [*Hint*: Table 2.2 may be useful.]

21. Give the Lewis structure for:
    (a) As_4 (b) S_x (chain polymer)   (c) Se_6 (cyclic)
    (d) Se_4^{2+} (cyclic)

22. Account for the difficulty of preparing $S_2(g)$ and $S_3(g)$ compared to the oxygen analogues.

23. Carbon was the first element to be studied by X-ray diffraction; the crystal structure of diamond was determined in 1913. The edge length of the unit cell is 356.7 pm. Show that the nonpolar covalent radius of carbon is 77 pm. [*Note: Not* the van der Waals radius.]

*24. Certain thermodynamic aspects of metallic bonding are considered in a brief article by R. J. Tykodi (*J. Chem. Educ.* **1989**, *66*, 306).
    (a) Explain in your own words what is meant by "effective bond number."
    (b) Why is that parameter meaningless for the Group 2 elements?
    (c) The effective bond numbers for the Group 1 elements are nearly constant. Does this contradict the data in Table 4.1? Why not?

*25. Experimental evidence has been obtained suggesting the existence of a metallic form of atomic hydrogen. Read the report by H.-K. Mao and R. J. Hemley (*Science* **1989**, *244*, 1462), and answer these questions:
    (a) Given that the prefix G (giga) = 10^9 and that 1 Pa = 9.87×10^{-6} atm, at about what pressure were these experiments performed? How was this achieved?
    (b) In what way is the evidence inconclusive?
    (c) Explain in your own words why high pressures should promote the conversion of H_2 to metallic hydrogen.
    (d) In a sense, this conversion can be explained via the delocalization of electrons. How?

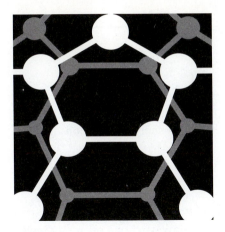

8

Secondary Chemical Interactions

The kinds of chemical associations that occur between nonbonded units—for example, between an ion and the dipole of a polar covalent species—are explored in this chapter. Such interactions are primarily electrostatic in nature, although there is often a directionality component more characteristic of covalency (orbital overlap) than of electrostatics. The term *van der Waals forces* is sometimes applied to certain interactions of this type.

As we have seen, chemical bonds (whether nonpolar covalent, ionic, or somewhere between these extremes) vary considerably in strength. There is no absolute division between true bonds and van der Waals forces, and some moderately strong attractions might be placed in either category. For example, many coordination compounds and ions (including those formed when metal salts are dissolved in water) can be considered to be held together either by dative covalent bonding or by ion–dipole interactions. Both interpretations have validity, and the explanation often depends on who does the explaining.

This chapter, then, should be considered an extension of the discussions of chemical bonding given in Chapters 3–6. Certain topics discussed in Chapter 7 provide useful background information.

8.1   Electrostatic Energies and Dipole Moments

Equation (6.6), repeated below, may be used to estimate the electrostatic energy of attraction between two isolated, oppositely charged ions.

$$E = \frac{(1.389 \times 10^5 \text{ kJ·pm/mol})(Z_+)(Z_-)}{d} \tag{8.1}$$

For ions, Z_+ and Z_- have values $(1+, 2-,$ etc.) that are integral multiples of what can be referred to as a *Z-unit*—that is, the charge of an electron $(1.602 \times 10^{-19}$ C). For interactions involving one or more dipoles, $Z_{+/-}$ has a fractional value less than 1 Z-unit.

Dipole moments are often expressed in debye (D) units, which can be converted to a form more amenable to use in equation (8.1) as follows:

$$1 \text{ D} \times \frac{3.336 \times 10^{-18} \text{ C·pm}}{1 \text{ D}} \times \frac{1 \text{ Z-unit}}{1.602 \times 10^{-19} \text{ C}} = 20.82 \text{ Z-unit·pm}$$

The resulting units (Z-unit·pm) indicate that a dipole moment has both a charge and a distance component. [Recall equation (5.12), $\mu = q \times d$.]

Consider chlorine monofluoride, which contains a polar covalent bond and has a dipole moment of 0.88 D, or 18 Z-unit·pm. The internuclear distance is 162.8 pm. The values of Z_+ and Z_- therefore can be estimated:

$$|Z| = \frac{18 \text{ Z-unit·pm}}{162.8 \text{ pm}} = 0.11$$

Thus, as a crude approximation we might consider ClF to consist of two point charges of $+0.11$ and -0.11 Z-unit, separated by a distance of 162.8 pm. Such a description is woefully inadequate for describing this molecule in the areas near and between the two nuclei. However, it can be useful for understanding the influence of the molecule on adjacent species.

8.2   Dipole–Dipole Interactions

Now, consider two molecules of ClF, adjacent to one another and oriented so as to maximize the intermolecular attraction:

$$^{\delta+}\text{Cl–F}^{\delta-}\cdots{}^{\delta+}\text{Cl–F}^{\delta-}$$

The energy of attraction between the dipoles can be estimated by inserting values of $+0.11$ and -0.11 for Z_+ and Z_-, respectively, into equation (8.1). The closest approach between the molecules must be greater than the Cl–F bond length of 162.8 pm. (The electron cloud of one molecule's chlorine does not overlap that of the second molecule's fluorine.) A reasonable estimate of the intermolecular distance is the sum of the van der Waals radii of F and Cl. Since fluorine is slightly larger than neon and chlorine is slightly larger than argon, the values for the noble gases given in Table 7.5 suggest an estimate of about 350 pm. Then the dipolar energy of interaction is roughly

$$E = \frac{(1.389 \times 10^5 \text{ kJ·pm/mol})(+0.11)(-0.11)}{350 \text{ pm}} \approx -5 \text{ kJ/mol}$$

The estimated energy is less than one-twentieth that of even a weak bond, so in this example the difference between a dipole–dipole attraction and a chemical bond is quite clear. For systems in which the dipole moment is greater and/or the intermolecular distance is smaller, however, the distinction becomes less clear-cut.

A more detailed treatment considers the influence of orientation (ie, directionality). This is illustrated schematically in Figure 8.1. The relevant equation is

$$E = \frac{3|Z_1 Z_2|l_1 l_2(2\cos\theta_1 \cdot \cos\theta_2 + \sin\theta_1 \cdot \sin\theta_2)}{d^3} \tag{8.2}$$

It can be shown that E is maximized if $\theta_1 = \theta_2 = 0$; that is, if the alignment is linear. Equation (8.2) is useful only if d is much greater than the molecular length l.

It is conceivable that two dipoles might double their energy of attraction by a head-to-tail alignment:

$$
\begin{array}{cc}
{}^{\delta+}\text{X} & {\cdots} & \text{Z}^{\delta-} \\
| & & | \\
{}^{\delta-}\text{Z} & {\cdots} & \text{X}^{\delta+}
\end{array}
$$

This arrangement is feasible only for species having rodlike shapes (much longer than they are wide). Such a head-to-tail interaction occurs when solid

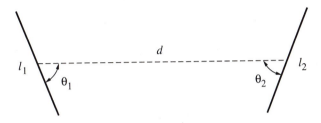

Figure 8.1 The influence of orientation on dipole–dipole interactions: l_1 and l_2 represent dipoles separated by distance d; θ_1 and θ_2 are the relevant angles. [Adapted with permission from Jaffe, H. H. *J. Chem. Educ.* **1963**, *40*, 649.]

HCl is condensed into a cold xenon matrix. (However, that particular situation might be better explained by hydrogen bonding; see Section 8.6.)

We should not expect a great deal of accuracy when estimating the magnitudes of dipole–dipole interactions. Three sources of uncertainty are:

1. Dipole moment data sometimes can be found in the chemical literature, but often must be estimated.

2. The intermolecular distance d usually can be estimated only roughly. If the solid-state structure of a compound has been determined, then d is probably known; otherwise, we must resort to gross estimates. In the liquid state, molecular motions cause such distances to have large variations with time, concentration, and temperature.

3. The effect of orientation on the energy of attraction is difficult or impossible to account for in the absence of detailed structural information for solids, and molecular motions make the orientation effect impossible to quantify for fluid states.

These three factors, coupled with the small energies generally obtained, often cause calculated dipolar energies to have uncertainties of the same magnitude as the results. Hence, the energies estimated here are rounded off, and must be recognized as nothing more than approximations. Their value lies in the comparisons they permit among different chemical systems.

Let us proceed with a comparison. The dipole moment of LiCl(g) is 7.13 D, or 148 Z-unit·pm; the internuclear distance is 202.1 pm; and $Z_{+/-}$ is then calculated to be 0.734 Z-unit. (This is equivalent to saying the bond has about 73% ionic character.) A reasonable estimate for the minimum intermolecular Li–Cl distance is about 340 pm.[1] The dipolar attraction between two adjacent LiCl molecules therefore can be estimated from equation (8.1):

$$E = \frac{(1.389 \times 10^5 \text{ kJ·pm/mol})(0.734)(-0.734)}{340 \text{ pm}} \approx -220 \text{ kJ/mol}$$

In contrast to ClF, the intermolecular energy of LiCl is quite large. Energies of this magnitude are consistent with "true" bonds. The point is that it is possible, though not necessarily desirable in all cases, to think of van der Waals forces as having the same basic nature as bonds; the difference is one of degree.

---

1. This value is obtained by summing the van der Waals radii of helium and argon. (The Li/Cl and He/Ar systems are isoelectronic.) If LiCl(g) is considered a true ion pair, an estimate of $r_{\text{Li}^+} + r_{\text{Cl}^-} = 90 + 167 = 257$ pm is obtained.

Regardless of the terminology, there is clearly a significant thermo- dynamic reason for LiCl to exist in a condensed phase—the extra stability from intermolecular interactions. The general relationship between van der Waals forces and boiling points should therefore be clear: Strong inter- molecular forces result in high boiling points. Energetically speaking, ClF has little to gain from condensation, so it is not surprising that it is a gas at room temperature and atmospheric pressure. In contrast, LiCl has a normal boiling point above 1300°C.

The condensation of a gas occurs when its thermal energy is overcome by intermolecular forces. The thermal energy is approximately equal to kT, where k is Boltzmann's constant (1.381×10^{-23} J/K, or 8.314×10^{-3} kJ/mol·K) and T is the Kelvin temperature. At the normal boiling point of ClF ($-101°C$, or 172 K), its thermal energy is

$$(172 \text{ K}) \times (8.314 \times 10^{-3} \text{ kJ/mol·K}) = 1.4 \text{ kJ/mol}$$

This also can be used as a rough estimate of the intermolecular energy of ClF. (However, this method is oversimplified; it neglects entropy, which plays an important role in phase changes.)

8.3 Ion–Dipole Interactions

It is reasonable to expect the energy of interaction between an ion and a dipole to be intermediate between ion–ion and dipole–dipole energies, and this is typically the case. For such interactions, equation (8.1) can be modified to

$$E = \frac{-(1.389 \times 10^5 \text{ kJ·pm/mol})|Z|\mu}{d^2} \tag{8.3}$$

where μ is again expressed in Z-unit·picometers.

Consider the energy of attraction between a fluoride ion and the positive end of a ClF molecule:

$$F^- \cdots {}^{\delta+}Cl-F^{\delta-}$$

A reasonable estimate for d is the van der Waals radius of chlorine plus the ionic radius of F^- (about 310 pm). The energy of attraction is about

$$E = \frac{-(1.389 \times 10^5)|1|(18)}{(310)^2} \approx -26 \text{ kJ/mol}$$

This is roughly five times the energy of the ClF dipolar interaction, but still small enough to clearly distinguish it from a covalent bond.

Interactions between water and metal cations, however, can be surprisingly strong. For the Ti^{3+}/H_2O system, the attraction may be represented as

$$Ti^{3+}\cdots^{\delta-}O\begin{array}{c}H^{\delta+}\\ \\H^{\delta+}\end{array}$$

The dipole moment of H_2O is 1.87 D, or 38.9 Z-unit·pm. If d is taken to be the sum of the van der Waals radius of oxygen and the ionic radius of Ti^{3+}, a value of about 250 pm is obtained. Then the energy of attraction is roughly

$$E = \frac{-(1.389 \times 10^5)|3|(38.9)}{(250)^2} \approx -260 \text{ kJ/mol}$$

which is of the same order as a moderate covalent bond.

8.4 The Hydration of Ions by Solvent Water

The interaction of Ti^{3+} with H_2O is a specific example of *hydration*. A general equation for the hydration of the ionic salt MX_n is

$$MX_n \xrightarrow{H_2O} M(H_2O)_y^{n+} + nX(H_2O)_z^- \tag{8.4}$$

Cation–anion interactions often remain intact upon dissolution into solvents of low polarity. The high dielectric constant of water, however, enables it to separate ion pairs because of the strong ion–dipole interactions with solvent molecules.[2] Numerous experimental and theoretical studies have been carried out in efforts to answer two questions about this process:

1. What are the hydration numbers [that is, the values of y and z in equation (8.4)]?

2. What are the quantitative thermodynamic aspects (changes in enthalpy, entropy, and free energy) of hydration?

Hydration Numbers[3]

Determining the number of coordinated water molecules is experimentally difficult, in part because coordination numbers sometimes vary with ion

2. The significance of the dielectric constant in this regard will be discussed in Chapter 11.

3. Hunt, J. P.; Friedman, H. L. *Prog. Inorg. Chem.* **1983**, *30*, 359.

Table 8.1 Hydration numbers for selected cations, as determined from NMR peak areas

Ion	Hydration Number	Ion	Hydration Number
Be^{2+}	4	Ga^{3+}	6
Mg^{2+}	6	In^{3+}	6
Fe^{2+}	6	Sc^{3+}	4, 5
Co^{2+}	6	V^{3+}	6
Ni^{2+}	6	Cr^{3+}	6
Zn^{2+}	6	Y^{3+}	2.5
Al^{3+}	6	Th^{4+}	9

Source: Values compiled by Burgess, J. *Metal Ions in Solution*; Ellis Horwood: Chichester, 1978; Chapter 5.

concentration. Beyond that, however, different experimental techniques lead to very different conclusions. For example, the solvation number for Mg^{2+} in water has variously been reported to be 3.8, 5.1, 6, 8, 9, 10.5–13, and 12–14![4] One of the most reliable methods is the measurement of peak areas for coordinated and free water molecules by the integration of ^1H and ^{17}O NMR spectra. Data obtained by this approach for various ions are given in Table 8.1. The dominance of coordination numbers 4 and 6 is suggestive of tetrahedral and octahedral geometries, which in turn is consistent with the notion of directed orbital overlap (covalency).

The Thermodynamics of Hydration

The enthalpy, entropy, and free energy of hydration of an ionic salt are calculated for the process described by equation (8.4). The enthalpy change depends on the lattice energy and the individual hydration energies of the cation and anion, as diagrammed in Figure 8.2 (p. 248).

Therefore, for a univalent salt MX,

$$\Delta H^\circ_{solv}(MX) = U(MX) + \Delta H^\circ_{solv}(M^+) + \Delta H^\circ_{solv}(X^-) \tag{8.5}$$

The hydration enthalpies of salts can be determined by calorimetry, and lattice energies are known from methods described in Chapter 6. However, two unknowns (the solvation enthalpies of the individual ions) remain. There are numerous ways to estimate ion hydration enthalpies. All involve the use of some reference ion, and H^+ is often chosen for this purpose. If the hydration enthalpy of the hydrogen ion is taken to be -1091 kJ/mol, then the data of Table 8.2 (p. 249) result.

---

4. See Burgess, J. *Metal Ions in Solution*; Ellis Horwood: Chichester; 1978, p. 144.

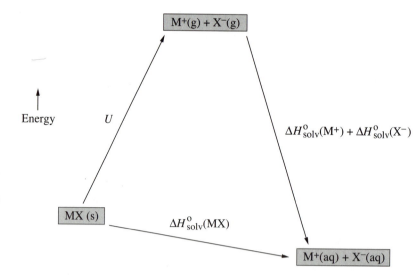

Figure 8.2
Qualitative relationships among the lattice and solvation energies of a 1:1 salt and its component monovalent ions. See equation (8.5).

The trends observed for the main group metals are generally in accord with expectations based on charge and size. This is less true for divalent cations of the first transition series. Since the charge densities of these ions increase in an approximately linear manner, we might also expect their hydration enthalpies to increase linearly. However, Figure 8.3 indicates otherwise. The failure to obey the expected trend suggests that these interactions are not simply of the ion–dipole type; rather, covalency is important in such systems. An explanation based on crystal field stabilization energies will be given in Chapter 16.

Figure 8.3
The variation of hydration enthalpies of 2+ ions of the first transition series. The dashed line represents the expected linear trend based on charge density.

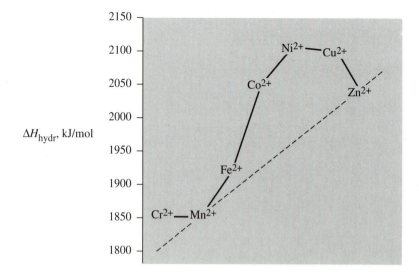

Table 8.2 Hydration enthalpies of metal cations

Li$^+$	Be^{2+}										
−515	−2487										
Na$^+$	Mg^{2+}	Al^{3+}									
−405	−1922	−4660									
K$^+$	Ca^{2+}	Sc^{3+}	Cr^{2+}	Mn^{2+}	Fe^{2+}	Co^{2+}	Ni^{2+}	Cu^{2+}	Zn^{2+}		Ga^{3+}
−321	−1592	−3960	−1850	−1845	−1920	−2054	−2106	−2100	−2044		−4685
			Cr^{3+}		Fe^{3+}						
			−4402		−4376						
Rb$^+$	Sr^{2+}	Y^{3+}						Ag$^+$	Cd^{2+}		In^{3+}	Sn^{2+}
−296	−1445	−3620						−475	−1806		−4109	−1554
Cs$^+$	Ba^{2+}	La^{3+}									Tl$^+$	Pb^{2+}
−263	−1304	−3283									−326	−1480

Note: Values are in kilojoules per mole, referenced to $\Delta H_{\text{hydr}} = -1091$ kJ/mol for H$^+$.
Source: Taken from values compiled by Burgess, J. *Metal Ions in Solution*; Ellis Horwood: Chichester, 1978; pp. 182–183.

The Solubilities of Ionic Salts

Ion–dipole attractions play an important role in determining the solubilities of ionic compounds in any polar solvent. An ionic lattice will be destroyed only if the ion–dipole attractions that result from dissolution are at least close in magnitude to the lattice energy. Thus, for a 1:1 salt in which both ions become coordinated to six water molecules, 1 mole of M^{n+}/X^{n-} interactions in the lattice is "challenged" by 6 moles of M^{n+}/H_2O plus 6 moles of X^{n-}/H_2O ion–dipole interactions. These two states are often close in energy. For example, ΔH°_{hydr} for the Group 1 halides ranges only from -63 kJ/mol for LiI to $+33$ kJ/mol for CsI, with the average value being about 0 and the majority between -20 and $+20$ kJ/mol.

The "competition principle" of Fajans states that hydration is favored when water shows a strong preference for one type of ion, thereby enabling it to displace the counter ion.[5] Thus, the hydration enthalpies for CsF and LiI are the most strongly negative of the Group 1 halides. These ion combinations have the largest size differentials, so one ion (the smaller of the two) is strongly hydrated. Conversely, the least favorable hydration enthalpies belong to salts with relatively similar sizes—LiF and CsI.

The enthalpy change is only part of the story, however. Entropy also must be considered, and entropy generally favors solubilization. Thus, most ionic solids have at least some solubility in water. The entropy effect is particularly important for the Group 1 halides because ΔH°_{hydr} is close to 0. Also, since entropy becomes more important as temperature increases, the solubilities of these salts tend to increase with increasing temperature.

For less polar solvents, of course, the ion–dipole attractions are reduced. The hydration enthalpy then becomes small compared to U, and insolubility is often observed. Solubilities in nonaqueous solutions will be discussed in Chapter 11.

8.5 Induced Dipoles

Ion and Dipolar Induction

When a nonpolar molecule comes in contact with either an ion or some other highly polar species, its electron cloud may momentarily become distorted. This leads to an instantaneous attraction between one end of the induced dipole and the polarizing species. If the polarizing species is an ion, then the relationship

$$E = \frac{-Z^2 \rho}{2d^4} \tag{8.6}$$

---

5. Fajans, K. *Struct. Bonding* **1969**, *6*, 157; *Naturwissenschaften* **1921**, *9*, 729.

Table 8.3 Polarizabilities of selected atoms and molecules

Species	ρ, pm^3	Species	ρ, pm^3
He	2.0×10^5	H_2O	1.5×10^6
Ne	3.9×10^5	H_2S	3.6×10^6
Ar	1.6×10^6	NH_3	2.2×10^6
H_2	8.0×10^5	CH_4	2.6×10^6
N_2	1.7×10^6	CCl_4	2.6×10^7
HCl	2.6×10^6	C_6H_6	2.5×10^7
HBr	3.6×10^6		

applies, where ρ represents the *polarizability* (usually given in volume units such as cubic picometers). Polarizability can loosely be described as the "softness" of the electron cloud; it increases with increasing size. This is apparent from Table 8.3, where values are listed for selected atoms and molecules.

Induced dipole energies are usually weak because the large exponent of d causes them to fall off very rapidly with increasing distance. Nevertheless, in extreme cases, they approach the magnitude of a weak bond. Induced dipoles are important factors in determining the solubilities of polar solutes in nonpolar solvents. Predicting solubility trends using equation (8.6) is not simple, however, because the size of the solvent molecule influences both ρ (in the numerator) and d (in the denominator).

London Forces

The electron cloud of an atom can become distorted under certain circumstances (eg, through collisions with other atoms or with the boundaries of the container), creating an instantaneous dipole. That momentary dipole may then induce a similar distortion in an adjacent species, resulting in an attractive force. Such interactions have variously been named *dispersion*, *instantaneous dipole*, or *London forces*; the latter is in reference to Fritz London, a pioneer of this area of chemistry.

Although a single attraction of this type is usually quite weak, the sum total of all such interactions may result in a considerable accumulation of attractive energy. This is especially likely in condensed states, where a given species can undergo more than one interaction at a time. An equation commonly used to estimate London energies is

$$E = \frac{-3IE_A \cdot IE_B \rho_A \rho_B}{2d^6(IE_A + IE_B)} \tag{8.7}$$

where IE_A and IE_B represent the ionization energies of the interacting particles.

The increase in London energies with increasing size can be illustrated by comparing two noble gases, neon and argon. The polarizability of neon is 3.9×10^5 pm^3 (Table 8.3); its van der Waals radius is 158 pm (Table 7.5); and its first ionization energy is 21.56 eV = 2080 kJ/mol (Table 1.7). The London energy per mole of neon atoms is therefore

$$E = \frac{-3(2080)^2(3.9 \times 10^5)^2}{2(158)^6(4160)} = -15 \text{ kJ/mol}$$

You should verify that the calculated value for argon is -68 kJ/mol, a difference greater than a factor of four.

It is sometimes stated that London forces are only important for nonpolar species (ie, in the absence of dipoles). For the solid state in particular, however, this does not appear to be true. The internuclear distances in crystals are relatively small. In addition, London forces are nondirectional, and so can operate in three dimensions simultaneously. They therefore make surprisingly large contributions to the lattice energies of covalent compounds, as demonstrated in Table 8.4. Even for so polar a molecule as HCl, nearly 80% of the lattice energy is due to London forces. If hydrogen bonds are present, however, as in H_2O and NH_3, then that factor (included in the dipole–dipole category in Table 8.4) is generally dominant.

London forces are the only source of intermolecular attractions for nonpolar molecules. This is true whether a species has no covalent bonds, as in Ar; only nonpolar bonds, as in Cl_2; or polar bonds canceled by

Table 8.4 Estimated percentage contributions to the lattice energies of certain elements and compounds

Species	E_{D-D}, %	E_{I-D}, %	E_L, %
Ar	0	0	100
HCl	16	5	80
HBr	3	2	95
H_2O	77	4	19
NH_3	45	4	50

Note: E_{D-D} = dipole–dipole (including hydrogen bonding); E_{I-D} = dipole–induced dipole; E_L = London contribution. Values are rounded to the nearest integer, and so do not always sum to 100%.
Source: Percentages calculated from data given by Wells, A. F. *Structural Inorganic Chemistry*, 5th ed.; Clarendon: Oxford, 1984; p. 304.

geometry, as in SiF_4. These forces are therefore the determining factor for certain properties (eg, boiling temperatures) of such compounds.

We demonstrated earlier that it is not always possible to distinguish a true bond from a van der Waals interaction on the basis of energy considerations. This is also the case when internuclear distance is used as the criterion, and the term *secondary bonding* has been applied to certain intermediate attractions.[6] Consider the I_8^{2-} ion of the salt Cs_2I_8 (Figure 8.4).

We might think of I_8^{2-} as arising from ion–dipole interactions between two triiodide (I_3^-) ions and an I_2 molecule:

$$[I–I–I\cdots I–I\cdots I–I–I]^{2-}$$

The nonpolar covalent radius of iodine is 133 pm. By analogy to xenon, the van der Waals radius is about 220 pm, which coincidentally equals the ionic radius of I^-. Therefore, the shortest I–I distance (283 pm, between the fourth and fifth atoms in the chain) reasonably corresponds to a covalent bond length. Note that this is consistent with the ion–dipole interpretation. The nonbonded I–I interactions should be about 440 pm (either $2r_{VDW}$ or $r_{I^-} + r_{VDW}$). However, the experimental values are 300 and 342 pm, seemingly too long for a covalent bond, but much too short to be of the van der Waals type.

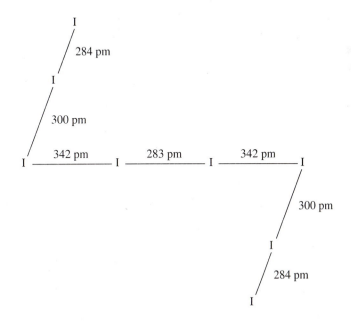

Figure 8.4
The approximate geometry and internuclear distances of the I_8^{2-} anion of Cs_2I_8.

6. Alcock, N. W. *Adv. Inorg. Chem. Radiochem.* **1972**, *15*, 1.

8.6 Hydrogen Bonding

Up to this point, compounds in which hydrogen is bonded to a highly electronegative element have generally been avoided. The dominant intermolecular attractions in many such species are *hydrogen bonds*.

To a first approximation, hydrogen bonds can be considered strong dipolar interactions ("superdipoles") between a partially positive hydrogen atom of one molecule and a highly electronegative atom of another species. However, several features distinguish hydrogen bonds from dipolar forces.

The general type of interaction can be symbolized as

$$^{\delta^-}X-H^{\delta^+}\cdots:Z$$

where the H$\cdots$:Z linkage is, at least in the formal sense, intermolecular. The X might be fluorine, oxygen, or nitrogen (in the most energetic cases), or even atoms such as chlorine, sulfur, or carbon; :Z is typically F, O, N, or an anion. A lone pair on Z appears to be required. The energy of interaction is mainly electrostatic; however, other factors, including delocalization/resonance and London forces, are operative as well.

At least as early as 1912, a difference between hydrogen bonds and dipolar forces was recognized, although the term *hydrogen bond* appears to have been used for the first time in 1920.[7] The strength of hydrogen bonding (typical energies ranging from about 10–60 kJ/mol, with values of up to 250 kJ/mol in exceptional cases) is related to the fact that a hydrogen atom has only one electron. The loss of electron density upon bonding to a highly electronegative element leaves an exposed nucleus of very small size and highly concentrated positive charge. That nucleus can embed itself in a lone-pair orbital from another molecule or ion (:Z from above); see Figure 8.5.

Figure 8.5 Orbital overlap in hydrogen bonding. The partially positive hydrogen atom of compound H–X becomes embedded in a lone-pair orbital of Z, resulting in a short X–Z internuclear distance.

7. Moore, T. S.; Winmill, T. F. *J. Chem. Soc.* **1912**, *101*, 1635; Latimer, W. M.; Rodebush, W. H. *J. Am. Chem. Soc.* **1920**, *42*, 1419.

The Lengths and Energies of Hydrogen Bonds

One consequence of the situation illustrated in Figure 8.5 is the remarkably short distance between atoms X and Z (often less than the sum of their van der Waals radii). One of the most conclusive kinds of evidence for hydrogen bonding is therefore a small internuclear distance. For example, consider the interaction of chloride ion with water:

$$Cl^- \cdots H—O \diagup^H$$

The predicted distance between the oxygen and chlorine nuclei is the sum of the van der Waals radius of oxygen and the ionic radius of Cl^-, plus some extra distance due to the hydrogen; a reasonable estimate is 330–350 pm. However, the O–Cl distance in $H_3O^+Cl^-$(s) is only 295 pm, more than 10% shorter than expected. This suggests that oxygen and chlorine orbitals overlap in space! The degree of shortening is often taken as a rough indication of the strength of a hydrogen bond. Some examples of intra- and intermolecular distances in hydrogen bonded systems are given in Table 8.5.

What does a "hydrogen bond energy" represent? One interpretation is as a dissociation energy; that is, ΔH for the reaction below is equated to the hydrogen bond energy:

$$X–H \cdots :Z \longrightarrow X–H + :Z \tag{8.8}$$

Table 8.5 Intra- and intermolecular distances in selected species that exhibit hydrogen bonding

Species	Bond Type	d, pm	
		Intramolecular	Intermolecular
HF	H–F$\cdots$H	92	256
$K^+HF_2^-$	H–F$\cdots$H	114	114
$K^+H_2F_3^-$	H–F$\cdots$H	116	116
H_2O	H–O$\cdots$H	101	175
$H_5O_2^+Cl^-$	H–O$\cdots$H	120	120
$Na^+HCO_3^-$	H–O$\cdots$H	107	154
$K^+H_2PO_4^-$	H–O$\cdots$H	107	142
ND_3	H–N$\cdots$H	100	237
DCl	H–Cl$\cdots$H	128	184

Note: Values are for crystalline solids as determined by X-ray diffraction, except for HF (vapor, electron diffraction) and ND_3 and DCl (neutron diffraction).
Source: Data taken from Wells, A. F. *Structural Inorganic Chemistry*, 5th ed.; Clarendon: Oxford, 1984.

However, calorimetric measurements are complicated by solvent, entropy, and other effects; hence, theoretical calculations are often used as a substitute for experimental data.

Certain instrumental tools (notably infrared, microwave, and NMR spectroscopy) are commonly used as semiquantitative or qualitative indicators of hydrogen bonding.[8] The following generalizations appear to be valid: Upon the hydrogen bonding of species H–X to :Z, the H–X bond weakens and lengthens, and the nucleus of the hydrogen atom becomes deshielded. The H–X infrared stretching frequency therefore shifts to lower energy, and is often broadened as well. The proton resonance moves to lower field in the NMR spectrum.

Since fluorine is the most electronegative of the bond-forming elements, it is not surprising that the strongest hydrogen bonds are found in systems containing fluorine. The simplest example is HF(s), with HF units interconnected by zigzag chains and H–F–H angles of about 120°:

The intermolecular H–F distance of 156 pm is about 64% greater than the intramolecular distance. Because those distances are unequal, the hydrogen bond is said to be *unsymmetrical*.

Hydrogen bonding is primarily a condensed phase phenomenon, and is usually destroyed by vaporization. Species containing strong hydrogen bonds tend to have high enthalpies of vaporization for that reason. Remarkably, the hydrogen bonding in HF is so strong that it persists even in the gas phase. Electron diffraction studies show that "HF(g)" consists of various oligomers formulated as $(HF)_x$, where x ranges from at least 2 to 6. The average F–F internuclear distance is about 256 pm. (Twice the van der Waals radius would be about 310 pm.)

The Covalent Model of Hydrogen Bonding

An even stronger hydrogen bond is found in the bifluoride ion, HF_2^-. This species is quite common, being commercially available in such compounds as potassium bifluoride (usually formulated as KF·HF, but more properly $K^+HF_2^-$). The bifluoride ion is especially interesting because of the central location of the hydrogen atom:

$$[F \cdots H \cdots F]^-$$

---

8. For a summary, see Schuster, P. *Top. Curr. Chem.* **1984**, *120*, 1.

Figure 8.6   Sketch of the bonding molecular orbital of HF_2^- (Σ_g^+ symmetry), one orbital of a three-center, four-electron bonding arrangement.

We might consider this to be an ion–dipole interaction between HF and F^-. However, in addition to the great hydrogen bond energy (estimated to be somewhere between 155 and 252 kJ/mol), the linkage is both linear and *symmetric*, with two equal H–F distances of 114 pm. Such symmetry is inconsistent with an ion–dipole argument.

Molecular orbital theory provides a different explanation—a three-center, four-electron bond (analogous to H_3^-; see Chapter 3). The H ($1s$) orbital overlaps with the $2p_z$ orbitals of both fluorines simultaneously, as shown in Figure 8.6. The MO diagram contains one bonding, one nonbonding, and one antibonding orbital, with the first two of these being filled. This amounts to one-half bond per H–F linkage.

Hydrogen Bonding in Water

The most common structural form of ice consists of puckered six-membered rings in which hydrogen bonds play an important role (Figure 8.7). Each oxygen lies at the center of a distorted tetrahedron, with the vertices occupied by four hydrogen atoms. Two are covalently bonded (at a distance of 101 pm), and two are hydrogen bonded (at 175 pm) to the oxygen. The

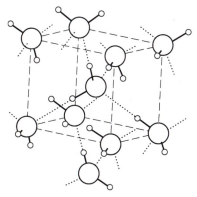

Figure 8.7   The structure of $H_2O(s)$. Large and small circles represent oxygen and hydrogen atoms, respectively; dotted lines are hydrogen bonds. [Reproduced with permission from Wells, A. F. *Structural Inorganic Chemistry*, 5th ed.; Clarendon: Oxford, 1984; p. 655.]

resulting crystal has a considerable amount of empty space, which accounts for the low density (compared to liquid water) of ice. The hydrogen bond energy is 20–25 kJ/mol.

The ionic conductances of H^+ and OH^- in aqueous solution, 350 and 199 $cm^2/\Omega \cdot mol$, respectively, are remarkably high. (Compare to 50 $cm^2/\Omega \cdot mol$ for Na^+, 39 $cm^2/\Omega \cdot mol$ for Li^+, 76 $cm^2/\Omega \cdot mol$ for Cl^-, etc.) This is explained by the *Grotthus–Huckel mechanism*, in which the intermolecular hydrogen bonding permits facile transfer of charge from one molecule to the next throughout the network (see Figure 8.8).[9] A similar mechanism is thought to be operative for OH^-. Thus, aqueous solutions of either high or low pH have much higher conductivities than does pure water.

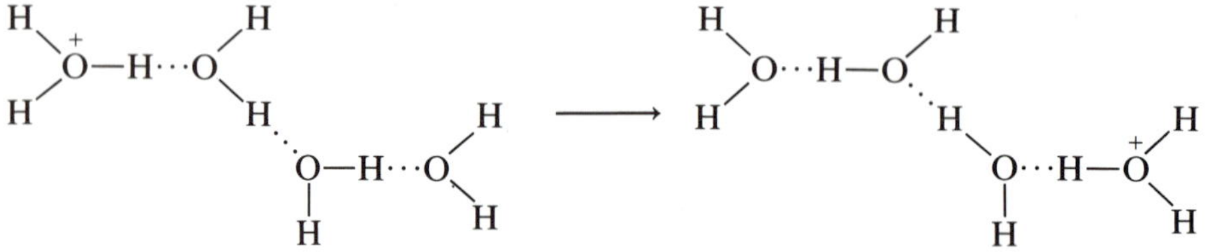

Figure 8.8 The Grotthus–Huckel mechanism. Positive charge is transferred via the hydrogen bonded network.

Hydrated Hydrogen and Hydroxide Ions in Crystals

The active species of acidic aqueous solutions is often formulated as H^+. However, this ion is highly solvated by water molecules; hence, it is better represented as $H(H_2O)_x^+$, where x ranges from 1 to at least 6. It is possible to obtain crystals containing cations of these formulations by the crystallization of appropriate salts from aqueous solutions.

The *hydronium* ion, H_3O^+, is found in the monohydrates of strong acids such as in hydrogen chloride monohydrate ("$HCl \cdot H_2O$," but more accurately, $H_3O^+Cl^-$). The cation contains three equivalent O–H linkages with bond angles of 117°. A chloride ion is hydrogen bonded to each hydrogen, with nonbonded O–Cl distances of only 295 pm. The hydronium ion is also found in $HNO_3 \cdot H_2O$, $HClO_4 \cdot H_2O$, and $H_2SeO_4 \cdot H_2O$.

The species $H_5O_2^+$ is present in many 2:1 acid hydrates, including $HCl \cdot 2H_2O$, $HClO_4 \cdot 2H_2O$, and $H_2SO_4 \cdot 4H_2O$. The hydrogen bonding is symmetric in the chloride; the O–H–O distances are 241 pm, and the linkage is nearly linear (175°). Such linearity is common in solid-state hydrogen

---

9. Grotthus, C. J. T. *Ann. Chem.* **1806**, *58*, 54; Huckel, E. *Z. Elektrochem.* **1928**, *34*, 546. See also, Hertz, H. G.; Braun, B. M.; Müller, K. J.; Maurer, R. *J. Chem. Educ.* **1987**, *64*, 777.

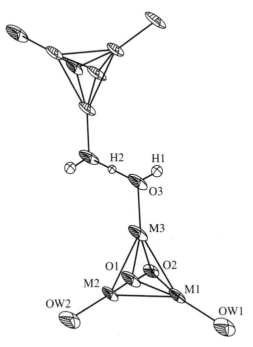

Figure 8.9 Skeletal structure of the cations $\{[M_3O_2(O_2CEt)_6(H_2O)_2]_2(H_3O_2)\}^{3+}$ (M = Mo and W). The $H_3O_2^-$ anion acts as a bridging ligand between the two metal centers.
[Reproduced with permission from Bino, A.; Gibson, D. *J. Am. Chem. Soc.* **1981**, *103*, 6742.]

bonding, since it reduces the steric interactions between groups about the tiny central hydrogen. (There is no directional requirement to the orbital overlap, since the $1s$ orbital is spherical.) The cations $H_7O_3^+$, $H_9O_4^+$, and $H_{13}O_6^+$ have also been observed in the solid state.

Similarly, the hydroxide ion occurs in the hydrated state $OH(H_2O)_x^-$. One example is $H_3O_2^-$, in which the O–H–O linkage is linear and the O–O internuclear distance is very short (229 pm). (Note the similarity to HF_2^-.) The $H_3O_2^-$ anion can act as a bridging ligand between metal centers; a representative example is shown in Figure 8.9.

8.7 Melting and Boiling Temperatures

We indicated earlier that a relationship exists between melting and boiling points and the strengths of intermolecular forces. The phase transition temperatures for several series of compounds will be examined below with that notion in mind.

Group 18 Elements

Element	mp, °C	bp, °C
He	−272	−269
Ne	−249	−246
Ar	−189	−186
Kr	−157	−152
Xe	−112	−107
Rn	−71	−62

The trend is clear, and the explanation should be equally clear. Since no true bonds are present, the only intermolecular interactions are London forces. These forces increase rapidly with increasing size, so the melting and boiling points do likewise. It is interesting to theorize about the next Group 18 element, which will have atomic number 118. Some believe that "noble gas" will actually be a "noble liquid" at or near room temperature!

Molecular Halogens

Compound	mp, °C	bp, °C
F_2	−220	−188
Cl_2	−101	−35
Br_2	−7	59
I_2	114	184

Here, covalent bonds are present, but they are of the nonpolar type. London forces are again the operative intermolecular attractions, so the melting and boiling temperatures increase with increasing size.

Silicon Tetrahalides

Compound	mp, °C	bp, °C
SiF_4	−90	−86
$SiCl_4$	−70	58
$SiBr_4$	5	154
SiI_4	120	288

These compounds all have polar bonds whose individual moments are canceled by the tetrahedral geometry. Again, London forces dominate, so the phase change temperatures increase with increasing size of the terminal atom. It is interesting to note that the melting and boiling points of silane, SiH_4, are $-185°C$ and $-112°C$, respectively; the values fit nicely with the above, even though silane does not formally belong to the series.

Hydrogen Halides

Compound	mp, °C	bp, °C
HF	−83	20
HCl	−115	−85
HBr	−88	−67
HI	−51	−35

Here, things become a bit more interesting. Since the H–X electronegativity difference is greatest for HF and least for HI, molecular polarity decreases going down the series. The melting and boiling temperatures of HCl, HBr, and HI are internally consistent. But HCl should have the highest phase change temperatures of the three if dipolar forces are dominant, and this is clearly not the case. London forces, which increase in going from HCl to HI, are the dominant factor.

It is obvious that HF is out of place relative to the others. The importance of hydrogen bonding in HF has already been discussed, and the resulting intermolecular energies are sufficient to cause the melting and boiling temperatures to be the highest of the series.

Group 15 Hydrides, Chlorides, and Fluorides

Compound	mp, °C	bp, °C
NH_3	−78	−33
PH_3	−133	−88
NF_3	−207	−129
PF_3	−152	−102
NCl_3	−40	71
PCl_3	−112	76

This group allows for several comparisons. The melting and boiling temperatures of NH_3 are each 55°C higher than those of PH_3. This is due to intermolecular hydrogen bonding in ammonia. (Phosphorus is not sufficiently electronegative to form hydrogen bonds in PH_3.)

The relative values for NF_3 and PF_3 can be explained in either of two ways. The P–F linkage is more polar than N–F, so PF_3 has the greater dipole moment.[10] In addition, the London forces are greater in PF_3 because of the larger central atom. With both factors in agreement, it is surprising that the differences in melting and boiling temperatures are not greater than they are.

The chlorides are included to demonstrate that this model is not always successful. There is no obvious way to rationalize the fact that NCl_3 melts at a higher temperature and boils at a lower temperature than its phosphorus analogue.

Tin Chlorides and Bromides

Compound	mp, °C	bp, °C
$SnCl_2$	246	652
$SnBr_2$	216	620
$SnCl_4$	−33	114
$SnBr_4$	31	202

These data emphasize that the dominant forces in an ionic lattice are true bonds. The strong electrostatic interactions in the ionic lattices of the tin(II) halides lead to high phase change temperatures. Comparing $SnCl_2$ to $SnBr_2$, the smaller size of Cl^- yields greater lattice energy (and, hence, a slightly higher melting point) for $SnCl_2$.

In contrast, the tetrahalides are covalent. Being symmetric molecules with tetrahedral electron geometries, the dominant intermolecular interactions are London forces; hence, the larger halogen forms the compound with the higher phase change temperatures.

8.8   Inclusion Compounds

Under the proper circumstances, molecules or ions may become trapped in the solid-state lattice of a chemically different species. In such situations, the trapped species are *guests* in a *host* lattice; the product is often called an *inclusion compound*. Such compounds are stabilized by van der Waals interactions between host and guest molecules. As we have seen for the case

---

10.   As it happens, the bond angles in these compounds are also different: 98° for PF_3 versus 102° for NF_3. The smaller angle of the former also contributes to its greater dipole moment.

of hydrogen atoms in Ti(s) (Section 6.7), the stoichiometry may vary up to some limit.

In an *intercalation compound* the host lattice is of a layered type, with guest molecules trapped between the layers. Two common intercalation hosts are graphite and the isostructural boron nitride. The layer structure of graphite (Figure 7.12) is an effective host for various metal atoms.[11] The structure of the intercalation compound KC_8 is shown in Figure 8.10.

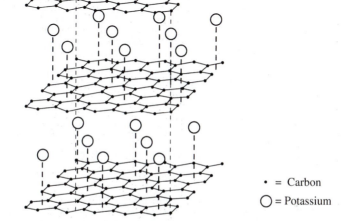

• = Carbon

◯ = Potassium

Figure 8.10
The structure of the intercalation compound KC_8.

When the host lattice forms three-dimensional cages (as in ionic and network crystals), the product is a *clathrate*. For example, elemental boron functions as a clathrate host by forming lattices related to the CsCl structure. Octahedral clusters of six boron atoms form a simple cubic sublattice; the interstices of the sublattice are then occupied by metals (see Figure 8.11, p. 264).

One of the best-known clathrate hosts is water. The ice lattice, being an open one, contains roughly spherical holes where such species as noble gas atoms, halogen molecules, and small organic compounds can be trapped.

Bibliography

Wulfsberg, G. *Principles of Descriptive Inorganic Chemistry*; Brooks/Cole: Pacific Grove, CA, 1987.

Lockhart, J. C., Ed. *Host–Guest Complex Chemistry*; Topics in Current Chemistry 121; Springer-Verlag: New York, 1984.

11. See Selig, H.; Ebert, L. B. *Adv. Inorg. Chem. Radiochem.* **1980**, *23*, 281; and references cited.

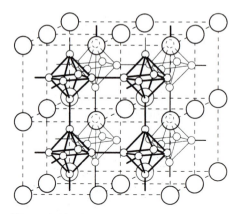

Figure 8.11 The structure of the clathrate compound CaB_6. Octahedra of boron atoms form a simple cubic sublattice, with the interstices occupied by calciums. [Reproduced with permission from Wells, A. F. *Structural Inorganic Chemistry*, 5th ed.; Clarendon: Oxford, 1984; p. 1056.]

Wells, A. F. *Structural Inorganic Chemistry*, 5th ed.; Clarendon: Oxford, 1984.

Dasent, W. E. *Inorganic Energetics*, 2nd ed.; Cambridge University: London, 1982.

Burgess, J. *Metal Ions in Solution*; Ellis Horwood: Chichester, 1978.

Schuster, P.; Zundel, G.; Sandorfy, C., Eds. *The Hydrogen Bond*; North-Holland: Amsterdam, 1976; Volumes 1–3.

Joesten, M. D.; Schaad, L. J. *Hydrogen Bonding*; Marcel Dekker: New York, 1974.

Hamilton, W. C.; Ibers, J. A. *Hydrogen Bonding in Solids*; Benjamin: New York, 1968.

Pimentel, G. C.; McClellan, A. L. *The Hydrogen Bond;* Freeman: San Francisco, 1959.

Questions and Problems

1. The dipole moment of the hydroxyl radical, $\cdot OH$, is 1.66 D; the O–H bond length is 97.1 pm in the gas phase.
   (a) Calculate the dipole moment in Z-unit·picometers.
   (b) Estimate the percent ionic character of $\cdot OH$.
   (c) Estimate the dipolar energy of attraction in OH(g).

2. Using data from appropriate tables of this book, estimate the energy of attraction between a molecule of ClF and a potassium ion.

3. Although the hydration enthalpy of Mg^{2+} is much greater than that of Na^+, the molar solubility of $MgCl_2$ is less than that of NaCl. Rationalize.

4. Explain in your own words why:
   (a) Ion–dipole energies are typically greater than dipole–dipole energies.

(b) London forces increase with increasing size.

(c) Boiling temperatures tend to increase as intermolecular forces increase.

5. The hydration enthalpy of NaOH is more exothermic than that of NaCl. Discuss.

6. Predict which member of each of the following pairs has the greater molar solubility in water; briefly explain your answers:

(a) LiF or LiBr (b) CsF or CsBr

7. As it is purchased for routine laboratory use, "nickel chloride" has a formula weight of 237.7 g/mol.

(a) Explain the "anomalous" formula weight.

(b) Suggest an appropriate structure.

8. Certain experimental methods for molecular weight determination (eg, freezing point depression) yield twice the expected value for certain organic acids. This is due to intermolecular hydrogen bonding. Explain this by drawing a structure for the dimer.

9. One of the most highly hydrated compounds for which the structure is known is $MgCl_2 \cdot 12H_2O$. Speculate on the composition of its solid-state lattice. (For the answer, see Wells, A. F. *Structural Inorganic Chemistry*, 5th ed.; Clarendon: Oxford, 1984, p. 673.)

10. Use appropriate data from Tables 1.7, 7.5, and 8.3 to estimate the London energy of elemental helium. Compare your answer to the values given in this chapter for neon and argon. Does the expected trend hold?

11. Weak acid dissociation equilibria usually take the form

$$HA \rightleftharpoons H^+ + A^-$$

However, the equilibrium for hydrofluoric acid is often written as

$$H_2F_2 \rightleftharpoons H^+ + HF_2^-$$

Why?

12. Decide which two of the following species are most likely to exhibit hydrogen bonding, and defend your answer: BeH_2, CHF_3, NH_4I, HCO_2H, NH_2OH, PH_2F.

13. In question 23 of Chapter 6 it was noted that Shannon's ionic radius for H^+ is −18 pm. Here's a second chance to explain the negative value.

14. The covalent radius of xenon is about 130 pm, while its van der Waals radius is 217 pm. In the solid-state lattice of XeF_4, each xenon is surrounded by four fluorines at a distance of 323 pm. In XeF_2, each xenon has eight fluorine neighbors at 341 pm.

(a) Is the Xe–F distance in XeF_4 compatible with a true bond? Defend your answer.

(b) Discuss the significance of the bond distance in $XeF_2(s)$.

15. The melting point of ammonium hydrogen fluoride (also called ammonium bifluoride) is only 126°C—too low for a lattice constructed of ionic bonds. Suggest another possibility.

16. Imagine that you are asked to determine whether the intermolecular interactions in the species HQ qualify as hydrogen bonds. List three specific types of data you would collect, and briefly explain how you would interpret the data to arrive at a conclusion.

17. For each of the following pairs, predict the higher-melting compound; briefly explain your reasoning:
    (a) BrCl or ICl   (b) CF_4 or CHF_3 (c) CH_4 or SiH_4
    (d) NCl_3 or $BiCl_3$ (e) PF_3 or PF_5 (f) H_2O or H_2S
    (g) H_2Se or H_2Te

18. Consider the two compounds Br_2 and ICl. Which do you feel has the higher boiling point? Why?

19. The distance between parallel layers in pure graphite is 335 pm. The corresponding distance in KC_8 is about 540 pm.
    (a) Explain the difference.
    (b) What specific type of intermolecular forces must be weakened upon the incorporation of potassium atoms into the graphite lattice?
    (c) What intermolecular energies are gained in compensation? Do you believe the gain exceeds the loss? (Is the addition of potassium to graphite an exothermic process?) Explain.

20. The open lattice of ice allows it to serve as a host for atmospheric molecules such as N_2, O_2, and CO_2.
    (a) There is a thermodynamic preference for CO_2 over O_2 or N_2. Explain.
    (b) A similar preference is shown for Xe compared to the other Group 18 elements. Rationalize.

*21. Bino and Gibson reported the structure of a cation having metal centers bridged by the $H_3O_2^-$ anion (see Figure 8.9 and the accompanying caption). Answer these questions after examining their article.
    (a) What is the nonbonded O–O distance? How is this indicative of hydrogen bonding?
    (b) What is the bond angle about the central hydrogen? Rationalize.

*22. J. Emsley, O. P. A. Hoyte, and R. E. Overill discuss what they believe to be "the strongest hydrogen bond" in *J. Chem. Soc. Chem. Commun.* **1977**, 225.
    (a) What species are involved?
    (b) Is their claim based primarily on experimental or theoretical work?
    (c) Do the authors believe that the linkage is better represented as O–H$\cdots$F$^-$ or O$\cdots$H–F? Why?

PART III

Reactions of Inorganic Compounds

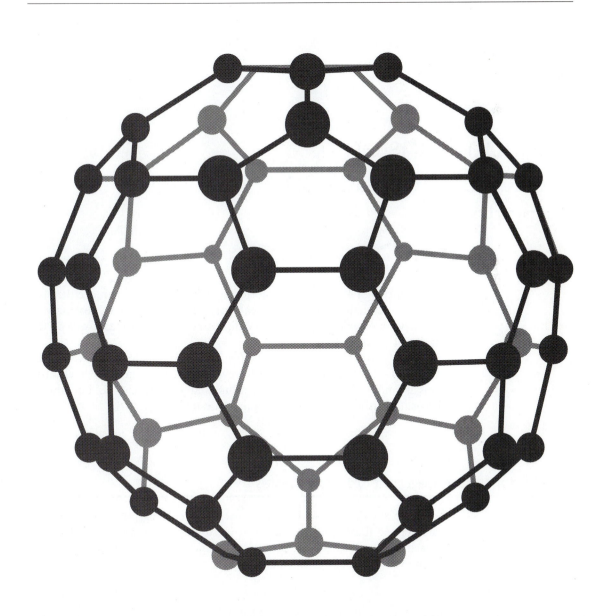

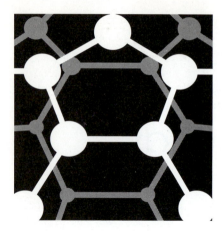

9

Electron Transfer: Oxidation–Reduction Reactions

The majority of known inorganic reactions can be placed into either the oxidation–reduction or acid–base category. In this chapter we discuss various aspects of oxidation–reduction chemistry. Acids and bases will be the subject of Chapter 10.

Simply stated, *oxidation–reduction* (or *redox*) reactions involve the transfer of one or more electrons. However, the distinction between electron transfer and two other processes, atom and ion transfer, is not always clear-cut. This leads to some intriguing questions!

9.1   Oxidation Numbers and Oxidation States

From the standpoint of Lewis structures, conventional oxidation numbers are obtained if all bonds are taken to be totally ionic, with both electrons of each bond allocated to the more electronegative of the involved atoms. As an example, consider the iodate ion:

$$\left[\begin{array}{c} :\ddot{O}-\ddot{I}-\ddot{O}: \\ | \\ :\ddot{O}: \end{array}\right]^{-}$$

The I–O bonds are polar covalent, polarized toward the more electronegative oxygens. If these bonds were ionic, then the "shared" electron density would reside entirely with the oxygens:

$$\left[\begin{array}{c} :\ddot{O}:\{\ddot{I}\}:\ddot{O}: \\ :\ddot{O}: \end{array}\right]^{-}$$

This leads to an oxidation number of -2 for each oxygen (eight valence electrons versus six in the free atom) and $+5$ for iodine (two valence electrons versus seven). As is the case for formal charges, the algebraic sum of all oxidation numbers equals the charge on the species.

Lewis structures are useful for assigning oxidation numbers in species having like atoms in nonequivalent environments. Consider the N_2O molecule. The major contributor to the resonance hybrid is

$$:\ddot{N}=N=\ddot{O}:$$

Since all homonuclear bonds are split equally (one electron of each bond to each atom), the oxidation number of the terminal nitrogen is seen to be -1, while that of the central nitrogen is $+3$. The average oxidation number of nitrogen is therefore $+1$. Similarly, you should be able to show that in carbon suboxide, $O{=}C{=}C{=}C{=}O$, the oxidation numbers of the carbon atoms are 0 (central) and $+2$ (end carbons), giving an average of $\frac{4}{3}$.

Given that redox reactions are characterized by changes in the oxidation numbers of at least two atoms (one increasing and another decreasing) in going from reactants to products, the following are readily recognized as redox processes:

$$2\,Na + 2\,H_2O \longrightarrow 2\,NaOH + H_2 \tag{9.1}$$

$$H_2 + I_2 \longrightarrow 2\,HI \tag{9.2}$$

$$2\,OCl^- \longrightarrow Cl^- + ClO_2^- \tag{9.3}$$

Maximum Oxidation States

With the electron configuration $1s^2 2s^2 2p^6 3s^2$, magnesium has two valence electrons. Consequently, the first two ionization energies of Mg are small relative to IE_3 and beyond. Thus, it is reasonable to think of a Mg^{2+} ion, but no cations of higher charge exist (except under highly energetic conditions). Applying this notion to oxidation numbers, the maximum oxidation state is Mg^{II}. Some other maximum oxidation states are B^{III}, La^{III}, C^{IV}, Hf^{IV}, Se^{VI}, and Cr^{VI}.

The transition and posttransition elements are often found in oxidation states two units less than the maximum; examples include Ge^{II}, Ti^{II}, P^{III}, V^{III}, and S^{IV}. This results from the relative inertness of the valence s electrons in the ground-state configurations ns^2np^x and ns^2nd^x, and is sometimes called the *inert pair effect*; some ramifications of this effect are described in Chapter 13. For certain other (especially transition) elements, less systematic states are known as well; examples include Fe^{III}, Pt^{IV}, and Mn^{VI}.

9.2 Electron versus Atom Transfer

The simplest possible redox reaction involves gas-phase electron transfer between a hydrogen atom and a hydrogen ion:

$$H + H^+ \longrightarrow H^+ + H \tag{9.4}$$

It is clear that equation (9.4) represents an oxidation–reduction process; however, for more complex systems the distinction between electron and atom transfer may be lost. Consider the reaction

$$O^- + O_2 \longrightarrow O + O_2^- \tag{9.5}$$

The products might arise either by electron transfer from O^- to O_2, or by oxygen atom transfer from O_2 to O^-. Experimental evidence (using isotopic labeling) suggests that both of these occur at low energies, but that the electron transfer mechanism dominates under more energetic conditions.

9.3 Electron Transfer in Aqueous Solution

Some General Relationships

Recall the basic equation relating enthalpy, entropy, and free energy:

$$\Delta G° = \Delta H° - T\Delta S° \tag{9.6}$$

The free energy change is usually given in kilojoules per mole, while T is in Kelvin. The superscript °, of course, refers to standard conditions: 25°C, 1 atm pressure for gases, and unit activities for solutes.[1]

---

1. It is customary in most books (including this one) to use concentrations in place of activities.

For aqueous solutions, it is convenient to use the electrochemical potential in volts as a substitute for free energy. The conversion equation is

$$\Delta G^{\circ} = -n\mathscr{F}\mathscr{E}^{0} \tag{9.7}$$

where $\mathscr{F}$ is the Faraday constant (96.487 kJ/mol·V) and n is the number of electrons transferred. This equation also can be applied to nonstandard conditions. Then the superscripts are deleted: $\Delta G = -n\mathscr{F}\mathscr{E}$.

Standard potential can be related to the equilibrium constant by the equation

$$\log K = \frac{n\mathscr{E}^{0}}{0.05916} \tag{9.8}$$

Standard Reduction Potentials

Reduction potentials in aqueous solution are conventionally referenced to the standard hydrogen electrode:

$$2H^{+} + 2e^{-} \longrightarrow H_{2} \qquad \mathscr{E}^{0} = 0.000 \text{ V} \tag{9.9}$$

Standard reduction potentials for the common elements are given later in this chapter (see Table 9.1). For oxidation half-reactions, the sign of the corresponding reduction potential is reversed.

Potentials under nonstandard conditions can be obtained using the *Nernst equation*:

$$\mathscr{E} = \mathscr{E}^{0} - \frac{RT(\ln Q)}{n} = \mathscr{E}^{0} - \frac{0.05916(\log Q)}{n} \quad \text{at } 25^{\circ}\text{C} \tag{9.10}$$

Here, n is again the number of electrons transferred; Q derives from the activities (concentrations) of the products and reactants, arranged as in the mass action expression. Thus, for the H^{+}/H_{2} couple, standard conditions refer to a pH of 0.00 (H^{+} concentration of 1.00 M). At a pH of 14.00 and a pressure of 1.00 atm H_{2}, Q is therefore

$$Q = \frac{P_{H_{2}}}{[H^{+}]^{2}} = \frac{1.00 \text{ atm}}{(1.0 \times 10^{-14})^{2}} = 1.00 \times 10^{28}$$

and the potential is

$$\mathscr{E} = 0.00 - \frac{0.05916[\log(1.0 \times 10^{28})]}{2} = -0.828 \text{ V}$$

The corresponding free energy change is therefore about $+160$ kJ/mol [from equation (9.7)]. Notice that the thermodynamic conventions are such that a negative ΔG always corresponds to a positive $\mathscr{E}$, and that either implies a spontaneous process.

Of course, in water at 25°C, $[H^+] \cdot [OH^-] = K_w = 1.0 \times 10^{-14}$. For any pH greater than 7.0, $[OH^-] > [H^+]$; hence, it becomes reasonable to express the "H^+/H_2" couple in terms of OH^-. The equivalent half-reaction in base can be obtained by adding two hydroxides to both sides of equation (9.9):

$$2H^+ + 2OH^- + 2e^- \longrightarrow H_2 + 2OH^- \tag{9.11}$$

or

$$2H_2O + 2e^- \longrightarrow H_2 + 2OH^- \qquad \mathscr{E}^0 = -0.828 \text{ V} \tag{9.12}$$

We can think of the potentials given for equations (9.9) and (9.12) as two points on a continuum of potential versus pH. The plot in Figure 9.1 shows that as the pH increases, the tendency for H_2 to be liberated from aqueous solution decreases. That is, H^I (H^+ or OH^-) is less easily reduced—is a poorer *oxidizing agent*—in basic than in acidic solution. In general, potent oxidizing agents have strongly positive reduction potentials.

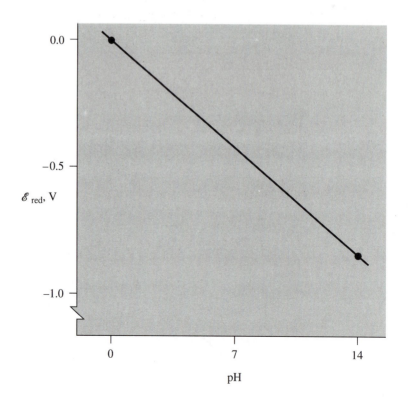

Figure 9.1

Plot of potential for the H^I/H^0 couple (H^+/H_2 in acidic and H_2O/H_2 in basic solution) as a function of pH.

Oxygen commonly exhibits three oxidation states in or above aqueous solution: 0 (O_2), -1 (H_2O_2 or HO_2^-), and -2 (H_2O or OH^-).[2] The relevant half-reactions in acidic solution are

$$O_2 + 4H^+ + 4e^- \longrightarrow 2H_2O \qquad \mathscr{E}^0 = +1.229 \text{ V} \qquad \textbf{(9.13)}$$

$$O_2 + 2H^+ + 2e^- \longrightarrow H_2O_2 \qquad \mathscr{E}^0 = +0.682 \text{ V} \qquad \textbf{(9.14)}$$

$$H_2O_2 + 2H^+ + 2e^- \longrightarrow 2H_2O \qquad \mathscr{E}^0 = +1.776 \text{ V} \qquad \textbf{(9.15)}$$

This information can be abridged into a *potential diagram*, a format popularized by Latimer:[3]

$$O_2 \xrightarrow{\ +0.682\ } H_2O_2 \xrightarrow{\ +1.776\ } H_2O$$
$$\underset{+1.229}{\underline{\hspace{6cm}}}$$

Potential diagrams have the disadvantage of omitting species (H^+, OH^-, and H_2O) necessary for mass balance. They do, however, condense a large amount of information into a small space. Table 9.1, on the following 7 pages, provides a set of potential diagrams for reference.

Volt-Equivalents

The rearrangement of equation (9.7) to $-\Delta G^\circ / \mathscr{F} = n\mathscr{E}^0$ demonstrates the direct relationship between ΔG° and $n\mathscr{E}^0$. Thus, it is the *volt-equivalent* ($n\mathscr{E}^0$), rather than the volt, that relates directly to free energy. Like free energies, volt-equivalents are additive. One use for this parameter can be illustrated using the potential diagram for iron in aqueous acid:

$$FeO_4^{2-} \xrightarrow[\mathscr{E}_1^0]{\ +2.20\ } Fe^{3+} \xrightarrow[\mathscr{E}_2^0]{\ +0.771\ } Fe^{2+} \xrightarrow[\mathscr{E}_3^0]{\ -0.440\ } Fe$$
$$\underset{\mathscr{E}_4^0}{\underline{\hspace{5cm}}_?}$$

---

2.   A fourth oxidation state, $-\frac{1}{2}$, is found in the superoxide ion (O_2^-).

3.   Latimer, W. *The Oxidation Potentials of the Elements and Their Values in Aqueous Solutions*, 2nd ed.; Prentice-Hall: Englewood Cliffs, NJ, 1952.

Table 9.1 Standard reduction potential diagrams (after Latimer) for the common elements

Acidic Solution

$$AgO^+ \xrightarrow{\ 2.1\ } Ag^{2+} \xrightarrow{\ 1.980\ } Ag^+ \xrightarrow{\ 0.799\ } Ag$$

$$Al^{3+} \xrightarrow{\ -1.676\ } Al$$

$$Au^{3+} \xrightarrow{\ 1.36\ } Au^+ \xrightarrow{\ 1.83\ } Au$$

$$H_3AsO_4 \xrightarrow{\ 0.560\ } H_3AsO_3 \xrightarrow{\ 0.248\ } As \xrightarrow{\ -0.607\ } AsH_3$$

$$H_3BO_3 \xrightarrow{\ -0.890\ } B$$

$$Ba^{2+} \xrightarrow{\ -2.92\ } Ba$$

$$Be^{2+} \xrightarrow{\ -1.97\ } Be$$

$$Bi_2O_5 \xrightarrow{\ 1.59\ } BiO^+ \xrightarrow{\ 0.320\ } Bi$$

$$BrO_4^- \xrightarrow{\ 1.763\ } BrO_3^- \xrightarrow{\ 1.505\ } HOBr \xrightarrow{\ 1.595\ } Br_2 \xrightarrow{\ 1.065\ } Br^-$$
$$BrO_3^- \xrightarrow{\ 1.52\ } Br_2$$

$$CO_2 \xrightarrow{\ -0.20\ } HCO_2H \xrightarrow{\ 0.034\ } HCHO \xrightarrow{\ 0.232\ } CH_3OH \xrightarrow{\ 0.59\ } CH_4$$

$$Ca^{2+} \xrightarrow{\ -2.84\ } Ca$$

$$Cd^{2+} \xrightarrow{\ -0.402\ } Cd$$

$$ClO_4^- \xrightarrow{\ 1.230\ } ClO_3^- \xrightarrow{\ 1.21\ } HClO_2 \xrightarrow{\ 1.645\ } HOCl \xrightarrow{\ 1.63\ } Cl_2 \xrightarrow{\ 1.360\ } Cl^-$$
$$HClO_2 \xrightarrow{\ 1.468\ } Cl_2$$

$$Co^{3+} \xrightarrow{\ 1.808\ } Co^{2+} \xrightarrow{\ -0.277\ } Co$$

$$Cr_2O_7^{2-} \xrightarrow{\ 1.33\ } Cr^{3+} \xrightarrow{\ -0.408\ } Cr^{2+} \xrightarrow{\ -0.912\ } Cr$$
$$Cr^{3+} \xrightarrow{\ -0.744\ } Cr$$

Latimer (reduction potential) diagrams:

$Cs^+ \xrightarrow{-2.923} Cs$

$Cu^{2+} \xrightarrow{0.153} Cu^+ \xrightarrow{0.521} Cu$
$Cu^{2+} \xrightarrow{0.337} Cu$

$F_2 \xrightarrow{3.05} HF$

$FeO_4^{2-} \xrightarrow{2.20} Fe^{3+} \xrightarrow{0.771} Fe^{2+} \xrightarrow{-0.440} Fe$
$Fe^{3+} \xrightarrow{-0.036} Fe$

$Ga^{3+} \xrightarrow{-0.529} Ga$

$Hg^{2+} \xrightarrow{0.920} Hg_2^{2+} \xrightarrow{0.788} Hg$

$H_5IO_6 \xrightarrow{1.644} IO_3^- \xrightarrow{1.133} HOI \xrightarrow{1.45} I_2 \xrightarrow{0.536} I^-$
$IO_3^- \xrightarrow{1.196} I_2$

$In^{3+} \xrightarrow{-0.49} In^{2+} \xrightarrow{-0.40} In^+ \xrightarrow{-0.14} In$
$In^{3+} \xrightarrow{-0.343} In$

$K^+ \xrightarrow{-2.925} K$

$Li^+ \xrightarrow{-3.040} Li$

$Mg^{2+} \xrightarrow{-2.36} Mg$

$MnO_4^- \xrightarrow{0.564} MnO_4^{2-} \xrightarrow{2.261} MnO_2 \xrightarrow{0.95} Mn^{3+} \xrightarrow{1.51} Mn^{2+} \xrightarrow{-1.180} Mn$
$MnO_4^- \xrightarrow{1.695} MnO_2$
$MnO_2 \xrightarrow{1.23} Mn^{2+}$

$NO_3^- \xrightarrow{0.803} N_2O_4 \xrightarrow{1.07} HNO_2 \xrightarrow{1.00} NO \xrightarrow{1.59} N_2O \xrightarrow{1.77} N_2 \xrightarrow{-3.09} HN_3 \xrightarrow{0.34} N_2H_5^+ \xrightarrow{1.275} NH_4^+$
$NO_3^- \xrightarrow{0.94} HNO_2$
$HNO_2 \xrightarrow{1.29} N_2O$
$HNO_2 \xrightarrow{1.45} N_2$
$N_2 \xrightarrow{-0.23} N_2H_5^+$
$HN_3 \xrightarrow{1.96} N_2H_5^+$
$N_2 \xrightarrow{0.27} NH_4^+$

(continued)

Table 9.1 (continued)

$$Na^+ \xrightarrow{-2.713} Na$$

$$NiO_2 \xrightarrow{1.678} Ni^{2+} \xrightarrow{-0.250} Ni$$

$$O_3 \xrightarrow{2.07} O_2 \xrightarrow{0.682} H_2O_2 \xrightarrow{1.776} H_2O$$
$$O_2 \xrightarrow{1.229} H_2O$$

$$OsO_4 \xrightarrow{1.005} OsO_2 \xrightarrow{0.687} Os$$

$$H_3PO_4 \xrightarrow{-0.276} H_3PO_3 \xrightarrow{-0.499} H_3PO_2 \xrightarrow{-0.508} P_4 \xrightarrow{-0.063} PH_3$$
$$H_3PO_3 \xrightarrow{-0.174} P_4$$

$$PbO_2 \xrightarrow{1.455} Pb^{2+} \xrightarrow{-0.126} Pb$$

$$PdO_3 \xrightarrow{\sim 2} Pd^{4+} \xrightarrow{\sim 1.6} Pd^{2+} \xrightarrow{0.987} Pd$$

$$PtO_2 \xrightarrow{1.045} PtO \xrightarrow{0.98} Pt$$

$$Rb^+ \xrightarrow{-2.924} Rb$$

$$ReO_4^- \xrightarrow{0.73} ReO_3 \xrightarrow{0.40} ReO_2 \xrightarrow{0.251} Re \xrightarrow{-0.4} Re^-$$
$$ReO_3 \xrightarrow{0.362} Re$$

$$RuO_4 \xrightarrow{0.9} RuO_4^- \xrightarrow{1.6} RuO_4^{2-} \xrightarrow{1.3} Ru^{2+} \xrightarrow{0.45} Ru$$
$$RuO_4^{2-} \xrightarrow{0.450} Ru^{2+}$$

$$S_2O_8^{2-} \xrightarrow{2.01} SO_4^{2-} \xrightarrow{0.172} SO_2 \xrightarrow{0.51} S_4O_6^{2-} \xrightarrow{0.08} S_2O_3^{2-} \xrightarrow{0.50} S_8 \xrightarrow{0.142} H_2S$$
$$SO_2 \xrightarrow{-0.082} S_2O_4^{2-}$$
$$S_2O_3^{2-} \xrightarrow{0.88} S_8$$

$$Sb_2O_5 \xrightarrow{0.581} SbO^+ \xrightarrow{0.152} Sb \xrightarrow{-0.510} SbH_3$$

$$Sc^{3+} \xrightarrow{-2.03} Sc$$

$$SeO_4^{2-} \xrightarrow{1.15} H_2SeO_3 \xrightarrow{0.740} Se \xrightarrow{-0.399} H_2Se$$

$$SiO_2 \xrightarrow{-0.909} Si$$

$$Sn^{4+} \xrightarrow{0.15} Sn^{2+} \xrightarrow{-0.136} Sn$$

$$Sr^{2+} \xrightarrow{-2.89} Sr$$

$$H_6TeO_6 \xrightarrow{1.02} H_2TeO_3 \xrightarrow{0.529} Te \xrightarrow{-0.739} H_2Te$$

$$TiO^{2+} \xrightarrow{0.099} Ti^{3+} \xrightarrow{-0.369} Ti^{2+} \xrightarrow{-1.628} Ti$$

$$Tl^{3+} \xrightarrow{1.25} Tl^{+} \xrightarrow{-0.336} Tl$$

$$VO_2^{+} \xrightarrow{1.00} VO^{2+} \xrightarrow{0.359} V^{3+} \xrightarrow{-0.256} V^{2+} \xrightarrow{-1.186} V$$

$$WO_3 \xrightarrow{-0.029} W_2O_5 \xrightarrow{-0.031} WO_2 \xrightarrow{-0.119} W$$

$$H_4XeO_6 \xrightarrow{2.3} XeO_3 \xrightarrow{1.8} Xe$$

$$Zn^{2+} \xrightarrow{-0.76} Zn$$

Basic Solution

$$Ag_2O_3 \xrightarrow{0.739} AgO \xrightarrow{0.607} Ag_2O \xrightarrow{0.345} Ag$$

$$Al(OH)_4^{-} \xrightarrow{-2.310} Al$$

$$AsO_4^{3-} \xrightarrow{-0.68} H_2AsO_3^{-} \xrightarrow{-0.675} As \xrightarrow{-1.21} AsH_3$$

$$B(OH)_4^{-} \xrightarrow{-1.24} B$$

$$Ba^{2+} \xrightarrow{-2.92} Ba$$

$$Bi_2O_5 \xrightarrow{0.6} Bi_2O_3 \xrightarrow{-0.46} Bi$$

(continued)

Table 9.1 *(continued)*

Basic Solution

$BrO_4^- \xrightarrow{0.99} BrO_3^- \xrightarrow{0.54} BrO^- \xrightarrow{0.45} Br_2 \xrightarrow{1.07} Br^-$
(with 0.761 from BrO^- to Br^-)

$CO_3^{2-} \xrightarrow{-1.01} HCO_2^- \xrightarrow{-1.07} HCHO \xrightarrow{-0.59} CH_3OH \xrightarrow{-0.2} CH_4$

$Ca^{2+} \xrightarrow{-2.84} Ca$

$Cd(OH)_2 \xrightarrow{-0.824} Cd$

$ClO_4^- \xrightarrow{0.36} ClO_3^- \xrightarrow{0.33} ClO_2^- \xrightarrow{0.66} OCl^- \xrightarrow{0.40} Cl_2 \xrightarrow{1.360} Cl^-$
(with 0.50 from ClO_3^-; 0.89 from OCl^- to Cl^-)

$Co(OH)_3 \xrightarrow{0.17} Co(OH)_2 \xrightarrow{-0.73} Co$

$CrO_4^{2-} \xrightarrow{-0.13} Cr(OH)_3 \xrightarrow{-1.1} Cr(OH)_2 \xrightarrow{-1.4} Cr$
(with -1.34 from $Cr(OH)_3$ to Cr)

$Cs^+ \xrightarrow{-2.923} Cs$

$Cu(OH)_2 \xrightarrow{-0.08} Cu_2O \xrightarrow{-0.358} Cu$

$F_2 \xrightarrow{2.87} F^-$

$FeO_4^{2-} \xrightarrow{0.72} Fe(OH)_3 \xrightarrow{-0.56} Fe(OH)_2 \xrightarrow{-0.877} Fe$

$IO_4^- \xrightarrow{0.7} IO_3^- \xrightarrow{0.14} IO^- \xrightarrow{0.45} I_2 \xrightarrow{0.54} I^-$
(with 0.26 from IO^- to I^-)

Potassium

$K^+ \xrightarrow{-2.925} K$

$Li^+ \xrightarrow{-3.040} Li$

$Mg(OH)_2 \xrightarrow{-2.687} Mg$

Nitrogen

$NO_3^- \xrightarrow{-0.86} N_2O_4 \xrightarrow{0.88} NO_2^- \xrightarrow{-0.46} NO \xrightarrow{0.76} N_2O \xrightarrow{0.94} N_2 \xrightarrow{-3.04} NH_2OH \xrightarrow{0.73} N_2H_4 \xrightarrow{0.11} NH_3$

$N_2O_4 \; (0.01) \qquad NO \; (0.15) \qquad 0.41 \qquad N_2 \; (-1.05) \qquad NH_2OH \; (0.42)$

Sodium

$Na^+ \xrightarrow{-2.713} Na$

$NiO_2 \xrightarrow{0.490} Ni(OH)_2 \xrightarrow{-0.72} Ni$

$O_3 \xrightarrow{1.24} O_2 \xrightarrow{0.365} O_2^- \xrightarrow{-0.517} \; \; 0.401 \xrightarrow{} HO_2^- \xrightarrow{0.878} OH^-$

$PO_4^{3-} \xrightarrow{-1.12} HPO_3^{2-} \xrightarrow{-1.565} H_2PO_2^- \xrightarrow{-2.05} P_4 \xrightarrow{-0.89} PH_3$

$PbO_2 \xrightarrow{0.247} PbO \xrightarrow{-0.580} Pb$

$Pd(OH)_4 \xrightarrow{0.73} Pd(OH)_2 \xrightarrow{0.07} Pd$

$Rb^+ \xrightarrow{-2.924} Rb$

$ReO_4^- \xrightarrow{-0.7} ReO_4^{2-} \xrightarrow{-0.4} ReO_2 \xrightarrow{-0.577} Re \xrightarrow{-0.4} Re^-$

-0.66

$SO_4^{2-} \xrightarrow{-0.93} SO_3^{2-} \xrightarrow{-0.80} S_4O_6^{2-} \xrightarrow{0.08} S_2O_3^{2-} \xrightarrow{-0.74} S_8 \xrightarrow{-0.447} S^{2-}$

$SO_3^{2-} \xrightarrow{-1.12} S_2O_4^{2-} \xrightarrow{-0.04}$

(continued)

Table 9.1 (*continued*)

Basic Solution

$$Sb(OH)_6^- \xrightarrow{-0.4} H_2SbO_3^- \xrightarrow{-0.66} Sb \xrightarrow{-1.34} SbH_3$$

$$Sc(OH)_3 \xrightarrow{-2.60} Sc$$

$$SeO_4^{2-} \xrightarrow{0.05} SeO_3^{2-} \xrightarrow{-0.366} Se \xrightarrow{-0.92} Se^{2-}$$

$$SiO_3^{2-} \xrightarrow{-1.69} Si$$

$$Sn(OH)_6^{2-} \xrightarrow{-0.93} Sn(OH)_3^- \xrightarrow{-0.909} Sn$$

$$Sr^{2+} \xrightarrow{-2.89} Sr$$

$$TeO_4^{2-} \xrightarrow{0.07} TeO_3^{2-} \xrightarrow{-0.57} Te \xrightarrow{-1.143} Te^{2-}$$

$$Tl(OH)_3 \xrightarrow{-0.05} TlOH \xrightarrow{-0.343} Tl$$

$$VO_4^{3-} \xrightarrow{2.19} HV_2O_5^- \xrightarrow{0.542} V_2O_3 \xrightarrow{-0.486} VO \xrightarrow{-0.820} V$$

$$WO_4^{2-} \xrightarrow{-1.259} WO_2 \xrightarrow{-0.982} W$$

$$HXeO_6^{3-} \xrightarrow{0.9} HXeO_4^- \xrightarrow{0.9} Xe$$

$$Zn(OH)_2 \xrightarrow{-1.25} Zn$$

Sources: Bard, A. J.; Parsons, R.; Jordan, J., Eds. *Standard Potentials in Aqueous Solution*; Marcel Dekker: New York, 1985; Milazzo, G.; Caroli, S. *Tables of Standard Electrode Potentials*; Wiley: New York, 1978; Latimer, W. M.; Hildebrand, J. L. *Reference Book of Inorganic Chemistry*, 3rd ed.; Macmillan: New York, 1952.

What is the standard potential for the reduction of Fe^{3+} to elemental iron ($\mathscr{E}_4^0$)? Using volt-equivalents,

$$n_4\mathscr{E}_4^0 = n_2\mathscr{E}_2^0 + n_3\mathscr{E}_3^0 \tag{9.16}$$

$$3\mathscr{E}_4^0 = 1(+0.771) + 2(-0.440)$$

$$\mathscr{E}_4^0 = -0.036 \text{ V}$$

Volt-equivalents are useful for evaluating the relative stabilities of the oxidation states of a given element in solution. In Figure 9.2 we have plotted volt-equivalents (relative to $n\mathscr{E}^0 = 0$ for the free element) as a function of oxidation state for iron.[4] The ionization energies of Fe are shown for

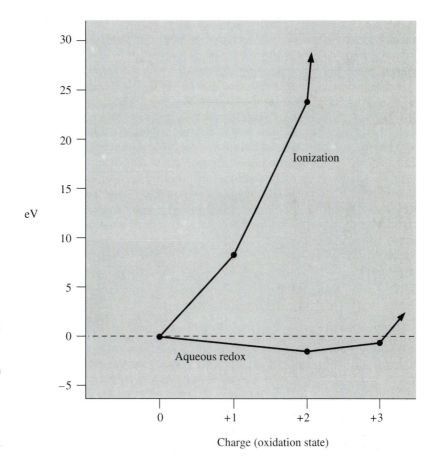

Figure 9.2

The relative stability of iron as a function of charge (referenced to $n\mathscr{E}^0 = 0$ for Fe^0) with respect to gas-phase ionization and aqueous oxidation–reduction.

4. Such presentations are sometimes called *Frost diagrams*. See Frost, A. A. *J. Am. Chem. Soc.* **1951**, *73*, 2680.

comparison. Since ionization energies are always positive, gas-phase stability decreases with increasing positive charge. In solution, however, the most stable species is Fe^{2+}. The correlation between the two plots is weak; this is the rule rather than the exception. The reasons for this are explored next.

9.4 Factors Influencing the Magnitude of Standard Potentials

A variety of factors play a role in determining redox potentials, including ionization energies and electron affinities, atomization energies, solvation energies, covalent bond energies, and the presence of nonaqueous ligands.

Consider the oxidation of a Group 1 element from its standard state to the monovalent, hydrated cation:

$$M(s) \longrightarrow M^+(aq) + 1e^- \tag{9.17}$$

A Born–Haber treatment can be used to express the qualitative relationships among ionization, atomization, and solvation energies, as shown in Figure 9.3.

Both ionization and atomization energies decrease going down Group 1. We might therefore expect the oxidation potentials corresponding to equation (9.17) to increase in the same manner.[5] That is the case for sodium through cesium, but lithium is out of order. The small size (more properly, the high charge density) of Li^+ allows for strong ion–dipole interactions with water molecules, producing large solvation energies. This causes the reduc-

Figure 9.3
The qualitative relationships among atomization, ionization, and hydration enthalpies in influencing $\mathscr{E}^0$ for the M^+/M redox couple. Entropy effects are ignored.

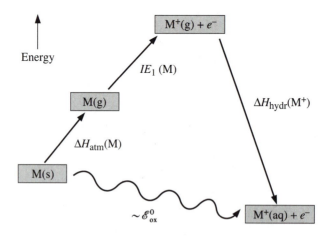

---

5. Provided that entropy effects are not great. Recall that electrochemical potential relates directly to free energy, not enthalpy.

tion potential of Li to be the most negative [ie, the aqueous oxidation of Li(s) is the most energetically favored] of the Group 1 elements.

There are well-defined periodic trends in redox potentials for the cations of the most electropositive metals. The oxidation process $M \rightarrow M^{n+} + ne^-$ generally becomes more favorable (the reduction potential of the cation becomes more negative) going down Groups 1, 2, and 3. (Lithium is an exception, as explained above.) The reverse is observed going across a period; that is, oxidation of the free metals becomes less favored. This is apparent from Figure 9.4, where we have plotted reduction potentials for selected M^{n+}/M couples for main group metals. The situation is more complex for transition metals, where ligand field effects also must be considered (see below).

For the gain of one or more electrons by neutral species, favorable reduction potentials should result from large electron affinities, low atomization energies, and large hydration energies for the resulting anions. Reduction potentials should therefore decrease going down a family of nonmetals and

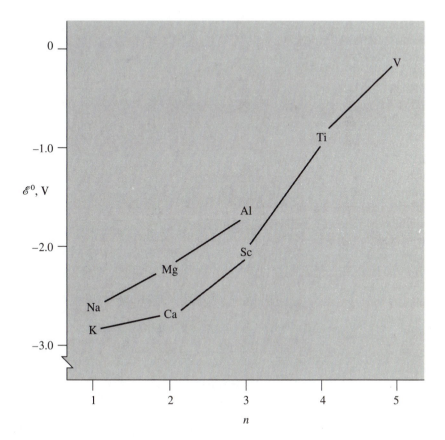

Figure 9.4

Periodic trends in reduction potentials for M^{n+}/M redox couples for selected metals of Groups 1–5. Values are for acidic solutions.

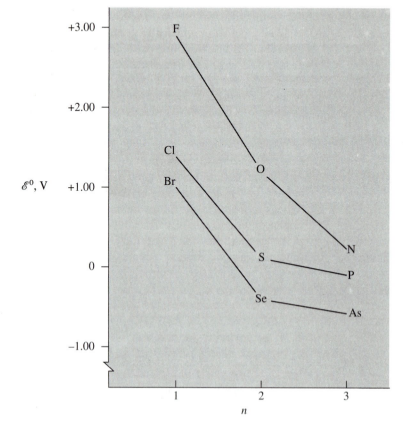

Figure 9.5
Periodic trends in reduction potentials for X^0/X^{n-} redox couples, where X belongs to Group 15, 16, or 17. Values are for acidic solutions.

increase going across a period. Figure 9.5 demonstrates that, at least for the upper right quadrant of the periodic table, this is the case. Thus, F_2 is among the most potent of all oxidizing agents.

Covalent Bond Energies

The effect of covalent bond energies can be discerned from Figure 9.6, which gives volt-equivalent plots for several nonmetals. In bond energy, S–O > Cl–O > Br–O.

Let us compare the oxidation of chlorate to perchlorate to that of sulfite to sulfate in basic solution. The half-reactions are very similar:

$$ClO_3^- + 2OH^- \longrightarrow ClO_4^- + H_2O + 2e^- \qquad \mathscr{E}^0 = -0.36 \text{ V} \tag{9.18}$$

$$SO_3^{2-} + 2OH^- \longrightarrow SO_4^{2-} + H_2O + 2e^- \qquad \mathscr{E}^0 = +0.93 \text{ V} \tag{9.19}$$

Note that in both cases an extra bond is formed between the heteroatom

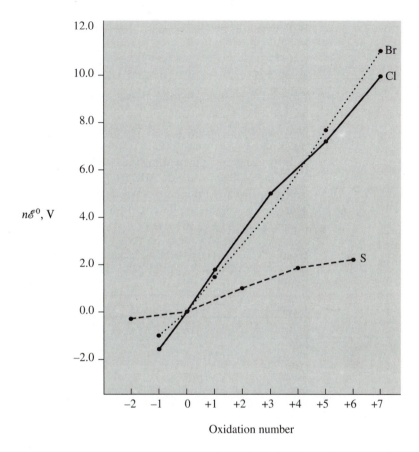

Figure 9.6
Plots of relative energy (in volt-equivalents, referenced to $n\mathscr{E}^0 = 0$ for the neutral species) for bromine, chlorine, and sulfur in acidic aqueous solution.

and oxygen. The greater energy of the S–O bond (a difference of roughly 80 kJ/mol,[6] which corresponds to $n\mathscr{E}^0 = 0.8$ V) accounts for most of the difference.

Bond energies also can be used to explain the spontaneity of certain oxygen atom transfer reactions, such as those described below:

$$SO_3^{2-} + XO^- \longrightarrow SO_4^{2-} + X^- \qquad X = Cl, \text{ Br, or I} \qquad (9.20)$$

$$SO_3^{2-} + NO_3^- \longrightarrow SO_4^{2-} + NO_2^- \qquad (9.21)$$

$$PO_3^{3-} + ClO_4^- \longrightarrow PO_4^{3-} + ClO_3^- \qquad (9.22)$$

In each case, the number of the stronger type of heteroatom–oxygen bonds is maximized as reaction proceeds from left to right. The importance of the bond energy effect is clear from the equilibrium constants of these

6. Based on data of Sanderson, R. T. *Chemical Bonds and Bond Energy*; Academic: New York, 1976; p. 139.

reactions—the smallest K_{eq} (for the oxidation of sulfite by NO_3^-) is 6×10^{31} at 25°C!

The Influence of Nonaqueous Ligands

The presence of coordinating ligands can play an important role in stabilizing or destabilizing the oxidation states of transition metals. Considerable data are available for the Fe^{III}/Fe^{II} couple, as summarized in Table 9.2. Thus, it is possible to fix iron in the Fe^{II} state using nitrogen-containing ligands such as bipyridine and phenanthroline, while oxygen-containing ligands tend to favor Fe^{III}. (See Appendix III for the structures of these ligands.) The $Fe(OH)_3/Fe(OH)_2$ couple is not strictly comparable because of a solubility effect (see below). The difference in behavior can be explained by crystal field theory (Chapter 16).

Table 9.2 The influence of nonaqueous ligands on the standard reduction potential of the Fe^{III}/Fe^{II} couple

Ligand	$\mathscr{E}^0$, V	Ligand	$\mathscr{E}^0$, V
H_2O	$+0.77$	CN^-	$+0.36$
OH^-	-0.56	bipy	$+1.10$
$C_2O_4^{2-}$	$+0.02$	phen	$+1.12$

Note: The value for OH^- is not strictly comparable to the others because of solubility and pH effects (see text); bipy = bipyridine, phen = *o*-phenanthroline (see Appendix III for structures).

A dramatic example of this effect is provided by aqueous cobalt; Co^{III} oxidizes water under standard conditions:

$$4Co^{3+} + 2H_2O \longrightarrow 4Co^{2+} + 4H^+ + O_2 \qquad \mathscr{E}^0 = +0.579 \text{ V}$$

$$(9.23)$$

However, certain ligands stabilize the Co^{III} state. The presence of excess ammonia lowers the reduction potential to $+0.11$ V, while cyanide ion causes the $Co^{III} \rightarrow Co^{II}$ conversion to be strongly disfavored:[7]

$$[Co(NH_3)_6]^{3+} + 1e^- \longrightarrow [Co(NH_3)_6]^{2+} \qquad \mathscr{E}^0 = +0.11 \text{ V}$$

$$(9.24)$$

$$[Co(CN)_6]^{3-} + 1e^- \longrightarrow [Co(CN)_6]^{4-} \qquad \mathscr{E}^0 = -0.83 \text{ V}$$

$$(9.25)$$

---

7. Actually, $[Co(CN)_6]^{4-}$ dissociates to $[Co(CN)_5]^{3-} + CN^-$, so the process is more complex than indicated by equation (9.25). The conclusion holds, however.

An interesting application of a combined oxidation–nonaqueous ligand effect concerns elemental gold. The resistance of gold to corrosion by atmospheric oxygen, water, and acids has long been known. Aqueous acids "dissolve" most free metals by oxidation, with the driving force for reaction being hydration energy. For gold, however, the process is strongly disfavored:

$$2\,Au + 6\,H^+ \longrightarrow 2\,Au^{3+} + 3\,H_2 \qquad \mathscr{E}^0 = -1.52 \text{ V} \qquad \textbf{(9.26)}$$

Though none of the common laboratory acids are individually able to dissolve metallic gold, a combination of 3 parts HCl to 1 part HNO_3 (known for centuries as *aqua regia*) does so. Each acid in this mixture performs a specific function. The HNO_3 is an *oxidizing acid*, since nitrate ion is a better oxidizing agent than H^+:

$$NO_3^- + 4\,H^+ + 3e^- \longrightarrow NO + 2\,H_2O \qquad \mathscr{E}^0 = +0.96 \text{ V}$$
$$\textbf{(9.27)}$$

The HCl supplies the nonaqueous ligand Cl^-, changing the oxidation potential of the Au/Au^{III} couple from -1.52 to -0.99 V. Thus, both the reduction and oxidation portions of the overall process are improved. The net equation is

$$Au + NO_3^- + 4\,Cl^- + 4\,H^+ \longrightarrow AuCl_4^- + NO + 2\,H_2O \quad \textbf{(9.28)}$$

9.5 Overall Cell Potentials and Thermodynamic Stability

Is the reaction below spontaneous under standard conditions?

$$2\,MnO_4^{2-} + 4\,H^+ \longrightarrow 2\,MnO_2 + 2\,H_2O + O_2 \qquad \textbf{(9.29)}$$

The answer can be obtained by writing two half-reactions (one oxidation and one reduction) and taking the algebraic sum of their potentials:

$$
\begin{array}{lr}
2 \times [MnO_4^{2-} + 4\,H^+ + 2e^- \longrightarrow MnO_2 + 2\,H_2O] & \mathscr{E}^0 = +2.26 \text{ V} \\
\hline
2\,H_2O \longrightarrow O_2 + 4\,H^+ + 4e^- & \mathscr{E}^0 = -1.23 \text{ V} \\
\hline
2\,MnO_4^{2-} + 4\,H^+ \longrightarrow 2\,MnO_2 + 2\,H_2O + O_2 & \mathscr{E}^0 = +1.03 \text{ V}
\end{array}
$$

Notice that the reduction half-reaction was multiplied by two for purposes of conservation of charge (the electrons gained in the reduction must equal the electrons lost in the oxidation). Because potential is independent of stoichiometry, however, the reduction potential is not doubled. Since the potential for the overall process is positive, the reaction is spontaneous under standard conditions; the manganate ion, MnO_4^{2-}, is thermodynamically unstable in acidic aqueous solution.

However, certain "unstable" solutions can be prepared and studied, because there is a difference between *thermodynamic* and *kinetic* stability. If a spontaneous reaction has a high activation energy (E_a), it will occur only slowly. Such reactions can be accelerated by a catalyst that provides an alternative reaction pathway of lower E_a, or by supplying extra energy (often by increasing the temperature). Also, reactions having only slightly favorable potentials may occur very slowly or not at all. The *overvoltage* is the extra potential (beyond $\mathscr{E} = 0$) required to cause a thermodynamically spontaneous process to occur at a detectable rate. For aqueous reactions involving the H^+/H_2 and the O_2/H_2O couples, the overvoltage is typically about 0.5 V. Thus, water is kinetically stable toward oxidizing and reducing agents over about a 2.2 V interval under standard conditions. The practical potential needed to oxidize H_2O to O_2 is $1.23 + 0.5$, or about 1.7 V, while the reduction of H^+ to H_2 requires about 0.5 V.

Disproportionation

Some species (always occupying intermediate positions in potential diagrams) are capable of self-oxidation–reduction, or *disproportionation*. A case in point is the hypochlorite ion. The relevant half-reactions in basic solution are

$$2 \times [OCl^- + H_2O + 2e^- \longrightarrow Cl^- + 2OH^-] \qquad \mathscr{E}^0 = +0.89 \text{ V}$$
$$\underline{OCl^- + 4OH^- \longrightarrow ClO_3^- + 2H_2O + 4e^- \qquad \mathscr{E}^0 = -0.50 \text{ V}}$$
$$3OCl^- \longrightarrow 2Cl^- + ClO_3^- \qquad \mathscr{E}^0 = +0.39 \text{ V}$$

$$\textbf{(9.30)}$$

Thus, OCl^- solutions are thermodynamically unstable under standard conditions, although they do exhibit some kinetic stability.

Potential diagrams are useful for identifying species having a tendency to disproportionate. If the potential listed immediately to the right of the species in question is greater than that immediately to its left, then disproportionation is indicated.

9.6 Nonstandard Conditions

Laboratory chemists are often able to promote or inhibit chemical reactions by controlling the appropriate conditions. Adjustments to concentration, temperature, pH, and solubility are commonly used in this regard. In the discussion below, we will use the example of nickel in acidic solution for illustration:

$$NiO_2 \xrightarrow{+1.678} Ni^{2+} \xrightarrow{-0.250} Ni$$

Nonstandard Concentrations

The Nernst equation describes the dependence of potential on concentration. For the Ni^{2+}/Ni couple, the specific relationship is

$$\mathscr{E} = -0.250 - \frac{0.05916 \cdot \log(1/[Ni^{2+}])}{2} \tag{9.31}$$

$$= -0.250 + 0.02958 \cdot \log[Ni^{2+}] \tag{9.32}$$

Therefore, a plot of reduction potential versus the logarithm of the nickel ion concentration is linear, with a slope equal to $0.05916/n$. Consistent with Le Châtelier's principle, the potential increases with increasing concentration of the reactant (Ni^{2+}).

Temperature Effects

A more general form of the Nernst equation is

$$\mathscr{E} = \mathscr{E}^0 - \frac{(1.985 \times 10^{-4})T \cdot \log Q}{n} \tag{9.33}$$

For the Ni^{2+}/Ni couple, this equates to

$$\mathscr{E} = -0.250 + (9.925 \times 10^{-5})T \cdot \log[Ni^{2+}] \tag{9.34}$$

The potential changes very slowly with changing temperature. Of much greater importance is the kinetic effect of a temperature increase. Recall that one way to enhance reaction rates is to increase T. The magnitude of this effect depends on the activation energy of the reaction. Typical E_a's for aqueous redox reactions are on the order of 40–50 kJ/mol; this corresponds to a doubling of the reaction rate for each 10–15°C increase in temperature.

pH Effects

If H^+ or OH^- appears in a redox equation, then the reduction potential can be influenced by pH. Consider the NiO_2/Ni^{2+} couple, for which the half-reaction is

$$NiO_2(s) + 4H^+ + 2e^- \longrightarrow Ni^{2+} + 2H_2O \quad \mathscr{E}^0 = +1.678 \text{ V} \tag{9.35}$$

The Nernst equation gives

$$\mathscr{E} = \mathscr{E}^0 - \frac{0.05916}{2} \log\left(\frac{[Ni^{2+}]}{[H^+]^4}\right)$$

Therefore, if the nickel ion concentration is 1.00 M,

$$\mathscr{E} = +1.678 - 0.1183 \cdot pH \qquad (9.36)$$

The resulting dependence of the potential on pH is shown in Figure 9.7.

The pH effect is a powerful one. Because the coefficients for H^+ and OH^- in redox equations are often large, conditions of acidity or basicity typically play a major role in electron transfer processes. Nickel(IV) oxide is reduced by water in acidic media, but becomes thermodynamically stable at a pH of about 7.6.[8] This is another demonstration of Le Châtelier's

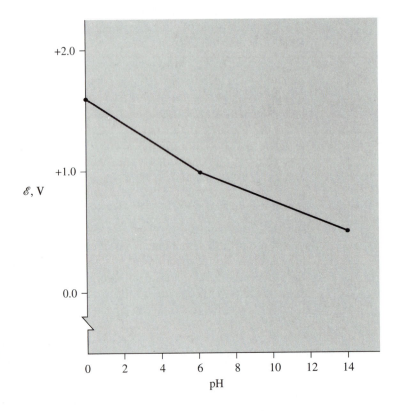

Figure 9.7

The dependence of the Ni^{IV}/Ni^{II} couple on pH. The slope changes near pH = 6.1 because of a solubility effect (see text).

$\mathscr{E}$, V

pH

---

8. The species NiO_2 is not well-characterized. Under the conditions described, it appears that several oxides in intermediate oxidation states, including Ni_2O_3 and Ni_3O_4, are also present.

principle—increasing the concentration of a reactant (H^+) favors the forward reaction.

As a further illustration, consider the interaction of nickel with aqueous acid. Under standard conditions, the metal reduces hydrogen ions to H_2:

$$Ni + 2H^+ \longrightarrow Ni^{2+} + H_2 \qquad \mathscr{E}^0 = +0.250 \text{ V} \qquad \textbf{(9.37)}$$

However, you should be able to show that if the pH is increased to 5.00, $\mathscr{E} = -0.046$ V and the reaction is nonspontaneous.

Solubility Effects

Because the half-reaction for the Ni^{2+}/Ni couple does not contain either H^+ or OH^-, we might expect its potential to be independent of pH. That is the case for acidic solutions; however, the solubility product of $Ni(OH)_2$ plays a role at higher $[OH^-]$:

$$Ni(OH)_2 \rightleftharpoons Ni^{2+} + 2OH^- \qquad K_{sp} = 1.5 \times 10^{-16} \qquad \textbf{(9.38)}$$

Equation (9.38) can be combined with the Ni^{2+}/Ni half-reaction to give the appropriate half-reaction for basic solution:

$$
\begin{array}{r}
Ni^{2+} + 2e^- \longrightarrow Ni \\
\underline{Ni(OH)_2 \longrightarrow Ni^{2+} + 2OH^-} \\
Ni(OH)_2 + 2e^- \longrightarrow Ni + 2OH^- \qquad \textbf{(9.39)}
\end{array}
$$

What is $\mathscr{E}^0$ for equation (9.39)? The change in volt-equivalents for the Ni^{2+}/Ni couple is $2(-0.250) = -0.500$ eV. The volt-equivalency for K_{sp} can be calculated using the equation

$$\log K = \frac{n\mathscr{E}^0}{0.05916} \qquad \textbf{(9.40)}$$

Then $n\mathscr{E}^0 = 0.05916(\log 1.5 \times 10^{-16}) = -0.936$ eV. Therefore, $\mathscr{E}^0$ for the $Ni(OH)_2/Ni$ couple is

$$\frac{-0.500 + (-0.936)}{2} = -0.72 \text{ V}$$

It follows that a plot of $\mathscr{E}$ versus pH for the reduction of Ni^{II} to Ni shows a change in slope at the point at which $Ni(OH)_2$ begins to precipitate (for 1.00 M Ni^{2+}, near a pH of 6.1; see Figure 9.8). This might be viewed as a concentration effect, of course, since the precipitation of $Ni(OH)_2$ reduces the $[Ni^{2+}]$ concentration of the solution.

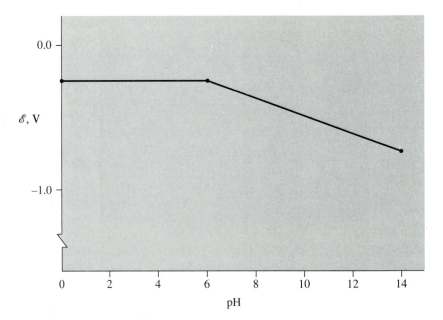

Figure 9.8
The dependence of the Ni^{II}/Ni^0 couple on pH. The slope changes near pH = 6.1 because of a solubility effect (see text). Potentials correspond to $[Ni^{2+}] = 1.00$ M except as limited by solubility.

Predominance Area Diagrams

The behavior of aqueous nickel-containing solutions as a function of pH is conveniently depicted by a *predominance area diagram*—a plot of $\mathscr{E}_{red}$ versus pH, blocked into domains of thermodynamic stability. In such diagrams it is customary for the pH to range from 0 to 14, although more extreme pH's are, of course, possible. Because species having very positive oxidation or reduction potentials are unstable toward reaction with water, the lines representing the O_2/H_2O and H^+/H_2 couples are boundaries of stability.

The predominance area diagram for nickel is shown in Figure 9.9. You should convince yourself that it is consistent with the conclusions drawn earlier. Predominance area diagrams for all of the common elements are available in the literature.[9]

Laboratory Oxidizing Agents and Kinetic Stability

Recall that water is kinetically stable over intervals of about 0.5 V beyond both its oxidation and reduction potentials. Solutes are stabilized toward oxidation or reduction by H_2O for that reason. This is especially important for two of the most common oxidizing agents used in the laboratory—the permanganate and dichromate ions. It should be impossible to prepare

---

9. See, for example, Campbell, J. A.; Whiteker, R. A. *J. Chem. Educ.* **1969**, *46*, 90; Delahay, P.; Pourbaix, M.; van Rysselberghe, P. *J. Chem. Educ.* **1950**, *27*, 683.

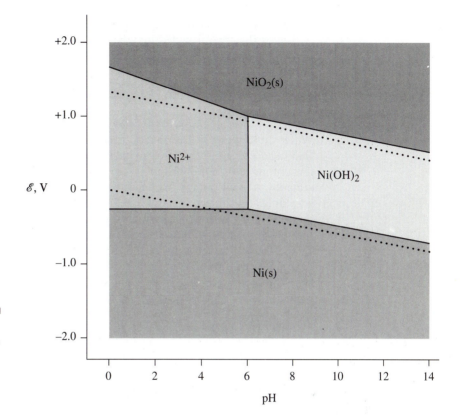

Figure 9.9
Predominance area
diagram for nickel
in aqueous solution.
Dotted lines
represent the
O_2/H_2O and H^+/H_2
couples.

aqueous solutions of these ions under standard conditions:

$$4MnO_4^- + 12H^+ \longrightarrow 4Mn^{2+} + 5O_2 + 6H_2O \qquad \mathscr{E}^0 = +0.28 \text{ V}$$
$$\text{(9.41)}$$

$$2Cr_2O_7^{2-} + 16H^+ \longrightarrow 4Cr^{3+} + 3O_2 + 8H_2O \qquad \mathscr{E}^0 = +0.10 \text{ V}$$
$$\text{(9.42)}$$

However, their kinetic stability permits their use as aqueous oxidizing agents. (Note that $\mathscr{E}^0$ is less than $+0.5$ V in each case.) Thus, acidic solutions of these oxyanions can be prepared and, if necessary, stored for brief periods of time.

9.7   Some Mechanistic Aspects of Aqueous Redox Reactions

How do electron transfer reactions occur in solution? The mechanistic possibilities are more complex than in the gas phase because of the greater numbers and types of species present. This has long been an exciting area of research. Some of the major subtopics are outlined in this section. The references that are provided may be consulted for more detail. In addition, the mechanisms of oxidation–reduction reactions at transition metal centers are discussed in Chapter 17.

Hydrated Electrons[10]

Water molecules can be ionized when bombarded with high-energy radiation. The electrons that are so generated then cause further (secondary) ionizations until equilibrium is reached, normally in about 10 ps (picoseconds). At that point, "free" electrons—actually, electrons associated with solvent water molecules—exist in solution. This state can be pictured as approximating ion–dipole attractions in which the electron lies in a *cavity* formed by surrounding waters. Alternatively, such an electron might be considered to reside in an antibonding molecular orbital of H_2O or some oligomer of H_2O.[11]

The lifetime of the equilibrium state is typically less than 1 ms (millisecond); decay occurs through processes such as

$$H^+ + e^-(aq) \longrightarrow H \tag{9.43}$$

$$H_2O + e^-(aq) \longrightarrow H + OH^- \tag{9.44}$$

Although brief, the lifetime of the hydrated electron is sufficient to permit spectroscopic and other kinds of experiments to be performed. It absorbs visible and infrared energy, with a maximum absorbance near 720 nm. As expected, solutions containing hydrated electrons are powerful reducing agents; the standard reduction potential for equation (9.43) is +2.7 V. The most characteristic reactions are one-electron reductions and bond rupture to hydrogen.

$$X^q + e^-(aq) \longrightarrow X^{q-1}$$
$$X^q = Ag^+, CO_2, MnO_4^-, OH, Fe^{3+}, Cu^{2+}, \text{ etc.} \tag{9.45}$$

$$HX + e^-(aq) \longrightarrow H + X^- \qquad X = Cl, ClO, \text{ etc.} \tag{9.46}$$

It is possible that lithium and the other strongly reducing metals (those having oxidation potentials greater than +2.7 V) may react with

---

10. Salem, L. *Electrons in Chemical Reactions*; Wiley: New York, 1982; pp. 222–226; Hart, E. J.; Anbar, M. *The Hydrated Electron*; Wiley: New York, 1970; Hart, E. J. *Acc. Chem. Res.* **1969**, 2, 161.

11. See Dye, J. L. *Adv. Inorg. Chem. Radiochem.* **1982**, 25, 135; and/or Symons, M. R. C. *Chem. Soc. Rev.* **1976**, 5, 337; and references cited therein.

water by ejecting electrons into the solvent. This also appears to be the case in liquid ammonia. The lifetimes of solvated electrons are considerably longer in that solvent than in water, so a great deal more is known about the chemistry of e^-(am) than e^-(aq) (see Chapter 11).

Inner-Sphere versus Outer-Sphere Mechanisms[12]

Oxidation–reduction reactions in solution are conveniently categorized as occurring through either inner- or outer-sphere processes. In an *outer-sphere* reaction electrons are transferred directly from the reducing agent to the oxidizing agent. This is exemplified by equation (9.47), where Q^+ represents the oxidant, R is the reductant, and the charges are relative rather than absolute:

$$Q^+ + R \longrightarrow [Q\text{---}R]^+ \longrightarrow Q + R^+ \tag{9.47}$$

Inner-sphere reactions require that some substituent ligand, bonded to either the oxidant or reductant, act as an intermediary by forming a bridge between the reacting centers. Transferred electrons then "hop across the bridge":

$$Q\text{--}L^+ + R \longrightarrow [Q\text{---}L\text{---}R]^+ \left\langle\begin{array}{l} \longrightarrow Q + L\text{--}R^+ \\ \longrightarrow Q\text{--}L + R^+ \end{array}\right. \tag{9.48}$$

As is shown in equation (9.48), the bridging ligand (L) may or may not be transferred in the process.

The factors that determine whether a given reaction follows an inner- or outer-sphere mechanism are complex, particularly when transition metal ions are involved. A more detailed account of such mechanisms is deferred until Chapter 17.

Single versus Multiple Electron Transfer[13]

Reactions in which the oxidant and reductant react with 1:1 stoichiometry are *complementary reactions*; those having other stoichiometries are *noncomplementary*. This distinction is significant primarily because noncomplementary processes have inherently more complex mechanisms. Thus, for the 1:1 process

$$Fe^{3+} + Cr^{2+} \longrightarrow Fe^{2+} + Cr^{3+} \tag{9.49}$$

12. Haim, A. *Prog. Inorg. Chem.* **1983**, *30*, 273; Sykes, A. G. *Adv. Inorg. Chem. Radiochem.* **1967**, *10*, 153; Taube, H. *Adv. Inorg. Chem. Radiochem.* **1959**, *1*, 1; *Chem. Rev.* **1952**, *50*, 69.

13. Cannon, R. D. *Electron Transfer Reactions*; Butterworths: London, 1980, Chapter 3; Sykes, A. G. *Adv. Inorg. Chem. Radiochem.* **1967**, *10*, 153.

the mechanism, as well as the stoichiometry, is expressed fully by equation (9.49). However, a reaction such as

$$Tl^{3+} + 2Fe^{2+} \longrightarrow Tl^{+} + 2Fe^{3+} \tag{9.50}$$

might proceed by either of two pathways. The first involves two one-electron transfers, and requires the relatively unstable Tl^{2+} intermediate:

(1) $Tl^{3+} + Fe^{2+} \longrightarrow Tl^{2+} + Fe^{3+}$

(2) $Tl^{2+} + Fe^{2+} \longrightarrow Tl^{+} + Fe^{3+}$

The second possibility involves initial two-electron transfer, forming another intermediate (Fe^{4+}) of low stability:

(1) $Tl^{3+} + Fe^{2+} \longrightarrow Tl^{+} + Fe^{4+}$

(2) $Fe^{4+} + Fe^{2+} \longrightarrow 2Fe^{3+}$

Experimental studies favor the first alternative.[14]

Another interesting possibility is complementary two-electron transfer. An example is

$$Tl^{3+} + *Tl^{+} \longrightarrow Tl^{+} + *Tl^{3+} \tag{9.51}$$

where the asterisk marks an isotopically distinct ion. The overall process can result either from a single two-electron transfer or from two sequential one-electron transfers. (The latter amounts to the formation of Tl^{2+} as an intermediate, which then disproportionates.) Data from numerous studies are suggestive of two-electron transfer.

Many other reports of simultaneous multiple electron transfer can be found in the literature, including a postulated four-electron transfer between two chromium centers ($Cr^{+} \leftrightarrow Cr^{5+}$).[15]

Electron Transfer in Biological Systems[16]

An especially active area of research concerns electron transfer processes of importance to life. For example, the cytochromes are known to catalyze the transport of electrons over unusually large distances. It has long been believed that these iron-containing enzymes catalyze reversible electron

---

14.   See Falcinella, B.; Falgate, P. D.; Laurence, G. S. *J. Chem. Soc. Dalton* **1975**, 1; and references cited.

15.   Rajasekar, N.; Subramaniam, R.; Gould, E. S. *Inorg. Chem.* **1982**, *21*, 4110.

16.   Isied, S. S. *Prog. Inorg. Chem.* **1984**, *32*, 443; Bennett, L. F. *Prog. Inorg. Chem.* **1973**, *18*, 1.

transfer reactions; however, a study of chemically modified cytochrome c suggests irreversible transfer.[17] Other recent advances have been reported on oxidation–reduction chemistry related to the hemoglobins.[18]

Bibliography

Katokis, D.; Gordon, G. *Mechanisms of Inorganic Reactions*; Wiley: New York, 1987.

Atwood, J. D. *Inorganic and Organometallic Reaction Mechanisms*; Brooks/Cole: Pacific Grove, CA, 1985.

Bard, A. J.; Parsons, R.; Jordan, J., Eds. *Standard Potentials in Aqueous Solution*; Marcel Dekker: New York, 1985.

Johnson, D. A. *Some Thermodynamic Aspects of Inorganic Chemistry*; Cambridge University: London, 1982.

Salem, L. *Electrons in Chemical Reactions*; Wiley: New York, 1982.

Meites, L. *An Introduction to Chemical Equilibrium and Kinetics*; Pergamon: Oxford, 1981.

Cannon, R. D. *Electron Transfer Reactions*; Butterworths: London, 1980.

Milazzo, G.; Caroli, S. *Tables of Standard Electrode Potentials*; Wiley: New York, 1978.

Hart, E. J.; Anbar, M. *The Hydrated Electron*; Wiley: New York, 1970.

Questions and Problems

1. Use Lewis structures to assign an oxidation number to each atom of:
   (a) Ketene, $H_2C{=}C{=}O$ (b) Thiosulfate ion, $S_2O_3^{2-}$
   (c) N_3^- (d) Pyruvate ion, $H–C(O)–C(O)–O^-$

2. Give the maximum oxidation state for:
   (a) Sn   (b) Sb   (c) In   (d) Tc   (e) Y
   (f) Mo   (g) Pr   (h) U

3. Suggest secondary oxidation states for the following elements:
   (a) Sn   (b) Sb   (c) In   (d) Nb   (e) Te

4. Use equations (9.8), (9.9), and (9.12) to verify that $K_w = 1.0 \times 10^{-14}$ at 25°C.

5. Use data for the NO_2^-/N_2O_4 (base) and N_2O_4/HNO_2 (acid) couples to determine K_a for HNO_2.

17. Bechtold, R.; Kuehn, C.; Lepre, C.; Isied, S. S. *Nature* **1986**, *322*, 286.

18. See Chapter 22. Also, see Joran, A. D.; Leland, B. A.; Felker, P. M.; Zewail, A. H.; Hopfield, J. J.; Dervan, P. B. *Nature* **1987**, *327*, 508; and references therein.

6. Write a balanced half-reaction for each potential in the Latimer diagram (Table 9.1) of:
   (a) Ru in acid   (b) P in acid   (c) I in base

7. Construct a volt-equivalent versus oxidation state diagram for:
   (a) Ru in acid   (b) P in base

8. (a) Use values from Table 9.1 to determine $\mathscr{E}^0$ and ΔG^0 for the aqueous reaction

   $$Li + Na^+ \longrightarrow Li^+ + Na$$

   (b) The atomization energies of Li and Na are 159 and 107 kJ/mol, respectively. Analyze this system from the standpoint of ionization energy (Table 1.7), atomization energy, and hydration energy (Table 8.2). What is the predicted enthalpy change? How does it compare to your answer to part (a)?

9. Explain in your own words why the order of reduction potentials for the reaction

   $$M^{n+} + ne^- \longrightarrow M \quad \text{is} \quad Li^+ < Na^+ < Mg^{2+} < Al^{3+}$$

10. Predict the order of ease of conversion of the Group 16 elements to H_2X according to the half-reaction

    $$X + 2H^+ + 2e^- \longrightarrow H_2X$$

11. Gold is a "noble metal" in part because of its resistance to acids (ie, because $\mathscr{E}^0$ for the oxidation of Au by H^+ is less than 0). Why is this the case? [*Hint*: Examine Figure 9.3.]

12. Determine the reduction potential for the ClO_4^-/ClO_3^- couple at $pH = 7.0$.

13. Write the oxidation and reduction half-reactions for the disproportionation of H_2O_2 in acidic solution, and demonstrate that this reaction is independent of $[H^+]$ for $pH = 1$ to 7.

14. Use the reduction potential for the H_4XeO_6/XeO_3 couple ($+2.3$ V) and that of the XeO_3/Xe couple ($+1.8$ V) to estimate $\mathscr{E}^0$ for the half-reaction

    $$8H^+ + H_4XeO_6 + 8e^- \longrightarrow Xe + 6H_2O$$

15. Carefully examine the potential diagram for nitrogen in acidic solution (Table 9.1) and identify all species prone to disproportionation. Write the balanced equation for each case.

16. Predict the products for each of the following. (Water is the solvent in all cases.)
    (a) The addition of Ni(s) to excess acidic Fe^{3+}
    (b) The addition of In(s) to excess acidic Fe^{3+}
    (c) The addition of $NaRuO_4(s)$ to excess acidic Br^-
    (d) The addition of MnO_4^- to excess acidic I^-
    (e) The addition of excess MnO_4^- to acidic I^-

17. The perbromate ion, BrO_4^-, is unstable in 1.00 M acid. Write the balanced equation for its reaction with H^+/H_2O.

18. Over what pH range is a 1.00 M solution of Tl^+ stable toward reduction by H_2 (1.00 atm)?

19. The hypothetical element Z has the following potential diagram in acid solution:

$$ZO_3 \xrightarrow{+0.50} ZO_2 \xrightarrow{+0.40} Z^{3+} \xrightarrow{-0.20} Z^{2+} \xrightarrow{0.00} Z$$

(a) Identify any species prone to disproportionation, and write the appropriate equation.

(b) Give the Z-containing product and the overall potential for the reaction of ZO_2 with excess tin.

(c) Calculate the pH below which the reaction $2Z^{2+} + 2H^+ \rightarrow 2Z^{3+} + H_2$ is spontaneous. (Assume standard conditions except for $[H^+]$.)

20. Use data from Table 9.1 to calculate the K_{sp} of palladium hydroxide, $Pd(OH)_2$.

21. Construct a predominance area diagram for:
    (a) Sn   (b) Pb   (c) Co

22. Sulfuric and nitric acids are commonly referred to as "oxidizing acids," because their anions (SO_4^{2-} and NO_3^-) have more positive reduction potentials than H^+. Using this criterion, which of the following are oxidizing acids?

    $HCl, \quad H_2Te, \quad H_3AsO_4, \quad H_3AsO_3, \quad HIO_3$

23. It has been suggested that the "ammoniated electron," $e(NH_3)_x^-$, is more stable than the hydrated electron, because the hydronium ion, H_3O^+, is relatively more stable in water than is NH_4^+ in ammonia. Explain.

24. Suggest two possible mechanisms for the noncomplementary reaction $V^{4+} + 2Cr^{2+} \rightarrow V^{2+} + 2Cr^{3+}$. Which of the two mechanisms do you favor? Defend your answer.

*25. For many years the compound FeI_3 was considered to be unstable. This notion was changed by K. B. Yoon and J. K. Kochi, who achieved its synthesis (*Inorg. Chem.* **1990**, *29*, 869). Answer these questions after reading their report.

(a) What redox reaction is responsible for the "instability"? What is $\mathscr{E}^0$ for that reaction?

(b) How was the synthesis achieved? What reagents and solvent were used?

(c) The extreme reactivity of FeI_3 made elemental analysis impossible, so the authors demonstrated its presence by preparing two derivatives. What were they? Write the appropriate equations and discuss.

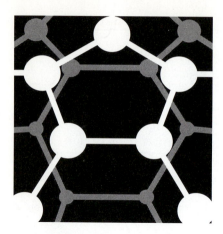

10

Acid–Base Chemistry

Reactions in which cations, anions, or electron pairs are transferred from one species to another are all considered (at least in this book) to be acid–base reactions. The logic behind this definition will be developed shortly, but it should be stated at the very beginning of the chapter that there is no consensus as to the proper definitions for the terms *acid* and *base*; nor is any single definition best for all situations.

First, we give a brief survey of some of the significant historical approaches toward understanding acids and bases. Later, two especially useful models—the Brønsted–Lowry and Lewis theories—will be examined in detail.

10.1   Definitions of Acids and Bases

Among the earliest definitions of the terms *acid* and *base* were those suggested by Arrhenius in 1887, as a consequence of his general notions of ionization. Arrhenius defined an acid as a hydrogen-containing species that dissociates in water to give H^+, and a base as a hydroxide-containing species that dissociates in water to give OH^-. While an important advance, these definitions limit acid–base chemistry to aqueous solutions; reactions in the gas and solid phases and in solvents other than water are excluded. Moreover, species such as boron trichloride do not contain H^+, but clearly behave as acids (eg, BCl_3 interacts with aqueous solutions to lower the pH). Also, NH_3 does not contain hydroxide ion, and yet is surely a base.

Such problems led to a modification of the Arrhenius theory. In the *solvent system* definitions, acids are defined as solutes that increase the concentration of the solvent cation (eg, H_3O^+ in water and NH_4^+ in ammonia), while bases increase the concentration of the solvent anion (OH^-, NH_2^-) upon dissolution.[1] *Neutralization*, the reaction of an acid with a base to produce a salt plus one or more solvent molecules, is emphasized in this approach:

$$\text{Acid} + \text{Base} \longrightarrow \text{Solvent} + \text{Salt} \qquad \textbf{(10.1)}$$

Thus, nitric acid is neutralized by potassium hydroxide in water,

$$HNO_3 + KOH \longrightarrow H_2O + KNO_3 \qquad \textbf{(10.2)}$$

while the parallel reaction in ammonia is

$$NH_4NO_3 + KNH_2 \longrightarrow 2NH_3 + KNO_3 \qquad \textbf{(10.3)}$$

Another generalization of the Arrhenius concept was made independently by Brønsted and Lowry,[2] who defined acids as hydrogen ion donors and bases as hydrogen ion acceptors. In this system, every acid and base has a *conjugate* to which it is related by the equation

$$\text{Acid} \rightleftharpoons \text{Base} + H^+ \qquad \textbf{(10.4)}$$

Brønsted–Lowry theory treats acidity and basicity as inverse properties—the stronger the acid, the weaker its conjugate base, and vice versa. Water can act as either an acid or a base (ie, can either donate or accept hydrogen ions), and so is *amphoteric*. There is a natural tendency for acidity and basicity to be reduced via chemical reactions; that is, a *leveling effect* is operative, and the strongest acid that can exist in water is its conjugate acid H_3O^+:

$$HA + H_2O \rightleftharpoons H_3O^+ + A^- \qquad \textbf{(10.5)}$$

Similarly, water levels all bases stronger than the hydroxide ion to OH^-.

1. Cady, H. P.; Elsey, H. M. *Science* **1922**, *56*, 27; Franklin, E. C. *The Nitrogen System of Compounds*; Reinhold: New York, 1935.

2. Brønsted, J. N. *Rec. Trav. Chim.* **1923**, *42*, 718; Lowry, T. M. *Chem. Ind.* **1923**, *42*, 43.

Theories Based on Anion Transfer

Consider the reaction of magnesium hydroxide with sulfuric acid in aqueous solution. This reaction takes the form of equation (10.1), and is unquestionably an acid–base process according to any of the *protonic* (hydrogen ion) theories.

$$Mg(OH)_2 + H_2SO_4 \longrightarrow 2H_2O + MgSO_4 \qquad (10.6)$$

The reaction of magnesium oxide [the *anhydride* of $Mg(OH)_2$; see Section 10.4] with sulfuric acid logically belongs to the same reaction category.

$$MgO + H_2SO_4 \longrightarrow H_2O + MgSO_4 \qquad (10.7)$$

Next, consider the interaction of MgO with SO_3, the anhydride of H_2SO_4:

$$MgO + SO_3 \longrightarrow MgSO_4 \qquad (10.8)$$

Like its two precursors, equation (10.8) surely describes an acid–base process; however, no hydrogen ions are present. Reactions of this type are common in molten salt solutions, and the failure of protonic theories to include them led to new definitions. In the *Lux–Flood* system, acids are defined as oxide ion acceptors, and bases as oxide ion donors.[3] The reaction described by equation (10.8) is viewed as the transfer of O^{2-} from the base MgO to the acid SO_3, producing a salt ($Mg^{2+}SO_4^{2-}$). The notion of conjugates is again important, but in this system the relationship is

$$Acid + O^{2-} \rightleftharpoons Base \qquad (10.9)$$

However, acids commonly bind to anions other than oxide. For example, CO_2 adds hydroxide ion, while in other systems the transfer of halide ions is observed:

$$CO_2 + OH^- \longrightarrow HCO_3^- \qquad (10.10)$$

$$CsCl + FeCl_3 \longrightarrow Cs^+ + FeCl_4^- \qquad (10.11)$$

In the *anionotropic* definitions, an acid is defined as an anion acceptor, and a base as an anion donor;[4] hence, the Lux–Flood acids represent a subset

---

3.   Lux, H. *Z. Elektrochem.* **1939**, *45*, 303; Flood, H.; Forland, T.; Roald, B. *Acta Chem. Scand.* **1947**, *1*, 790; and references cited therein.

4.   Gutmann, V.; Lindquist, I. *Z. Phys. Chem.* **1954**, *203*, 250.

of the anionotropic acids. Similarly, in the *cationotropic* system, an acid is a cation donor, and a base is a cation acceptor. This incorporates the Brønsted–Lowry theory.

The Lewis Theory—Electron Pair Donors and Acceptors

Another early approach, quite different from those described above, was proposed by Lewis in 1923.[5] He defined acids as electron pair acceptors and bases as electron pair donors. An acid reacts with a base with the formation of a *coordinate covalent* (or *dative*) bond; the reaction product is an *adduct*:

$$A + :B \longrightarrow A:B \qquad (10.12)$$

The Lewis definitions can be shown to incorporate all the previously discussed theories. For example, H^+ is a Lewis acid, since its vacant $1s$ orbital is capable of accommodating an added pair of electrons. Hydroxide, oxide, and other anions all have one or more lone pairs, and so can function as Lewis bases. The Lewis system also embraces certain reactions that do not qualify as acid–base processes according to other definitions. Some examples follow:

$$BCl_3 + NMe_3 \longrightarrow Me_3N \cdot BCl_3 \qquad (10.13)$$

$$AlH_3 + H^- \longrightarrow AlH_4^- \qquad (10.14)$$

$$Ag^+ + 2CN^- \longrightarrow [Ag(CN)_2]^- \qquad (10.15)$$

The Usanovich Definition

The most inclusive of all the acid–base theories is that of Usanovich, who suggested that any salt-forming reaction be considered an acid–base reaction.[6] More explicitly, an acid is any substance that forms salts by reacting with bases, by donating a cation, or by adding anions or electrons; a base forms salts by reacting with acids, accepting a cation, or donating an anion or electrons. Not only are the Arrhenius, Brønsted–Lowry, Lux–Flood, and Lewis theories incorporated into these definitions, but redox reactions are included as well—an oxidizing agent is considered to be an acid, and a reducing agent a base.

5.   Lewis, G. N. *Valence and the Structure of Atoms and Molecules*; Chemical Catalog: New York, 1923.

6.   Usanovich, M. *Zhur. Obschei. Khim.* **1939**, *9*, 182.

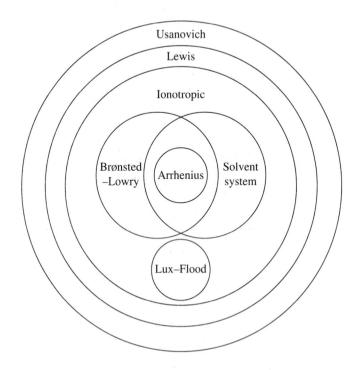

Figure 10.1
Interrelationships among the commonly used acid–base theories. [Adapted with permission from Jensen, W. B. *The Lewis Acid–Base Concepts*; Wiley: New York, 1980; p. 65.]

A diagram illustrating the interrelationships among the various ionic and electronic acid–base definitions is given in Figure 10.1.

10.2 The Relationship Between Acid–Base and Oxidation–Reduction Reactions

It has been argued that the Usanovich system is so broad as to lose usefulness, since it places the vast majority of all inorganic reactions into the acid–base category. The inclusion of electron transfer reactions within the acid–base domain is intriguing, however, and seems to have at least some merit. Consider that strong reducing agents such as lithium metal react with water to increase the pH, while strong oxidizing agents such as Cl_2 react with water to decrease the pH:

$$2Li + 2H_2O \longrightarrow H_2 + 2Li^+ + 2OH^- \tag{10.16}$$

$$2Cl_2 + 2H_2O \longrightarrow O_2 + 4H^+ + 4Cl^- \tag{10.17}$$

Furthermore, equation (10.18) represents an oxidation–reduction process by the criterion given in Chapter 9, since oxidation numbers change in going from reactants to products:

$$Ca + S \longrightarrow CaS \qquad (10.18)$$

This might also be viewed as a Lewis acid–base reaction, however, since an electron pair is transferred from the base Ca: to the acid :S̈:, and an adduct ($Ca^{2+}S^{2-}$) is formed.

Now consider the very similar process

$$2K + S \longrightarrow K_2S \qquad (10.19)$$

Again, this is unquestionably a redox reaction. But Lewis theory does not allow it as an acid–base reaction, because it requires two single electron transfers rather than the transfer of one electron pair. Common sense suggests, however, that the processes represented by equations (10.18) and (10.19) are so similar (in terms of their rates, positions of the equilibria, exothermicity, etc.) that it is unreasonable to call one an acid–base reaction and not the other.

Another aspect of the relationship between acid–base and redox reactions was discussed by Finston and Rychtman, who determined the pH's of mixed $HCl/K_2Cr_2O_7$ solutions for various acid/oxidant ratios.[7] They found that the addition of dichromate ion to HCl solutions increases their acidity. This was explained through the Grotthus–Huckel mechanism for proton transfer (see Section 8.6); $Cr_2O_7^{2-}$ coordinates to aquated hydrogen ions, forming large species of low mobility:

$$Cr_2O_7^{2-} + nH_3O^+ \longrightarrow [Cr_2O_7 \cdot nH_3O]^{n-2} \qquad (10.20)$$

The proposed structure of such an aggregate is diagrammed in Figure 10.2 (p. 306).

The interaction of hydrogen atoms from solvent H_2O with the oxygens of $Cr_2O_7^{2-}$ creates an inductive effect. This loosens the bonds to hydrogen, rendering them more available for rapid proton transfer and thereby increasing the apparent acidity of the solution. This is suggestive of a general correlation between oxidizing agents and acids. For example, the addition of species such as H_2SO_4, HNO_3, P_2O_5, or CrO_4^{2-} to water leads to both a decrease in pH and an increase in the oxidizing ability of the system.

---

7. Finston, H. L.; Rychtman, A. C. *A New View of Current Acid–Base Theories*; Wiley: New York, 1982; Chapter 6.

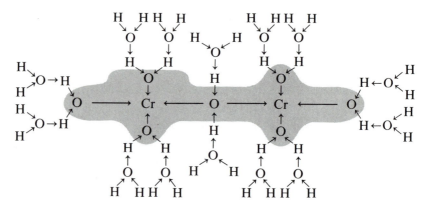

Figure 10.2 The coordination of solvent water molecules to dichromate ions. [Reproduced with permission from Finston, H. L.; Rychtman, A. C. *A New View of Current Acid–Base Theories*; Wiley: New York, 1982; p. 194.]

10.3   Hydrogen Ion Donors and Acceptors

What determines whether a chemical species behaves as a Brønsted–Lowry acid or base?[8] At least two factors must be considered to answer this question: First, something must be known about the inherent properties of the species itself; and second, the chemical environment must be specified. Experimental information relevant to the first factor is best obtained in the gas phase, while the latter normally relates to solution chemistry.

Proton Affinities

The best way to quantify inherent Brønsted–Lowry basicity is through gas-phase proton affinity. Analogous to electron affinity, the proton affinity (PA) of species M is defined as the energy released by the process

$$M(g) + H^+(g) \longrightarrow MH^+(g) \qquad \Delta H = -PA \qquad (10.21)$$

Because of the nature of this definition, a positive proton affinity corresponds to the exothermic addition of H^+.

A Born–Haber treatment suggests that the proton affinities of neutral molecules depend on three factors: the first ionization energy of M, the

---

8. Or, for that matter, fits either of two other classes: *aprotic* (neither acidic nor basic) or *amphoteric* (capable of both acidic and basic behavior)?

ionization energy of hydrogen, and the M^+–H bond energy:

$$M(g) \longrightarrow M^+(g) + 1e^- \qquad IE_1 (M)$$
$$H^+(g) + 1e^- \longrightarrow H(g) \qquad -IE (H)$$
$$\underline{M^+(g) + H(g) \longrightarrow MH^+(g) \qquad -E (M^+\text{–}H)}$$
$$M(g) + H^+(g) \longrightarrow MH^+(g) \qquad -PA$$

Therefore, for the neutral atom or molecule M,

$$PA (M) = IE (H) - IE_1 (M) + E (M^+\text{–}H) \qquad (10.22)$$

Equation (10.22) is useful for the trends that can be predicted or rationalized from it. Basicity is expected to be greatest for species that form strong bonds to hydrogen and have low ionization energies. Comparing binary hydrides of the nonmetals and metalloids, the ionization energies span a greater range of values than do M–H bond energies. (Compare Table 1.7 to Table 4.6.) As a result, the ionization energy term generally determines the relative order of proton affinities. For example, the order of ionization energy is

$$C > N < O < F$$

while the order of proton affinity is

$$CH_4 < NH_3 > H_2O > HF$$

Also, ionization energy decreases going down Group 17,

$$F > Cl > Br > I$$

while the proton affinities of the hydrogen halides increase in the order

$$HF < HCl < HBr < HI$$

Proton affinities have been measured by various instrumental techniques, but most often by ion cyclotron resonance and high-pressure mass spectroscopies. Unfortunately, different methods yield slightly different results. For example, reported proton affinities for ammonia range from at least 845 to 858 kJ/mol.[9] This should be remembered when using Table 10.1, which lists values for a variety of species.

9. Aue, D. H.; Bowers, M. T. In *Gas-Phase Ion Chemistry*; Bowers, M. T., Ed.; Academic: New York, 1979; Volume 2; Kebarle, P. *Ann. Rev. Phys. Chem.* **1977**, *28*, 445.

Table 10.1 Gas-phase proton affinities of selected atoms, molecules, and ions

(a) Binary Hydrides			
H_2 423			
CH_4 527	NH_3 845	H_2O 711	HF 481
SiH_4 640	PH_3 774	H_2S 728	HCl 565
	AsH_3 711	H_2Se 743	HBr 590
			HI 611

(b) Anions			
H^- 1674			
CH_3^- 1743	NH_2^- 1689	OH^- 1632	F^- 1554
SiH_3^- 1554	PH_2^- 1548	HS^- 1474	Cl^- 1395
GeH_3^- 1509	AsH_2^- 1500	HSe^- 1420	Br^- 1355
			I^- 1315

(c) Other Atoms and Molecules

Species	PA	Species	PA
Kr	424	Me_3N	884
Xe	478	Et_3N	895
N_2	476	CH_3OH	762
O_2	423	CH_3CH_2OH	782
CO	582	CF_3CH_2OH	720
LiOH	1007	CCl_3CH_2OH	749
NaOH	1036	CH_3CN	782
KOH	1099		

Note: Values are in kilojoules per mole, and are generally accurate to ± 5 kJ/mol.

Sources: Data taken primarily from compilations by Bartmess, J. E.; McIver, R. T. In *Gas-Phase Ion Chemistry*; Bowers, M. T., Ed.; Academic: New York, 1978; Kebarle, P. *Ann. Rev. Phys. Chem.* **1977**, *28*, 445.

The proton affinities of conjugate bases provide a useful measure of inherent acidity, since the relevant equation corresponds to the reverse of an acid dissociation:

$$M^-(g) + H^+(g) \longrightarrow MH(g) \qquad \Delta H = -PA\ (M^-) \qquad \textbf{(10.23)}$$

Such PA's can be considered to result from contributions from the electron affinity of M, the ionization energy of H, and the M–H bond energy:

$$
\begin{array}{lll}
M^-(g) & \longrightarrow M(g) + 1e^- & EA_1\ (M) \\
H^+(g) + 1e^- & \longrightarrow H(g) & -IE\ (H) \\
\underline{M(g) + H(g)} & \underline{\longrightarrow MH(g)} & \underline{-E\ (M\text{–}H)} \\
M^-(g) + H^+(g) & \longrightarrow MH(g) & -PA\ (M^-)
\end{array}
$$

Thus, $PA = IE\ (H) - EA_1(M) + E\ (M\text{–}H)$.

Comparing the hydrogen halides, the primary differentiating factor is the M–H bond energy.[10] Thus, the order of inherent acidity,

$$HF < HCl < HBr < HI$$

is explained by the increasing ease of H–X bond cleavage with increasing size. However, the observed increase in acidity going across a period results from a different factor. For the series below, where proton affinity increases from left to right,

$$F^- < OH^- < NH_2^- < CH_3^-$$

the major differentiating factor is not the M–H bond energy, but rather the electron affinity of M. That is, the order of acidity,

$$CH_4 < NH_3 < H_2O < HF$$

correlates with the electron affinities of the heteroatoms.

Charge, Inductive, and Resonance Effects

Table 10.1 provides numerous examples of the effects of charge, induction, and resonance on proton affinity. For a series in which charge varies in a regular way, such as

$$H_2O < OH^- < O^{2-}$$

---

10. Since the range of electron affinities is very small (only 54 kJ/mol) among these four halogens.

where proton affinity increases from left to right, the difference in gas-phase basicity is typically about 600–800 kJ/mol per step. Inductive effects are apparent in series such as

$$H_2O < LiOH < NaOH < KOH$$

$$CF_3CH_2OH < CCl_3CH_2OH < CH_3CH_2OH$$

where, again, proton affinity increases from left to right.

The effect of resonance is most evident when the negative charge of a conjugate base is delocalized over two or more atoms. For example, the proton affinities of ions having the general formula RCH_2^- are small when the R group can stabilize the negative charge, as is the case when R = CN or NO_2:

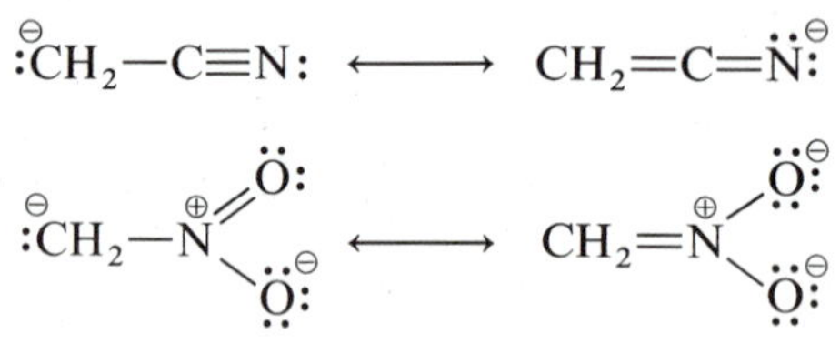

10.4 Proton Donors and Acceptors in Aqueous Solution

Experimental Determination of Acid–Base Strength in Water

Recall that solvent water levels any acid stronger than the hydronium ion to H_3O^+, as shown in equation (10.5). Similarly, any base stronger than OH^- is consumed by the solvent:

$$B + H_2O \longrightarrow BH^+ + OH^- \tag{10.24}$$

Therefore, only a limited range of acid and base strengths (ie, K_a and K_b) can be experimentally determined by direct methods. If the arbitrary limit of 99.99% dissociation for a 1.00 M solution of acid HA is chosen as the limit of accurate measurement, then the corresponding K_a is calculated to be

$$K_a = \frac{[H^+] \cdot [A^-]}{[HA]} = \frac{(0.9999)^2}{0.0001} = 1 \times 10^4$$

Similarly, if 0.01% dissociation for a 1.00 M solution is taken to be the weakest directly measurable acid, then $K_a = 1.0 \times 10^{-8}$. Thus, the most readily detected range of acid strengths is over the pK_a range of -4 to $+8$.

The situation is actually somewhat better than that, however. Indirect methods permit the measurement of pK for weaker acids and bases. One approach utilizes oxidation–reduction data. For example, the potentials of two half-reactions can be used to estimate pK_a for H_2:

$$\frac{1}{2}H_2 + 1e^- \longrightarrow H^- \qquad \mathscr{E}^0_{red} = -2.25 \text{ V}$$

$$\frac{1}{2}H_2 \longrightarrow H^+ + 1e^- \qquad \mathscr{E}^0_{ox} = 0.00 \text{ V}$$

$$H_2 \longrightarrow H^+ + H^- \qquad \mathscr{E}^0 = -2.25 \text{ V} \qquad \textbf{(10.25)}$$

Then, using equation (9.8),

$$\log K = \frac{n\mathscr{E}^0}{0.05916} = \frac{1(-2.25 \text{ V})}{0.05916} = -38.0$$

$$K_a = 1 \times 10^{-38} \qquad pK_a = 38$$

Another common method involves determination of the rate constants for forward and reverse reactions, followed by application of the relationship

$$K_{eq} = \frac{k_f}{k_r} \qquad \textbf{(10.26)}$$

For example, the forward and reverse rate constants for the deprotonation of acetone are $4.7 \times 10^{-10} \text{ s}^{-1}$ and 5×10^{10} L/mol·s, respectively, at 25°C. Therefore,

$$K_a = \frac{k_f}{k_r} = \frac{4.7 \times 10^{-10}}{5 \times 10^{10}} = 1 \times 10^{-20} \qquad pK_a = 20$$

The Strengths of Binary Acids

The pK_a's of a variety of binary hydrides are given in Table 10.2 (p. 312).

There is general (but not perfect) agreement between the trends for aqueous dissociation constants and those for proton affinities. Differences between the two scales can be attributed to hydration and/or entropy effects. The hydration factor can be understood through a Born–Haber cycle:

$$HA(g) \longrightarrow H^+(g) + A^-(g) \qquad PA \text{ } (A^-)$$

$$H^+(g) \longrightarrow H^+(aq) \qquad \Delta H_{hydr} \text{ } (H^+)$$

$$A^-(g) \longrightarrow A^-(aq) \qquad \Delta H_{hydr} \text{ } (A^-)$$

$$HA(aq) \longrightarrow HA(g) \qquad -\Delta H_{hydr} \text{ } (HA)$$

$$HA(aq) \longrightarrow H^+(aq) + A^-(g) \qquad \Delta H \text{ } (K_a)$$

Table 10.2 The aqueous pK_a's of binary hydrides

CH$_4$	NH$_3$	H$_2$O	HF
49	39	15.74	3.18
C$_2$H$_4$		H$_2$O$_2$	
44		11.65	
C$_2$H$_2$			
25			
SiH$_4$	PH$_3$	H$_2$S	HCl
35	27	7.05	−6.1
		14.0*	
GeH$_4$	AsH$_3$	H$_2$Se	HBr
25	23	3.8	−8
		H$_2$Te	HI
		2.6	−9

Note: Values are for 25°C. * = pK_{a_2}.
Sources: Albert, A.; Serjeant, E. P. *The Determination of Ionization Constants*, 3rd ed.; Chapman and Hall: London, 1984; pp. 162–163; Jolly, W. L. *Modern Inorganic Chemistry*, 2nd ed.; McGraw-Hill: New York, 1991; p. 207.

Therefore, the enthalpy change for aqueous dissociation is given by the equation

$$\Delta H\ (K_a) = PA\ (A^-) + \Delta H_{hydr}\ (H^+) + \Delta H_{hydr}\ (A^-) - \Delta H_{hydr}\ (HA) \quad \textbf{(10.27)}$$

and gas-phase acidity correlates with $\Delta H\ (K_a)$ only to the extent that the $\Delta H_{hydr}\ (HA)$ term balances the terms $\Delta H_{hydr}\ (H^+) + \Delta H_{hydr}\ (A^-)$.

Ternary (Oxy-) Acids

Many important compounds contain hydrogen–oxygen–heteroatom linkages and conform to the general formula H$_m$XO$_n$. Such species tend to be acidic when X is a nonmetal or a metalloid. With very few exceptions (notably the phosphorus system; see below), oxyacids contain hydrogen bound exclusively to oxygen rather than to the heteroatom. This is exemplified by the structures for the major oxyacids of nitrogen, as listed in Table 10.3.

Several factors influence the strength of an oxyacid:

1. If there is more than one ionizable hydrogen, $K_{a_1} > K_{a_2} > K_{a_3}$. (This is true for binary polyprotics as well.) As a rough rule of thumb, the successive acidity constants typically differ by about 5 powers of 10. Thus, for H$_3$PO$_4$, p$K_{a_1} = 2.1$, p$K_{a_2} = 7.2$, and p$K_{a_3} = 12.4$.

2. Inductive and resonance effects are important. Acidity is greatest when the heteroatom is highly electronegative, and is increased by the in-

Table 10.3 The primary oxyacids of nitrogren

Formula	Name	Oxidation Number(s)
(structure)	Hyponitrous	+1
(structure)	Hyponitric	+3, +1
(structure)	Hydronitrous	+2
(structure)	Nitrous	+3
(structure)	Peroxonitrous	+3
(structure)	Nitric	+5
(structure)	Peroxonitric	+5

corporation of electron-withdrawing groups (eg, –F or –CF$_3$) on the heteroatom. Thus, in the following series, K_a increases from left to right:

$$H_3PO_4 < H_2SO_4 < HClO_4$$

$$HIO_3 < HBrO_3 < HClO_3$$

$$CH_3CO_2H < FCH_2CO_2H < F_2CHCO_2H < CF_3CO_2H$$

$$H_2SO_4 < HSO_3F$$

3. Acidity increases with increasing number of oxygens. (This is also an inductive effect.) For the oxyacids of chlorine, K_a increases in the following order:

$$HClO < HClO_2 < HClO_3 < HClO_4$$

Here again, it is found that the difference between successive ionization constants is generally about 10^5.

Table 10.4 Acidity constants for some ternary (oxy-) acids

Acid		pK_{a_1}	pK_{a_2}	pK_{a_3}
Formula	Name			
Boron				
$(HO)_3B$	Boric	9.234		
Carbon				
$(HO)_2C{=}O$	Carbonic	3.60	10.32	
$HOC(H){=}O$	Formic	3.75		
$HOCH_3$	Methanol	16		
$[HOC({=}O)]_2$	Oxalic	1.271	4.27	
$HOC(O)CH_3$	Acetic	4.757		
Silicon				
$(HO)_4Si$	Silicic	9.77		
Germanium				
$(HO)_4Ge$	Germanic	8.73	11.7	
Nitrogen				
$HONO$	Nitrous	3.20		
$HONO_2$	Nitric	-1.44		
$HO{-}N{=}N{-}OH$	Hyponitrous	7.15	11.54	
Phosphorus				
$HOP(H)_2{=}O$	Hypophosphorous	1.23		
$(HO)_2P(H)O$	Phosphorous	1.43	6.67	
$(HO)_2P(H)O_2$	Phosphoric	2.148	7.198	12.375
Arsenic				
$(HO)_3As$	Arsenious	9.18		
$(HO)_3AsO$	Arsenic	2.21	6.98	11.5
Oxygen				
HOH	Water	15.7		
$HOOH$	Hydrogen peroxide	11.65		
Sulfur				
$(HO)_2SO$	Sulfurous	1.89	7.21	
$(HO)_2SO_2$	Sulfuric	-3	1.96	
$(HO)_2S(S)O$	Thiosulfuric	1.74		
Selenium				
$(HO)_2SeO$	Selenious	2.62	8.32	
$(HO)_2SeO_2$	Selenic	1.66		
Tellurium				
$(HO)_6Te$	Telluric	7.70	10.95	
Chlorine				
$HOCl$	Hypochlorous	7.54		
$HOClO$	Chlorous	1.94		

(continued)

Table 10.4 (*continued*)

Formula	Name	pK_{a_1}	pK_{a_2}	pK_{a_3}
$HOClO_2$	Chloric	-2.0		
$HOClO_3$	Perchloric	-7.3		
Bromine				
HOBr	Hypobromous	8.49		
Iodine				
$HOIO_2$	Iodic	0.80		
$(HO)_5IO$	Periodic	1.64	8.36	

Note: Values are for aqueous solution at 25°C.
Sources: Where available, data are taken from Albert, A.; Serjeant, E. P. *The Determination of Ionization Constants*, 3rd ed.; Chapman and Hall: London, 1984; pp. 162–163. Others are from Manahan, S. E. *Quantitative Chemical Analysis*; Brooks/Cole: Pacific Grove, CA, 1986; p. 682.

These factors allow us not only to predict acidity trends, but also to estimate actual K_a values with reasonable accuracy. Toward that end Ricci and Pauling developed similar equations.[11] The Ricci version for compounds having the general formula H_mXO_n is

$$pK_a = 8 - 9f + 4(n - m) \qquad (10.28)$$

Here, f is the formal charge of heteroatom X (calculated assuming that the X–O bonds have no multiple character) and $(n - m)$ is the number of oxygen atoms not bonded to hydrogen. The application of equation (10.28) to arsenic acid, H_3AsO_4, gives

$$pK_{a_1} = 8 - 9(+1) + 4(4 - 3) = \ 3$$

$$pK_{a_2} = 8 - 9(+1) + 4(4 - 2) = \ 7$$

$$pK_{a_3} = 8 - 9(+1) + 4(4 - 1) = 11$$

These predictions compare favorably to the accepted values of 2.21, 6.98, and 11.5, respectively. The accuracy of Ricci's equation can be tested further by comparing its predictions to the experimental pK_a values for other oxyacids given in Table 10.4.

The Ricci equation seems to fail for phosphorous (H_3PO_3) and hypophosphorous (H_3PO_2) acids, both of which are much stronger than predicted. This discrepancy results from their molecular structures. As shown in

11. Ricci, J. E. *J. Am. Chem. Soc.* **1948**, *70*, 109; Pauling, L. *College Chemistry*, 3rd ed.; Freeman: San Francisco, 1964; p. 540.

Expected structure
Calculated $pK_{a_1} = 8$

Actual structure
Calculated $pK_{a_1} = 3$

(a) Hypophosphorous acid, H_3PO_2, experimental $pK_{a_1} = 1.23$

Figure 10.3
The structures and calculated (Ricci equation) pK_a's of hypophosphorous and phosphorous acids (see text).

Expected structure
Calculated $pK_{a_1} = 8$

Actual structure
Calculated $pK_{a_1} = 3$

(b) Phosphorous acid, H_3PO_3, experimental $pK_{a_1} = 1.43$

Figure 10.3, these species contain P–H and P=O bonds rather than the expected P–O–H linkages only. When this is taken into account, the Ricci prediction is $pK_a = 8 - 9(+1) + 4(1) = 3$ for each of these acids, in reasonable agreement with their experimental values.

Anhydrides

Oxyacids are produced from the action of water on nonmetal oxides, while basic metal hydroxides are formed from reactions between water and metal oxides. The reagent oxides are *acidic* and *basic anhydrides*, respectively. The relationship between an anhydride and its corresponding acid or base is described by the following equations:

$$\text{Acid anhydride} + \text{Water} \longrightarrow \text{Oxyacid} \tag{10.29}$$

$$\text{Basic anhydride} + \text{Water} \longrightarrow \text{Base (metal hydroxide)} \tag{10.30}$$

For example, the conversions of three nitrogen oxides to their oxyacids are given by equations (10.31)–(10.33):[12]

$$2NO + H_2O \longrightarrow H_2N_2O_3 \tag{10.31}$$

$$N_2O_3 + H_2O \longrightarrow 2(HO)N{=}O \tag{10.32}$$

$$N_2O_5 + H_2O \longrightarrow 2HONO_2 \tag{10.33}$$

---

12. Oxyhyponitrous acid, $H_2N_2O_3$, has not been isolated because of low thermal stability. However, certain of its salts (eg, the hydrate $Na_2N_2O_3 \cdot H_2O$) are known.

Note that these reactions do not involve changes in oxidation number. This provides a facile way to correlate acids with their anhydrides.

There is, of course, no absolute distinction between metals and non-metals, and certain elements (the metalloids) exhibit intermediate chemical properties. The oxides of such elements are generally amphoteric. Silicon dioxide is one example:

$$SiO_2 + 2H_2O \quad \overbrace{??}^{\qquad} \quad \begin{array}{l} \longrightarrow H_4SiO_4 \qquad\qquad \textbf{(10.34)} \\ \longrightarrow Si(OH)_4 \qquad\qquad \textbf{(10.35)} \end{array}$$

There is only one product, of course; it is written as an acid in equation (10.34) and as a base in equation (10.35). It might be named either silicic acid or silicon hydroxide, but the former is preferable because it is more acidic than basic.

The periodic trends in oxide acidity and basicity parallel those of electronegativity—the most basic oxides lie at the bottom left of the periodic table, while the most acidic oxides lie at the upper right. These trends are apparent from predominance area diagrams for selected elements in fixed oxidation states, as shown in Figure 10.4.

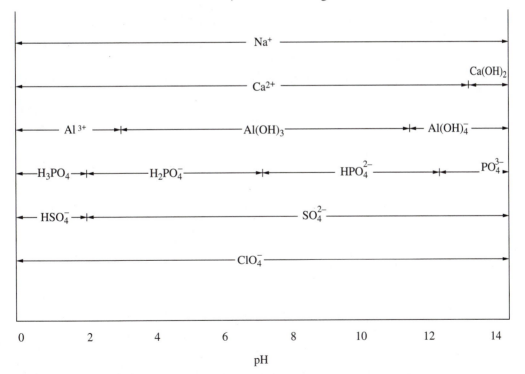

Figure 10.4 Abridged predominance area diagrams for selected elements in fixed oxidation states.

Pyroacids

The reaction of an acid anhydride with water is sometimes reversible. Heating an oxyacid may simply result in the elimination of $H_2O(g)$ and reversion to the anhydride. In other cases, however, heating causes the coupling of two fragments to form the X–O–X linkage characteristic of a *pyroacid*. Two examples are

$$SO_2 + H_2O \longrightarrow H_2SO_3 \xrightarrow{\Delta} \tfrac{1}{2}H_2O + \tfrac{1}{2}H_2S_2O_5 \quad (10.36)$$

$$P_2O_5 + 3H_2O \longrightarrow 2H_3PO_4 \xrightarrow{\Delta} H_2O + H_4P_2O_7 \quad (10.37)$$

The products are named pyrosulfurous and pyrophosphoric (or disulfurous and diphosphoric) acids. Can you write their Lewis structures?

Acidity Resulting from Metal Ion Hydrolysis

Recall that the dissolution of metal salts into water is promoted by ion–dipole attractions (and, in some cases, by covalent M–O interactions as well). The resulting flow of electron density from solvent molecules to the metal centers increases the partial positive charges of the hydrogens of the co-ordinated water, and thereby increases the Brønsted acidity of the system. Thus, the aqueous solutions of certain metal ions are acidic because of the sequential processes

$$M^{n+} + xH_2O \longrightarrow M(H_2O)_x^{n+} \quad (10.38)$$

$$M(H_2O)_x^{n+} + H_2O \rightleftharpoons [M(H_2O)_{x-1}OH]^{(n-1)+} + H_3O^+ \quad (10.39)$$

Equation (10.38) describes the process known as *hydration*, while equation (10.39) represents *hydrolysis*. Equilibrium constants for the latter are symbolized by either K_h or K_a. Hydrolysis constants for the common metal cations are given in Table 10.5.

For purely electrostatic (ion–dipole) interactions, acidity is a function of the charge-to-size ratio of the cation. Thus, hydrolysis of the hydrated Group 1 ions decreases in the order $Li^+ > Na^+ > K^+$, Rb^+, Cs^+; the charge densities of the latter three are so small that hydrolysis is negligible. Hydrolysis is more extensive in Group 2; in fact, the K_a of $Be^{2+}(aq)$ is greater than that of many oxyacids.

Table 10.5 : Hydrolysis constants (pK_h) for cations at 25°C

Li^+ 13.9	Be^{2+} 6.2											Al^{3+} 5.0
Na^+ 14.7	Mg^{2+} 11.4											
K^+ —	Ca^{2+} 12.6	Sc^{3+} 4.7	Ti^{3+} 2.3	V^{2+} 6.5	Cr^{2+} 8.7	Mn^{2+} 10.6	Fe^{2+} 7.1	Co^{2+} 9.6	Ni^{2+} 10.0	Cu^{2+} 7.6	Zn^{2+} 9.5	Ga^{3+} 2.6
				V^{3+} 2.6	Cr^{3+} 3.9		Fe^{3+} 2.0	Co^{3+} 3.2				
Rb^+ —	Sr^{2+} 13.1	Y^{3+} 8.0						Rh^{3+} 3.2	Pd^{2+} 1.6	Ag^+ 11.8	Cd^{2+} 7.9	In^{3+} 3.2
Cs^+ —	Ba^{2+} 13.3	La^{3+} 9.5						Ir^{3+} 4.8	Pt^{2+} >2.5		Hg^{2+} 2.5	Tl^+ 13.3

Note: Values are those reported for ionic strengths of 0 when available, and in many cases are averages of data from two or more sources; — = negligible hydrolysis.

Source: Taken from data compiled by Burgess, J. *Metal Ions in Solution*; Ellis Horwood: Chichester, 1978; pp. 264–265.

Hydrolysis often leads to precipitation for metal ions having charges of $2+$ or greater. An example is aqueous Fe^{III}:[13]

$$[Fe(H_2O)_6]^{3+} + H_2O \;\rightleftharpoons\; H_3O^+ + [Fe(H_2O)_5(OH)]^{2+} \quad \textbf{(10.40)}$$

$$[Fe(H_2O)_5(OH)]^{2+} + H_2O \;\rightleftharpoons\; H_3O^+ + [Fe(H_2O)_4(OH)_2]^+ \quad \textbf{(10.41)}$$

$$[Fe(H_2O)_4(OH)_2]^+ \;\rightleftharpoons\; H_3O^+ + 2H_2O + Fe(OH)_3(s) \quad \textbf{(10.42)}$$

The amphoteric nature of metal hydroxides such as $Zn(OH)_2$ and $Al(OH)_3$ is apparent from the fact that they redissolve upon the addition of either acid or base.

$$Zn(OH)_2(s) + 2H^+ \;\longrightarrow\; Zn^{2+} + 2H_2O \quad \textbf{(10.43)}$$

$$Zn(OH)_2(s) + 2OH^- \;\longrightarrow\; [Zn(OH)_4]^{2-} \quad \textbf{(10.44)}$$

10.5 Lewis Acids and Bases

Next, we will examine the other acid–base model of primary importance to inorganic chemists—the Lewis concept of electron pair donors and acceptors.

To act as a Lewis base, a species must have an electron pair available for donation. These are normally nonbonded electrons, and so the best-known Lewis bases are from Groups 14–17 of the periodic table. They include:

1. Carbanions, $:CR_3^{\ominus}$

2. NH_3, PH_3, AsH_3, SbH_3, their conjugate bases, and their derivatives

3. H_2O, H_2S, H_2Se, H_2Te, their conjugate bases, and their derivatives

4. Halide anions (but neutral or covalently bonded halogens are very weak bases)

In addition, certain doubly and triply bonded molecules and ions (eg, alkenes) can donate π electrons.

Lewis acids have an empty orbital in which to house an "extra" electron pair; this is most often a nonbonding orbital, but also might be of the antibonding type. Categories of Lewis acids include:

---

13. Equations (10.40)–(10.42) should not be taken too literally, as they involve more water molecules than are indicated from the stoichiometries. For example, $Fe(OH)_3$ is a hydrated material of complex (and variable) composition.

1. H^+, because of its empty $1s$ orbital (thus, any Brønsted acid also exhibits Lewis acid behavior)

2. Compounds lacking octets, such as those of divalent beryllium and trivalent boron and aluminum

3. Species capable of expanding their octets, such as PR_3 and SR_2

4. Species containing polar multiple bonds, which may serve as acid sites at the less electronegative atom (eg, carbon in $R_2C{=}O$ and $O{=}C{=}O$, and sulfur in $O{=}S{-}O$)

5. Metal cations

The Strengths of Lewis Acids and Bases

An obvious way to compare the relative strengths of different Lewis bases is to measure the enthalpy changes for their reactions with some common (reference) acid. Proton affinities are a natural choice in this regard, the reference being H^+. Table 10.1 is therefore consistent with the following order of gas-phase basicity toward the proton:

$$CH_3^- > NH_2^- > H^- > OH^- > F^- \approx SiH_3^- > PH_2^- > HS^- > Cl^-$$

$$> Br^- > I^- >> NH_3 > PH_3 > H_2S > H_2O > HI > \text{etc.}$$

We might think that this order should hold regardless of the reference acid chosen, and for many acids it does. However, such situations are less common than might be expected. There are many cases in which different reference acids have different relative affinities for bases. Consider that when bonding to halide ions, the Lewis acids Sc^{3+} and Hg^{2+} show exactly opposite preferences:

$$Sc^{3+}: \quad F^- > Cl^- > Br^- > I^-$$

$$Hg^{2+}: \quad I^- > Br^- > Cl^- > F^-$$

Ahrland, Chatt, and Davies used data of this type to divide the periodic table into class (*a*), class (*b*), and borderline regions (see Figure 10.5).[14] Metals having a greater affinity for F^- than for I^- (such as Sc^{3+}) were placed in class (*a*), while those preferring iodide belong to class (*b*).

---

14. Ahrland, S.; Chatt, J.; Davies, N. R. *Quart. Rev. Chem. Soc.* **1958**, *12*, 265.

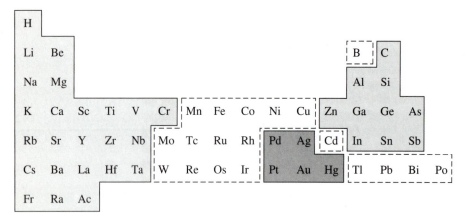

Figure 10.5 The classification of Lewis acid elements by their acceptor preferences, according to Ahrland, Chatt, and Davies: □ = class (*a*), ■ = class (*b*), and ⬚ = border region.

10.6 Hard and Soft Acids and Bases

R. G. Pearson extended the approach of Ahrland, Chatt, and Davies by categorizing acids and bases in terms of their *hardness* and *softness*.[15] The hard description is applied to species having relatively small sizes, high charge densities, and low polarizabilities. Conversely, soft acids and bases are relatively large in size, and have low charge densities and high polarizabilities.

Reactions between matched acids and bases (both hard or both soft) are thermodynamically favored over those involving hard–soft combinations. Moreover, displacement reactions are often governed by the tendency for acids and bases to be matched in hardness. Thus, a hard acid such as Al^{3+} readily displaces a softer acid from a hard base,

$$Al^{3+} + HgF^+ \rightleftharpoons AlF^{2+} + Hg^{2+} \qquad K_{eq} > 10^5 \qquad (10.45)$$

while the reverse is true for softer bases:

$$AlI^{2+} + Hg^{2+} \rightleftharpoons Al^{3+} + HgI^+ \qquad K_{eq} > 10^{12} \qquad (10.46)$$

15. Pearson, R. G. *J. Am. Chem. Soc.* **1963**, *85*, 3533. See also, Pearson, R. G. *J. Chem. Educ.* **1987**, *64*, 561; Pearson, R. G., Ed. *Hard and Soft Acids and Bases*; Dowden, Hutchinson, and Ross: East Stroudsburg, PA, 1973.

A large number of double exchange (metathesis) reactions are nicely explained by this theory:

$$LiI \quad + \quad AgF \quad \longrightarrow \quad LiF \quad + \quad AgI \qquad \Delta H° = -72\,kJ/mol$$

Hard–Soft   Soft–Hard   Hard–Hard   Soft–Soft

$$(10.47)$$

Many of Pearson's categorizations are given in Table 10.6.[16]

The natural preferences for hard–hard and soft–soft combinations can be explained using certain principles of molecular orbital theory. As shown in Figure 10.6, there is a correlation with orbital energies—the harder the acid, the more stable its available orbital (its LUMO), and the harder the

Table 10.6 Some hard and soft acids and bases

Acids	
Hard:	Soft:
H^+, Li^+, Na^+, K^+	Cu^+, Ag^+, Au^+, Hg^+, CH_3Hg^+, Tl^+
Be^{2+}, Mg^{2+}, Ca^{2+}, Sr^{2+}	Pd^{2+}, Pt^{2+}, Cd^{2+}, Hg^{2+}
BF_3, $B(OR)_3$, AlH_3, $AlCl_3$, $AlMe_3$	BH_3, $GaMe_3$, $GaCl_3$, GaI_3, $InCl_3$
CO_2, RCO^+, NC^+, Si^{4+}, CH_3Sn^{3+},	CH_2, carbenes
$\quad N^{3+}$, Cl^{3+}, I^{5+}, I^{7+}	Br_2, I_2, Br^+, I^+
Al^{3+}, Sc^{3+}, Ga^{3+}, In^{3+}, La^{3+}	Metal atoms
Cr^{3+}, Fe^{3+}, Co^{3+}	
Ti^{4+}, Zr^{4+}, Hf^{4+}	
Borderline:	
Fe^{2+}, Ru^{2+}, Os^{2+}, Co^{2+}, Rh^{3+}, Ir^{3+}, Ni^{2+}, Cu^{2+}, Zn^{2+}, BMe_3, GaH_3, R_3C^+, $C_6H_5^+$, Sn^{2+}, Pb^{2+}, NO^+, Sb^{3+}, Bi^{3+}, SO_2	

Bases	
Hard:	Soft:
CO_3^{2-}, $CH_3CO_2^-$	CO, CN^-, RNC, C_2H_4, C_6H_6
NH_3, RNH_2, N_2H_4	R_3P, $(RO)_3P$, R_3As
H_2O, OH^-	R_2S, RSH
ROH, RO^-, R_2O	H^-, R^-
F^-, Cl^-	I^-, SCN^-
NO_3^-, PO_4^{3-}, SO_4^{2-}, ClO_4^-	$S_2O_3^{2-}$
Borderline:	
N_2, N_3^-, NO_2^-, C_5H_5N, $C_6H_5NH_2$, Br^-	

Source: Categorizations are from Pearson, R. G. *J. Chem. Educ.* **1968**, *45*, 581.

16. It should be stressed that *hard* and *soft* are relative terms. The essence of this theory is comparative; thus, when we say that an acid is hard, we nearly always mean "harder than some other acid."

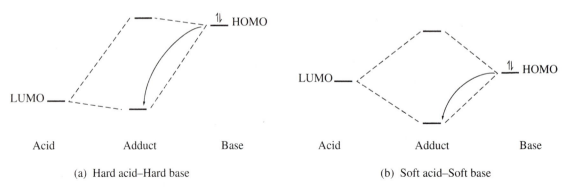

(a) Hard acid–Hard base   (b) Soft acid–Soft base

Figure 10.6 The molecular orbital interpretation of hard–soft acid–base theory. (a) The LUMO's of hard acids lie at relatively low energies, while the HOMO's of hard bases are at high energies; this gives rise to ionic-type interactions. (b) Soft acids and bases have frontier orbitals of intermediate (more closely matched) energies, producing strong covalent interactions. Arrows indicate the direction of electron flow upon adduct formation.

base, the less stable its HOMO. The large energy difference between hard orbitals causes base-to-acid charge transfer to be exothermic (ie, the interaction is primarily ionic). In contrast, soft acids and bases have intermediate—and, therefore, more closely matched—frontier orbital energies; this promotes strong covalent bonding. (Recall that overlap is most effective for orbitals having approximately equal energies.) Generalizing, the strong interaction between a hard acid and a hard base is due to a large electrostatic contribution to the bond energy. Soft–soft interactions benefit from a large covalent contribution. Mismatches (hard–soft and soft–hard combinations) do not produce either.

Inherent Acid–Base Strength versus the Hard–Soft Model

The hard–soft effect is important, but must be tempered with a consideration of inherent acidities and basicities. Consider Figure 10.7, where we have plotted the calculated enthalpy changes for the gas-phase reactions of five divalent metal cations with halide ions. Notice that there is no direct relationship between inherent acidity and hardness or softness. Among the acids, the strongest (Be^{2+}, based on ΔH) and the weakest (Ca^{2+}) are both hard. Fluoride has the greatest inherent basicity of the halide ions; hence, the reaction with F^- is the most exothermic for each of the cations, regardless of hard–soft character.

The slope of each line in Figure 10.7 is a measure of the hardness of the acid. The three hardest acids (Be^{2+}, Mg^{2+}, Ca^{2+}) give the steepest slopes, while the two soft acids (Hg^{2+}, Cd^{2+}) have the smallest slopes.

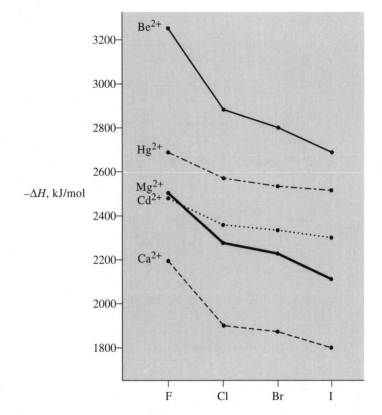

Figure 10.7
Calculated enthalpies for gas-phase reactions of selected divalent metal cations with halide ions. The enthalpy is greatest for interaction with F$^-$ in each case, but the lines vary in slope. [Enthalpy data compiled by Pearson, R. G.; Mawby, R. J. In *Halogen Chemistry*; Gutmann, V., Ed.; Academic: New York, 1967; Volume 3, pp. 55*ff.*]

Ambidentate Ligands

Certain species are capable of electron pair donation through two or more different atoms; such bases are said to be *ambidentate*. Reactions between acids and ambidentate bases provide interesting tests of the hard–soft model, since the acid can "choose" between a harder or softer site. For example, the thiosulfate ion has both hard (oxygen) and soft (sulfur) electron pairs:

$$\begin{bmatrix} \ddot{S}-\overset{\displaystyle \overset{..}{O}}{\underset{\displaystyle \underset{..}{O}}{\overset{\|}{\underset{\|}{S}}}}-\ddot{O} \end{bmatrix}^{2-}$$

Another ambidentate ligand is the thiocyanate ion. Hard acids such as Cr^{3+} preferentially bind to SCN^- through the nitrogen atom (the difference in Cr–N versus Cr–S isomer stability is about 30 kJ/mol), while the soft acids Cd^{2+}, Hg^{2+}, and Pt^{2+} all bond to sulfur.

Symbiosis

For certain acids and bases (especially those in the borderline region), hard–soft behavior preferences are influenced by substituents. For example, the carbocations CH_3^+ and CF_3^+ show very different reactivities as Lewis acids, with CF_3^+ having a greater affinity for hard bases. This results from the influence of the fluorines, whose high electronegativities increase the positive charge density at the acidic site (carbon). Thus, there is a general tendency for acid–base substrates to "take on more of the same," a phenomenon that has been given the name *symbiosis*. Symbiosis provides an explanation for the large equilibrium constants observed for reactions such as

$$CH_3F + CF_3I \rightleftharpoons CH_3I + CF_4 \tag{10.48}$$

Similarly, the reactivity of Co^{3+} toward donors is influenced by its other ligands. The species $[Co(NH_3)_5F]^{2+}$ is much more stable in aqueous solution than $[Co(NH_3)_5I]^{2+}$. This is because Co^{III} is "hardened" by the presence of five hard NH_3 substituents, causing it to prefer another hard base for the sixth ligand. Thus, $[Co(NH_3)_5I]^{2+}$ reacts readily with water to form $[Co(NH_3)_5(H_2O)]^{3+}$, while $[Co(NH_3)_5F]^{2+}$ is hydrolyzed more slowly and to a lesser extent.

In contrast, $[Co(CN)_5I]^{3-}$ and $[Co(CN)_5H]^{3-}$ are both more stable in water than $[Co(CN)_5F]^{3-}$. This is due to the softening influence of the CN^- ligands, which favors a soft base (I^- or H^- rather than F^-) at the sixth site.

Solvent Water as a Hard Base

The solubilities of metal salts depend on the hard–soft qualities of both the solute and the solvent. Aqueous solubilities tend to be higher for salts in which the cation and anion are mismatched (hard–soft or soft–hard) than when matched. Consider the following orders of molar solubility:

$$AgF \gg AgCl > AgBr > AgI$$

$$NaF < NaCl < NaBr < NaI$$

In hardness, $F^- > H_2O > Cl^- > Br^- > I^-$. The soft acid Ag^+ prefers the larger halides, so solvent molecules are effective only at displacing F^- from the crystalline silver halides. In contrast, Na^+ is relatively hard; hence, displacement by H_2O—and therefore solubility—increases with increasing size of the halide.[17] Quantitative data supporting these generalizations are given in Table 10.7.

---

17. Hard–soft theory is not the whole story, of course. Consistent with discussions in Chapters 6 and 8, other factors are also important in determining solubilities.

Table 10.7 Molar solubilities of some halide salts in water at 25°C

	F^-	Cl^-	Br^-	I^-
Hard acids				
Li^+	0.10	15.1	16.7	12.3
Na^+	1.0	6.1	11.3	12.3
Mg^{2+}	0.0012	5.7	5.5	5.3
Al^{3+}	0.067	5.2	Decomposes	Decomposes
Soft acids				
Ag^+	14.3	1.3×10^{-5}	7.2×10^{-7}	9.1×10^{-9}
Cu^+	—	1.1×10^{-3}	7.2×10^{-5}	1.0×10^{-6}
Tl^+	3.52	1.2×10^{-2}	1.8×10^{-3}	1.8×10^{-5}
Cs^+	24.2	9.6	5.8	1.7

10.7 Quantitative Aspects of Lewis Acid–Base Theory

The hard–soft concept as originally presented by Pearson was qualitative in nature. However, various attempts have been made to quantify acid–base strengths via thermodynamic data.

Donor and Acceptor Numbers

A common approach is to measure the reaction enthalpies for a series of bases with some common acid; proton affinities represent one example. Antimony pentachloride also has been used as a reference acid. In this system, the reaction enthalpies define the *donicities*, or *donor numbers*, of the bases (Table 10.8, p. 328).[18] A scale of *acceptor numbers* for acids also has been developed. Donor and acceptor numbers are especially useful for evaluating the Lewis acid–base tendencies of solvents.

E and C Parameters: The Drago–Wayland Equation

The equation

$$-\Delta H_{AB} = E_A E_B + C_A C_B \tag{10.49}$$

has been used to evaluate acid–base pairs from the standpoint of hardness and softness. In equation (10.49), E_A and E_B reflect the ability of the acid and base, respectively, to participate in electrostatic interactions, while C_A and C_B do likewise for their covalent (orbital overlap) tendencies. These

18. Gutmann, V.; Steininger, A.; Wychera, E. *Monatsh. Chem.* **1966**, *97*, 460; Mayer, U.; Gutmann, V. *Adv. Inorg. Chem. Radiochem.* **1975**, *17*, 189.

Table 10.8 Donor and acceptor numbers for selected solvents

Solvent	DN	AN
Acetic acid	—	52.9
Acetic anhydride	10.5	—
Acetone	17.0	12.5
Acetonitrile	14.1	19.3
Acetyl chloride	0.7	—
Antimony pentachloride	—	100.0
Benzene	0.1	8.2
Carbon tetrachloride	—	8.6
Chloroform	—	23.1
Dichloromethane	—	20.4
Diethyl ether	19.2	3.9
Dimethylsulfoxide (DMSO)	29.8	19.3
Dioxane	14.8	10.8
Ethanol	19.0	37.1
Ethyl acetate	17.1	—
Formamide	—	39.8
Hexamethylphosphorotriamide (HMPA)	38.8	10.6
Methanol	20.0	41.3
Methyl acetate	16.5	—
Methylsulfonic acid	—	126.1
Nitrobenzene	4.4	14.8
Nitromethane	2.7	20.5
Phosphorus oxychloride	11.7	—
Propanol	18.0	33.5
Pyridine	33.1	14.2
Tetrahydrofuran	20.0	8.0

Note: DN = Donor number; AN = Acceptor number. Donor numbers correspond to reaction enthalpies (in kilocalories per mole) with SbF_5. Acceptor numbers are based on ^{31}P chemical shifts on an arbitrary scale setting hexane at 0 and $SbCl_5$ at 100.
Source: Taken from data of Gutmann, V. *The Donor–Acceptor Approach to Molecular Interactions*; Plenum: New York, 1978.

constants have been determined from experimental data via least-squares computer analysis.[19] The results are consistent with Pearson's original premise, in that large energies of interaction (strongly negative ΔH_{AB}'s) result in two ways: through reactions in which both the acid and base have large

19. Drago, R. S.; Wayland, B. *J. Am. Chem. Soc.* **1965**, *87*, 3571.

Table 10.9   Parameters *E* and *C* for selected acids and bases

Acid	*E*	*C*	Base	*E*	*C*
I_2	1.00	1.00	NH_3	1.15	4.75
ICl	5.10	0.830	$MeNH_2$	1.30	5.88
IBr	1.56	2.41	Me_2NH	1.09	8.73
H_2O	1.64	0.571	Me_3N	0.808	11.54
$BF_3(g)$	9.88	1.62	Me_3P	0.838	6.55
$BMe_3(g)$	6.14	1.70	C_5H_5N	1.17	6.40
$AlMe_3$	16.9	1.43	MeCN	0.886	1.34
SO_2	0.920	0.808	Me_2SO	1.34	2.85
EtOH	3.88	0.451	Me_2S	0.343	7.46
Me_3COH	2.04	0.300	Et_2S	0.339	7.40
C_6H_5OH	4.33	0.422	Me_2Se	0.217	8.33
$p\text{-}MeC_6H_4OH$	4.18	0.404	$Me_2C{=}O$	0.937	2.33
C_6H_5SH	0.99	0.198	Et_2O	0.936	3.25
			Tetrahydrofuran (THF)	0.978	4.27
			C_6H_6	0.280	0.590

Note: When used in equation (10.49), these values give reaction enthalpies in kilocalories per mole.
Source: Taken from a compilation by Drago, R. S. *Coord. Chem. Rev.* **1980**, *33*, 251.

E values (hard combinations such as Me_3Al with NH_3) or those in which both have large *C* values (soft combinations such as $SbCl_5$ and Me_3N).

Equation (10.49) works best for uncharged acids and bases. Published *E* and *C* constants for such species are given in Table 10.9. Consistent with the original literature, they yield reaction enthalpies per mole of adduct via equation (10.49).

10.8   Electronic, Steric, and Leaving-Group Effects

Lewis acid–base reactions in which the acidic centers are nonmetal or metalloid atoms are especially sensitive to steric and electronic effects. We will illustrate this with some of the considerable amount of information that has been collected for amine–borane adducts.

The boron trihalides have the relative acidities

$$BF_3 < BCl_3 < BBr_3 < BI_3$$

which is precisely the opposite order from that predicted by inductive effects. (The strongly electronegative fluorines should withdraw electron density from boron and thereby increase its acidity.) The explanation for this anomaly involves dative π bonding:

The acidity of the central boron is diminished to whatever extent these secondary resonance structures contribute. Since π bonding decreases with increasing internuclear distance, the effect is greatest for BF_3; thus, it is the least basic member of the series.

Steric effects also play an important role. This is apparent from Table 10.10, which gives thermodynamic data for amine adducts of trimethylborane. It is clear from the table that reaction enthalpies do not correlate well with amine basicity. For example, although trimethylamine is more basic toward H^+ than is $MeNH_2$ (proton affinities of 884 versus 856 kJ/mol) and much more basic toward water than is pyridine ($pK_b = 4.2$ versus 8.7), the formation enthalpies of the three trimethylborane adducts are roughly equal. This is due in part to steric hindrance, which reduces the B–N bond energy in $Me_3N·BMe_3$. It may seem surprising that steric hindrance is important in a system where no substituents are larger than methyl groups, since the reagents, Me_3N and Me_3B, are not considered to be hindered molecules. However, upon $N \rightarrow B$ bond formation, the CH_3 groups on the opposing atoms come sufficiently close to reduce the stability. This general phenomenon, in which unhindered acids and bases interact to create a hindered adduct, has been termed *F-strain* (for frontal strain).

This effect also is apparent for adducts derived from pyridine and its methyl derivatives. Substitution at one ortho position of C_5H_5N increases the Brønsted basicity, but decreases the stability of the Me_3B adduct.

Table 10.10 Reaction enthalpies (vapor phase) for the formation of some amine adducts of trimethylborane

Adduct	ΔH, kJ/mol	Adduct	ΔH, kJ/mol
$Me_3B·NH_3$	-58	$Me_3B·Et_2NH$	-68
$Me_3B·MeNH_2$	-74	$Me_3B·NEt_3$	-42
$Me_3B·EtNH_2$	-75	$Me_3B·C_5H_5N$	-71
$Me_3B·i\text{-}PrNH_2$	-73	$Me_3B·3\text{-}MeC_5H_4N$	-75
$Me_3B·t\text{-}BuNH_2$	-54	$Me_3B·2\text{-}MeC_5H_4N$	-42
$Me_3B·Me_2NH$	-81	$Me_3B·2,6\text{-}Me_2C_5H_3N$	0
$Me_3B·NMe_3$	-74		

Source: Brown, H. C. *J. Chem. Soc.* **1956**, 1248.

Substitution at both ortho positions (to give lutidene, $2,6\text{-}Me_2C_5H_3N$) creates such strong destabilization that the adduct does not form.

Addition versus Displacement Reactions

The interaction of a Lewis acid and base often results in a substitution rather than addition reaction. In such cases, it is conventional to refer to the base as a *nucleophile* and to the acid as an *electrophile*. The ultimate reaction products often depend on whether a suitable leaving group is present. Consider the interaction of boron trichloride with amines. The tertiary amine adduct $Me_3N \cdot BCl_3$ is stable to high temperatures, melting without decomposition at $243°C$. For BCl_3 adducts of secondary amines, however, decomposition to an aminoborane often occurs at or below room temperature:

$$BCl_3 + R_2NH \longrightarrow [Cl_3B \cdot NHR_2] \xrightarrow{-HCl} Cl_2BNR_2 \qquad (10.50)$$

The BCl_3 adducts of primary amines often undergo a second elimination of HCl to form cyclic *borazines* (see Chapter 13):

$$BCl_3 + RNH_2 \longrightarrow [Cl_3B \cdot NH_2R] \xrightarrow{-2HCl} \tfrac{1}{3}(ClBNR)_3 \qquad (10.51)$$

In a similar manner, the nucleophile OH^- reacts with CO_2 (a Lewis acid because of its polar double bonds) to form bicarbonate ion, which is stabilized by the lack of a good leaving group:

$$OH^- + O{=}C{=}O \longrightarrow \left[HO-C\overset{\displaystyle O}{\underset{\displaystyle O}{\Big\backslash}} \right]^- \qquad (10.52)$$

In contrast, the reaction of hydroxide ion with acetyl halides leads to substitution because of the facile loss of X^-:

$$OH^- + Me-C\overset{\displaystyle O}{\underset{\displaystyle X}{\Big\backslash}} \longrightarrow Me-C\overset{\displaystyle O}{\underset{\displaystyle OH}{\Big\backslash}} + X^- \qquad (10.53)$$

Bibliography

Wulfsberg, G. *Principles of Descriptive Inorganic Chemistry*; Brooks/Cole: Pacific Grove, CA, 1987.

Albert, A.; Serjeant, E. P. *The Determination of Ionization Constants*, 3rd ed.; Chapman and Hall: London, 1984.

Finston, H. L.; Rychtman, A. C. *A New View of Current Acid–Base Theories*: Wiley: New York, 1982.

Jensen, W. B. *The Lewis Acid–Base Concepts*; Wiley: New York, 1980.

Gutmann, V. *The Donor–Acceptor Approach to Molecular Interactions*; Plenum: New York, 1978.

Baes, C. F.; Mesmer, R. E. *The Hydrolysis of Cations*; Wiley: New York, 1976.

Pearson, R. G., Ed. *Hard and Soft Acids and Bases*; Dowden, Hutchinson, and Ross: East Stroudsburg, PA, 1973.

Bell, R. P. *The Proton in Chemistry*, 2nd ed.; Chapman and Hall: London, 1972.

Questions and Problems

1. Write equations that show HCl to behave as:
   (a) An Arrhenius acid
   (b) A Brønsted acid toward ammonia
   (c) A Brønsted base toward H_2SO_4
   (d) A Lewis acid (electrophile) toward hydride ion in a displacement reaction

2. Write equations to demonstrate that NH_3 is amphoteric in the Brønsted sense.

3. Explain why each of the following can be considered an acid–base reaction:
   (a) $CaO + H^+ \longrightarrow Ca^{2+} + OH^-$
   (b) $SnCl_4 + 2Cl^- \longrightarrow SnCl_6^{2-}$
   (c) $CO_2 + Na_2O \longrightarrow Na_2CO_3$
   (d) $HF + KF \longrightarrow K^+ + HF_2^-$
   (e) $AgCN + CN^- \longrightarrow [Ag(CN)_2]^-$
   (f) $BaCl_2 + TlCl \longrightarrow Ba^{2+} + TlCl_3^{2-}$
   (g) $2\,PCl_5 \longrightarrow PCl_4^+ + PCl_6^-$

4. Write equations showing PF_3 acting as:
   (a) A Lewis acid toward F^-
   (b) A Lewis base toward BH_3

5. Explain why PH_3 behaves as both a Lewis acid and a Lewis base, while NH_3 is a Lewis base only.

6. Each of the following equations qualify NO_2^- as a base according to the Usanovich definition. Explain.
   (a) $NO_2^- + K^+ \longrightarrow KNO_2$
   (b) $NO_2^- + :\ddot{O}: \longrightarrow NO_3^-$
   (c) $NO_2^- + HCl \longrightarrow HNO_2 + Cl^-$
   (d) $NO_2^- + H_2O \longrightarrow NO_3^- + 2H^+ + 2e^-$

7. Tin(II) chloride exhibits both acidic and basic properties, although it is neither a Brønsted acid nor base.

(a) Write an equation in which $SnCl_2$ undergoes the anionotropic equivalent to autoprotolysis.

(b) Explain the formation of Sn_2Cl_4 as the product of the Lewis equivalent to autoprotolysis.

8. Use equation (10.22) to rationalize the fact that in PA, Xe > Kr.

9. Given that the bond dissociation energy of KH is 180 kJ/mol, estimate the proton affinity of elemental potassium.

10. Use the Ricci equation (10.28) to estimate pK_a for each of the following acids; compare your predictions to the actual values given in Table 10.4.

   (a) H_2SeO_4 (b) H_2SeO_3 (c) $HSeO_3^-$ (d) HNO_3

   (e) H_2O_2 (f) H_3BO_3

11. Explain the following observations about oxyacids using ideas discussed in Chapters 4 and 5:

   (a) The bond linkage is nearly always H–O–X, rather than H–X–O.

   (b) For a given heteroatom, K_a increases with increasing number of oxygens.

   (c) For a given number of oxygens, K_a increases with increasing electronegativity of the heteroatom.

12. For each of the given pairs, identify the stronger Brønsted base. Your answers should be based on the acidity of the conjugate acid.

   (a) NH_2^- or OH^- (b) HCO_2^- or $MeCO_2^-$

   (c) O^{2-} or S^{2-} (d) ClO_2^- or ClO_3^-

   (e) SeO_4^{2-} or $HSeO_4^-$ (f) $[Fe(H_2O)_5OH]^+$ or $[Fe(H_2O)_5OH]^{2+}$

13. Identify the anhydrides of the following acids and bases:

   (a) KOH   (b) $Al(OH)_3$ (c) $H_4P_2O_7$ (d) H_2CrO_4

   (e) HNO_3 (f) H_5IO_6

14. Sketch the structures of the following pyroacids:

   (a) $H_2Cr_2O_7$ (b) $H_2S_4O_5$ (c) $H_4P_2O_7$

15. Write chemical equations corresponding to these observations.

   (a) When $Cr(NO_3)_3$ is dissolved in water, the pH decreases.

   (b) When the pH of a 1.0 M solution of Mg^{2+} is raised above 8.5, precipitation occurs.

   (c) A precipitate of aluminum hydroxide can be redissolved with 6 M KOH.

   (d) A precipitate of aluminum hydroxide can be redissolved with 6 M HNO_3.

16. Place the following species in order of increasing Brønsted acidity, and briefly explain your reasoning: H_3AsO_3, $H_2AsO_3^-$, H_3AsO_4, H_2SO_4, H_2SeO_4.

17. The pH of a 1.00 M aqueous solution of $NiCl_2$ is about 4.9.

   (a) Explain the acidity through one or more chemical equations.

   (b) Calculate K_h for Ni^{2+}.

18. The following species can be considered to be Lewis adducts. Identify a reasonable precursor acid–base pair for each.

   (a) I_3^- (b) PCl_6^- (c) HNO_2 (d) $CsPF_6$

   (e) H_2SO_3 (f) Me_2SO (g) F_3NO (h) $LiBH_4$

19. Use Pearson's hard–soft theory to predict whether the following equilibria favor the reactants or products, and explain:
    (a) $Me_3N \cdot BH_3 + Me_3As \cdot BF_3 \rightleftharpoons Me_3As \cdot BH_3 + Me_3N \cdot BF_3$
    (b) $TiBr_4 + 4HgF \rightleftharpoons TiF_4 + 4HgBr$
    (c) $2CuCN + CuI_2 \rightleftharpoons 2CuI + Cu(CN)_2$
    (d) $CsI + NH_3 \rightleftharpoons HI + CsNH_2$

20. When dimethylsulfoxide (Me_2SO, DMSO) acts as a ligand toward ferric ion, coordination occurs through the oxygen. When it binds to Hg^{2+}, it normally does so through sulfur.
    (a) Explain these observations.
    (b) Write a Lewis structure (including formal charges) for the major contributor to the hybrid for free DMSO, for the ligand bound to a metal via the oxygen atom, and for the ligand bound to a metal via the sulfur atom.

21. Imagine that you wish to prepare compounds in which the selenocyanate ion, $SeCN^-$, acts as a bridging ligand between two metal centers (M–SeCN–M').
    (a) For your first synthesis, would you choose one kind of metal ion (M = M') or two different metals? Why?
    (b) What specific metal(s) would you choose? Why?

22. The aminophosphine Me_2NPF_2 is an ambidentate ligand. Experimental evidence shows that this base binds to BH_3 through the phosphorus atom, but to BF_3 through the nitrogen. Rationalize.

23. Consider the compound 1,1,1-trifluoroethane to be an adduct formed from charged $CH_3^{+/-}$ and $CF_3^{+/-}$ units.
    (a) Which of the possible bases, CH_3^- or CF_3^-, is softer? Why?
    (b) Which should be the least stable adduct: CH_3CH_3, CH_3CF_3, or CF_3CF_3? Why?
    (c) It has been calculated that the equilibrium $CH_3CH_3 + CF_3CF_3 \rightleftharpoons 2CH_3CF_3$ lies well to the left. Is this consistent with your answer to part (b)? Explain.

24. The molar solubility of $CaCl_2$ is roughly four times that of $BaCl_2$. Explain this on the basis of hard–soft acid–base theory.

25. Zinc nitrate is dissolved in water. Slow addition of 2 M NH_3 leads to the formation of a blue-gray precipitate. Continued addition causes the precipitate to redissolve. Write equations to describe what has occurred.

26. What properties of $SbCl_5$ do you believe prompted Gutmann to choose it as his standard for the generation of donor numbers? Discuss.

27. Use the Drago–Wayland parameters to calculate the approximate enthalpy change for:
    (a) $SO_2 + C_5H_5N \longrightarrow C_5H_5N \cdot SO_2$
    (b) $BF_3 + NMe_3 \longrightarrow Me_3N \cdot BF_3$
    (c) $BF_3 + PMe_3 \longrightarrow Me_3P \cdot BF_3$
    (d) $BF_3 + THF \longrightarrow THF \cdot BF_3$

28. Explain this observed order of adduct stability:
    $Me_3N \cdot BF_3 > Me_3N \cdot BMe_3 > Me_3P \cdot BMe_3$

29. Explain in your own words why BF_3 is a weaker Lewis acid than BI_3.

30. Predict products for the following displacement reactions:
    (a) $Me_3SiCl + Et_2NH \longrightarrow$
    (b) $Me_3SiCl + MeOH \longrightarrow$
    (c) $Me_2SiCl_2 + 2Me_2NH \longrightarrow$

*31. A quantitative acidity scale for oxides based on the Lux–Flood definitions has been proposed by D. W. Smith (*J. Chem. Educ.* **1987**, *64*, 480). Answer these questions after reading Smith's article.
    (a) How were the values for the acidity parameter *a* assigned?
    (b) What is the approximate enthalpy change for the reaction $Na_2O + CO_2 \rightarrow Na_2CO_3$?
    (c) Do you expect the displacement reaction $PbO + SO_2 \rightarrow Pb^{2+} + SO_3^{2-}$ to be exothermic or not? Explain.

*32. The proton affinities of several metal carbonyls have been reported (Miller, A. E. S.; Kawamura, A. R. *J. Am. Chem. Soc.* **1990**, *112*, 457).
    (a) Describe how these experiments were performed.
    (b) The reaction $HMn(CO)_5 + X^- \rightarrow HX + Mn(CO)_5^-$ is exothermic if X = F, Cl, or Br, but not I. What did the authors of the article learn from this?
    (c) It was concluded that $HRe(CO)_5$ is a weaker acid than $HMn(CO)_5$. To what factor was this attributed?

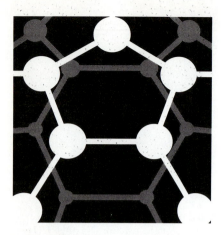

11

Inorganic Chemistry in Nonaqueous Solvents

What we call chemical reactivity is influenced to a remarkable degree by environment. For example, a compound may be stable in the gas phase but unstable in solution, or it may behave as an acid in one solvent but as a base in another.

What are the important factors to consider when choosing a solvent for an inorganic reaction? Why is AgCl insoluble and $BaCl_2$ soluble in water, while AgCl is soluble and $BaCl_2$ insoluble in NH_3? Under what circumstances do molten salts make useful solvents? These types of questions are considered in this chapter.

11.1 Solvent Classifications

One consequence of Brønsted's studies of acids and bases was the organization of solvents into eight classes. The groupings were based on properties such as dielectric constant and acidity/basicity. Water and methanol were placed in the same class because of their similarities in reactions such as those described by equations (11.1)–(11.3):

$$2HQ \rightleftharpoons H_2Q^+ + Q^- \tag{11.1}$$

$$HQ + NH_3 \rightleftharpoons NH_4^+ + Q^- \tag{11.2}$$

$$HQ + HClO_4 \rightleftharpoons H_2Q^+ + ClO_4^- \tag{11.3}$$

where $Q = OH$ and OCH_3.

Though not in common use today, Brønsted's classes represent an important first effort in organizing solvent chemistry. A more modern approach might consider five parameters: (*1*) dielectric constant, which relates to the ability to solvate, and thereby separate, ion pairs; (*2*) tendency toward self-ionization; (*3*) acidic and/or basic properties; (*4*) complexation; and (*5*) oxidation–reduction tendencies. Each of these factors will be considered individually.

Dielectric Constant

The *dielectric constant* (ϵ) of a medium is defined in terms of the equation

$$\epsilon = \frac{kq_+q_-}{Ed} \tag{11.4}$$

where E is the energy of attraction between two opposite charges q_+ and q_- separated by distance d, and k is a constant. It can be described as a measure of the attenuation of the q_+/q_- electrostatic energy by a medium. The dielectric constants of solvents are referenced to that of a vacuum—that is,

$$\frac{\epsilon}{\epsilon_0} = \frac{\epsilon_{solvent}}{\epsilon_{vacuum}}$$

The value for water is 81 (no units), meaning that H_2O is 81 times more effective than a vacuum at separating charges.

Charge separation may be achieved through the destruction of an ionic lattice, as, for example, in the dissolution of crystalline sodium chloride in water:

$$NaCl(s) + (x + y)H_2O \longrightarrow Na(H_2O)_x^+ + Cl(H_2O)_y^- \tag{11.5}$$

Or it may be achieved by the heterolytic cleavage of polar covalent bonds, for example, the dissolution of HCl(g) in water:[1]

$$HCl(g) + (x + y)H_2O \longrightarrow H(H_2O)_x^+ + Cl(H_2O)_y^- \tag{11.6}$$

---

1. Hydrogen chloride forms nonconducting, nonionic solutions in solvents of low polarity such as benzene.

In general, the higher the dielectric constant, the more polar the solvent and the greater its interaction with ions. Thus, in ϵ,

$$H_2O > CH_3CN > CH_3OH > C_2H_5OH > CH_2Cl_2 > C_6H_{14}$$

Liquids having similar dielectric constants (eg, water and acetonitrile) are likely to be miscible, while liquids having very different ϵ's (water and hexane) are normally immiscible.

Self-Ionization

Protic solvents (primarily, those in which hydrogen is bonded to a highly electronegative element such as oxygen, nitrogen, phosphorus, sulfur, or a halogen) normally undergo autoprotolysis to one extent or another. That is, they demonstrate amphoterism in the Brønsted sense, according to equation (11.1). Self-ionization does not require the presence of hydrogen, however, since other acid–base theories also allow for amphoteric behavior.

$$2\,PBr_5 \; \rightleftharpoons \; PBr_4^+ + PBr_6^- \tag{11.7}$$

$$3\,HF \; \rightleftharpoons \; H_2F^+ + HF_2^- \tag{11.8}$$

$$POCl_3 \; \rightleftharpoons \; POCl_2^+ + Cl^- \tag{11.9}$$

Brønsted-based equivalencies can be established among molecular solvents, as well as among their conjugate acids and bases (Table 11.1). Thus, the H_3O^+ equivalent (and the strongest acid that can exist) in acetic acid is $CH_3C(OH)_2^+$, while the hydroxide ion equivalent (and the strongest base that can exist) in methanol is CH_3O^-.

Table 11.1 Brønsted acid–base equivalents in various solvents

Solvent	Conjugate Acid	Conjugate Base	Dibase
H_2O	H_3O^+	OH^-	O^{2-}
NH_3	NH_4^+	NH_2^-	NH^{2-}
CH_3COOH	$CH_3C(OH)_2^+$	CH_3COO^-	—
H_2SO_4	$H_3SO_4^+$	HSO_4^-	SO_4^{2-}
HF	H_2F^+	F^- (HF_2^-)	—
CH_3OH	$CH_3OH_2^+$	CH_3O^-	—
CH_3CN	CH_3CNH^+	CH_2CN^- [a]	—

[a] Unstable in host solvent.

Acidic and Basic Tendencies

Whether a solute is protonated or deprotonated in solution depends on both its own acid–base properties and those of the solvent. Consider the dissolution of hydrogen chlorite (chlorous acid) into three different solvents, ammonia, water, and sulfuric acid:

$$HOClO + NH_3 \longrightarrow NH_4^+ + ClO_2^- \qquad (11.10)$$

$$HOClO + H_2O \rightleftharpoons H_3O^+ + ClO_2^- \qquad (11.11)$$

$$HOClO + H_2SO_4 \longrightarrow Cl(OH)_2^+ + HSO_4^- \qquad (11.12)$$

Thus, HOClO is a strong Brønsted acid in NH_3, a weak acid in H_2O, and a strong base in H_2SO_4.

One way to compare the relative acid–base properties of solvents is through a plot, such as Figure 11.1, of the effective pH range of selected solvents along the aqueous pH scale. This diagram was constructed in the following way. The pK_a of water is about 14 at room temperature, while the pK_a for its conjugate acid H_3O^+ is 0. The H_2O line is therefore 14 units long and runs from pH = 0 through 14. The line for NH_3 runs from about 9 to 39, since the pK_a's of NH_3 and NH_4^+ are 39 and 9.2, respectively; and so on.

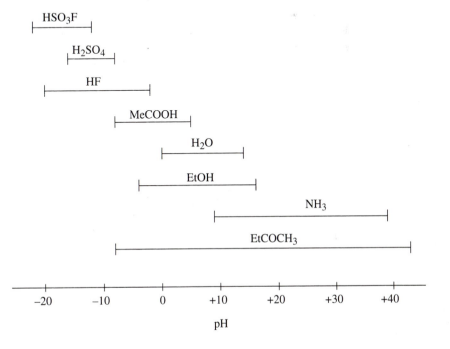

Figure 11.1
The effective pH ranges of some common solvents. The two ends of each line correspond to the pK_a of the solvent and its conjugate acid (see text).

Any solute that is a stronger acid than the conjugate acid of its solvent will react with that solvent; that is, the solute is leveled to its conjugate base. Since HOClO ($pK_a = 1.94$) is a stronger acid than NH_4^+, it is leveled to the chlorite ion in ammonia. However, since its pK_a lies between those of H_2O and H_3O^+, an equilibrium is established in water. The ratio in a 1.0 M solution is about 90% HOClO/10% ClO_2^-.

This relates to the *differentiation* of solute acidity or basicity. The acids HCl, HBr, and HI are all completely dissociated (they are equally strong acids) in water. A less basic solvent such as glacial acetic acid permits their differentiation, since the pK_a's of these acids lie between those of CH_3COOH and $CH_3C(OH)_2^+$. The experimental order of acid strength is then found to be HI > HBr > HCl.

A common reason for using a nonaqueous solvent is to stabilize a solute that is reactive toward water. For example, the nitryl ion is such a potent (Lewis) acid that it is instantly consumed by H_2O:

$$NO_2^+ + H_2O \longrightarrow [H_2NO_3^+] \longrightarrow 2H^+ + NO_3^- \quad \textbf{(11.13)}$$

However, it can be used as a reactant in sulfuric acid (and is, in fact, often prepared as an in situ reagent by dissolving nitric acid in anhydrous H_2SO_4).

$$HNO_3 + 2H_2SO_4 \longrightarrow NO_2^+ + H_3O^+ + 2HSO_4^- \quad \textbf{(11.14)}$$

Complexation

The tendency for stable stoichiometric complexes to form upon the dissolution of a given solute varies greatly with the solvent, and the formation of complexes can lead to a number of secondary effects. An example is the remarkably high solubility of $AgNO_3$ in acetonitrile—solutions having concentrations of over 9 M can be prepared. This is due to the formation of stable complexes having the formula $Ag(NCMe)_x^+$, where x is 4 in dilute solutions (<0.5 M) and either 1 or 2 at higher concentrations.

Solute acidity and basicity are also affected by complexation. Solvents that strongly coordinate to H^+ enhance Brønsted acidity. An example is dimethylsulfoxide, DMSO. The transfer of H^+ from H_2O to DMSO is exothermic:

$$H(H_2O)_x^+ + y\text{DMSO} \rightleftharpoons H(\text{DMSO})_y^+ + xH_2O$$

$$\Delta H^\circ = -25 \text{ kJ/mol} \quad \textbf{(11.15)}$$

This indicates greater solvation of H^+ by DMSO than by H_2O. All else being equal, then, acidities should be greater in DMSO than in water. That is not usually the case, however; solvation of the (anionic) conjugate base is also an important factor, and anion solvation is greater in H_2O because of hydrogen bonding.[2] For species such as picric acid (2,4,6-trinitrophenol), in which the anion is poorly solvated by both solvents, the acidity in DMSO is equal to or greater than that in water.

Oxidation–Reduction Tendencies

Standard reduction potentials typically exhibit only small variations from one solvent to the next. Exceptions often can be explained by complexation effects. For example, the aforementioned complexation of Ag^+ by acetonitrile has a large effect on its ease of reduction. The lack of strong interaction between Ag^+ and H_2O (soft acid–hard base) contributes to the favorable Ag^+/Ag reduction potential in aqueous solution ($\mathscr{E}^0 = +0.80$ V). In CH_3CN (a softer base), complexation stabilizes Ag^+ and makes it more difficult to reduce ($\mathscr{E}^0 = +0.23$ V). On the other hand, the harder acid Zn^{2+} is more stabilized (and hence, more difficult to reduce) in H_2O.

The ease or difficulty of solvent oxidation–reduction is another important aspect of solution chemistry. Compared to water, HF is very difficult to oxidize. As a result, certain extremely strong oxidizing agents can be studied in HF. An example is xenon difluoride:

$$2XeF_2 + 2H_2O \longrightarrow 2Xe + O_2 + 4HF \tag{11.16}$$

$$XeF_2 + HF \longrightarrow (\text{No reaction}) \tag{11.17}$$

Table 11.2 (p. 342) lists 20 solvents commonly used by inorganic chemists, along with numerical values for some relevant physical and chemical properties.

We will now turn our attention to some specific chemistry in selected solvents. A basic (NH_3), an acidic (HF), and a polar aprotic solvent (CH_3CN) will be discussed in the next three sections.

11.2 Chemistry in Liquid Ammonia

Of all the common nonaqueous solvents, ammonia probably bears the greatest similarity to water. This should be remembered when reading the

---

2. Some important aspects of water's solvation of anions have been discussed by Sharpe, A. G. *J. Chem. Educ.* **1990**, *67*, 304.

Table 11.2 Important physical and chemical properties of some common solvents

Solvent	ϵ/ϵ_0	T_f, °C	T_b, °C	pK_s	d, g/mL
H_2O	81	0	100	14	0.998
HF	84	−83	20	10.7^a	1.002
H_2SO_4	101	10	270	3.6	1.83
CH_3COOH	6.2	−17	118	14.4	1.04
HI	3.6^a	−51	−35	—	—
NH_3	23^b	−78	−33	33	0.69^b
$HCONH_2$	110	3	193	—	1.13
CH_3OH	33	−94	65	16.7	0.791
C_2H_5OH	24	−117	78	19	0.790
$(CH_3)_2O$	21	−95	56	—	0.792
DMF	37	61	152	—	0.945
DMSO	47	18	189	—	1.101
THF	7.6	−65	66	—	0.889
CH_3CN	36	−45	82	32	0.786
HMPA	30	7	235	—	1.03
CCl_4	2.2	−23	77	—	1.594
DCE	9.9	−97	57	—	1.18
CH_2Cl_2	9	−97	40	—	1.424
C_6H_{12}	2.2	7	81	—	0.779
$n\text{-}C_6H_{14}$	2	−95	69	—	0.660

Note: Dielectric constants, pK_s's and densities are for 25°C unless otherwise indicated. DMF = N,N-dimethylformamide; DMSO = dimethylsulfoxide; THF = tetrahydrofuran; HMPA = hexamethylphosphoramide; DCE = 1,1-dichloroethane.
[a] At −45°C. [b] At −40°C.

following statements, which emphasize the *differences* in solvent properties between NH_3 and H_2O. The primary differences are:

1. Ammonia has a lower boiling temperature (-33°C) and a smaller liquid range (only 45°C at atmospheric pressure).

2. Ammonia has a smaller dielectric constant, and so is less able to dissolve ionic compounds. Also, ionic solutes exhibit more ion pairing in NH_3 than in water. (For saturated KCl, conductivity data indicate that only about 35% of the K^+Cl^- ion pairs are dissociated in liquid ammonia, compared to virtually 100% dissociation in H_2O.)

3. Ammonia is a weaker acid, and is less liable to protonate solutes than is H_2O. (Brønsted basicities are repressed.)

4. Ammonia is considerably more basic, and enhances solute acidities compared to H_2O.

Solubilities

Table 11.3 gives solubility data for liquid ammonia and, for comparison, the corresponding values in water.[3]

From the table, it is apparent that ionic solubilities increase with the softness of ions (eg, NaF < NaCl < NaBr). Ammonium salts have high

Table 11.3 Molal solubilities of selected solutes in water and ammonia

	H_2O	NH_3
Inorganic solutes		
NaF	1.0	0.082
NaCl	6.2	0.52
NaBr	11.3	13.4
NaI	12.3	10.8
NH_4Cl	5.6	19.2
NH_4NO_3	14.8	48.7
AgCl	10^{-5}	0.06
AgI	10^{-8}	8.8
$BaCl_2$	1.8	0
CaI_2	0.71	0.14
$NaNH_2$	Reacts with solvent	0.001
KNH_2	Reacts with solvent	0.65
Organic solutes		
n-C_4H_9OH	1.1	Miscible
$C_6H_5CH_2OH$	0.38	Miscible
$(C_2H_5)_3N$	2.4	Miscible
$C_6H_5NH_2$	0.40	Miscible
C_2H_5COOH	Miscible	Miscible
$(COOH)_2$	1.0	0

Note: Solubilities are in molal units (m, moles of solute per kilogram of solvent), taken in water at 25°C; values for NH_3 at −33°C.

Source: Converted from data compiled by Lagowski, J. J.; Moczygemba, G. A. In *The Chemistry of Non-aqueous Solvents*; Lagowski, J. J., Ed.; Academic: New York, 1967; Volume 2, pp. 323–330.

---

3. Throughout this chapter solubilities are tabulated in molal units (m, moles of solute per kilogram of solvent).

solubilities, and, in fact, saturated NH_4NO_3 is so concentrated that its normal boiling temperature is over 30°C higher than that of pure NH_3. Nitrates and thiocyanates are also quite soluble. Salts of divalent and trivalent ions such as Ba^{2+}, Sr^{2+}, CO_3^{2-}, and PO_4^{3-} have low solubilities, except for those cations that form complexes:

$$CuX_2 + 4NH_3 \longrightarrow [Cu(NH_3)_4]^{2+} + 2X^- \tag{11.18}$$

These solubility tendencies make for some interesting contrasts to aqueous chemistry. For example, for the reaction

$$2AgNO_3 + CaCl_2 \rightleftharpoons 2AgCl + Ca(NO_3)_2 \tag{11.19}$$

the equilibrium lies far to the right in H_2O (because of the insolubility of AgCl), but far to the left in NH_3 (because of the insolubility of $CaCl_2$).

Salts of the amide ion (NH_2^-) have low solubilities. This is in contrast to aqueous chemistry, where certain salts of the conjugate base OH^- are very soluble (eg, over 15 M for NaOH).

Moderately polar organic compounds (alcohols, aldehydes, ketones, amines, and amides) are generally more soluble in ammonia than in water. Low molecular weight carboxylic acids are soluble, but longer-chain mono-acids and diacids can be dissolved only as their ammonium salts.

Brønsted Acid–Base Chemistry

As mentioned above, the high basicity of ammonia causes certain species that produce neutral or even basic aqueous solutions to be proton donors in NH_3. However, strong bases (those that are 100% protonated) are rare in ammonia. The two most common examples are the oxide and hydride ions:

$$O^{2-} + NH_3 \longrightarrow OH^- + NH_2^- \tag{11.20}$$

$$H^- + NH_3 \longrightarrow H_2 + NH_2^- \tag{11.21}$$

We stated above that acidities are enhanced, but because of ion pairing, this increased acidity is not always reflected in the K_a. That is, for the sequence

$$HA + NH_3 \xrightarrow{\;\;K_1\;\;} [NH_4^+A^-] \tag{11.22}$$

$$[NH_4^+A^-] + (x + y - 1)NH_3 \xrightarrow{\;\;K_2\;\;} H(NH_3)_x^+ + A(NH_3)_y^- \tag{11.23}$$

the second equilibrium constant K_2 is often small. (Dissociation is so complete in water that K_2 is normally disregarded.) The pairing of NH_4^+ and Cl^- ions is so extensive that the experimental K_a ($K_a = K_1 \cdot K_2$) of HCl in liquid ammonia is only about 10^{-4}!

The high basicity of ammonia can be used to synthetic advantage, since it stabilizes reactants that would be destroyed by water. For example, the acetylide ion and its derivatives are rapidly protonated in aqueous solution:

$$RC{\equiv}C^- + H_2O \longrightarrow RC{\equiv}CH + OH^- \qquad (11.24)$$

However, these anions can be prepared in liquid ammonia and then reacted with metal cations to produce metal acetylides such as $CuC{\equiv}CR$, $Ni(C{\equiv}CR)_2 \cdot 4NH_3$, and $K_2[Zn(C{\equiv}CR)_4]$.[4]

Several cations demonstrate amphoteric behavior in liquid NH_3. For example, zinc ion can be precipitated and then redissolved by the addition of excess conjugate base in either water or ammonia:

$$Zn^{2+} \xrightarrow{\ 2OH^-\ } Zn(OH)_2(s) \xrightarrow{\ 2OH^-\ } [Zn(OH)_4]^{2-} \qquad (11.25)$$

$$Zn^{2+} \xrightarrow{\ 2NH_2^-\ } Zn(NH_2)_2(s) \xrightarrow{\ 2NH_2^-\ } [Zn(NH_2)_4]^{2-} \qquad (11.26)$$

Complexation (Lewis Acid–Base) Chemistry

The cations of ammonia-soluble metal salts are solvated by NH_3, with solvation numbers similar to those in H_2O. Such complexes are often acidic because of solvolysis (*ammoniolysis*):

$$M(NH_3)_x^{n+} \xrightarrow{\ NH_3\ } NH_4^+ + [M(NH_3)_{x-1}(NH_2)]^{(n-1)+} \qquad (11.27)$$

The extent of solvolysis is generally less than in water. For example, the K_h of Au^{3+} is about 3×10^{-8} in NH_3 and 3×10^{-2} in H_2O. As in water, the addition of strong base often causes precipitation; see equations (11.25) and (11.26), and the following:

$$[Cr(H_2O)_6]^{3+} + 3OH^- \longrightarrow Cr(OH)_3(s) + 6H_2O \qquad (11.28)$$

$$[Cr(NH_3)_6]^{3+} + 3NH_2^- \longrightarrow Cr(NH_2)_3(s) + 6NH_3 \qquad (11.29)$$

---

4. But great care must be exercised because of the explosive nature of such species!

Also as in water, solvent coordination to positive centers often leads to nucleophilic displacement if appropriate leaving groups are present.[5]

$$R_3SiCl + 2H_2O \longrightarrow R_3SiOH + H_3O^+ + Cl^- \tag{11.30}$$

$$R_3SiCl + 2NH_3 \longrightarrow R_3SiNH_2 + NH_4^+ + Cl^- \tag{11.31}$$

Oxidation–Reduction Chemistry

Standard reduction potentials in NH_3 are referenced to the standard hydrogen electrode (SHE) equivalent:

$$2NH_4^+ + 2e^- \longrightarrow 2NH_3 + H_2 \qquad \mathscr{E}^0 = 0.00 \text{ V} \tag{11.32}$$

The reduction of N_2 to NH_3 occurs at $+0.04$ V with respect to that reference:

$$N_2 + 6NH_4^+ + 6e^- \longrightarrow 8NH_3 \qquad \mathscr{E}^0 = +0.04 \text{ V} \tag{11.33}$$

The corresponding potentials in basic solution are

$$2NH_3 + 2e^- \longrightarrow 2NH_2^- + H_2 \qquad \mathscr{E}^0 = -1.59 \text{ V} \tag{11.34}$$

$$N_2 + 4NH_3 + 6e^- \longrightarrow 6NH_2^- \qquad \mathscr{E}^0 = -1.55 \text{ V} \tag{11.35}$$

The range of thermodynamic stability toward oxidation or reduction in acidic or basic ammonia is therefore only ± 0.04 V! Fortunately, redox reactions in NH_3 tend to be slow, and there is considerable overvoltage (typically about 1 V, compared to roughly 0.5 V in water). Hence, there is a reasonable domain of kinetic stability (about 2.0 V, compared to 2.3 V in water; see Figure 11.2).

A sampling of reduction potentials in liquid ammonia are plotted in Figure 11.3 (p. 348). There is a noticeable correlation to aqueous potentials, but differences can be observed as well. The more positive potentials exhibited for metal cations reflect their reduced stability (because of the smaller solvation energies) in NH_3. As a consequence, some of the less active free metals are stabilized toward oxidation by acids.

$$Pb + 2H^+ \longrightarrow Pb^{2+} + H_2 \qquad \mathscr{E}^0 = +0.13 \text{ V } (H_2O) \tag{11.36}$$

$$Pb + 2NH_4^+ \longrightarrow Pb^{2+} + H_2 + 2NH_3 \qquad \mathscr{E}^0 = -0.32 \text{ V } (NH_3) \tag{11.37}$$

---

5. The products R_3SiOH and R_3SiNH_2 may undergo condensation reactions to give $(R_3Si)_2O$ and $(R_3Si)_2NH$, respectively.

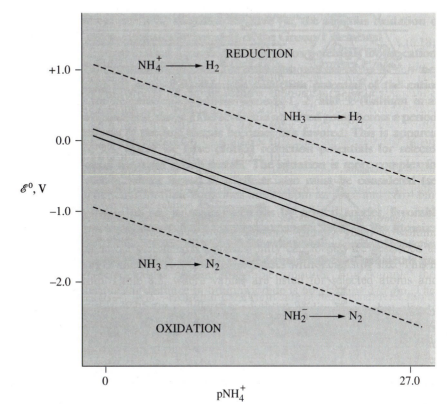

Figure 11.2
The stability domain of the solvent toward oxidation–reduction as a function of "pH" (ie, pNH_4^+) in liquid ammonia. The solid and dashed lines represent boundaries of thermodynamic and kinetic stability, respectively.

The preference of ammonia for softer ions is also apparent from Figure 11.3. For example, there is a clear preference for Cu^+ over Cu^{2+}. This has the effect of reversing the aqueous tendency of Cu^+ to disproportionate.

$$2Cu^+ \longrightarrow Cu + Cu^{2+} \qquad \mathscr{E}^0 = +0.18 \text{ V } (H_2O) \qquad \textbf{(11.38)}$$
$$= -0.02 \text{ V } (NH_3)$$

Solutions of Metals in Liquid Ammonia[6]

The solvated electron was discussed briefly in Chapter 9. It is most often prepared in the laboratory by dissolving active metals in liquid NH_3.

6. Edwards, P. P. *Adv. Inorg. Chem. Radiochem.* **1982**, *25*, 135; Dye, J. L. *J. Chem Educ.* **1977**, *54*, 332.

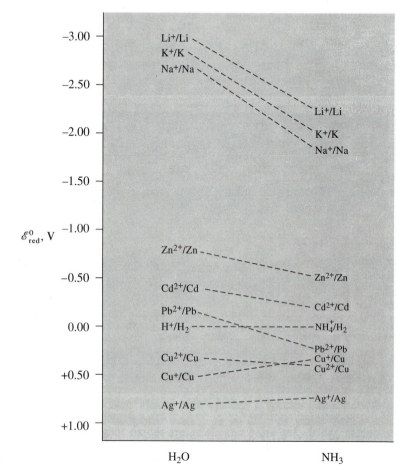

Figure 11.3
Standard reduction potentials (acidic solution) for selected cations in ammonia and water. [Aqueous potentials are taken from Table 9.1; those for NH₃ solutions are from Jolly, W. L. *J. Chem. Educ.* **1956**, *33*, 512.]

When a Group 1 metal dissolves in ammonia under rigorously anhydrous conditions, a deep blue solution is formed initially. Such solutions have been the subject of much study. The first of a complex sequence of reactions is thought to be the dissociation of an electron into the solvent, with the resulting cation and the electron both becoming solvated:

$$\text{M(am)} \xrightarrow{\text{NH}_3} \text{M}^+\text{(am)} + e^-\text{(am)} \tag{11.39}$$

This reaction is remarkable in that gas-phase ionization requires at least 375 kJ/mol, and is suggestive of considerable solvation energy for one or both of the charged species.

Some of the other equilibria established in metal–ammonia solutions are described by equations (11.40)–(11.44):

$$M(s) + e^- \rightleftharpoons M^- \tag{11.40}$$

$$e^- + e^- \rightleftharpoons e_2^{2-} \tag{11.41}$$

$$M^+ + e_2^{2-} \rightleftharpoons M^- \tag{11.42}$$

$$M + M \rightleftharpoons M_2 \tag{11.43}$$

$$M^+ + M^- \rightleftharpoons M_2 \tag{11.44}$$

For sodium–ammonia solutions near the boiling point (240 K), the equilibrium constants are 1.0×10^{-4} for equation (11.39), 1.0×10^3 for equation (11.40), and 5.2×10^3 for equation (11.43). A plot of the distribution of various species as a function of metal concentration is given in Figure 11.4.

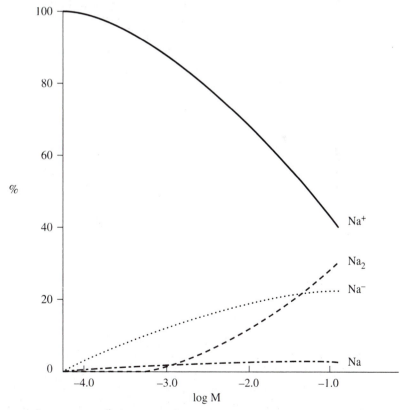

Figure 11.4 The distribution of sodium-containing species as a function of concentration of sodium–ammonia solutions at 240 K. The logarithm of the molarity of dissolved sodium is plotted on the horizontal axis. [Values calculated from data compiled by Thompson, J. In *The Chemistry of Non-aqueous Solvents*; Lagowski, J. J., Ed.; Academic: New York, 1967; Volume 2.]

The blue solution described above is sometimes referred to as the "dilute phase," because it persists only at relatively low concentrations (up to about 2 mol% Na at 220 K). Above 8 mol% Na, a bronze-colored, metallic luster is observed, and the electrical conductivities and magnetic susceptibilities of such solutions are comparable to those of pure metals. Under certain conditions (eg, the cooling of 4 mol% Na in NH_3 to below 231 K), the two solutions coexist as immiscible phases. Both the blue and bronze solutions have enormous equivalent conductances. The conductance drops to a minimum near 0.05 M (probably because of ion pairing) and then increases with increasing concentration.

The high solubilities of the alkali metals in ammonia are also noteworthy. Saturated Cs in NH_3 has a concentration of about 25 M at 220 K, and saturated Na in NH_3 is about 11 M at that temperature.

The blue color characteristic of metal–ammonia solutions arises from an absorption having a wavelength maximum at about 1500 nm and tailing into the visible region. Since the metals themselves show no comparable absorption, this must be due to the solvated electrons. The prevailing models of electron solvation place the electron in a "cavity" of solvent molecules, and calculations indicate that the cavity is about 600–680 pm in diameter.[7] This suggests considerable unoccupied volume (see Figure 11.5), and is consistent with the very low densities of these solutions (lower than those of either the free metals or liquid ammonia itself).

Reactions

The reactions that occur in metal–ammonia solutions can be organized into several categories. We give some specific examples below.

In some cases, simple reduction occurs:

$$O_2 \xrightarrow{e^-(am)} O_2^- \xrightarrow{e^-(am)} O_2^{2-} \tag{11.45}$$

$$[Pt(NH_3)_4]^{2+} + 2e^-(am) \longrightarrow Pt(NH_3)_4 \tag{11.46}$$

$$[Ni(CN)_4]^{2-} + 2e^-(am) \longrightarrow [Ni(CN)_4]^{4-} \tag{11.47}$$

Certain metals can be stabilized in unusually low oxidation states (such as Pt^0 and Ni^0) in this manner.

---

7.   See Dye, J. L. *Adv. Inorg. Chem. Radiochem.* **1982**, *25*, 135; and/or Symons, M. R. C. *Chem. Soc. Rev.* **1976**, *5*, 337; and references cited.

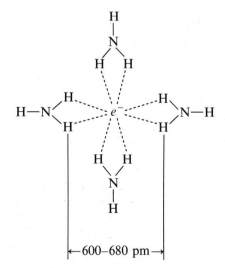

Figure 11.5
Idealized representation of the cavity model for the solvation of electrons in liquid ammonia.

Two-electron reduction may lead to homonuclear bond rupture and the formation of anions:

$$R_3Si–SiR_3 + 2e^-(am) \longrightarrow 2R_3Si^- \tag{11.48}$$

$$Mn_2(CO)_{10} + 2e^-(am) \longrightarrow 2[Mn(CO)_5]^- \tag{11.49}$$

$$C_6H_5NH–NH_2 + 2e^-(am) \longrightarrow C_6H_5NH^- + NH_2^- \tag{11.50}$$

Nonmetal hydrides often react via the evolution of H_2:

$$2SiH_4 + 2e^-(am) \longrightarrow 2SiH_3^- + H_2 \tag{11.51}$$

The cleavage of covalent halides to produce halide ions is common:

$$CH_3Br + 2e^-(am) + NH_3 \longrightarrow CH_4 + Br^- + NH_2^- \tag{11.52}$$

$$2C_6H_5I + 2e^-(am) \longrightarrow C_6H_5–C_6H_5 + 2I^- \tag{11.53}$$

Carbon–carbon double bonds can be hydrogenated:

$$R_2C{=}CR_2 + 2e^-(am) + 2NH_3 \longrightarrow R_2CH–CHR_2 + 2NH_2^- \tag{11.54}$$

$$\bigcirc\!\!\!\!\| + 4e^-(am) + 4NH_3 \longrightarrow \bigcirc + 4NH_2^- \tag{11.55}$$

Some remarkable polynuclear anions of metalloids and representative metals have been synthesized:

$$Pb \xrightarrow{e^-(am)} Pb_7^{4-}, Pb_9^{4-}, etc. \tag{11.56}$$

$$As_4 \xrightarrow{e^-(am)} As_3^{3-}, As_5^{3-}, As_7^{3-}, etc. \tag{11.57}$$

$$S_8 \xrightarrow{e^-(am)} S_x^{2-} \qquad (x = 4, 6, 7, 8, etc.) \tag{11.58}$$

(Homonuclear anions such as Pb_9^{4-} and As_7^{3-} will be discussed in Chapter 19.) The elemental sulfur–liquid ammonia system is particularly complex.[8] If metallic sodium is present, then the least reduced sulfur-containing anion in solution is S_6^{2-}, which is in equilibrium with the S_3^- radical anion:

$$S_6^{2-} \rightleftharpoons 2S_3^- \qquad K_{eq} = 0.043\ (20°C) \tag{11.59}$$

There is also spectroscopic evidence pointing to the existence of heteronuclear anions such as S_4N^-. This species is thought to arise from the reaction

$$\tfrac{5}{4}S_8 + 4NH_3 \longrightarrow S_4N^- + S_6^{2-} + 3NH_4^+ \tag{11.60}$$

11.3 Hydrogen Fluoride as a Solvent

The dominant characteristics of HF as a solvent can be deduced from the periodic relationship $H_3N–H_2O–HF$. The same factors that cause ammonia to be generally similar to water, but more basic, cause hydrogen fluoride to be generally similar to H_2O, but more acidic. Some of the trends that can be discerned from Table 11.2 are listed below.

$$\epsilon: \qquad HF \approx H_2O > NH_3$$

Boiling point: $\quad HF < H_2O > NH_3$

Liquid range: $\quad HF \approx H_2O > NH_3$

$$K_s: \qquad HF > H_2O >> NH_3$$

Many of the chemical and physical properties of liquid hydrogen fluoride can be traced to its very strong hydrogen bonding. The molecular structure consists of a variety of chains and rings, with an average molecular

---

8. Dubois, P.; Lelieur, J. P.; Lepoutre, G. *Inorg. Chem.* **1988**, *27*, 73; *Inorg. Chem.* **1987**, *26*, 1897.

weight of about 70 g/mol (roughly 3.5 times the formula weight). The fact that HF forms strong hydrogen bonds to both of its autoprotolysis products ($H^+\cdots F$–H and $F^-\cdots H$–F) leads to a large K_s (2.1×10^{-11} at $0°C$); in fact, hydrogen bonding is so strong that the self-ionization equilibrium is more correctly written as

$$3HF \rightleftharpoons H_2F^+ + HF_2^- \tag{11.61}$$

Solubilities

Solubility data are largely unavailable for liquid hydrogen fluoride, but a sampling is given in Table 11.4. Many of the same tendencies observed for H_2O and NH_3 also apply for HF. Monovalent salts are generally more soluble than those of divalent or trivalent ions, because it is easier to overcome the smaller lattice energies of 1:1 salts. We would expect ionic solutes to be dissociated because of the large dielectric constant, and the high experimental conductivities of such solutions show that this is indeed the case. Many metal fluorides (especially those of Group 1) simply dissolve in HF with dissociation. In that respect they are formally equivalent to the corresponding hydroxides in H_2O,

$$MOH \xrightarrow{\ H_2O\ } M^+ + OH^- \tag{11.62}$$

$$MF \xrightarrow{\ HF\ } M^+ + F^- \tag{11.63}$$

although the previously mentioned hydrogen bonding between F^- and HF converts equation (11.63) to

$$MF + HF \longrightarrow M^+ + HF_2^- \tag{11.64}$$

Table 11.4 Molal solubilities of selected fluorides in HF

Compound	Solubility, m	Compound	Solubility, m
LiF	3.96	CaF_2	0.11 (14°C)
NaF	7.17	SrF_2	1.18
KF	6.28	BaF_2	0.32
RbF	10.53 (20°C)	ZnF_2	0.0023 (14°C)
CsF	13.09	CdF_2	0.013
NH_4F	8.81	HgF_2	0.023
AgF	6.56	AlF_3	0.0002
MgF_2	0.004	CrF_3	0.022

Note: All at temperatures of $10 \pm 2°C$ unless otherwise noted.
Source: Jache, A. W.; Cady, G. W. *J. Chem. Phys.* **1952**, *56*, 1109.

Reaction Chemistry in HF

Most nonionic solutes undergo some type of chemical reaction with liquid HF. Common types of behavior are described below.

Brønsted Acid–Base Reactions The acidity of hydrogen fluoride causes many species to be protonated upon dissolution:

$$H_2O + 2HF \longrightarrow H_3O^+ + HF_2^- \tag{11.65}$$

$$Fe(CO)_5 + 2HF \longrightarrow HFe(CO)_5^+ + HF_2^- \tag{11.66}$$

$$CH_3CO_2H + 2HF \longrightarrow CH_3C(OH)_2^+ + HF_2^- \tag{11.67}$$

$$\text{\Large$\searrow$}C{=}O + 2HF \longrightarrow \text{\Large$\searrow$}C{=}OH^+ + HF_2^- \tag{11.68}$$

$$-\overset{|}{\underset{|}{C}}-O-\overset{|}{\underset{|}{C}}- + 2HF \longrightarrow -\overset{|}{\underset{|}{C}}-\overset{H}{\underset{+}{O}}-\overset{|}{\underset{|}{C}}- + HF_2^- \tag{11.69}$$

Unsaturated carbon–carbon bonds are also protonated, with the end result often being the addition of HF across the double bond:

$$\text{\Large$\searrow$}C{=}C\text{\Large$\nwarrow$} \xrightarrow{\ H^+\ } \left[H-\overset{|}{\underset{|}{C}}-\overset{|}{\underset{|}{C}}^+ \right] \xrightarrow{\ F^-\ } H-\overset{|}{\underset{|}{C}}-\overset{|}{\underset{|}{C}}-F \tag{11.70}$$

Lewis Acid–Base Reactions: Fluoride Ion Transfer A few fluorine-containing solutes are sufficiently strong Lewis acids to accept F^- from the solvent; for example,

$$SbF_5 + 2HF \longrightarrow H_2F^+ + SbF_6^- \tag{11.71}$$

$$BF_3 + 2HF \longrightarrow H_2F^+ + BF_4^- \tag{11.72}$$

However, a greater number behave as Lewis bases:

$$BrF_3 + HF \longrightarrow BrF_2^+ + HF_2^- \tag{11.73}$$

$$IF_5 + HF \longrightarrow IF_4^+ + HF_2^- \tag{11.74}$$

Fluorination and Redox Reactions Inorganic oxides and oxyanions are prone to fluorination in liquid hydrogen fluoride, often producing H_2O (H_3O^+) as a by-product:

$$HSO_4^- + 3HF \longrightarrow SO_3F^- + H_3O^+ + HF_2^- \tag{11.75}$$

$$SiO_2 + 8HF \longrightarrow SiF_4 + 2H_3O^+ + 2HF_2^- \tag{11.76}$$

Equation (11.76) explains the destruction of laboratory glassware by HF. To circumvent this reaction, a variety of special techniques and apparatus have been developed for preparative work involving hydrogen fluoride. They include glassware coated with paraffin wax and the use of polyethylene, polypropylene, and teflon reaction vessels.[9]

The fact that HF and F^- are very difficult to oxidize makes this a nonreducing solvent, and many species that are reduced by H_2O or NH_3 are stable in HF. An example is the permanganate ion, which is partly fluorinated but remains in the Mn^{VII} state:

$$MnO_4^- + 5HF \longrightarrow MnO_3F + H_3O^+ + 2HF_2^- \qquad (11.77)$$

Superacids[10]

Antimony pentafluoride behaves as a Lewis acid in HF, as indicated by equation (11.71). Its addition to liquid HF produces a mixture of remarkable acidity, capable of protonating species that normally do not act as bases. For example,

$$Xe \xrightarrow{\;HF/SbF_5\;} HXe^+ \qquad (11.78)$$

$$Br_2 \xrightarrow{\;HF/SbF_5\;} HBr_2^+ \qquad (11.79)$$

$$(C_6H_5)_3COH \xrightarrow{\;HF/SbF_5\;} (C_6H_5)_3CH^+ + H_3O^+ \qquad (11.80)$$

The HF/SbF_5 system has been described as a *superacid* because of this impressive Brønsted acidity. Another superacid is derived from HF, H_2SO_4, and SbF_5:

$$H_2SO_4 + 3HF \longrightarrow HSO_3F + H_3O^+ + HF_2^- \qquad (11.81)$$

$$2HSO_3F + SbF_5 \longrightarrow H_2SO_3F^+ + SbF_5(SO_3F)^- \qquad (11.82)$$

The HSO_3F/SbF_5 mixture has many practical advantages as an acidic solvent. It has a long and convenient liquid range (from $-89°C$ to $+163°C$); its viscosity and self-ionization constants are low compared to H_2SO_4; and unlike hydrogen fluoride, it is inert toward laboratory glassware (provided that care is taken to remove traces of HF). In addition to protonation

9.   See Hyman, H. H.; Katz, J. J. In *Non-aqueous Solvent Systems*; Waddington, T. C., Ed.; Academic: New York, 1965; pp. 50–52.

10.   Olah, G. A.; Prakash, G. K. S.; Sommer, J. *Science* **1979**, *206*, 13; Jache, A. W. *Adv. Inorg. Chem. Radiochem.* **1974**, *16*, 177; Gillespie, R. J.; Passmore, J. *Acc. Chem. Res.* **1971**, *4*, 413.

reactions, such mixtures have been used for fluorinations (eg, the conversions of ClO_4^- to ClO_3F, Br_2O_3 to BrF_3, and CrO_4^{2-} to CrO_2F_2) and also for the formation of some unusual polyatomic cations such as I_3^+, I_5^+, S_8^{2+}, and Se_4^{2+}.

11.4   Acetonitrile: An Aprotic Solvent

Certain liquids are popular solvents by virtue of their relatively high polarities and lack of strong Brønsted acidity or basicity. Acetonitrile provides a representative example.

Acetonitrile has a moderate dielectric constant ($\epsilon/\epsilon_0 = 36$ at room temperature), a large dipole moment, and very low inherent acidity. Its autoprotolysis constant is difficult to determine (partly because the CH_2CN^- anion promotes solvent polymerization), but is believed to be about 10^{-32}. Thus, its effective pH scale is very long, making it an excellent differentiating solvent.

Solubilities

Acetonitrile is a mediocre solvent for ionic compounds. Salts of small anions tend to be less soluble than those of large, polarizable anions, so the solubilities of salts increase in the series

$$MOH, MF << MCl < MBr < MI < MSCN < MB(C_6H_5)_4$$

For certain metal halides, the addition of excess X^- increases the solubility through complex formation. This is the case for mercuric iodide:

$$HgI_2(s) + I^- \xrightarrow{\quad CH_3CN \quad} [HgI_3]^-(\text{acetonitrile}) \tag{11.83}$$

Alternatively, it is sometimes possible to increase solubility by adding a hydrogen bonding agent (eg, acetic acid) to improve the solvation of anions.

The solubilities of selected inorganic compounds in acetonitrile are given in Table 11.5.

Acid–Base Chemistry

Since most conjugate bases are anions, the poor solvation of such species by acetonitrile tends to inhibit Brønsted acidities. As is the case for ammonia, an acid may be ionized but not necessarily dissociated:

$$HA + CH_3CN \; \rightleftharpoons \; CH_3CNH^+\cdots A^- \; \rightleftharpoons \; CH_3NH^+ + A^-$$
$$\tag{11.84}$$

Table 11.5 Molal solubilities of selected inorganic compounds in acetonitrile

Compound	Solubility, m	Compound	Solubility, m
LiCl	3×10^{-2}	CsBr	6×10^{-3}
NaF	7×10^{-4}	CsI	3×10^{-2}
NaCl	3×10^{-5}	AgCl	4×10^{-7}
KCl	3×10^{-4}	AgBr	4×10^{-7}
KBr	2×10^{-5}	AgI	8×10^{-8}
KI	1×10^{-1}	AgSCN	1×10^{-5}
CsCl	4×10^{-4}	AgB(C_6H_5)$_4$	3×10^{-4}

Sources: Alexander, R.; Ko, E. C. F.; Mac, Y. C.; Parker, A. J. *J. Am. Chem. Soc.* **1967**, *89*, 3703; Luehrs, D. C.; Iwamoto, R. T.; Kleinberg, J. *Inorg. Chem.* **1966**, *5*, 201; Price, E. In *The Chemistry of Non-aqueous Solvents*; Lagowski, J. J., Ed.; Academic: New York, 1967; Volume 1, p. 70.

Because of the poor solvation, the phenomenon of *homoconjugation*— that is, "solvation" of a solute anion by a solute molecule—is important in CH_3CN.

$$A^- + HA \longrightarrow HA_2^- \tag{11.85}$$

The "K_a" reaction then amounts to

$$2HA + CH_3CN \longrightarrow CH_3CNH^+ + HA_2^- \tag{11.86}$$

Acid dissociation constants in acetonitrile must be interpreted with caution because of homoconjugation. Some literature values are given in Table 11.6.

Table 11.6 Selected acid and base dissociation constants in acetonitrile

Acid	pK_a	Base	pK_b
HCl	8.94	NH_3	16.46
HBr	5.51	CH_3NH_2	18.37
HNO_3	8.89	$(CH_3)_2NH$	18.73
H_2SO_4	7.29 (K_1)	$(C_2H_5)_3N$	18.46
C_6H_5COOH	20.7	$C_6H_5NH_2$	10.56
CH_3COOH	22.3	C_5H_5N	12.33
C_6H_5OH	26.6		
NH_4^+	16.5		
$HClO_4$	<1		

Source: Taken from data compiled by Popovych, O.; Tompkins, R. P. T. *Non-aqueous Solution Chemistry*; Wiley: New York, 1981; pp. 61–65.

The nitrogen lone pair makes acetonitrile somewhat basic (although both its Lewis and Brønsted basicities are low compared to most other nitrogen-containing species). However, it displays virtually no acid activity. Basicity constants (K_b's) are therefore very small compared to those in H_2O. For example, the pK_b of NH_3 is 4.7 in H_2O but 16.5 in CH_3CN.

Oxidation–Reduction Potentials

It can be observed from the tabulated potentials in Table 11.7 that oxidation–reduction chemistry in CH_3CN is not appreciably different from that in water. For example, the Na^+/Na couple has a standard potential of -2.71 V in water and -2.87 V in CH_3CN; the values for Zn^{2+}/Zn are -0.76 and -0.74 V, respectively. The few cases in which there are significant differences (eg, the Ag^+/Ag couple) can be explained by complexation or hard–soft effects.

Table 11.7   Standard reduction potentials in CH_3CN and H_2O

Couple	$\mathscr{E}^0_{red}$, CH_3CN	$\mathscr{E}^0_{red}$, H_2O	$\mathscr{E}^0(CH_3CN) - \mathscr{E}^0(H_2O)$
Li^+/Li	-3.23	-3.04	-0.19
Na^+/Na	-2.87	-2.71	-0.16
K^+/K	-3.16	-2.92	-0.24
Cs^+/Cs	-3.16	-2.92	-0.24
Ca^{2+}/Ca	-2.75	-2.84	$+0.09$
Zn^{2+}/Zn	-0.74	-0.76	$+0.02$
Cd^{2+}/Cd	-0.47	-0.40	-0.07
Cu^+/Cu	$+0.38$	$+0.52$	-0.14
Ag^+/Ag	$+0.23$	$+0.80$	-0.57

Note: Values are in volts.
Source: Strehlow, H. In *The Chemistry of Non-aqueous Solvents*; Lagowski, J. J., Ed.; Academic: New York, 1967; Volume 1, Chapter 4.

11.5   Chemistry in Molten Salts

The use of molten salts as solvents has become common over the past 30 years, and considerable information has been accumulated concerning such systems. There are several reasons why molten salts are useful media for chemical reactions, the most important of which are the following:

1. Molten salts dissolve a broad range of solutes, including those in which the dominant forces are ionic, polar and nonpolar covalent, and metallic bonds.

2. A wide range of liquid temperatures, ranging from below room temperature to over 1000°C, is available. This can be important for influencing solubility, in kinetic studies, etc.

3. The entire gamut of chemical reactions (acid–base, oxidation–reduction, complexation, substitution, etc.) has been shown to occur in these media.

4. The use of mixed solvents often allows for fine-tuning to a medium having the characteristics desired for a given purpose. For example, mixed melts can be prepared that are highly acidic but virtually inert to oxidation or reduction. Other mixed salts promote oxidation–reduction but not acid–base reactions.

General Properties: Correlations with Traditional Solvents

The same properties that were used to categorize other solvents (notably, polarity, complexing ability, and acidity) also can be applied to molten salts. Such solvents are divided into two broad categories—ionic or covalent—depending on the extent of self-ionization. Liquid NaCl is an ionic solvent because of its considerable dissociation, as evidenced by its very high electrical conductivity ($8000 \ \Omega^{-1} \ cm^{-1}$):

$$NaCl(l) \longrightarrow Na^+ + Cl^- \tag{11.87}$$

Liquid $AsCl_3$ is a covalent solvent because of its low conductivity ($10^{-3} \ \Omega^{-1} \ cm^{-1}$). The self-ionization of $AsCl_3$ involves chloride ion transfer:

$$2AsCl_3 \rightleftharpoons AsCl_2^+ + AsCl_4^- \tag{11.88}$$

Equation (11.88) is useful for defining acids and bases in $AsCl_3$ melts. A solute that increases the concentration of the solvent-derived cation, $AsCl_2^+$, is an acid:

$$HClO_4 + AsCl_3 \longrightarrow HCl + ClO_4^- + AsCl_2^+ \tag{11.89}$$

A solute that increases the concentration of the solvent-derived anion, $AsCl_4^-$, is a base:

$$KCl + AsCl_3 \longrightarrow K^+ + AsCl_4^- \tag{11.90}$$

The Eutectic Mixture of LiCl and KCl

We will illustrate the chemistry of molten salts using one particularly well-studied system—the eutectic mixture containing 59 mol% LiCl and 41 mol% KCl, which melts at 440°C and is most often used at 450–480°C.

Solutes in this binary solvent system fall into any of several categories. Ionic compounds are generally soluble and dissolve with dissociation. Oxides having appreciable covalent character (eg, Al_2O_3 and SiO_2) are often insoluble. It is frequently possible to solubilize a metal cation through the addition of either an oxidizing agent or an acid:

$$2FeO + 3Cl_2 \longrightarrow 2Fe^{3+} + 6Cl^- + O_2 \tag{11.91}$$

$$NiO + S_2O_7^{2-} \longrightarrow Ni^{2+} + 2SO_4^{2-} \tag{11.92}$$

$$SnO_2 + 4HCl \longrightarrow SnCl_4 + 2H_2O \tag{11.93}$$

Acid–base chemistry is dominated by ion-transfer reactions:

$$FeCl_2 + 2Cl^- \longrightarrow [FeCl_4]^{2-} \tag{11.94}$$

$$MgO + SO_3 \longrightarrow Mg^{2+} + SO_4^{2-} \tag{11.95}$$

A considerable amount of oxidation–reduction chemistry has been studied. Oxides of metals in high oxidation states are often reduced; for example, CuO to Cu^+, PbO_2 to Pb^{2+}, etc. The sulfite ion tends to be unstable, with the reaction products depending on their solubilities. The evolution of $SO_2(g)$ occurs for metal cations that form insoluble oxides:

$$PdSO_3 \longrightarrow PdO(s) + SO_2 \tag{11.96}$$

If the metal sulfide is insoluble, then disproportionation of SO_3^{2-} to sulfide and sulfate is observed:

$$4BaSO_3 \longrightarrow BaS(s) + 3BaSO_4 \tag{11.97}$$

Table 11.8 Standard reduction potentials for the LiCl + KCl eutectic

Half-Reaction	$\mathcal{E}^0$, V	Half-Reaction	$\mathcal{E}^0$, V
$Li^+ + 1e^- \longrightarrow Li$	-3.30	$Ni^{2+} + 2e^- \longrightarrow Ni$	-0.80
$Mg^{2+} + 2e^- \longrightarrow Mg$	-2.58	$Ag^+ + 1e^- \longrightarrow Ag$	-0.74
$Al^{3+} + 3e^- \longrightarrow Al$	-1.76	$Cr^{3+} + 1e^- \longrightarrow Cr^{2+}$	-0.52
$Zn^{2+} + 2e^- \longrightarrow Zn$	-1.57	$I_2 + 2e^- \longrightarrow 2I^-$	-0.21
$Cr^{2+} + 2e^- \longrightarrow Cr$	-1.42	$Pt^{2+} + 2e^- \longrightarrow Pt$	0.00
$Cd^{2+} + 2e^- \longrightarrow Cd$	-1.32	$Cu^{2+} + 1e^- \longrightarrow Cu^+$	$+0.06$
$Fe^{2+} + 2e^- \longrightarrow Fe$	-1.17	$Fe^{3+} + 1e^- \longrightarrow Fe^{2+}$	$+0.09$
$Sn^{2+} + 2e^- \longrightarrow Sn$	-1.08	$Br_2 + 2e^- \longrightarrow 2Br^-$	$+0.18$
$Cu^+ + 1e^- \longrightarrow Cu$	-0.96	$Cl_2 + 2e^- \longrightarrow 2Cl^-$	$+0.32$

Note: Values are referenced to the half-reaction Pt^{2+}/Pt.
Source: Laitinen, H. A.; Panky, J. W. *J. Inorg. Nucl. Chem.* **1977**, *39*, 255; *J. Am. Chem. Soc.* **1959**, *81*, 1053.

Oxidation–reduction potentials are given in Table 11.8. Since the table reference is the Pt^{II}/Pt half-reaction, direct comparisons to aqueous potentials are difficult. However, comparisons can be made for overall cells. For example, for the reaction between zinc metal and silver ion,

$$Zn + 2Ag^+ \longrightarrow Zn^{2+} + 2Ag \tag{11.98}$$

$\mathscr{E}^0_{cell} = +1.57 - 0.74 = +0.83$ V in $LiCl + KCl$, compared to $+0.763 + 0.799 = +1.562$ V in H_2O.

Bibliography

Olah, G. A.; Surya Prakash, G. K.; Sommer, J. *Superacids*; Wiley: New York, 1985.

Burger, K. *Ionic Solvation and Complex Formation Reactions in Non-aqueous Solvents*; Elsevier: Amsterdam, 1983.

Popovych, O.; Tompkins, R. P. T. *Non-aqueous Solution Chemistry*; Wiley: New York, 1981.

Nicholls, D. *Inorganic Chemistry in Liquid Ammonia*; Elsevier: Amsterdam, 1979.

Burgess, J. *Metal Ions in Solution*; Ellis Horwood: Chichester, 1978.

Jolly, W. L., Ed. *Metal–Ammonia Solutions*; Dowden, Hutchinson, and Ross: East Stroudsburg, PA, 1972.

Hills, G. J.; Kerridge, D. *Fused Salts*; Elsevier: New York, 1971.

Gutmann, V. *Coordination Chemistry in Non-aqueous Solutions*; Springer-Verlag: Berlin, 1968.

Bloom, H. *The Chemistry of Molten Salts*; Benjamin: New York, 1967.

Questions and Problems

1. Organize the solvents listed in Table 11.2 into four categories: acidic, basic, polar aprotic, and nonpolar. Justify your assignments.

2. Classify each of the following solutes as an acid or base in the solvent given:
   (a) $LiPBr_6$ in PBr_5 (b) $AsBr_2(NO_3)$ in $AsBr_3$ (c) KHF_2 in HF
   (d) H_2O in HF   (e) $CaCl_2$ in $POCl_3$ (f) BF_3 in HF

3. Identify a solvent that would permit the differentiation of acids having pK_a's of:
   (a) 2.0 and 7.0   (b) 10.0 and 16.0   (c) 20.0 and 28.0

4. (a) In which (if any) of the following solvents is water essentially 100% protonated?

   $$NH_3, \quad CH_3OH, \quad H_2SO_4$$

(b) In which (if any) of the following solvents is HF essentially 100% deprotonated?

$$NH_3, \quad CH_3CH_2OH, \quad HSO_3F$$

5. The enthalpies of transfer from water to acetonitrile for Na^+, K^+, Rb^+, and Ag^+ are -3.1, -5.4, -5.5, and -12.6 kJ/mol, respectively. Rationalize this order.

6. Write a self-ionization equation for each of the following solvents, and use it to identify the prototype acid and base of each system:
   (a) $CH_3CH_2NH_2$ (b) BrF_3 (c) $NaNO_3$

7. Calcium oxide is a basic anhydride with respect to aqueous acid–base chemistry.
   (a) Write the formula for the equivalent species in NH_3.
   (b) Write the equation for NH_3 solution that corresponds to the aqueous solution equation $CaO + H_2O \rightarrow Ca(OH)_2$.

8. For each of the following aqueous reactions, write the equation for the parallel reaction in liquid ammonia:
   (a) $HCl + RbOH \longrightarrow RbCl + H_2O$
   (b) $H_2SO_4 + 2KOH \longrightarrow K_2SO_4 + 2H_2O$
   (c) $Fe + 2H^+ \longrightarrow Fe^{2+} + H_2$
   (d) $O^{2-} + H_2O \longrightarrow 2OH^-$
   (e) $(C_4H_9)_2BCl + H_2O \longrightarrow (C_4H_9)_2BOH + HCl$
   (f) $AlCl_3 + 3OH^- \longrightarrow Al(OH)_3(s) + 3Cl^-$

9. Consider the species $As(NH_2)_3$ in liquid ammonia.
   (a) Suggest a plausible synthesis for this compound, and write the corresponding equation.
   (b) Of the reactions given below, which has the larger equilibrium constant?

$$As(NH_2)_3 + NH_3 \rightleftharpoons NH_4^+ + [As(NH_2)_2(NH)]^-$$

$$As(NH_2)_3 + NH_3 \rightleftharpoons [As(NH_2)_2(NH_3)]^+ + NH_2^-$$

     Explain your answer.

10. The pK_s for liquid ammonia is 33.
    (a) What is the molarity of NH_4^+ in neutral ammonia?
    (b) What is the molarity of NH_4^+ in strongly basic ($[NH_2^-] = 1.0$ M) ammonia?

11. Which solvent would be preferred for the titration of $HAsO_4^{2-}$ with strong base: H_2O, NH_3, or CH_3CO_2H? Defend your answer.

12. Determine the thermodynamically stable products for each of the following mixtures in liquid ammonia (assume standard conditions):
    (a) Cd and NH_4^+ (b) Cu and NH_4^+ (c) Zn, Cu^+, and NH_4^+

13. Complete and balance the following equations, assuming they occur in sodium–ammonia solution:

     (a) $SnH_4 + e^-(am) \longrightarrow$ _____ $+$ _____

     (b) $NO_2 + e^-(am) \longrightarrow$ _____

     (c) $NH_4^+ + e^-(am) \longrightarrow$ _____ $+$ _____

(d) $Me_3SnBr + e^-(am) \longrightarrow \underline{\hspace{1.5cm}} + \underline{\hspace{1.5cm}}$

(e) $OCl^- + e^-(am) \longrightarrow \underline{\hspace{1.5cm}} + \underline{\hspace{1.5cm}}$

(f) $HI + e^-(am) \longrightarrow \underline{\hspace{1.5cm}} + \underline{\hspace{1.5cm}}$

14. For each of the following aqueous reactions, write the equation for the parallel reaction in HF solution. If the position of equilibrium is significantly different in the two solvents, explain.

(a) $HClO_4 + H_2O \longrightarrow H_3O^+ + ClO_4^-$

(b) $CH_3NH_2 + H_2O \rightleftharpoons CH_3NH_3^+ + OH^-$

(c) $KCN + H_2O \rightleftharpoons KOH + HCN$

(d) $Mg(OH)_2 \rightleftharpoons Mg^{2+} + 2OH^-$

15. Write products for the following reactions in HF and balance the equations:

(a) $CaCl_2 + HF \longrightarrow \underline{\hspace{1.5cm}} + \underline{\hspace{1.5cm}}$

(b) $NO_2^- + HF \longrightarrow \underline{\hspace{1.5cm}} + \underline{\hspace{1.5cm}}$

(c) $CH_3CHO + HF \longrightarrow \underline{\hspace{1.5cm}} + \underline{\hspace{1.5cm}}$

(d) $C_3H_7COOH + HF \longrightarrow \underline{\hspace{1.5cm}} + \underline{\hspace{1.5cm}}$

(e) $CrO_4^{2-} + HF \longrightarrow CrO_2F_2 + \underline{\hspace{1.5cm}}$

(f) $AlF_3 + HF \longrightarrow \underline{\hspace{1.5cm}} + \underline{\hspace{1.5cm}}$

(g) $F_2C{=}CF_2 + HF \longrightarrow \underline{\hspace{1.5cm}}$

16. Identify the stepwise intermediates that link the given reactants and products in superacid media. Also, list any products that are not given.

(a) $KNO_3 \longrightarrow K^+, NO_2^+, H_3O^+$

(b) $XeF_6 \longrightarrow XeF_5^+$

(c) $CO_3^{2-} \longrightarrow CO_2, H_3O^+$

(d) $(Me_3Si)_2O \longrightarrow Me_3SiF$

17. The acidity of HF can be increased by the addition of BF_3. Explain, using words and an equation.

18. Draw Lewis structures for:

(a) $H_2SO_3F^+$ (b) $SbF_5(SO_3F)^-$

19. Write an equation to describe the acid dissociation of methanol in CH_3CN solution. Take homoconjugation into account.

20. The equilibrium constant for the disproportionation of cuprous ion in DMSO has a numerical value of about 2 at 25°C:

$$2Cu^+ \longrightarrow Cu + Cu^{2+}$$

The corresponding equilibrium constant in water is about 10^6. Rationalize this difference.

21. Complete and balance the following reactions occurring in molten salt (chloride ion-rich) media:

(a) $CuO + Cl^- \longrightarrow CuCl_2 + \underline{\hspace{1.5cm}}$

(b) $SO_3^{2-} + Hg^{2+} \longrightarrow Hg + SO_4^{2-} + \underline{\hspace{1.5cm}}$

(c) $Ca(OH)_2 + Cl^- \longrightarrow \underline{\hspace{1.5cm}} + \underline{\hspace{1.5cm}}$

(d) $HClO_4 + Cl^- \longrightarrow$ _____ + _____

(e) $TiCl_4 + Cl^- \longrightarrow$ _____

22. Estimate the equilibrium constant for the disproportionation of Cu^+ in the LiCl + KCl eutectic.

23. Arrange the following reactions in the LiCl + KCl eutectic in increasing order of completion:
    (a) $Sn^{2+} + Ni \longrightarrow Ni^{2+} + Sn$
    (b) $3Cr^{3+} + Al \longrightarrow Al^{3+} + 3Cr^{2+}$
    (c) $Cd + I_2 \longrightarrow Cd^{2+} + 2I^-$

24. The following are statements from the text of this chapter. Explain in your own words why each is true.
    (a) The extent of solvolysis is generally less in NH_3 than in water.
    (b) The fact that HF forms strong hydrogen bonds to both of its autoprotolysis products leads to a large K_s.
    (c) The K_a values in acetonitrile must be interpreted with caution.
    (d) Acetonitrile's basicity is low compared to most other nitrogen-containing species.

*25. Electrochemical evidence has been obtained for the existence of Cs^{III} (Moock, K.; Seppelt, K. *Angew. Chem. Int. Ed. Engl.* **1989**, *28*, 1676). Answer these questions after reading the cited article.
    (a) Acetonitrile is not a particularly good solvent for cesium salts, but cesium teflate ($Cs^+OTeF_5^-$) is sufficiently soluble for study. Why?
    (b) What property of acetonitrile made it a good solvent for this work? [*Hint*: What is the potential of the Cs^{III}/Cs^I couple?]
    (c) The authors discount the possibility that the oxidized state is Cs^{II}. What is their logic?

The Main Group Elements

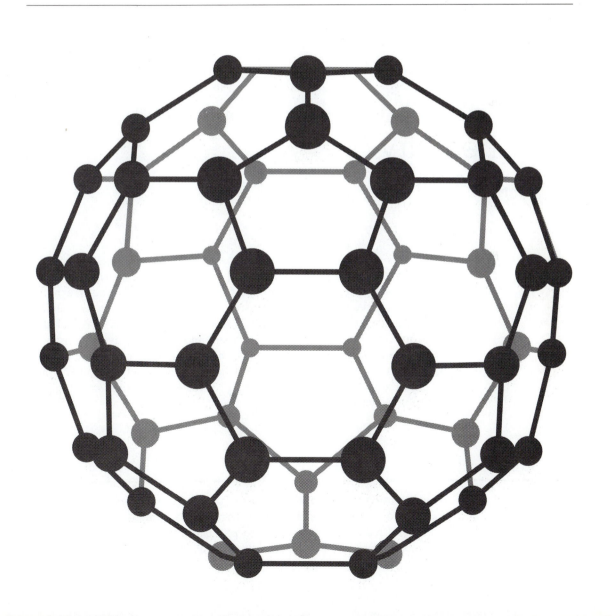

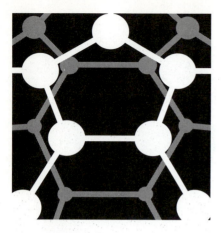

12

The Chemistry of
Groups 1 and 2

The metals of Groups 1 and 2 are often referred to as the *s-block* elements. The placement of hydrogen into Group 1 or Group 17 is somewhat arbitrary, and depends at least in part on personal preference. Hydrogen exhibits certain properties that are characteristic of each group. However, its chemical properties are unique, so it requires a discussion of its own. It is included in Group 1 for the purposes of this book, and its descriptive chemistry is considered in Sections 12.1 and 12.2. Other important aspects have already been described (hydrogen bonding in Chapter 8, Brønsted acids in Chapter 10, etc.). Later in the chapter (Sections 12.3–12.7), the Group 1 and 2 metals will be considered.

12.1   Elemental Hydrogen

As we mentioned in Chapter 7, most of the hydrogen found on Earth is in the H^I oxidation state, as H_2O and other hydroxy compounds. Chemical or electrolytic reduction is necessary to produce the free element as H_2. The best method of synthesis depends on the quantity needed. For small-scale laboratory preparations, H_2 is usually generated either by the electrolysis of water or through reactions of active metals with acids:

$$2H_2O \xrightarrow{\text{Electrolysis}} 2H_2(g) + O_2(g) \tag{12.1}$$

$$Zn + 2HCl \longrightarrow ZnCl_2 + H_2(g) \tag{12.2}$$

It is prepared industrially by high-temperature reactions of steam with either methane or coal.

$$CH_4 + H_2O(g) \xrightarrow{\Delta} 3H_2(g) + CO(g) \tag{12.3}$$

$$C + H_2O(g) \xrightarrow{\Delta} H_2(g) + CO(g) \tag{12.4}$$

$$C + 2H_2O(g) \xrightarrow{\Delta} 2H_2(g) + CO_2(g) \tag{12.5}$$

The major industrial uses for H_2 are for the synthesis of ammonia and as a fuel (often mixed with CO in *synthesis gas* or *water gas*).

The three known isotopes of hydrogen are ^1H (protium, 99.98% natural abundance), ^2H (deuterium, D, 0.016% abundance), and ^3H (tritium, T, about $10^{-16}\%$ abundance). Tritium is radioactive, undergoing decay by beta emission with a half-life of 12.3 years. Its presence in the atmosphere results from cosmic rays, which promote the nuclear reaction

$$^{14}_{7}N + ^{1}_{0}n \longrightarrow ^{12}_{6}C + ^{3}_{1}H \tag{12.6}$$

The presence of either or both of the heavier isotopes may alter the chemical properties of compounds. For example, the autoprotolysis constant of D_2O at room temperature is 3×10^{-15} (about three times less than that of normal water), and its dielectric constant is about 2% smaller than that of H_2O. The physical properties of isotopically different hydrides are sometimes sufficiently distinct to permit their separation. For example, the boiling points of H_2, D_2, and T_2 are 20.4, 23.7, and 25.0 K, respectively.

All six of the possible diatomic combinations, H_2, D_2, T_2, HD, HT, and DT, are known. Diprotium, H_2, is found in either of two isomeric forms, ortho or para, in which the nuclear spins of the bonded atoms are aligned with or in opposition to one another. The equilibrium mixture is close to 100% para near absolute zero, but the distribution is close to the theoretical ratio of 3:1 ortho to para at room temperature. The ortho–para interconversion is catalyzed by paramagnetic materials, so the two forms can be separated by chromatography.

12.2 Compounds of Hydrogen

Protonic and Hydridic Compounds

Hydrogen forms stable compounds with virtually every element in the periodic table, except those of Group 18. Because its electronegativity is close to the median (about 2.2 on a scale of 0.7 to 4.0), both positive and negative oxidation states are common. The oxidation number is commonly +1 when bonded to nonmetals and −1 when bonded to metals. Hydrogen is said to be *protonic* in the former case and *hydridic* in the latter. Certain nonmetals and metalloids (eg, C, P, As, and Ge) have electronegativities so close to hydrogen that neither label is appropriate. These hydrides are neither acidic nor basic, except under forcing conditions.

The preparation of compounds containing bonds between hydrogen and nonmetals or metalloids usually involves either water or some hydridic reagent:

$$Ca_3P_2 + 6H_2O \longrightarrow 2PH_3 + 3Ca(OH)_2 \qquad (12.7)$$

$$2BCl_3 + 6NaH \longrightarrow B_2H_6 + 6NaCl \qquad (12.8)$$

$$GeCl_4 + LiAlH_4 \longrightarrow GeH_4 + LiCl + AlCl_3 \qquad (12.9)$$

Saline Hydrides

Binary compounds in which hydrogen is bonded to a Group 1 or Group 2 element (except beryllium) are called *saline hydrides*. They conform to the ionic bonding model for the most part, although their experimental lattice energies are often less than predicted (see Chapter 6). They are much less thermally stable than the corresponding halides, with all except LiH decomposing to the free elements below their melting points. Most saline hydrides are best prepared by direct union at 300°C or above:

$$2M + H_2 \xrightarrow{\Delta} 2MH \qquad M = Li, Na, K, Rb, Cs \qquad (12.10)$$

$$M + H_2 \xrightarrow{\Delta} MH_2 \qquad M = Mg, Ca, Sr, Ba \qquad (12.11)$$

The saline hydrides are potent Brønsted bases and reducing agents:

$$HCl + H^- \longrightarrow H_2 + Cl^- \qquad (12.12)$$

$$SF_2 + 2H^- \longrightarrow H_2S + 2F^- \qquad (12.13)$$

$$Fe^{2+} + 2H^- \longrightarrow Fe + H_2 \qquad (12.14)$$

Hydride ion is a powerful reducing agent, with a standard oxidation potential of $+2.25$ V in water:

$$2H^- \longrightarrow H_2 + 2e^- \qquad \mathcal{E}^0_{ox} = +2.25 \text{ V} \qquad (12.15)$$

A major chemical property of H^- is its nucleophilicity, and equations (12.12) and (12.13) might be described as nucleophilic displacements. The relative reactivities of the saline hydrides are as expected based on electronegativity. As the metal becomes less electronegative, the M–H bond becomes more polar; thus, nucleophilicity decreases in the order

$$CsH > RbH > KH > NaH > CaH_2 > MgH_2 \approx LiH$$

For example, the cesium and rubidium hydrides burn spontaneously in dry air. The others are less reactive toward O_2, but are decomposed by water and water vapor.

Another reaction common to saline hydrides is nucleophilic addition (Lewis basicity). Some important ternary hydrides are prepared in this manner.

$$LiH + MH_3 \longrightarrow LiMH_4 \qquad M = Al, B \qquad (12.16)$$

These species, along with their many derivatives and analogues such as $LiBH_3R$ and $NaAlH_4$, are widely used reducing agents in inorganic and organometallic synthesis.

Transition Metal Hydrides

The bonding in compounds containing hydrogen linked to a transition metal is difficult to characterize. There appears to be a combination of ionic, polar covalent, and/or metallic bonding, with the relative proportions varying with the situation. Related to this, the stoichiometries of binary metal–hydrogen systems have been the subject of much debate. It is now known that many such systems form lattices that are typical of ionic or metallic bonding, but which are nonstoichiometric because of systematic or nonsystematic vacancies. For example, "TiH_2" crystallizes in the fluorite lattice. However, the observed stoichiometry is typically about $TiH_{1.8}$, because roughly 10% of the anionic sites are vacant. Such situations are intermediate between ionic and metallic bonding. The "vacant" lattice sites are presumably occupied by delocalized electrons. Thus, in this interpretation, there is 1 mol of M^{x+} cations, $x - n$ moles of hydride ions, and n moles of delocalized electrons per formula unit of MH_x.

Of all transition metal–hydrogen interactions, that with palladium may be the most interesting.[1] When placed under a hydrogen atmosphere, metallic palladium first adsorbs H_2 onto its surface. If the gas pressure is sufficiently high, the incorporation of hydrogen into the lattice itself then occurs, forming either of two phases (α or β); up to about 0.8 mol of hydrogen atoms can be added per mole of Pd. The lattice structure is not significantly altered, although there is some expansion (about 10% of the total volume). The nature of the bonding is not completely understood. Hydrogen gas is recovered quantitatively upon heating.

This system and others like it are potentially useful for the storage of H_2. Molecular hydrogen has been suggested as an ideal alternative to fossil fuels—it is abundant, it burns exothermically, and its combustion product is nonpolluting.[2] A major problem, however, is the great volume required for the storage of $H_2(g)$. In theory, the reversible absorption of H_2 by palladium can be used to overcome this difficulty.

There are at least two remaining types of hydrogen compounds. Hydride ion serves as a ligand in many transition metal complexes, and as such is an important component of many catalysts. This aspect of its chemistry will be discussed in Chapter 18. Finally, bridging hydrogens are found in the boron hydrides, metalloboranes, and other cluster systems. These species are unusual in that the hydrogen is, if not truly divalent, "connected" to two or more atoms simultaneously. The structures and reactions of such compounds will be described in Chapter 19.

12.3 The Free Metals

The Group 1 elements are the *alkali metals*, while those of Group 2 are the *alkaline earths*. The chemical similarities of all these elements, except beryllium, are such that they can be described together. Beryllium is almost as distinctive among the elements as hydrogen, so the primary aspects of its chemistry are considered separately in Section 12.6.

The heaviest members of these families, francium and radium, have no stable isotopes. Francium has at least 21 known isotopes, with the longest-lived (^{212}Fr) having a half-life of only 19 min. Thus, little is known about its reaction chemistry. Radium also has an abundance of isotopes, but since at least one has a very long lifetime (^{226}Ra, with $t_{1/2} = 1600$ years), it is more readily studied.

---

1. The deuterium–palladium system was the focal point of the "cold fusion" brouhaha. For a summary, see the June 1989 issue of *Scientific American*, pp. 22–24.

2. For discussions of the use of hydrogen as a fuel, see Dinga, G. P. *J. Chem. Educ.* **1988**, *65*, 688; and/or Cohen, R. L.; Wernick, J. H. *Science* **1981**, *214*, 1081.

Periodic Trends

The periodic trends exhibited by the Group 1 and 2 elements are satisfying in that the variations in chemical and physical properties are perhaps the most logical. That is, they obey the theoretical models we have developed better than any of the other families. For example, the regular variations in ionization energy and size are shown in Figures 12.1 and 12.2.

Some other important group trends and the underlying principles that explain them are listed below:

1. ***Melting Temperatures*** Recall from Chapter 7 that the melting points of the free elements decrease going down both families (except for magnesium), and that the Group 2 elements all melt at higher temperatures than the corresponding alkali metals. This is satisfactorily explained by the metallic bonding model.

2. ***Bond Energies of*** M_2 ***Compounds*** We noted in Chapter 3 that the Group 1 metals are diatomic in the vapor state, with bond orders of 1.0. Dissociation energies decrease in regular progression going down the

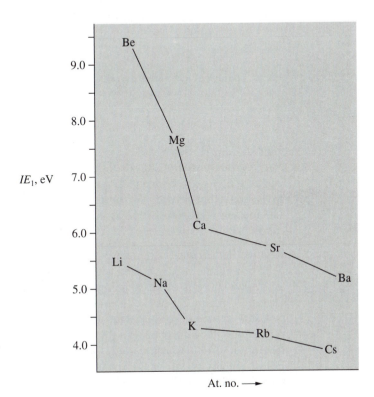

Figure 12.1
Trends in first ionization energies for the Group 1 and Group 2 elements. (Data taken from Table 1.7.)

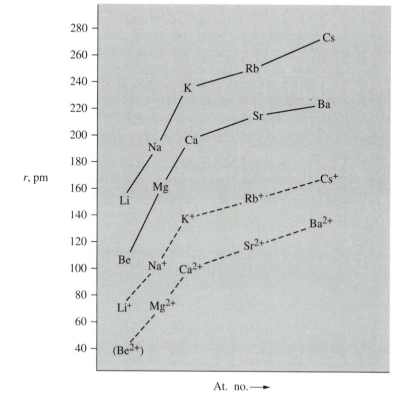

Figure 12.2
Trends in metallic and ionic radii for the Group 1 and Group 2 elements. Metallic radii are taken from Table 7.6. Ionic radii are for hexacoordination (monovalent for Group 1 and divalent for Group 2), as given in Table 6.7.

family. This is a size effect—increased size leads to decreased orbital overlap and weaker bonds.

3. **Hydrolysis Constants**   Recall that the extent of metal ion hydrolysis is much greater for the Group 2 cations than for their Group 1 counterparts, and that hydrolysis decreases going down each group. The explanation (Chapter 10) is based on the charge-to-size ratios of the cations.

4. **Relative Reactivities**   The relative reactivities of the free metals toward oxidizing agents vary inversely with their ionization energies. Recall that ionization energies decrease going down each family, and are greater in Group 2 than in Group 1. Visual observations of the reactions of these metals with water are consistent with these trends. Lithium pellets react very slowly when dropped into water at room temperature. Sodium tends to sputter on the surface and fizz because of the evolution of H_2. Potassium reacts with sufficient heat to cause the evolved hydrogen to burst into flame, while the oxidations of rubidium and cesium are nearly

instantaneous. The Group 2 metals are considerably less reactive. The two having the most endothermic ionizations, beryllium and magnesium, do not react with H_2O at room temperature.

Generalizing, the chemistry of the alkali metals is dominated by one-electron oxidations. Since the first ionization energy is greatest for lithium and smallest for cesium, reactivity increases going down Group 1. Furthermore, the reaction chemistry of lithium marks it as the most covalent member of the family, since it forms the strongest covalent bonds (because of its small size) and is also the most reluctant to lose its valence electron.

Two-electron oxidation is the rule for the alkaline earths, so it is useful to consider the sum of the first two ionization energies. These sums are 27.5 eV for beryllium, 22.7 eV for magnesium, and 18.0 eV or less for the remaining elements. The four heaviest members of the family, Ca, Sr, Ba, and Ra, all require relatively little energy for oxidation, and their reactions commonly involve ionic bond formation. For beryllium, the sum of IE_1 and IE_2 is so great that the loss of two electrons is less feasible; hence, covalency is the normal bonding pathway. Magnesium is intermediate between these extremes, and this is reflected in its chemical reactions.

Preparation and Uses

The Group 1 and 2 metals are found in a variety of minerals and ores. Sodium, potassium, magnesium, and calcium are present in huge quantities in the lithosphere; the others occur in lower abundance. The best-known minerals of these elements are listed in Table 12.1.

Table 12.1 Some common minerals of the Group 1 and 2 metals

Formula	Mineral (Common Name)	Formula	Mineral (Common Name)
$LiAlSi_2O_6$	Spodumene	$Mg_3Si_4O_{10}(OH)_2$	Talc
$NaCl$	Rock salt	$MgCO_3$	Magnesite
$Na_2CO_3 \cdot NaHCO_3$	Trona	$MgCO_3 \cdot CaCO_3$	Dolomite
$NaNO_3$	Chile saltpeter	$CaCO_3$	Calcite (limestone, marble, etc.)
$Na_2B_4O_7 \cdot 10H_2O$	Borax		
KCl	Sylvite	$CaSO_4 \cdot 2H_2O$	Gypsum, alabaster
$KCl \cdot MgCl_2 \cdot 6H_2O$	Carnallite	CaF_2	Fluorspar, fluorite
$KCl \cdot NaCl$	Sylvinite	$SrSO_4$	Celestite
KNO_3	Saltpeter	$SrCO_3$	Strontianite
$Be_3Al_2Si_6O_{18}$	Beryl, emerald	$BaSO_4$	Barite

The free metals are generally prepared by electrolytic reduction, usually of the molten chloride:

$$2\,MCl(l) \xrightarrow{\text{Electrolysis}} 2\,M(l) + Cl_2(g) \qquad \text{(Group 1)} \qquad \textbf{(12.17)}$$

$$MCl_2(l) \xrightarrow{\text{Electrolysis}} M(l) + Cl_2(g) \qquad \text{(Group 2)} \qquad \textbf{(12.18)}$$

Water must be rigorously excluded from the system; otherwise, H_2O is reduced instead of the metal. In mixtures of metal halides, the cation having the least negative reduction potential is deposited. Magnesium chloride is often found mixed with $NaCl$ and $CaCl_2$ in nature, yet magnesium metal that is sufficiently pure for many commercial purposes can be obtained by the electrolysis of $MgCl_2/NaCl/CaCl_2$ melts without preliminary separation.

The metals that occur primarily as carbonates or sulfates in nature (see Table 12.1) must be pretreated, since these oxyanions are reduced more easily than the metal cations. The metal salts are usually converted to the chlorides by reaction with HCl and then electrolyzed:

$$CaCO_3 + 2\,HCl \longrightarrow CaCl_2 + H_2O + CO_2 \qquad \textbf{(12.19)}$$

Magnesium and sodium are the metals produced in greatest quantity. Magnesium has many uses in construction because of its low density and (especially in certain alloys) considerable strength. Molten sodium (liquid range 98–883°C) is used as a coolant in nuclear reactors. Elemental sodium is a useful reducing agent:

$$4\,Na + TiCl_4 \longrightarrow Ti + 4\,NaCl \qquad \textbf{(12.20)}$$

$$2\,Na + 2\,Me_3SiBr \longrightarrow Me_3Si\text{–}SiMe_3 + 2\,NaBr \qquad \textbf{(12.21)}$$

Alloys of these metals are sometimes better reducing agents than the pure metals themselves. Sodium, potassium, rubidium, and cesium are miscible in all combinations when liquefied. The alloys have lower melting points than the free metals; hence, mixtures such as 10–50% Na in K are liquid at room temperature, and as such are especially convenient reducing agents in the laboratory.

Interestingly, sodium–potassium alloy dissolves in certain solvents (especially ethers), in which the pure metals are only slightly soluble. This is believed to be due to the influence of solvation on the rather surprising equilibrium

$$K + Na \rightleftharpoons K^+ + Na^- \qquad \textbf{(12.22)}$$

(See the discussion of metal anions in Section 12.5.) Consequently, an ether is often the solvent of choice for reductions employing sodium–potassium alloy.

12.4   Some Important Binary and Ternary Salts

The Group 1 and Group 2 cations (except Rb^+ and Cs^+) are relatively hard Lewis acids, and show preferences for hard bases. Thus, fluorides tend to be more stable than iodides (based on decomposition temperatures, bond energies, and solubilities), and oxides are favored over sulfides. The metal salts are more soluble in water, a hard base, than in most other solvents.

Halides

The structures of the Group 1 and 2 halides were discussed in Chapter 6. A characteristic reaction of these compounds is halogen atom or halide ion exchange, which relates to the hard–soft theory. Thus, the equilibrium lies to the right for each of the following reactions:

$$NaBr + F^- \longrightarrow NaF + Br^- \qquad (12.23)$$

$$2\,LiI + F_2 \longrightarrow 2\,LiF + I_2 \qquad (12.24)$$

$$CsF + KI \longrightarrow CsI + KF \qquad (12.25)$$

The binary halides have enormous commercial importance. About 50 million tons of NaCl are produced in the United States each year, most of which is mined (rock salt) or obtained by evaporation of salt water (sea salt). Sodium chloride has a wide variety of uses. About 15–20% goes for melting ice and snow on highways, and about 10% is consumed by humans and animals. Among the technical uses, roughly 50% is consumed in industrial reactions known collectively as the *chlor–alkali process*. The end products are Na(s) and Cl_2(g) (from the electrolysis of molten NaCl) and NaOH, Cl_2, and H_2 (from the electrolysis of brine solutions):

Anode: $2\,Cl^- \longrightarrow Cl_2 + 2e^-$

Cathode: $2\,H_2O + 2e^- \longrightarrow 2\,OH^- + H_2$

$$2\,Cl^- + 2\,H_2O \longrightarrow Cl_2 + 2\,OH^- + H_2 \qquad (12.26)$$

Or

$$2\,NaCl + 2\,H_2O \xrightarrow{\text{Electrolysis}} Cl_2 + 2\,NaOH + H_2 \qquad (12.27)$$

The products are used primarily for the synthesis of other materials (eg, NH_3 from H_2, and Na_2CO_3 and $NaHCO_3$ from NaOH).

Oxides and Related Compounds

Burning these metals in oxygen-rich environments leads to some intriguing contrasts. The alkaline earths form oxides, as does lithium:

$$2M + O_2 \longrightarrow 2MO \qquad M = Be, Mg, Ca, Sr \qquad \textbf{(12.28)}$$

$$4Li + O_2 \longrightarrow 2Li_2O \qquad \textbf{(12.29)}$$

However, sodium and barium give the peroxides, and the heavier Group 1 elements form superoxides:

$$2Na + O_2 \longrightarrow Na_2O_2 \qquad \textbf{(12.30)}$$

$$Ba + O_2 \longrightarrow BaO_2 \qquad \textbf{(12.31)}$$

$$M + O_2 \longrightarrow MO_2 \qquad M = K, Rb, Cs \qquad \textbf{(12.32)}$$

This is the reverse of what might be expected on the basis of metal reactivity. The O_2^- and O_2^{2-} anions contain oxygen in less reduced states than in the oxide ion; thus, the most active metals (K, Rb, and Cs) achieve less reduction of O_2 than do Li and Be. The explanation relates to size. There is a general tendency for oppositely charged ions of about the same size to form stable ionic lattices.[3] Thus, the larger O_2^- and O_2^{2-} ions stabilize the larger cations, while the small cations prefer the monatomic oxide ion. However, under the proper circumstances the superoxides and peroxides of $Na^+, K^+, Rb^+, Cs^+, Ca^{2+}, Sr^{2+}$, and Ba^{2+} all can be prepared.

Because of this tendency, the best way to prepare several of the oxides is by heating the corresponding carbonates:

$$Li_2CO_3 \xrightarrow{\Delta} Li_2O + CO_2 \qquad \textbf{(12.33)}$$

$$MgCO_3 \xrightarrow{\Delta} MgO + CO_2 \qquad \textbf{(12.34)}$$

A number of other, less stable binary oxides are known. Examples include potassium sesquoxide, K_2O_3, which is obtained by carefully heating the superoxide, and cesium ozonide, made by the reaction of O_3 with CsOH.

$$4KO_2 \xrightarrow{\Delta} 2K_2O_3 + O_2 \qquad \textbf{(12.35)}$$

$$6CsOH + 4O_3 \longrightarrow 4CsO_3 + 2CsOH \cdot H_2O + O_2 \qquad \textbf{(12.36)}$$

---

3. For a discussion of this phenomenon and a variety of examples, see Basolo, F. *Coord. Chem. Rev.* **1968**, *3*, 168.

The sesquoxides are believed to contain oxygen in both the peroxide and superoxide ions [ie, $(M^+)_4(O_2^-)_2(O_2^{2-})$], while ozonides are simple salts of the paramagnetic O_3^- ion.

Carbonates, Nitrates, and Sulfates

All the Group 1 and 2 cations form salts of various oxyanions. Only those having particular chemical or industrial importance are discussed here.

Many of the carbonates are found in nature. Soda ash, Na_2CO_3, is used in the soap and paper industries and, in large-scale chemical processes, as an inexpensive reagent for acid neutralization. The combination of Na^+, CO_3^{2-}, and H_2O is found in a variety of chemical systems, including $NaHCO_3$; the mono-, hepta-, and decahydrates of Na_2CO_3; and various hydrated carbonate–bicarbonate mixed salts. Calcium carbonate occurs in several natural forms (see Table 12.1), and huge quantities of $CaCO_3$ are consumed in the steel and fertilizer industries, for water and sewage treatment, and in the manufacture of acetylene. The other Group 2 carbonates have lesser importance except for the mixed salt dolomite, $CaMg(CO_3)_2$, which is used as a fertilizer and as a "scrubber" for the removal of $SO_2(g)$ from smokestacks:

$$CO_3^{2-} + SO_2 \longrightarrow SO_3^{2-} + CO_2 \qquad (12.37)$$

The industrial chemistries of sodium and calcium are interwoven by the *Solvay process*, in which Na_2CO_3 and $CaCl_2$ are produced through a series of reactions that can be summarized as

$$CaCO_3 + 2\,NaCl \longrightarrow Na_2CO_3 + CaCl_2 \qquad (12.38)$$

The Solvay process has long been important because of the great abundance (and low cost) of the bulk starting materials, limestone and rock salt.

Among the nitrates, the best known are KNO_3 (saltpeter, the oxidizing agent in gunpowder) and $NaNO_3$ (Chile saltpeter). The most important sulfates are those of Mg^{2+} and Ca^{2+}. Anhydrous magnesium sulfate absorbs large amounts of water (forming the heptahydrate), and is a common drying agent in the laboratory. Magnesia, $MgSO_4 \cdot 7H_2O$, is an antacid and laxative. Hydrates of calcium sulfate are best known for their uses as building materials. Gypsum, $CaSO_4 \cdot 2H_2O$, is used in plaster and wallboard, while $CaSO_4 \cdot \frac{1}{2}H_2O$ is plaster of Paris. The sodium, potassium, and calcium compounds of major importance to consumers are listed in Table 12.2 (at the top of the next page).

Table 12.2 Important industrial chemicals of sodium, potassium, and calcium

Formula	Common Name(s)	Formula	Common Name(s)
NaOH	Lye, caustic soda	KOH	Caustic potash
Na_2CO_3	Soda ash	KNO_3	Saltpeter
$Na_2CO_3 \cdot 7H_2O$	Washing soda	K_2CO_3	Potash, pearl ash
Na_2SO_4	Salt cake	CaO	Lime, quicklime, burnt lime, calx
$NaNO_3$	Chile saltpeter, soda niter	$Ca(OH)_2$	Slaked lime, caustic lime
$NaHCO_3$	Bicarbonate of soda, baking soda	$CaSO_4 \cdot 2H_2O$	Gypsum
		$CaSO_4 \cdot \frac{1}{2}H_2O$	Plaster of Paris

12.5 Complexation Reactions: Lewis Acid–Base Chemistry[4]

The Group 1 (and, to a lesser extent, Group 2) cations are rather poor Lewis acids because of their low charge densities. The dative interactions in complexes such as $M(H_2O)_x^+$ and $M(NH_3)_x^+$ are relatively weak. This is evidenced by the rapid exchange of ligands in solution (see Chapter 17), and also by the difficulty of isolating the complexes in crystalline salts. Thus, there is relatively little to be said about their chemistry.

Complex stabilities can be improved by the use of chelates. For example, complexes with the bidentate ligands ethylenediamine and salicylaldehyde are known. Also, the unlikely ligand 2,3-di-(4-aminophenyl)-butane reacts with aqueous sodium salts to form the complex salts $[Na(L)_3]^+X^-$ (X = Cl, NO_3, and CN), in which Na^+ has approximately octahedral coordination.

Polydentate ligands are even more effective. An obvious example is the well-known analytical reagent EDTA (ethylenediaminetetraacetic acid):

$$\begin{array}{c} HOOCCH_2 \\ \\ HOOCCH_2 \end{array} \diagdown \hspace{-0.5em} \diagup \hspace{-1em} N - CH_2 - CH_2 - N \hspace{-1em} \diagup \hspace{-0.5em} \diagdown \begin{array}{c} CH_2COOH \\ \\ CH_2COOH \end{array}$$

This hexadentate ligand complexes strongly (particularly as its dianion) to Mg^{2+}, Ca^{2+}, Sr^{2+}, and Ba^{2+}:

$$M^{2+} + EDTA^{2-} \longrightarrow M(EDTA) \qquad (12.39)$$

The large equilibrium constants for these reactions (K_f ranging from 5.8×10^7 for Ba^{2+} to 5.0×10^{10} for Ca^{2+}) make EDTA valuable for the quantitatively analysis of the Group 2 cations.

4. Poonia, N. S.; Bajaj, A. V. *Chem. Rev.* **1979**, *79*, 389.

Crown Ethers and Cryptands[5]

An important class of polydentate ligands is the *crown ethers*, which are cyclic polyethers in the general shape of a crown.[6] Such compounds show different affinities for cations, depending on their size. For example, the ether commonly named 12-crown-4 (12 total ring atoms, 4 of which are oxygens) contains a cavity of the appropriate size for Li^+, and therefore complexes strongly to that cation. Similarly, 16-crown-5 has a large formation constant with Na^+, while 18-crown-6 favors potassium. This is useful for chemical separations, and also for improving solubilities. For example, potassium metal is nearly insoluble in Et_2O ($\sim 10^{-3}$ M at room temperature), but the addition of 18-crown-6 increases its solubility by a factor of over a thousand.

Complexation is also useful for the activation of reagent anions, as it disrupts ion pairing in less polar solvents. The structures of several crown ethers are pictured in Figure 12.3a–d.

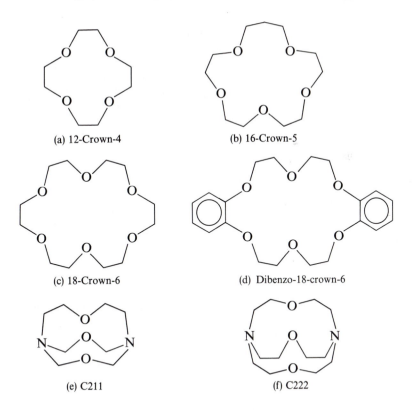

(a) 12-Crown-4

(b) 16-Crown-5

(c) 18-Crown-6

(d) Dibenzo-18-crown-6

(e) C211

(f) C222

Figure 12.3
The structures of some commonly used crown ethers and cryptands.

5. Dietrich, B. *J. Chem. Educ.* **1985**, *62*, 954; Parker, D. *Adv. Inorg. Chem. Radiochem.* **1983**, *27*, 1; Knipe, A. C. *J. Chem. Educ.* **1976**, *53*, 618.

6. Pederson, C. J. *J. Am. Chem. Soc.* **1970**, *92*, 391; **1967**, *89*, 7017.

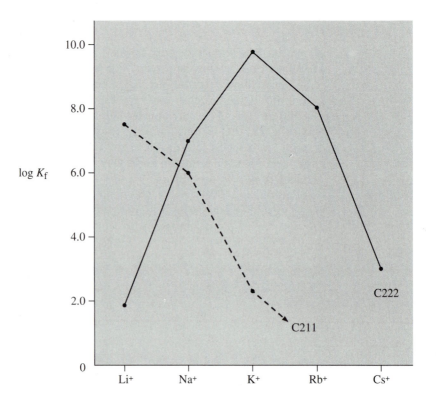

Figure 12.4
The relative affinities of two cryptands for Group 1 cations; the smaller C211 prefers Li^+, while the larger C222 favors K^+. (See Figure 12.3e–f for structures.) Values calculated from data compiled by Parker, D. *Adv. Inorg. Chem. Radiochem.* **1983**, *27*, 1.

Coordination in three dimensions can be achieved by incorporating a bridging group containing a donor atom into a crown ether. Such ligands are *cryptands*, and their complexes are *cryptates* (Figure 12.3e–f).[7] The sequestration of cations by such species is so great that they have been described as "inclusion ligands." Plots of the relative affinities of the Group 1 cations for two cryptands are given in Figure 12.4.

Cryptands have been instrumental in the isolation of some unusual polyatomic anions, including Sb_7^{3-}, Te_3^{2-}, Sn_5^{2-}, Pb_5^{2-}, and Ge_9^{n-} ($n = 2, 4$), from solution by complexation of the sodium counterion:[8]

$$Na_3Sb_7 + 3L \longrightarrow 3NaL^+ + Sb_7^{3-} \tag{12.40}$$

$$Na_2Te_3 + 2L \longrightarrow 2NaL^+ + Te_3^{2-} \tag{12.41}$$

7. Pederson, C. J.; Frensdorff, H. K. *Angew. Chem. Int. Ed.* **1972**, *11*, 16; *Angew. Chem.* **1972**, *84*, 16.

8. Cisar, A.; Corbett, J. D. *Inorg. Chem.* **1977**, *16*, 632; Edwards, P. A.; Corbett, J. D. *Inorg. Chem.* **1977**, *16*, 903; Belin, C. H. E.; Corbett, J. D.; Cisar, A. *J. Am. Chem. Soc.* **1977**, *99*, 7163; Adolphson, D. G.; Corbett, J. D.; Merryman, D. J. *J. Am. Chem. Soc.* **1976**, *98*, 7234.

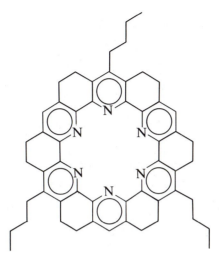

Figure 12.5
The structure of a torand ligand. [Reproduced with permission from Bell, T. W.; Firestone, A. *J. Am. Chem. Soc.* **1986**, *108*, 8109.]

Another group of three-dimensional polydentate ligands is the *torands*. These are nitrogen analogues of crown ethers. An example is the cyclic hexamine shown in Figure 12.5, which forms a stable complex with Ca^{2+}.[9]

Metal Anions and Electrides[10]

Another remarkable development resulting from the use of macrocyclic ligands is the isolation of salts of monatomic metal anions. Although electron transfer between two sodium atoms is strongly disfavored,

$$2\,Na \longrightarrow Na^+ + Na^- \qquad \Delta H = IE_1 - EA_1 = +440 \text{ kJ/mol}$$

$$\textbf{(12.42)}$$

the stabilization of Na^+ by the C222 cryptand (Figure 12.3f) made possible the isolation of $Na(C222)^+Na^-$.[11] This bright gold, metallic-like salt consists of hexagonal close-packed cations with Na^- (sodide) anions in the octahedral interstices. The radius of Na^- has been estimated to be about 220 pm, making it roughly the same size as I^-. Salts containing K^-, Rb^-, Cs^-, and Au^- also have been identified.

It is also possible to prepare compounds in which the "anion" is an electron; such species are called *electrides*. In electron-rich media (eg,

  9. Bell, T. W.; Firestone, A. *J. Am. Chem. Soc.* **1986**, *108*, 8109.

10. Dye, J. L. *Prog. Inorg. Chem.* **1984**, *32*, 327.

11. Dye, J. L.; Ceraso, J. M.; Lok, M. T.; Barnett, B. L.; Tehan, F. J. *J. Am. Chem. Soc.* **1974**, *96*, 608.

metal–amine solutions), the mole ratio of metal to cryptand may determine whether a metal anion or a metal electride is formed. For example, for mixtures of sodium with C222, spectroscopic evidence indicates that both equations below are obeyed:

$$\text{Na} + \text{L} \longrightarrow (\text{NaL})^+ e^- \tag{12.43}$$

$$2\,\text{Na} + \text{L} \longrightarrow (\text{NaL})^+ \text{Na}^- \tag{12.44}$$

However, the results are different in the reaction of cesium metal with 18-crown-6. The large Cs^+ ion can coordinate to two crown ligands simultaneously. The 1:1 stoichiometry produces the ceside salt, but the electride is formed when the metal/crown ratio is 1:2.[12]

$$2\,\text{Cs} + 2\,\text{L} \longrightarrow (\text{CsL}_2)^+ \text{Cs}^- \tag{12.45}$$

$$\text{Cs} + 2\,\text{L} \longrightarrow (\text{CsL}_2)^+ e^- \tag{12.46}$$

12.6   Variations from the Group Norms

Now that we have discussed the major similarities and trends exhibited by the Group 1 and 2 metals, it is appropriate to consider some differences. The exceptions are most often found in the behavior of the three smallest elements—lithium, beryllium, and magnesium. Hence, the focus of this section is on these elements and their compounds.

Beryllium

Beryllium's compact valence orbitals promote strong covalent bonding. With only two valence electrons, a maximum of two covalent bonds normally can be formed. This leaves Be four electrons short of an octet in its divalent compounds. It is not surprising, then, that Lewis acidity and complex formation are important aspects of its chemistry.

Beryllium chloride is useful for illustration. Unlike the other Group 1 and Group 2 chlorides, $\text{BeCl}_2(s)$ is clearly covalent. It is a linear polymer in the solid state (Figure 12.6), a structure that allows each atom to achieve an electron octet. The average hybridization is sp^3.

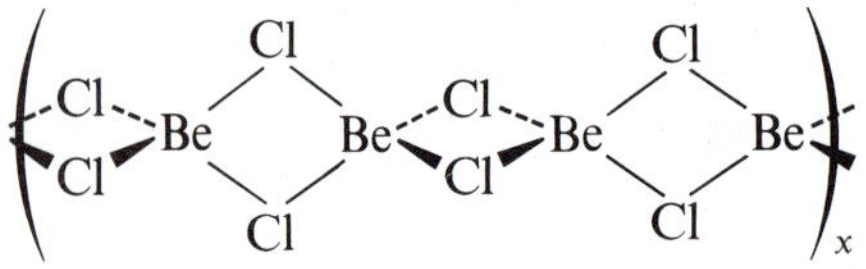

Figure 12.6   The solid-state structure of BeCl_2, a linear polymer. Beryllium has four-coordinate, approximately tetrahedral coordination.

12.   Ellaboudy, A.; Dye, J. L.; Smith, P. B. *J. Am. Chem. Soc.* **1983**, *105*, 6490.

The reaction chemistry of $BeCl_2$ is dominated by Lewis acid–base interactions, particularly with fluorine, chlorine, oxygen, and nitrogen donors.

$$BeCl_2 + 2Cl^- \longrightarrow [BeCl_4]^{2-} \tag{12.47}$$

$$BeCl_2 + 4NH_3 \longrightarrow [Be(NH_3)_4]^{2+} + 2Cl^- \tag{12.48}$$

$$BeCl_2 + 2Et_2O \longrightarrow Be(OEt_2)_2Cl_2 \tag{12.49}$$

The affinity for these hard bases is not surprising, since the small size and high charge density of Be^{II} make it an extremely hard acid. As we would expect, the analogous reactions with softer bases such as Br^- and PH_3 are much less exothermic. It is also significant that the stronger base, NH_3, is able to displace chloride ion (ie, cause substitution as well as addition), while the weaker base, Et_2O, gives only the adduct.

Soluble beryllium salts dissolve in water to initially form the cation $[Be(H_2O)_4]^{2+}$. This species is stable only in strongly acidic solution, since the high charge density of Be^{2+} makes it prone to hydrolysis. (With a pK_h of 6.5, Be^{2+} is hydrolyzed nearly 100,000 times more extensively than any of the other Group 2 ions.) The result is a complex mixture of products. Two equilibria that are established are

$$[Be(H_2O)_4]^{2+} + H_2O \rightleftharpoons [Be(H_2O)_3OH]^+ + H_3O^+ \tag{12.50}$$

$$2[Be(H_2O)_4]^{2+} + H_2O \rightleftharpoons [(H_2O)_3Be-O-Be(H_2O)_3]^{2+} + 2H_3O^+ \tag{12.51}$$

Precipitation occurs if the pH is increased,

$$[Be(H_2O)_3OH]^+ + OH^- \longrightarrow Be(OH)_2(s) + 3H_2O \tag{12.52}$$

followed by redissolution if excess base is added:

$$Be(OH)_2(s) + 2OH^- \longrightarrow [Be(OH)_4]^{2-} \tag{12.53}$$

Except for valency, the chemistry of beryllium is more similar to that of aluminum than to the other members of its own family. This is an example of a *diagonal relationship*. Some specific similarities between this pair of elements will be discussed in the next chapter.

Lithium and Magnesium

The similarities between lithium and magnesium, and especially between the Li^I and Mg^{II} oxidation states, is striking. This results from the fact that these diagonal elements have roughly the same electronegativity, charge density, and hardness as acids. The major similarities are:

Solubility Trends Lithium fluoride is the least soluble of the lithium halides, and MgF_2 is the least soluble of the magnesium halides. This reflects the preference of these hard cations for hard bases. (The fluorides have high solubilities for the largest members of each family.)

Affinity for Nitrogen In contrast to the heavier metals, complexes of Li^+ and Mg^{2+} can be precipitated from liquid ammonia:

$$LiI + 4NH_3 \longrightarrow [Li(NH_3)_4]I(s) \qquad (12.54)$$

$$MgCl_2 + 6NH_3 \longrightarrow [Mg(NH_3)_6]Cl_2(s) \qquad (12.55)$$

These metals also are remarkable in their reactivity toward molecular nitrogen. Metallic lithium is one of the few species known to react with N_2 at room temperature. Magnesium is oxidized by N_2 above 300°C.

$$6Li + N_2 \longrightarrow 2Li_3N \qquad (12.56)$$

$$3Mg + N_2 \xrightarrow{\Delta} Mg_3N_2 \qquad (12.57)$$

The remaining Group 1 and 2 metals do not form nitrides by direct reaction with N_2.

Decompositions of Oxyanion Salts Lithium and magnesium carbonate have the lowest decomposition temperatures of their respective families.

$$Li_2CO_3 \xrightarrow{\Delta} Li_2O + CO_2 \qquad (12.58)$$

$$MgCO_3 \xrightarrow{\Delta} MgO + CO_2 \qquad (12.59)$$

The comparative ease of these decompositions is a reflection of the relative preference of the small cations for O^{2-} compared to the larger CO_3^{2-}.

The nitrates of lithium and magnesium undergo a different type of thermal decomposition than the heavier family members. The normal mode is elimination of O_2 to produce the nitrite:

$$2NaNO_3 \xrightarrow{\Delta} 2NaNO_2 + O_2 \qquad (12.60)$$

However, lithium and magnesium give oxides.

$$4LiNO_3 \xrightarrow{\Delta} 2Li_2O + 2N_2O_4 + O_2 \qquad (12.61)$$

$$2Mg(NO_3)_2 \xrightarrow{\Delta} 2MgO + 2N_2O_4 + O_2 \qquad (12.62)$$

Behavior Toward Molecular Oxygen Recall that lithium and magnesium burn in air to give oxides, rather than peroxides or superoxides.

Organometallic Compounds The structures and reactions of the organometallic compounds of these elements show marked similarities, as described next.

12.7 Organometallic Compounds

Although organometallics of all the Group 1 and Group 2 elements have been synthesized, the best known and most important are those of magnesium, lithium, sodium, and potassium (probably in that order). This is primarily because of their value as reagents in organic and organometallic synthesis. The emphasis in this section is therefore on these four elements.

Synthetic Methods

Metal–carbon bonds are formed in a variety of ways. The most important are described below.

Direct Union or Addition It is sometimes possible to add metal atoms across a $C=C$ double bond. However, insertion into a carbon–halogen bond is more common. The best-known example of the latter is the preparation of Grignard reagents, as described by equation (12.64).

$$2Cs + R_2C=CR_2 \longrightarrow R_2(Cs)C-C(Cs)R_2 \tag{12.63}$$

$$R-I + Mg \longrightarrow RMgI \tag{12.64}$$

Metal–Metal Exchange It is possible to replace certain carbon-bonded metals with others, with the general replacement order being

$$Cs < Na, K < Hg \approx Zn < Li < Mg$$

This is especially true when halogens or other groups of high electronegativity are present, and is consistent with Pauling's generalization that there is a tendency for bonds to form between the elements of highest and lowest electronegativity in any chemical system (Chapter 4). Thus, the position of equilibrium favors the products for each of the following equations:

$$RMgX + LiR \rightleftharpoons MgR_2 + LiX \tag{12.65}$$

$$KR + LiX \rightleftharpoons LiR + KX \tag{12.66}$$

$$HgR_2 + 2Li \rightleftharpoons 2LiR + Hg \tag{12.67}$$

Reduction of Acidic Hydrogen   If a C–H bond is sufficiently polar (ie, if the hydrogen is sufficiently acidic), it is often possible to induce metal–carbon bond formation by the elimination of H_2.

$$2Na + 2HC{\equiv}CR \longrightarrow 2NaC{\equiv}CR + H_2 \tag{12.68}$$

Metal–Halogen Exchange   Probably the best-known method for the preparation of organolithium compounds involves cleavage of a carbon–halogen bond.

$$R{-}Br + 2Li \longrightarrow LiR + LiBr \tag{12.69}$$

Cleavage of Ethers   Some organolithiums cannot be prepared in reasonable yield via equation (12.69) because of carbon–carbon bond formation. For example,

$$2C_6H_5CH_2Br + 2Li \longrightarrow C_6H_5CH_2CH_2C_6H_5 + 2LiBr \tag{12.70}$$

In such cases, the desired compound often can be prepared by the cleavage of an appropriate ether:

$$H_3C{-}O{-}CH_2C_6H_5 + 2Li \longrightarrow LiCH_2C_6H_5 + LiOCH_3 \tag{12.71}$$

The Structures of Organometallic Compounds

The polarities of metal alkyls are as expected from electronegativities:

$$BeR_2 < MgR_2 < LiR < NaR < KR$$

Covalent $\longleftarrow\!\!\!\longrightarrow$ Ionic

The "covalent" and "ionic" labels are arbitrary (as usual!), but are supported by the structures of these compounds. Dimethylberyllium and dimethylmagnesium are similar to $BeCl_2$, with bridging CH_3 groups. Methyllithium is also covalent; it consists of tetramers in the solid state (Figure 12.7). The

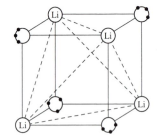

Figure 12.7
The solid-state structure of the tetramer of methyllithium.

very short Li–Li distance (256 pm, compared to 267 pm in Li_2 vapor) seems to suggest strong metal–metal bonding interactions. This is not expected based on the formula $LiCH_3$, however, and there is controversy on this point.

Methylsodium and methylpotassium behave as ionic compounds (low solubilities in nonpolar solvents, etc.), and have low thermal stabilities compared to $LiCH_3$.

Much effort has been expended toward elucidating the structures of Grignard reagents. When crystallized from diethyl ether, such compounds conform to the formula $RMgX(OEt_2)_2$. Magnesium is in a tetrahedral environment, and the bonding is of the polar covalent type. The situation in solution is more complex; a large number of species, including $RMgX$, MgR_2, MgX_2, and various dimers, are known to be present.

Chemical Reactions

The replacement order given above is, for most purposes, the order of reactivity. Thus, Grignard reagents are less reactive than the corresponding organolithium compounds, which in turn are less reactive than the organo-sodiums or organopotassiums. The last two behave as true carbanions. Some characteristic reactions of these compounds are described below.

Reactions as Brønsted Bases   Since the pK_a of CH_4 and other alkanes is about 40, their conjugates are very strong bases. They are therefore capable of deprotonating even extremely weak acids:

$$R_2NH + n\text{-}C_4H_9Li \longrightarrow R_2N^-Li^+ + C_4H_{10} \qquad (12.72)$$

$$R_2C{=}CHR + CH_3Na \longrightarrow R_2C{=}CR^-Na^+ + CH_4 \qquad (12.73)$$

Alkylation   This is the best known and perhaps the most common type of reaction of these compounds. Two examples are

$$BBr_3 + 3\,RMgBr \longrightarrow BR_3 + 3\,MgBr_2 \qquad (12.74)$$

$$(CH_3)_2SiCl_2 + 2\,LiCH_3 \longrightarrow Si(CH_3)_4 + 2\,LiCl \qquad (12.75)$$

Formation of Alkenes by Thermal Decomposition   Many alkyllithiums form LiH and an alkene upon heating.

$$RCH_2CH_2Li \xrightarrow{\Delta} RCH{=}CH_2 + LiH \qquad (12.76)$$

Addition and Insertion Reactions   These polar reagents add across numerous types of double bonds, including $C{=}C$, $C{=}O$, and $C{=}N$.

$$O{=}C{=}O + RMgX \longrightarrow R\text{–}\overset{\overset{\displaystyle O}{\|}}{C}\text{–}OMgX \qquad (12.77)$$

A slightly different reaction [analogous to the addition of metal hydrides to Lewis acids, equation (12.16)] occurs between alkyllithiums and alkyl-coppers.

$$LiR + CuR \longrightarrow LiCuR_2 \qquad\qquad (12.78)$$

Redistribution Reactions   The reactions described by the equations below can be influenced by temperature, mole ratio, and/or removal of a product from the reaction medium by precipitation or gas evolution.

Such redistribution reactions are often useful for synthetic purposes. For example, equation (12.80) describes an efficient synthesis for dichloro-phenylborane.[13]

$$MgR_2 + MgCl_2 \rightleftharpoons 2\,RMgCl \qquad\qquad (12.79)$$

$$Sn(C_6H_5)_4 + 4\,BCl_3 \rightleftharpoons 4\,C_6H_5BCl_2 + SnCl_4 \qquad\qquad (12.80)$$

Bibliography

Thayer, J. S. *Organometallic Chemistry, an Overview*; VCH: Weinheim, 1988.

Puddephatt, R. T.; Monaghan, P. K. *The Periodic Table of the Elements*, 2nd ed.; Clarendon: Oxford, 1986.

Hajtos, A. *Complex Hydrides*; Elsevier: Amsterdam, 1979.

Parish, R. V. *The Metallic Elements*; Longman: New York, 1977.

Wakefield, B. S. *The Chemistry of Organolithium Compounds*; Pergamon: Oxford, 1976.

Brown, H. C. *Organic Synthesis via Boranes*; Wiley: New York, 1975.

Muetterties, E. L. *Transition Metal Hydrides*; Marcel Dekker: New York, 1971.

Wiberg, E.; Amberger, E. *Hydrides of the Elements of Main Groups I–IV*; Elsevier: Amsterdam, 1971.

Lewis, F. A. *The Palladium–Hydrogen System*; Academic: New York, 1967.

Questions and Problems

1.   From the perspective of hydrogen, explain why:
     (a) KH is a base, but HBr is an acid.
     (b) HOCl, which has an O–H bond, is an acid, but NaOH, which also has an O–H bond, is a base.

13. For details of the procedure, see Jolly, W. L. *The Synthesis and Characterization of Inorganic Compounds*; Prentice-Hall: Englewood Cliffs, NJ, 1970; pp. 481–483.

(c) H_2Se is a stronger acid than H_2O in spite of the fact that oxygen is more electronegative than selenium.

2. Complete and balance the following equations:
   (a) $Ca + HNO_3 \longrightarrow$ _____ + _____
   (b) $LiH + HOCl \longrightarrow$ _____ + _____
   (c) $I_2 + H^- \longrightarrow$ _____ + _____
   (d) $C_2H_5BH_2 + KH \longrightarrow$ _____
   (e) $RbH \xrightarrow{\Delta}$ _____ + _____
   (f) $C_3H_7OH + CsH \longrightarrow$ _____ + _____

3. The density of metallic palladium is 12.0 g/cm^3. About what volume of Pd is needed to store 1 mol of H_2? Compare your answer to the volume occupied by 1 mol of H_2 gas, 22.4 L at STP.

4. It is difficult to assign a specific ionic radius to the hydride ion because it appears to change from compound to compound. An average value is about 150 pm.
   (a) Why is the ionic radius of H^- especially prone to variation?
   (b) Use the average value given above to predict the lattice type (zinc blende, rock salt, or CsCl) for LiH.
   (c) Predict whether MgH_2 crystallizes in the rutile or fluorite lattice.
   (d) Based on lattice dimensions, the radius of H^- is 130 pm in LiH and 154 pm in CsH. Would you have expected the value for CsH to be the larger of the two? Explain.

5. We indicated in the text that PH_3 is neither a Brønsted acid nor base, except under forcing conditions. Suggest reagents that would cause PH_3 to behave in each manner, and write the corresponding equations.

6. Explain what happens (using both words and an equation) when strontium metal is added to water at room temperature.

7. Write the two half-reactions for the electrolysis of:
   (a) Molten $MgCl_2$ \qquad (b) A molten mixture of LiCl and NaCl
   (c) An acidic aqueous solution of $CaBr_2$

8. Beryllium hydride was not discussed in this chapter. It is known to be polymeric. Speculate on its structure, and explain why it is different from those of the other Group 2 hydrides.

9. Write the products for the following reactions and briefly explain why each occurs:

   (a) $LiI + BaF_2 \longrightarrow$
   (b) $Na_2CO_3 + SO_2 \longrightarrow$
   (c) $Na_2O_2 \xrightarrow{\Delta}$
   (d) $BeS + \text{Excess } F^- \longrightarrow$
   (e) $LiNO_3 \xrightarrow{\Delta}$

10. The synthesis of cesium ozonide, as described by equation (12.36), is experimentally demanding. Discuss the reaction $Cs + O_3 \rightarrow CsO_3$ in terms of the reduction potential diagrams for Cs and O.

11. The anions of Na, K, Rb, Cs, and Au are all known. Suggest other metals that are good candidates for addition to this list, and defend your answers.

12. The basicity of KOH in moderately polar solvents is greatly enhanced by the addition of an appropriate crown ether. Explain.

13. Certain lithium salts (eg, LiBr) are hydrated, while the corresponding sodium and potassium salts are generally anhydrous. Rationalize.

14. The experimental mobilities (ie, the relative rates of migration) of the Group 1 cations in water follow the order $Li^+ < Na^+ < \cdots < Cs^+$.
    (a) Is this what you would have expected? Why or why not?
    (b) Rationalize this experimental order.

15. The complex formed between Li^+ and NH_3 is more stable than that between K^+ and NH_3. However, the complex formed between Li^+ and $S_2O_3^{2-}$ is less stable than the corresponding K^+ complex. Explain.

16. The standard reduction potentials for the Group 1 cations do not follow a smooth trend. Why not?

17. Consider the three elements K, Ca, and Sr:
    (a) In general, which one forms the most stable complexes?
    (b) Which is the most likely to undergo a reaction in which it exchanges chloride for fluoride?
    (c) Which forms the highest-melting binary fluoride?
    (d) Which one is not strongly complexed by EDTA?

18. Use electronegativities, ionization energies, and redox potentials to decide which Group 1 element is the most similar to strontium.

19. Explain in your own words why:
    (a) The ionization energies for the Group 1 metals are low.
    (b) The aqueous molar solubility of LiCl is greater than that of LiF.
    (c) The aqueous molar solubility of BaS is less than that of $BaCl_2$.
    (d) The melting point of Ca is higher than that of K.
    (e) The melting point of Ca is higher than that of Sr.

20. Complete and balance (no stoichiometries are implied):

    (a) $C_2H_5Br + Mg \longrightarrow$ _____

    (b) $CH_3Br + Li \longrightarrow$ _____ + _____

    (c) $C_6H_5OH + n\text{-BuLi} \longrightarrow$ _____ + _____

    (d) $SnCl_4 + \text{Excess } NaC_2H_5 \longrightarrow$ _____ + _____

    (e) $CH_3COCH_3 + CH_3MgI \longrightarrow$ _____

*21. In a report by N. A. Bell, G. E. Coates, and J. A. Heslop (*J. Organomet.*

Chem. **1987**, *329*, 287), a tetrameric hydroborane was reacted with diethyl-beryllium:

$$[NaBHMe_3]_4 + Et_2Be \longrightarrow ??$$

The beryllium was recovered as the trimethylamine complex of ethyl-beryllium hydride.

(a) Explain the structure of the reactant hydroborane.

(b) Predict the formula of the $Me_3N–Be$ complex, and defend your answer.

*22. The validity of the lithium–magnesium diagonal relationship was challenged by T. P. Hanusa (*J. Chem. Educ.* **1987,** *64,* 686).

(a) What element does Hanusa feel shows the greatest similarity to lithium? On what criteria does he base his argument?

(b) List other information you would want to consider before deciding which relationship is stronger, lithium–magnesium or lithium–??

(c) Using any data given in the text, argue whichever side of the case you choose.

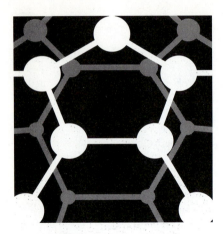

13

The Chemistry of Groups 13 and 14

The *p-block* elements—that is, those of Groups 13–18—are considered in this chapter and the next. The first two of these families (headed by boron and carbon) are discussed here, with the remaining four covered in Chapter 14. There are several reasons for making the division in this manner. The elements of Groups 13 and 14 are all metals or metalloids except carbon, while most of the elements of Groups 15–18 are nonmetals. Also, the elements of Groups 13 and 14 behave as electrophiles in their most characteristic chemical reactions, and their hydrogen compounds tend to be hydridic rather than protonic. Atoms of elements of Groups 15–18 have one or more lone pairs in the majority of their compounds, and act as nucleophilic centers. Their hydrides are usually protonic.

A trend observed for Groups 1 and 2 also is apparent in these families. The chemistry of the first element of a group is markedly different from that of the heavier members. Just as hydrogen[1] and beryllium are dissimilar from the other Group 1 and 2 elements, boron and carbon differ in their physical and chemical properties from the remaining members of their respective families.

1. Or lithium, if hydrogen is placed in Group 17.

13.1 Trends within the Families

The chemical properties of compounds derive, of course, from the properties of their constituent atoms. Thus, it is useful to begin a discussion of the descriptive chemistry of a given family by listing some relevant atomic properties for its elements, as we have done in Table 13.1 for Group 13.

As we discussed in Chapter 7, boron has numerous allotropes, all of which are three-dimensional networks held together by delocalized covalent bonding. The remaining elements of Group 13 form metallic lattices. This difference is not surprising, given the large ionization energies and small covalent radius of boron relative to the other members of the family.

The melting points of the free elements are, by and large, bizarre. The very high melting temperature of boron is consistent with its network structure, but the low values for indium and especially gallium are not easily explained. The boiling points are more consistent, and follow the order expected for metallic-type interactions:

Al > Ga > In > Tl

The electronegativities and ionization energies are also worthy of comment. The values for gallium are anomalously high. This can be understood from consideration of effective nuclear charges. In going from boron to aluminum, one electron shell is added, and the difference in nuclear charge is 8 protons. Comparing aluminum to gallium, one electron shell is again added; however, the nuclear charge differs by 18 protons. The 10 extra electrons do not shield the extra protons on a one-for-one basis. As a result, the valence electrons of gallium experience a greater effective nuclear charge than would otherwise be expected, and therefore require relatively more energy for ionization. This also leads to a greater electronegativity. (Recall the relationship between valence orbital energy and electronegativity described in Chapter 1.)

Size is also affected by this higher-than-expected effective nuclear charge. The nonpolar covalent radius of aluminum is about 68% larger than that

Table 13.1 Selected properties of the Group 13 elements

Element	mp, °C	bp, °C	χ_{A-R}	IE_1, eV	r_{cov}, pm	r_{3+}, pm
B	2300	2550	2.01	8.3	85	—
Al	660	2467	1.47	6.0	143	54
Ga	30	2403	1.82	6.0	135	62
In	157	2080	1.49	5.8	162	80
Tl	303	1457	1.44	6.1	170	88

Note: χ_{A-R} = Allred–Rochow electronegativity.

Table 13.2   Selected properties of the elements of Group 14

Element	mp, °C	bp, °C	χ_{A-R}	IE_1, eV	r_{cov}, pm
C	3550	4827	2.50	11.3	77
Si	1410	2355	1.74	8.1	118
Ge	937	2830	2.02	7.9	120
Sn	232	2270	1.72	7.3	140
Pb	328	1740	1.55	7.4	144

Note: χ_{A-R} = Allred–Rochow electronegativity.

of boron, but gallium and aluminum are about the same size. The incomplete shielding by *d* electrons influences the properties of many of the posttransition elements, but particularly those of the first long period—that is, Ga, Ge, and As.

A similar situation arises when indium and thallium are compared. Here, the difference in nuclear charge is 32 protons, as the $4f$ subshell comes into play. Shielding by the extra electrons again does not compensate. Thallium therefore has a smaller size and greater ionization energies and electronegativity than expected by comparison with the other family members.

In a different presentation of the periodic table the "boron family" might contain B, Al, Sc, Y, La, and Ac. There is some logic to such an organization. Each of these elements has three valence electrons, two of which are in an *s*-type orbital. The observed trends in electronegativity, ionization energy, and size are in excellent agreement with expectations for this alternative grouping.

Physical and chemical properties of the Group 14 elements are given in Table 13.2. The melting points in this family were discussed in Chapter 7; the anomaly (Ge > Sn < Pb) coincides with the transition from network covalent to metallic-type bonding in the free elements. As for Group 13, the boiling points more closely follow the expected trend.

The effects of incomplete shielding are again apparent. The electronegativity of germanium is greater than that of silicon, and the covalent radii and first ionization energies of the two are nearly equal.

We will discuss the specific members of these families in Sections 13.2–13.4. The subdivisions are apparent from the section titles: first, the only nonmetal (carbon) will be considered; the three metalloids follow; and then the chemistry of the metals will be described.

13.2   The Inorganic Chemistry of Carbon

The vast majority of known carbon chemistry involves hydrocarbons and their derivatives, of course, and is treated (in excruciating detail, according to some students!) in organic chemistry courses. Inorganic aspects are emphasized here.

Isotopes and Natural Occurrence

Natural carbon exists in three isotopes: ^{12}C (98.89% abundance), ^{13}C (1.11%), and radioactive ^{14}C (trace). Carbon-14 is formed in the atmosphere from collisions of high-energy neutrons with ^{14}N:[2]

$$^{14}_{7}N + ^{1}_{0}n \longrightarrow ^{14}_{6}C + ^{1}_{1}H \tag{13.1}$$

Radioactive decay occurs by beta emission:

$$^{14}_{6}C \longrightarrow ^{14}_{7}N + ^{0}_{-1}e \tag{13.2}$$

Because these reactions occur at constant rates, a steady-state $^{14}C/^{12}C$ ratio of about 10^{-12} exists in the atmosphere. This ratio is eventually established in all living things through respiration, and provides the basis for *carbon-14 dating*. Dead organisms no longer ingest radioactive carbon (as $^{14}CO_2$) from the atmosphere. Therefore, as the reaction described by equation (13.2) occurs, the $^{14}C/^{12}C$ ratio decreases at a predictable rate, enabling the determination of the elapsed time since death. The half-life of ^{14}C (about 5730 years) makes it useful for determining the age of once living matter over a span of roughly 500–50,000 years.

Carbon occurs as the free element in either of two allotropes (graphite and diamond; see Chapter 7), but is found in much greater quantities in oxidized forms in the lithosphere and hydrosphere (CO_3^{2-} and HCO_3^- ions) and in the atmosphere (CO_2). The majority is in the form of inorganic carbonates.

Carbides[3]

Binary compounds in which carbon is bonded to a metal or metalloid are called *carbides*. The subclasses of carbides are reminiscent of the hydrides: saline (ionic), interstitial, and covalent. The best-known saline carbides contain C_2^{2-} units, and may be considered salts of the acetylide ion. They are normally prepared either by the reaction of an active metal with C_2H_2

---

2. A competing process is the formation of $^{12}_{6}C + ^{3}_{1}H$ [equation (12.6)].

3. Johansen, H. A. *Surv. Prog. Chem.* **1977**, *8*, 57; Frad, W. A. *Adv. Inorg. Chem. Radiochem.* **1968**, *11*, 153.

in a solvent such as liquid ammonia, or by the high-temperature reaction of a metal oxide with graphite:

$$2\,Na + C_2H_2 \xrightarrow{\ NH_3\ } Na_2C_2 + H_2 \tag{13.3}$$

$$CaO + 3\,C \xrightarrow{\ >2000°C\ } CaC_2 + CO \tag{13.4}$$

The best-known ionic carbide is CaC_2. This compound is a convenient laboratory source of acetylene, and in the past was used to generate the fuel in miners' lamps. The acetylene was produced by slow hydrolysis:

$$CaC_2 + 2\,H_2O \longrightarrow Ca(OH)_2 + C_2H_2 \tag{13.5}$$

The CaC_2 lattice is of the rock salt type, with distortion in one dimension to accommodate the C_2^{2-} units (Figure 13.1). This is typical of several carbides, including SrC_2 and LaC_2. However, structure type is not necessarily indicative of bond type in these compounds. Calcium carbide has the properties expected for an ionic species, while LaC_2 has a metallic luster and is an electrical conductor; this clearly places it in the interstitial category.

Interstitial carbides contain close-packed metal atoms, with carbon occupying some or all of the octahedral interstices. In La_3C, one-third of the available octahedral holes of a cubic close-packed lattice are filled in a random manner; in V_2C, there is random occupancy of half the octahedral sites of hexagonal close-packed metal atoms; and in TiC and ZrC, all the octahedral sites are occupied, giving a true (but based on chemical evidence, nonionic) NaCl type of lattice.

Figure 13.1
The crystal structure of CaC_2. [Reproduced with permission from Wells, A. F. *Structural Inorganic Chemistry*, 5th ed.; Clarendon: Oxford, 1984; p. 948.]

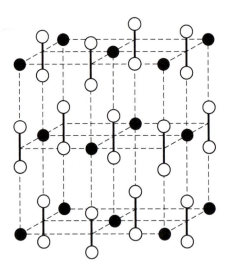

$\bullet = Ca^{2+}$

$\circ\!-\!\circ = C_2^{2-}$

The covalent carbides are network polymers. The best known is carborundum, SiC, with a diamondlike structure in which both silicon and carbon have sp^3 hybridization. Carborundum is very hard, making it useful for industrial drills, sharpening tools, etc. Beryllium carbide, Be_2C, has an antifluorite structure in which the beryllium atoms are four-coordinate (tetrahedral) and the carbons occupy cubic sites. The bonding is polar covalent.

Carbon–Oxygen Systems

The best-characterized binary compounds of carbon and oxygen are CO, CO_2, and C_3O_2. The latter, commonly named carbon suboxide, is the anhydride of malonic acid. It can be prepared by heating the acid in the presence of a dehydrating agent:

$$HOOC–CH_2–COOH \xrightarrow{\Delta} C_3O_2 + 2H_2O \qquad (13.6)$$

The C_3O_2 molecule is linear, with the Lewis structure

$$:\ddot{O}=C=C=C=\ddot{O}:$$

It is stable only when kept cold, undergoing polymerization near to and above its boiling point of 7°C. The polymer consists of sheets of fused six-membered rings (five carbons and one oxygen per ring), and is structurally related to graphite (Figure 13.2).

The great stability of CO and CO_2 (which can be traced to the strength of C–O π bonding) has an important influence on the chemical reactivities of many carbon-containing species. It is for this reason that C(s) and CO are excellent high-temperature reducing agents, being particularly useful for the preparation of certain free metals from their ores:

$$ZnO + C \xrightarrow{\Delta} Zn + CO \qquad (13.7)$$

$$Fe_2O_3 + 3CO \xrightarrow{\Delta} 2Fe + 3CO_2 \qquad (13.8)$$

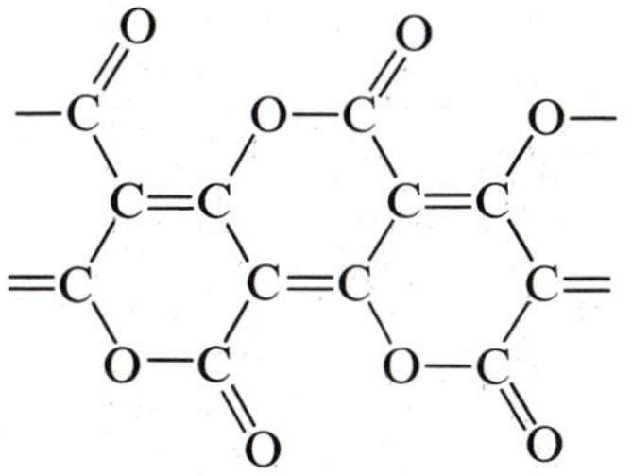

Figure 13.2
The polymer of carbon suboxide, C_3O_2.

The formation of CO or CO_2 also provides a driving force for certain thermal decomposition reactions. Compounds that have the general formula X–C(O)–Y tend to be unstable when X and Y are small (eg, a halogen, OH, or NH_2).

$$Cl-C(O)-Cl \underset{25°C}{\rightleftharpoons} CO + Cl_2 \tag{13.9}$$

$$Cl-C(O)-OH \longrightarrow CO_2 + HCl \tag{13.10}$$

$$H_2N-C(O)-OH \longrightarrow CO_2 + NH_3 \tag{13.11}$$

However, the stability of the –C(O)– system is increased by the incorporation of larger substituents; thus, many species that have the RO–C(O)–OR′ (carbonate), R_2N–C(O)–OR′ (carbamate or urethane), and R_2N–C(O)–NR′$_2$ (urea) linkages are known.

Aqueous carbon–oxygen chemistry is dominated by CO_2 and its derivatives. Unlike CO, carbon dioxide is reasonably soluble in water (about 0.003 M at room temperature). At least three acid–base equilibria are established among four carbon-containing species upon dissolution:

$$CO_2 + H_2O \rightleftharpoons (HO)_2CO \qquad K_f = 2 \times 10^{-3} \tag{13.12}$$

$$(HO)_2CO \rightleftharpoons H^+ + HOCO_2^- \qquad K_a = 2 \times 10^{-4} \tag{13.13}$$

$$HOCO_2^- \rightleftharpoons H^+ + CO_3^{2-} \qquad K_a = 5 \times 10^{-11} \tag{13.14}$$

Calculations suggest that in saturated solutions at room temperature over 99% of the carbon is present as CO_2.

13.3 The Metalloids: Boron, Silicon, and Germanium

The remaining members of Groups 13 and 14 are either metals or metalloids; boron, silicon, and germanium fall into the latter category. We will discuss these three elements together because of both their chemical similarities and their pronounced differences from the metals of the two families.

The boron/silicon pair represents one of the strongest diagonal relationships in the periodic table. Germanium might not be expected to fit into this grouping as well as it does; however, the effective nuclear charge effect discussed in Section 13.1 causes it to be comparable to boron and silicon in properties such as size and electronegativity.

Some of the reasons for segregating these three elements from the heavier members of Groups 13 and 14 are listed below:

1. The solid-state structures of elemental boron, silicon, and germanium are three-dimensional network polymers. The heavier elements utilize metallic lattices.

2. Boron, silicon, and germanium do not form true ionic bonds; even their bonds to fluorine and oxygen are polar covalent.

3. The binary halides BCl_3, $SiCl_4$, and $GeCl_4$ dissolve in water with solvolysis, producing acidic solutions. The products (H_3BO_3, H_4SiO_4, and H_4GeO_4) are amphoteric, but more acidic than basic. The heavier elements of both families form basic hydroxides.

4. All three of these elements form stronger covalent bonds than their heavier congeners. In many cases, the bonding is enhanced by partial double bonding (π-type Lewis acidity).

5. These three elements are more prone to *catenation* (the formation of rings or chains) than the others.

Bond Energies and Their Ramifications

Boron, silicon, and germanium all show a preference for bonding to the elements of the upper right corner of the periodic table—more specifically, to small elements having one or more lone pairs. In such cases, dative $(p \rightarrow p)\pi$ or $(p \rightarrow d)\pi$ bonding often contributes to very large bond energies. As might be expected, this tendency decreases in the order B > Si > Ge; it is most obvious for B–N, B–O, and B–F bonds in compounds in which the boron formally lacks an electron octet (Figure 13.3).

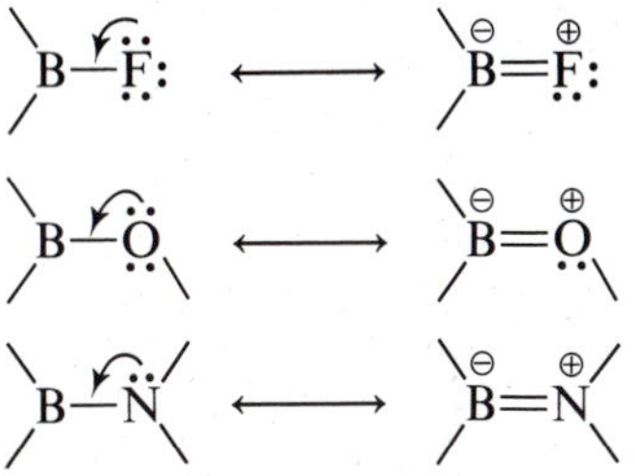

Figure 13.3
Dative $(p \rightarrow p)\pi$ bonding to trivalent boron in species containing B–F, B–N, and B–O bonds.

The resulting effect on bond energy can be seen in Table 13.3 (p. 400), where we give data for selected compounds in which boron is linked to atoms that have one or more lone pairs. The strongest bonds to atoms without lone pairs are listed for comparison.

A size effect is also evident from Table 13.3. The B–F bond is about 45% stronger than B–Cl, and B–O is about 64% stronger than B–S. This is expected if π bonding is an important factor, since π overlap decreases rapidly with increasing size (Chapter 4). It is interesting to compare boron, silicon, germanium, and carbon in this regard (Figure 13.4). Among these elements, carbon forms strong bonds to itself and to hydrogen, while boron and silicon favor fluorine and oxygen. As expected, this influences chemical reactivities. For example, B–H (and also Si–H and Ge–H) bonds are destroyed by water, while C–H bonds are generally inert. Thus, the

Table 13.3 Thermochemical energies of some covalent bonds involving boron

Bond	Compound	Energy, kJ/mol
To atoms with lone pairs:		
B–F	BF_3	649
B–Cl	BCl_3	447
B–Br	BBr_3	372
B–N	$B_3N_3H_6$	586
B–O	B_2O_3	686
B–S	B_2S_3	418
To atoms without lone pairs:		
B–B	$B_2(g)$	274
B–C	BMe_3	371
B–H	B_2H_6	372

Source: Values taken from Table 4.6 and from Muetterties, E. L., Ed. *The Chemistry of Boron and Its Compounds*; Wiley: New York, 1967; Chapter 1.

following reactions go rapidly to completion at or well below room temperature, but ethane does not react with H_2O even upon moderate heating:

$$B_2H_6 + 6H_2O \longrightarrow 2H_3BO_3 + 6H_2 \tag{13.15}$$

$$Si_2H_6 + 8H_2O \longrightarrow 2H_4SiO_4 + 7H_2 \tag{13.16}$$

The diatomic halogens differ in their reactivities toward boron compounds. The reaction

$$B_2S_3 + 3X_2 \longrightarrow 2BX_3 + \tfrac{3}{8}S_8 \tag{13.17}$$

is strongly favored for X = F, but not for X = Cl, Br, or I. The same is true for the corresponding reaction of the oxide.

$$B_2O_3 + 3X_2 \longrightarrow 2BX_3 + \tfrac{3}{2}O_2 \tag{13.18}$$

Borates and Silicates[4]

Silicon is one of the most abundant elements in the Earth's crust, boron is of moderate abundance, and germanium is quite rare. All occur in nature in

4. Liebau, F. *Structural Chemistry of Silicates*; Springer-Verlag: New York, 1985; Farmer, J. B. *Adv. Inorg. Chem. Radiochem.* **1982**, *25*, 187.

Figure 13.4
Thermochemical energies of selected single bonds involving boron, carbon, silicon, and germanium. Where available, values are taken from Table 4.6; others are from Cottrell, T. L. *The Strengths of Chemical Bonds*, 2nd ed.; Butterworths: London, 1958.

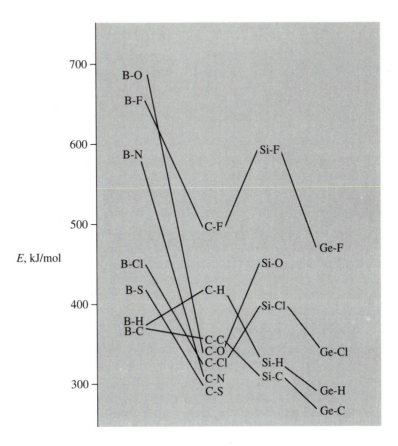

positive oxidation states, covalently bonded to oxygen. The structural diversity of the naturally occurring boron–oxygen and silicon–oxygen compounds and ions is astonishing, and only a brief sample is provided here.

The borates and silicates are anions of either ternary acids or their anhydrides. For example, H_3BO_3 is often called orthoboric acid. It can be partly dehydrated to metaboric acid, $(HOBO)_x$, or completely dehydrated to boric oxide:

$$H_3BO_3 \xrightarrow{\Delta} \frac{1}{x}(HOBO)_x + H_2O \tag{13.19}$$

$$2H_3BO_3 \xrightarrow{\Delta} B_2O_3 + 3H_2O \tag{13.20}$$

Both H_3BO_3 and $(HOBO)_x$ can be deprotonated to give a variety of anions, and metal salts of these anions are known.

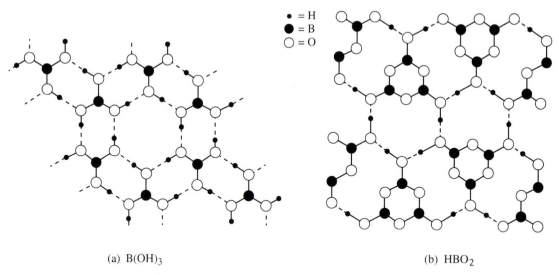

● = H
● = B
○ = O

(a) B(OH)$_3$ (b) HBO$_2$

Figure 13.5 The structures of (a) orthoboric and (b) metaboric acids.

The crystalline forms of both orthoboric and metaboric acid contain BO$_3$ units linked by hydrogen bonds, forming rings that interconnect to create planar sheets (Figure 13.5).

The planar BO$_3$ units persist in borate anions, but the rings may or may not be destroyed. In Na$_3$BO$_3$ and Mg$_3$(BO$_3$)$_2$, the anion has a simple, trigonal planar geometry. In Fe$_2$B$_2$O$_5$, one of the five oxygens of the formula unit is in a bridging position, $(O_2B-O-BO_2)^{4-}$; such salts are called *pyroborates*. Six-membered rings with alternating boron and oxygen atoms are also common. There is an additional (terminal) oxygen on each boron, so the formula unit is B$_3$O$_6^{3-}$.

The metaborate ions have a boron/oxygen ratio of 1:2. This is achieved either through cyclization or by the formation of infinite chains; the latter is found in Ca(BO$_2$)$_2$ (see Figure 13.6).

Figure 13.6
The structures of
(a) the cyclic
metaborate anion
of Na$_3$B$_3$O$_6$ and
(b) the polymeric
anion of Ca(BO$_2$)$_2$.

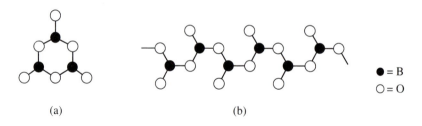

● = B
○ = O

(a)   (b)

Metal borates also exist in which boron has a tetrahedral geometry; that is, the parent anion is BO_4^{5-} rather than BO_3^{3-}. One example is $TaBO_4$. Borax, the major naturally occurring source of boron, is a hydrate containing the $H_4B_4O_9^{2-}$ ion. This anion contains an eight-membered ring of alternating boron and oxygen atoms, with two opposite borons connected by an oxygen bridge. Both trivalent and tetravalent borons are present.

The silicate system is similar, but even more complex. The "parent" is silicic acid, H_4SiO_4, and its anhydride is silicon dioxide:

$$H_4SiO_4 \longrightarrow \frac{1}{x}(SiO_2)_x + 2H_2O \qquad (13.21)$$

Silicic acid is thermally unstable toward the elimination of water; thus, SiO_2 (silica) and a multitude of partly hydrated species are the naturally occurring forms.

The simplest of the anions derived from silicic acid is orthosilicate (SiO_4^{4-}), which is found in Be_2SiO_4 (phenacite) and Na_4SiO_4. The pyrosilicate ion, $(O_3Si-O-SiO_3)^{6-}$, is known as well. Metasilicate ions have the general formula $(SiO_3)_x^{2x-}$. Like the metaborates, they may be either linear or cyclic. The linear chains have alternating silicon and oxygen atoms in which each tetrahedral unit shares a corner with the next (see Figure 13.7, p. 404). Cyclic metasilicates exist in various ring sizes. The best-known members of the series are the six- ($x = 3$), eight- ($x = 4$), twelve- ($x = 6$), and sixteen-($x = 8$) membered rings.

More complex silicates are formed by the cross-linking of two chains (to produce polymeric anions having empirical formulas such as $Si_2O_5^{2-}$ and $Si_4O_{11}^{6-}$), or by the fusion of rings to form infinite sheets (again giving $Si_2O_5^{2-}$). The mineral mica and an assortment of clays fall into this category.

Most complex of all are extended structures in which all the oxygens are shared between two units. This description fits silicon dioxide itself. Silica is polymorphic, with at least six known solid-state structures. The stable form at room temperature is α-quartz. (A β form is also known.) High-temperature variations include tridymite and cristobalite (α and β forms of both). β-Cristobalite is structurally related to elemental silicon, and can be derived from that diamondlike lattice by the insertion of an oxygen into each Si–Si bond. The structures of the β forms of quartz, tridymite, and cristobalite are pictured in Figure 13.8 (p. 405).

The structural characteristics of the most common borate and silicate anions are summarized in Table 13.4 (p. 405).

Germanium–oxygen chemistry is somewhat less developed, but as expected, there are close parallels to silicon–oxygen systems. For example, one form of GeO_2 has the α-quartz structure, and anions such as GeO_4^{4-} and $Ge_2O_5^{2-}$ are known. A notable difference is the existence of the monoxide GeO. When elemental germanium and germanium dioxide are

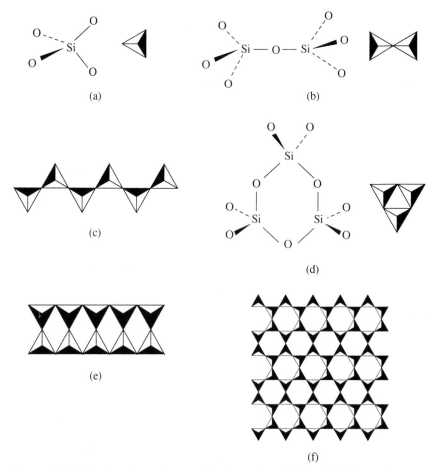

Figure 13.7   The structures of some silicate anions: (a) the SiO_4^{4-} ion of Be_2SiO_4; (b) the $Si_2O_7^{6-}$ ion of $Sc_2Si_2O_7$, with a linear Si–O–Si linkage; (c) the chain metasilicate ion $(SiO_3)_x^{2x-}$ of Na_2SiO_3; (d) the cyclic metasilicate ion $Si_3O_9^{6-}$ of $BaTiSi_3O_9$; (e) the cross-linked chain polymeric anion $(Si_2O_5)_x^{2x-}$; (f) a sheet polysilicate $(Si_2O_5)_x^{2x-}$ ion of the type found in micas.

mixed together and heated, a conproportionation reaction (the opposite of a disproportionation) occurs:

$$Ge + GeO_2 \xrightarrow{\Delta} 2GeO \tag{13.22}$$

The product contains Ge^{II}. The silicon analogue does not have comparable stability.

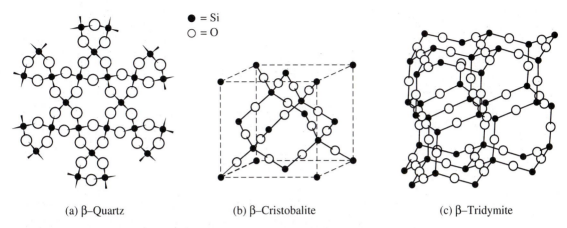

● = Si
○ = O

(a) β–Quartz   (b) β–Cristobalite   (c) β–Tridymite

Figure 13.8 Three crystalline forms of silica, SiO_2. [Reproduced with permission from Wells, A. F. *Structural Inorganic Chemistry*, 5th ed.; Clarendon: Oxford, 1984; pp. 1006–1007.]

Table 13.4 Relationships among borates and silicates

	Borate	Silicate	Description
Discrete ions:			
Ortho	BO_3^{3-}	SiO_4^{4-}	Triangular or
	BO_4^{5-}		tetrahedral
Pyro	$B_2O_5^{4-}$	$Si_2O_7^{6-}$	Two units sharing
			an oxygen
Meta cyclic	$B_3O_6^{3-}$	$(SiO_3^{2-})_n$,	Cyclic units
		$n = 3, 4, 6, 8$	
Extended structures:			
Single chains	$(BO_2^-)_x$	$(SiO_3^{2-})_x$	Edge-sharing units
Double chains		$(Si_4O_{11}^{6-})_x$	
Infinite sheets		$(Si_2O_5^{2-})_x$	Micas, clay minerals
Framework	B_2O_3	SiO_2	Each O shared
			between two units;
			quartz, zeolites

Nucleophilic Addition and Displacement Reactions

The reaction chemistry of trivalent, sp^2 hybridized boron is dominated by nucleophilic addition (the formation of Lewis acid–base adducts) and displacement reactions. The same is true for silicon and germanium, but stable adducts are somewhat less common because the central atoms possess valence electron octets in Si^{IV} and Ge^{IV} compounds.

A few of the many reactions in which BF_3 adducts are produced are described by equations (13.23)–(13.26). The hybridization of boron changes from sp^2 to sp^3 in each case.

$$BF_3 + F^- \longrightarrow BF_4^- \tag{13.23}$$

$$BF_3 + H^- \longrightarrow BHF_3^- \tag{13.24}$$

$$BF_3 + Et_2O \longrightarrow Et_2O{\cdot}BF_3 \tag{13.25}$$

$$BF_3 + H_2O \longrightarrow H_2O{\cdot}BF_3 \tag{13.26}$$

The stability of the water adduct toward elimination of HF is remarkable [compare to BCl_3; see equation (13.31), below]. Solid $H_2O{\cdot}BF_3$ is stable below its melting point of 10°C. The pure liquid undergoes a self-ionization reaction in which the OBF_3 unit is maintained:

$$2H_2O{\cdot}BF_3 \rightleftharpoons H_3O^+ + BF_3OH^- + BF_3 \tag{13.27}$$

Recall that whether nucleophilic attack leads to an addition or substitution reaction depends in large measure on the presence of appropriate leaving groups. For example, the reaction between trimethylamine and boron trichloride gives an adduct. However, if dimethylamine or another secondary amine is used, elimination of HCl either occurs spontaneously or can be induced by heating [equation (13.28)] or with excess base [equation (13.29)]:

$$Me_2NH + BCl_3 \longrightarrow \overset{\oplus}{Me_2N}\!\!-\!\!\overset{\ominus}{BCl_2} \xrightarrow[\Delta]{-HCl} Me_2NBCl_2 \tag{13.28}$$
$$\underset{\text{H \ Cl}}{\big| \ \big|}$$

$$6\,Me_2NH + BCl_3 \longrightarrow B(NMe_2)_3 + 3\,Me_2NH_2^+Cl^- \tag{13.29}$$

(If ammonia or a primary amine is used, the aminoborane itself may be thermally unstable toward further elimination; see below.) Similar reactions occur with other nucleophiles.

$$BCl_3 + 3\,LiMe \longrightarrow Me_3B + 3\,LiCl \tag{13.30}$$

$$BCl_3 + 3\,H_2O \longrightarrow H_3BO_3 + 3\,HCl \tag{13.31}$$

$$BCl_3 + \underset{\text{HO}}{\overset{\text{HO}}{\bigcirc\!\!\!\bigcirc}} \longrightarrow Cl\!-\!B\underset{\text{O}}{\overset{\text{O}}{\bigcirc\!\!\!\bigcirc}} + 2HCl \tag{13.32}$$

The situation for silicon and germanium is similar, but the adducts are often unstable at room temperature. Thus, there is no apparent reaction when, for example, trimethylamine and tetrachlorosilane are mixed.[5] However, adducts with fluoride ion (and for germanium, with chloride as well) have been isolated:

$$SiF_4 + Et_4N^+F^- \longrightarrow Et_4N^+ + SiF_5^- \tag{13.33}$$

$$GeCl_4 + 2HCl \longrightarrow 2H^+ + GeCl_6^{2-} \tag{13.34}$$

If a good leaving group is present, the reactions often parallel those described by equations (13.28)–(13.32), above:

$$Me_3SiCl + 2Me_2NH \longrightarrow Me_3SiNMe_2 + Me_2NH_2^+Cl^- \tag{13.35}$$

$$GeBr_4 + 4MeMgBr \longrightarrow GeMe_4 + 4MgBr_2 \tag{13.36}$$

$$R_2GeCl_2 + 2MeOH \longrightarrow R_2Ge(OMe)_2 + 2HCl \tag{13.37}$$

A different type of adduct formation is sometimes observed, especially with boron. If the substituents are small and a donor atom having sufficient basicity is present, head-to-tail dimerization can occur. An example involves the compound dichloro(dimethylamino)borane. Upon standing for 1–2 weeks under an inert atmosphere, the colorless liquid monomer crystallizes as the solid dimer:

$$2Me_2NBCl_2 \longrightarrow \begin{matrix} \overset{\oplus}{Me_2N}-\overset{\ominus}{BCl_2} \\ | \quad | \\ \underset{\ominus}{Cl_2B}-\underset{\oplus}{NMe_2} \end{matrix} \tag{13.38}$$

Similar $N \rightarrow Si$ oligomers can be produced if the nitrogen is sufficiently basic. The best-known example is the anion $(Me_3Si)_2N^-$:

$$\begin{bmatrix} Me_3Si-\overset{\ominus}{N}-\overset{\ominus}{Si}Me_3 \\ | \quad | \\ Me_3\underset{\ominus}{Si}-N-SiMe_3 \end{bmatrix}^{2-}$$

More unusual are the silatranes, which have the general formula $RSi(OCH_2CH_2N)_3$. Silicon is pentacoordinate in these compounds, with the fifth bond arising from electron pair donation by the nitrogen (Figure 13.9). Many silatranes are potent poisons, acting on the central nervous system of mammals in an unknown manner.

---

5. Spectroscopic evidence demonstrating adduct formation can be obtained at low temperatures.

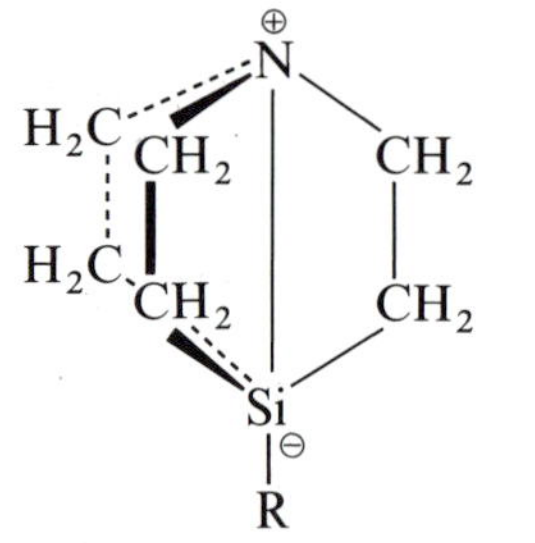

Figure 13.9
The silatrane
structure, with a
dative N → Si
bond.

Catenation: Rings and Chains

The most extensive catenation chemistry of any element belongs to carbon. This is a natural result of its valence (four bonds per atom) and its large homonuclear bond energies (because of its small size and lack of adjacent atom lone-pair repulsions). Silicon–silicon and germanium–germanium bonds are much weaker. (Comparing the diamond-type lattices of the free elements, the thermochemical C–C bond energy is about 58% greater than that of Si–Si and about 89% greater than for Ge–Ge.) Therefore, *homocatenation* (the formation of chains or rings composed of only one kind of atom), in which the framework is constructed from Si–Si or Ge–Ge bonds, produces compounds of relatively low stability, although a variety of such species have been characterized.

The boron–boron σ bond is intermediate in strength between C–C and Si–Si. However, homocatenation by boron is limited by its electron deficiency, since a chain such as

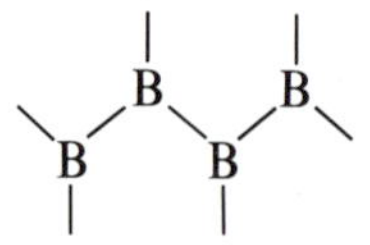

is susceptible to attack by nucleophiles at each link. As a result, the isolation of homocatenated boron compounds is difficult. One prominent example is the six-membered ring $(BNMe_2)_6$ shown in Figure 13.10.

An alternative to catenation is the formation of three-dimensional clusters, which have extensive electron delocalization. The chemistry of these *polyhedral boranes* and their derivatives will be discussed in Chapter 19.

Boron, silicon, and germanium engage in extensive *heterocatenation* (more than one type of atom in the chain or ring). The great bond energies to nitrogen and oxygen are important in this regard, as structures containing M–N–M or M–O–M linkages are extremely common. The simplest examples involve the formation of short chains through reactions in which a small molecule (typically, NH_3, H_2O, or HCl) is eliminated. Most primary amino-

Figure 13.10
The structure of the hexamer $(BNMe_2)_6$, as determined by X-ray diffraction. [Reproduced with permission from Nöth, H.; Pommerening, H. *Angew. Chem. Int. Ed. Engl.* **1980**, *19*, 482.]

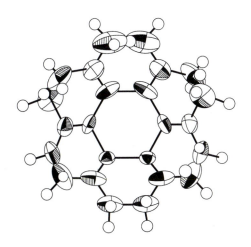

boranes, aminosilanes, and aminogermanes are thermally unstable:

$$3H_3SiNH_2 \xrightarrow{-20°C} (H_3Si)_3N + 2NH_3 \qquad \textbf{(13.39)}$$

$$2Et_3GeNH_2 \longrightarrow (Et_3Ge)_2NH + NH_3 \qquad \textbf{(13.40)}$$

Similarly, unhindered silanols and germanols eliminate water. The presence of sterically bulky groups such as *t*-butyl inhibits condensation.

Condensation reactions sometimes produce rings. Consider the reaction of BCl_3 with ammonia. The initial product is an adduct, which undergoes elimination of HCl to form the aminoborane:

$$BCl_3 + NH_3 \longrightarrow H_3N \cdot BCl_3 \xrightarrow{-HCl} [H_2NBCl_2] \qquad \textbf{(13.41)}$$

Good leaving groups remain on both boron and nitrogen, so H_2NBCl_2 is unstable toward elimination of a second mole of HCl. By analogy to organic chemistry, we might expect a B=N double bond to be formed, giving an *iminoborane*. However, the

$$\diagdown B{=}N\diagup$$

linkage is energetically inferior to

Thus, the actual product is a six-membered ring composed of alternating boron and nitrogen atoms:

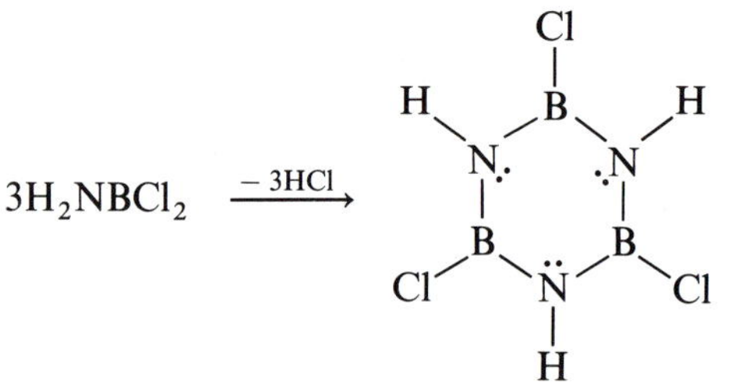

$$3H_2NBCl_2 \xrightarrow{-3HCl} \qquad\qquad\qquad (13.42)$$

Compounds containing such rings are called *borazines*.

The parent compound, $B_3N_3H_6$, is borazine itself. It can be prepared by the reduction of *B*-trichloroborazine or, more directly, by heating ammonia with diborane.

$$2Cl_3B_3N_3H_3 + 6NaBH_4 \longrightarrow 2B_3N_3H_6 + 6NaCl + 3B_2H_6 \quad (13.43)$$

$$3B_2H_6 + 6NH_3 \xrightarrow{\Delta} 2B_3N_3H_6 + 12H_2 \qquad (13.44)$$

The similarity between borazine and benzene has intrigued chemists for many years, and in fact borazine has been called "inorganic benzene." Their physical properties are quite comparable. Both are oily, colorless liquids with similar densities, viscosities, and refractive indices. They are also isoelectronic, and borazine is considered to be aromatic via two Kekulé-type resonance structures:

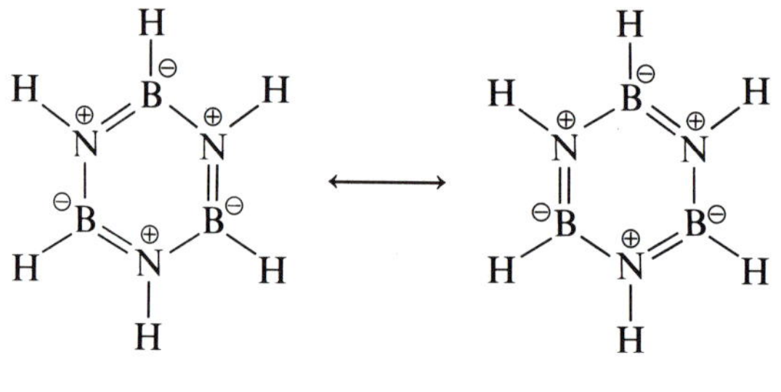

However, the polar boron–nitrogen bonds make borazine more chemically reactive than benzene. Some characteristic reactions are described below.

1. *Hydrogenation*

$$B_3N_3H_6 + 3H_2 \xrightarrow{\Delta} B_3N_3H_{12} \qquad (13.45)$$

The hydrogenation product is named *borazane*; it is structurally similar to cyclohexane.

2. ***Adduct Formation***

$$B_3N_3H_6 + 3HCl \longrightarrow Cl_3(H)_3B_3N_3H_6 \qquad (13.46)$$

This reaction demonstrates that the usual polarity of the boron–nitrogen bond (boron relatively positive) is maintained in spite of the formal charges indicated in the resonance structures above. This might also be considered an addition across a polar double bond, as demonstrated by the similar behavior toward methanol:

$$B_3N_3H_6 + 3MeOH \longrightarrow (MeO)_3(H)_3B_3N_3H_6 \qquad (13.47)$$

3. ***Nucleophilic Displacement at Boron***

$$H_3B_3N_3R_3 \xrightarrow{\text{MeMgBr}} Me_3B_3N_3R_3 \qquad R \neq H \qquad (13.48)$$

$$B_3N_3H_6 \xrightarrow{\text{NaBD}_4} D_3B_3N_3H_3 \qquad (13.49)$$

4. ***Deprotonation at Nitrogen***

$$R_3B_3N_3H_3 + 3CH_3Li \longrightarrow R_3B_3N_3Li_3 + 3CH_4 \qquad (13.50)$$

Nucleophiles tend to attack either at boron or at a nitrogen-bonded hydrogen. The reaction of excess methyllithium with borazine gives alkylation at boron and deprotonation at nitrogen.

Boron–nitrogen cyclization reactions occasionally give other than six-membered rings, especially if substituents having unusual steric requirements are present. In that event, eight-membered *borazocine* rings (tetramers) may result:

$$4t\text{-BuNH}_2 \cdot BCl_3 \xrightarrow{\Delta} (t\text{-BuNBCl})_4 + 8HCl \qquad (13.51)$$

These B_4N_4 rings exhibit a boat conformation (Figure 13.11), with alternating short and long B–N bonds. This is suggestive of localized $N \rightarrow B$ π interactions, rather than delocalization as in borazine. (Hückel's $4n + 2$ rule for aromaticity is obeyed for B_3N_3, but not for B_4N_4 rings.)

Why does steric hindrance influence the ring size? Borazine rings must be planar to maintain aromaticity. The substituents therefore lie at 60° angles (the internal angles of a hexagon) with respect to one another. Much larger

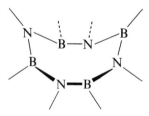

Figure 13.11 Generalized structure of the eight-membered B_4N_4 (borazocine) ring system. The B–N bonds alternate in length (see text).

angles are achieved for the nonplanar tetramer. In particular, the boat conformation maximizes separation if the large groups are on alternating ring atoms (see Figure 13.11).

For silicon–nitrogen and germanium–nitrogen systems, cyclic trimers are again common. In the condensation reaction

$$Me_2SiCl_2 + 3NH_3 \longrightarrow \frac{1}{x}(Me_2SiNH)_x + 2NH_4Cl \qquad (13.52)$$

the major product is the six-membered cyclotrisilazane; some tetramer and small amounts of other oligomers are obtained as well. This tendency to form six-membered rings carries over to B–O, Si–O, and Ge–O combinations.

$$RBCl_2 + H_2O \longrightarrow \tfrac{1}{3}(RBO)_3 + 2HCl \qquad (13.53)$$

$$R_2GeCl_2 + H_2O \longrightarrow \tfrac{1}{3}(R_2GeO)_3 + 2HCl \qquad (13.54)$$

Under the proper conditions it is possible to obtain linear polymers rather than rings; this is especially true for the Si–O system. For example, the hydrolysis of Me_2SiCl_2 produces polydimethylsilicone, $(Me_2SiO)_x$, in addition to cyclic compounds. Polymer yield can be maximized by heating the hydrolysis mixture in the presence of dilute sulfuric acid to promote ring-opening reactions.

Polydimethylsilicone is a hydrophobic material of high thermal stability and low viscosity; thus, it is useful as a waterproofing agent and in lubricants. The physical properties can be altered by changing the substituents (to larger alkyl or aryl moieties, and/or by the incorporation of functional groups) or in other ways. For example, the addition of $MeSiCl_3$ to the $Me_2SiCl_2 + H_2O$ reaction mixture promotes cross-linking between the polymer chains. The resulting material is either a rubber ("silly putty") or a resin, depending on the degree of cross-linking. It is also possible to prepare polymers having Si–N, Ge–O, and Ge–N frameworks, but they have less commercial importance.

An enormous number of less symmetric ring systems have been produced through various synthetic techniques. A few examples are given in equations (13.55)–(13.58).

$$\text{Me}_2\text{Si(NHMe)}_2 \xrightarrow[\text{(2) } \text{C}_6\text{H}_5\text{BCl}]{\text{(1) } 2n\text{-BuLi}} \text{(ring: Me}_2\text{Si, N-Me, N-Me, BC}_6\text{H}_5\text{)} \tag{13.55}$$

$$\text{R}_2\text{GeCl}_2 + 4\text{Na} + \text{ClCH}_2\text{CH}_2\text{CH}_2\text{Cl} \xrightarrow{-4\text{NaCl}} \text{R}_2\text{Ge} \tag{13.56}$$

$$8\,\text{R}_2\text{Ge} + \text{S}_8 \longrightarrow 8\,\text{R}_2\text{Ge} \tag{13.57}$$

$$\text{BCl}_3 + \begin{array}{c}\text{HO} \\ \text{Me(H)N}\end{array} \xrightarrow{-2\text{HCl}} \text{Cl-B} \tag{13.58}$$

Organometallic Compounds

Boron, silicon, and germanium all form strong bonds to carbon, and so each has extensive organometallic chemistry. The order of relative reactivity does not parallel bond energies:

Bond energy: B–C > Si–C > Ge–C

Reactivity: Ge–C, B–C > Si–C

This is due to a combination of factors. The Ge–C bond is longer than the others, which makes it both weaker and less sterically hindered. This is reflected in the relative ease of redistribution reactions such as

$$\text{R}_4\text{Ge} + \text{GeCl}_4 \xrightarrow{\Delta} 2\,\text{R}_2\text{GeCl}_2 \tag{13.59}$$

which occur only under forcing conditions (or not at all) with organoboranes and organosilanes. Trivalent organoboranes are highly reactive toward nucleophiles, especially if the organic substituents are small. Also, in contrast to Me_4Si and Me_4Ge, which are stable in air, trimethylborane is spontaneously flammable in air.

Table 13.5 The major carbosilanes produced from the pyrolysis of tetramethylsilane

Compound	Yield, wt %
$Me_3Si–CH_2–SiMe_3$	6.70
$Me_2Si\underset{\triangle}{\triangledown}SiMe_2$	5.33
$Me_2Si{\underset{\text{—}SiMe_2}{\overset{\text{—}SiMe_2}{\Big\langle}}}$	4.57
$Me_3Si–CH_2–SiMe_2–CH_2–SiMe_3$	3.54
$Me_3Si–CH_2–SiHMe_2$	3.52

Source: Fritz, G.; Matern, E. *Carbosilanes*; Springer-Verlag: Berlin, 1986; p. 11.

The Si–C bond is as strong as or slightly stronger than the C–C single bond, depending on the environment. As a result, the chemistry of *carbosilanes* (compounds containing alternating carbon and silicon atoms) is remarkably extensive. Even so "simple" a reaction as the pyrolysis of tetramethylsilane yields at least 45 identifiable products, including chains of up to seven members (C–Si–C–Si–C–Si–C), an assortment of four-, five-, six-, and eight-membered rings, and a number of bicyclic and multicyclic compounds. The structures of some of these species are given in Table 13.5.

Organoboron and organosilicon reagents are useful in organic synthesis, and there is a considerable body of literature in that regard.[6]

13.4 The Metallic Elements: Al, Ga, In, Tl, Sn, and Pb

Occurrence and Uses

The metals of Groups 13 and 14 vary considerably in natural abundance. Aluminum is the third most abundant element by mass in the lithosphere (behind oxygen and silicon), while indium and thallium are rare. The general order is Al $\gg$ Ga > Pb > Sn > Tl, In. The abundance of lead is very high in comparison to its neighbors in the periodic table. This probably derives

---

6. For boron, see Brown, H. C. *Organic Synthesis via Boranes*; Wiley: New York, 1975; Cragg, G. M. L. *Organoboranes in Organic Synthesis*; Marcel Dekker: New York, 1973. For silicon, see Walton, D. R. M., Ed. *Organometallic Compounds of Silicon*; Chapman and Hall: London, 1985; Weber, W. P. *Silicon Reagents for Organic Synthesis*; Springer-Verlag: Berlin, 1983; Colvin, E. *Silicon in Organic Synthesis*; Butterworths: London, 1981.

from its magic number of protons (see Chapter 7), and the fact that certain lead isotopes are the end products of radioactive decay chains of heavier elements.

As expected, all these elements occur in nature in positive oxidation states. Aluminum, gallium, and tin are most often found bonded to oxygen (ie, they are lithophiles), while lead, indium, and thallium occur more often as sulfides (ie, they are chalcophilic). For example, the most important ore of tin is casserite, SnO_2, while that of lead is galena, PbS. The difference in counterions can be explained by the relative hardness of Sn^{IV} compared to Pb^{II}. The difference in oxidation states is an example of a general tendency that will be discussed shortly.

The aluminum–oxygen system has many natural variations, including a number of well-known minerals. Aluminosilicates can be considered to be close-packed arrays of oxide ions, with Al^{III} and Si^{IV} occupying the interstices. Feldspar is a prominent aluminosilicate. One polymorph of anhydrous aluminum oxide, Al_2O_3, is the colorless mineral corundum. Colored gemstones result when certain transition metal ions are present in the lattice as impurities; ruby contains chromium, and sapphire has titanium.

Another variant is spinel, $MgAl_2O_4$, which is the parent of a large class of mixed oxides having the general formula AB_2O_4. In spinel the oxygens form a cubic close-packed lattice, with magnesium occupying tetrahedral and aluminum occupying octahedral interstices.

The most important natural source of aluminum is bauxite, Al_2O_3 containing between one and three waters per formula unit. The formula of bauxite is variously written as AlOOH, $Al(OH)_3$, and $Al_2O_3 \cdot xH_2O$; each can be correct depending on the Al/H_2O ratio. For example, $Al_2O_3 \cdot H_2O$ corresponds to the empirical formula $HAlO_2$ (or AlOOH), while $Al_2O_3 \cdot 3H_2O$ has the empirical formula H_3AlO_3 [or $Al(OH)_3$].

The free metal is usually obtained from bauxite ore through two sequential industrial processes. The *Bayer method* is a prepurification in which the aluminum is dissolved (along with some impurities) by addition of excess aqueous NaOH:

$$Al_2O_3 \cdot xH_2O \xrightarrow{OH^-} 2\,Al(OH)_4^- \tag{13.60}$$

After filtration and heating under pressure to remove impurities, the mixture is dissolved in molten cryolite (Na_3AlF_6) at 900–1000°C and subjected to electrolysis (the *Hall–Herault process*) to obtain the pure metal.

Its physical and chemical properties make aluminum a valuable commodity. It has a low density, and is extremely strong on a per-weight basis compared to other metals. It has a high malleability. Though nonmagnetic, it is an excellent electrical conductor. Aluminum forms a nonpermeable oxide surface coating, making it resistant to destructive oxidation by air or water.

Finally, it can be mixed into alloys to alter specific properties; for example, it can be mixed with copper, manganese, and/or other elements to increase its strength. The primary uses for aluminum are in construction materials for buildings, automobiles, and aircraft. Other uses include packaging materials, electrical power lines, and, as the oxide, in paint bases.

The other metals of commercial importance from among this group are lead and tin. Both are obtained from their ores by chemical reduction at high temperature.

$$2\,PbS + 3\,O_2 \xrightarrow{\Delta} 2\,PbO + 2\,SO_2 \tag{13.61}$$

$$PbO + C \xrightarrow{\Delta} Pb + CO \tag{13.62}$$

$$SnO_2 + 2\,C \xrightarrow{\Delta} Sn + 2\,CO \tag{13.63}$$

Tin is most often used in alloys. Most solders contain a mixture of tin and lead; bronze is 5–10% tin dissolved in copper; pewter is an alloy of tin, copper, and antimony; etc. The use of lead is declining because of its toxicity. At present, over 50% of all lead refined in the United States is used in automobile batteries ($Pb/PbO_2/PbSO_4$), while lesser (and rapidly decreasing) amounts are used for making tetraethyllead (an antiknocking gasoline additive) and paint pigments.

Aluminum and Beryllium: Another Diagonal Relationship

We stated in Chapter 12 that the chemistry of beryllium parallels that of aluminum. The major similarities are described here.

1. The free metals are noteworthy for their low densities, and both have high mechanical strength. Aluminum is used to a much greater extent than beryllium because of its higher natural abundance and relative ease of purification (and hence, lower cost).

2. Both metals form nonpermeable surface oxides, making them resistant to air oxidation.

3. The chlorides of both metals form oligomers containing M–Cl–M bridges; $(BeCl_2)_x$ is a linear polymer (Figure 12.6), while Al_2Cl_6 is dimeric. The metals achieve electron octets in both cases.

4. The hydrides of both metals are polymeric. The structure of $BeH_2(s)$ is believed to be similar to $BeCl_2$. Aluminum is larger and has valence level d orbitals; it can thus accommodate six substituents. Therefore, $(AlH_3)_x$ forms a three-dimensional structure in which each metal atom has an octahedral environment.

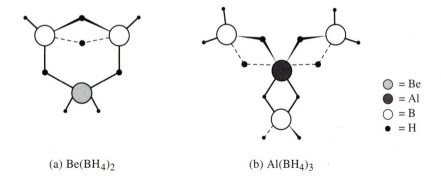

Figure 13.12
The structures of two hydrogen-bridged borohydrides.

(a) Be(BH$_4$)$_2$ (b) Al(BH$_4$)$_3$

○ = Be
● = Al
○ = B
● = H

5. Both elements form covalent borohydrides. In the best-known MBH$_4$ compounds (eg, NaBH$_4$), the BH$_4^-$ ion is present as a discrete anion. However, Be(BH$_4$)$_2$ and Al(BH$_4$)$_3$ contain M–H–B covalent linkages (see Figure 13.12).

6. The cations exhibit similar aqueous behavior. Recall that the charge density of Be^{2+} is so great that it hydrolyzes in water, producing a variety of species. The same is true for Al^{3+}; in fact, K_h of Al^{3+} is about 25 times greater than that of Be^{2+}. Both the metals and their hydroxides behave similarly toward excess base:

$$Be + 2OH^- + 2H_2O \longrightarrow [Be(OH)_4]^{2-} + H_2 \qquad \textbf{(13.64)}$$

$$2Al + 2OH^- + 6H_2O \longrightarrow 2[Al(OH)_4]^- + 3H_2 \qquad \textbf{(13.65)}$$

$$Be(OH)_2(s) + 2OH^- \longrightarrow [Be(OH)_4]^{2-} \qquad \textbf{(13.66)}$$

$$Al(OH)_3(s) + OH^- \longrightarrow [Al(OH)_4]^{2-} \qquad \textbf{(13.67)}$$

Dehydration of the hydroxy anions gives the beryllate and aluminate anions, respectively:

$$[Be(OH)_4]^{2-} \longrightarrow BeO_2^{2-} + 2H_2O \qquad \textbf{(13.68)}$$

$$[Al(OH)_4]^- \longrightarrow AlO_2^- + 2H_2O \qquad \textbf{(13.69)}$$

7. The metal cations have strongly negative standard reduction potentials (-1.97 V for the Be^{2+}/Be couple and -1.68 V for Al^{3+}/Al), and each element has only one common positive oxidation state.

Lower Oxidation States: The "Inert Pair" Effect

The metals discussed up to this point (Groups 1 and 2 and aluminum) have only one important positive oxidation state. The heavier members of·Groups 13 and 14 provide examples of a rather general phenomenon—the existence of a second oxidation state, two units less than the maximum. Gallium, indium, and thallium all form stable compounds in which their oxidation number is +1, and tin and lead (and less commonly germanium) are sometimes +2. This tendency increases going down each family (ie, $Ga \ll In < Tl$ and $Ge \ll Sn < Pb$). Because the valence electron configurations of these species are ... $ns^2 np^x$ and the lower oxidation state is M^{x+}, this behavior is sometimes attributed to the relative inertness of the ns orbital toward chemical bonding. Although theoretical and stereochemical aspects of this notion have been examined, the "inert pair" description is probably misleading.[7] A key to understanding these lower oxidation states is the realization that they are usually found in compounds having significant ionic character.

Going down a family, the increase in size has two important ramifications: Covalent bonds become weaker and ionization energies decrease. The former disfavors covalency, and the latter favors ionic interactions. However, the ionization of three or four electrons requires a prohibitive amount of energy (eg, the sum of the first three ionization energies of Tl is 56.4 eV, or about 5440 kJ/mol). The loss of only the valence p electron(s) is more feasible (6.1 eV for Tl, about the same as for Ca). Thus, TlCl and TlBr form ionic (specifically, CsCl-type) lattices and have the high melting and boiling temperatures characteristic of ionic species. In contrast, $TlCl_3$ has a low melting temperature (25°C), and $TlBr_3$ decomposes below its melting point:

$$2\,TlBr_3 \longrightarrow Tl[TlBr_4] + Br_2 \qquad\qquad (13.70)$$

The situation is similar for lead: $PbCl_2$ (ionic-type lattice) melts at 501°C, and $PbCl_4$ (covalent) at −15°C. The latter is easily decomposed (eg, by cold water), with Cl_2 being evolved:

$$PbCl_4 \longrightarrow PbCl_2 + Cl_2 \qquad\qquad (13.71)$$

Because of their relative softness, these lower oxidation states are stabilized by soft anions such as Br^- and I^-. Evidence for this can be found in Table 13.6, which lists the lower-valent halides of Groups 13 and 14 that have measurable melting points.

The inert pair effect is manifested by aqueous reduction potentials. In Group 13, the M^{III}/M^I potential can be measured only for indium and

7. Ng, S.-W.; Zuckerman, J. J. *Adv. Inorg. Chem. Radiochem.* **1985**, *29*, 297; Pitzer, K. S. *Acc. Chem. Res.* **1979**, *12*, 271; Pyykko, P.; Desclaux, J.-P. *Acc. Chem. Res.* **1979**, *12*, 276.

Table 13.6 The melting points of the lower-valent halides of Groups 13 and 14

Group 13	mp, °C	Group 14	mp, °C
InCl	225	$GeBr_2$	122
InBr	220	$SnCl_2$	146
TlF	327	$SnBr_2$	216
TlCl	430	SnI_2	320
TlBr	480	PbF_2	855
TlI	440	$PbCl_2$	501
		$PbBr_2$	373
		PbI_2	402

thallium:

$$In^{3+} + 2e^- \longrightarrow In^+ \qquad \mathscr{E}^0 = -0.44 \text{ V} \qquad \text{(13.72)}$$

$$Tl^{3+} + 2e^- \longrightarrow Tl^+ \qquad \mathscr{E}^0 = +1.25 \text{ V} \qquad \text{(13.73)}$$

In Group 14, M^{IV}/M^{II} potentials are measurable for tin and lead.

$$Sn^{4+} + 2e^- \longrightarrow Sn^{2+} \qquad \mathscr{E}^0 = +0.15 \text{ V} \quad \text{(13.74)}$$

$$PbO_2 + 4H^+ + 2e^- \longrightarrow Pb^{2+} + 2H_2O \quad \mathscr{E}^0 = +1.455 \text{ V} \quad \text{(13.75)}$$

Another measure of the stability of the lower state is its ease of reduction:

$$In^+ + 1e^- \longrightarrow In \qquad \mathscr{E}^0 = -0.14 \text{ V} \qquad \text{(13.76)}$$

$$Tl^+ + 1e^- \longrightarrow Tl \qquad \mathscr{E}^0 = -0.336 \text{ V} \qquad \text{(13.77)}$$

$$Sn^{2+} + 2e^- \longrightarrow Sn \qquad \mathscr{E}^0 = -0.136 \text{ V} \qquad \text{(13.78)}$$

$$Pb^{2+} + 2e^- \longrightarrow Pb \qquad \mathscr{E}^0 = -0.126 \text{ V} \qquad \text{(13.79)}$$

Note that In^+ is unstable with respect to disproportionation while Tl^+ is stable, and that Sn^{2+} is a better reducing agent than Pb^{2+} by 1.3 V.

There is also an apparent, though false, intermediate oxidation state in certain compounds. For example, solid materials that have the empirical formulas $GaCl_2$ and $TlBr_2$ are known, but in both cases, +2 is an average oxidation number. The individual metal atoms are in the M^I and M^{III} states, so the true formulations are $M^+[MX_4]^-$. A similar case is that of $KPbO_2$, which superficially appears to contain Pb^{III}, but its solid-state lattice contains K^+, Pb^{2+}, and PbO_4^{4-} ions.

Table 13.7 Calculated radius ratios for the trivalent halides of the Group 13 metals

	F	Cl	Br	I
Al	0.41	0.30	0.28	0.25
Ga	0.47	0.34	0.36	0.28
In	0.60	0.44	0.41	0.36
Tl	0.66	0.49	—	—

Note: Compounds to the right of the line have lattices in which the metal has coordination number 4; the others have hexacoordinate metals. Radius ratios are calculated from data given in Table 6.7 assuming hexavalent M^{3+} and X^- ions. No value is given for $TlBr_3$ because it decomposes below its melting point. No value is given for TlI_3 because its composition ($Tl^+I_3^-$) differs from the others (see text).

Lewis Acidity and Complexation[8]

Trivalent aluminum, gallium, indium, and thallium are electron-deficient. In addition, all these metals (as well as tin and lead) are capable of octet expansion. These elements therefore form a wide variety of complexes.

Self-complexation through the formation of dimers occurs for Al_2X_6 (X = Cl, Br, and I), Ga_2X_6 (X = Cl, Br, and I), In_2I_6, and Tl_2Br_6. Most of the remaining Group 13 metal halides (all the fluorides plus $InCl_3$, $TlCl_3$, and $InBr_3$) form ionic-type lattices in the solid state. The anions occupy a close-packed sublattice, with metals in the octahedral interstices. These differences can be understood by considering the relative sizes of the metal and halogen. Recall that the theoretical requirement for a cation to fit into the octahedral holes of a close-packed lattice is a radius ratio above 0.41 (Chapter 6). As can be seen from Table 13.7, all combinations having ratios greater than 0.41 have hexacoordinate metals.

The above discussion does not include TlI_3 because of its very different lattice, which consists of Tl^+ and I_3^- ions. This structure is favored by at least two factors: the proclivity of thallium for the Tl^I oxidation state and stabilization of the Tl^+ cation by the large triiodide ion.

Many other acid–base adducts are known. They may be tetra-, penta-, or hexavalent, depending on the situation. Some examples are described by equations (13.80)–(13.83).

$$AlF_3 + 3NaF \longrightarrow 3Na^+ + AlF_6^{3-} \tag{13.80}$$

$$InCl_3 + 2py \longrightarrow In(py)_2Cl_3 \qquad (py = pyridine) \tag{13.81}$$

8. Carty, A. J.; Tuck, D. G. *Prog. Inorg. Chem.* **1975**, *19*, 243; Lee, A. L. *Coord. Chem. Rev.* **1972**, *8*, 289; Walton, R. A. *Coord. Chem. Rev.* **1971**, *6*, 1.

$$Al_2Cl_6 + 2CsCl \longrightarrow 2Cs^+ + 2AlCl_4^- \tag{13.82}$$

$$Ga_2Cl_6 + 2Me_3N \longrightarrow 2Me_3N{\cdot}GaCl_3 \tag{13.83}$$

Nucleophilic displacement also can occur, either with or without addition, under the proper conditions.

$$InCl_3 \xrightarrow{NH_3(l)} [In(NH_3)_6]^{3+} + 3Cl^- \tag{13.84}$$

$$SnCl_4 + 8Me_2NH \longrightarrow Sn(NMe_2)_4 + 4Me_2NH_2^+Cl^- \tag{13.85}$$

$$GaCl_3 + 3NH_3 \longrightarrow (H_3N)_3GaCl_3 \tag{13.86}$$

$$GaCl_3 + 6NH_3 \longrightarrow [Ga(NH_3)_6]^{3+} + 3Cl^- \tag{13.87}$$

Catenation

Homocatenation is even less common for the Group 13 and 14 metals than for boron, silicon, and germanium. Tin forms the strongest of the homonuclear metal–metal σ bonds (about 146 kJ/mol). Homocatenated Sn chains of up to nine atoms, and rings (cyclostannanes) of at least four, five, six, and nine members are known. Such species appear to be stabilized by moderately sized organic substituents such as $-C_2H_5$ and $-C_6H_5$. They are usually prepared either by the reduction of dihalides or by dehydrogenation.

$$R_2SnCl_2 + 2Na \longrightarrow \frac{1}{x}(R_2Sn)_x + 2NaCl \tag{13.88}$$

$$R_2SnH_2 \xrightarrow{\Delta} \frac{1}{x}(R_2Sn)_x + H_2 \tag{13.89}$$

Heterocatenation is much more common, and usually arises through one of three synthetic pathways. The first two, condensation and metallation reactions, are similar to those discussed earlier for B, Si, and Ge.

$$3R_2SnCl_2 + 9NH_3 \longrightarrow (R_2SnNH)_3 + 6NH_4Cl \tag{13.90}$$

$$2t\text{-}Bu_2SnCl_2 + 2Na_2S \xrightarrow{-4NaCl} (t\text{-}Bu_2SnS)_2 \tag{13.91}$$

$$R_2SnCl_2 + LiNR(SiR_2)_2NRLi \xrightarrow{-2LiCl} \begin{array}{c} R \\ \diagdown \\ N{-}SiR_2 \\ R_2Sn \diagup \quad | \\ \diagdown \\ N{-}SiR_2 \\ \diagup \\ R \end{array} \tag{13.92}$$

Many noncyclic adducts contain bridging halides. One well-known example is $Sn_2F_5^-$ (F_2Sn–F–SnF_2^-), which can be considered an adduct of the acid SnF_2 and the base SnF_3^-. Also known is $Sn_3F_{10}^{4-}$, which

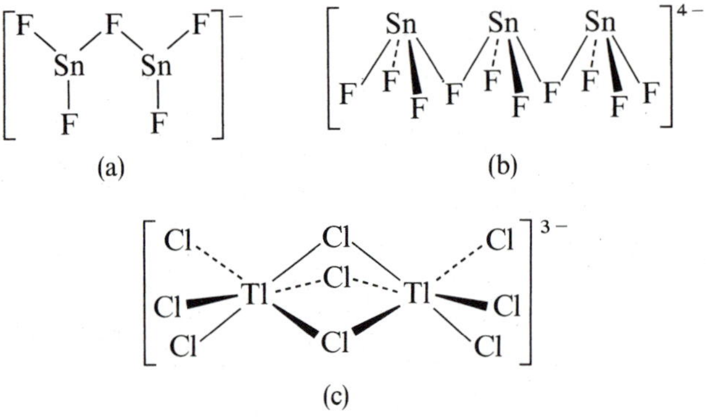

Figure 13.13
The structures of
some halide-
bridged adducts:
(a) the $Sn_2F_5^-$
anion of $NaSn_2F_5$;
(b) the $Sn_3F_{10}^{4-}$
anion of
$Na_4Sn_3F_{10}$;
(c) the $Tl_2Cl_9^{3-}$
anion of $Cs_3Tl_2Cl_9$.

has an odd structure consisting of square pyramidal SnF_4 groups with two bridging fluorines. The anions $In_2Cl_9^{3-}$ and $Tl_2Cl_9^{3-}$ contain hexacoordinate metals; in each case, the chlorines form two octahedra sharing a common face. Several of these complex anions are shown in Figure 13.13.

Organometallic Compounds

Metal–carbon bond energies follow the order Al–C > Ga–C > Sn–C > In–C > Pb–C > Tl–C; in all cases the M–C bond is weaker than the corresponding M–F, M–Cl, M–O, and M–N bonds. Thus, the organometallic compounds of these elements tend to be stable, but chemically reactive. Several are useful reagents for industrial-scale syntheses (see below).

The metal alkyls SnR_4 and PbR_4 are generally similar in structure and behavior to their silicon and germanium analogues, but are more reactive toward redistribution reactions and nucleophiles. In Group 13, trivalent compounds such as AlR_3 and GaR_3 are electron-deficient, and like their boron congeners are often pyrophoric. They are also powerful Lewis acids. Dimerization via bridging carbon atoms is common. For example, many triorganoaluminums are dimers in both the solid and vapor states:

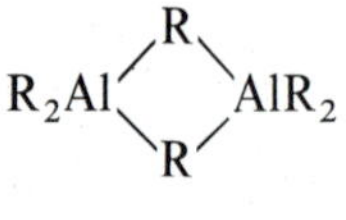

$$R_2Al \diagdown_{R}^{R} AlR_2$$

The tendency to dimerize decreases going down the family.

Metal alkyls are usually synthesized by metal–metal exchange reactions, often using Grignard or alkyllithium reagents:

$$2Ga + 3HgR_2 \longrightarrow 2GaR_3 + 3Hg \tag{13.93}$$

$$2AlX_3 + 6RMgX \longrightarrow Al_2R_6 + 6MgX_2 \tag{13.94}$$

$$3SnX_4 + 4AlR_3 + 4NaX \longrightarrow 3SnR_4 + 4NaAlX_4 \tag{13.95}$$

Direct synthesis is possible for many dialkyltin dihalides:

$$Sn + 2RX \xrightarrow[\Delta]{Catalyst} R_2SnX_2 \qquad (13.96)$$

Tetraethyllead, which has long been used as a gasoline additive, is prepared commercially by either of two electrolysis reactions in which lead metal is the anode:

$$Pb + 4EtI \xrightarrow{Electrolysis} PbEt_4 + 2I_2 \qquad (13.97)$$

$$Pb + 2EtMgCl + 2EtCl \xrightarrow{Electrolysis} PbEt_4 + 2MgCl_2 \qquad (13.98)$$

Certain organoaluminum and organotin compounds have commercial importance. Triethylaluminum and related compounds readily add across $C{=}C$ double bonds:

$$AlEt_3 + C_2H_4 \longrightarrow Et_2AlCH_2CH_2Et \qquad (13.99)$$

Sequential reactions of this type lead to polymerization under the proper conditions:

$$AlEt_3 + nC_2H_4 \longrightarrow Et_2Al(CH_2CH_2)_nEt \qquad (13.100)$$

One such set of conditions uses a mixture of $TiCl_4$ and Al_2Cl_6 in heptane (the *Ziegler–Natta catalyst*[9]) at 50–150°C to produce high-density polyethylene. Alternatively, the process can be terminated at $n = 5$–10 units. Hydrolysis of the product mixture then gives long-chain alcohols, which are used for the production of biodegradable detergents. The mechanism of insertion into the $C{=}C$ double bond is believed to involve a π-type complex, and will be discussed in Chapter 18.

Various tin(IV) organometallics, especially SnR_4, $HSnR_3$, and H_2SnR_2, are useful in organic synthesis on both the laboratory and industrial scale. Here again, insertion into double bonds is common. A few examples include the following:

$$HSnR_3 + H_2C{=}CHR' \longrightarrow R_3SnCH_2CH_2R' \qquad (13.101)$$

$$HSnR_3 + R_2'C{=}O \longrightarrow R_3Sn{-}O{-}CHR_2' \qquad (13.102)$$

$$HSnR_3 + C_6H_5N{=}C{=}S \longrightarrow R_3Sn{-}S{-}CH{=}NC_6H_5 \qquad (13.103)$$

9. Sinn, H. *Adv. Organomet. Chem.* **1980**, *18*, 99; Boor, J. *Ziegler–Natta Catalysts and Polymerizations*; Academic: New York, 1978.

The inert pair effect appears to be lessened by organic substituents (consistent with the increased tendency exhibited by carbon-bonded atoms toward covalency). Thus, most compounds containing Tl–C bonds are trivalent. Compounds with the formulas SnR_2 and PbR_2 are also rather rare; those that are known contain very large groups such as $-CH(SiMe_3)_2$.

Bibliography

Urry, G. *Elementary Equilibrium Chemistry of Carbon*; Wiley: New York, 1988.

Woolins, J. D. *Non-metal Rings, Cages, and Clusters*; Wiley: New York, 1988.

Haiduc, I.; Sowerby, D. B., Eds. *The Chemistry of Inorganic Homo- and Heterocycles*; Academic: New York, 1987.

Fritz, G.; Matern, E. *Carbosilanes*; Springer-Verlag: Berlin, 1986.

Haiduc, I.; Zuckerman, J. J. *Basic Organometallic Chemistry*; Walter de Gruyter: New York, 1985.

Liebau, F. *Structural Chemistry of Silicates*; Springer-Verlag: New York, 1985.

Walton, D. R. M., Ed. *Organometallic Compounds of Silicon*; Chapman and Hall: London, 1985.

Cowley, A. H., Ed. *Rings, Clusters, and Polymers of the Main-Group Elements*; ACS Symposium Series, No. 232: American Chemical Society: Washington, DC, 1983.

Rheingold, A. L., Ed. *Homoatomic Rings, Chains, and Macromolecules of the Main-Group Elements*; Elsevier: Amsterdam, 1977.

Muetterties, E. L., Ed. *Boron Hydride Chemistry*; Academic: New York, 1975.

Jeffrey, E. A. *Organoaluminum Compounds*; Elsevier: Amsterdam, 1972.

Questions and Problems

1. In the chapter introduction we stated that the first member of most chemical families is quite different from the others. With that in mind, compare and contrast:
   (a) CH_4 versus the other Group 14 hydrides
   (b) BMe_3 versus the other Group 13 methyls
   (c) The B–B bond versus the homonuclear bonds of the other Group 13 elements (in terms of strength and reactivity)
   (d) $C=C$ and $C\equiv C$ bonds versus the other Group 14 homonuclear multiple bonds

2. Explain in your own words why:
   (a) In IE_1, B > Al   (b) In IE_1, B ≈ Si
   (c) In melting point, C > Si   (d) In χ, Al < Ga

3. We described Be_2C as polar covalent even though it forms an antifluorite lattice. What aspects of this structure are consistent with covalency? What physical properties might be indicative of covalent rather than ionic bonding?

4. (a) Give the formula for the anhydride of pyruvic acid, $O{=}CH{-}COOH$. Write an equation describing the most probable pathway for its thermal decomposition.
   (b) Give the formula for the anhydride of oxalic acid.
   (c) Suggest a structure for your answer to part (b), and rationalize its instability.

5. Write the most probable products for the following reactions and then balance:
   (a) $HCO_2H \xrightarrow{\Delta}$
   (b) $C_3O_2 + H_2O \longrightarrow$
   (c) $C_3O_2 + R_2NH \longrightarrow$

6. Explain the low thermal stability of each of the following compounds by writing an equation for its decomposition:
   (a) $Se{=}C{=}O$ (b) $Cl{-}CO{-}COOH$ (c) $(NH_4)HCO_3$

7. Because of its explosive reaction with atmospheric oxygen, SiH_4 is highly dangerous and difficult to work with. In contrast, methane is kinetically stable in air.
   (a) Write the balanced equation for the burning of SiH_4.
   (b) Explain the difference in behavior between SiH_4 and CH_4.
   (c) Do you expect stannane, SnH_4, to be stable in air? Why or why not?

8. Complete and balance the following equations:

   (a) $B_2H_6 + $ Excess $O_2 \longrightarrow$ _____ + _____
   (b) $SiO_2 + $ Excess $HF \longrightarrow$ _____ + _____
   (c) $H_4GeO_4 \longrightarrow$ _____ + _____
   (d) $B_2S_3 + $ Excess $H_2O \longrightarrow$ _____ + _____

9. Use Figure 13.4 to evaluate the favorability of each of the following general types of reactions. (That is, which ones might be useful for synthetic purposes?)

   (a) $\diagdown B{-}Cl + {-}\underset{|}{\overset{|}{Si}}{-}O\diagdown \longrightarrow {-}\underset{|}{\overset{|}{Si}}{-}Cl + \diagup B{-}O\diagdown$

   (b) $-\underset{|}{\overset{|}{C}}{-}H + -\underset{|}{\overset{|}{Ge}}{-}F \longrightarrow -\underset{|}{\overset{|}{C}}{-}F + -\underset{|}{\overset{|}{Ge}}{-}H$

   (c) $\diagdown B{-}H + -\underset{|}{\overset{|}{C}}{-}O\diagdown \longrightarrow \diagdown B{-}O\diagdown + -\underset{|}{\overset{|}{C}}{-}H$

   (d) $-\underset{|}{\overset{|}{C}}{-}Cl + -\underset{|}{\overset{|}{Si}}{-}H \longrightarrow -\underset{|}{\overset{|}{C}}{-}H + -\underset{|}{\overset{|}{Si}}{-}Cl$

10. Use Lewis structures and normal valences to rationalize the following observations:
    (a) For the cyclic $B_3O_6^{n-}$ ion, $n = 3$.
    (b) For the cyclic $Si_3O_9^{n-}$ ion, $n = 6$.
    (c) For the acyclic $B_2O_5^{n-}$ ion, $n = 4$.

11. Sketch the linear silicate anion $Si_3O_8^{n-}$, and determine its charge.

12. Make a sketch of a portion of the $(Si_2O_5^{2-})_x$ double chain and use it to:
    (a) Describe this structure in terms of the rings that are formed.
    (b) Demonstrate that the limiting mole ratio is two silicons per five oxygens.
    (c) Demonstrate that the charge is $2-$ per formula unit, and identify the atoms on which these charges are localized.

13. Rationalize the following observations concerning silicon and germanium:
    (a) The inert pair effect is more evident for Ge than for Si.
    (b) The Si–H bond is slightly stronger than Ge–H, but the Si–F bond is much stronger than Ge–F.

14. The compound Me_3SiNH_2 undergoes condensation (elimination of NH_3) below $-25°C$. In contrast, t-$BuMe_2SiNH_2$ is stable above $100°C$. What does this suggest about the reaction mechanism? Is it more likely to be inter- or intramolecular? Defend your answer.

15. A sample of trichloroborane is dissolved in toluene and cooled to $0°C$. An equimolar amount of methylamine is added, producing a white solid that melts at $126°C$ after purification. Elemental analysis of this solid shows it to contain 71.8% chlorine by mass. The compound is then redissolved in hot toluene, and excess triethylamine is added. Two products result: One is water-soluble and has 25.8% chlorine, while the other is soluble in less polar solvents and contains 47.1% chlorine. Identify all three of the materials described, and write a pair of balanced equations to represent their formation.

16. Complete and balance the following equations, in which borazine or one of its derivatives is either a reactant or product.

    (a) _____ + $MeBCl_2$ $\longrightarrow$ $(MeBNMe)_3$ + _____
    (b) $B_3N_3H_6 + 3Me_3SiCl$ $\xrightarrow{\Delta}$ _____ + $3HCl$

    (c) $B_3N_3H_6$ + Excess C_2H_5Li $\longrightarrow$ _____ + _____

17. Review the description of the Bayer/Hall–Herault process for the industrial production of aluminum (Section 13.4). Speculate on:
    (a) The identities of some of the impurities removed by precipitation with OH^-
    (b) The purpose of adding Na_3AlF_6 prior to electrolysis

18. We stated in the text that the spinel lattice has close-packed oxygens, with magnesium occupying tetrahedral and aluminum occupying octahedral interstices. What fraction of each type of interstice is filled?

19. Both $BeCl_2$ and $AlCl_3$ are oligomeric, with bridging chlorine atoms. However, one is a dimer while the other is a linear polymer. Tell which is which, and explain the different behavior.

20. Many Ga^{III} salts dissolve in water to initially give the complex ion $Ga(H_2O)_6^{3+}$. Upon standing at room temperature, this ion slowly decomposes to give a material formulated as GaOOH. Write a balanced equation to describe this process. How might the decomposition be inhibited?

21. The reaction of $AlMe_3$ with $MeNH_2$ at 70°C gives a product having a molecular weight of about 260 g/mol. Suggest its structure and write the balanced equation.

22. Based on its empirical formula, the linear polymer TlSe appears to contain Tl^{II}. Suggest a more likely structure.

23. There are two different Al–C bond distances in Al_2Me_6, 197 pm (terminal) and 214 pm (bridging). Explain why the distances are unequal.

24. Suggest structures for:
    (a) $MgAl_2Me_8$ (b) $Al_2Me_5NR_2$

25. In the solid-state lattice of GaS, all the gallium atoms have coordination number 4 and are in equivalent positions. In TlS, there are two chemically different kinds of thallium. Speculate on each structure.

26. The Tl^+ ion is similar in chemical behavior to Ag^+.
    (a) Explain why this is the case; support your argument with whatever tabulated data you choose from anywhere in the text.
    (b) These two ions are markedly different in their oxidation–reduction chemistry. Identify two such differences.

27. Thallium(III) chloride forms a 1:2 adduct with DMSO.
    (a) The adduct contains Tl–O (not Tl–S) bonds. Why?
    (b) The electron geometry is trigonal bipyramidal. Three isomers are theoretically possible, but only one is observed. Which one do you predict?
    (c) Assign a point group to this molecule that is consistent with your answer to part (b). Disregard the sulfurs and methyl groups (ie, give the local symmetry about thallium).

*28. For many years the synthesis of gallane (GaH_3) eluded inorganic chemists; however, it has now been achieved (Downs, A. J.; Goode, M. J.; Pulham, C. R. *J. Am. Chem. Soc.* **1989**, *111*, 1936). Answer these questions after reading the article cited.
    (a) What properties of GaH_3 make it difficult to isolate?
    (b) What synthetic route was used?
    (c) What experimental evidence did the authors present?

*29. The preparation of the first "alumazine" has been claimed by K. M. Waggoner, H. Hope, and P. P. Power (*Angew. Chem. Int. Ed. Engl.* **1988**, *27*, 1699).
    (a) What in the world is an alumazine? Where does the name come from?
    (b) The substituents on nitrogen are sterically bulky. What is the significance of this?
    (c) Why was the determination of the Al–N bond distance important to the authors?

*30.   S. Adams and M. Dräger reported experimental results for a series of reactions conforming to the general equation

$$2(C_6H_5)_3SnLi + I(t\text{-}Bu_2Sn)_nI \longrightarrow \ ??$$

for $n = 1\text{--}4$ (*Angew. Chem. Int. Ed. Engl.* **1987**, *26*, 1255).

(a) Sketch the primary tin-containing product for each of these reactions.

(b) In what way are these reactions more complex than is suggested by the stoichiometry?

(c) The authors state that two types of substituents tend to increase Sn–Sn bond distances. What are they? Can you explain why?

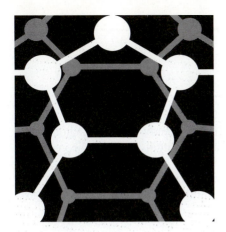

14

The Chemistry of
Groups 15–18

Our survey of the main group elements will be completed in this chapter with discussion of Groups 15–18. Only 15 of these elements are considered in any significant depth. Relatively little descriptive chemistry is known for radioactive polonium and astatine.[1] Among the noble gases, well-characterized compounds of only krypton and xenon have been reported to date.

Although their physical and chemical properties are quite varied, the elements of Groups 15–18 may be grouped together because of several general similarities. All except bismuth are either nonmetals or metalloids, and as such they have relatively high ionization energies, electron affinities, and electronegativities. As noted in the introduction to Chapter 13, their hydrogen compounds tend to be more acidic than basic. Their compounds and ions usually contain one or more lone pairs, so nucleophilic behavior is common.

Many compounds of these elements (especially the halogens) have already been discussed. The emphasis in this chapter is therefore on systems not yet covered—primarily, species containing bonds between nonmetals.

---

1.   However, the known chemistry of these two elements has been reviewed. See Brown, I. *Adv. Inorg. Chem. Radiochem.* **1987**, *31*, 43; Bagnall, K. W. *Adv. Inorg. Chem. Radiochem.* **1962**, *4*, 198.

14.1 Elemental Properties: Trends, Comparisons, and Contrasts

The normal periodic variations generally apply to the Group 15–18 families. For example, within each family there is a steady decrease in ionization energy with increasing atomic number. Horizontal trends (for consecutive elements, as in the series phosphorus, sulfur, chlorine, argon) are also as expected, save for certain exceptions explained in Chapter 1 (eg, in IE_1, P > S < Cl). Both vertical and horizontal trends are apparent in first ionization energies, electronegativities, and nonpolar covalent radii (Figures 14.1–14.3).

Following the trend of Al/Ga and Si/Ge (Chapter 13), electronegativity reversals occur for P/As and S/Se. The inert pair effect persists: the (III) oxidation state (two less than the maximum) is common for P, As, Sb, and

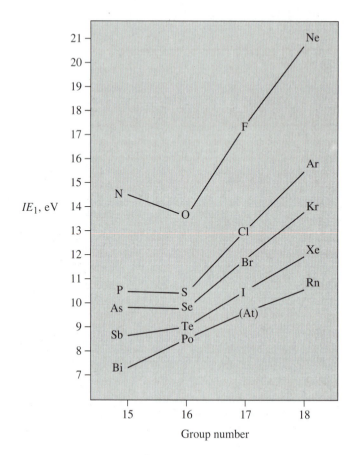

Figure 14.1
Trends in the first ionization energies of the Group 15–18 elements. The value for astatine is approximate.

IE_1, eV

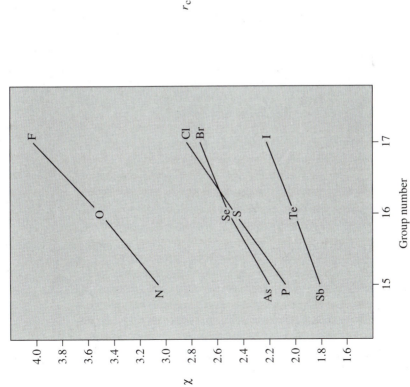

Figure 14.3 Nonpolar covalent (single-bond) radii of the Group 15–17 elements. The radius of tellurium is approximate.

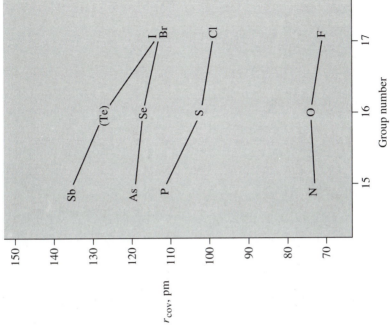

Figure 14.2 Electronegativities (Allred–Rochow scale) of the Group 15–17 elements.

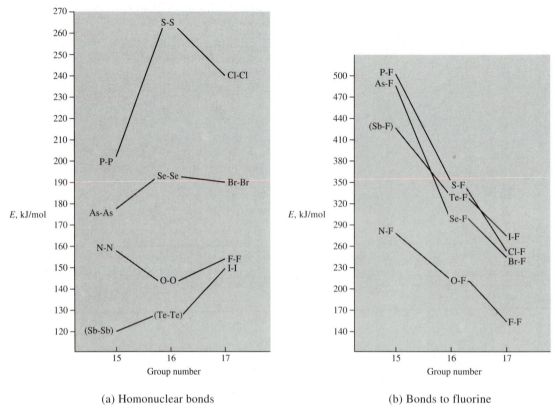

(a) Homonuclear bonds (b) Bonds to fluorine

Figure 14.4 Single-bond energies among the Group 15 17 elements. Data taken from Table 4.6; values in parentheses are estimates.

Bi; (IV) is common for S, Se, and Te; and (V) is well-established for Cl, Br, and I.

Covalent bond energies are strongly influenced by adjacent atom lone-pair repulsions. As can be seen in Figure 14.4, the N–N, O–O, F–F, N–F, and O–F single bonds all are much weaker than would be expected from the small sizes of the involved atoms. Hence, there are several exceptions to the usual tendency for bond energies to decrease going down a family. For example, in bond energy,

$$N–N < P–P > As–As > Sb–Sb$$

$$O–O < S–S > Se–Se > Te–Te$$

$$N–F < P–F > As–F > Sb–F$$

14.2 Preparation and Uses of the Free Elements

The Group 15–18 elements found in the zero oxidation state in nature are nitrogen, oxygen, sulfur, and the noble gases. All except sulfur are present in the atmosphere, and are isolated by the fractional distillation of air.[2] In that process, air is first condensed at a temperature close to absolute zero and then gradually warmed. The fractions that are collected are summarized in Table 14.1.

Table 14.1 Fractions from the distillation of dry, liquid air

Fraction	bp, K	Volume %
He	4	<0.001
H_2	20	<0.001
Ne	27	0.002
N_2	77	78.0
Ar	87	0.93
O_2	90	21.0
CH_4	109	<0.001
Kr	120	<0.001
Xe	165	<0.001
CO_2	195	0.03
		100.0%

The product obtained in largest volume, N_2, is used mainly for the synthesis of ammonia. Oxygen is an oxidant, of course, and more than half its total consumption is in steel-making and other metallurgical purposes. Its ability to promote burning also leads to smaller-scale uses, such as in oxyacetylene torches and rocket fuels. It is also used for sewage treatment and medical purposes.

Sulfur is found in deposits of the free element, as well as in oxidized (SO_4^{2-}) and reduced (S^{2-}) forms. Since the sulfide ion is a soft base, it is bonded to soft metals (chalcophiles) in most of its ores (see Table 14.2). Elemental sulfur is extracted from the earth by the *Frasch process*, in which superheated water is used to melt the sulfur (mp 113°C). The molten material is forced to the surface by high pressure and then collected.

The dominant sulfur-containing industrial chemical (and the most important of all synthetic chemicals in terms of quantity produced) is sulfuric

---

2. Significant quantities of helium are obtained from natural gas as well.

Table 14.2 Some common sulfur-containing minerals

Formula	Name	Formula	Name
FeS	Pyrrhotite	HgS	Cinnabar
FeS_2	Pyrite	PbS	Galena
CuS	Covellite	As_2S_3	Orpiment
Cu_2S	Chalcocite	Sb_2S_3	Stibnite
ZnS	Zinc blende	$CuSbS_3$	Tetrahedrite
	Wurtzite	Bi_2S_3	Bismuthinite
CdS	Greenockite		

acid, which is prepared from the free element by oxidation to SO_3, followed by reaction with water:

$$\tfrac{1}{8}S_8 + \tfrac{3}{2}O_2 \xrightarrow[V_2O_5]{500°C} SO_3 \xrightarrow{H_2O} H_2SO_4 \tag{14.1}$$

The remaining elements exist in nature in nonzero oxidation states, so their isolation as free elements requires either oxidation or reduction. Fluorine, chlorine, and bromine occur as the halide ions. The oxidation potentials of F^- and Cl^- are so negative that electrolysis is the only efficient method for the large-scale production of F_2 and Cl_2. Elemental bromine can be produced by chemical reduction; Cl_2 is the usual oxidizing agent:

$$2\,Br^- + Cl_2 \longrightarrow Br_2 + 2\,Cl^- \qquad K_{eq} \approx 10^{10} \tag{14.2}$$

The major industrial uses for Cl_2 and Br_2 involve the halogenation of organic compounds. Prominent examples include the synthesis of CH_3Cl, CH_2Cl_2, $CHCl_3$, CCl_4, $H_2C{=}CHCl$, $C_6H_4Cl_2$, $CHBr_3$, and $C_2H_4Br_2$. Certain bromide salts, particularly AgBr, are important in photography.

Iodine occurs as both the iodide and iodate anions. The free element is prepared from I^- by oxidation with Cl_2, and from IO_3^- by reduction (conveniently, with I^-):

$$5I^- + IO_3^- + 6H^+ \longrightarrow 3I_2 + 3H_2O \tag{14.3}$$

Elemental phosphorus is made by a high-temperature process involving three bulk starting materials: phosphate rock, sand, and coal. An idealized equation is

$$2Ca_3(PO_4)_2 + 6SiO_2 + 10C \xrightarrow{1000°C} P_4 + 6CaSiO_3 + 10CO \tag{14.4}$$

The product is condensed as white phosphorus using cold water. (It is also stored under water, since P_4 is rapidly oxidized by atmospheric oxygen.) The other allotropes (see Section 7.11) are obtained from white phosphorus by the appropriate treatment, which may include high temperature, high pressure, and/or photolysis.

The most important phosphorus-containing chemical is phosphoric acid, which is used for making fertilizers and, on a smaller scale, detergents and food additives. Because the oxidation state of phosphorus in H_3PO_4 is the same as in its naturally occurring form, the free element is not required as an intermediate. Phosphoric acid is made directly by acidifying phosphate rock:

$$Ca_3(PO_4)_2 + 3H_2SO_4 \longrightarrow 2H_3PO_4 + 3CaSO_4 \qquad (14.5)$$

The remaining elements are produced by chemical reduction. Example reactions are

$$2As_2S_3 + 9O_2 \xrightarrow{\Delta} 2As_2O_3 + 6SO_2 \qquad (14.6)$$

$$2As_2O_3 + 6C \xrightarrow{\Delta} As_4 + 6CO \qquad (14.7)$$

$$2Bi_2S_3 + 9O_2 \xrightarrow{\Delta} 2Bi_2O_3 + 6SO_2 \qquad (14.8)$$

$$Bi_2O_3 + 3H_2 \xrightarrow{\Delta} 2Bi + 3H_2O \qquad (14.9)$$

14.3 The Hydrides of Groups 15–17

Most of the Group 15–17 elements have only one stable binary hydride. Exceptions are those elements that form strong homonuclear bonds: S, N, P, and O. Sulfur forms a series of compounds having the general formula HS_nH ($n = 1$–6 and perhaps others). Other hydrides that show reasonable stability include H_2N-NH_2, H_2P-PH_2, and $HO-OH$. Nitrogen also forms two hydrides containing N–N π bonds, diazine ($HN{=}NH$) and hydrogen azide (HN_3). Each has low thermal stability, readily decomposing to the free elements. This is due to the large energies of $N{\equiv}N$ and H–H bonds.

The thermal stabilities of the remaining hydrides are directly related to the strengths of their bonds to hydrogen (the decompositions are to the free elements), and therefore decrease going down each family. Bismuthine, BiH_3, is difficult to isolate. Hydrogen telluride is unstable above about 0°C, and is also light-sensitive. The common names and phase transition temperatures of the best-characterized hydrides of Groups 15–17 are given in Table 14.3.

These compounds are synthesized in various ways. Those having sufficient thermal stability (NH_3, H_2S, and the hydrogen halides) are made by

Table 14.3 Formulas, common names, and phase change temperatures for the binary hydrides of Groups 15–17

Formula	Common Name	mp, °C	bp, °C
NH_3	Ammonia	-78	-33
N_2H_2	Diazine	—	d
N_2H_4	Hydrazine	2	114
HN_3	Hydrogen azide	-80	36
PH_3	Phosphine	-134	-88
P_2H_4	Diphosphine	-99	d
AsH_3	Arsine	-116	-62
SbH_3	Stibine	-88	-18
BiH_3	Bismuthine	—	d
H_2O	Water	0	100
H_2O_2	Hydrogen peroxide	0	d
H_2S	Hydrogen sulfide	-86	-60
H_2S_2	Sulfane	-40	70
H_2Se	Hydrogen selenide	-66	-41
H_2Te	Hydrogen telluride	-51	-4
HF	Hydrogen fluoride	-83	20
HCl	Hydrogen chloride	-115	-84
HBr	Hydrogen bromide	-89	-67
HI	Hydrogen iodide	-51	-35

Note: d = decomposes.

direct reaction at high temperature:

$$H_2 + Br_2 \xrightarrow[\text{Pt}]{300°C} 2\,HBr \tag{14.10}$$

$$8\,H_2 + S_8 \xrightarrow{600°C} 8\,H_2S \tag{14.11}$$

Less stable hydrides are prepared by reducing the corresponding chlorides:

$$4\,SbCl_3 + 3\,LiAlH_4 \longrightarrow 4\,SbH_3 + 3\,AlCl_3 + 3\,LiCl \tag{14.12}$$

Hydrogen fluoride and HCl are made on an industrial scale by the action of sulfuric acid on naturally occurring minerals:[3]

$$CaF_2 + H_2SO_4 \xrightarrow{\Delta} 2\,HF + CaSO_4 \tag{14.13}$$

$$2\,NaCl + H_2SO_4 \xrightarrow{\Delta} 2\,HCl + Na_2SO_4 \tag{14.14}$$

---

3. Hydrogen chloride is a by-product of many industrial organic reactions (especially the chlorination of hydrocarbons), and a great deal of commercial HCl is produced in that manner.

This method is not viable for HBr or HI, because they are oxidized by H_2SO_4.

Acid–Base and Oxidation–Reduction Tendencies

Recall that acid–base and oxidation–reduction tendencies are often inter-related (Chapter 10). Acidic hydrides are relatively easy to reduce, and many basic hydrides are prone to oxidation. This can be shown in a number of ways. For example, Figure 14.5 is a plot of the proton affinities of the conjugate bases of some nonmetal hydrides versus their reduction potentials. It can be seen that the lines connecting family members move in the same general direction. This indicates that as the basicity of a hydride's conjugate increases, its tendency to be oxidized to the free element also increases. The same is true for the oxyanions of these elements (see Section 14.4).

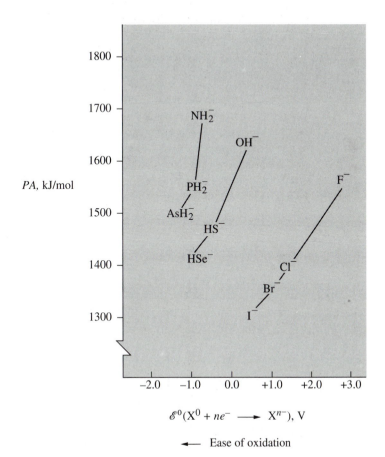

Figure 14.5
The relationship between proton affinity and tendency toward oxidation for the conjugate bases of the Group 15–17 hydrides. Data taken from Tables 9.1 (basic solution) and 10.1.

$\mathscr{E}^0(X^0 + ne^- \longrightarrow X^{n-})$, V

$\longleftarrow$ Ease of oxidation

Brønsted basicity generally decreases going down each family. Hence, there is considerable known chemistry for NH_4^+, less for PH_4^+, and little or none for the other Group 15 analogues. Similarly, H_2O is the most easily protonated of the Group 16 hydrides.

Most of these compounds act as Lewis bases as well. One result of this is the high toxicity of several nonmetal hydrides, most notoriously PH_3, AsH_3, and H_2S. The toxicity arises from the formation of strong Lewis adducts with biochemical iron, as in hemoglobin and the cytochromes. This will be discussed more completely in Chapter 22.

14.4   Oxides and Other Oxygen-Containing Species

The single bonds formed by oxygen to the Group 14–17 elements generally follow expectations; that is, there is a decrease in bond energy and an increase in polarity going down each family. (However, N–O, O–O, and O–F single bonds are exceptionally weak, as noted above.) Several of these elements form two common oxides (eg, Bi_2O_3 and Bi_2O_5; SeO_2 and SeO_3), a reflection of the inert pair effect.

Oxides of Nitrogen[4]

Nitrogen forms an unusual number of binary oxides—between 6 and 12 (several have been only incompletely characterized). This is a result of strong N–O π bonding. As shown in Table 14.4, all these compounds have at least one multiple bond.

At least five different oxidation states, N^I to N^V, are evidenced. Some interconversions among the oxides and/or their anions are shown in Figure 14.6. The three of greatest commercial importance are N_2O, NO, and NO_2.

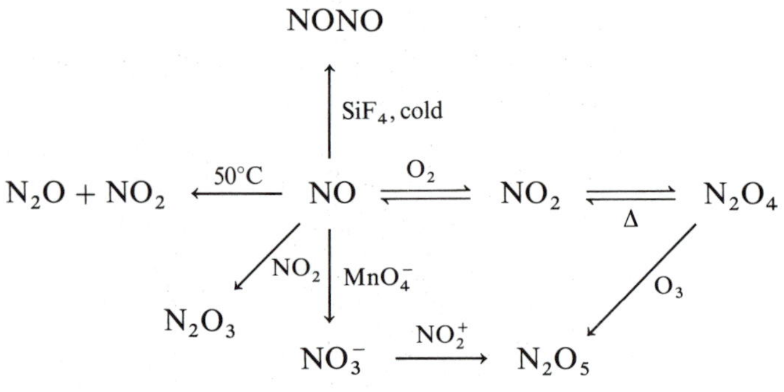

Figure 14.6
Some interconversions among the oxides of nitrogen.

4.   Laane, J.; Ohlsen, J. R. *Prog. Inorg. Chem.* **1980**, *27*, 465.

Table 14.4 The structures and properties of the common oxides of nitrogen

Formula	Structure	mp, °C	bp, °C	Disproportionation Products
N_2O	$\overset{\ominus}{:}\ddot{N}=\overset{\oplus}{N}=\ddot{O}:$	-91	-88	$N_2 + NO$
NO	$:\dot{N}=\ddot{O}:$	-164	-152	$N_2O + NO_2$
NO_2	$:\ddot{O}=\overset{\oplus}{N}-\ddot{O}:^{\ominus}$	a	a	$NO^+ + NO_3^-$
N_2O_3		-102	4 (d)	$NO + NO_2$
N_2O_4		-11	21	$NO^+ + NO_3^-$
N_2O_5		32	47	—

Note: All these species are resonance hybrids; only one resonance structure is shown for each.
d = decomposes.
^{a} The NO_2/N_2O_4 equilibrium favors N_2O_4 at the melting and boiling points.

Nitrous oxide is produced by the thermal decomposition of molten ammonium nitrate:[5]

$$NH_4NO_3(l) \xrightarrow{250°C} N_2O + 2H_2O \tag{14.15}$$

Nitrous oxide is a colorless gas that liquefies at $-88°C$. It is quite soluble in water, but thermodynamically unstable toward disproportionation in aqueous solution:

$$2N_2O \longrightarrow N_2 + 2NO \qquad \mathscr{E}^0 = +0.18 \text{ V} \tag{14.16}$$

Because of its low toxicity, N_2O has long been used as a general anesthetic ("laughing gas"). More important industrial uses are as a propellant in aerosol cans and as an oxidant for welding. The $N_2O + C_2H_2$ flame is nearly as hot (about 2950°C) as that of $O_2 + C_2H_2$.

Nitric oxide can be prepared by direct union, but only in poor yields and at very high temperatures. Unlike most reactions of the $X_2 + Z_2 \rightarrow 2XZ$

---

5. Ammonium nitrate is explosive when heated; hence, this reaction requires special apparatus and extreme caution!

type, that between N_2 and O_2 is endothermic:

$$N_2 + O_2 \longrightarrow 2NO \qquad \Delta H = +180 \text{ kJ/mol} \tag{14.17}$$

Nitric oxide is unstable not only with respect to decomposition to the free elements, but also toward disproportionation. The latter reaction dominates at mild temperatures (30–70°C), especially under high pressure:

$$3NO(g) \longrightarrow N_2O(g) + NO_2(g) \tag{14.18}$$

As an odd-electron molecule, NO is necessarily paramagnetic. The bond length (115 pm) and energy (628 kJ/mol) are consistent with the bond order of 2.5 predicted from MO theory. Because its HOMO is an antibonding π^* orbital, nitric oxide is easily ionized to the nitrosonium (sometimes called nitrosyl) cation, NO^+. In its chemical behavior, however, NO shows a greater tendency toward reduction to either N_2 or N_2O than toward oxidation. Thus, it can serve as a moderate oxidizing agent; for example,

$$2NO + Sn^{II} + 2H^+ \longrightarrow N_2O + Sn^{IV} + H_2O \tag{14.19}$$

$$2NO + CH_3CH_2OH \longrightarrow N_2O + CH_3CHO + H_2O \tag{14.20}$$

Nitric oxide has some tendency to dimerize, but less than might be expected given its unpaired electron. In the solid state, the molecules are paired to give apparent N–N interactions ($:\ddot{O}{=}\ddot{N}\cdots\ddot{N}{=}\ddot{O}:$). However, the N–N distance of 218 pm far exceeds that expected for a single bond. (Compare to 145 pm in N_2H_4.) An asymmetric dimer $\left(:\ddot{O}{=}\ddot{N}{-}\overset{\oplus}{\ddot{O}}{=}\overset{\ominus}{\ddot{N}}:\right)$ can be prepared at low temperatures in the presence of a Lewis acid, but this red-colored species decomposes upon warming.

Nitrogen dioxide, NO_2, is best prepared by the thermal decomposition of metal nitrates such as $Mg(NO_3)_2$, although the product must be rapidly cooled to avoid decomposition:[6]

$$2Mg(NO_3)_2 \xrightarrow{330°C} 2MgO + 4NO_2 + O_2 \tag{14.21}$$

As would be expected for N^{IV}, NO_2 is a strong oxidizing agent:

$$NO_2 + 2HBr \longrightarrow NO + H_2O + Br_2 \tag{14.22}$$

$$NO_2 + SO_2 \longrightarrow NO + SO_3 \tag{14.23}$$

---

6. Most other metal nitrates decompose to the nitrites upon heating (see Chapter 12).

An interesting aspect of the chemistry of NO_2 is its dimerization to N_2O_4. The equilibrium

$$2NO_2 \rightleftharpoons N_2O_4 \tag{14.24}$$

is exothermic ($\Delta H = -57\,\text{kJ/mol}$), but is disfavored by entropy. At low temperatures (below the boiling point of 21°C), more than 99% of all molecules are dimeric, but the monomer dominates above 135°C. This temperature-dependence is visually obvious, since the free radical NO_2 is red-brown in color, while the dimer is colorless. (This makes an excellent classroom demonstration of equilibrium.)

The dimer is planar (D_{2h} point group), and contains a weak N–N bond (175 pm). Although its thermal decomposition leads simply to NO_2, much of the chemistry of N_2O_4 is suggestive of asymmetrical cleavage to NO^+ and NO_3^-. It is virtually 100% dissociated to the nitrosonium and nitrate ions in anhydrous HNO_3. Further evidence is provided by chemical reactions such as the following:

$$KCl + N_2O_4 \longrightarrow NOCl + KNO_3 \tag{14.25}$$

$$SnCl_4 + N_2O_4 \longrightarrow NO^+[Cl_4SnONO_2]^- \tag{14.26}$$

$$Na + N_2O_4 \longrightarrow NaNO_3 + NO \tag{14.27}$$

Note that equation (14.27) amounts to simple electron transfer from Na to NO^+.

Nitric oxide and NO_2 are intermediates in the *Ostwald process*, by which nitric acid is produced on an industrial scale. The sequence of reactions is

$$(1) \quad 4NH_3 + 5O_2 \xrightarrow[\text{Pt}]{900°C} 4NO + 6H_2O \tag{14.28}$$

$$(2) \quad 2NO + O_2 \longrightarrow 2NO_2 \tag{14.29}$$

$$(3) \quad 3NO_2 + H_2O \longrightarrow 2HNO_3 + NO \tag{14.30}$$

These gases are dangerous pollutants, and their presence in the atmosphere (primarily arising from the combustion of fossil fuels having nitrogen-containing impurities) is a continuing concern. They also have been implicated in the destruction of the ozone layer.[7]

The anhydrides of nitrous and nitric acid are N_2O_3 and N_2O_5, respectively. Both can be prepared from HNO_3, the trioxide by reduction and the

---

7. See Donovan, R. J. *Educ. Chem.* **1978**, *15*, 110.

pentoxide by dehydration:

$$2HNO_3 + H_2O + 2SO_2 \longrightarrow N_2O_3 + 2H_2SO_4 \qquad (14.31)$$

$$12HNO_3 + P_4O_{10} \xrightarrow{\Delta} 6N_2O_5 + 4H_3PO_4 \qquad (14.32)$$

Nitrogen(III) oxide is a blue solid below its melting point of $-102°C$. It has low thermal stability, decomposing slowly in the liquid phase and rapidly (but incompletely) above about $10°C$ to produce equilibrium mixtures of NO, NO_2, and N_2O_3. Because of this instability, relatively little is known about its reaction chemistry. It undergoes self-ionization,

$$N_2O_3 \rightleftharpoons NO^+ + NO_2^- \qquad (14.33)$$

and is a good source of NO^+. Nitrosonium salts result when N_2O_5 is dissolved in strong acids:

$$N_2O_3 + 3H_2SO_4 \longrightarrow 2NO^+HSO_4^- + H_3O^+ + HSO_4^- \quad (14.34)$$

$$N_2O_3 + 3HClO_4 \longrightarrow 2NO^+ClO_4^- + H_3O^+ + ClO_4^- \quad (14.35)$$

Nitrogen(V) oxide is ionic in the solid state, with a structure corresponding to "nitronium nitrate," $NO_2^+NO_3^-$ (Figure 14.7), but is molecular in the gas phase (O_2N–O–NO_2), with C_{2v} symmetry. It has low thermal stability;

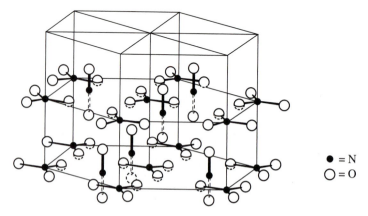

Figure 14.7   The solid-state structure of N_2O_5. Note the NO_2^+ and NO_3^- ions. [Reproduced with permission from Addison, C. C.; Logan, N. In *Developments in Inorganic Nitrogen Chemistry*; Colburn, C. B., Ed.; Elsevier: Amsterdam, 1973; Volume 2.]

the half-life for its first-order decomposition is only about 10 h at room temperature.

$$2N_2O_5 \longrightarrow 4NO_2 + O_2 \qquad (14.36)$$

Nonredox reactions of N_2O_5 produce either nitryl (NO_2X) compounds or the nitrate ion:

$$SiCl_4 + 2N_2O_5 \longrightarrow SiO_2 + 4NO_2Cl \qquad (14.37)$$

$$Cr(CO)_6 + 3N_2O_5 \longrightarrow Cr(NO_3)_3 + 6CO + 3NO_2 \qquad (14.38)$$

Pure N_2O_5 is remarkable for its ability to dissolve gold. The product is the nitronium salt $NO_2^+ \, [Au(NO_3)_4]^-$.

The Oxides of Phosphorus, Arsenic, and Bismuth

Phosphorus forms two well-known oxides, one for each of its common oxidation states. The empirical formulas are P_2O_3 and P_2O_5; however, each undergoes dimerization to give P_4O_6 and P_4O_{10}, respectively. Both have structures directly related to that of white phosphorus, with the P_4 tetrahedron maintained in each case (see Figure 14.8). The P_4O_6 molecule is figuratively derived from P_4 by separating the atoms of the latter by about 32% and inserting an oxygen between each adjacent phosphorus pair. (More precisely, the oxygens are centered above the six edges of the tetrahedron.) Thus, six P–P bonds in P_4 are replaced by twelve P–O bonds in P_4O_6. The

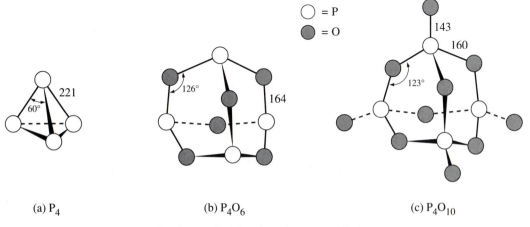

(a) P_4 (b) P_4O_6 (c) P_4O_{10}

Figure 14.8 The molecular structures of white phosphorus and its two best-known oxide derivatives. Bond distances in picometers.

bond angles are improved from $60°$ ($\angle$ P–P–P of P_4) to $126°$ ($\angle$ P–O–P of P_4O_6).

The P_4O_6 molecule is susceptible to oxidation, and is readily attacked by O_2. Under controlled conditions the intermediate compounds P_4O_7, P_4O_8, and P_4O_9 can be isolated. Complete oxidation gives P_4O_{10}, which has the same structure as P_4O_6, except that each phosphorus has an extra (terminal) oxygen replacing the nonbonded electron pair. The terminal linkages have significant double bond character. (Compare the terminal bond lengths to those for the bridging oxygens.)

The anhydrides of H_3PO_3 and H_3PO_4 are P_4O_6 and P_4O_{10}, respectively, and each has a great affinity for water. As evidenced by equations (14.32), (14.49), and (14.56), P_4O_{10} is a common agent for laboratory dehydration reactions. Much of the reaction chemistry of P_4O_6 also involves dehydration; for example,

$$6HCOOH + P_4O_6 \longrightarrow 4H_3PO_3 + 6CO \qquad (14.39)$$

Arsenic and antimony form M_4O_6 and M_4O_{10} compounds that are structurally similar to those of phosphorus. However, Bi_2O_3 is different; its structure is more characteristic of an ionic lattice. Following the usual trend, the anhydrides become increasingly basic going down the family; Sb_2O_3 is amphoteric, while Bi_2O_3 is clearly basic. The sequential equilibria described by the following equations apply in aqueous solution:

$$Bi_2O_3 + 3H_2O \rightleftharpoons 2Bi(OH)_3 \qquad (14.40)$$

$$Bi(OH)_3 \rightleftharpoons BiOOH + H_2O \qquad (14.41)$$

$$BiOOH \rightleftharpoons BiO^+ + OH^- \qquad (14.42)$$

Recall that in Groups 13 and 14 the maximum oxidation state becomes increasingly less stable going down the family, partly because covalent bond energies decrease with increasing size. The same is true in Groups 15 and 16. Antimony(V) oxide is more difficult to prepare than its phosphorus and arsenic analogues, and Bi_2O_5 has not yet been isolated.

The Oxides of Sulfur, Selenium, and Tellurium

The stable oxides of sulfur are SO_2 and SO_3. There is also evidence for sulfur monoxide (SO), disulfur monoxide (SSO), and higher oxides such as S_7O and S_7O_2, but these species have been less characterized and/or have low thermal stabilities.

Sulfur dioxide is the primary product of the burning of sulfur. The S–O bond energy (530 kJ/mol) and length (143 pm) are compatible with bonds intermediate between single and double, as predicted by both the MO and

Lewis theories:

$$:\overset{\oplus}{\ddot{O}}=\overset{}{\ddot{S}}-\overset{\ominus}{\ddot{O}}: \quad \longleftrightarrow \quad \overset{\ominus}{:\ddot{O}}-\overset{\oplus}{\ddot{S}}=\ddot{O}:$$

Sulfur dioxide has extensive acid–base and redox chemistry, and is a useful nonaqueous solvent for both types of reactions. It can act as a Lewis base through either the sulfur or oxygen lone pair, or as a Lewis acid:

$$SO_2 + BF_3 \longrightarrow OSO \cdot BF_3 \tag{14.43}$$

$$SO_2 + SnMe_4 \longrightarrow O_2S \cdot SnMe_4 \tag{14.44}$$

$$SO_2 + NMe_3 \longrightarrow Me_3N \cdot SO_2 \tag{14.45}$$

Sulfur dioxide is a serious air pollutant.[8] As with the nitrogen oxides, the problem arises because sulfur-containing compounds are present as impurities in coal and other fossil fuels.

This species is very soluble in water; saturated solutions contain 53% SO_2 by mass at 0°C. The acidity of such solutions usually has been attributed to "sulfurous acid."

$$SO_2 + H_2O \overset{?}{\rightleftharpoons} H_2SO_3 \rightleftharpoons H^+ + HSO_3^- \tag{14.46}$$

However, the dominant S^{IV} species in acidic aqueous solution is actually the molecular hydrate $SO_2 \cdot nH_2O$; hence, the acidity is really due to hydrolysis:

$$SO_2 \cdot nH_2O + H_2O \rightleftharpoons H_3O^+ + [SO_2(OH)(H_2O)_{n-1}]^- \tag{14.47}$$

Sulfur trioxide is produced by the vigorous oxidation of elemental sulfur or SO_2 in the presence of a catalyst [see equation (14.1)]. Two solid forms are known. The first, β-SO_3, is a helical chain polymer, while γ-SO_3 consists of six-membered S_3O_3 rings (see Figure 14.9b, p. 477). The S–O bond distances in the trimer are 167 (ring) and 140 (terminal) pm, which suggests multiple bonding to the terminal oxygens. Monomeric SO_3 (D_{3h} point group) exists in the gas phase.

Sulfur trioxide dissolves in water to give H_2SO_4. Further addition of SO_3 produces pyrosulfuric acid, $H_2S_2O_7$, which is an even stronger acid than sulfuric:

$$SO_3 + H_2SO_4 \longrightarrow H_2S_2O_7 \tag{14.48}$$

8. Husar, R. B.; Lodge, J. P.; Moore, D. J. *Sulfur in the Atmosphere*; Pergamon: London, 1978; Meyer, B. *Sulfur, Energy, and the Environment*; Elsevier: Amsterdam, 1977.

Solutions of S^{VI} in H_2O are both strongly acidic and strongly oxidizing. The affinity of SO_3 for water makes it a good dehydrating agent.

The aqueous chemistry of sulfur is complex. There are seven different oxidation states, and the situation is further complicated by catenation. For example, both H_2SO_4 and $H_2S_2O_7$ contain S^{VI}. The names and structures of the best-known ternary acids of sulfur are given in Table 14.5.

Selenium dioxide and trioxide are oligomers—infinite chains for SeO_2 and eight-membered rings for SeO_3 (Figure 14.9c–d). The π bonding is weaker than in the sulfur–oxygen analogues, so there is a smaller difference between the bridging and terminal bond distances. The bond lengths in $(SeO_2)_x$ are 173 and 178 pm, a difference of only 3%.

Table 14.5 Formulas, names, and structures for the common oxyacids of sulfur

Formula	Name	Structure
H_2SO_4	Sulfuric	$\begin{array}{c} O \\ \parallel \\ HO-S-OH \\ \parallel \\ O \end{array}$
H_2SO_5	Peroxosulfuric	$\begin{array}{c} O \\ \parallel \\ HO-S-O-OH \\ \parallel \\ O \end{array}$
$H_2S_2O_3$	Thiosulfuric	$\begin{array}{c} S \\ \parallel \\ HO-S-OH \\ \parallel \\ O \end{array}$
$H_2S_2O_7$	Disulfuric (pyrosulfuric)	$\begin{array}{c} O \quad\quad O \\ \parallel \quad\quad \parallel \\ HO-S-O-S-OH \\ \parallel \quad\quad \parallel \\ O \quad\quad O \end{array}$
$H_2S_2O_8$	Peroxodisulfuric	$\begin{array}{c} O \quad\quad\quad\quad O \\ \parallel \quad\quad\quad\quad \parallel \\ HO-S-O-O-S-OH \\ \parallel \quad\quad\quad\quad \parallel \\ O \quad\quad\quad\quad O \end{array}$
H_2SO_3	Sulfurous[a]	$\begin{array}{c} O \\ \parallel \\ HO-S-OH \end{array}$
$H_2S_2O_5$	Disulfurous	$\begin{array}{c} O \quad\quad O \\ \parallel \quad\quad \parallel \\ HO-S-O-S-OH \end{array}$

[a] The structure of sulfurous acid is better written as $SO_2 \cdot nH_2O$ (see text).

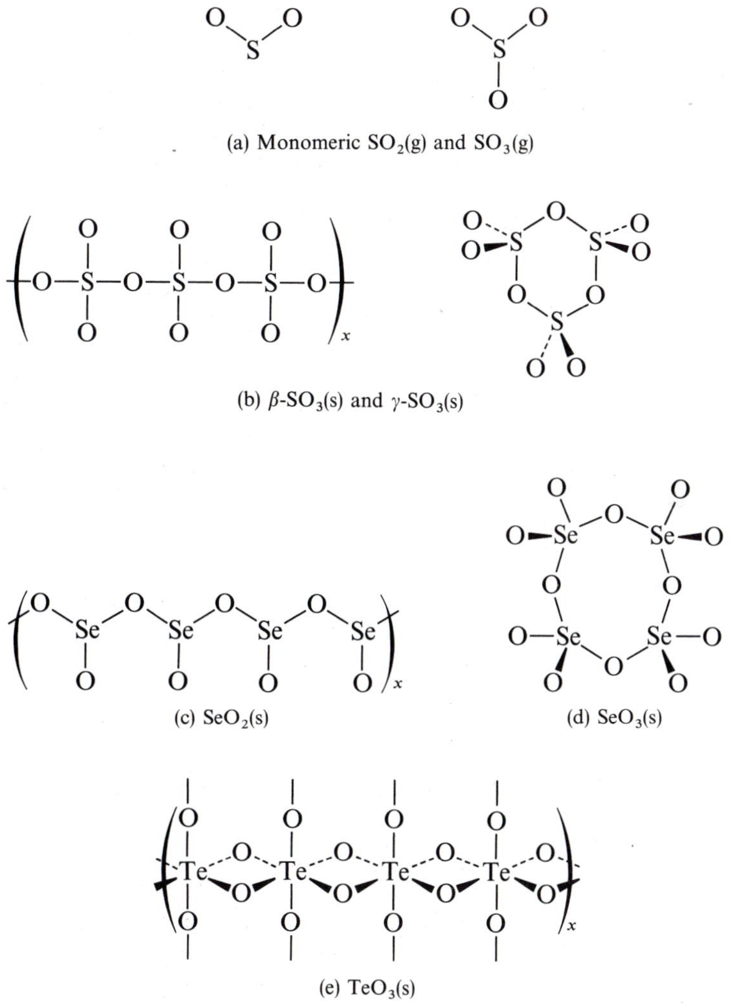

(a) Monomeric $SO_2(g)$ and $SO_3(g)$

(b) β-$SO_3(s)$ and γ-$SO_3(s)$

(c) $SeO_2(s)$ (d) $SeO_3(s)$

Figure 14.9
The structures of
some Group 16
oxides.

(e) $TeO_3(s)$

Selenium trioxide is more difficult to prepare than SO_3, and must be made by dehydration rather than oxidation.

$$H_2SeO_4 \xrightarrow{P_4O_{10}} SeO_3 \tag{14.49}$$

Tellurium dioxide occurs in nature as the mineral tellurite. It forms a lattice in which tellurium has an approximately trigonal bipyramidal electron geometry (four sites occupied by oxygens and the fifth by a lone pair). The trioxide also utilizes a three-dimensional lattice, with tellurium in an octahedral environment (Figure 14.9e).

Halogen Oxides[9]

Binary oxygen–halogen compounds have two dominant chemical properties: they are potent oxidizing agents, and they have low thermal stabilities. The latter results from the weak X–O bond energies compared to the homonuclear diatomics, O_2 and X_2.

There are two well-characterized oxygen fluorides, oxygen difluoride, OF_2, and the peroxide, O_2F_2. Neither is sufficiently stable to be prepared by heating the free elements, but O_2F_2 can be made by electrical discharge of mixtures of O_2 and F_2 at low temperature and pressure. Oxygen difluoride is usually obtained via the reaction

$$2F_2 + 2NaOH \longrightarrow OF_2 + 2NaF + H_2O \tag{14.50}$$

Both OF_2 and O_2F_2 decompose vigorously to the free elements; the decomposition is rapid above about $-100°C$ for O_2F_2. Their tremendous oxidizing abilities can be appreciated from reactions such as the following:

$$S_8 + 8OF_2 \longrightarrow 4SO_2 + 4SF_4 \tag{14.51}$$

$$H_2S + 4O_2F_2 \longrightarrow SF_6 + 2HF + 4O_2 \tag{14.52}$$

$$HCl + 2O_2F_2 \longrightarrow ClF_3 + HF + 2O_2 \tag{14.53}$$

The best-known binary oxides of chlorine are Cl_2O and ClO_2. Dichlorine oxide can be produced from sodium carbonate, chlorine gas, and water, while chlorine dioxide is made by partial reduction of ClO_3^-:

$$2CO_3^{2-} + 2Cl_2 + H_2O \longrightarrow Cl_2O + 2HCO_3^- + 2Cl^- \tag{14.54}$$

$$2ClO_3^- + 2Cl^- + 4H^+ \longrightarrow 2ClO_2 + Cl_2 + 2H_2O \tag{14.55}$$

Both compounds are bleaching agents. Solutions of Cl_2O in aqueous base (in which chlorine occurs as OCl^-) are used as household bleaches. Chlorine dioxide is a bactericide, and is sometimes used in the purification of drinking water. It is also a whitening agent for wood pulp in paper manufacture.

Like nitrogen dioxide, ClO_2 is a free radical compound. Unlike NO_2, however, it has little tendency to dimerize (Cl_2O_4 is known, but is thermally unstable and has the asymmetrical structure $ClOClO_3$). The difference in behavior can be rationalized by a resonance argument. The two most

9. Renard, J. J.; Bolker, H. I. *Chem. Rev.* **1976**, *76*, 487; Gordon, G. *Prog. Inorg. Chem.* **1972**, *15*, 201; Schmeisser, M.; Brandle, K. *Adv. Inorg. Chem. Radiochem.* **1963**, *5*, 42.

important resonance structures for NO_2 are

$$:\ddot{O}=\overset{\oplus}{\overset{\cdot}{N}}-\overset{\cdots}{\underset{\cdot\cdot}{O}}:^{\ominus} \longleftrightarrow {}^{\ominus}:\overset{\cdots}{\underset{\cdot\cdot}{O}}-\overset{\oplus}{\overset{\cdot}{N}}=\ddot{O}:$$

in which nitrogen lacks an octet; this electron deficiency is alleviated by dimerization. Because chlorine can violate the octet rule, three different types of resonance structures are possible for ClO_2:

$$^{\ominus}:\overset{\cdots}{\underset{\cdot\cdot}{O}}-\overset{\overset{\textcircled{+2}}{\cdot\cdot}}{\underset{}{Cl}}-\overset{\cdots}{\underset{\cdot\cdot}{O}}:^{\ominus} \longleftrightarrow :\underset{\cdot\cdot}{O}=\overset{\oplus}{\overset{\cdot}{Cl}}-\overset{\cdots}{\underset{\cdot\cdot}{O}}:^{\ominus} \longleftrightarrow :\underset{\cdot\cdot}{O}=\overset{\cdot}{Cl}=\underset{\cdot\cdot}{O}:$$

The second and third structures suggest that chlorine is stabilized (with regard to both electron deficiency and formal charge) by dative $(p \rightarrow d)\pi$ bonding. Hence, it shows no tendency to dimerize. This explanation is supported by experimental data, which show that the Cl–O bond length in ClO_2 is only 147 pm, compared to 170 pm in Cl_2O.

Less stable chlorine oxides are known, including Cl_2O_7, ClO_3, and Cl_2O_6. Chlorine(VII) oxide has the pyro-type structure O_3Cl–O–ClO_3, and is the anhydride of perchloric acid. Like many anhydrides, it is obtained by dehydration of the acid with P_4O_{10}:

$$12\,HClO_4 + P_4O_{10} \xrightarrow{-10°C} 6\,Cl_2O_7 + 4\,H_3PO_4 \qquad (14.56)$$

Considerable care must be taken when working with Cl_2O_7. Like some ClO_4^- salts, it is shock- and heat-sensitive, and undergoes very vigorous redox reactions.

The bromine oxides parallel their chlorine analogues, but the compounds are even less stable and have little commercial importance.

For iodine, IO_2 and I_2O are scarcely known. The most stable iodine–oxygen compound is I_2O_5, the anhydride of HIO_3. It can be obtained either by the thermal dehydration of that acid,

$$2\,HIO_3 \xrightarrow{>200°C} I_2O_5 + H_2O \qquad (14.57)$$

or by direct reaction of the elements in the presence of electrical discharge. It is the most stable of all binary halogen oxides, decomposing at about 300°C.

The structure of I_2O_5 is symmetric (O_2I–O–IO_2). The bond angle about the central oxygen atom is quite large (139°). There are intermolecular I–O interactions in the solid state, giving iodine an approximately octahedral electron geometry—three covalent and two van der Waals interactions to oxygens, plus one lone pair.

14.5 Some Chemistry Related to Halogen–Halogen Bonding

The Stabilities of Halides and Hydrides

For virtually any comparable series of halogen-containing compounds, it is the fluoride that has the greatest thermal stability. There are several reasons for this. For ionic compounds, the relatively small size of F^- yields large lattice energies. For covalent species, fluorine's small size promotes effective orbital overlap, as well as efficient π back-bonding if the conditions are appropriate. Another reason is somewhat less obvious—the relative weakness of the homonuclear bond in F_2. The bond energies for the homonuclear diatomics F_2, Cl_2, H_2, O_2, and N_2 are listed below:

	D, kJ/mol
F_2	154
Cl_2	240
H_2	432
O_2	494
N_2	942

Why is this important? Consider the decomposition of some hypothetical compound MX_n (X = F, Cl, H, O, or N) to its free elements. For most M, the strongest heteronuclear bond is to fluorine (see Table 4.6); these bonds must be cleaved during the decomposition reaction. Among the X_2 products listed above, the fluorine–fluorine bond is the weakest of the five. Thus, in terms of the relative thermal stability of MX_n, both bond breaking and bond formation favor the fluorides.[10]

Bond energies for trivalent phosphorus compounds can be used as a quantitative example. Consider the reaction

$$4PX_3 \longrightarrow P_4 + 6X_2 \tag{14.58}$$

This process requires the rupture of 12 P–X bonds for every 6 P–P and 6 X–X bonds formed. Using bond energies from Table 4.6, the estimated enthalpy change for the three cases X = H, F, and Cl are calculated as follows:

For H: $\Delta H = 12(321) - 6(200) - 6(432) = +60$ kJ

For F: $\Delta H = 12(503) - 6(200) - 6(154) = +3912$ kJ

For Cl: $\Delta H = 12(322) - 6(200) - 6(240) = +1224$ kJ

---

10. Of course, all thermal decompositions do not produce the free elements. For example, IF is thermally unstable but does not decompose to I_2 and F_2 [see equation (14.61)].

The fluoride is estimated to be more stable than the chloride by about $(3912 - 1224)/4$ or 672 kJ/mol of PX_3, and more stable than the hydride by 963 kJ/mol!

Many such examples can be cited. Among the trihalides of nitrogen, NCl_3 is shock-sensitive and extremely difficult to work with; NBr_3 is explosive at temperatures as low as $-100°C$; and NI_3 has been isolated only in adducts. However, NF_3 is a relatively unreactive compound of high thermal stability. Other nitrogen fluorides such as N_2F_2 and N_2F_4 are more stable than their heavier halogen analogues.

Similarly, hexacoordinate sulfur is rather rare, and only those cases in which most or all of its substituents are fluorines have reasonable stability (eg, SF_6, SF_5Cl, and SF_5CF_3).

At least for some purposes, hydrogen fits about as well into Group 17 as into Group 1. It is therefore interesting to compare the stabilities of binary hydrides to the corresponding halides. The bond in H_2 is over 180 kJ/mol stronger than for any of the diatomic halogens, and this tends to destabilize hydrides. However, heteronuclear bonds to hydrogen are also very strong. Thus, in thermal stability (as measured by ΔH_f°) for MX_n compounds, there is a nearly invariant order $F > Cl > Br > I$; hydrogen falls anywhere between the most and least stable. Consider that in stability:

$$KF > KCl > KBr > KI > KH \quad \text{(The same order is observed for } CaX_2, SiX_4, PX_3, PX_5, \text{ etc.)}$$

$$HF > HCl > HBr > H_2 > HI$$

$$SF_2 > SCl_2 > H_2S > SBr_2$$

$$NF_3 > NH_3 > NCl_3 > NBr_3 > NI_3$$

$$H_2O > OF_2 > Cl_2O > Br_2O$$

The hydrides are the least stable when bonded to heteroatoms of low electronegativity, but the most stable with heteroatoms of very high electronegativity. Intermediate cases give intermediate stabilities.

The Diatomic Cations, X_2^{+} [11]

Recall that the diatomic halogens have bond orders of 1.0, with the HOMO being an antibonding π^* orbital. Hence, the bond order is increased by removal of an electron. Not surprisingly, salts of X_2^+ cations can be produced in strongly acidic, strongly oxidizing media. For example, Br_2^+ is obtained by dissolving elemental bromine in superacid ($HSO_3F + SbF_5 + SO_3$)

11.   Shamir, J. *Struct. Bonding* **1979**, *37*, 141; Gillespie, R. J.; Passmore, J. *Adv. Inorg. Chem. Radiochem.* **1975**, *17*, 49.

solution. Similarly, I_2 reacts with the potent acid/oxidizing agent $S_2O_6F_2$, peroxydisulfuryl difluoride:

$$2I_2 + S_2O_6F_2 \longrightarrow 2I_2^+ + 2SO_3F^- \tag{14.59}$$

Salts such as $Br_2^+AsF_6^-$, $I_2^+Ta_2F_{11}^-$, and $Br_2^+Sb_3F_{16}^-$ have been isolated. The internuclear distances in these cations are as expected. Removal of an electron increases the bond order to 1.5, causing d_{X-X} to decrease by about 5%. For example, the chlorine–chlorine bond length is 189 pm in Cl_2^+ versus 199 pm in Cl_2.

These ions are colored: Cl_2^+ is yellow, Br_2^+ red, and I_2^+ blue. When dissolved in superacid media and cooled to below $-60°C$, the blue color of I_2^+ gives way to a deep red color due to the formation of the dimer I_4^{2+}. The equilibrium constant for dimerization is about 23 at $-86°C$, which corresponds to a reaction enthalpy of about -40 kJ/mol.

Interhalogen Compounds and Ions

The well-characterized interhalogen compounds are listed in Table 14.6. All but one of the polyatomics have fluorine as the terminal atom, and iodine

Table 14.6 Physical properties of the binary interhalogen compounds

	Color	mp, °C	bp, °C
Diatomics			
ClF	Colorless	-157	-100
BrF	Orange-brown	-33	20
BrCl	Red		5
ICl	Red	27	100
IBr	Black	40	116
Tetraatomics			
ClF_3	Colorless	-76	12
BrF_3	Yellow	9	126
IF_3	Yellow	-28 (d)	d
Hexaatomics			
ClF_5	Colorless	-103	-14
BrF_5	Colorless	-60	41
IF_5	Colorless	10	100
Octaatomics			
IF_7	Colorless	6	s
$(ICl_3)_2$	Yellow-orange	101	d

Note: The phase-change temperatures are approximate in several cases because of disproportionation or other decomposition reactions. d = decomposes; s = sublimes.

is the predominant central atom. The thermal stabilities of fluorides come into play here, and there is also a steric effect. For example, only the combination of the largest central atom with the smallest terminal atoms gives an isolable MX_7 compound.

If astatine is excluded, then there are six possible diatomic combinations: ClF, BrF, IF, BrCl, ICl, and IBr. All except IF can be prepared by direct reaction at or above room temperature:

$$X_2 + X_2' \longrightarrow 2XX' \tag{14.60}$$

The properties of iodine monofluoride are not well-established, because IF undergoes disproportionation to I_2 and IF_5. This is surprising until we consider the bonds broken and formed during this process:

$$5\,IF \longrightarrow 2\,I_2 + IF_5 \tag{14.61}$$
$$\text{5 I–F bonds} \qquad \text{2 I–I + 5 I–F bonds}$$

Hence, there is a net gain of two I–I bonds. Although this is somewhat misleading (the I–F bonds in IF_5 are weaker than those in IF because of steric crowding and reduced orbital overlap), the reaction is exothermic.

Bromine monofluoride also undergoes disproportionation, but to BrF_3:

$$3\,BrF \xrightarrow{\,<25°C\,} Br_2 + BrF_3 \tag{14.62}$$

The tri- and pentafluoride compounds, as well as IF_7, are usually prepared by the fluorination of either the diatomic halogens or metal salts:

$$\tfrac{1}{2}Br_2 \xrightarrow{\tfrac{1}{2}F_2} BrF \xrightarrow[25°C]{F_2} BrF_3 \xrightarrow[>150°C]{F_2} BrF_5 \tag{14.63}$$

$$KCl + 3F_2 \longrightarrow KF + ClF_5 \tag{14.64}$$

$$KI + 4F_2 \xrightarrow{250°C} KF + IF_7 \tag{14.65}$$

The structures of interhalogen molecules are effectively rationalized by the VSEPR model. For example, BrF_3 (with two lone pairs on bromine) is T-shaped, IF_5 (with one lone pair on the central iodine) is square pyramidal, and IF_7 has a pentagonal bipyramidal geometry. Structural parameters for several of the polyatomic interhalogens are given in Figure 14.10.

Several interhalogens undergo dimerization. In the solid state, "ICl_3" has two bridging chlorines (Figure 14.10c), and there is evidence for a monomer–dimer equilibrium in ClF_3. To whatever extent halogen bridging

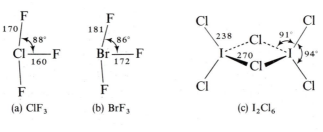

Figure 14.10
The structures of some interhalogen compounds. Bond lengths in picometers.

(a) ClF_3 (b) BrF_3 (c) I_2Cl_6 (d) IF_5 (e) IF_7

occurs, halide exchange reactions should be facile. This notion has been tested using isotopically labeled reagents, and it has been found that ClF_3, BrF_3, BrF_5, IF_5, and IF_7 all undergo rapid exchange with deuterium fluoride at room temperature.

Given this knowledge, it should not be surprising that ion transfer reactions (including self-ionization) are common. For example, ClF_3 can either donate or accept F^-:

$$2ClF_3 \;\rightleftharpoons\; ClF_2^+ + ClF_4^- \tag{14.66}$$

$$ClF_3 + MF_5 \longrightarrow ClF_2^+ + MF_6^- \qquad M = P, As, Sb, Pt, \text{ etc.} \tag{14.67}$$

$$ClF_3 + MF \longrightarrow M^+ + ClF_4^- \qquad M = Rb, K, NO, \text{ etc.} \tag{14.68}$$

The higher fluorides are potent fluorinating and oxidizing agents; some examples are

$$3SiO_2 + 4BrF_3 \longrightarrow 2Br_2 + 3SiF_4 + 3O_2 \tag{14.69}$$

$$H_2O + IF_7 \longrightarrow IOF_5 + 2HF \tag{14.70}$$

$$5Sn + 4ClF_5 \longrightarrow 5SnF_4 + 2Cl_2 \tag{14.71}$$

Such reactions tend to be highly exothermic. In fact, many of the interhalogens explode on contact with water or certain organic compounds, and BrF_3 is known to set fire to asbestos! They must therefore be handled with great caution in grease-free, inert-atmosphere apparatus.

Chlorine monofluoride is commercially available. Its most characteristic reaction is addition across multiple bonds:

$$CO + ClF \longrightarrow ClC(O)F \tag{14.72}$$

$$SO_2 + ClF \longrightarrow OS(Cl)OF \tag{14.73}$$

$$RCN + 2ClF \longrightarrow RC(F)_2NCl_2 \tag{14.74}$$

Many interhalogen ions are known, and they are conveniently categorized as anionotropic conjugate acids or bases. For example, the conjugate acid of ClF_3 is ClF_2^+, and its conjugate base is ClF_4^-. All conjugate bases of the stable interhalogens are known, with the exception of ClF_6^-. The conjugate of the unstable IF, IF_2^-, also has been reported. Among the acids, all nine cations of the XF_2^+, XF_4^+, and XF_6^+ (X = Cl, Br, and I) series have been characterized. Like IF_2^-, ClF_6^+ and BrF_6^+ derive from unstable interhalogens.

These ions are generally prepared by ion transfer reactions. However, in many cases an alternative synthesis is available as well—the fluorination of a metal or nonmetal halide:

$$CsX + 2F_2 \longrightarrow Cs^+XF_4^- \qquad X = Cl, Br, I \tag{14.75}$$

$$2ClF_5 + 2PtF_6 \longrightarrow ClF_6^+PtF_6^- + ClF_4^+PtF_6^- \tag{14.76}$$

Since PtF_6 is a very strong oxidizing agent, equation (14.76) may be understood as a one-electron oxidation to produce ClF_5^+, followed by its disproportionation.

14.6 Compounds and Ions of the Noble Gases[12]

For many years it was believed that the Group 18 elements were incapable of forming chemical compounds. That myth was shattered in 1962 by Bartlett, who reported that the oxidation of xenon by PtF_6 yields a yellow-orange solid formulated as $XePtF_6$:[13]

$$Xe + PtF_6 \longrightarrow Xe^+PtF_6^- \quad (??) \tag{14.77}$$

It now appears that this species is not a simple salt. The actual structure is uncertain, with one possibility being a linear polymer with bridging fluorines, $(PtF_4–F–Xe–F)_x$.

12. Selig, H.; Holloway, J. H. *Top. Curr. Chem.* **1984**, *124*, 33; Seppelt, K.; Lentz, D. *Prog. Inorg. Chem.* **1982**, *29*, 167; Moody, G. J. *J. Chem. Educ.* **1974**, *51*, 628.
13. Bartlett, N. *Proc. Chem. Soc. (London)* **1962**, 218.

Table 14.7 Experimental conditions for the synthesis of xenon fluorides from Xe + F_2 mixtures

Compound	Mole Ratio Xe:F_2	Conditions	Product
XeF_2	1:1	Room temperature, hv	Colorless solid, mp = 140°C
XeF_4	1:5	400°C, 1 h, 6 atm	Colorless solid, mp = 114°C
XeF_6	1:20	300°C, 16 h, 60 atm	Colorless solid, mp = 46°C

Source: Chernick, C. L. In *Noble Gas Compounds*; Hyman, H. H., Ed.; University of Chicago: Chicago, 1963; pp. 35–38.

The Binary Xenon Fluorides

Shortly after Bartlett's work, three covalent xenon fluorides, XeF_2, XeF_4, and XeF_6, were synthesized by direct reaction. The products and yields depend on the reaction conditions, as described in Table 14.7.

Like the interhalogens, the structures of XeF_2 and XeF_4 are as predicted by the VSEPR model; XeF_2 is linear, while XeF_4 is square planar.

Four different solid-state structures of XeF_6 are known. The lowest-temperature version is an ionic-type lattice ($XeF_5^+F^-$) in which the cation has square pyramidal geometry. Gas-phase electron diffraction data indicate that the shape is distorted octahedral (see Figure 5.2). Structural details for several xenon-containing compounds and ions are given in Table 14.8.

Table 14.8 Bond distances and angles in some xenon-containing molecules and ions

Species	d(Xe–F), pm	d(Xe–O), pm	Bond Angle(s), °
XeF_2	198		180
XeF_4	195		90
XeF_5^+	188 (eq)		79
	181 (ax)		
XeF_6	189		?
XeO_3		176	103
XeO_4		174	109
XeO_6^{2-}		186	90
XeO_2F_2	190	171	106, 175
$XeOF_4$	190	170	∼90

Source: Compiled from data given by Wells, A. F. *Structural Inorganic Chemistry*, 5th ed.; Clarendon: Oxford, 1984; Chapter 8.

The xenon fluorides undergo a variety of ion transfer reactions. For example,

$$XeF_2 + IrF_5 \longrightarrow XeF^+IrF_6^- \tag{14.78}$$

$$XeF_2 + 2PtF_5 \longrightarrow XeF^+Pt_2F_{11}^- \tag{14.79}$$

$$2XeF_2 + AsF_5 \longrightarrow Xe_2F_3^+AsF_6^- \tag{14.80}$$

$$XeF_6 + RbF \longrightarrow Rb^+XeF_7^- \tag{14.81}$$

$$2NOF + XeF_6 \longrightarrow (NO^+)_2XeF_8^{2-} \tag{14.82}$$

Additional reactions involving xenon fluorides are summarized in Figure 14.11.

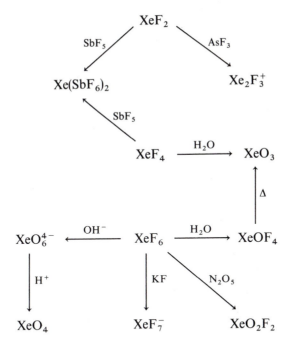

Figure 14.11
Some interconversions among xenon compounds and ions.

Other Xenon-Containing Species

The careful, partial hydrolysis of XeF_6 yields a compound containing both Xe–F and Xe–O bonds,

$$XeF_6 + H_2O \longrightarrow XeOF_4 + 2HF \tag{14.83}$$

The square pyramidal geometry of $XeOF_4$ is again consistent with the VSEPR model. The Xe–O bond length of 170 pm is surprisingly short, and is suggestive of partial double bond character:

$$:\overset{\ominus}{\overset{\cdot\cdot}{O}}: \qquad\qquad :O:$$

Upon mild heating, $XeOF_4$ undergoes a redistribution reaction to produce XeO_3:

$$3XeOF_4 \xrightarrow{\Delta} XeO_3 + 2XeF_6 \tag{14.84}$$

But XeO_3 is more readily obtained by the complete hydrolysis of XeF_6 [compare equation (14.85) to (14.83)]:

$$XeF_6 + 3H_2O \longrightarrow XeO_3 + 6HF \tag{14.85}$$

Xenon trioxide is pyramidal, with bond angles of 103°. It is notoriously dangerous to work with, and quantities of less than 3 mg have been known to detonate without apparent provocation, producing the free elements. The exothermicity of the reaction ($\Delta H = -490$ kJ/mol) is primarily due to the high O=O bond energy. This behavior contrasts with the relative stabilities of the xenon fluorides.

If the hydrolysis of XeF_6 is carried out under basic conditions, then the octahedral perxenate ion, XeO_6^{4-}, is produced. Note that this is a redox process, and that the product contains xenon in its maximum oxidation state (Xe^{VIII}).

$$XeF_6 + 10OH^- \longrightarrow XeO_6^{4-} + 6F^- + 4H_2O + H_2 \tag{14.86}$$

Salts such as K_4XeO_6 have been isolated and studied. The perxenate anion reacts with acids to produce xenon tetroxide:

$$XeO_6^{4-} + 4H^+ \longrightarrow XeO_4 + 2H_2O \tag{14.87}$$

The first reported compound containing a xenon–nitrogen bond was $FXeN(SO_2F_2)_2$, which was prepared from XeF_2 according to the following equation:[14]

$$XeF_2 + HN(SO_2F_2)_2 \longrightarrow FXeN(SO_2F_2)_2 + HF \tag{14.88}$$

14. DesMarteau, D. D.; LeBlond, R. D.; Hossain, S. F.; Nothe, D. *J. Am. Chem. Soc.* **1981**, *103*, 7734; DesMarteau, D. D. *J. Am. Chem. Soc.* **1978**, *100*, 6270.

This was followed by a report of a xenon–carbon compound:[15]

$$XeF_2 + C_2F_6 \longrightarrow Xe(CF_3)_2 + F_2 \tag{14.89}$$

This reaction is thought to involve the attack of $\cdot CF_3$ free radicals at xenon. The product is unstable, decomposing at room temperature within $\frac{1}{2}$ h after its preparation.

Krypton-Containing Compounds and Ions

The chemistry of the other Group 18 elements is extremely limited. Radon should form covalent bonds that are approximately as strong as xenon,[16] but the study of that element is difficult because it has no stable isotopes.

The first krypton compound to be reported was KrF_2, which was prepared by electric discharge of $Kr + F_2$ mixtures at 90 K. It is thermodynamically unstable ($\Delta H_f^\circ = +60$ kJ/mol), and decomposes slowly at room temperature. As would be expected, KrF_2 is an awesome fluorinating agent.

$$Xe + 3\,KrF_2 \longrightarrow XeF_6 + 3\,Kr \tag{14.90}$$

$$I_2 + 7\,KrF_2 \longrightarrow 2\,IF_7 + 7\,Kr \tag{14.91}$$

$$2\,Au + 7\,KrF_2 \longrightarrow 2\,AuF_5 + 7\,Kr + 2\,F_2 \tag{14.92}$$

The KrF^+ and $Kr_2F_3^+$ cations can be obtained from the interaction of KrF_2 with SbF_5:[17]

$$KrF_2 + 2\,SbF_5 \longrightarrow KrF^+Sb_2F_{11}^- \tag{14.93}$$

$$KrF_2 + KrF^+Sb_2F_{11}^- \longrightarrow Kr_2F_3^+Sb_2F_{11}^- \tag{14.94}$$

A different kind of interaction occurs between KrF_2 and graphite. There is some evidence for chemical reaction (fluorination of carbon), but it also appears that an intercalation species is produced when krypton difluoride is stored over graphite at 0°C for long periods of time. Heating to 200°C causes the evolution of $Kr(g)$.

The first species containing a krypton–nitrogen bond was prepared according to the reaction

$$KrF_2 + HC{\equiv}NH^+AsF_6^- \xrightarrow{-60°C} HC{\equiv}N{-}KrF^+AsF_6^- + HF \tag{14.95}$$

15. Turbini, L. J.; Aikman, R. E.; Lagow, R. J. *J. Am. Chem. Soc.* **1979**, *101*, 5833.
16. Seppelt, K.; Lentz, D. *Prog. Inorg. Chem.* **1982**, *29*, 167.
17. Gillespie, R. J.; Schrobilgen, G. J. *Inorg. Chem.* **1976**, *15*, 22.

The product is stable only below about $-50°C$. It was prepared in both HF and BrF_5, but appears to be more stable in the latter.[18]

14.7 Homocatenation in Groups 15–17

As might be expected from bond energies, sulfur has a great tendency to form homonuclear rings and chains. (The S–S σ bond is the strongest of any of the Group 15–17 elements.) The most obvious example, of course, is the free element itself. We have already discussed the crown-shaped S_8 molecule (Chapters 2, 5, and 7), and rings of other sizes are also known (see Figure 14.12).

A variety of homocatenated ions can be synthesized. When S_8 is dissolved in an oxidizing solvent system such as $SbF_5 + SO_2$ or $SO_3 + H_2SO_4$, cations having the general formula S_n^{2+} ($n = 4, 8, 19$, and probably others) are formed. Consistent with Hückel aromaticity (14π electrons), the S_4^{2+} cation is a square, planar ring.

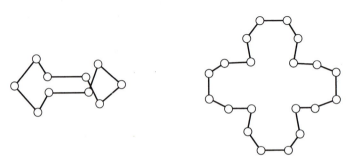

Figure 14.12
The structures of some of the known homoatomic sulfur rings: S_6, S_7, S_8, S_{10}, and S_{20}.

---

18. Schrobilgen, G. J. *J. Chem. Soc. Chem. Commun.* **1988**, 863.

The S_8^{2+} cation is derived from neutral S_8 by the formation of a transannular bond, producing two fused five-membered rings:

$$\left(\begin{array}{c} S-S-S \\ S \quad | \quad S \\ S-S-S \end{array}\right)^{2+}$$

The S_{19}^{2+} cation [found in $S_{19}(AsF_6)_2$] consists of two seven-membered rings connected by a chain of five sulfur atoms.[19]

A number of S_n^{2-} anions are known as well. They can be prepared in reducing media such as metal–ammonia solutions. Examples include K_2S_3, K_2S_4, K_2S_5, and K_2S_6. In all cases, the sulfurs form chains rather than rings. The S–S–S bond angles are typically between 96° and 109°, in the range expected for sp^3 hybridized central atoms having lone pairs:

$$\left[\ddot{\underset{..}{S}} \diagdown \overset{..}{\underset{..}{S}} \diagdown \left(\overset{..}{\underset{..}{S}} \right)_x \overset{..}{\underset{..}{S}} \diagdown \ddot{\underset{..}{S}} \right]^{2-}$$

Selenium and tellurium also exhibit homocatenation.[20] Three different allotropes of elemental selenium contain eight-membered rings, and the cations Se_4^{2+}, Te_4^{2+}, Te_6^{2+}, Se_8^{2+}, and Se_{10}^{2+} are known. There are also examples of mixed rings, including $(Te_2Se_2)^{2+}$ (both cis and trans isomers), $(Te_2Se_4)^{2+}$, $(Te_3Se_3)^{2+}$, and $(Te_2Se_8)^{2+}$.

Homocatenation is less common in Group 15. Nitrogen–nitrogen σ bonds are weakened by adjacent atom lone-pair repulsions, and the great bond energy of N_2 increases the likelihood of an exothermic thermal decomposition pathway. The P–P and As–As σ bonds are sufficiently strong to permit the isolation of stable entities, but these elements tend to form three-dimensional cages rather than rings or chains (recall P_4, P_4O_6, and P_4O_{10}). Several such compounds will be described in Chapter 19.

A few examples of Group 15 homocycles can be cited, however. Five-membered nitrogen rings (pentazoles) are obtained from reactions of diazonium salts with metal azides:

$$R-\overset{\oplus}{N}\equiv N: + N_3^- \longrightarrow R-N\underset{N=N}{\overset{N=N}{\diagup}}\Big| \tag{14.96}$$

In contrast to RN_5, the known P_5 rings have five substituents and are therefore saturated. The compound $P_5(C_6H_5)_5$ contains a puckered ring in

19. Burns, R. C.; Gillespie, R. J.; Sawyer, J. F. *Inorg. Chem.* **1980**, *19*, 423; Low, H. S.; Beaudet, R. A. *J. Am. Chem. Soc.* **1976**, *98*, 3849.

20. Steudel, R.; Strauss, E. M. *Adv. Inorg. Chem. Radiochem.* **1984**, *78*, 135.

which one phosphorus atom lies above the plane of the other four. Syntheses of P_5R_5 compounds in low yields are accomplished by several routes, usually using monomeric dichlorophosphines:

$$5\,RPCl_2 + 5\,RPH_2 \longrightarrow 2\,R_5P_5 + 10\,HCl \qquad \textbf{(14.97)}$$

$$5\,RPCl_2 + 5\,Mg \longrightarrow R_5P_5 + 5\,MgCl_2 \qquad \textbf{(14.98)}$$

Other known ring systems include P_3R_3, P_4R_4, and P_6R_6. It is also possible to prepare bicyclic and multicyclic species. An example is $P_4[N(SiMe_3)_2]_2$, which contains two fused three-membered rings:[21]

$$R-P \underset{\underset{\displaystyle P}{\diagdown}}{\overset{\overset{\displaystyle P}{\diagup}}{<}} P-R \qquad R = N(SiMe_3)_2$$

This rather strange ring system is best understood through its relationship to white phosphorus. The rupture of one of the six bonds of the P_4 tetrahedron leads to the bicyclic framework. The bonding in this compound will be considered from another perspective in Chapter 19.

The chemistry of the polyiodides is quite extensive. This is due in large measure to the Lewis acidity of I_2 and the basicity of I^-, which promote chain-forming reactions:

$$I_2 + I^- \rightleftharpoons I_3^- \qquad K_{eq} = 80 \text{ at } 25°C \qquad \textbf{(14.99)}$$

$$I_3^- + n I_2 \rightleftharpoons I_{(2n+3)}^- \qquad \textbf{(14.100)}$$

The I_5^- anion is V-shaped (idealized C_{2v} symmetry). This structure can be rationalized by Lewis/VSEPR theory, but also might be considered to be an example of either an ion–dipole interaction or secondary bonding (see Chapter 8).

Catenation by chlorine and bromine is more limited. The Cl_3^- and Br_3^- anions have been characterized, as have triatomic anions such as ICl_2^-, $IBrF^-$, and $IBrCl^-$.

14.8   Heterocatenation

Heterocatenation involving Group 15 and 16 elements is very common. (Recall the borazines, boroxines, and polysiloxanes.) The magnitude of the

21.   Niecke, R.; Rüger, R.; Krebs, B. *Angew. Chem. Int. Ed. Engl.* **1982**, *21*, 544; *Angew. Chem.* **1982**, *94*, 553.

known chemistry in this area is intimidating; a review of cyclic phosphorus–nitrogen chemistry contains 1057 references![22] The primary emphasis here is on three particularly well-studied systems—those in which the ring or chain frameworks contain phosphorus–nitrogen, phosphorus–oxygen, or sulfur–nitrogen bonds.

Phosphorus–Nitrogen Heterocycles and Polymers[22,23]

The best-known cyclic P–N systems are *phosphazanes* (saturated rings and chains) and *phosphazenes* (unsaturated systems with alternating P–N single and double bonds). Cyclic dimers, trimers, tetramers, etc. (up to a total of 34 ring atoms), as well as linear polymers, all are known. As is the case for B–N and Si–N rings, cyclic P–N chemistry is dominated by the six-membered trimers.

The interaction of ammonium chloride with PCl_5 is one of the most intriguing and complex reactions in inorganic chemistry. It can be used to produce, in sequential order, short-chain, cyclic, and polymeric phosphazenes. Upon mixing the two reagents in a solvent such as chlorobenzene, the initial product is an ionic compound in which the cation has a —P=N—P— framework:

$$3\,PCl_5 + NH_4Cl \xrightarrow[\Delta]{-4HCl} (Cl_3P{=}N{-}PCl_3)^+\,PCl_6^- \qquad (14.101)$$

More HCl is evolved as the reaction continues; this leads to chain lengthening and then cyclization. The distribution of the product mixture depends on the reaction conditions, but the trimer (hexachlorocyclotriphosphazene) is the major product. The overall equation for its formation is

$$3\,PCl_5 + 3\,NH_4Cl \xrightarrow{\Delta} (Cl_2PN)_3 + 12\,HCl \qquad (14.102)$$

As for the formation of borazine (Section 13.3), a multiply bonded species ($Cl_2P{\equiv}N$) that undergoes spontaneous oligomerization can be visualized.[24] This gains one P–N σ bond for each P–N π bond lost, and since the σ bond is the stronger of the two, the cyclization process is favored.

---

22. Allen, C. W. In *The Chemistry of Inorganic Homo- and Heterocycles*; Haiduc, I.; Sowerby, D. B., Eds.; Academic: New York, 1987; Volume 2, pp. 501–616.

23. Krishnamurthy, S. S.; Sau, A. C.; Woods, M. *Adv. Inorg. Chem. Radiochem.* **1978**, *21*, 41; Allcock, H. R. *Chem. Rev.* **1972**, *72*, 315.

24. This does not imply a reaction mechanism, but rather a rationalization of the thermal instability of this monomer.

There are significant differences between borazines and phosphazenes, however. The latter has two substituents on each phosphorus and none on nitrogen, rather than one on each ring atom. This requires multiple bonding to produce reasonable valences:

Thus, two Kekulé-type resonance structures can be drawn for the cyclo-triphosphazene framework. However, note that each phosphorus has apparent sp^3 hybridization; this precludes p orbital involvement by phosphorus in the π bonding scheme. Although there is controversy about the nature of the multiple bonding, a common interpretation is that an unhybridized p orbital of nitrogen overlaps with one or more phosphorus d orbitals. Thus, the classic model of aromaticity—a rigid, planar ring structure dictated by the π overlap requirements of a set of parallel p orbitals—does not apply to the phosphazenes. This helps explain why some cyclotriphosphazenes are planar, but others are not. Also, spectroscopic data suggest that those P_3N_3 rings that are planar are more flexible toward out-of-plane distortion than are substituted benzenes and borazines.

The two substituents on phosphorus create the possibility of isomerism for compounds having more than one kind of substituent. Thus, the known isomers of $(Me_2N)_3(Cl)_3P_3N_3$ include one in which the chlorines are all on the same side of the ring (the *syn* isomer), and another having two chlorines on one side and one on the other (*anti*).

The chemical reactions of cyclotriphosphazenes can be divided into those that alter the ring structure and those that do not; the second category amounts to exocyclic substitution reactions. One or more chlorines of $Cl_6P_3N_3$ can be replaced by groups such as $-R$, $-F$, $-OR$, and $-NR_2$. For double displacements [giving $R_2(Cl)_4P_3N_3$], *geminal* (both replacements on the same phosphorus) and *viscinal* (reactions at two different phosphorus atoms) isomers are possible.

The most important ring-opening reaction of the cyclotriphosphazenes is thermal polymerization. Heating to about 250°C gives *polyphosphazenes*, which are linear polymers of high molecular weight (10^4–10^5 g/mol):

These polymers also can be formed in other ways. For example, $(Me_2PN)_x$ can be synthesized by elimination of chlorotrimethylsilane from the appropriate silyl-substituted aminophosphorane.[25]

$$Me_2(Cl)_2P-N(SiMe_3)_2 \xrightarrow[\Delta]{-2Me_3SiCl} \frac{1}{x}(Me_2PN)_x \qquad (14.103)$$

The polyphosphazenes are superseded only by the silicones in commercial importance among inorganic polymers. Depending on the reaction conditions and the substituents, they can be formed as thin sheets, fibers, or elastomers. A precursor for many of the others is $(Cl_2PN)_x$, since it can be produced inexpensively and in high yield. However, it is hydrolyzed in moist air to the P–OH derivative. It is therefore customary to react the polydichloro substrate with appropriate nucleophiles to give alkyl, aryl, alkoxy, amino, or other derivatives. This not only improves the hydrolytic stability, but also can be used to impart the desired mechanical properties and/or to cross-link adjacent chains.

A large number of saturated P–N rings are known. They are generally prepared from reactions of chlorophosphines (trivalent phosphorus) or chlorophosphoranes (pentavalent phosphorus) with weak nitrogen bases such as aniline:

$$PCl_3 + C_6H_5NH_2 \xrightarrow{-2HCl} ClP{=}NC_6H_5 \longrightarrow \frac{1}{x}(ClPNC_6H_5)_x \qquad (14.104)$$

$$PCl_5 + C_6H_5NH_2 \xrightarrow{-2HCl} Cl_3P{=}NC_6H_5 \longrightarrow \frac{1}{x}(Cl_3PNC_6H_5)_x \qquad (14.105)$$

The most common ring size, especially for sterically bulky amines, is four $(x = 2)$, but larger rings are sometimes observed as well.

Phosphorus–Oxygen Chains and Heterocycles[26]

Many species having structures based on PO_4 tetrahedra are known. These compounds show relationships to silicon–oxygen systems, with many of the phosphate anions being isostructural with silicates. (The anionic charges are different, of course.) Just as H_4SiO_4 and its anions may be considered the "parents" of the silicates, H_3PO_4 and its anions serve as starting materials for phosphates.

Like most ternary acids, H_3PO_4 eliminates water upon heating. The product is diphosphoric (or pyrophosphoric) acid, $H_4P_2O_7$, which contains

25. Neilson, R. H.; Wisian-Neilson, P. *Inorg. Chem.* **1980**, *19*, 1875.

26. Kalliney, S. Y. *Top. Phosph. Chem.* **1972**, *7*, 255.

the P–O–P structural unit. Further elimination of water produces longer chains. The limiting empirical formula is $(HPO_3)_n$.

Heterocycles also can be obtained. The pyrolysis of sodium dihydrogen phosphate gives $P_3O_9^{3-}$:

$$3\,NaH_2PO_4 \xrightarrow[\Delta]{} Na_3P_3O_9 + 3\,H_2O \qquad\qquad (14.106)$$

This is the first member of a series of cyclic anions having the general formula $(P_nO_{3n})^{n-}$; they are sometimes called *metaphosphates*. Rings having 6, 8, 10, 12, 16, 20, and 24 members are all known. The P_3O_3 ring has a nonplanar chair conformation, as shown in Figure 14.13a.

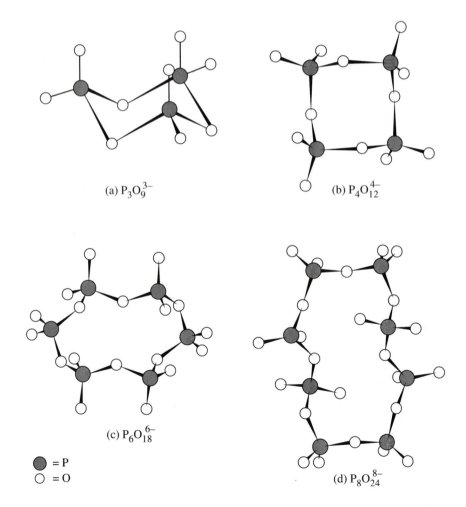

(a) $P_3O_9^{3-}$

(b) $P_4O_{12}^{4-}$

(c) $P_6O_{18}^{6-}$

(d) $P_8O_{24}^{8-}$

Figure 14.13
The structures of some cyclic phosphate anions.

● = P
○ = O

The eight-membered cyclotetraphosphate ring can be prepared directly from P_4O_{10} by careful hydrolysis using ice water:

$$P_4O_{10} + 2H_2O \xrightarrow{0°C} H_4P_4O_{12} \tag{14.107}$$

Anions are then obtained by reacting $H_4P_4O_{12}$ with base. These eight-membered rings are nonplanar, with bond angles and lengths similar to those of the trimer (Figure 14.13b).

Sulfur–Nitrogen Compounds[27]

Although sulfur–nitrogen chemistry might be expected to parallel the nitrogen–oxygen system, there is remarkably little similarity between the two. There are at least two reasons for this. First, the bond polarities are different—nitrogen is relatively positive in its oxides, but negative when bonded to sulfur. Second, the relative σ and π bond energies are quite different. Sulfur–nitrogen multiple bonds are weaker than their N–O counterparts, while S–S single bonds are considerably stronger than O–O. The effects of these differences will be evident in the following discussion.

A case in point is sulfur nitride, SN, the analogue of nitric oxide. The monomer is much less stable than NO because of the weaker π bonding; hence, SN undergoes spontaneous oligomerization. The dimer (a planar ring) is known, but it is unstable toward polymerization. Polymeric $(SN)_x$ consists of long, nonlinear chains. The average bond distance (150 pm) and energy (463 kJ/mol) are indicative of multiple bonding. The chains are oriented to allow interactions between sulfurs of adjacent chains (see Figure 14.14).

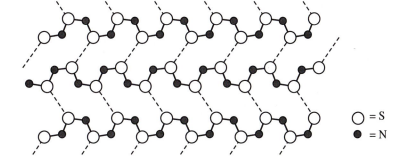

Figure 14.14
The structure of the polymer of sulfur nitride, $(SN)_x$.

$\bigcirc$ = S
$\bullet$ = N

27. Roesky, H. W. *Adv. Inorg. Chem. Radiochem.* **1979**, *22*, 239; Heal, H. G. *Adv. Inorg. Chem. Radiochem.* **1972**, *15*, 375.

Pure samples of $(SN)_x$ have electrical conductivities of about 2×10^3 $\Omega^{-1}\ cm^{-1}$ at room temperature—nearly as large as many metals. Superconductivity is observed at very low temperatures. This is not totally understood, but is certainly related to the presence of antibonding electrons.[28]

The NH group is valence isoelectronic with S, so we might expect the formal replacement of one or more sulfur atoms of S_8 with NH to give stable heterocycles. This is the case, and four members of the series $S_{8-n}(NH)_n$ (those having $n = 1$–4) are known. These compounds are usually prepared by the reaction of S_2Cl_2 with ammonia in a polar aprotic solvent such as DMF. The product mixture includes S_7NH, three isomers of $S_6(NH)_2$, and 1,3,6-$S_5(NH)_3$. [The last member of the series, $S_4(NH)_4$, also is obtained, but can be made more easily by the chemical or electrolytic reduction of S_4N_4.]

All these compounds retain the crown shape of S_8. The N–S bond lengths are shorter than expected from the covalent radii, probably because of dative $(p \rightarrow d)\pi$ bonding:

$$HN\underset{\ddot{S}-\ddot{N}H-\ddot{S}}{\overset{\ddot{S}-\ddot{N}H-\ddot{S}}{\diagup\diagdown}}NH \longleftrightarrow HN\underset{\ominus\ddot{S}=\ddot{N}H-\ddot{S}}{\overset{\ddot{S}-\ddot{N}H-\ddot{S}}{\diagup\diagdown}}NH \longleftrightarrow \text{etc.}$$

This idea is supported by X-ray diffraction studies, which show approximate planarity (ie, sp^2 rather than sp^3 hybridization) about the nitrogens.

Most of the chemical reactions of the cyclic sulfur imides fall into three categories: acid–base (nucleophilic addition and displacement) reactions, condensations, and ring contractions. Some examples follow:

$$S_4(NH)_4 + AlCl_3 \longrightarrow S_4(NH)_4 \cdot AlCl_3 \tag{14.108}$$

$$S_7NH + BCl_3 \longrightarrow S_7NBCl_2 + HCl \tag{14.109}$$

$$2S_7NH + SCl_2 \xrightarrow{\Delta} (S_7N)_2S + 2HCl \tag{14.110}$$

$$3S_4(NH)_4 + 12Cl_2 \longrightarrow 4Cl_3S_3N_3 + 12HCl \tag{14.111}$$

Unlike hydrogen, halogen substituents bond to sulfur rather than to nitrogen. For example, $(FSN)_4$ and $Cl_2S_4N_4$ have quite different structures from $S_4(NH)_4$. The fluoride has alternating single and double bonds around

28. For further information about poly(sulfur nitride), see Labes, M. M.; Love, P.; Nichols, L. F. *Chem. Rev.* **1979**, *79*, 1; and/or King, R. B., Ed. *Inorganic Compounds with Unusual Properties*; Advances in Chemistry 150; American Chemical Society: Washington, DC, 1976.

the ring, with S–N bond distances of 154 and 166 pm:

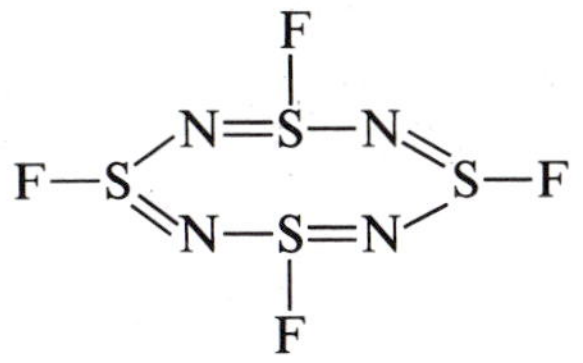

$$
\begin{array}{c}
\text{F} \\
\text{|} \\
\text{N=S—N} \\
\text{F—S} \qquad \text{S—F} \\
\text{N—S=N} \\
\text{|} \\
\text{F}
\end{array}
$$

On the other hand, $Cl_2S_4N_4$ contains a transannular bond and two fused five-membered rings. Many resonance structures can be drawn for that molecule.

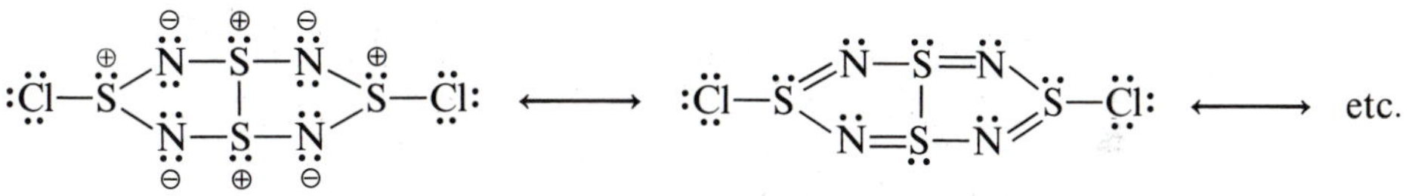

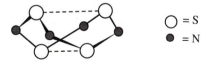

The most easily prepared and thoroughly studied of the cyclic sulfur–nitrogen compounds is tetrasulfur tetranitride, S_4N_4. It can be synthesized either by dissolving elemental sulfur in liquid ammonia or by reacting S_2Cl_2 with NH_3 in carbon tetrachloride.

$$5\,S_8 + 16\,NH_3 \longrightarrow 4\,S_4N_4 + 24\,H_2S \qquad (14.112)$$

$$6\,S_2Cl_2 + 16\,NH_3 \longrightarrow S_4N_4 + 12\,NH_4Cl + S_8 \qquad (14.113)$$

A large number of resonance structures can be written for S_4N_4, including some that contain transannular S–S bonds; S–S bonding is also predicted by molecular orbital calculations.[29] The solid-state structure contains alternating sulfur and nitrogen atoms, with two short S–S distances (see Figure 14.15).

$$
\bigcirc = S \qquad \bullet = N
$$

Figure 14.15 The structure of tetrasulfur tetranitride, S_4N_4. The short S–S distances are indicated by dashed lines (see text).

29. See Findlay, R. H.; Palmer, M. H.; Downs, A. J.; Egdell, R. G.; Evans, R. *Inorg. Chem.* **1980**, *19*, 1307; and references cited therein.

Table 14.9 Structural data for some sulfur–nitrogen compounds and ions

	d(S–N), pm	d(S–S), pm	∠NSN, °	∠SNS, °
S_4N_4	162	258	104	113
$S_4N_4^{2+}$	155		119	151
$S_4N_4Cl_2$	159	248	109	119
S_2N_2	165		90	90
$(SN)_x$	159		106	120
	163			
S_4N_2	156	206	123	127
	167			
$S_4N_3^+$	155 (av)	206		149
				155
$S_3N_3^-$	160 (av)		117	123
$S_4(NH)_4$	165		110	129
S_7NH	168	205	103	113
			110	
$S_3N_3Cl_3$	160		114	129

Source: Data compiled mainly from that given in Wells, A. F. *Structural Inorganic Chemistry*, 5th ed.; Clarendon: Oxford, 1984; Chapter 18.

It is interesting to compare molecular structures in the sequence $S_4N_4^{2+} \rightarrow S_4N_4 \rightarrow S_4(NH)_4$, which are related by oxidation–reduction:

$$S_4N_4^{2+} \xrightarrow{+2e^-} S_4N_4 \xrightarrow{+4e^-, 4H^+} S_4(NH)_4 \qquad \textbf{(14.114)}$$

The oxidation of S_4N_4 to its dication increases the total number of S–N bonds, while its reduction to $S_4(NH)_4$ has the opposite effect. This is consistent with the data compiled in Table 14.9. The average S–N bond distances vary from about 155 pm in the cation to 165 pm in $S_4(NH)_4$.

Representative reactions of S_4N_4 are given in Figure 14.16.

Mixed Systems[30]

An imposing number of mixed chains and rings (ie, those containing more than two kinds of framework atoms) are known. Only a very few examples are given here.

30. Burford, N.; Chivers, T.; Rao, M. N. S.; Richardson, J. F. *Adv. Chem. Ser.* **1983**, *232*, 81; van de Grampel, J. C. *Rev. Inorg. Chem.* **1981**, *3*, 1; Heal, H. G. *The Inorganic Heterocyclic Chemistry of Sulfur, Nitrogen, and Phosphorus*; Academic: New York, 1980.

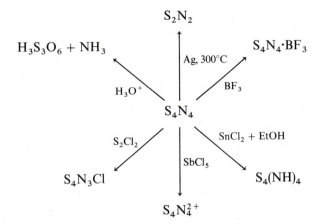

Figure 14.16
Some reactions of S_4N_4.

The reaction of sulfuric acid or one of its anions with secondary amines can be used to produce chains of five or more atoms. The general equation is

$$(HO)_2SO_2 + 2R_2NH \xrightarrow{-2H_2O} R_2N-\overset{\overset{O}{\|}}{\underset{\underset{O}{\|}}{S}}-NR_2 \qquad (14.115)$$

Mixed chains having the $-Si-N-S-N-Si-$ framework can be prepared by using silyl-substituted amines. An example is bis[methyl(trimethylsilyl)-amino]sulfate:

$$\begin{array}{c} Me_3Si \\ Me \end{array}\!\!\!\!\!\!\!\!>\!N-\overset{\overset{O}{\|}}{\underset{\underset{O}{\|}}{S}}-N\!\!<\!\!\!\!\!\!\!\!\begin{array}{c} Me \\ SiMe_3 \end{array}$$

This is a useful reagent for the formation of mixed-ring compounds. For example, its reaction with PF_3Cl_2 gives a four-membered SN_2P ring:

$$SO_2[N(Me)SiMe_3]_2 + PF_3Cl_2 \xrightarrow{-2Me_3SiCl} \underset{Me}{\overset{Me}{O_2S\diagdown\!\!\!\underset{N}{\overset{N}{\diagup}}\!\!\!PF_3}} \qquad (14.116)$$

Other mixed rings can be synthesized from S_4N_4. One example is its reaction with diphenylphosphine:[31]

$$S_4N_4 + \phi_2PH \longrightarrow \text{[ring structure]} \qquad (14.117)$$

(Seven- and eight-membered rings are obtained as well.) Another example is the formation of a six-membered S_3N_2O ring by reaction with excess SO_3:[32]

$$S_4N_4 + 6SO_3 \longrightarrow 2 \text{[ring structure]} + 4SO_2 \qquad (14.118)$$

Bibliography

Woolins, J. D. *Non-metal Rings, Cages, and Clusters*; Wiley: New York, 1988.

Haiduc, I.; Sowerby, D. B., Eds. *The Chemistry of Inorganic Homo- and Heterocycles*: Academic: New York, 1987.

Goldwhite, H. *Introduction to Phosphorus Chemistry*; Cambridge University: Cambridge, 1981.

Heal, H. G. *The Inorganic Heterocyclic Chemistry of Sulfur, Nitrogen, and Phosphorus*; Academic: New York, 1980.

Husar, R. B.; Lodge, J. P.; Moore, D. J. *Sulfur in the Atmosphere*; Pergamon: London, 1978.

Ring, N. H. *Inorganic Polymers*; Academic: New York, 1978.

Meyer, B. *Sulfur, Energy, and the Environment*; Elsevier: Amsterdam, 1977.

Steudel, R. *Chemistry of the Non-metals*; Walter de Gruyter: Berlin, 1977.

Emsley, J.; Hall, D. *The Chemistry of Phosphorus*; Harper & Row: New York, 1976.

Downs, A. J.; Adams, C. J. *The Chemistry of Chlorine, Bromine, Iodine, and Astatine*; Pergamon: New York, 1975.

31. Chivers, T.; Rao, M. N. S.; Richardson, J. F. *J. Chem. Soc. Chem. Commun.* **1982**, 982.
32. Rodek, E.; Amin, N.; Roesky, H. W. *Z. Anorg. Allg. Chem.* **1979**, *457*, 123.

O'Donnell, T. A. *The Chemistry of Fluorine*; Pergamon: Oxford, 1975.

Corbridge, D. E. C. *The Structural Chemistry of Phosphorus*; Elsevier: Amsterdam, 1974.

Powell, P.; Timms, P. L. *The Chemistry of the Non-metals*; Chapman and Hall: London, 1974.

Colburn, C. B., Ed. *Developments in Inorganic Nitrogen Chemistry*; Elsevier: Amsterdam, 1973.

Allcock, H. R. *Phosphorus–Nitrogen Compounds*; Academic: New York, 1972.

Bartlett, N. *The Chemistry of the Noble Gases*; Elsevier: Amsterdam, 1971.

Questions and Problems

1. Use Figures 14.1–14.3 to answer these questions about astatine.
   (a) Estimate the first ionization energy and the electronegativity of At.
   (b) Can a case be made for a diagonal relationship between Te and At? Defend your answer.
   (c) Estimate the energy of the At–F bond; refine your estimate as much as possible, and explain.

2. We might expect argon to form stronger covalent bonds than xenon because of its smaller size, and ArF_2 is a logical compound from this perspective. This species has never been synthesized, however, and evidence suggests that it is unstable. Use Figures 14.1 and 14.2 to develop an explanation.

3. The higher sulfanes, H_2S_n ($n > 1$), are thermodynamically unstable toward disproportionation to $S_8 + H_2S$.
   (a) Can a simple bond energy argument be used to explain this observation? Answer quantitatively.
   (b) If disproportionation is not based on ΔH, then a logical alternative is ΔS. How might entropy provide the explanation?
   (c) Discuss the probable cause for the spontaneous decomposition of P_2H_4 to $P_4 + PH_3$.

4. Given the four hydrides NH_3, AsH_3, H_2O, and H_2Se, identify the one that is the:
   (a) Strongest Brønsted acid   (b) Weakest Brønsted acid
   (c) Most prone to oxidation   (d) Least thermally stable

5. Consider the industrial preparations of HF and HCl described by equations (14.13) and (14.14). Do you believe that these reactions are strongly exothermic? How is the equilibrium driven in the desired direction?

6. Complete and balance the following equations:

   (a) $OF_2 + PF_3 \longrightarrow$ _____ + _____
   (b) $SO_2 + PCl_5 \longrightarrow SOCl_2 +$ _____
   (c) $Cl_2 + I^- \longrightarrow$ _____ + _____

(d) $P_2H_4 \xrightarrow{\Delta}$ _____ + _____

(e) $Bi_2O_3 + C \xrightarrow{\Delta}$ _____ + _____

(f) $AsCl_3 + NaBH_4 \longrightarrow$ _____ + _____ + _____

(g) $NaF + H_3PO_4 \longrightarrow$ _____ + _____

7. The first ionization energy of NO is considerably less than that of O_2, in spite of the fact that an electron is removed from a π^* orbital in both cases. Explain.

8. Write Lewis structures for the symmetric and asymmetric dimers of NO, and use them to answer the following questions:
   (a) Which of the two structures is favored on the basis of formal charge?
   (b) Compare the total number of bonds in $(NO)_2$ (either isomer) to the number in two NO molecules. Do you expect dimerization to be highly exothermic?
   (c) Why is the asymmetric dimer favored in the presence of Lewis acids?

9. Complete and balance the following reactions involving nitrogen oxides:

   (a) $N_2O \xrightarrow{\Delta}$ _____ + _____

   (b) $NO + CH_3OH \longrightarrow$ _____ + _____

   (c) $NO_2 + F_2 \longrightarrow$ _____

   (d) $2 N_2O_4 +$ _____ $\longrightarrow NOF + NaNO_3$

10. Elemental analysis of the product of the reaction between N_2O_5 and excess SO_3 is consistent with the formulation $N_2O_5 \cdot 2SO_3$. However, this "adduct" behaves like an ionic compound. Suggest a structure.

11. The addition of small amounts of water to P_4O_{10} gives a product that analyzes as HPO_3. Write Lewis structures for this product as both a monomer and a linear polymer.

12. The Arbusov reaction involves the following transformation:

$$P(OR)_3 + R'CH_2X \longrightarrow (RO)_2 \overset{\displaystyle O}{\overset{\displaystyle \|}{P}} CH_2R' + RX$$

The initial step of the reaction is thought to involve nucleophilic attack by the phosphorus lone pair at the carbon–halogen bond. Complete the mechanism.

13. Given your answer to problem 12, predict the products of the reaction of triethylphosphite, $(EtO)_3P$, with Br_2.

14. Suggest the molecular framework and write reasonable Lewis structures for:
   (a) S_2O (b) S_3O_2 (c) $S_2O_3^{2-}$ (d) S_7O (cyclic)

15. The adduct of SO_2 with $SnMe_4$ contains a tin–sulfur bond. The adduct of SO_2 with BF_3 contains a boron–oxygen bond. Explain the difference in behavior.

16. Each of the following can undergo disproportionation in acidic aqueous solution. Write the balanced equation for each case.
   (a) NO (b) N_2O (c) N_2O_4 (d) SO_2 (e) $S_2O_3^{2-}$

17. Quantitatively explain why the thermal stability of O_2F_2 greatly exceeds that of O_2Cl_2.

18. Write the product(s) for the reaction of PCl_3 with:
    (a) F_2 (b) Excess methanol   (c) S_8
    (d) 1 molar equivalent of t-BuLi
    (e) 1 molar equivalent of Me_3SiNEt_2
    (f) Excess Me_3SiNEt_2

19. The compound $S_2O_6F_2$ is a potent oxidizing agent.
    (a) Write the most reasonable Lewis structure for $S_2O_6F_2$.
    (b) What is the weakest bond in this molecule?
    (c) Given your answer to part (b), explain why this compound is a strong, two-electron oxidizing agent.

20. Suggest structures for the anions of $Br_2^+Sb_2F_{11}^-$ and $Br_2^+Sb_3F_{16}^-$.

21. Consider the two compounds NO_2X (X = H and F). One contains an X–O bond, and the other an N–X bond. Which is which? Why the difference?

22. Phosphorus pentachloride is partly dissociated to $PCl_3 + Cl_2$ in the vapor phase. This is not the case for PF_5. Give an explanation based on bond energies.

23. Give your best estimate of the bond angles in:
    (a) IF_2^+ (b) IF_2^- (c) IF_4^- (d) IF_5

24. Which of the species in question 23 contains the strongest I–F bond? Defend your answer.

25. The $IBrCl_3^-$ ion has been observed, but its structure is not yet known. Consider two frameworks, one in which the iodine is three-coordinate and one in which it is four-coordinate.
    (a) Write a Lewis structure for each.
    (b) Give the predicted geometries according to the VSEPR model.
    (c) Which of these possibilities do you prefer? Why?

26. Review the bond energy analysis for PX_3 compounds, and then carry out a parallel analysis for SX_2. To what extent do the results differ?

27. The compound XeO_2F_2 might exist in either of two frameworks: one in which xenon is two-coordinate and one in which it is four-coordinate.
    (a) Write a Lewis structure for each.
    (b) The compound actually contains two Xe–O and two Xe–F bonds. Rationalize.

28. Use VSEPR theory to help you assign a point group to:
    (a) IF_2^- (b) IF_3 (c) IF_4^- (d) IF_7 (e) XeF_2
    (f) XeF_4 (g) XeF_5^+ (h) XeO_3

29. Consider the gas-phase reactions below, where Z is a Group 18 element.

$$Z + 3X_2 \longrightarrow ZX_6 \qquad (X = F \text{ and } Cl)$$

$$2Z + 3O_2 \longrightarrow 2ZO_3$$

(a) Assume that a compound might be isolable if its standard enthalpy of formation is more negative than $+100 \, \text{kJ/mol}$. What would be the lower limit for the Z–F and Z–Cl bond energies of ZX_6 to permit the preparation of these compounds?

(b) Repeat part (a) for the Z–O bond of ZO_3.

30. Construct a molecular orbital diagram for the hypothetical compound XeH_2. This species should be linear, so the diagram for BeH_2 (Figure 3.18) can be used as a basis. What is the predicted bond order? How is the diagram for XeF_2 different from that of BeH_2?

31. Use group theory to identify the specific xenon orbitals involved in the hybridization scheme for:

(a) XeF_4 (b) XeO_3

Be sure to include vectors for the central atom lone pairs. (If a review is needed, see Chapter 5.)

32. In what way does Lewis acid–base theory explain the existence of the triiodide ion? Include both an equation and an explanation for why I_3^- is more stable than Cl_3^- and Br_3^- in your answer.

33. Sketch the $(Cl_3P{=}N{-}PCl_3)^+$ cation [equation (14.101)]; be careful to show the proper geometry about each atom.

34. Sketch the three isomeric cyclotriphosphazenes having the formula $Me_2(Cl)_4P_3N_3$.

35. Consider the series $S_{8-n}(NH)_n$.

(a) Only three of the four possible isomers of $S_6(NH)_2$ are known. Predict which is unknown, and explain.

(b) None of the compounds having $n = 5$–8 are known. What factor mitigates against their stability?

36. The dimer S_2N_2 is known to be a four-membered ring. Describe the bonding using Lewis theory. Do you consider this compound to be aromatic or not? Defend your answer.

37. The cation $S_3N_2Cl^+$ contains a five-membered ring. Predict its probable framework, and draw at least three Lewis structures.

38. The acyclic molecule $S_3N_2O_2$ has C_{2v} symmetry. Sketch its most important resonance structure.

39. The reaction of $P(NMe_2)_3$ with phosphorus trichloride in a 1:2 molar ratio produces product A, which has a molecular weight of approximately 145 g/mol. Warming A in the presence of elemental sulfur gives B, which contains 39.8% chlorine by mass. Give chemical formulas for A and B.

40. When Se_8 reacts with excess AsF_5 in SO_2 solution at 80°C, two products result. One is the salt $[Se_4]^{2+}(AsF_6)_2$. What is the other? Explain.

*41. Several ways to prepare XeO_2F_2 from XeF_6 have been reported by K. O. Christie and W. W. Wilson (*Inorg. Chem.* **1988**, *27*, 3763). Answer these questions after reading their article.

(a) In what way are these preparations of XeO_2F_2 more convenient (ie, safer) than those previously used?

(b) The $XeF_6 \rightarrow XeO_2F_2$ conversion can be accomplished using either $CsNO_3$ or N_2O_5. What do the two reagents have in common? Why is N_2O_5 more convenient?

(c) Predict the products of the reaction of BrF_5 with $CsNO_3$.

*42. The rather surprising transformation

$$CHF{=}CH_2 + N_2F_4 \longrightarrow FN{=}C(F)C{\equiv}N$$

has been discussed by H. M. Marsden and J. M. Shreeve (*Inorg. Chem.* **1987**, *26*, 169).

(a) A sequence of several reactions is obviously required. Suggest such a sequence. Your answer should make clear what happens to the extra fluorines.

(b) The formula $FN{=}C(F)C{\equiv}N$ actually indicates two products. How so?

The Transition Elements

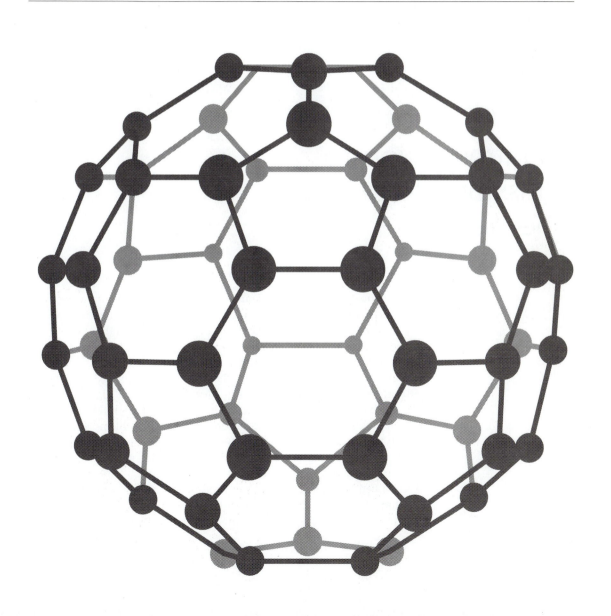

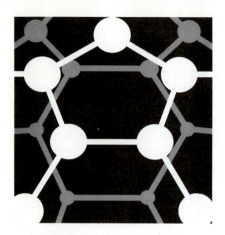

15

An Introduction to Transition Metal Chemistry

Part V is devoted to the structural and reaction chemistry of the transition metals, with particular emphasis on their coordination complexes. A logical place to begin is to ask which elements of the periodic table should be considered transitional. This apparently innocent question becomes surprisingly difficult to answer when examined in depth, and has been a source of considerable disagreement. If quantum theory is used as a basis, it might be argued that the transition elements are those in which the "last" electron (per the Aufbau principle) occupies an inner (d or f) orbital. However, irregularities in ground-state configurations then cause problems. For example, in going from copper ($[Ar]4s^13d^{10}$) to zinc ($[Ar]4s^23d^{10}$), that electron enters an s orbital; the same is true for cadmium and mercury. Are the Group 12 metals then transition elements?

A modified approach uses this definition: A transition metal is an element that has a partly filled d or f subshell in any of its common oxidation states.[1] Zinc, cadmium, and mercury are excluded by this definition, and, in

1. Cotton, F. A.; Wilkinson, G. *Advanced Inorganic Chemistry*, 5th ed.; Wiley: New York, 1988; p. 625.

fact, there is some justification for placing them in the main group category (see below). However, this results in only nine members (scandium through copper) for the first transition series, in spite of the fact that quantum theory allows for ten d electrons in a given subshell.

For the organizational purposes of this book, the transition elements are taken to be those of Groups 3–12, plus the lanthanides and actinides. Thus, the first transition series comprises ten elements, scandium through zinc. Keep in mind, however, that this is to a large extent an arbitrary division—periodicity is continuous throughout the periodic table (eg, for the duration of the first long period, potassium through krypton).[2]

15.1   Properties Common to the Transition Elements

According to the above definition, 59 of the first 103 elements are transitional. Those having sufficient nuclear stability to allow for systematic study demonstrate several common properties. Five of the most significant are the following:

1.   The free elements conform to the metallic bonding model. Thus, their lattices are typically either close-packed or body-centered cubic. They generally have high thermal and electrical conductivities, and they are malleable and ductile.

2.   Nearly all have more than one positive oxidation state. (Given the discussion above, it is interesting that zinc and cadmium are two exceptions.)

3.   Nearly all have one or more unpaired electrons in their atomic ground states, and form paramagnetic compounds and ions. (Zinc is again an exception.) Because of this, studies of magnetic properties using experimental tools such as NMR, ESR (electron spin resonance), and magnetic susceptibility are often employed in this area of chemistry.

4.   Low-energy electron transitions are observed for the free elements and their compounds and complexes. The transitions may fall into the infrared, visible, or ultraviolet region; for the visible cases, of course, colored species are the result.

5.   The cations of these elements (and often the neutral atoms as well) behave as Lewis acids, and there is a strong tendency to form complexes.

---

2.   Why is it necessary to distinguish between main group and transition elements at all? Nearly all chemists would agree that there are significant differences in the properties of transition and nontransition elements. These differences result primarily from the presence or absence of chemically active d or f electrons.

Complexes having 2–6 bases (ligands) are common, and species with as many as 14 ligands about a single metal are known.[3]

15.2 Oxidation State Tendencies and Their Causes

A key to understanding the chemistry of the transition elements is knowledge of their oxidation state preferences. These preferences can be rationalized through analysis of such factors as ionization energy and bond energy. In this section, both the observed trends in oxidation states and the underlying reasons for these trends are discussed.

It is useful to view oxidation states from two perspectives: One is the maximum (the most oxidized) state, and the second is the preferred (most stable) state. For the first half of the $3d$ series (scandium through manganese), the maximum state corresponds to the "loss" (or, more accurately, the participation in chemical bonding) of all valence electrons. Examples are the oxygen-containing species Sc_2O_3, TiO_2, VO_2^+, CrO_3, and MnO_4^-. Beyond manganese there is a general decrease in the highest state observed, as demonstrated by FeO_4^{2-}, $Co_2O_3 \cdot nH_2O$, NiO_2, CuO^+, and ZnO. The decrease from Fe^{VI} to Zn^{II} correlates better with the number of vacancies than with the number of occupancies in the valence orbitals of these elements.

Maximum oxidation states also can be examined through binary oxides and fluorides (Table 15.1). For chromium and manganese, a higher state is achieved with oxide than with fluoride. This is probably a coordination effect; for example, Mn^{VII} would require seven fluorines but only four oxygens. The most common state is M^{II}, with the difluorides of all these elements except scandium and titanium being known. Since the electron configurations of the M^{2+} cations are given by $[Ar]4s^0 3d^x$, this might be considered a variation of the inert pair effect in which the s orbital is empty (rather than the s orbital containing a nonbonding electron pair).

The preferred state is the one that is the most stable (toward disproportionation or thermal dissociation) and/or the least reactive. Such states are highly dependent on the types of ligands present. Metals in high oxidation states are relatively hard, and so are stabilized by hard bases (eg, F and O donors). Lower oxidation states are stabilized by softer bases such as S^{2-} and I^-.

Table 15.1 shows that, when bonded only to oxygen or fluorine, higher states (M^{III} and/or M^{IV}) are preferred by the early elements; M^{II} is favored later in the series. This correlates with ionization energies. The energy requirements for the removal of two and three electrons from the

---

3. For example, in $U(BH_4)_4$ the uranium takes part in 14 different U–H–B linkages.

Table 15.1   The common binary fluorides and oxides
of the elements of the first transition series

					Fluorides					
M^{II}			VF_2	CrF_2	**MnF_2**	FeF_2	**CoF_2**	NiF_2	CuF_2	ZnF_2
M^{III}	ScF_3	TiF_3	VF_3	**CrF_3**	MnF_3	**FeF_3**	CoF_3			
M^{IV}		**TiF_4**	VF_4	CrF_4	MnF_4					
M^{V}			VF_5	CrF_5						

					Oxides					
M^{I}									Cu_2O	
M^{II}		TiO	VO	CrO	**MnO**	FeO	**CoO**	NiO	CuO	ZnO
M^{III}	Sc_2O_3	Ti_2O_3	V_2O_3	**Cr_2O_3**	Mn_2O_3	Fe_2O_3	Co_2O_3			
M^{IV}		**TiO_2**	VO_2	CrO_2	MnO_2			NiO_2		
M^{V}			V_2O_5							
M^{VI}				CrO_3						
M^{VII}					Mn_2O_7					

Note: The most thermodynamically stable states are shown in boldface type. The most stable oxide of iron is the mixed oxide Fe_3O_4.

$3d$ elements are plotted in Figure 15.1 (p. 484). As expected, there is a steady increase going across the period in both cases. The removal of three electrons requires 44.10 eV for scandium and 56.75 eV for manganese, the first element for which the M^{II} state is preferred. Hence, Mn^{3+} is destabilized relative to its neutral atom by 12.65 eV (about 1220 kJ/mol) more than is Sc^{3+} versus Sc. The relative destabilization is even greater for cobalt, nickel, copper, and zinc; iron is an exception because of a pairing energy effect (see Chapter 1).

Volt-equivalents provide information about oxidation state preferences in aqueous solution. (See Chapter 9 for a background discussion.) Table 15.2 gives data for the $3d$ metals. The overall trends are similar to those observed for binary fluorides and oxides: M^{III} is favored by the early elements, while M^{II} dominates later in the series.

Comparisons of the 3*d*, 4*d*, and 5*d* Elements

Ionization energies tend to be slightly lower for the $4d$ elements than for their $3d$ congeners. For example, the sum of the first three energies of technetium is 52.08 eV, about 8% less than for manganese. These lower ionization energies and reduced steric interactions (because of the longer bonds) both favor high oxidation states. Thus, the most stable fluorides of niobium, molybdenum, technetium, and ruthenium are NbF_5, MoF_6, TcF_6, and RuF_5, respectively. (Compare these to the corresponding $3d$ elements in Table 15.1.) Silver is a notable exception, since it shows a marked preference for the Ag^I state. The $5d$ elements are similar to their

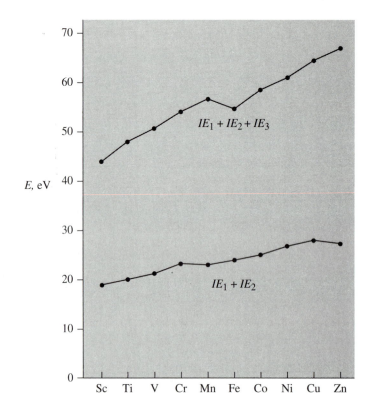

Figure 15.1
Ionization energies for the elements of the first transition series. Data taken from Table 1.7.

Table 15.2 Volt-equivalents as a function of oxidation state for the 3d elements

	M^I	M^{II}	M^{III}	M^{IV}	M^V	M^{VI}	M^{VII}
Sc			**−6.09**				
Ti		−3.26	**−3.62**	−3.53			
V		−2.37	**−2.63**	−2.27	−1.27		
Cr		−1.82	**−2.23**			+1.76	
Mn		**−2.36**	−0.85	+0.10		+4.62	+5.19
Fe		**−0.88**	−0.11			+6.49	
Co		**−0.55**	+1.25				
Ni		**−0.50**		+2.86			
Cu	+0.52	+0.67					
Zn		**−1.52**					

Note: Values are calculated from the standard potentials (acidic solution) of Table 9.1, and referenced to 0.00 V for the M^0 state. The most stable oxidation states are shown in boldface type. The most stable state for copper is Cu^0.

$4d$ congeners (TaF_5, WF_6, etc.), and in some cases still higher oxidation states are stabilized (ReF_7, OsF_6, IrF_4, and PtF_4).

Aqueous reduction potentials are known with less certainty for the $4d$ and $5d$ elements, but the expected trends appear to hold. There is a tendency toward decreasing oxidation state going across each period and toward increasing oxidation state going down each group.

15.3 Occurrences, Isolation, and Uses of the Free Elements

The natural abundances and naturally occurring forms of the transition metals follow the patterns established by the main group elements. More specifically:

1. Abundances tend to decrease going down each family.

2. The lighter elements are normally bonded to oxygen in their natural deposits. The heavier, softer metals generally prefer sulfur.

3. The oxidation state preferences discussed in the previous section are reflected in the ores and minerals of these metals.

Going across the first transition series, the relative natural abundances follow the order

$$Sc \ll Ti \gg V \approx Cr < Mn \ll Fe \gg Co < Ni > Cu \approx Zn$$

(Here, $\ll$ and $\gg$ represent differences greater than a power of 10.) This order is consistent with comments made in Chapter 7 concerning nuclear stabilities; even atomic numbers are favored over odd, and binding energy per nucleon reaches a maximum at iron. The $4d$ and $5d$ elements are rare, with the exception of zirconium (about 160 ppm in the lithosphere, roughly the same as vanadium). The most abundant element of each family is the lightest one. The major naturally occurring minerals of the d-block elements are listed in Table 15.3 (p. 486). Several elements, including silver, gold, and platinum, have such positive M^{n+}/M reduction potentials that they occur as native metals (see Figure 7.5).

The remaining free elements are obtained by chemical reduction. A few specific examples are described below.

Titanium (the Kroll Method):

$$TiO_2 + 2Cl_2 + 2C \xrightarrow{\Delta} TiCl_4 + 2CO \tag{15.1}$$

$$TiCl_4 + 2Mg \xrightarrow{1000° C} Ti + 2MgCl_2 \tag{15.2}$$

Table 15.3 Natural occurrences and primary end uses of the commercially important d-block elements

Element	Common Ore(s)	Major Commercial Uses
Sc	Thortveitite, $Sc_2Si_2O_7$	
Ti	Rutile, TiO_2 Ilmenite, $FeTiO_3$	Paint base (as TiO_2); alloy with Al for low density, high-strength uses (aircraft engines and bodies, etc.)
Zr	Zircon, $ZrSiO_4$	Photoflash bulbs; gemstones
V	Vanadinate, $Pb_{10}(VO_4)_6Cl_2$	Steel additive (increases strength for springs, etc.)
Cr	Chromite, $FeCr_2O_4$ Crocoite, $PbCrO_4$	Stainless steels (may be up to 25% Cr)
W	Wolframite, $(Fe, Mn)WO_4$	Filaments in electric lights; industrial sharpening tools (WC)
Mn	Pyrolusite, MnO_2 Hausmannite, Mn_3O_4 Rhodochrosite, $MnCO_3$	Steel manufacture and additive
Fe	Limonite, $2Fe_2O_3 \cdot 3H_2O$ Magnetite, Fe_3O_4 Pyrite, FeS_2 Siderite, $FeCO_3$	Steels
Co	Cobaltite, CoAsS	Pigments (glass, paints, inks)
Ni	Garnierite, $(Ni,Mg)_6Si_4O_{10}(OH)_8$	Ferrous and nonferrous alloys (nichrome, alnico, etc.)
Pd	Native deposits	White gold (Au/Pd); resistance wires; catalysis
Pt	Native deposits	Catalysts; thermocouples
Cu	Chalcopyrite, $CuFeS_2$ Malachite, $Cu_2(OH)_2CO_3$	Electrical conduction purposes; coinage
Ag	Argentite, Ag_2S Native deposits	Photographic chemicals; jewelry
Au	Native deposits	Jewelry; electrical contacts
Zn	Zinc blende, sphalerite, wurtzite, ZnS Calamine, $ZnCO_3$	Anticorrosion coatings; nonferrous alloys (brass, bronze); roofing
Cd	Greenockite, CdS	Batteries (Ni/Cd); ceramic glazes
Hg	Cinnabar, HgS	Liquid electrical conductor; pressure and temperature gauges; germicide, fungicide

Since $TiCl_4$ is volatile (bp $= 136°C$), it is easily separated from impurities such as $FeCl_3$ by distillation. High-temperature chemical reduction then yields the free metal.

Iron (the Bessemer Blast Furnace):

$$CaCO_3 \xrightarrow{\Delta} CaO + CO_2 \tag{15.3}$$

$$SiO_2 + CaO \xrightarrow{\Delta} CaSiO_3 \tag{15.4}$$

$$2C + O_2 \xrightarrow{\Delta} 2CO \tag{15.5}$$

$$Fe_3O_4 + 4CO \xrightarrow{\Delta} 3Fe(l) + 4CO_2 \tag{15.6}$$

Iron ore typically contains siliceous impurities, which are removed using limestone as an inexpensive precipitating agent [equations (15.3) and (15.4)]. Coke is both a fuel (generating temperatures in excess of $2000°C$) and a source of CO, the reducing agent. The gaseous by-products are vented through the top of the furnace, while molten iron is removed from the bottom. Slag (calcium metasilicate and other impurities) floats to the top of the iron layer, and so is easily removed.

Nickel (the Mond Method):

$$2NiS + 3O_2 \xrightarrow{\Delta} 2NiO + 2SO_2 \tag{15.7}$$

$$NiO + H_2 \xrightarrow{\Delta} Ni + H_2O \tag{15.8}$$

$$Ni + 4CO \xrightarrow{50°C} Ni(CO)_4(g) \tag{15.9}$$

$$Ni(CO)_4(g) \xrightarrow{>200°C} Ni + 4CO \tag{15.10}$$

Nickel generally occurs in ores in the presence of a variety of other metals. Heating in an oxygen atmosphere followed by reduction with H_2 or CO therefore gives a mixture of free metals. The nickel is isolated by conversion to the highly toxic $Ni(CO)_4$ (bp $= 43°C$), which is then decomposed at higher temperatures.

The commercial uses of the free metals are dominated by the steel industry. Steel (of which there are many classifications) can be loosely defined as an alloy for which iron is the major component. The addition of other metals is useful for improving certain desirable properties such as strength, hardness, and ductility. For example, the addition of vanadium or nickel increases mechanical strength. Vanadium steel is used in automobile springs, and nickel steel is used for armor plating. Chromium adds luster and general attractiveness (stainless steel). Zinc coating (galvanization) improves resistance to rust.

Some of the major commercial uses of the d-block elements are summarized in Table 15.3.

15.4 Coordination Compounds and Complex Ions

The chemistry of the transition elements is dominated by species in which the metal simultaneously engages in Lewis acid–base interactions with two or more donor ligands. These have long been called *coordination compounds* or *complexes*; if charged, they are *complex ions*.[4] A general equation for complex formation is

$$M^{m+} + nL^{x-} \longrightarrow [ML_n]^{m-nx} \tag{15.11}$$

The variable metal oxidation states, great variety of ligands available, and different possible coordination numbers combine to create an enormous number of possibilities for study. However, among the known, stable complexes the number of ligands and the geometric structure about a given metal are far from random. The structural aspects of coordination chemistry are considered next, with the discussion organized according to coordination number.

Low-Coordinate Geometries[5]

Two ligands might be arranged about a central metal in either a linear or an angular manner; the linear case is much more common. For such complexes, the idealized symmetry (ie, the point group of highest possible order) is either $D_{\infty h}$ or $C_{\infty v}$, depending on whether the ligands are identical or different.

The monovalent cations of Group 11 form numerous dicoordinate complexes, including $CuCl_2^-$, $[Ag(NH_3)_2]^+$, and $[Au(CN)_2]^-$, all of which are linear. Complexes of Hg^{II} (which is isoelectronic with Au^I), such as $Hg(CH_3)_2$ and $Hg(Br)SCN$, are known as well. The metal has ten valence

---

4. Recent advances, especially in organometallic chemistry, have led to a broadening of this definition. Many species are now known that have metal–metal or metal–nonmetal bonds that are not dative in nature and yet are referred to as coordination compounds or complex ions.

5. Eller, P. G.; Bradley, D. C.; Hursthouse, M. B.; Meek, D. W. *Coord. Chem. Rev.* **1977**, *24*, 1.

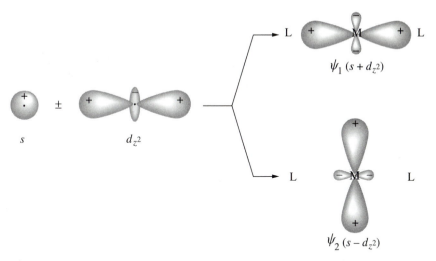

Figure 15.2   The formation of *sd* hybrids for linear complexes. The hybrid perpendicular to the bond axis (ψ_2) permits the transfer of charge from the L–M–L region into the *xy* plane, leading to stabilization. (See Orgel, L. E. *An Introduction to Transition Metal Chemistry: Ligand Field Theory*, 2nd ed.; Wiley: New York, 1966; pp. 69–70.)

electrons in every case. The configurations $\ldots ns^0(n-1)d^{10}$ and $\ldots ns^1(n-1)d^9$ are energetically close to one another. This permits at least two plausible hybridization schemes. For the $s^0 d^{10}$ case, the electron pairs donated by the ligands might reside in some linear combination of the *ns* and np_z orbitals. This corresponds to *sp* hybridization, of course, and is consistent with linear geometry. However, sd_{z^2} hybridization is also feasible. It has been argued that such hybridization is actually favored, because it removes charge from the region between the metal and ligands (see Figure 15.2).[6]

Keep in mind that stoichiometry is not predictive of coordination, especially in the solid state. For example, you might assume the coordination number of iron in FeF_2 is 2, but in both crystalline FeF_2 and FeF_3, the metals occupy 6-coordinate, octahedral interstices of fluoride ion sublattices.

Complexes in which the metal has a coordination number of 3 are relatively uncommon. Several have Cu^+ as the metal ion.[7] In

6.   Orgel, L. E. *J. Chem. Soc.* **1958**, 4816; see also, Orgel, L. E. *An Introduction to Transition Metal Chemistry: Ligand Field Theory*, 2nd ed.; Wiley: New York, 1966; pp. 69–71.

7.   For a review of Cu^I complexes, see Jardine, F. H. *Adv. Inorg. Chem. Radiochem.* **1975**, *17*, 116.

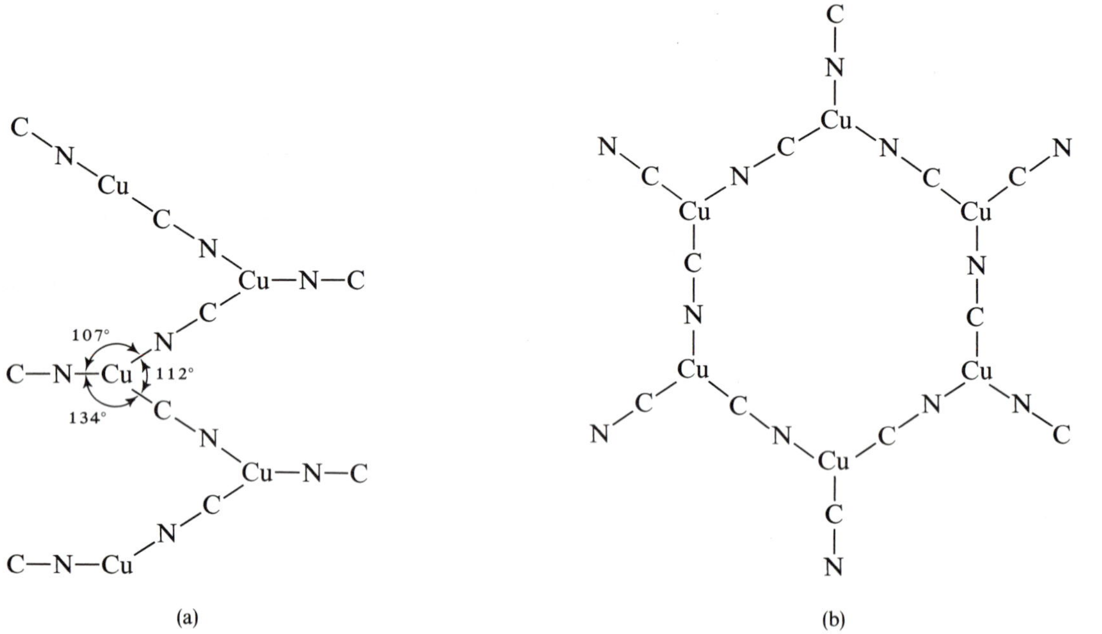

Figure 15.3 The structures of two 3-coordinate Cu^I complexes: (a) the [Cu(CN)$_2$]$^-$ anion of K[Cu(CN)$_2$]; (b) the [Cu(CN)$_3$]$^{2-}$ anion of K$_2$[Cu(CN)$_3$] · H$_2$O.

K$^+$[Cu(CN)$_2$]$^-$ (note the 1:2 metal:ligand stoichiometry), there are two bridging and one terminal cyanide ligands, with the –Cu–C–N– framework creating a helical chain. As shown in Figure 15.3a, the bond angles about copper do not sum to 360°. This creates puckering, which in turn produces the helical structure. The hydrated salt K$_2$[Cu(CN)$_3$] · H$_2$O contains –Cu–N–C– links that close to form 18-membered, planar rings; the interconnection of these rings creates a planar sheet.

Sterically large ligands often promote trigonal coordination. Two well-known examples are (Me$_3$Si)$_2$N$^-$ and (Me$_3$Si)$_3$C$^-$, which form trigonal complexes with a majority of the 3d elements. Apparently, even four such bulky groups surrounding a metal create steric repulsions sufficient to destabilize the system. In most such complexes the metal lies in the plane of the three donor atoms, but in Sc[N(SiMe$_3$)$_2$]$_3$ the scandium is above the plane (C_{3v} local symmetry).

It is possible (but not common) for three ligands to form an approximately T-shaped geometry about the metal. This is the case in the cation [Rh(Pϕ_3)$_3$]$^+$, which has a nearly linear P–Rh–P linkage.

Tetrahedral, Square Planar, and Intermediate Geometries[8]

not necessary
e.g. Ni(CN)₄

Four-coordinate complexes are usually either tetrahedral (for four equivalent ligands, giving local T_d symmetry and L–M–L bond angles of 109.5°) or square planar (ideal symmetry D_{4h}, 90° angles). Many complexes of each type will be encountered in this chapter and in Chapters 16–18. There are an infinite number of possible geometries between these limiting cases; all have idealized D_{2d} symmetry and bond angles between 90° and 109.5°. The situation can be visualized either by considering the dihedral angle between two intersecting planes, or through the perspective of a cubic superstructure (see Figure 15.4).

The intermediate geometries (variously referred to as flattened, squashed, or distorted tetrahedra) are less common than either the tetrahedral or square planar structures. Two examples are the complex anions in $Cs_2[CuX_4]$ (X = Cl and Br), which have X–Cu–X bond angles between 100° and 103°.

The geometry of a given tetracoordinate complex depends on both steric and electronic factors. The tetrahedron is superior from a steric standpoint, because it maximizes the distance between ligands. As will be explained in

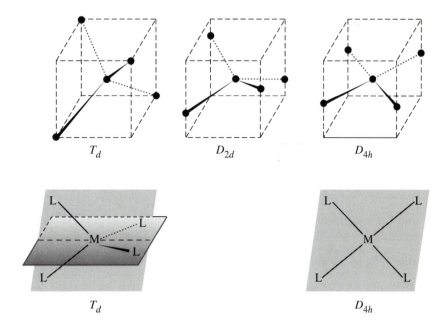

Figure 15.4
Comparisons of the tetrahedral (idealized T_d point group), square planar (D_{4h}), and intermediate (D_{2d}) geometries.

8. Favas, M. C.; Kepert, D. L. *Prog. Inorg. Chem.* **1980**, *27*, 325; Miller, J. R. *Adv. Inorg. Chem. Radiochem.* **1962**, *4*, 133.

the next chapter, the number of d electrons determines which possibility is favored electronically.

Square planar geometry is favored by metal ions having eight d electrons, especially Pd^{II}, Pt^{II}, and Au^{III}. Both geometries are common for Ni^{II}. The steric effect is more important for nickel than for its $4d$ and $5d$ congeners because of its smaller size. As a result, it is possible to prepare nickel complexes for which the difference in stability between the tetrahedral and square planar geometries is very small. For example, compounds having the general formula $NiX_2(P\phi_2R)_2$ ($X = Cl^-$, Br^-, and I^-; R = alkyl) show a gradual transition from square planar to tetrahedral geometry with increasing steric bulk (see Table 15.4). Several of the intermediate species, such as $NiBr_2(P\phi_2Et)_2$, can be produced in either of the two forms.

Table 15.4 The preferred structures of some tetracoordinate Ni^{II} complexes

Complex	Geometry	Complex	Geometry
$NiBr_4^{2-}$	T-4	$NiBr_2(P\phi_3)_2$	T-4
NiI_4^{2-}	T-4	$NiCl_2(P\phi_2Me)_2$	SP
$[Ni(CN)_4]^{2-}$	SP	$NiBr_2(P\phi_2Me)_2$	T-4
$[Ni(NCO)_4]^{2-}$	T-4	$NiI_2(P\phi_2Me)_2$	T-4
$[Ni(P\phi_3)_4]^{2+}$	SP	$NiBr_2(P\phi_2Et)_2$	Either
$[Ni(py)_4]^{2+}$	SP		

Note: T-4 = tetrahedral; SP = square planar.

Certain ligands dictate (or at least strongly favor) square planar geometry. One such donor is terpyridine, a tridentate ligand containing three aromatic rings (see Figure 15.5a). Even better examples are the tetradentate porphine molecule and its derivatives, called *porphyrins*. The conjugated π system must remain planar to maintain effective $p\pi$ overlap. The result is a natural "hole" for the metal ion, created by the planar arrangement of donor nitrogens. The metal may be either in the ligand plane (D_{4h} site symmetry) or out of the plane (C_{4v}). Porphyrins are found in many biologically important compounds, several of which will be discussed in Chapter 22.

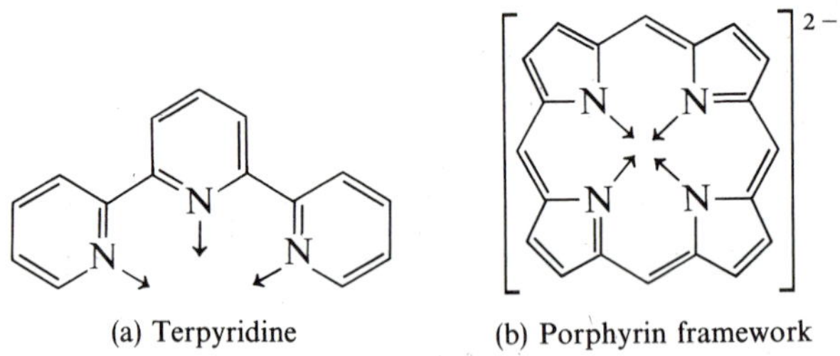

Figure 15.5
Ligands that promote square planar geometries.

(a) Terpyridine

(b) Porphyrin framework

Geometric and Optical Isomerism

Stereoisomers differ in their spatial arrangement of atoms. The two broad types of stereoisomerism are *geometric* and *optical* isomerism. Stereoisomerism is possible for certain 4-coordinate complexes having more than one kind of ligand. For example, geometric (cis–trans) isomers exist for square planar complexes of the type Ma_2b_2, where a and b represent nonequivalent donors. No isomerism is possible for tetrahedral Ma_2b_2 compounds, a fact of historical importance. The existence of two different compounds having the molecular formula $Pt(NH_3)_2Cl_2$ was used by the father of coordination chemistry, Alfred Werner, to prove that (unlike carbon) transition metals often have square planar geometries.[9] A fascinating sidelight is that the cis isomer of $Pt(NH_3)_2Cl_2$ (marketed under the name cisplatin) is an antitumor agent, while the trans isomer is inactive.

The possibilities for isomerism are affected by the presence of chelating ligands. The bidentate ethylenediamine molecule (en, $H_2N-CH_2-CH_2-NH_2$) can substitute for two NH_3 ligands to give $Pt(en)Cl_2$. However, only the cis isomer can be isolated, because en is not long enough for its nitrogens to occupy trans positions. This is generally the case regardless of geometry; that is, complexes in which bidentate ligands span non-cis coordination sites are rare.[10]

Although the tetrahedral geometry does not give rise to geometric isomers, the presence of four different ligands creates an asymmetric metal, thereby leading to optical isomerism. (Convince yourself that square planar Mabcd complexes have three geometric but no optical isomers.) Compared to carbon chemistry, there are few tetracoordinate metal complexes for which optically pure enantiomers have been isolated; the racemization of such species is usually rapid at or below room temperature.

It is possible to introduce stereoisomerism through the ligands themselves. One way to do this is via trivalent donor atoms having three different substituents, since they become asymmetric upon coordination to a metal. This is the case for ligands such as $PH(Me)\phi$ and the *N*-monomethyl derivative of ethylenediamine, $MeNH-CH_2-CH_2-NH_2$.

Trigonal Bipyramidal, Square Pyramidal, and Intermediate Geometries[11]

Five-coordinate complexes were once rather rare. This led to directed efforts toward their synthesis, and in recent years a large number of such species

9.   Werner, A. *Z. Anorg. Chem.* **1893**, *3*, 267. See also, Kauffman, G. B., Ed. *Classics in Coordination Chemistry*; Dover: New York, 1968.

10.   A specific example was reported by Kapoor, P. N. *J. Organomet. Chem.* **1986**, *315*, 383.

11.   Holmes, R. R. *Prog. Inorg. Chem.* **1984**, *32*, 119; Favas, M. C.; Kepert, D. C. *Prog. Inorg. Chem.* **1980**, *27*, 325; Hoskins, B. F.; Williams, F. D. *Coord. Chem. Rev.* **1972–1973**, *9*, 305; Wood, J. S. *Prog. Inorg. Chem.* **1972**, *16*, 227.

have been reported. This coordination number appears to produce less stable complexes than similar systems having four or six ligands. As a result, either of two types of decomposition may be observed. One (common in solution chemistry) is simple dissociation to give a square planar or tetrahedral product:

$$ML_5 \longrightarrow ML_4 + L \qquad\qquad (15.12)$$

The other mode of "decomposition" involves an increase in coordination number through oligomerization. This is especially common for complexes containing halide ligands, which can bridge two metal centers:

$$M \leftarrow L \rightarrow M$$

Thus, "$MoCl_5$" is a dimer in the solid state, with 6-coordinate molybdenums. The structure can be described as two octahedra sharing a common edge (ie, the metals share two bridging chlorides; see Figure 15.6a). Niobium(V) fluoride is a tetramer, again with 6-coordinate metals. The core of this structure is a square, eight-membered unit having a metal at each corner and a fluorine centering each side (Figure 15.6b).

For complexes that are truly 5-coordinate, there are again two limiting structures and an infinite number of intermediate cases. The limiting geometries are the trigonal bipyramid (TBP) and square pyramid (SPY), with maximum local symmetries of D_{3h} and C_{4v}, respectively. Main group atoms with ten valence electrons usually have TBP electron geometries. The situation is less clear-cut for metal complexes, however; the energy difference

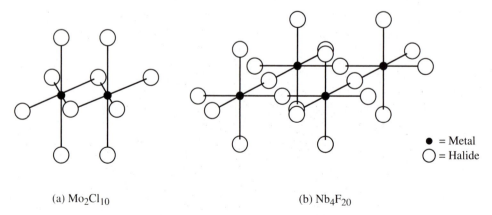

(a) Mo_2Cl_{10} (b) Nb_4F_{20}

● = Metal
○ = Halide

Figure 15.6 The structures of two halide-bridged oligomers. [Reproduced with permission from Wells, A. F. *Structural Inorganic Chemistry*, 5th ed.; Clarendon: Oxford, 1984; p. 427.]

between the TBP and SPY structures is often small. A classic example is the complex salt $2[Cr(en)_3][Ni(CN)_5] \cdot 3H_2O$. Its solid-state structure contains two different $[Ni(CN)_5]^{3-}$ anions, one having a slightly distorted TBP geometry, and the other, SPY geometry.[12]

Two isomers of the copper complex CuL_3^{2+} (where L is a derivative of imidazole) can be isolated, depending on the mode of purification. The green salt $[Cu(C_7H_{12}N_2S)_3](BF_4)_2$ is obtained when recrystallized from ethanol. The copper ion of this isomer is surrounded by three nitrogen and two sulfur ligands in a TBP array. When recrystallized by vapor diffusion into nitromethane, however, a blue isomer is obtained. The copper is again pentacoordinate, but the geometry is square pyramidal.[13]

The ideal SPY geometry is essentially never encountered. In square pyramidal complexes, the metal ion is normally raised out of the basal plane (typically, by 30–50 pm) toward the axial group:

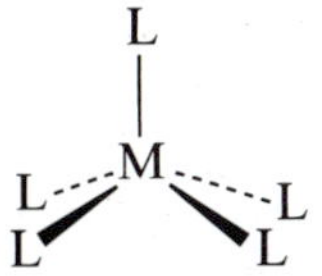

This changes the SPY bond angles to (calculated) optimum values of about 104° (axial–basal) and 87° (basal–basal) from the 90° ideal. Bond angles for the ideal TBP and SPY and for the optimized SPY geometries are summarized in Figure 15.7.

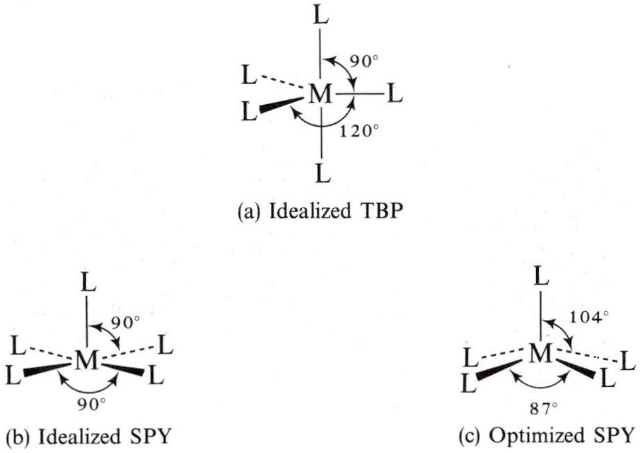

Figure 15.7
Shapes and bond angles for the idealized trigonal bipyramidal and idealized and optimized square pyramidal geometries (see text).

(a) Idealized TBP

(b) Idealized SPY

(c) Optimized SPY

12. Raymond, K. N.; Corfield, P. W. R.; Ibers, J. A. *Inorg. Chem.* **1968**, *7*, 1362.
13. Glass, R. S.; Sabahi, M.; Hajjatie, M.; Wilson, G. S. *Inorg. Chem.* **1987**, *26*, 2194.

Another way the TBP and SPY geometries become indistinct is through their interconversion. This is thought to occur either via the Berry pseudo-rotation (Figure 5.5) or a similar process. Such interconversion is sometimes facile, resulting in fluxional behavior in solution. For example, all members of the series $Fe(CO)_x(PF_3)_{5-x}$ have been prepared and exhibit a TBP geometry. Isomerism is possible when $x = 1-4$, since there are two non-equivalent ligand positions (axial and equatorial). In solution it is possible to detect all three geometric isomers of $Fe(CO)_3(PF_3)_2$ (and determine their relative concentrations) by infrared spectroscopy. However, no individual isomer can be separated from the others (at least at room temperature) because of their rapid interconversions.

$$
\begin{array}{ccc}
\underset{|}{\overset{PF_3}{\text{OC}-\text{Fe}}}\overset{\text{-CO}}{\underset{\text{CO}}{}} & \rightleftharpoons & \underset{|}{\overset{CO}{\text{OC}-\text{Fe}}}\overset{\text{-PF}_3}{\underset{\text{CO}}{}} & \rightleftharpoons & \underset{|}{\overset{CO}{\text{OC}-\text{Fe}}}\overset{\text{-PF}_3}{\underset{\text{PF}_3}{}}
\end{array} \quad (15.13)
$$

The axial and equatorial bond distances in trigonal bipyramidal complexes are of interest. For most cases studied to date, the equatorial bonds are longer than the axial linkages (Table 15.5). Recall from Chapter 5 that in main group compounds having a 5-coordinate central atom, the axial bonds are invariably the longer of the two types; for example, the distances in $PCl_5(g)$ are 202 pm (equatorial) and 214 pm (axial). This is consistent with the subhybridizations sp^2 (equatorial) and pd (axial), since

Table 15.5 Experimental distances for the axial and equatorial bonds in some trigonal bipyramidal complexes

Complex	Number of d Electrons	$d(M-L)$, ax	$d(M-L)$, eq
$[Fe(N_3)_5]^{2-}$	5	204	200
$[Co(C_6H_7NO)_5]^{2+}$	7	210	198
$[Mn(CO)_5]^-$	8	182	180
$Fe(CO)_5$	8	181	183
$[Co(NCCH_3)_5]^+$	8	184	188
$[Ni(CN)_5]^{3-}$	8	184 (av)	193 (av)
$[Pt(GeCl_3)_5]^{3-}$	8	240	243
$[Pt(SnCl_3)_5]^{3-}$	8	254	254
$[CuCl_5]^{3-}$	9	230	239
$[CuBr_5]^{3-}$	9	245	252
$[CdCl_5]^{3-}$	10	253	256

Note: Bond distances in picometers; av = average values (distorted geometry).

the more diffuse $3d$ orbitals cause the axial hybrids (50% d character) to extend further from the nucleus. The same overall sp^3d hybridization applies to TBP metal complexes. However, the bonding orbitals are from the ns, np, and $(n-1)d$ subshells. Being from an inner shell, the hybridized d_{z^2} orbital lies closer to the nucleus than do the others, causing the axial bonds to be relatively short.

As we observed earlier for 4-coordinate species, the presence of certain kinds of ligands can dictate whether a pentacoordinate complex has a TBP, a SPY, or an intermediate geometry. *Tren* [2,2′,2″-triaminotriethylamine, $N(CH_2CH_2NH_2)_3$] is a well-known chelate with a shape ideally suited to TBP geometry. (It is described as a *tripod* ligand for obvious reasons; see Figure 15.8a.) In complexes having the formula [M(tren)Br]$^+$ (M = Co, Ni, and Cu), the four nitrogens occupy one of the axial sites and the three equatorial sites of a trigonal bipyramid. Another example is *tap* [tris(3-dimethylarsinopropyl)phosphine]. In [Ni(tap)CN]$^+$, the arsenic donors lie in the equatorial plane, with the phosphorus and cyano groups in the axial positions (Figure 15.8b).

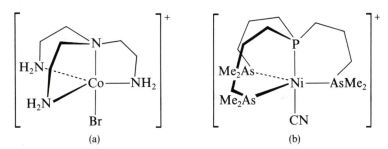

Figure 15.8 Two trigonal bipyramidal species containing tripod ligands: (a) the [Co(tren)Br]$^+$ cation; (b) the [Ni(tap)CN]$^+$ cation; tren = 2,2′,2″-triaminotriethylamine; tap = tris(3-dimethylarsinopropyl)phosphine.

Octahedral and Distorted Octahedral Complexes[14]

Among the metal complexes reported to date, the largest percentage contain 6-coordinate metal atoms or ions. All are either octahedral or distorted in some manner from that geometry. We will first consider stereoisomerism in true octahedral complexes, and then examine the three most common types of distortion.

14. Kepert, D. L. *Prog. Inorg. Chem.* **1977**, *23*, 1; Serpone, N.; Bickley, D. G. *Prog. Inorg. Chem.* **1972**, *17*, 391.

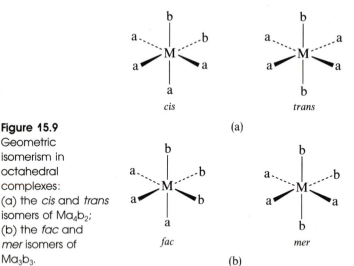

Figure 15.9
Geometric
isomerism in
octahedral
complexes:
(a) the *cis* and *trans*
isomers of Ma_4b_2;
(b) the *fac* and
mer isomers of
Ma_3b_3.

As shown in Figure 15.9, geometric isomerism is possible for species conforming to the formulas Ma_4b_2 and Ma_3b_3. The familiar designations *cis* and *trans* are used in the former case, depending on whether the b ligands are adjacent to or opposite one another (said another way, depending on whether the b–M–b bond angle is 90° or 180°). The Ma_3b_3 isomers are named *fac* (for *facial*) and *mer* (for *meridional*). In the facial isomer, each set of three common ligands forms a triangular face of the octahedron, and the a–M–a and b–M–b bond angles are all 90°. The meridional structure has 180° a–M–a and b–M–b linkages.

The possibilities for geometric isomerism increase dramatically as the number of different kinds of ligands increases. For octahedral $Ma_2b_2c_2$ systems, 5 geometric isomers (and a total of 6 stereoisomers, including an enantiomeric pair) are possible. For the totally unsymmetric (and rarely encountered!) Mabcdef case, 30 stereoisomers (15 geometric, each with two enantiomers) are possible.

Optical isomerism is especially common for octahedral complexes having chelating ligands. Examples include the hexadentate EDTA ligand and tris(ethylenediamine) complexes. Thus, $[Co(en)_3]^{3+}$ belongs to the D_3 point group; it has neither a mirror plane nor an inversion center, and so is optically active.[15] The two enantiomers are shown in Figure 15.10; the Greek

---

15. Actually, optical activity results from the lack of an improper axis of rotation. Since $S_1 = \sigma$ and $S_2 = i$, the mirror plane–inversion center guide identifies nearly all such cases, and so is often used in place of the more rigorous test. See Cotton, F. A. *Chemical Applications of Group Theory*, 3rd ed.; Wiley: New York, 1990; Chapter 3.

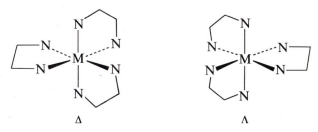

Figure 15.10   Enantiomers of a tris(ethylenediamine) complex. The isomers are assigned the Δ (right-handed) and Λ (left-handed) labels, as shown.

letters Δ and Λ are used to designate right- and left-handed spirals looking down the threefold axis, respectively.

Werner's resolution of *cis*-$[Co(en)_2(NH_3)Cl]^{2+}$ into its enantiomers in 1911 was an important step in the history of coordination chemistry. Until that time most of the known structural chemistry involved organic compounds, and there was doubt about whether coordination numbers greater than 4 could lead to stable species.[16]

A compilation of the possible geometric and optical isomers for selected categories of octahedral complexes is given in Table 15.6 (p. 500).

There are three common types of distortion from octahedral geometry. One results from metal–ligand bond shortening or lengthening to two opposite ligands. This is *tetragonal* (or *Jahn–Teller*) distortion (Figure 15.11a), and destroys the threefold axes of the original octahedron. Thus, the ideal point group symmetry changes from O_h to D_{4h} (or to C_{4v} if the change

Figure 15.11
The three common types of distortion from octahedral geometry.

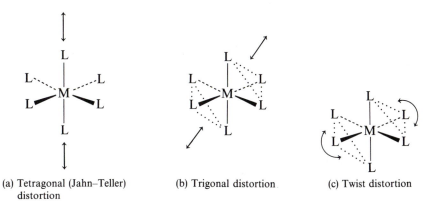

(a) Tetragonal (Jahn–Teller) distortion

(b) Trigonal distortion

(c) Twist distortion

16.   Werner, A.; King, V. L. *Chem. Ber.* **1911**, *44*, 1887; also, Kauffman, G. B. *Coord. Chem. Rev.* **1973**, *11*, 161.

Table 15.6 Possible stereoisomers in selected types of octahedral complexes

Type	Geometric	Optical Pairs	Total
Complexes having only monodentate ligands:			
Ma_5b	1	0	1
Ma_4b_2	2	0	2
Ma_3b_3	2	0	2
Ma_4bc	2	0	2
Ma_3b_2c	3	0	3
Ma_3bcd	4	1	5
$Ma_2b_2c_2$	5	1	6
Ma_2b_2cd	6	2	8
Mabcdef	15	15	30
Complexes containing chelate ligands:			
$M(aa)_3$	1	1	2
$M(aa)_2b_2$	2	0	2
$M(aa)_2bc$	2	1	3
$M(ab)_3$	2	2	4
$M(ab)_2cd$	6	5	11
$M(abc)_2$	6	5	11

is unequal on the two sides). As we will explain in the next chapter, such distortion improves the overall quality of the metal–ligand bonding (and thereby increases the stability) of certain complexes. Among the $3d$ metals, Cu^{II} complexes are particularly prone to tetragonal distortion. For example, four of the Cu–N bonds of $[Cu(NH_3)_6]^{2+}$ have lengths of 207 pm, while the other two distances are 262 pm (a difference of over 26%).

A second type of distortion involves bond lengthening or shortening to three facial ligands (see Figure 15.11b). This *trigonal* distortion destroys the C_4 axes of the octahedron, reducing the idealized symmetry to D_{3d}. The result is a *trigonal antiprism*. Such structures are rare for metal complexes, but are found in certain ionic lattices. One polymorph of CdBrI consists of alternating layers of close-packed bromide and iodide ions. The cadmiums occupy the octahedral interstices, and are therefore coordinated to three Br^- "ligands" on one side and to three iodides on the other. The Cd–Br and Cd–I distances are unequal, of course, and D_{3d} site symmetry results.

A third mode of distortion involves the twisting of a triangle of three ligands (ie, one face of the octahedron). In a regular octahedron these triangles are skewed by 60° with respect to one another (Figure 15.11c). Any twist angle is possible between the limiting cases of 60° and 0°, with the latter corresponding to the eclipsed (trigonal prismatic) structure.[17] Like the

---

17.   For a detailed comparison of trigonal prismatic and octahedral coordination, see Wentworth, R. A. D. *Coord. Chem. Rev.* **1972–1973**, *9*, 171.

trigonal antiprism, this geometry is more common in extended lattices than in discrete complexes. It is found, for example, in structures in which the metal ion is sufficiently large to separate adjacent layers of the anion sublattice, thereby reducing the repulsive interactions. Such is the case for MoS_2. There are also complexes having twist angles between $0°$ and $60°$, and in particular tris(dithiolene) complexes have been thoroughly studied. The twist angle in $V(S_2C_2\phi_2)_3$ is about $8.5°$, while in $Mo(S_2C_2(CN)_2)_3$ it is $27°$. For $Mo(S_2C_2H_2)_3$ it is $0°$ (ie, a regular trigonal prism).

High-Coordinate Geometries[18]

Complexes in which a metal coordinates to seven or more ligands are less common than hexacoordinate species, but can no longer be considered rare. As a general rule, structural diversity increases with increasing coordination. There are three well-established geometries for 7-coordinate complexes: the pentagonal bipyramid (idealized D_{5h} symmetry), capped octahedron (C_{3v}), and capped trigonal prism (C_{2v}).[19] These are shown in Figure 15.12.

At least two of these geometries are likely to lie close to one another in energy for any given complex, and theory suggests (and experiments verify) that they are readily interconvertible. Such conversions can be achieved by "stretching" the edges of the polyhedron—that is, by deformations of the L–M–L bond angles. Fluxional behavior is therefore common. For example, $W(PMe_3)F_6$ undergoes rapid intramolecular interconversion in solution at temperatures as low as $-85°C$.

Heptacoordination sometimes arises from the oligomerization of lower-coordinate species. The ZrF_6^{2-} anion forms a dimer in which each metal lies at the center of a pentagonal bipyramid; the two units share a common edge. An even more interesting case is that of UBr_4, which is a linear polymer in the solid state. Its structure is based on fused pentagons of bromines,

Figure 15.12
The three common 7-coordinate geometries.

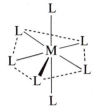

(a) Pentagonal bipyramid

(b) Capped octahedron

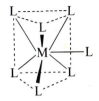

(c) Capped trigonal prism

18.   Favas, M. C.; Kepert, D. L. *Prog. Inorg. Chem.* **1979**, *25*, 41; Kepert, D. L. *Prog. Inorg. Chem.* **1978**, *24*, 179; Drew, M. G. B. *Prog. Inorg. Chem.* **1977**, *23*, 67.

19.   An atom can "cap" a trigonal prism in two ways: It can center either a triangular or a square face. The latter is the normal mode, and the one that yields C_{2v} symmetry.

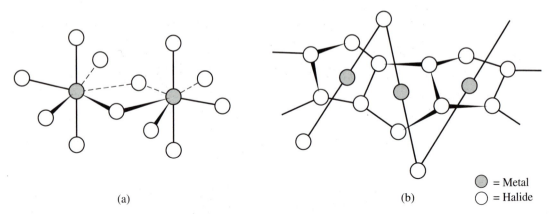

= Metal

= Halide

(a)   (b)

Figure 15.13 The solid-state structures of two halide-bridged oligomers: (a) the $[ZrF_6^{2-}]_2$ anion in $K_2Cu(ZrF_6)_2 \cdot 6H_2O$; (b) the chain polymeric structure of UBr_4.

which provide the five equatorial ligands about a given uranium; this requires three bromines per metal. The remaining ligands lie in bridging positions between the polymer chains and occupy the axial positions, completing the pentagonal bipyramids (Figure 15.13).

The isolation of 7-coordinate complexes is sometimes facilitated by the incorporation of polydentate ligands. For example, a heptadentate cryptand was used to prepare the salt $[Co(crypt)]^{2+}[Co(SCN)_4]^{2-}$, and several pentagonal bipyramidal complexes containing the planar, pentadentate ligand $pal_{2,2,2}$ are known (Figure 15.14).[20]

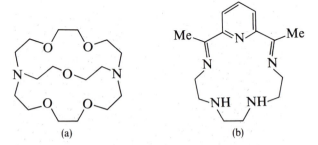

(a)   (b)

Figure 15.14 Two polydentate ligands that stabilize high-coordinate species: (a) the cryptand ligand of the complex salt $[Co(crypt)]^{2+}[Co(SCN)_4]^{2-}$; (b) the pentadentate ligand $pal_{2,2,2}$.

20. Drew, M. G. B.; bin Othman, A. H.; McIlroy, P. D. A.; Nelson, S. M. *J. Chem. Soc. Dalton* **1975**, 2507; Mathieu, F.; Weiss, R. *J. Chem. Soc. Chem. Commun.* **1973**, 816.

Figure 15.15
The structures and modes of interconversion of the common 8-coordinate geometries.

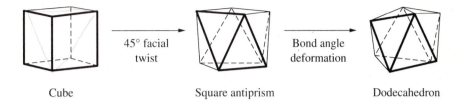

Cube   45° facial twist   Square antiprism   Bond angle deformation   Dodecahedron

The best-known 8-coordinate geometries are the cube (like the trigonal prism and antiprism, common in extended lattices but rare in discrete complexes), the square antiprism, and the dodecahedron (see Figure 15.15). The relationship between the square antiprism and the cube is the same as between the octahedron and the trigonal prism—twisting one face (for the 8-coordinate case, by 45°) generates the alternative structure. As might be expected, the energy differences among these geometries are small, inter-conversions are usually facile, fluxionality is common, and the separation of isomers is often impossible. Most of the known 8-coordinate complexes have large metal ions ($4d$, $5d$, lanthanide, or actinide elements) and/or either small or macrocyclic ligands.

Oxyanions such as NO_2^-, NO_3^-, IO_3^-, and SO_4^{2-} are often found in octacoordinate complexes. These ions act as bidentate ligands, and stabilize high coordination through what might be called "bond angle compatibility." As coordination number increases, the average L–M–L bond angle necessarily decreases. Chelate linkages such as

$$-X\overset{\displaystyle O}{\underset{\displaystyle O}{\diamond}}M$$

produce angles of 90° or less. For example, $Ti(NO_3)_4$ has dodecahedral geometry about the metal, with O–M–O angles of about 62°.

Species having more than eight ligands surrounding a metal are comparatively rare. The best-known geometries are the tricapped trigonal prism (9-coordinate), bicapped square antiprism (10), and icosahedron (12). These structures are important in molecular clusters, and will be examined from that perspective in Chapter 19.

15.5 Structural Isomerism

The term *structural isomerism* is loosely applied to include all kinds of isomerism other than geometric or optical.[21] The types of structural isomerism most common to transition metal complexes are described in this section.

21. Kauffman, G. B. *Coord. Chem. Rev.* **1973**, *11*, 161.

Coordination Isomerism

Several kinds of *coordination isomerism* were described by Werner. A common type arises when a complex cation and complex anion form a salt. Thus, $[Co(NH_3)_6]^{3+}[Cr(CN)_6]^{3-}$ and $[Cr(NH_3)_6]^{3+}[Co(CN)_6]^{3-}$ both have the molecular formula $CoCrC_6H_{18}N_{12}$. Other salts having the same formula are possible as well—for example, $[Co(NH_3)_5(CN)]^{2+}[Cr(NH_3)(CN)_5]^{2-}$. A variation has only one kind of metal, but in two different oxidation states. Two compounds that are analyzed as $Pt_2H_{12}Cl_6N_4$ are the salts $[Pt(NH_3)_4]^{2+}[PtCl_6]^{2-}$ and $[Pt(NH_3)_4Cl_2]^{2+}[PtCl_4]^{2-}$. The former contains Pt^{II} in the cation and Pt^{IV} in the anion, while the reverse is true in the latter. (As we will explain in the next chapter, Pt^{II} prefers square planar coordination, while Pt^{IV} prefers octahedral coordination.)

Coordination polymerization isomerism (or, simply, *polymerization isomerism*) is not true isomerism in that it involves species having different molecular weights. Such "isomers" have identical empirical but different molecular formulas. Consider the empirical formula $CoH_9N_6O_6$. One possibility is the neutral, octahedral complex $Co(NH_3)_3(NO_2)_3$ (with *fac* and *mer* geometric isomers).[22] The same empirical formula results from the complex salt $[Co(NH_3)_6]^{3+}[Co(NO_2)_6]^{3-}$; hence, these two species are polymerization isomers. A third isomer is $[Co(NH_3)_4(NO_2)_2]^+[Co(NH_3)_2(NO_2)_4]^-$, and $[Co(NH_3)_5NO_2]^{2+}\{[Co(NH_3)_2(NO_2)_4]^-\}_2$, with a molecular weight three times that indicated by the empirical formula, is yet another.

Coordination position isomerism can arise for species having two or more metals connected by bridging ligands. An example is the pair of cations

$$\left[(H_3N)_3(Cl)Co\diagup\!\!\!\!\!\!\!\!\overset{OH}{\underset{OH}{}}\!\!\!\!\!\!\!\diagdown Co(NH_3)_3Cl\right]^{2+}$$

and

$$\left[(H_3N)_4Co\diagup\!\!\!\!\!\!\!\!\overset{OH}{\underset{OH}{}}\!\!\!\!\!\!\!\diagdown Co(NH_3)_2Cl_2\right]^{2+}$$

The two cobalts can[23] occupy symmetrically equivalent positions in the first structure, but not in the second.

Ionization and Hydrate Isomerism

Ionization isomerism occurs for certain salts containing complex cations. Consider the formula $Pt(NH_3)_3(Br)(NO_2)$. (Recall that Pt^{II} favors 4-coordinate complexes.) Three of the four donor sites must be occupied by

---

22. Linkage isomerism (see below) is also possible.
23. But must they? Consider geometric isomerism.

ammonia molecules, with either Br^- or NO_2^- as the fourth ligand. Both possibilities produce a complex having a $1+$ charge; the second anion serves as the counterion rather than as a ligand. Thus, two ionization isomers are known for this formula: $[(H_3N)_3PtBr]^+NO_2^-$ and $[(H_3N)_3PtNO_2]^+Br^-$. A second example involves the two Pt^{IV} salts $[Pt(NH_3)_4Cl_2]Br_2$ and $[Pt(NH_3)_4Br_2]Cl_2$.[24] Any salt of a complex cation containing two or more different anionic groups is susceptible, at least in theory, to this type of isomerism.

A related case involves water. Complexes crystallized from aqueous solution can incorporate H_2O molecules in at least two ways: as a ligand or through hydrogen bonding to one or more ligands. *Hydrate isomerism* can result. The $Cr^{III} + Cl^- + H_2O$ system includes at least three compounds having the molecular formula $CrH_{12}O_6Cl_3$. One is the violet $[Cr(H_2O)_6]Cl_3$, in which all six waters are coordinated to the metal. In a second (blue-green) isomer, one chloride ion acts as a ligand and the other two serve as counterions; a water molecule is hydrogen bonded to Cl^-, so the formulation is $[Cr(H_2O)_5Cl]Cl_2 \cdot H_2O$. A third isomer (green in color) is *trans*-$[Cr(H_2O)_4Cl_2]Cl \cdot 2H_2O$, with two chloride ligands and two waters of hydration.

Linkage Isomerism

Recall from Chapter 10 that ambidentate ligands can act as Lewis bases through two or more kinds of donor atoms. The potential for isomerism exists in any complex containing ambidentate ligands, and many examples of *linkage isomerism* are known. The earliest and perhaps most thoroughly studied examples contain the NO_2^- ion, as in the pair $[(H_3N)_5CoNO_2]^{2+}$ and $[(H_3N)_5CoONO]^{2+}$. The former is the nitro and the latter the nitrito isomer. (An alternate mode of nomenclature uses the names nitro-*N* and nitro-*O*; see Appendix II.) Whether NO_2^- binds to a metal through nitrogen or oxygen depends on several factors, including the relative hardness or softness of the metal and steric and statistical effects.

Two other ambidentate ligands that have received considerable attention are SCN^- and $SeCN^-$. In these cases, the hard–soft factor appears to be the major determinant of structure; hard acids such as Fe^{III} usually bind through nitrogen, while Hg^{II} and other soft acids prefer the Group 16 donor. The intermediate Cd^{II} binds through N, S, or Se, but appears to prefer nitrogen. This is evidenced by the structure of the bridged complex anion $[Cd_2(SeCN)_6]^{2-}$:[25]

24. Geometric isomerism is possible as well.
25. Norbury, A. H. *Adv. Inorg. Chem. Radiochem.* **1975**, *17*, 232.

Other ligands known to form linkage isomers include SO_3^{2-}, $S_2O_3^{2-}$, urea, thiourea, and dimethylsulfoxide.

The fact that linkage isomerism is possible for a given system does not necessarily mean that more than one isomer actually can be isolated. One isomer may have such high stability relative to the other(s) that only that form can be obtained. For example, carbon monoxide theoretically might interact with metals through either the carbon or oxygen atom; yet, when unidentate (ie, not in a bridging position between two or more metals), the donor atom is invariably found to be carbon. This can be rationalized through either Lewis or MO theory (see Chapter 16).

Bibliography

Basolo, F.; Johnson, R. *Coordination Chemistry*; Science Reviews: Northwood, 1987.

Wulfsberg, G. *Principles of Descriptive Inorganic Chemistry*; Brooks/Cole: Pacific Grove, CA, 1987.

Parish, R. V. *The Metallic Elements*; Longman: New York, 1977.

Cotton, S. A.; Hart, F. A. *The Heavy Transition Elements*; Wiley: New York, 1975.

Fergusson, J. E. *Stereochemistry and Bonding in Inorganic Chemistry*; Prentice-Hall: Englewood Cliffs, NJ, 1974.

Hartley, F. R. *The Chemistry of Palladium and Platinum*; Wiley: New York, 1973.

Kepert, D. L. *The Early Transition Metals*; Academic: New York, 1972.

Fackler, J. P. *Symmetry in Coordination Chemistry*; Academic: New York, 1971.

Orgel, L. E. *An Introduction to Transition Metal Chemistry: Ligand Field Theory*, 2nd ed.; Wiley: New York, 1966.

Questions and Problems

1. The most common organization of the periodic table sets the 14 elements between Ce and Lu apart as the $4f$, or lanthanide, elements. Note that lanthanum is not included in this group.
   (a) On the basis of quantum theory, what do you consider to be the $4f$ elements?
   (b) What reason(s) can you give for organizing that section of the periodic table as it is?

2. In footnote 2 we stated that transition elements differ from main group elements because of the effects of d or f orbitals. Explain the following observations with this in mind.
   (a) Most transition metals have two or more common positive oxidation states.
   (b) Low-energy electron transitions are common for transition metals and their compounds.

3. The first ionization energy of zinc is greater than that of copper, as expected based on normal periodic trends. However, the IE_2 of copper is greater than that of zinc. Why is this the case? How does it lead to the observation of different oxidation states for the two elements?

4. Table 15.2 gives the relative stabilities of metal oxidation states in aqueous solution. How might these values differ for NH_3 solutions? Specifically, would high oxidation states be stabilized or destabilized relative to aqueous media? Defend your answer.

5. The compound FeI_3 has very low stability (see problem 25 in Chapter 9), so the complex ion FeI_4^- is better characterized.
   (a) With the aid of aqueous reduction potentials, write the equation for the decomposition of FeI_3.
   (b) Speculate on why FeI_4^- has greater stability than FeI_3.

6. The transition elements given in this chapter as occurring as native metals are silver, gold, and platinum. We stated that this is due to their very negative oxidation potentials.
   (a) Provide a deeper explanation. What underlying factors cause the oxidation potentials of these elements to be so negative? (A review of appropriate sections of Chapter 9 may be useful.)
   (b) Actually, several other metals are sometimes found in native deposits. Considering your answer to part (a), suggest probable candidates.

7. Consider the Kroll method for the production of titanium metal outlined in equations (15.1) and (15.2).
   (a) Do you expect Ti^{IV} to bond preferentially to oxygen or to chlorine? Why?
   (b) Given your answer to part (a), what factor(s) cause equation (15.1) to occur as written?
   (c) The reaction described by equation (15.2) is exothermic. Equilibrium constants for exothermic reactions decrease with increasing temperature. Why, then, is a high temperature used in that reaction?

8. In main group chemistry, species of the XY_3 type exhibit any of three geometries, corresponding to the idealized point groups D_{3h}, C_{3v}, and C_{2v}.
   (a) Give a specific example for each case.
   (b) What factor determines the choice of geometry?
   (c) Speculate on additional factors that might influence the geometries of transition metal complexes having the general formula ML_3.

9. Sketch all possible stereoisomers for square planar species having each of the following general formulas:
   (a) M(aa)bc, where aa is a symmetrical bidentate ligand such as ethylene-diamine
   (b) M(ab)cd, where ab is an asymmetrical bidentate ligand such as $H_2P-CH_2-CH_2-NH_2$
   (c) M(ab)$_2$

10. The anion $[Ni_2Cl_8]^{2-}$ belongs to the C_{2h} point group. Each nickel has a square pyramidal arrangement of ligands, and there are no Ni–Ni bonds. Sketch the structure.

11. Compare the number of geometric isomers expected for octahedral and trigonal prismatic complexes having each of the following formulas:
    (a) Ma_4b_2 (b) Ma_3b_3

12. Sketch the six possible stereoisomers of the octahedral complex $Ma_2b_2c_2$.

13. Sketch all possible geometric isomers for the following octahedral complexes:
    (a) $[Co(NH_3)_3(H_2O)(Br)(Cl)]^+$ (b) $[Co(en)_2(Br)(Cl)]^+$
    (c) $Co(OCH_2CH_2NH_2)_3$

14. Indicate which of the species in problem 13 have optical isomers.

15. It has been reported that solutions of $VCl_3(NCMe)_3$ give two ^1H NMR signals at room temperature, but only one above 50°C. Suggest an explanation.

16. Give the number of stereoisomers expected for each of the three 7-coordinate geometries discussed in this chapter for complexes having the general formula Ma_6b.

17. There are four possible stereoisomers for Ma_5b_2 complexes having pentagonal bipyramidal geometry about the metal. Sketch each.

18. The $[Zr_2F_{13}]^{5-}$ ion can be described as two monocapped trigonal prisms sharing the capped vertex. Sketch this anion and assign a point group.

19. (a) Sketch two isomers having the molecular formula $Co(en)_3Cl_3$. (Co^{III} is 6-coordinate!)
    (b) Identify four isomers having the molecular formula $Co_2(en)_3Cl_6$. Classify pairs of isomers according to type.

20. Consider the empirical formula $Pt(NH_3)_2Cl_3$. Recall that platinum has two common oxidation states: Pt^{II} (square planar geometry) and Pt^{IV} (octahedral). At least six isomers that correspond to the given empirical formula and have a molecular weight of about 670 g/mol are possible. Describe at least four.

21. Sketch four structures that correspond to the formula $Cr(NH_3)_4(Cl)_3(H_2O)_2$. Each isomer should contain an octahedral cation.

22. Use the empirical formula $CoPt(NH_3)_5F_7$ to generate at least five coordination isomers.

23. Use the empirical formula $Co(H_2O)_4(SCN)_2Cl$ to illustrate each of the following types of isomerism:
    (a) Geometric   (b) Ionization   (c) Hydrate   (d) Linkage

24. The $[Cu(NO_2)_5]^{3-}$ ion has approximate pentagonal bipyramidal geometry (read carefully!) about the copper. There are three Cu–N and four Cu–O bonds in this anion, which belongs to the C_{2v} point group. Make a sketch consistent with the above description.

25. The nitrito isomer of $[Co(NH_3)_5ONO]^{2+}$ is converted to the nitro isomer upon heating. An ^{18}O-labeling study showed this rearrangement to be intramolecular. Speculate on the structure of the intermediate.

*26. The complex $Zn[Si(SiMe_3)_3]_2$ has been discussed by J. Arnold, T. D. Tilley, A. L. Rheingold, and S. J. Geib (*Inorg. Chem.* **1987**, *26*, 2106). Answer these questions after reading their paper.

(a) How was the complex synthesized? What experimental method was used to determine its structure?

(b) Nearly all zinc complexes are tetrahedral. Why is this complex stable, given its coordination number of 2?

*27. F. A. Cotton and R. L. Luck have studied two isomers with the formula $[Mo_2(PMe_3)_2Cl_7]^-$ (*Inorg. Chem.* **1989**, *28*, 182).

(a) How were these species synthesized? What is the structure of the molybdenum-containing reagent?

(b) Sketch the two isomers. Are they geometric, optical, or structural?

(c) Describe the geometry about the molybdenum. What phrase applies to the ligand positions?

*28. A study of several compounds having the general formula $Fe(PR_3)_nCl_3$ ($n = 1$, 2, or 3) was carried out by J. D. Walker and R. Poli (*Inorg. Chem.* **1989**, *28*, 1793). After reading their report, describe the structures of the various species observed. What are the relative stabilities of these species?

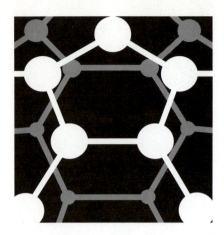

16

Bonding Models for Transition Metal Complexes

Having surveyed the most important structure types for coordination compounds and complex ions, we will now undertake the task of explaining the bonding in such species. Beginning about 1930, two very different approaches were developed. One, crystal field theory, is based on electrostatics. The other, molecular orbital theory, derives from the covalent model. We now realize that these seemingly diverse theories are related. The bonding interactions in metal complexes are neither strictly electrostatic nor totally covalent in nature. As you will see, the MO approach is the more powerful (but also the more complex to use) of the two, since it incorporates the major features of crystal field theory.

16.1 The Crystal and Ligand Field Theories

As originally developed, *crystal field theory* (CFT) was concerned with the electrostatic interactions about a given ion in a solid-state ionic lattice.[1] This

---

1. Bethe, H. *Ann. Phys.* **1929**, *3*, 133; Van Vleck, J. H. *Phys. Rev.* **1932**, *41*, 208; *J. Chem. Phys.* **1935**, *3*, 803, 807.

model was later applied to coordination compounds and complex ions—for example, for evaluating the repulsions that arise when the electron(s) in a d subshell of a metal cation are surrounded by ligand electron pairs in a specific geometric array.

Consider an isolated transition metal ion that has one d electron. The five d orbitals are equivalent in energy, so the electron is equally likely to reside in any of the five. Next, visualize an "invasion" of this system by two negative charges. The electron experiences electrostatic repulsion from these charges. If the charge distribution is spherically symmetric (as in the form of a diffuse, spherical electron cloud), the five d orbitals are destabilized by equal amounts and thus remain degenerate. For any distribution other than spherical, however, the repulsion is different for different orbitals; that is, the d orbital equivalence is destroyed. For example, two point charges on opposite sides of the metal would occupy positions along the z axis according to group theory convention. In this situation the orbitals having lobal density along z (d_{z^2} and, to a lesser extent, d_{xz} and d_{yz}) are destabilized relative to the others. The resulting energy-level diagram is shown in Figure 16.1.

The d^1 system therefore has a doubly degenerate ground state in a linear crystal field ($D_{\infty h}$ point group), with the symmetry label Δ_g assigned to the $d_{xy}/d_{x^2-y^2}$ level.[2] The excited states (corresponding to occupancy of the Π_g and Σ_g^+ levels, respectively) are accessible through $d \rightarrow d$ electron transitions of relatively low energy.

Octahedral Complexes

Recall that the single most common geometry observed for transition metal complexes is octahedral. Thus, we will now consider the effect of an octahedral crystal field on the d subshell of a metal cation.

Figure 16.1
Qualitative crystal field theory diagram for linear ML$_2$ complexes. The group theory symbols for the relevant irreducible representations ($D_{\infty h}$ point group) are indicated.

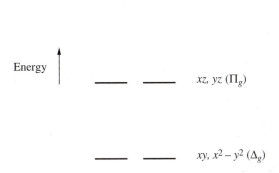

$$z^2 \ (\Sigma_g^+)$$

Energy

$$xz, yz \ (\Pi_g)$$

$$xy, x^2 - y^2 \ (\Delta_g)$$

2. The use of capital and lowercase letters in this chapter (Δ_g versus δ_g, T_{2g} versus t_{2g}, etc.) follows convention. Recall from Chapters 2 and 3 that symmetry labels for irreducible representations include capital letters (Greek for groups of infinite order, roman for all others), while molecular orbitals are identified by lowercase symbols.

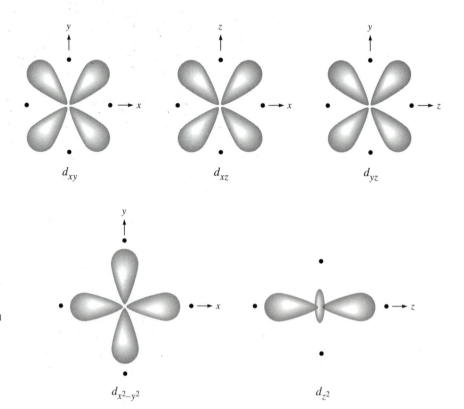

Figure 16.2
The orientations of a set of five d orbitals in an octahedral crystal field;
● = point charge (ligand).

In forming a regular octahedron, the six point charges (ligands) identify the x, y, and z axes of a Cartesian coordinate system; that is, they can be assigned the positions $(x, 0, 0)$, $(-x, 0, 0)$, $(0, y, 0)$, $(0, -y, 0)$, $(0, 0, z)$, and $(0, 0, -z)$. Therefore, the metal d orbitals having lobal density directed along any of the three axes are destabilized. As can be seen from Figure 16.2, $d_{x^2-y^2}$ and d_{z^2} are oriented toward the ligand point charges; the others have lobes directed between, rather than along, the axes.

The resulting energy-level diagram (Figure 16.3) shows two levels. The d_{xy}, d_{xz}, and d_{yz} orbitals (which belong to the T_{2g} representation in the O_h point group; see the character table in Appendix IV) lie at relatively low energy, while $d_{x^2-y^2}$ and d_{z^2} (E_g symmetry) are equivalent to one another and lie at higher energy.[3]

By convention, the energy gap between the two levels is symbolized by either Δ_o or $10\,Dq$ [equation (16.1)]. In fact, the two conditions below are

---

3. It is not apparent from Figure 16.2 that d_{z^2} and $d_{x^2-y^2}$ are equivalent because d_{z^2} is a linear combination of two orbitals (see Section 1.3).

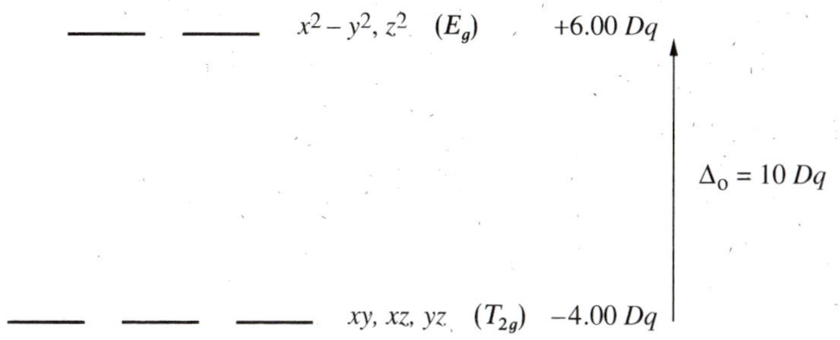

$$x^2 - y^2, z^2 \quad (E_g) \qquad +6.00 \, Dq$$

Figure 16.3
The crystal field theory diagram for octahedral complexes.

$$\Delta_o = 10 \, Dq$$

$$xy, xz, yz \quad (T_{2g}) \quad -4.00 \, Dq$$

generally set:

$$E(e_g) - E(t_{2g}) = \Delta_o = 10 \, Dq \tag{16.1}$$

$$\sum E(d) = 0 \, Dq \tag{16.2}$$

These two equations contain two unknowns, and so can be uniquely solved to give energies of $-4 \, Dq$ for the t_{2g} level and $+6 \, Dq$ for e_g. Note that $0 \, Dq$ is both the sum and the average of the five orbital energies; it is called the *baricenter*. For an octahedral complex having one d electron, the ground state is stabilized by $4 \, Dq$ with respect to the baricenter; hence, $4 \, Dq$ is the *crystal field stabilization energy* (CFSE).[4] Systems with two d electrons have the ground-state configuration $(t_{2g})^2$ and a CFSE of $8 \, Dq$; for d^3 the configuration is $(t_{2g})^3$ and CFSE $= 12 \, Dq$. In general, the CFSE is taken to be the absolute value of the sum of all the d electron energies; that is, CFSE's are always positive in sign.

The Dq unit is flexible, with its magnitude changing from complex to complex. Thus, $1 \, Dq$ has different values in $CrCl_6^{3-}$ and $MoCl_6^{3-}$ (about 16 kJ/mol in the former and 23 kJ/mol in the latter). This is a consequence of how Dq is defined [see equation (16.1)], as well as the theoretical relationship[5]

$$Dq = \frac{Ze^2 \langle r \rangle^4}{6d^5} \tag{16.3}$$

Here, Z is the charge on the metal ion, e is the charge of an electron, $\langle r \rangle$ is an integral related to the mean radius of the orbital in question, and d is the metal–ligand distance. This equation indicates that Dq increases with increasing metal charge and with closeness of approach by the ligands, which

---

4. Or *ligand field stabilization energy* (LFSE).

5. For a discussion of this equation, see Douglas, B. E.; Hollingsworth, C. A. *Symmetry in Bonding and Spectra*; Academic: Orlando, FL, 1985; p. 223; and/or Burdett, J. K. *Molecular Shapes*; Wiley: New York, 1980; pp. 128ff.

is intuitively what we would expect. This is consistent with the following two series, in which Dq decreases from left to right:

$$[Cr(NH_3)_6]^{3+} > [Cr(NH_3)_6]^{2+}$$

$$CrF_6^{3-} > CrCl_6^{3-} > CrBr_6^{3-}$$

However, equation (16.3) can be applied only qualitatively (see below).

Strong–Weak Field and High–Low Spin Complexes

An interesting situation arises for octahedral complexes having four d electrons. The fourth electron might enter one of the previously occupied t_{2g} orbitals. This would maximize the CFSE (16 Dq). However, there would be a cost in pairing energy (Chapter 1); hence, the stabilization energy for this system would be given as 16 $Dq - P$ (where P = pairing energy). An alternative configuration is $(t_{2g})^3(e_g)^1$, for which the CFSE is only 6 Dq, but there is no pairing energy requirement. Clearly, the latter option is favored if $P > 10 Dq$. These two factors are rather closely balanced for octahedral complexes of the first transition series, and examples of both $(t_{2g})^4$ and $(t_{2g})^3(e_g)^1$ ground states are known. Theoretical pairing energies have been calculated, with typical values for divalent cations of the 3d elements being 230–280 kJ/mol.[6] The t_{2g}/e_g energy separation is a function of the strength of the metal–ligand interaction and shows large variation depending on the complex. Cases having 10 $Dq > P$ result from large crystal field splittings and are described as *strong field* complexes. The alternative is the *weak field* case.

The number of unpaired electrons (and hence, the spin quantum number of the system) is always maximized by the weak field alignment. As a result, two other terms are often used. *Low spin* and strong field complexes are synonymous, and *high spin* is functionally equivalent to weak field.

Strong and weak field possibilities exist for octahedral systems having between four and seven d electrons. For d^{1-3} and d^{8-10}, the two cases merge into one. The situation is summarized in Table 16.1; you should be able to verify each value given in the table.

A notable success of the CFSE concept concerns the experimental hydration enthalpies of the first-row transition metal cations (Figure 8.3), and that diagram should be reexamined with the values of Table 16.1 in mind.

---

6. Orgel, L. E. *J. Chem. Phys.* **1955**, *23*, 1819. These calculated values are for free ions, and are typically 10–30% larger than for complexes.

Table 16.1 Crystal field stabilization energies for strong and weak field (low and high spin) octahedral complexes

d^n, $n =$	Configuration		CFSE, Dq	
	Strong field	Weak field	Strong field	Weak field
1	$(t_{2g})^1$		4	
2	$(t_{2g})^2$		8	
3	$(t_{2g})^3$		12	
4	$(t_{2g})^4$	$(t_{2g})^3(e_g)^1$	16	6
5	$(t_{2g})^5$	$(t_{2g})^3(e_g)^2$	20	0
6	$(t_{2g})^6$	$(t_{2g})^4(e_g)^2$	24	4
7	$(t_{2g})^6(e_g)^1$	$(t_{2g})^5(e_g)^2$	18	8
8	$(t_{2g})^6(e_g)^2$		12	
9	$(t_{2g})^6(e_g)^3$		6	
10	$(t_{2g})^6(e_g)^4$		0	

Note: The strong and weak field configurations are equivalent for d^{1-3} and d^{8-10}. In calculating the CFSE's, the pairing energies have been neglected (see text).

The Adjustment for Covalency: Ligand Field Theory

The complex anion $FeCl_6^{3-}$ has five unpaired electrons, which identifies it as a high spin (weak field) complex. On the other hand, $[Fe(CN)_6]^{3-}$ has only one unpaired electron, indicative of low spin d^5:

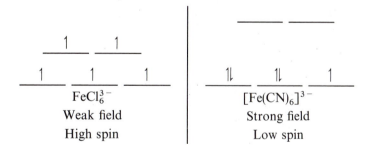

Since the configurations differ, these complexes must have different Dq's. (In fact, the difference is almost a factor of 2.) This cannot be explained by electrostatic CFT. Consider the situation from the standpoint of equation (16.3). The ionic radii of Cl^- and CN^- are approximately the same, so the metal–ligand distances should be approximately equal. All the other terms are identical for the two cases; hence, their Dq values are predicted to be very similar.

This is one example of a general phenomenon. Upon close examination the CFT model fails for many complexes, because it treats metal–ligand interactions as being purely electrostatic in nature, when in fact they have a covalent component. Some relevant observations in this regard are the following:

1. The theoretical relationship given by equation (16.3) does not closely match experimental results. (It is not possible to satisfactorily calculate values of Dq from theory alone.)

2. There is a discrepancy between the pairing energies of free ions and their complexes (footnote 6). This indicates that complexes do not result solely from simple electrostatic interactions involving free ions.

3. Many experimental bond distances are inconsistent with purely electrostatic interactions. Based on ionic radii, the Cr^{2+} and F^- "ions" of CrF_2 should approach no closer than about 210 pm; however, there are Cr–F bonds of only 200 pm.

4. Certain other types of experimental data clearly point to metal–ligand orbital overlap. As an example, the electron spin resonance (ESR) spectrum of $IrCl_6^{2-}$ suggests that about 30% of the unpaired electron density of Ir^{IV} resides in the chloride ion orbitals.[7]

Ligand field theory (LFT) results from the incorporation of covalency considerations into CFT. Such adjustments are often made on an ad hoc basis; that is, covalency is added to the basic model as is reasonable and necessary to explain experimental observations.

We are now ready to discuss the question of what factors determine the value of $\Delta_o = 10\,Dq$. The influences of metal charge (oxidation state) and metal–ligand distance were mentioned in conjunction with equation (16.3). A more complete discussion includes two other factors:

1. ***The Period of the Metal Is Important.*** For the octahedral complexes of a given *d*-block family, Δ_o increases in the order $3d < 4d < 5d$. Consider the following data:

Complex	Δ_o, kJ/mol
$[Co(NH_3)_6]^{3+}$	274
$[Rh(NH_3)_6]^{3+}$	408
$[Ir(NH_3)_6]^{3+}$	490

The increase is generally on the order of 40–50% between the corresponding 3*d* and 4*d* elements, and 20–25% between the 4*d* and 5*d*

---

7. A discussion of ESR spectroscopy is given in Chapter 21.

elements. There are two important consequences. First, complexes of the $4d$ and $5d$ metals often have such large splittings that electron transitions of the $t_{2g} \rightarrow e_g$ type occur in the ultraviolet rather than the visible region; hence, such species are often colorless. Second, the large splittings cause $10\,Dq$ to be greater than the pairing energy. Thus, octahedral complexes of the cations of the second and third transition series are virtually always of the low spin (strong field) variety.

2. ***Different Ligands Create Very Different Splittings.*** This is evident from the data for octahedral complexes of Cr^{III} compiled in Table 16.2. An ordering of ligands based on the ligand field splittings they cause is often referred to as the *spectrochemical series*, because it is obtained from spectroscopic determinations of the energies of $d \rightarrow d$ electron transitions. The spectrochemical series will be discussed in more detail in Section 16.2.

Distorted Octahedral Complexes: The Jahn–Teller Effect[8]

Recall from Chapter 15 that tetragonal distortion from octahedral geometry involves the lengthening or shortening of two trans M–L bonds. This reduces the idealized point group symmetry from O_h to D_{4h} or C_{4v}. In both cases, the major rotational axis corresponds to the direction of distortion, which

Table 16.2 Approximate energies and wavelengths for the $t_{2g} \rightarrow e_g$ transition for some octahedral chromium(III) complexes

Complex	E, kJ/mol	λ, nm
$CrCl_6^{3-}$	158	760
CrF_6^{3-}	190	630
$[Cr(H_2O)_6]^{3+}$	209	575
$[Cr(C_2O_4)_3]^{3-}$	211	570
$[Cr(NH_3)_5Cl]^{2+}$	233	515
$[Cr(NH_3)_5(H_2O)]^{3+}$	250	480
$[Cr(NH_3)_6]^{3+}$	258	465
$[Cr(en)_3]^{3+}$	264	455
$[Cr(CN)_6]^{3-}$	320	375

Source: Taken from data compiled by Forster, L. S. In *Transition Metal Chemistry*; Carlin, R. L., Ed.; Marcel Dekker: New York, 1969; Volume 5, p. 7.

8. Bersuker, I. B. *Coord. Chem. Rev.* **1975**, *14*, 357; Engleman, R. *The Jahn–Teller Effect in Molecules and Crystals*; Wiley: New York, 1972.

is therefore defined as the z axis. Thus, the terms z-in and z-out are sometimes used to describe tetragonal compression and elongation, respectively.

This type of distortion was predicted on theoretical grounds by Jahn and Teller shortly after the introduction of crystal field theory.[9] A general statement of the Jahn–Teller theorem is as follows: *A nonlinear system having a degenerate ground state will become nondegenerate, and thereby stabilized, by distortion.*

A rigorous proof of the theorem is beyond the scope of this book, but its origin can be understood through an example. The best-known cases of Jahn–Teller distortion involve hexacoordinate d^9 (particularly, Cu^{II}) complexes. In a regular octahedral array, the half-filled orbital might be either d_{z^2} or $d_{x^2-y^2}$; the ground state is doubly degenerate. Assume that the vacancy is in the $d_{x^2-y^2}$ orbital. The shielding of the protons of the copper nucleus is therefore asymmetric, being greater in the z direction than along x and y. Said another way, the electronegativity of Cu^{II} along z is less than in the other directions. Thus, the two z ligands experience less "pull" than the other four, causing the equilibrium distance in that direction to be greater than along x and y; that is, the complex becomes elongated (z-out).

As shown in Figure 16.4, either elongation or contraction increases the CFSE by reducing the energy of whichever orbital is filled. However, the theorem does not predict whether "octahedral" d^9 species are elongated (two relatively long and four shorter bonds) or compressed (four long and two short).

The Jahn–Teller effect applies to a number of d^n systems. High spin d^4 complexes (eg, Cr^{II}) are also doubly degenerate at the e_g level, and show the predicted behavior. For example, CrF_2 forms a distorted rutile lattice, with four Cr–F bond lengths of 200 pm and two of 243 pm (Figure 16.5). Chromium(II) chloride has a very different solid-state structure, but again exhibits a $4 + 2$ distribution of ligands.

Degeneracy also occurs at the t_{2g} level. The Jahn–Teller theorem predicts that octahedral d^1 and d^2 complexes should be distorted. Compression, but not elongation, gives a nondegenerate ground state for d^1, while the reverse is true for d^2. Metal ions with three d electrons have nondegenerate ground states, and therefore form regular octahedral complexes.

The Jahn–Teller predictions for octahedral complexes are summarized in Table 16.3, p. 520. (You should be able to independently arrive at any of the entries in this table.)

Although there are experimental data in support of these predictions, the effect is much weaker for t_{2g} distortions than for the e_g cases (d^9 and high spin d^4). This is because the t_{2g} orbitals are only indirectly influenced by the ligands, while the e_g orbitals interact directly. It is interesting to

---

9. Jahn, H. A.; Teller, E. *Proc. Roy. Soc. London A* **1937**, *161*, 220.

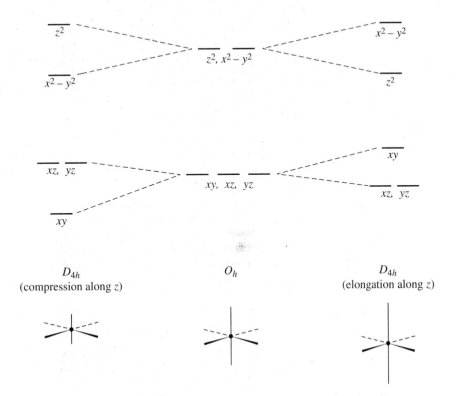

Figure 16.4
Tetragonal distortion in "octahedral" d^9 complexes.

compare solid-state internuclear distances for a series of divalent metal fluorides, MF_2 (M = Cr, Mn, Fe, Co, Ni, Cu, and Zn), in this regard. These salts are isostructural, with approximately octahedral geometry about the metal; the d^4–d^7 cases have high spin configurations. As we can see in Table 16.4 (p. 520), those species that are subject to Jahn–Teller distortion at the e_g level (CrF_2 and CuF_2) exhibit large differences in bond lengths (roughly

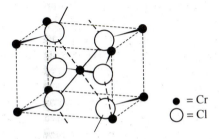

Figure 16.5 The solid-state structure of chromium(II) fluoride, a Jahn–Teller distorted rutile lattice. Dashed lines represent the two longer Cr–F bonds. [Reproduced with permission from Wells, A. F. *Structural Inorganic Chemistry*, 5th ed.; Clarendon: Oxford, 1984; p. 249.]

Table 16.3 Jahn–Teller predictions for "octahedral" d^n complexes

d^n, $n =$	Prediction	d^n, $n =$	Prediction
0	N	5 (high spin)	N
		(low spin)	E
1	C	6 (high spin)	C
		(low spin)	N
2	E	7 (high spin)	E
		(low spin)	C or E
3	N	8	N
4 (high spin)	C or E	9	C or E
(low spin)	C	10	N

Note: C = compression; E = elongation; N = no distortion predicted.

20%). However, the differences for the t_{2g} cases are so small that the data cannot reasonably be argued as supportive of Jahn–Teller splitting at the t_{2g} level.

Some complexes show distortion only at low temperature. This is attributed to a dynamic Jahn–Teller effect. Recall that for d^9 species the predicted energy difference between z-in and z-out complexes is 0, or at least very small. A macrosample of such a complex would contain an equilibrium mixture of the two, and certain experimental probes would then detect only an average (undistorted) structure. The same result would arise if the interconversion of these "isomers" is rapid on the experimental time scale. Removing thermal energy (ie, lowering the temperature) may permit the observation of distortion in such cases.

Table 16.4 Bond distances in MF_2 salts utilizing (distorted) rutile lattices

Compound	d^n, $n =$	Predicted Distortion	d_{M-L}, pm x, y	z
CrF_2	4	C or E (e_g)	200	243
MnF_2	5	N	211	214
FeF_2	6	C (t_{2g})	210	203
CoF_2	7	E (t_{2g})	204	205
NiF_2	8	N	198	204
CuF_2	9	C or E (e_g)	193	227
ZnF_2	10	N	203	204

Note: CrF_2, MnF_2, FeF_2, and CoF_2 are all high spin species. C = compression; E = elongation; N = no Jahn–Teller distortion predicted.
Source: Data compiled from Wells, A. F. *Structural Inorganic Chemistry*, 5th ed.; Clarendon: Oxford, 1984.

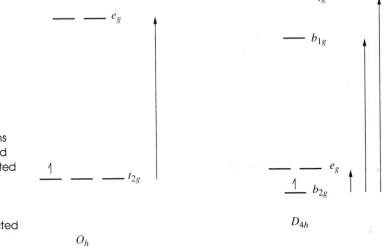

Figure 16.6
Electron transitions
in undistorted and
Jahn–Teller distorted
(compressed)
octahedral d^1
species. Three
bands are expected
for the latter.

Although the most concrete experimental evidence for the Jahn–Teller effect comes from X-ray diffraction (through precise measurements of bond lengths), electronic spectroscopy is useful as well. The spectrum of a 6-coordinate d^1 complex should show only one band for a regular octahedron, but up to three if it is tetragonally distorted (Figure 16.6). In fact, the spectra of Ti^{III} complexes do contain three bands—one ($b_{2g} \rightarrow e_g$) in the infrared and the others in the visible and/or ultraviolet regions. (Because of the small energy difference between the b_{1g} and a_{1g} levels, the two higher-energy absorptions are often not well-resolved.) The absorption spectra of metal complexes are discussed more fully in Chapter 21.

The Square Planar Geometry

Extreme tetragonal elongation of an octahedron ultimately leads to the complete removal of two opposite ligands, giving a square planar complex. The effect on the CFT splitting diagram is shown in Figure 16.7. It may seem surprising that d_{z^2} lies at higher energy than the $d_{xz,yz}$ pair. However, recall that d_{z^2} does not lie solely along the z axis; it contains a torus in the xy plane, causing it to be destabilized relative to d_{xz} and d_{yz}.

The tendency for d^8 metal ions to form square planar complexes, first discussed in Chapter 15, can be readily understood from Figure 16.7. Four of the five orbitals are relatively stable. The fifth ($d_{x^2-y^2}$) lies very high in energy, and its occupancy is strongly destabilizing. It follows that only those d^8 complexes with sufficient crystal field splitting to be low spin exhibit square planar geometry. For example, the cyanide ion creates a larger Dq

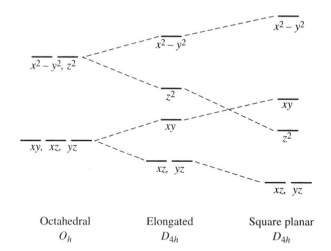

Figure 16.7
Comparison of the relative d orbital energies for octahedral, Jahn–Teller distorted (elongated), and square planar complexes.

unit than does fluoride (by nearly a factor of 2). Hence, the equilibrium

$$NiL_4^{2-} + 2L^- \; \rightleftharpoons \; NiL_6^{4-} \tag{16.4}$$

favors the reactants when L = CN, but the product when L = F.

Cubic and Tetrahedral Geometries

The symmetry relationships among the cube, tetrahedron, and octahedron were described in Chapter 2. Recall that the eight corners of a cube can be made to correspond to the face centers of an octahedron, and that the reverse is also true; that is, the six vertices of an octahedron center the faces of a cube. In terms of crystal field theory, this opposite relationship causes an exact reversal of the relative d orbital energies. The symmetry labels T_{2g} and E_g still apply (the point group of a cube is O_h), and the d_{xy}, d_{xz}, d_{yz} and the $d_{x^2-y^2}$, d_{z^2} subsets remain but e_g is the more stable level in a cubic crystal field. Also, note from Figure 16.8 that the location of ligands at the corners of a cube allows for less direct interaction (from the CFT standpoint, creates less repulsion) than does an octahedral array. It can be shown that the splitting caused by each cubic ligand is exactly $\frac{2}{3}$ that per octahedral ligand. However, there are eight cubic versus six octahedral point charges. Thus, for any given metal–ligand combination, the cubic splitting parameter Δ_c is $(\frac{2}{3})(\frac{8}{6}) = \frac{8}{9}$ that of Δ_o.

A tetrahedron is, for purposes of crystal field theory, equivalent to half of a cube. (Four of the eight vertices of a cube form a tetrahedron; see Figure 2.7.) It follows that the CFT diagram for tetrahedral geometry is qualitatively identical to the cubic case, and that $\Delta_t = \frac{1}{2}\Delta_c = \frac{4}{9}\Delta_o$ (Figure 16.8).

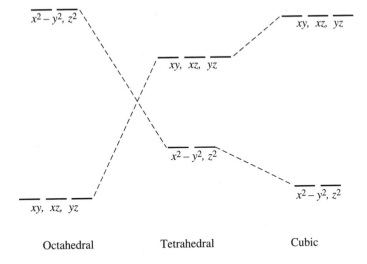

Figure 16.8
Comparison of the crystal field theory diagrams for cubic, octahedral, and tetrahedral geometries; $\Delta_t = \frac{1}{2}\Delta_c = \frac{4}{9}\Delta_o$.

Because of the small splitting, Δ_t is almost always less than the pairing energy; that is, nearly all tetrahedral complexes are high spin. It follows that tetrahedral crystal field stabilization energies are small. The maximum CFSE is obtained for the d^2 and d^7 configurations, but it is only 12 Dq [equivalent to $\frac{4}{9}(12.00) = 5.33\ Dq$ in octahedral units]. As a result, most d-block complexes prefer octahedral (or if d^8, square planar) rather than tetrahedral geometry. The most common exceptions occur for d^0 and d^{10} (for which CFSE = 0 regardless of ligand orientation), and for complexes in which steric effects are significant.

Quantitative Aspects of Crystal Field Theory

A unique CFT splitting diagram exists, of course, for each geometry. Qualitative diagrams usually can be constructed with little effort. An approach advocated by Krishnamurthy and Schaap is useful for quantitative comparisons.[10] Their method (slightly revised here) divides the ligands of a given complex into some combination of four different subsets:[11] a single ligand along the z axis (subset I), a pair at right angles in the xy plane (subset II), a trio lying at 120° angles in the xy plane (subset III), and a set of four in a tetrahedral array (subset IV). Each of these subsets is assigned a

---

10. Krishnamurthy, R.; Schaap, W. B. *J. Chem. Educ.* **1970**, *47*, 433; *J. Chem. Educ.* **1969**, *46*, 799.

11. Only three ligand groups are identified in the references cited in footnote 10, one of which is subdivided. The addition of a fourth subset simplifies the calculations for certain geometries, but is less general than the original method.

Table 16.5 Contributions to d orbital (ligand field) energies from ligand subsets

Subset	xy	xz	yz	$x^2 - y^2$	z^2
I	−3.14	+0.57	+0.57	−3.14	+5.14
II	+1.14	−2.57	−2.57	+6.14	−2.14
III	+5.46	−3.86	−3.86	+5.46	−3.21
IV	+1.78	+1.78	+1.78	−2.67	−2.67

Note: Subsets I–IV are identified in the text.

contribution to the overall energy of each d orbital (Table 16.5). Note that the five energies for each subset sum to 0; that is, the baricenter is maintained.

Most of the common geometries derive from some combination of these four ligand types. Since the values given in Table 16.5 are additive, this makes possible the determination of the d orbital energies. For example, since a cube is equivalent to two tetrahedra, the orbital energies (in octahedral Dq units) for a cubic crystal field can be obtained by doubling the values given for subset IV. The octahedral geometry is equivalent to two subset I plus two subset II types, yielding the expected values of −4.00 and +6.00 Dq.

This method can be combined with structural, spectroscopic, or other data with fruitful results. We noted earlier that the Cr–F bond lengths in CrF_2 are 200 pm (four, along the x and y axes) and 243 pm (two, along z). Equation (16.3) indicates that Dq is inversely related to the fifth power of the metal–ligand distance. We will therefore assume that $E(d_{x^2-y^2})/E(d_{z^2}) = (243/200)^5 = 2.648$. If the baricenter of these orbitals is set at +6.00 Dq, then the equations below can be solved:

$$E(d_{x^2-y^2}) = (2.648)E(d_{z^2}) \tag{16.5}$$

$$E(d_{x^2-y^2}) + E(d_{z^2}) = 12.00 \ Dq \tag{16.6}$$

Energies of +3.29 and +8.71 Dq are obtained for d_{z^2} and $d_{x^2-y^2}$, respectively. Hence, the Jahn–Teller splitting is estimated to be about 5.4 Dq units.

The $[Cr(NH_3)_6]^{3+}$ ion has three d electrons and is therefore a regular (undistorted) octahedron. Spectroscopic data (an absorbance maximum at 465 nm) indicate that Δ_o for this complex is 21,500 cm^{-1}, or 258 kJ/mol.[12] Let us consider the effect of replacing one NH_3 with water to give $[Cr(NH_3)_5(H_2O)]^{3+}$. The new idealized point group is C_{4v}, with the H_2O ligand oriented along the z axis. The ligand field splitting caused by water is about $\frac{1}{1.25} = 0.80$ times that caused by ammonia (see Table 16.6, later in

---

12. Using the equation $E = hc/\lambda$ and appropriate conversion factors from Appendix I, the relationship E (kJ/mol) = $(1.20 \times 10^5)/\lambda$ (nm) is obtained.

this section). Therefore, the replacement of NH_3 by H_2O slightly stabilizes the d_{z^2}, d_{xz}, and d_{yz} orbitals relative to those in $[Cr(NH_3)_6]^{3+}$, resulting in the qualitative CFT diagram

$$\underline{\hspace{1.5cm}}\quad x^2 - y^2$$

$$\underline{\hspace{1.5cm}}\quad z^2$$

$$\underset{\displaystyle 1}{\underline{\hspace{1.5cm}}}\quad xy$$

$$\underset{\displaystyle 1}{\underline{\hspace{1.5cm}}}\quad\underset{\displaystyle 1}{\underline{\hspace{1.5cm}}}\quad xz,\ yz$$

Quantitatively, the d orbital energies can be estimated as follows. The complex consists of one NH_3 along the z axis (subset I), one H_2O along z (0.80 · subset I), and four NH_3's in the xy plane (2 · subset II). Thus, for d_{z^2},

$$E(d_{z^2}) = +5.14 + 0.80(+5.14) + 2(-2.14) = +4.97\ Dq$$

Similarly, the other levels are calculated to be $d_{x^2-y^2} = +6.63\ Dq$, $d_{xy} = -3.37\ Dq$, and $d_{xz} = d_{yz} = -4.11\ Dq$. According to this method, then, the energies are as shown in Figure 16.9. Note that the baricenter remains at $0\ Dq$; that is, the five orbital energies sum to 0. The CFSE is 11.59 Dq in $[Cr(NH_3)_6]^{3+}$ units. The value of Δ_o' (computed as the energy difference between the averages for the $d_{x^2-y^2,\,z^2}$ and $d_{xy,\,xz,\,yz}$ subsets) in the same units is 9.66 Dq (249 kJ/mol), in excellent agreement with the experimental value (see Table 16.2).

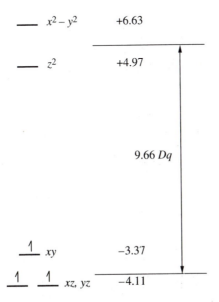

Figure 16.9
The quantitative ligand field theory diagram predicted for $[Cr(NH_3)_5(H_2O)]^{3+}$. Energies are in octahedral Dq units of $[Cr(NH_3)_6]^{3+}$ (see text).

16.2 Molecular Orbital Theory

The other primary bonding model for metal complexes is molecular orbital theory. We emphasize the LCAO approach in the following discussion, so a review of Section 3.1 may be in order before you proceed.

Initially, we will treat metal–ligand complexes as Lewis adducts in which each ligand donates a σ electron pair. This will produce MO diagrams that are approximately, but not entirely, consistent with experimental observations. Then we will obtain more accurate diagrams through the inclusion of π bonding.

Linear Complexes

The first polyatomic compound considered in Chapter 3 was beryllium hydride, BeH_2. The BeH_2 treatment can be used as a basis for linear transition metal complexes, but must be amended in two ways:

1. Beryllium has only s- and p-type valence orbitals, while transition metals have valence d orbitals as well. Thus, a total of nine metal atomic orbitals [from the ns, np, and $(n-1)d$ subshells] will be used for generating the molecular orbitals.

2. Ligands generally bond to metals via lone pairs located either in pure p orbitals or in hybrids having at least 50% p character, while the terminal atoms of BeH_2 use s orbitals. For the approximate, σ-only cases, one p orbital will be selected from each ligand to form the symmetry-adapted linear combination (SALC) orbitals.

In linear complexes, only those metal orbitals having lobal density along the L–M–L bond axis (defined as the z axis by group theory convention) can participate in σ bonding. The possible candidates are s, p_z, and d_{z^2}. Nonzero overlap can occur for each of these orbitals (Figure 16.10), as well as for their hybrid combinations.

Now we apply group theory to develop an MO diagram, taking the linear $CuCl_2^-$ ion as a specific example. The point group is $D_{\infty h}$. A total of eleven atomic orbitals—nine from the metal plus one from each ligand—will be mixed to produce eleven MO's. The two ligand SALC orbitals are

$$\psi_1 = \frac{1}{\sqrt{2}}(\phi_1 + \phi_2)$$

$$\psi_2 = \frac{1}{\sqrt{2}}(\phi_1 - \phi_2)$$

Using the $D_{\infty h}$ character table (Appendix IV) and the methodology described

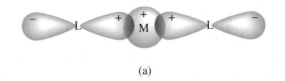

(a)

Figure 16.10
Possible modes of overlap for metal–ligand σ bonding in linear complexes:
(a) ligand p_z–metal s; (b) ligand p_z–metal p_z;
(c) ligand p_z–metal d_{z^2}.

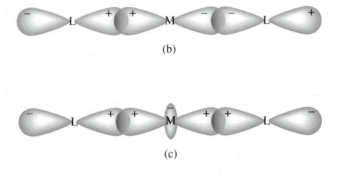

(b)

(c)

in Chapter 3, we find that ψ_1 and ψ_2 belong to the irreducible representations Σ_g^+ and Σ_u^+, respectively. The metal orbitals transform as follows:

$$
\begin{cases}
\Sigma_g^+: & s, d_{z^2} \\
\Pi_g: & d_{xz}, d_{yz} \\
\Delta_g: & d_{x^2-y^2}, d_{xy} \\
\Sigma_u^+: & p_z \\
\Pi_u: & p_x, p_y
\end{cases}
$$

The s and d_{z^2} orbitals each have the appropriate symmetry to interact with ψ_1, and the resulting molecular orbitals have contributions from both. The p_z orbital can interact with ψ_2. The six other metal orbitals are necessarily nonbonding.

Two important assumptions were made in constructing the σ-only MO diagram (Figure 16.11). First, the order of Cu^+ orbital energies was taken to be $3d < 4s < 4p$. (Recall that $3d$ is generally more stable than $4s$ for metal cations.) Second, the electronegativity of Cl^- was assumed to be greater than that of Cu^+; hence, the skewing of the atomic orbitals.[13]

Fourteen valence electrons are shown—ten from Cu^+ and four from the two ligand electron pairs. The HOMO is primarily metal in character.

---

13. Copper is certainly a less electronegative element than chlorine. However, it is less certain that Cu^+ has a lower electronegativity than Cl^-. The skewing shown in Figure 16.11 is consistent with general practice.

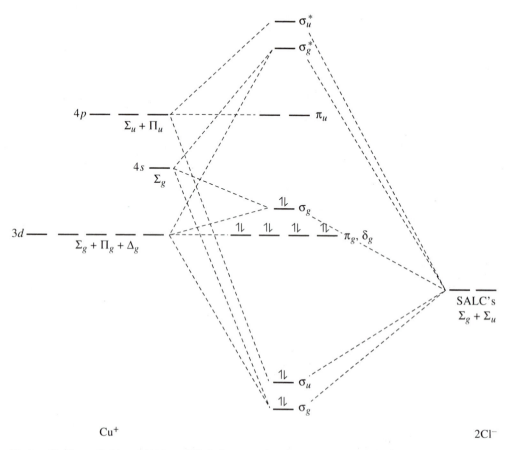

Figure 16.11 σ-Only molecular orbital diagram for the linear complex $CuCl_2^-$.

π-Type Interactions

The diagram in Figure 16.11 can be improved by considering the effect of π bonding between appropriate metal and ligand orbitals. It is apparent that such side-to-side interactions are possible from the relative orientations of the metal d_{xz} and d_{yz} orbitals and ligand p_x and p_y orbitals:

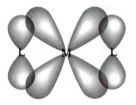

The $3p_x$ and $3p_y$ orbitals of the ligands can be used to construct two different kinds of SALC's (labeled ψ_3 and ψ_4; see Figure 16.12). Since these are filled (lone-pair) orbitals, it is necessary to add another eight electrons to the MO diagram.

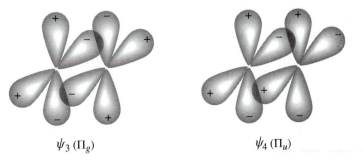

Figure 16.12
Π-Type SALC's for $CuCl_2^-$.

$\psi_3\ (\Pi_g)$ $\psi_4\ (\Pi_u)$

By following the procedure described in Chapter 3, the SALC symmetries are found to be Π_g for the ψ_3 set and Π_u for ψ_4. Since the metal $3d_{xz,yz}$ pair transforms as Π_g, those orbitals overlap with ψ_3. Similarly, the Cu ($4p_{x,y}$) pair overlaps with ψ_4. The resulting $\sigma + \pi$ diagram is shown in Figure 16.13 (p. 530).

Some pertinent observations are given below:

1. There are now fifteen atomic orbitals—nine from the metal plus the three $3p$ orbitals from each chloride—and fifteen molecular orbitals. The number of valence electrons is 22, so eleven of the MO's are filled.[14]

2. Twelve electrons reside in bonding and four in antibonding orbitals. The excess of eight bonding electrons corresponds to two σ and two (net) π bonds.

3. The three highest occupied orbital levels (δ_g, π_g^*, and σ_g) are primarily metal in character. Their relative energies are identical to those obtained from the crystal field theory analysis. (Compare that portion of Figure 16.13 to Figure 16.1.) This is a characteristic result, and in that respect a CFT diagram can be viewed as a section of a more complete MO diagram.

Octahedral Complexes

A generalized, σ-only molecular orbital diagram for octahedral complexes was developed in Chapter 3 (see Figure 3.22). The six SALC's are pictured in Figure 16.14 (p. 531), along with the metal orbitals to which they bond.

---

14.   The fourth lone pair from each Cl^- has been ignored. In this scheme it is presumed to reside in a nonbonding, s-type orbital.

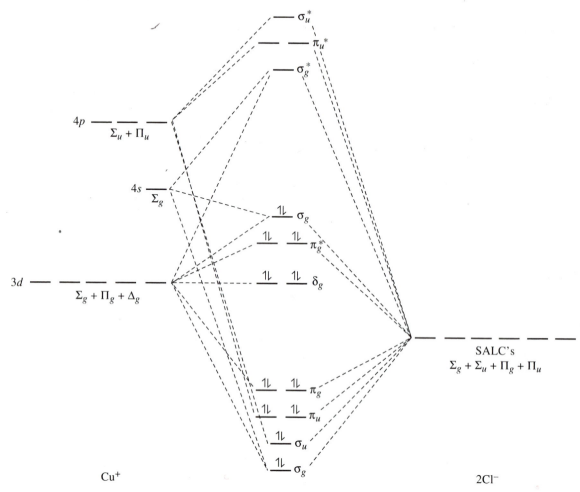

Figure 16.13 Complete $\sigma + \pi$ molecular orbital diagram for the $CuCl_2^-$ ion.

A diagram specific to the complex ion $[Cr(NH_3)_6]^{3+}$ is presented in Figure 16.15 (p. 532). It is recommended that you work through that example in its entirety.

The major conclusions to be drawn are the following:

1. The σ overlap involves one s-, three p-, and two d-type metal orbitals, and in that sense is consistent with sp^3d^2 hybridization. The $d_{x^2-y^2}$ and d_{z^2} orbitals overlap with σ-type SALC's to produce molecular orbitals of e_g symmetry.

2. There are three unpaired electrons, located in degenerate, nonbonding

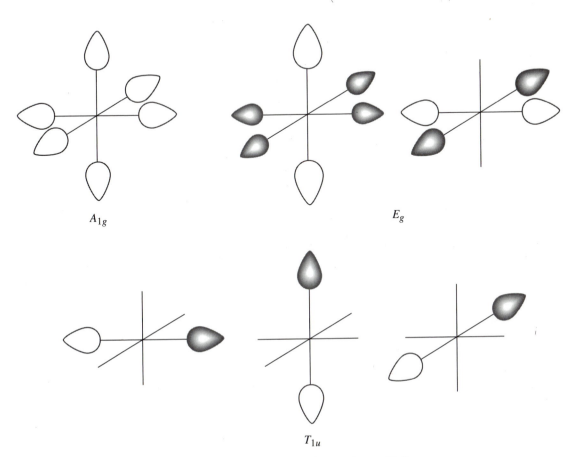

A_{1g}

E_g

T_{1u}

Figure 16.14 σ-Type SALC's for octahedral complexes. A_{1g} overlaps with the metal s-type orbital; the E_g pair overlap with the metal d_{z^2} and $d_{x^2-y^2}$ orbitals; and the T_{1u} set overlaps with the three p orbitals of the metal.

metal orbitals having t_{2g} symmetry. The prediction of paramagnetism has been verified experimentally.

3. A total of 18 electrons would be needed to completely fill all the bonding and nonbonding orbitals. However, since the t_{2g} level is nonbonding, the maximum possible number of bonding electrons is 12.

4. As for $CuCl_2^-$, the MO's derived from the metal $3d$ orbitals correlate in both relative energy and occupancy with the CFT diagram. (Compare the HOMO/LUMO section of Figure 16.15 to Figure 16.3.) The $t_{2g}-e_g^*$ energy gap corresponds to $\Delta_o = 10\,Dq$ in CFT terminology, and is a reflection of the strength of the bonds formed by the d orbitals.

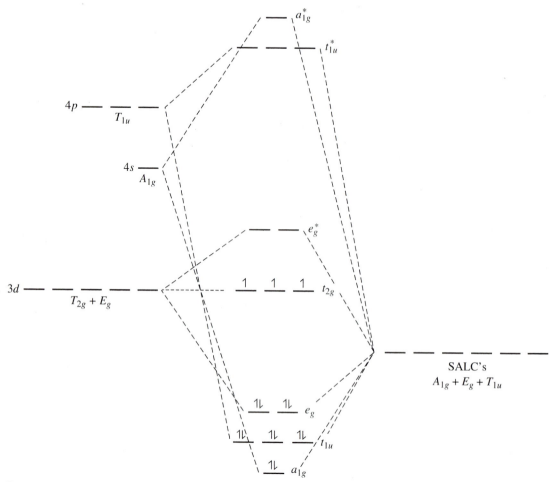

Figure 16.15 Molecular orbital diagram for the octahedral complex $[Cr(NH_3)_6]^{3+}$.

For $[Cr(NH_3)_6]^{3+}$, π-type interactions are not possible, because after the formation of the six $N \to Cr$ σ bonds the donor nitrogens are coordinately saturated. For certain other complexes, however, π bonding involving the t_{2g} subset does occur. This converts chromium's σ nonbonding d orbitals to π antibonding MO's of predominantly metal character:

Figure 16.16
Two modes of
metal–ligand π-type
interaction:
(a) ligand-to-metal π
bonding;
(b) metal-to-ligand
π bonding
(back-bonding).

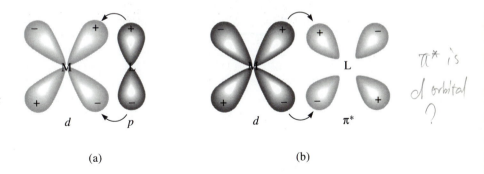

(a)   (b)

π^* is
d orbital
?

This is *ligand-to-metal π bonding*; because the ligand SALC's are filled, any transfer of π electron density must be from the ligand to the metal (see Figure 16.16a). In general, bases having more than one lone pair on the donor atom can act as π *donor* ligands; common examples include the halide ions, H_2O, OH^-, and NH_2^-.

A different type of π bonding is exhibited by $Cr(CO)_6$. The LUMO's of CO (two π^* orbitals) are capable of side-to-side interaction with the metal t_{2g} subset (Figure 16.16b). Since these π^* orbitals are empty, the flow of π electron density in $Cr(CO)_6$ is from the metal to the ligand. This is *metal-to-ligand π bonding* (or *back-bonding*). Carbon monoxide and related bases (CN^-, NO, and also PH_3 and its derivatives, which utilize empty $3d$ rather than π^* orbitals) are classified as π *acceptor* ligands. Because these antibonding ligand orbitals lie above the metal t_{2g} level in energy, the bonding MO's are mostly metal in character:

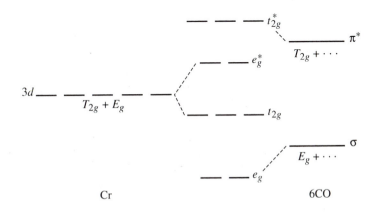

Back-bonding decreases the net transfer of negative charge from the ligand to the metal. Not surprisingly, it is most important for metals in low oxidation states (eg, Cr^0).

Table 16.6 Jørgensen's f and g constants for use in equation (16.7)

Ligand	f	Ligand	f	Metal	g	Metal	g
Br^-	0.72	NC^-	1.15	Mn^{II}	8.0	Mo^{III}	24.6
SCN^-	0.73	$MeNH_2$	1.17	Ni^{II}	8.7	Rh^{III}	27.0
Cl^-	0.78	MeCN	1.22	Co^{II}	9	Tc^{IV}	30
N_3^-	0.83	py	1.23	V^{II}	12.0	Ir^{III}	32
F^-	0.9	NH_3	1.25	Fe^{III}	14.0	Pt^{IV}	36
Me_2SO	0.91	en	1.28	Cr^{III}	17.4		
EtOH	0.97	NH_2OH	1.30	Co^{III}	18.2		
H_2O	1.00	CN^-	1.7	Ru^{II}	20		
NCS^-	1.02			Mn^{IV}	23		

Sources: Jørgensen, C. K. *Modern Aspects of Ligand Field Theory*; North-Holland: Amsterdam, 1971; pp. 347–348; *Absorption Spectra and Chemical Bonding in Complexes*; Pergamon: Oxford, 1962; p. 113.

The Spectrochemical Series

The spectrochemical series, first mentioned in Section 16.1, can now be understood. Jørgensen used spectroscopic data to quantify this series.[15] The basic premise of his method is that the CFT parameter Δ_o is the product of ligand and metal contributions (symbolized by f and g, respectively):

$$\Delta_o \text{ (in cm}^{-1}) = 10^3 \cdot f \cdot g \tag{16.7}$$

The ligand constants are normalized to $f = 1.00$ for H_2O. Jørgensen's values are given in Table 16.6.

Equation (16.7) predicts Δ_o for $[Cr(NH_3)_6]^{3+}$ to be 21,750 cm^{-1}, in good agreement with the actual value of 21,500 cm^{-1}. However, the accuracy of this method varies. Carbonyl and cyanide complexes are particularly prone to deviations from the relationship, since the extent of back-bonding varies with the metal.

Interestingly, infrared spectroscopy can be used to arrive at the same conclusion. We will reexamine π bonding in metal complexes using IR data in Chapter 21.

Tetrahedral Complexes

We will use the complex ion $CoCl_4^{2-}$ to illustrate the construction of an MO diagram for a tetrahedral complex. From the T_d character table

---

15. Jørgensen, C. K. *Absorption Spectra and Chemical Bonding in Complexes*; Pergamon: Oxford, 1962.

(Appendix IV) and the procedures used in Chapters 2 and 3, the reducible representation for the σ-type SALC's is found to be

	E	$8C_3$	$3C_2$	$6S_4$	$6\sigma_d$
Γ_{SALC}	4	1	0	0	2

This reduces to $A_1 + T_2$. The metal orbitals that have appropriate symmetry for overlap are $4s$ (A_1) plus either of the two subsets $4p_{x,y,z}$ or $3d_{xy,xz,yz}$, both of which transform as T_2; that is, either sp^3 or sd^3 hybridization is feasible. (In fact, group theory requires all molecular orbitals that have T_2 symmetry to have both metal p and d parentage.) The resulting energy levels are shown in Figure 16.17.

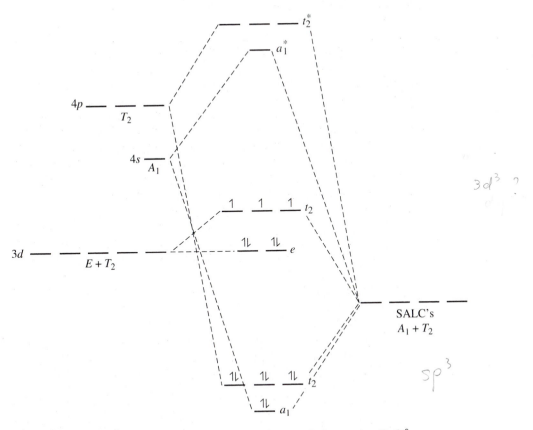

Figure 16.17   Molecular orbital diagram for the tetrahedral complex $CoCl_4^{2-}$; only σ-type overlap is considered.

The configurations that most often produce tetrahedral complexes are d^0, d^2, d^7, and d^{10}. It should be apparent why this is the case. There are four σ bonds regardless of the number of d electrons. The d^0 configuration causes only the bonding MO's to be filled; d^2 results in single and d^7 in double occupancy through the nonbonding e level; and for d^{10}, all the bonding and nonbonding orbitals are filled.

Metal–ligand π bonding occurs in many tetrahedral complexes. That topic is not discussed here, however, since the basic principles are the same as for octahedral species, and the end result does not significantly alter the qualitative MO diagram.

16.3   Binuclear Complexes

Many complexes are known, of course, in which more than one metal is present; a large number of these contain metal–metal bonds. The simplest are *binuclear* (two-metal) species.

A useful first example is $Mn_2(CO)_{10}$, which can be visualized as two $Mn(CO)_5$ fragments joined by a metal–metal bond. The environment around each manganese is pseudo-octahedral—five CO's plus one Mn "ligand." The bonding is similar to that in a regular octahedral complex, with the d_{z^2} orbitals used to form the Mn–Mn bond. The valence electron count is 34 (7 from each manganese plus 20 from the ligands). Recalling that 18 electrons fill the MO's of an octahedral complex through the t_{2g} level (Figure 16.15), 34 is precisely the number needed to occupy the analogous orbitals of $Mn_2(CO)_{10}$. (Two electrons are saved, since the octahedra share a common bond.[16]) Because CO is a π acceptor ligand, the t_{2g}-type orbitals are π bonding rather than nonbonding. This explains why $Mn_2(CO)_{10}$ is much more stable than either $Cr_2(CO)_{10}$ (with a vacancy in a bonding orbital) or $Fe_2(CO)_{10}$ (with occupancy of an antibonding orbital).

Figure 16.18a shows that $Mn_2(CO)_{10}$ has a staggered conformation; it belongs to the D_{4d} point group. This minimizes the nonbonded repulsions between the two sets of ligands.

A more complex analysis is required for $Re_2Cl_8^{2-}$. The structure of that anion is remarkable in two ways. First, the Re–Re distance is extremely short (224 pm, compared to 275 pm in the free element); this is suggestive of very strong metal–metal bonding. Second, the two sets of chloride ions lie in a perfectly eclipsed conformation (D_{4h} symmetry; see Figure 16.18b).

These results are consistent with quadruple bonding between the metal centers. Recall from Chapter 3 that a quadruple bond consists of one σ, two

---

16.   Note that each manganese atom is associated with 18 valence electrons. This is a prelude to the *18-electron*, or *effective atomic number*, *rule*, which will be discussed in Chapter 18.

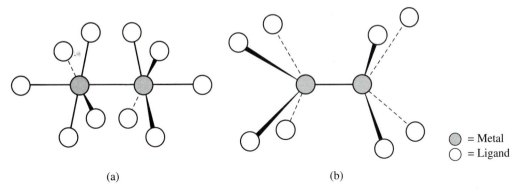

Figure 16.18 The structures of two species containing metal–metal bonds: (a) $Mn_2(CO)_{10}$; (b) $Re_2Cl_8^{2-}$. In $Mn_2(CO)_{10}$, the ligands are staggered, while in $Re_2Cl_8^{2-}$, they are eclipsed.

π, and one δ (four-lobed) interactions. A group theory approach is as follows: The local symmetry of each $ReCl_4^-$ unit is C_{4v} (square pyramidal). Using the character table for that point group, the reducible representation for the σ-type SALC's is:

	E	$2C_4$	C_2	$2\sigma_v$	$2\sigma_d$
Γ_{SALC}	4	0	0	2	0

This reduces to $A_1 + B_1 + E$, and indicates that the metal s (A_1 symmetry), $p_{x,y}$ (E), and $d_{x^2-y^2}$ (B_1) orbitals can be used to form the metal–ligand bonds. The metal–metal σ bond might result from overlap of either the two p_z or the two d_{z^2} orbitals (or a hybrid combination of the two). The d_{xz} and d_{yz} orbitals have the proper orientation for π bonding, while d_{xy}–d_{xy} overlap produces the δ bond (Figure 16.19, p. 538). Note that δ bonding would not be possible if the geometry were staggered instead of eclipsed.

A partial MO diagram (showing d–d overlap only) is given in Figure 16.20. The eight electrons occupy bonding MO's that are completely metal in character; that is, there are four Re–Re bonds.

A large number of species are known that contain quadruple bonds, and their chemistry has been summarized in the literature.[17]

17. Cotton, F. A.; Walton, R. A. *Multiple Bonds Between Metal Atoms*; Wiley: New York, 1983; Cotton, F. A. *Chem. Soc. Rev.* **1983**, *12*, 35; Templeton, J. L. *Prog. Inorg. Chem.* **1979**, *26*, 211.

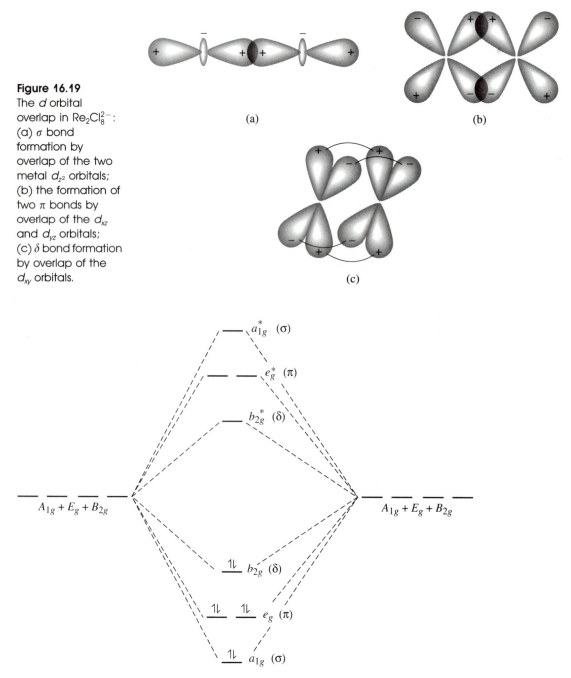

Figure 16.19
The d orbital overlap in $Re_2Cl_8^{2-}$: (a) σ bond formation by overlap of the two metal d_{z^2} orbitals; (b) the formation of two π bonds by overlap of the d_{xz} and d_{yz} orbitals; (c) δ bond formation by overlap of the d_{xy} orbitals.

(a)

(b)

(c)

Figure 16.20 Partial molecular orbital diagram (showing d-d overlap only) for $Re_2Cl_8^{2-}$. A quadruple bond is indicated.

Bibliography

Cotton, F. A. *Chemical Applications of Group Theory*, 3rd ed.; Wiley: New York, 1990.

Basolo, F.; Johnson, R. *Coordination Chemistry*; Science Reviews: Northwood, 1987.

Douglas, B. E.; Hollingsworth, C. A. *Symmetry in Bonding and Spectra*; Academic: Orlando, FL, 1985.

Murrell, J. N.; Kettle, S. F. A.; Tedder, J. M. *The Chemical Bond*, 2nd ed.; Wiley: New York, 1985.

Cotton, F. A.; Walton, R. A. *Multiple Bonds Between Metal Atoms*; Wiley: New York, 1983.

Ballhausen, C. J. *Molecular Electronic Structures of Transition Metal Complexes*; McGraw-Hill: New York, 1979.

Engleman, R. *The Jahn–Teller Effect in Molecules and Crystals*; Wiley: New York, 1972.

Jørgensen, C. K. *Modern Aspects of Ligand Field Theory*; North-Holland: Amsterdam, 1971.

Figgis, B. N. *Introduction to Ligand Fields*; Wiley: New York, 1966.

Orgel, L. E. *An Introduction to Transition Metal Chemistry: Ligand Field Theory*, 2nd ed.; Wiley: New York, 1966.

Questions and Problems

1. Construct a qualitative CFT diagram for the following cases. (The z axis should correspond to the major axis of rotation.)
   (a) ML_5, trigonal bipyramidal
   (b) ML_5, square pyramidal
   (c) $[Rh(NH_3)_5Cl]^{2+}$
   (d) $[Rh(NH_3)_5CN]^{2+}$

2. Calculate the CFSE in Dq units for octahedral complexes of the following metal ions (ignore Jahn–Teller effects). If both high and low spin complexes are possible, calculate the CFSE for each.
   (a) V^{2+}
   (b) Mn^{3+}
   (c) Mn^{2+}
   (d) Mo^{3+}
   (e) Rh^{3+}
   (f) Au^{3+}

3. The calculated free ion pairing energy for Fe^{2+} is 229 kJ/mol. The experimental value of 10 Dq for $[Fe(H_2O)_6]^{2+}$ is 10,400 cm^{-1}.
   (a) Assuming that pairing energies for octahedral complexes are about 80% of the free ion values, predict whether $[Fe(H_2O)_6]^{2+}$ is a high or low spin species.
   (b) With the aid of Table 16.6, predict whether the following are high or low spin: $FeCl_6^{4-}$, FeF_6^{4-}, $[Fe(NH_3)_6]^{2+}$, and $[Fe(CN)_6]^{4-}$.

4. (a) The pairing energy of Co^{III} is 283 kJ/mol, about 24% greater than for the isoelectronic Fe^{II}. Rationalize.

(b) For $[Co(H_2O)_6]^{3+}$, Δ_o is 20,700 cm^{-1}, much greater than for $[Fe(H_2O)_6]^{2+}$. Why?

(c) The complex $[Co(H_2O)_6]^{3+}$ is diamagnetic, but $[Fe(H_2O)_6]^{2+}$ has four unpaired electrons per ion. Explain.

5. Predict whether or not the following complexes undergo Jahn–Teller distortion, and defend your answers:
(a) VCl_6^{3-} (b) VCl_6^{4-} (c) $Cr(CO)_6$ (d) $OsCl_6^{3-}$
(e) $[Ni(en)_3]^{2+}$ (f) $[Rh(CN)_6]^{3-}$

6. The Jahn–Teller theorem applies to complexes of any geometry. For which of the d^0–d^{10} configurations is Jahn–Teller distortion expected for tetrahedral complexes?

7. For the two equilibria

$$[M(H_2O)_6]^{2+} + 4Cl^- \; \rightleftharpoons \; MCl_4^{2-} + 6H_2O$$

(M = Co and Ni), the value of K_{eq} is greater for M = Co. Rationalize.

8. Metal oxidation states that commonly yield octahedral complexes include Cr^{III}, Co^{III}, and Pt^{IV}. Those that favor tetrahedral complexes include Ti^{IV}, Co^{II}, and Zn^{II}. Use crystal field theory to explain.

9. Use the experimental data given in Table 16.2 to calculate f values for the various ligands; set $f(H_2O) = 1.00$. Compare your results to those of Jørgensen (Table 16.6).

10. Estimate the energies (in Dq units) of the d_{z^2} and $d_{x^2-y^2}$ orbitals for each of the following ligand geometries:
(a) Linear   (b) Square planar
(c) Compressed octahedral in which the distance along z is 95% of that along x and y

11. Calculate the d orbital energies (in Dq units) for a trigonal bipyramidal complex.

12. The electronic spectrum of Ti^{3+} in LiCl + KCl eutectic at 400°C shows maxima at about 10,000 and 13,000 cm^{-1}. These bands are believed to be due to the $TiCl_6^{3-}$ ion.
(a) Convert these energies to kilojoules per mole.
(b) Explain why two bands are observed.
(c) A third (low-energy) band also should be present. Identify its source.

13. Predict whether the following complexes are high or low spin. Assume that the pairing energies are approximately 220 kJ/mol.
(a) $[Fe(CN)_6]^{3-}$ (b) $CoCl_6^{3-}$ (c) $IrBr_6^{3-}$ (d) $[Rh(H_2O)_6]^{3+}$

14. The anion FeF_6^{3-} has five unpaired electrons per ion.
(a) Construct a CFT diagram that is consistent with this description.
(b) Construct a σ-only MO diagram that is consistent with this description.

15. Use group theory to construct a σ-only MO diagram for a trigonal planar complex. What (if any) d orbitals are involved in the bonding?

16. Predict whether $[Fe(CN)_6]^{3-}$ or $[Fe(CN)_6]^{4-}$ has the longer C–N bonds, and defend your answer.

17. Classify the following ligands as π donating, π accepting, or π nonbonding, and explain your answers: PH_3, NH_3, NH_2^-, en, H_2S, H_2O, and OH^-.

18. Predict the most probable value of n for the following reactions; also indicate the geometry expected for each product.
    (a) $Rh^{3+} + Excess\ CN^- \longrightarrow [Rh(CN)_n]^{3-n}$
    (b) $Pd^{2+} + Excess\ CN^- \longrightarrow [Pd(CN)_n]^{2-n}$
    (c) $Zn^{2+} + Excess\ CN^- \longrightarrow [Zn(CN)_n]^{2-n}$
    (d) $Ni^{2+} + Excess\ I^- \longrightarrow [NiI_n]^{2-n}$

19. Among mononuclear carbonyls, $V(CO)_6$ has an unusual tendency to undergo one-electron reduction. Explain.

20. (a) Sketch the structure of $Os_3(CO)_{12}$, in which the three atoms are in equivalent, pseudo-octahedral positions.
    (b) How many metal–metal bonds are present in this molecule?

21. Explain the following observations in your own words:
    (a) It is sometimes said that MO theory encompasses, and also goes beyond, CFT.
    (b) The ligand NH_2^- is a π donor, while NH_3 is π nonbonding.
    (c) Although OH^- is a stronger base than H_2O, the latter lies higher in the spectrochemical series.
    (d) There is a tendency for metals to be associated with exactly 18 valence electrons when bonded to six π acceptor ligands in an octahedral array.
    (e) Ligands that are π donors bond most strongly to metals in high oxidation states.

*22. Many dinuclear complexes having M≡M triple bonds are known. Two examples were reported by T. P. Blatchford, M. H. Chisholm, and J. C. Huffman (*Inorg. Chem.* **1987**, *26*, 1920).
    (a) Sketch the coordination compounds $M_2(MeNCH_2CH_2NMe)_3$ (M = Mo and W).
    (b) How were these compounds synthesized?
    (c) What are torsion angles? Why are they significant in these species?

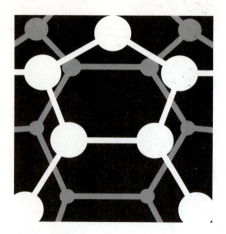

17

Reactions of Transition Metal Complexes

The theoretical groundwork has now been laid for a discussion of the chemical reactions of transition metal complexes. In the first part of this chapter we have made an attempt to categorize the types of reactions that are characteristic of these species. Later, in Sections 17.5–17.7, some of what we know about the mechanisms of such reactions will be described.

Most reactions of mononuclear complexes fall into one of three categories: those involving the formation and/or rupture of one or more metal–ligand bonds (generally, ligand substitution reactions); those in which the oxidation state of the metal changes (redox reactions); and those in which both of these occur (oxidative additions and reductive eliminations). A fourth category, isomerization, might also be cited. However, isomerizations amount to substitutions if they occur by a bimolecular mechanism. Unimolecular isomerization is related to fluxionality, which is discussed elsewhere in this text (Chapters 15 and 21).

17.1 Precursors: Metal Salts and Solvate Complexes

The ultimate source of the metal(s) present in any coordination compound is, in most cases, an ore. A typical (but far from universal) sequence is

$$\text{Ore} \longrightarrow \text{Free element} \longrightarrow \text{Metal salt} \longrightarrow$$

$$\text{Solvate complex} \longrightarrow \text{Other complexes}$$

The conversion of ores to free elements was discussed in Chapter 15. The second and third steps of the sequence are considered in this section.

Recall that reactions of the type

$$X_2 + Y_2 \longrightarrow 2XY \tag{17.1}$$

are normally exothermic, because heteronuclear bonds are stronger than their parent homonuclear bonds. Thus, it is not surprising that many metal salts can be prepared by simply mixing the free elements. (Heating is often required to overcome the enthalpy of activation.)

$$Ti + 2Cl_2 \xrightarrow{\Delta} TiCl_4 \tag{17.2}$$

$$Cu \xrightarrow[\Delta]{\frac{1}{2}F_2} CuF \xrightarrow[\Delta]{\frac{1}{2}F_2} CuF_2 \tag{17.3}$$

Metal salts often can be converted to *solvate complexes* (those in which solvent molecules act as ligands) by dissolving them in an appropriate (coordinating) solvent. Such reactions may or may not be favored by enthalpy, since lattice energy is lost in the process. They are usually favored by entropy. Nitrogen and oxygen donors (eg, water, ammonia, alcohols, and amines) are commonly used for this purpose. Thus, the following equations represent the solvation of $NiCl_2$ by ammonia and water, respectively:

$$NiCl_2(s) + 6NH_3 \xrightarrow{NH_3} [Ni(NH_3)_6]^{2+} + 2Cl^- \tag{17.4}$$

$$NiCl_2(s) + 6H_2O \xrightarrow{H_2O} [Ni(H_2O)_6]^{2+} + 2Cl^- \tag{17.5}$$

Such complexes can then be used for further transformations.

17.2 Ligand Substitution Reactions

Substitutions Based on Lewis Acid–Base Strength

A process related to that described by equation (17.5) (similar in that it involves the replacement of Cl^- ligands by H_2O) can be accomplished in noncoordinating or weakly coordinating solvents:

$$NiCl_4^{2-} + 6H_2O \longrightarrow [Ni(H_2O)_6]^{2+} + 4Cl^- \tag{17.6}$$

This reaction is exothermic because the Ni^{2+}–OH_2 interactions are stronger than those between Ni^{2+} and Cl^-. (Water is a stronger Lewis base than chloride ion.) In aqueous solution, coordinated water is displaced by stronger bases such as NH_3 and CN^-.

$$[Ni(H_2O)_6]^{2+} \xrightarrow{6NH_3} [Ni(NH_3)_6]^{2+} + 6H_2O \qquad K_{eq} \approx 10^9 \quad (17.7)$$

$$[Ni(H_2O)_6]^{2+} \xrightarrow{4CN^-} [Ni(CN)_4]^{2-} + 6H_2O \qquad K_{eq} \approx 10^{31} \quad (17.8)$$

It is also possible for one metal to displace another from a complex. For example, consider the reaction

$$Ca(EDTA) + Zn^{2+} \longrightarrow Ca^{2+} + Zn(EDTA) \qquad K_{eq} = 6 \times 10^5 \quad (17.9)$$

The large equilibrium constant can be rationalized on the basis of the small size of Zn^{2+} compared to Ca^{2+}, which produces stronger metal–ligand interactions. In fact, the general order of reactivity for divalent metals,

$$Ba^{II} < Sr^{II} < Ca^{II} < Mg^{II} < Mn^{II} < Fe^{II} < Co^{II} < Ni^{II} < Cu^{II} > Zn^{II}$$

known as the *Irving–Williams series*,[1] has long been used to predict the position of equilibrium for certain metal displacement reactions. This series results mainly from electrostatics (note the general relationship to size), but steric and crystal field stabilization energy effects also play a role. The high position of Cu^{II} is due in part to the special stabilization of its hexacoordinate complexes by Jahn–Teller distortion.

The reactivities of many ligands depend on the pH, since conjugates have different basicities from their parent acids and bases. It follows that the control of pH can be used to influence the reactivity of such ligands toward metal cations. Hence, acids and bases are often used to aid the synthesis of metal complexes. An example is the $Cr^{III} + NH_3$ system. Solvation might appear to be a logical way to prepare $[Cr(NH_3)_6]^{3+}$, but unfortunately, the dissolution of $CrCl_3$ into $NH_3(l)$ does not yield complete substitution. The dominant reaction is

$$CrCl_3 + 5NH_3 \xrightarrow{NH_3} [Cr(NH_3)_5Cl]Cl_2 \qquad (17.10)$$

However, NH_2^- is a stronger base than ammonia. Thus, if this reaction is carried out under basic conditions (ie, in the presence of $NaNH_2$), the

---

1.   Irving, H.; Williams, R. J. P. *J. Chem. Soc.* **1953**, 3192; *Nature* **1948**, *162*, 746.

hexaammine complex is rapidly produced:

$$CrCl_3 + 5NH_3 + NH_2^- \xrightarrow{-3Cl^-} [Cr(NH_3)_5(NH_2)]^{2+} \qquad (17.11)$$

$$[Cr(NH_3)_5NH_2]^{2+} + NH_3 \rightleftharpoons [Cr(NH_3)_6]^{3+} + NH_2^- \qquad (17.12)$$

Acids can extract certain ligands from complexes by protonating them. An example is the reaction

$$[Co(NH_3)_5F]^{2+} + H_3O^+ \longrightarrow [Co(NH_3)_5(H_2O)]^{3+} + HF \qquad (17.13)$$

A more dramatic case involves the hexaaqua and hexaammine complexes of nickel(II).[2] In the absence of acid, the greater σ donating ability of ammonia makes $[Ni(NH_3)_6]^{2+}$ the more stable complex [see equation (17.7)]. In acidic aqueous solution, however, the protonation of NH_3 by H_3O^+ promotes the reaction

$$[Ni(NH_3)_6]^{2+} + 6H_3O^+ \longrightarrow [Ni(H_2O)_6]^{2+} + 6NH_4^+ \qquad (17.14)$$

In fact, K_{eq} for equation (17.14) is on the order of 10^{45} at room temperature!

The Use of Le Châtelier's Principle

For reactions in which the equilibrium constant is insufficiently large to give complete conversion, product yields often can be improved by the application of Le Châtelier's principle—usually by control of temperature or concentration, and/or by gas evolution or precipitation.

Based on ligand basicity the reaction

$$CoCl_4^{2-} + 6H_2O \longrightarrow [Co(H_2O)_6]^{2+} + 4Cl^- \qquad (17.15)$$

should be exothermic as written, and in fact $CoCl_2$ dissolves in water to give the pink hexaaquacobalt(II) cation. However, addition of excess Cl^- (most conveniently as HCl) leads to a shift in the product distribution and hence in the color; the solution is blue (the color of $CoCl_4^{2-}$) in 12 M HCl. An intermediate purple color is observed at intermediate concentrations (~ 6 M).

The formation of a gaseous side product often causes substitution and elimination reactions to go to completion. This is especially common in metal

---

2. See Appendix II for a summary of the nomenclature of metal complexes.

carbonyl systems, as illustrated by the sequence

$$2\,Fe(CO)_5 \xrightarrow{\Delta} Fe_2(CO)_9 + CO(g) \qquad (17.16)$$

$$3\,Fe_2(CO)_9 \xrightarrow{\Delta} 2\,Fe_3(CO)_{12} + 3\,CO(g) \qquad (17.17)$$

Gas evolution also can be used to induce counterions to enter the ligand sphere. For example, iodide ion becomes a ligand in the following equations:

$$[Rh(NH_3)_5(H_2O)]I_3 \xrightarrow{100°C} [Rh(NH_3)_5I]I_2 + H_2O(g) \qquad (17.18)$$

$$[Rh(NH_3)_6]I_3 \xrightarrow{>150°C} [Rh(NH_3)_5I]I_2 + NH_3(g) \qquad (17.19)$$

The elimination of H_2O in preference to the more volatile NH_3 is consistent with the greater strength of the $Rh \leftarrow N$ bond. Ammonia is necessarily evolved (albeit at a higher temperature) when no water is present.

Another alternative is precipitation. The formation of an insoluble solid (either the complex itself or a side product) often controls the course of a reaction. The interaction of $PtCl_4^{2-}$ with pyridine is a case in point. Solvation—that is, simply dissolving $K_2[PtCl_4]$ in pyridine—leads to complete substitution:

$$K_2[PtCl_4] + 4py \longrightarrow [Pt(py)_4]Cl_2(s) + 2\,KCl \qquad (17.20)$$

Here, the ionic product precipitates from the moderately polar solvent. However, only two displacements are observed when excess pyridine is added to an aqueous solution of K_2PtCl_4:

$$K_2[PtCl_4] + 2py \xrightarrow{H_2O} cis\text{-}Pt(py)_2Cl_2(s) + 2\,K^+ + 2\,Cl^- \qquad (17.21)$$

The reaction is terminated by the formation of the nonionic product, which precipitates from the polar solvent before further substitution can occur.

The Chelate Effect[3]

We might reasonably expect ammonia and ethylenediamine to exhibit very similar characteristics as ligands, since their K_b's and positions in the

---

3. Chung, C.-S. *J. Chem. Educ.* **1984**, *61*, 1062; Simmons, E. L. *J. Chem. Educ.* **1979**, *56*, 578.

spectrochemical series are nearly identical. On that basis the equilibrium constant for the reaction

$$[Ni(NH_3)_6]^{2+} + 3en \rightleftharpoons [Ni(en)_3]^{2+} + 6NH_3 \qquad (17.22)$$

should be close to 1.0. In fact, this is far from the case; K_{eq} is 1.1×10^9 at 25°C.

This is an example of a general phenomenon known as the *chelate effect*, the natural preference for chelate complexes over those having similar but monodentate ligands. This effect has its basis in entropy. The total number of moles of species increases from 4 to 7 as equation (17.22) proceeds from left to right, and entropy normally favors processes that increase the number of particles. Since $\Delta S°$ is positive and $\Delta H°$ is very close to 0, it follows that $\Delta G°$ must be negative and the equilibrium constant must be greater than 1.

We can gain an intuitive understanding of why chelates are favored by considering the intermediate complex $[Ni(NH_3)_4(en)]^{2+}$. The rupture of an Ni–NH$_3$ bond may be sufficient to cause the loss of an ammonia molecule from the vicinity (by diffusion through the solvent medium). However, if cleavage of one of the Ni–N bonds to an ethylenediamine ligand occurs, that nitrogen is held in proximity to the metal via the covalent Ni–N–C–C–N framework. This greatly enhances the likelihood of bond reformation. Said another way, to lose ethylenediamine from the system requires two Ni–N bond cleavages, and therefore approximately twice the energy as does loss of NH$_3$.

The chelate effect explains a variety of reactions that are inexplicable on the basis of ligand basicity. For example,

$$[Co(NH_3)_6]^{2+} + EDTA^{2-} \xrightarrow{-6NH_3} Co(EDTA) \quad K_{eq} = 8 \times 10^{11} \quad (17.23)$$

$$[Ni(NH_3)_6]^{2+} + 2dmg^- \xrightarrow{-6NH_3} Ni(dmg)_2 \quad K_{eq} = 1 \times 10^8 \quad (17.24)$$

$$dmg^- = \text{dimethylglyoxime anion,} \qquad \underset{{}^-O N}{\overset{H_3C}{\diagdown}}C = C\underset{NOH}{\overset{CH_3}{\diagup}}$$

17.3 Oxidation–Reduction Reactions

The basic principles of electron transfer were discussed in Chapter 9. Only certain aspects specific to metal complexes are considered in this section.

Consider the general reaction

$$FeX_6^{3+} + CoX_6^{2+} \longrightarrow FeX_6^{2+} + CoX_6^{3+} \qquad (17.25)$$
$$\quad d^5 \qquad\quad d^7 \qquad\qquad\quad d^6 \qquad\quad d^6$$

If X is a weak field ligand (ie, if all four complexes are high spin), then the crystal field stabilization energies total $8\,Dq - 2P$ for both the reactants and the products. Conversely, for low spin complexes, the indicated electron transfer results in an energy gain of $10\,Dq - P$. Therefore, for very strong ligand fields ($10\,Dq \gg P$), CFSE's provide a driving force for electron transfer.

An analysis based on MO theory leads to a similar conclusion. In the strong field case, the transferred electron moves from an antibonding e_g^* to a σ nonbonding t_{2g} orbital, with a corresponding increase in stability. Hence, the ligands can exert an important influence on the oxidation and reduction tendencies of the metals to which they are bonded.

The effect of ligands on standard reduction potentials was mentioned earlier (see Table 9.2). It is especially significant for cases in which a high spin–low spin conversion occurs. Consider the Co^{II}/Co^{III} couple. The standard aqueous reduction potential of Co^{III} is very large ($+1.8$ V), making cobalt(III) complexes thermodynamically unstable toward reduction by H_2O. However, strong field ligands discourage this reduction (and in some cases actually promote the oxidation of Co^{II} to Co^{III}) because of the large CFSE of the low spin d^6 configuration. For example, aqueous solutions of the Co^{II}-containing complex anion $[Co(CN)_5(H_2O)]^{3-}$ must be prepared in an inert atmosphere because of its facile oxidation by molecular oxygen.

$$[Co(CN)_5(H_2O)]^{3-} \xrightarrow{\;O_2\;} [Co(CN)_5OH]^{3-} \qquad (17.26)$$

Crystal field stabilization energies also can play a role in determining the stability of metal oxidation states toward disproportionation. As an example, certain Au^{II} complexes are thermodynamically unstable:

$$2\,AuBr_4^{2-} \longrightarrow AuBr_4^{3-} + AuBr_4^{-} \qquad (17.27)$$
$$\quad d^9 \qquad\qquad\qquad d^{10} \qquad\quad d^8$$

The d^8 configuration is strongly favored for square planar geometry, while d^{10} is a common configuration for tetrahedral complexes. Neither geometry is particularly favored for d^9.

17.4 Oxidative Addition and Reductive Elimination Reactions[4]

Recall that d^8 systems show a preference for square planar geometry, while low spin d^6 complexes are virtually always octahedral—observations that are nicely explained by crystal field theory. As a result, it should not be

4. Millstein, D. *Acc. Chem. Res.* **1984**, *17*, 221; Halpern, J. *Acc. Chem. Res.* **1970**, *3*, 386; Collman, J. P.; Roper, W. R. *Adv. Organomet. Chem.* **1968**, *7*, 53.

surprising that the two-electron oxidation of a d^8 complex is often accompanied by an increase in coordination number:

$$\underset{d^8}{Ma_4} + b_2 \longrightarrow \underset{d^6}{Ma_4b_2} \tag{17.28}$$

Reactions of this type are *oxidative additions*. Although there are many variations, the most common cases are characterized by: (1) an increase in the formal oxidation state of the metal by exactly 2 units; and (2) an increase in coordination number by 2.

Typical oxidative addition reagents are small molecules such as X_2, HX, or RX. (Here, X is not necessarily a halogen, nor is R always a carbon-containing moiety.) The process normally involves the cleavage of one bond of the incoming group.

A species widely known for its propensity for oxidative addition reactions is *Vaska's compound*,[5] *trans*-Ir(Pϕ_3)$_2$(CO)Cl. Some examples are given in Figure 17.1.

The reverse of oxidative addition is *reductive elimination*. One way to induce such reactions is by heating to promote ligand dissociation and, in many cases, the evolution of some small molecule as a gas.

$$Ni(PMe_3)_2(CH_3)_2 \xrightarrow{\Delta} Ni(PMe_3)_2 + H_3C–CH_3(g) \tag{17.29}$$

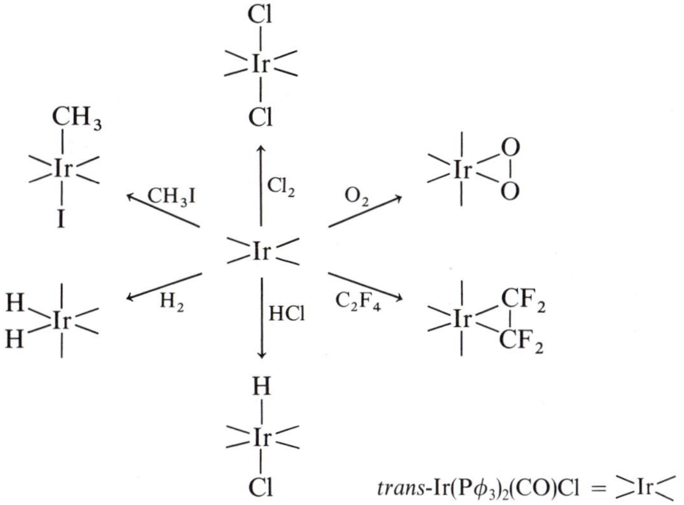

Figure 17.1
Some oxidative addition reactions of Vaska's compound.

5. Vaska, L.; Di Luzio, J. W. *J. Am. Chem. Soc.* **1961**, *83*, 2784.

An oxidative addition, reductive elimination sequence often amounts to the equivalent of a ligand substitution reaction. This is the case for the sequence below, in which the overall result is the substitution of a chloride ligand for CH_3^-:

$$Pt(PEt_3)_2(CH_3)Cl \xrightarrow{+HCl} Pt(PEt_3)_2(Cl)_2(H)CH_3 \qquad (17.30)$$

$$Pt(PEt_3)_2(Cl)_2(H)CH_3 \xrightarrow[\Delta]{-CH_4} Pt(PEt_3)_2Cl_2 \qquad (17.31)$$

The stereochemistry of oxidative addition and reductive elimination reactions is not straightforward. For example, the addition of H_2 to square planar complexes usually occurs in a cis fashion, but the addition of polar molecules is often trans. Other cases are known in which a mixture of isomers is produced. Reductive elimination should reasonably involve the loss of cis groups (to facilitate bond formation between the leaving ligands). But this logic applies only to intramolecular eliminations, and some such reactions are known to be bimolecular.[6]

17.5   Ligand Exchange and Substitution in Octahedral Complexes

Coordination Sphere Theory

Figure 17.2 describes the environment about a complexed metal cation dissolved in a polar solvent. Four types of species are shown. The first, of course, is the metal itself. A second (symbolized by L) represents the ligands; this set is often described as the *first coordination sphere*. This region is characterized by strong interactions (metal–ligand bonds), specific population (a fixed number of ligands), and a high degree of order (fixed bond distances and geometry).

The *second coordination sphere*, X, consists of solvent molecules held in proximity to the complex by hydrogen bonds, ion–dipole, and/or dipole–dipole interactions. These forces are weaker than in the first sphere. The population is variable, and while there is some order (recall that hydrogen bonds and dipoles have directionality), it is less than in the first sphere. The M–X distances are greater than those to the ligands, and unlike the latter they are not fixed.

The *third coordination sphere* is simply the balance of the solvent molecules.

---

6.   See, for example, Jones, W. D.; Bergman, R. G. *J. Am. Chem. Soc.* **1979**, *101*, 5447.

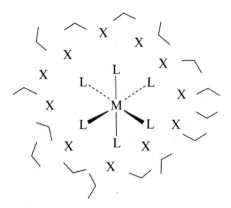

Figure 17.2 Schematic diagram of an octahedral metal complex dissolved in a polar solvent, according to coordination sphere theory. M = metal; L = ligands (first coordination sphere); X = solvent molecules (second coordination sphere); ⌒ = balance of solvent (third coordination sphere).

The relevance of Figure 17.2 to ligand substitution reactions is this: Substitution can be thought of as the exchange of a species between the first and second coordination spheres. That is, the general reaction

$$\text{ML}_n + \text{X} \longrightarrow \text{ML}_{n-1}\text{X} + \text{L} \tag{17.32}$$

represents the exchange of L and X moieties. If L and X have the same identity, the process is commonly described as a ligand exchange reaction; if L and X are chemically different, it is a substitution.

Associative, Dissociative, and Interchange Mechanisms

There are two limiting pathways by which the X ↔ L interchange described above might occur. In one mechanism, the initial step involves the introduction of a new species into the first coordination sphere—that is, the formation of an extra bond to the metal. This is followed by bond rupture and the movement of a (former) ligand away from the metal, returning the first sphere to its original population. This is *associative* activation, and is symbolized in the chemical literature as a.[7] It is essentially equivalent to the S_N2 mechanism of organic chemistry.

---

7. The lowercase, italic a and d (for dissociative) are conventional. A distinction is often made between *intimate* and *stoichiometric* mechanisms; the latter are assigned the capital letters A and D. See Langford, C. H.; Gray, H. B. *Ligand Substitution Processes*; Benjamin: New York, 1965; pp. 7–17.

At the limit of an associative process the transition state is actually an intermediate, since it has a sufficiently long lifetime to be detected experimentally. Said another way, in the two-step sequence

$$ML_n + X \underset{k_{-1}}{\overset{k_1}{\rightleftharpoons}} ML_nX \tag{17.33}$$

$$ML_nX \overset{k_2}{\longrightarrow} ML_{n-1}X + L \tag{17.34}$$

the second step is relatively slow ($k_2 < k_1$), so the expression below applies:

$$\text{Rate} = k_2[ML_nX] \tag{17.35}$$

However, the change in concentration of an intermediate such as ML_nX is not usually measurable in kinetic experiments. It is therefore useful to use the *steady-state approximation*,[8]

$$\frac{d[ML_nX]}{dt} \equiv 0 = k_1[ML_n][X] - k_{-1}[ML_nX] - k_2[ML_nX] \tag{17.36}$$

$$= k_1[ML_n][X] - (k_{-1} + k_2)[ML_nX] \tag{17.37}$$

By solving equation (17.37) for $[ML_nX]$ and substituting the result into equation (17.35), the rate law below is derived:

$$\text{Rate} = \frac{k_1 k_2[ML_n][X]}{k_{-1} + k_2} \tag{17.38}$$

Thus, an associative process exhibits a rate-dependence on both reactants.

Alternatively, the initial step might be bond cleavage and the creation of a vacancy in the first sphere; a species from the second sphere then fills that vacancy. This is *dissociative* (*d*) activation, and is analogous to the organic chemist's S_N1 process. A true dissociative mechanism again requires a relatively slow second step:

$$ML_n \underset{k_{-1}}{\overset{k_1}{\rightleftharpoons}} ML_{n-1} + L \tag{17.39}$$

$$ML_{n-1} + X \overset{k_2}{\longrightarrow} ML_{n-1}X \tag{17.40}$$

---

8. The steady-state approximation is discussed in numerous monographs on kinetics. See, for example, Espensen, J. H. *Chemical Kinetics and Reaction Mechanisms*; McGraw-Hill: New York, 1981; pp. 72*ff*.

In this case, the rate expression is

$$\text{Rate} = k_2[\text{ML}_{n-1}][\text{X}] \tag{17.41}$$

The steady-state approach then yields the rate law

$$\text{Rate} = \frac{k_1 k_2[\text{ML}_n][\text{X}]}{k_{-1}[\text{L}] + k_2[\text{X}]} \tag{17.42}$$

If $k_2[\text{X}] \gg k_{-1}[\text{L}]$, this simplifies to

$$\text{Rate} = k_1[\text{ML}_n] \tag{17.43}$$

indicating a first-order concentration dependence only. If $k_{-1}[\text{L}] \approx k_2[\text{X}]$, however, then the rate is influenced by both $[\text{ML}_n]$ and $[\text{X}]$. In that case, it becomes difficult to distinguish between a and d activation.

The above descriptions merely represent the two extremes of a continuum. Most reactions proceed by processes intermediate between these; that is, the bond forming and bond breaking steps are not completely sequential. A mechanism in which bond formation and bond rupture overlap in time is said to be of the *interchange (I)* type. Interchange mechanisms are divided into I_a, where the transition state is X–ML$_n \cdots$ L, with the bond formation leading the bond rupture; and I_d, primarily dissociative activation, with the transition state X $\cdots$ ML$_n$–L and bond rupture leading bond formation.

Inert and Labile Complexes

For an aqua complex in water, the exchange of H_2O between the first and second coordination spheres does not lead to an observable macroscopic change unless either the ligands or the solvent molecules are isotopically distinct. In that case, the kinetics of the exchange reaction

$$\text{M(H}_2\text{O)}_x^{n+} + \text{H}_2\text{O}^* \;\rightleftharpoons\; [\text{M(H}_2\text{O)}_{x-1}(\text{H}_2\text{O})^*]^{n+} + \text{H}_2\text{O} \tag{17.44}$$

can be experimentally measured. Isotopic labeling is therefore common in the study of such reactions, and H_2O exchange rates for many complexes have been determined in this manner. They show tremendous variation—the first-order rate constants[9] range from about 10^{10} to 10^{-8}/s at 25°C. It has long been conventional to describe complexes having a half-life toward exchange longer than 1 min as *inert*. (This corresponds to a maximum rate constant of roughly 10^{-2}/s.) Those with half-lives shorter than 1 min ($k > 10^{-2}$/s) are *labile*.[10]

---

9. The incoming group in such exchange reactions is a solvent molecule, so its concentration is invariant. Regardless of the mechanism, therefore, (pseudo-) first-order kinetics apply.

10. Taube, H. *Chem. Rev.* **1952**, *50*, 69.

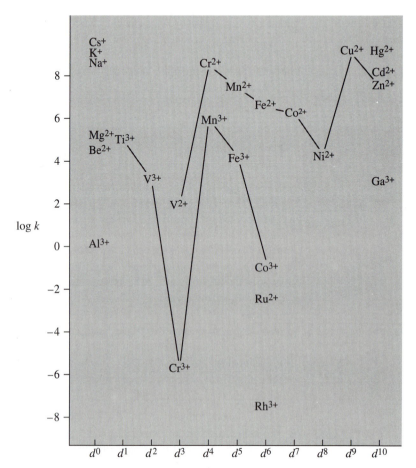

Figure 17.3 Rates of H_2O exchange reactions (plotted as log k versus ground-state configuration) for selected metal cations at 25°C. Lines connect ions that have the same charge and differ by one electron. Data taken from compilations by Wilkens, R. G. *The Study of Kinetics and Mechanism of Reactions of Transition Metal Complexes*; Allyn & Bacon: Boston, 1974; pp. 219, 221; Basolo, F.; Pearson, R. G. *Mechanisms of Inorganic Reactions*, 2nd ed.; Wiley: New York, 1967; p. 152; Eigen, M. *Pure Appl. Chem.* **1963**, 6, 105.

A plot of rate constants for exchange reactions of selected aqua complexes is given in Figure 17.3. Of the metal ions included, the slowest rates belong to Cr^{3+} (which has a d^3 configuration) and Rh^{3+}, Ru^{2+}, and Co^{3+} (d^6).

Two other noteworthy observations can be made:

1. For the main group metals, there is an obvious relationship between the exchange rate and the charge-to-size ratio: Those ions having the

greatest charge densities are the most inert. This is reasonable, since these species engage in the strongest ion–dipole interactions.

2. The rates for transition metal ions are influenced by both charge density and crystal field effects. The role of crystal field theory was first explained by Taube (see footnote 10), who argued that octahedral complexes in which the antibonding e_g^* level is occupied (high spin d^4–d^6 and all d^7–d^{10} species) should be labile, since such electrons readily participate in chemical reactions. Similarly, complexes having a vacant t_{2g} orbital (d^0–d^2) also should be labile. The remaining configurations (d^3 and low spin d^4–d^6) are relatively inert.

A more sophisticated approach is to quantitatively compare the crystal field stabilization energies of reactant complexes to those of their transition states. To understand this method, it must be recognized that inertness and lability are kinetic phenomena, and so depend on the enthalpy of activation (the energy difference between the reactant ground state and the transition state). If a dissociative mechanism is operative, then the most probable geometry of the pentacoordinate transition state is square pyramidal (since it results from the removal of one "leg" of an octahedron). The most likely transition state for associative activation is pentagonal bipyramidal. The CFSE's for octahedral, square pyramidal, and pentagonal bipyramidal complexes (in common, octahedral Dq units) are given in Table 17.1 (p. 556). It can be seen that the d^3, low spin d^4–d^6, and d^8 configurations experience a loss in CFSE for either transition-state geometry. Therefore, metal ions having these configurations should form relatively inert complexes.

Based on this analysis, the predicted order of exchange rates for the divalent metals of the first transition series is

$$V^{2+}, Ni^{2+} < Mn^{2+}, \quad Zn^{2+} < \quad Fe^{2+} \quad < Ti^{2+}, \quad Co^{2+} \quad < \quad Cr^{2+}, \quad Cu^{2+}$$
$$d^3, \quad d^8 \qquad d^5 \text{ (high spin)}, \quad d^{10} \qquad d^6 \text{ (high spin)} \qquad d^2, \quad d^7 \text{ (high spin)} \qquad d^4 \text{ (high spin)}, \quad d^9$$

The experimental order is

$$V^{2+} < Ni^{2+} < Co^{2+} < Fe^{2+} < Mn^{2+}, Zn^{2+} < Cr^{2+} < Cu^{2+}$$

The agreement is not perfect, but there is a definite correlation.

Whether an a or d mechanism is operative depends on several factors. Table 17.2 is Pearson's summary of eight of these, along with their predicted effects on associative and dissociative activation mechanisms.[11] We discuss some of these factors below.

---

11. Pearson, R. G. *J. Chem. Educ.* **1961**, *38*, 164.

Table 17.1 Predictions of relative lability for exchange reactions of octahedral complexes via dissociative and associative activation

d^n, $n =$	Crystal Field Stabilization Energy, Dq				Prediction
	O_h	C_{4v}	D_{5h}	Δ	
1	4.00	4.57	5.28*	+1.28	L, a
2	8.00	9.14	10.56*	+2.56	L, a
3	12.00	10.00*	7.74	−2.00	I
4 (high spin)	6.00	9.14*	4.93	+3.14	L, d
(low spin)	16.00	14.57*	13.02	−1.43	I
5 (high spin)	0.00	0.00	0.00	0.00	—
(low spin)	20.00	19.14*	18.30	−0.86	I
6 (high spin)	4.00	4.57	5.28*	+1.28	L, a
(low spin)	24.00	20.00*	15.48	−4.00	I
7 (high spin)	8.00	9.14	10.56*	+2.56	L, a
(low spin)	18.00	19.14*	12.66	+1.14	L, d
8	12.00	10.00*	7.74	−2.00	I
9	6.00	9.14*	4.93	+3.14	L, d
10	0.00	0.00	0.00	0.00	—

Note: O_h = octahedral reactant; C_{4v} = square pyramidal (dissociative) transition state; D_{5h} = pentagonal bipyramidal (associative) transition state; ΔCFSE = difference between the more favored of the two transition states (marked by *) and the ground state. L = labile; I = inert; a = associative; d = dissociative. See Basolo, F.; Pearson, R. G. *Mechanisms of Inorganic Reactions*, 2nd ed.; Wiley: New York, 1967; p. 146.

Table 17.2 Factors that influence the rates of ligand exchange and substitution in octahedral complexes

	d	a
Increased positive charge of central atom	Decrease	Opposing effects
Increased size of central atom	Increase	Increase
Increased negative charge of entering group	No effect	Increase
Increased size of entering group	No effect	Decrease
Increased negative charge of leaving group	Decrease	Decrease
Increased size of leaving group	Increase	Opposing effects
Increased negative charge of other ligands	Increase	Opposing effects
Increased size of other ligands	Increase	Decrease

Source: Pearson, R. G. *J. Chem. Educ.* **1961**, *38*, 164.

Table 17.3 The effect of the leaving group on ligand substitution reactions of octahedral complexes; rate constants for the acid hydrolysis of $[Co(NH_3)_5X]^{2+}$ [equation (17.45)]

X^-	k_f, s^{-1} (at 25°C)	X^-	k_f, s^{-1} (at 25°C)
NCS^-	5.0×10^{-10}	Cl^-	1.7×10^{-6}
N_3^-	2.1×10^{-9}	Br^-	6.3×10^{-6}
$HC_2O_4^-$	2.2×10^{-8}	I^-	8.3×10^{-6}
F^-	8.6×10^{-8}	NO_3^-	2.7×10^{-5}
$H_2PO_4^-$	2.6×10^{-7}	SO_3F^-	2.2×10^{-2}

Source: Data taken primarily from Basolo, F.; Pearson, R. G. *Mechanisms of Inorganic Reactions*, 2nd ed.; Wiley: New York, 1967; p. 164.

Entering- and Leaving-Group Effects

The role of the leaving group in substitution reactions has been studied for a variety of systems. A well-known example is the acid-catalyzed hydrolysis of a series of octahedral cobalt(III) complexes:

$$[Co(NH_3)_5X]^{2+} + H_2O \underset{k_r}{\overset{k_f}{\rightleftharpoons}} [Co(NH_3)_5(H_2O)]^{3+} + X^- \quad \textbf{(17.45)}$$

In the forward reaction, the entering group is always H_2O, while the leaving group varies. These reactions are amenable to experimental study, because the relative inertness of low spin Co^{III} complexes causes them to occur at rates that are conveniently measured. The rate constants (k_f) for ten such reactions are given in Table 17.3. The data demonstrate a huge leaving-group effect, with the slowest and fastest reactions differing by a factor of over 2×10^8.

The reverse process, in which H_2O is displaced by X^-, is an *anation* reaction. The rate constants [k_r in equation (17.45)] provide a test of the influence of the entering group, since that reactant is varied while the substrate complex remains unchanged. Data for such anations (Table 17.4, p. 558) show much less variation than do those for hydrolysis. The small entering-group and large leaving-group effects are indications that the mechanism is primarily dissociative in nature.

Notice that none of the anions listed in Table 17.4 react with the substrate complex faster than H_2O. This is to be expected from coordination sphere theory. The population of solvent water in the second sphere is greater than that of the reactant anion, so H_2O is statistically more likely to occupy the vacancy created by the initial ligand dissociation. As a result, the rate constants for substitution by X^- cannot be greater than for exchange of H_2O, *if* these reactions occur by a completely dissociative process.[12]

---

12. However, there are ligands that do substitute faster than H_2O. Hence, at least for those reactions, the mechanism is best described as I_d.

Table 17.4 The effect of the entering group on ligand substitution reactions of octahedral complexes; rate constants for anation reactions of $[Co(NH_3)_5(H_2O)]^{3+}$ [the reverse of equation (17.45)]

X^-	k_r, s^{-1} (at 25°C)	X^-	k_r, s^{-1} (at 25°C)
NCS^-	1.3×10^{-6}	NO_3^-	2.3×10^{-6}
$H_2PO_4^-$	2.0×10^{-6}	Br^-	2.5×10^{-6}
Cl^-	2.1×10^{-6}		

Note: For comparison, the rate constant for H_2O exchange by this complex is 6.6×10^{-6} s^{-1} at 25°C.
Source: Data taken from Basolo, F.; Pearson, R. G. *Mechanisms of Inorganic Reactions*, 2nd ed.; Wiley: New York, 1967; p. 203.

From these and other results (see below), it is generally agreed that substitution reactions of Co^{III} complexes proceed by d or I_d mechanisms. This should not be surprising for two reasons. First, the transition-state CFSE for the low spin d^6 configuration is greater for penta- than for heptacoordination (Table 17.1). Also, the relatively small size of Co^{III} disfavors 7-coordination.

The situation is less well-resolved for other metal ions. Rate constants for a series of anation reactions of hexaaquachromium(III) are given in Table 17.5. The entering-group effect is considerably greater than for the cobalt complexes discussed above. In addition, anation by SO_4^{2-} occurs more rapidly than does exchange of H_2O. These data are supportive of an I_a mechanism.

Increasing the size of the metal ion should facilitate 7-coordination and make associative activation more feasible. Thus, it is not surprising that

Table 17.5 Rate constants for some anation reactions of $[Cr(H_2O)_6]^{3+}$

X^-	k, s^{-1} (at 25°C)	X^-	k, s^{-1} (at 25°C)
I^-	8×10^{-10}	HSO_4^-	1.3×10^{-7}
SCN^-	4×10^{-9}	NO_3^-	7.3×10^{-7}
Br^-	9×10^{-9}	NCS^-	1.8×10^{-6}
Cl^-	2.9×10^{-8}	SO_4^{2-}	1.1×10^{-5}

Note: For comparison, the rate constant for H_2O exchange by this complex is 2.5×10^{-6} s^{-1} at 25°C.
Source: Taken from data compiled by Espenson, J. H. *Inorg. Chem.* **1969**, *8*, 1554.

complexes of the 4d and 5d elements (eg, Ru^{III}, Rh^{III}, and Ir^{III}) give evidence for associative behavior.[13]

The Influence of "Uninvolved" Ligands

The five ligands of an octahedral complex that are not directly involved in a substitution reaction also play a role in determining the rate of the reaction. A classic study by Pearson, Boston, and Basolo demonstrated a steric effect for hydrolysis of a series of Co^{III} complexes (Table 17.6):[14]

$$trans\text{-}[Co(en^*)_2Cl_2]^+ + H_2O \xrightarrow{-Cl^-} [Co(en^*)_2(H_2O)Cl]^{2+} \quad \textbf{(17.46)}$$

Here, en* represents a substituted ethylenediamine. The sequential replacement of hydrogen by methyl groups at the carbons of en increases its steric bulk. This should decrease the reaction rate if the mechanism is associative (by increasing the difficulty of incoming attack), but increase it for a dissociative process (by destabilizing the reactant complex and thereby lowering the activation enthalpy). As can be seen in Table 17.6, the data are consistent with dissociative activation.

Nonreacting ligands can influence rates by altering the electron density of the metal ion. The π donating and accepting properties of ligands adjacent to the leaving group are particularly important in this regard. Consider the acid hydrolysis of two Co^{III} species:

$$cis\text{-}[Co(en)_2(X)Cl]^+ + H_2O \xrightarrow{-Cl^-} [Co(en)_2(H_2O)X]^{2+} \quad \textbf{(17.47)}$$

$$X = Cl^-, \quad k = 2.4 \times 10^{-4}\ s^{-1}$$

$$X = CN^-, \quad k = 6.2 \times 10^{-7}\ s^{-1}$$

Table 17.6 Steric effects in acid hydrolysis reactions [Equation (17.46)] of substituted bis(ethylenediamine) complexes of Co^{III}

en*	k, s^{-1} (at 25°C)	en*	k, s^{-1} (at 25°C)
$-CH_2CH_2-$	3.2×10^{-5}	$-CH_2C(CH_3)_2-$	2.2×10^{-4}
$-CH_2CH(CH_3)-$	6.2×10^{-5}	$-C(CH_3)_2C(CH_3)_2-$	Rapid

Source: Pearson, R. G.; Boston, C. R.; Basolo, F. *J. Am. Chem. Soc.* **1953**, 75, 3089.

13. See, for example, Fairhurst, M. T.; Swaddle, T. W. *Inorg. Chem.* **1979**, 18, 3241; Borghi, E.; Monacelli, F.; Prosperi, T. *Inorg. Nucl. Chem. Lett.* **1970**, 6, 667; Monacelli, F. *Inorg. Chem. Acta* **1968**, 2, 263.

14. Pearson, R. G.; Boston, C. R.; Basolo, F. *J. Am. Chem. Soc.* **1953**, 75, 3089.

Since Cl^- is a π donor and CN^- a π acceptor ligand, the former causes the metal to be electron-rich relative to the latter. On that basis, an associative reaction should proceed faster for $X = CN^-$. However, the experimental evidence points in the opposite direction.

Let us therefore assume a dissociative process. How does Cl^- facilitate the reaction? The 5-coordinate transition state has a vacant orbital that lies parallel to the filled p_z orbitals of the cis chlorides (see Figure 17.4), permitting dative π interaction. The resulting stabilization of the transition state lowers the activation energy, and thereby increases the reaction rate. This type of stabilization is possible for a cis, but not a trans, π donating group. Thus, the conclusion is supported by additional data showing that the trans complexes $[Co(en)_2(X)Cl]^+$ ($X = Cl^-$ and CN^-) undergo hydrolysis at approximately equal rates.

Figure 17.4
The stabilization of a square pyramidal transition state (dissociative substitution mechanism) by cis Cl^- ligands (see text).

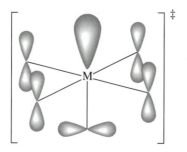

pH Effects: Acid and Base Hydrolysis

The rates of hydrolysis reactions are often accelerated by addition of either acid or base. For example, the rate law for the hydrolysis of the Cr^{III} complex $[Cr(H_2O)_5OAc)]^{2+}$ ($OAc^- =$ acetate ion) contains two independent rate constants:

$$k_{obs} = k_1 + k_2[H^+] \tag{17.48}$$

This suggests two different reaction pathways. The first-order rate constant k_1 is for "normal" behavior. The second term corresponds to a pathway beginning with the protonation of the acetate ligand; this weakens its Lewis basicity and facilitates dissociation. Studies with other substrates indicate that this is a general effect; the replacement by H_2O of ligands that are Brønsted bases is usually faster in acidic than in neutral solutions.

The base-catalyzed hydrolysis of $[Co(NH_3)_5Cl]^{2+}$ also follows second-order kinetics. It might be argued that this is a consequence of the superior nucleophilicity of OH^-, which causes the activation to be associative in nature. However, other data are consistent with a d-type mechanism.[15] It is generally believed that hydroxide ion serves to deprotonate one of the NH_3 ligands. This promotes dissociation by stabilizing the 5-coordinate transition state through ligand-to-metal π donation by NH_2^-:

$$\left[\begin{array}{c} NH_2 \\ \| \\ H_3N \cdots M \cdots NH_3 \\ H_3N \quad NH_3 \end{array} \right]^{\ddagger}$$

Since this mechanism involves the formation of the conjugate base of an "uninvolved" ligand, it is often symbolized S_N1CB.

17.6 Substitution in Square Planar Complexes[16]

In square planar complexes, the four ligands lie in the xy plane, and the metal p_z and d_{z^2} orbitals are σ nonbonding. As a result, such systems are susceptible to attack by Lewis bases along the z axis. Not surprisingly, various studies have shown that ligand substitution reactions occur by associative activation. The transition state is pentacoordinate, and the reaction rates generally depend upon both the reactant complex and the entering group.

A typical rate law for reactions of the type

$$ML_4 \xrightarrow{\;+X\;} [ML_4X] \xrightarrow{\;-L\;} ML_3X \tag{17.49}$$

is

$$\text{Rate} = k_1[ML_4] + k_2[ML_4][X] \tag{17.50}$$

This implies that there are two competing processes, as illustrated in Figure 17.5. The last term of equation (17.50) is consistent with associative activation for direct $X \leftrightarrow L$ interchange. The $k_1[ML_4]$ term arises from association by, and then rapid displacement of, a solvent molecule S:

$$ML_4 \xrightarrow{\;+S\;} [ML_4S] \xrightarrow{\;-L\;} ML_3S \tag{17.51}$$

$$ML_3S \xrightarrow{\;+X\;} [ML_3SX] \xrightarrow{\;-S\;} ML_3X \tag{17.52}$$

15. Buckingham, D. A.; Foxman, B. M.; Sargeson, A. M. *Inorg. Chem.* **1970**, *9*, 1790.

16. Cross, R. J. *Adv. Inorg. Chem.* **1989**, *34*, 219; Hartley, F. R. *Chem. Soc. Rev.* **1973**, *2*, 163; Cattalini, L. *Prog. Inorg. Chem.* **1970**, *13*, 263; Basolo, F.; Pearson, R. G. *Adv. Inorg. Chem. Radiochem.* **1961**, *3*, 1.

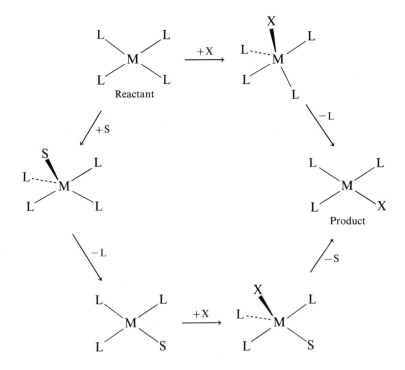

Figure 17.5
Competing mechanisms for ligand substitution by the square planar complex ML_4 to produce ML_3X.
X = entering group;
S = solvent molecule.

If the first step of this sequence is rate-determining, then the resulting rate expression is

$$\text{Rate} = k[ML_4][S] \tag{17.53}$$

Since the concentration of the solvent is constant, however, this amounts to

$$\text{Rate} = k_1[ML_4] \tag{17.54}$$

which is identical to the relevant term of the overall rate law given by equation (17.50).

Supporting evidence for solvent participation in these reactions is readily found. For example, the rate constants k_1 and k_2 for chloride ion exchange by the isotopically labeled complex *trans*-$Pt(py)_2Cl_2^*$ are shown in Table 17.7.

The data fall into two groups. For water and other nucleophilic solvents, $k_1 \gg k_2$, with only the former being measurable; thus, the rate is independent of chloride ion concentration. For poorly nucleophilic solvents (eg, CCl_4 and C_6H_6), $k_1 \ll k_2$. Only the second pathway is observed in such media, and a second-order rate law (rate dependence on both the reactant complex and on Cl^-) results.

Table 17.7 Rate constants for chloride ion exchange by *trans*-Pt(py)$_2$Cl$_2$ in various solvents at 25°C

Solvent	k_1, s^{-1}	k_2, L/mol·s
DMSO	3.8×10^{-4}	
H$_2$O	3.5×10^{-5}	
CH$_3$NO$_2$	3.2×10^{-5}	
EtOH	1.4×10^{-5}	
n-C$_3$H$_7$OH	4.2×10^{-6}	
CCl$_4$		10^4
C$_6$H$_6$		10^2
t-C$_4$H$_9$OH		10^{-1}
(CH$_3$)$_2$C=O		10^{-2}

Source: Taken from data compiled by Basolo, F. *Adv. Chem. Ser.* **1965**, *49*, 81.

The Trans Effect

Substitutions in square planar complexes are influenced to a remarkable extent by a moiety that might appear to be uninvolved in the process—the ligand trans to the eventual leaving group. This *trans effect* has been defined as "the effect of a coordinated group upon the rate of substitution reactions of ligands opposite to it."[17] Thus, for the square planar complex ML$_2$TY (where Y is the leaving group and T is the ligand positioned trans to it), cleavage of the M–Y bond is facilitated by certain T ligands:

$$\begin{array}{c} L\diagdown\quad\diagup Y \\ \quad M \\ T\diagup\quad\diagdown L \end{array} \xrightarrow[-Y]{+X} \begin{array}{c} L\diagdown\quad\diagup X \\ \quad M \\ T\diagup\quad\diagdown L \end{array} \qquad (17.55)$$

The experimental evidence for the trans effect is dramatic. Consider the reaction of [Pt(NH$_3$)Cl$_3$]$^-$ with NH$_3$:

$$[Pt(NH_3)Cl_3]^- + NH_3 \longrightarrow Pt(NH_3)_2Cl_2 + Cl^- \qquad (17.56)$$

Two geometric isomers are possible for the product. In the absence of other factors, a statistical distribution of $\frac{2}{3}$ cis and $\frac{1}{3}$ trans is expected; however, the product "mixture" actually comprises more than 99% of the cis isomer.

---

17. Basolo, F.; Pearson, R. G. *Prog. Inorg. Chem.* **1962**, *4*, 381.

Other examples are described by the following pairs of reactions, which illustrate the value of the trans effect for synthetic purposes:

$$PtBr_4^{2-} \quad \xrightarrow[-Br^-]{+CN^-} \quad [Pt(CN)Br_3]^{2-} \tag{17.57}$$

$$[Pt(CN)Br_3]^{2-} \quad \xrightarrow[-Br^-]{+NH_3} \quad trans\text{-}[Pt(NH_3)(CN)Br_2]^- \tag{17.58}$$

$$PtBr_4^{2-} \quad \xrightarrow[-Br^-]{+NH_3} \quad [Pt(NH_3)Br_3]^- \tag{17.59}$$

$$[Pt(NH_3)Br_3]^- \quad \xrightarrow[-Br^-]{+CN^-} \quad cis\text{-}[Pt(NH_3)(CN)Br_2]^- \tag{17.60}$$

In each sequence, only one of the two possible isomeric products is obtained, uncontaminated by the other. (The Pt–Br bond is cleaved in preference to either Pt–N or Pt–C for two reasons: that bond is the weakest of the three, and Br^- is the best available leaving group.)

Equations (17.58) and (17.60) suggest that the order of trans-directing ability is $CN^- > Br^- > NH_3$. A more complete series is

$$CN^- \approx CO \approx C_2H_4 > PR_3 \approx H^- > CH_3^- > NO_2^- \approx I^- \approx SCN^-$$

$$> Br^- \approx Cl^- > py > NH_3 \approx OH^- > H_2O$$

Approximately the same ordering of ligands has been encountered before in the spectrochemical series. As in the latter, reversals in the trans-directing series are occasionally observed (depending on the metal, etc.).

The trans effect is kinetic in nature.[18] The stereospecificity results from the fact that bond cleavage occurs much faster at one site than the others. Kinetic studies indicate that the difference in rates is often on the order of 10^6. (To appreciate what a large difference this is, consider that reaction half-lives of 1.0 s versus 12 days are in that ratio!)

A complete understanding of the trans-directing series requires consideration of both σ bonding (which primarily affects the ground-state stability) and π bonding (which may affect either the ground state or transition state). We will consider the σ contribution first.

Two trans ligands in a square planar complex must bond with the same metal orbital. If one of the bases is a significantly better σ donor than the opposite ligand, then the strength of the interaction of the latter with the

---

18.   As will shortly be seen, there is evidence for a thermodynamic (bond weakening) effect as well. The term *trans influence*, rather than *trans effect*, is often used when discussing the thermodynamic aspects of this subject.

Figure 17.6
The π interactions involving mutually trans ligands in a square planar complex; T = π acceptor trans ligand; Y = leaving group.

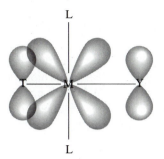

metal will be reduced. This simple notion explains the high positions of H$^-$ and CH$_3^-$ (both strong, σ-only bases) in the trans-directing series.

Metal–ligand π bonding generally involves the side-to-side overlap of ligand p with metal d orbitals. The situation with respect to one of the bond axes of a square planar complex (ie, for a trans T–M–Y linkage) is shown in Figure 17.6. For the different ligands T and Y, the π interactions with the metal orbitals are unequal—one or the other is preferred for overlap. If T is the stronger π bonding ligand, then the M–Y bond is weakened and more easily cleaved. In this way, the general similarity between the trans-directing series and the spectrochemical series can be rationalized, since both depend on the degree and type of metal–ligand π bonding.

Supporting evidence for this idea is found in bond distances. Bond lengths in some square planar complexes are given in Table 17.8. It can be

Table 17.8 Pt–Cl bond distances in some square planar complexes

Complex	Trans Ligand	d(Pt–Cl), pm
PtCl$_4^{2-}$	Cl$^-$	233
[Pt(NH$_3$)Cl$_3$]$^-$	NH$_3$	232
	Cl$^-$	235
[Pt(H)Cl$_3$]$^{2-}$	H$^-$	232
	Cl$^-$	230
trans-Pt(NH$_3$)$_2$Cl$_2$	Cl$^-$	230
trans-Pt(PEt$_3$)$_2$Cl$_2$	Cl$^-$	230
cis-Pt(PMe$_3$)$_2$Cl$_2$	PMe$_3$	238
Pt(C$_2$H$_4$)Cl$_3$	C$_2$H$_4$	234
	Cl$^-$	230

Note: In the trans-directing series, C$_2$H$_4$ > PMe$_3$ $\approx$ H$^-$ > Cl$^-$ > NH$_3$.
Source: Data compiled primarily from that given by Wells, A. F. *Structural Inorganic Chemistry*, 5th ed.; Clarendon: Oxford, 1984.

seen that Pt–Cl bonds tend to be longer (and presumably weaker) when trans to a π acceptor ligand such as PR_3 or C_2H_4.[19]

These ligands also play a role in stabilizing the trigonal bipyramidal transition state, where there are four nonbonding d orbitals capable of side-to-side overlap. Most or all of these orbitals are filled in the common square planar configurations (eg, d^8). Moreover, the metal is electron-rich in such a transition state (relative to the reactant ground state) because of the presence of five, rather than four, ligands. Therefore, ligands that accept π electron density stabilize the transition state, and thereby enhance the rate of ligand substitution.

17.7 The Mechanisms of Electron Transfer Reactions[20]

Consider the aqueous electron transfer process

$$[Fe^*(H_2O)_6]^{2+} + [Fe(H_2O)_6]^{3+} \rightleftharpoons$$
$$[Fe^*(H_2O)_6]^{3+} + [Fe(H_2O)_6]^{2+} \quad \textbf{(17.61)}$$

which can be followed using an isotopically labeled reactant. The rate of electron exchange varies markedly with the pH, being much faster in basic solution. More specifically, the rate equation below is obeyed:

$$k_{obs} = k_1 + \frac{k_2 K_h}{[H^+]} \quad \textbf{(17.62)}$$

Here, k_1 and k_2 are both second-order rate constants (first-order in each reagent), and K_h is the hydrolysis constant for $[Fe(H_2O)_6]^{3+}$ (see Section 10.4). This intriguing result is evidence for two competing processes. In the first (the *outer sphere* mechanism, with rate constant k_1), electron transfer is believed to occur directly from one metal center to another. The second (*inner sphere*) mechanism involves the formation of the bridged complex $[(H_2O)_5Fe–OH–Fe(OH_2)_6]^{4+}$, where a hydroxo ligand is simultaneously bound to both metals. The reason for the pH dependence then becomes obvious—the formation of the bridged intermediate requires the deprotonation of $[Fe(H_2O)_6]^{3+}$ to $[Fe(H_2O)_5OH]^{2+}$ (H_2O is not sufficiently basic to bind to two metal ions simultaneously). The transferred electron therefore "hops across the bridge" via the ligand orbitals.

---

19. Bond length data are more properly used when discussing the trans influence (see footnote 18.)

20. Bennett, L. E. *Prog. Inorg. Chem.* **1973**, *18*, 2; Taube, H. *Electron Transfer Reactions of Complex Ions in Solution*; Academic: New York, 1970; Sykes, A. G. *Adv. Inorg. Chem. Radiochem.* **1967**, *10*, 153.

Electron transfer appears to occur by an inner-sphere pathway whenever possible. That process is generally the faster of the two, provided, of course, that an appropriate bridging ligand is available. For example, consider the oxidation of Cr^{II} by two similar Cr^{III} complexes, $[Cr(H_2O)_6]^{3+}$ and $[Cr(H_2O)_5Br]^{2+}$. No bridging ligand is present in the hexaaqua complex (at least at low pH), so the electron transfer is necessarily of the outer-sphere type. The experimental rate constant is less than 2×10^{-5} L/mol·s at 25°C. Conversely, in $[Cr(H_2O)_5Br]^{2+}$, the bromide can act as a bridging ligand. Thus, the inner-sphere pathway is possible, and $k_{obs} = 6 \times 10^1$ L/mol·s, over 3 million times faster! These two mechanisms are considered in more depth below.

Outer-Sphere Reactions[21]

Outer-sphere reactions are thought to occur through the following sequence:

1. The formation of a *precursor complex*, in which the oxidant and reductant are momentarily trapped within a solvent cage

2. Activation to some type of excited state

3. Reorganization to a *successor complex*, in which the metals have acquired new oxidation states

4. Dissociation of the oxidized and reduced products

We will use the exchange reaction

$$[Cr(H_2O)_6]^{2+} + [Cr^*(H_2O)_6]^{3+} \rightleftharpoons$$
$$[Cr(H_2O)_6]^{3+} + [Cr^*(H_2O)_6]^{2+} \quad \textbf{(17.63)}$$

in which the reactants contain isotopically different chromiums, to illustrate this sequence. The precursor complex can be thought of as a loosely bound unit in which the oxidant is in the second coordination sphere of the reductant, and vice versa. Next, activation is required. The Cr–O equilibrium bond lengths are unequal in the two reactants (Cr^{II} is larger than Cr^{III}), and these distances must change before the electron can be transferred. (The movement of atoms is much slower than that of electrons, so the nuclei must move before, not during, the electron transfer.) The transition state presumably contains equal bond distances, since that represents the greatest degree of distortion necessary. After the electron is transferred, the bonds of the successor complex readjust to the new equilibrium distances. This is

---

21. Sutin, N. *Prog. Inorg. Chem.* **1982**, *30*, 441; Taube, H. *J. Chem. Educ.* **1968**, *45*, 452.

Table 17.9 Rate constants for some outer-sphere electron transfer reactions between metal complexes

Oxidant	Reductant	k, L/mol·s (at 25°C)
Exchange reactions		
$[V(H_2O)_6]^{3+}$	$[V(H_2O)_6]^{2+}$	0.01
$[Cr(H_2O)_6]^{3+}$	$[Cr(H_2O)_6]^{2+}$	$<2 \times 10^{-5}$
MnO_4^-	MnO_4^{2-}	3.6×10^{-3}
$[Fe(H_2O)_6]^{3+}$	$[Fe(H_2O)_6]^{2+}$	4.0
$[Fe(CN)_6]^{3-}$	$[Fe(CN)_6]^{4-}$	7.4×10^2
$[Fe(phen)_3]^{3+}$	$[Fe(phen)_3]^{2+}$	3×10^7
$[Co(NH_3)_6]^{3+}$	$[Co(NH_3)_6]^{2+}$	3×10^{-12}
$[Co(en)_3]^{3+}$	$[Co(en)_3]^{2+}$	2×10^{-5}
$[Co(phen)_3]^{3+}$	$[Co(phen)_3]^{2+}$	1.1
$[Ru(NH_3)_6]^{3+}$	$[Ru(NH_3)_6]^{2+}$	8.2×10^2
$[Ru(phen)_3]^{3+}$	$[Ru(phen)_3]^{2+}$	$>10^7$
Cross reactions		
$[Fe(H_2O)_6]^{3+}$	$[Cr(H_2O)_6]^{2+}$	2.3×10^3
$[Fe(H_2O)_6]^{3+}$	$[V(H_2O)_6]^{2+}$	1.8×10^4
$[Co(NH_3)_6]^{3+}$	$[Cr(H_2O)_6]^{2+}$	8.9×10^{-5}
$[Co(NH_3)_6]^{3+}$	$[Cr(bipy)_3]^{2+}$	2.5×10^2
$[Co(NH_3)_5(H_2O)]^{3+}$	$[Cr(H_2O)_6]^{2+}$	0.5

Sources: Data taken from compilations by Wilkens, R. G. *The Study of Kinetics and Mechanism of Reactions of Transition Metal Complexes*; Allyn & Bacon: Boston, 1974; Chapter 5; Taube, H. *Electron Transfer Reactions of Complex Ions in Solution*; Academic: New York, 1970; Basolo, F.; Pearson, R. G. *Mechanisms of Inorganic Reactions*, 2nd ed.; Wiley: New York, 1967; Chapter 6.

followed by separation (diffusion into the solution) of the oxidized and reduced products.

Experimental rate constants for a variety of outer-sphere reactions are given in Table 17.9.

Two observations can be made relevant to the data in the table:

1. The most rapid outer-sphere transfers occur when the electron in question originates from and enters an e_g^*, rather than t_{2g}, molecular orbital, because the e_g^* orbitals are the more accessible of the two types. They are partly ligand in character, and therefore extend further into space than the σ nonbonding t_{2g} orbitals.

2. The ligands affect the rate of electron transfer. (Compare the rates for $[Fe(H_2O)_6]^{2+/3+}$, $[Fe(CN)_6]^{3-/4-}$, and $[Fe(phen)_3]^{2+/3+}$.) Ligands might be involved in the transfer in any of at least three ways: *chemical exchange*, in which the ligand is temporarily reduced and then reoxidized; *double exchange*, in which the ligand accepts one electron from

the reducing agent and then donates a different electron to the oxidizing agent; or (for ligands with π systems) *superexchange*, in which a vacant π^* ligand orbital serves as a temporary repository for the electron.

Electron transfer from one kind of metal to another is a *cross reaction* (as opposed to an exchange). The rates of cross reactions are usually greater than those for the parent exchange processes. Marcus proposed the relationship[22]

$$k_{12} = (k_{11}k_{22}K_{12}f)^{1/2} \tag{17.64}$$

where k_{12} is the rate constant and K_{12} is the equilibrium constant for the cross reaction, and k_{11} and k_{22} are the rate constants for the parent exchange reactions. The f term is defined by the equation

$$\log f = \frac{(\log K_{12})^2}{4\log(k_{11}k_{22}/z^2)} \tag{17.65}$$

where z is the collision frequency. Since it often approaches a value of 1, f is sometimes neglected.

Equation (17.64) can be used to predict the rate constants for cross reactions. As an example, the relevant data for the one-electron oxidation of $[Fe(CN)_6]^{4-}$ by $IrCl_6^{2-}$ at 25°C are: $k_{11}[Fe^{II}/Fe^{III}] = 7.4 \times 10^2$ L/mol·s; $k_{22}[Ir^{III}/Ir^{IV}] = 2.3 \times 10^5$ L/mol·s; and $K_{12} = 1.2 \times 10^4$. Ignoring f, the predicted rate k_{12} is calculated to be 1×10^6 L/mol·s, compared to the experimental value of 4×10^5 L/mol·s. This is typical of the level of accuracy obtained.

Inner-Sphere Reactions[23]

The first unequivocal example of an inner-sphere process was provided by Taube and co-workers.[24] Their seminal study involved the reduction of a Co^{III} complex by hexaaquachromium(II):

$$[Co(NH_3)_5Cl]^{2+} + [Cr(H_2O)_6]^{2+} \xrightarrow{H^+, H_2O}$$

$$[Cr(H_2O)_5Cl]^{2+} + [Co(H_2O)_6]^{2+} + 5NH_4^+ \tag{17.66}$$

The complex $[Co(H_2O)_6]^{2+}$ is formed from the initial reduction product $[Co(NH_3)_5Cl]^+$, which is labile toward ligand substitution (high spin d^7) and therefore undergoes rapid acid-catalyzed hydrolysis.

22. Marcus, R. A. *J. Phys. Chem.* **1963**, *67*, 853. See also, Marcus, R. A. *Ann. Rev. Phys. Chem.* **1964**, *15*, 155.

23. Haim, A. *Prog. Inorg. Chem.* **1982**, *30*, 273; *Acc. Chem. Res.* **1975**, *8*, 265.

24. Taube, H.; Myers, H. *J. Am. Chem. Soc.* **1954**, *76*, 2103; Taube, H.; Myers, H.; Rich, R. L. *J. Am. Chem. Soc.* **1953**, *75*, 4118.

The transfer of Cl^- from cobalt to chromium is strongly suggestive of a Co–Cl–Cr bridged intermediate, and hence, of an inner-sphere process. However, it might be argued that ligand transfer occurs in an indirect manner. A sequence of outer-sphere electron transfer, followed by the hydrolysis of $[Co(NH_3)_5Cl]^+$ to give free Cl^- in solution, followed by anation of $[Cr(H_2O)_6]^{3+}$ by the freed Cl^- might seem feasible. The proof of the inner-sphere mechanism is that the experimental rate constant for the redox reaction (about 6×10^5 L/mol·s at 25°C) is many orders of magnitude greater than for the anation of $[Cr(H_2O)_6]^{3+}$ (3×10^{-8} s^{-1}). This rules out the outer-sphere transfer, hydrolysis, anation sequence.

An inner-sphere mechanism is generally indicated whenever an electron transfer reaction also gives ligand transfer. However, the converse is not true. The lack of ligand transfer does not obviate an inner-sphere process, since the bridging ligand might remain with its original metal. Consider the following two cases. After electron transfer has occurred, the successor complex of equation (17.66) is $[(H_2O)_5Cr^{III}–Cl–Co^{II}(NH_3)_5]^{4+}$. This successor complex contains a metal that is inert toward ligand substitution (Cr^{III}, d^3) and one that is labile (Co^{II}). Thus, the bridging chloride is held in the first coordination sphere of chromium upon rupture of the bridge. However, the situation is different for the reduction of $[Fe(CN)_6]^{3-}$ by Cr^{II}, where the successor complex is $Cr^{III}–NC–Fe^{II}$, and both metals are substitution inert (the π acceptor ligands cause Fe^{II} to be low spin). Therefore, ligand transfer may or may not occur.

Experimental rate constants for a variety of inner-sphere reactions are given in Table 17.10. The relative abilities of ligands to serve as bridges can be inferred from the table. For example, for the reduction of $[Co(H_2O)_5X]^{2+}$ by Cr^{II}, the relative bridging ability is

$$OH^- > Cl^- > F^- > SCN^- > NCS^-$$

It is not surprising that the best bridging ligand, OH^-, is also the strongest base. That Cl^- is more efficient than F^- is probably a steric effect. Hard–soft acid–base theory can be used to rationalize the difference between thiocyanate and isothiocyanate. These bridges are thought to form through *remote attack*; that is, the incoming Cr^{II} attacks at an atom not directly bonded to the cobalt. Thus, in one case the bridge is Co–S=C=N–Cr, while in the other it is Co–N=C=S–Cr. Since Cr^{II} is a hard acid, with a greater affinity for nitrogen than for sulfur, N-attack at Co–SCN is faster than S-attack at Co–NCS.

Bridged Binuclear Complexes

Before closing the discussion of bridging ligands, we will briefly consider compounds that contain permanent bridges (ie, isolable complexes in which

Table 17.10 Rate constants for some inner-sphere electron transfer reactions between metal complexes

Oxidant	Reductant	k, L/mol·s
Exchange reactions		
$[Cr(H_2O)_5OH]^{2+}$	$[Cr(H_2O)_6]^{2+}$	0.7
$[Cr(H_2O)_5F]^{2+}$	$[Cr(H_2O)_6]^{2+}$	2.7×10^{-4}
$[Cr(H_2O)_5Cl]^{2+}$	$[Cr(H_2O)_6]^{2+}$	5.1×10^{-2}
$[Cr(H_2O)_5Br]^{2+}$	$[Cr(H_2O)_6]^{2+}$	0.32
$[Cr(H_2O)_5I]^{2+}$	$[Cr(H_2O)_6]^{2+}$	5.5
$[Cr(H_2O)_5CN]^{2+}$	$[Cr(H_2O)_6]^{2+}$	7.7×10^{-2}
$[Fe(H_2O)_5F]^{2+}$	$[Fe(H_2O)_6]^{2+}$	9.7 (0°C)
$[Fe(H_2O)_5Cl]^{2+}$	$[Fe(H_2O)_6]^{2+}$	5.4 (0°C)
$[Fe(H_2O)_5Br]^{2+}$	$[Fe(H_2O)_6]^{2+}$	4.9 (0°C)
$[Fe(H_2O)_5NCS]^{2+}$	$[Fe(H_2O)_6]^{2+}$	4.2 (0°C)
$[Fe(H_2O)_5N_3]^{2+}$	$[Fe(H_2O)_6]^{2+}$	1.9×10^3 (0°C)
$[Fe(H_2O)_5OH]^{2+}$	$[Fe(H_2O)_6]^{2+}$	1.01×10^3
Cross reactions		
$[Co(H_2O)_5F]^{2+}$	$[Cr(H_2O)_6]^{2+}$	2.5×10^5
$[Co(H_2O)_5Cl]^{2+}$	$[Cr(H_2O)_6]^{2+}$	6×10^5
$[Co(H_2O)_5Br]^{2+}$	$[Cr(H_2O)_6]^{2+}$	1.4×10^6
$[Co(H_2O)_5N_3]^{2+}$	$[Cr(H_2O)_6]^{2+}$	3×10^5
$[Co(H_2O)_5NCS]^{2+}$	$[Cr(H_2O)_6]^{2+}$	19
$[Co(H_2O)_5SCN]^{2+}$	$[Cr(H_2O)_6]^{2+}$	61
$[Co(NH_3)_5F]^{2+}$	$[Cr(H_2O)_6]^{2+}$	2.5×10^5
$[Co(NH_3)_5Cl]^{2+}$	$[Cr(H_2O)_6]^{2+}$	6×10^5
$[Co(NH_3)_5Br]^{2+}$	$[Cr(H_2O)_6]^{2+}$	1.4×10^6
$[Co(NH_3)_5I]^{2+}$	$[Cr(H_2O)_6]^{2+}$	3×10^6
$[Co(NH_3)_5OH]^{2+}$	$[Cr(H_2O)_6]^{2+}$	1.5×10^6
$[Co(H_2O)_5F]^{2+}$	$[Fe(H_2O)_6]^{2+}$	6.6×10^{-3}
$[Co(H_2O)_5Cl]^{2+}$	$[Fe(H_2O)_6]^{2+}$	1.4×10^{-3}
$[Co(H_2O)_5Br]^{2+}$	$[Fe(H_2O)_6]^{2+}$	7.3×10^{-4}

Note: Rate constants measured at 25°C, unless otherwise indicated.
Source: Data taken from compilations by Wilkens, R. G. *The Study of Kinetics and Mechanism of Reactions of Transition Metal Complexes*; Allyn & Bacon: Boston, 1974; Chapter 5; Taube, H. *Electron Transfer Reactions of Complex Ions in Solution*; Academic: New York, 1970; Basolo, F.; Pearson, R. G. *Mechanisms of Inorganic Reactions*, 2nd ed.; Wiley: New York, 1967; Chapter 6.

two metal centers share one or more ligands). Many such species are known. The most common bridging anions include the halides and pseudohalides; O^{2-} and OH^-; and NH_2^-, NH^{2-}, and N^{3-}.[25]

---

25. Certain carbon-containing groups might be included as well. Such species will be discussed in Chapter 18.

The synthesis of bridged complexes might involve ligand replacement, oxidation–reduction, or some combination of the two. For example, dissolution of $RuCl_3(s)$ into concentrated ammonia water gives, among other products, a doubly bridged species:[26]

$$RuCl_3 \xrightarrow{NH_3/H_2O} \left[(H_3N)_4Ru \underset{NH_2}{\overset{NH_2}{<>}} Ru(NH_3)_4 \right] Cl_4 \qquad (17.67)$$

Here, the two RuN_6 subunits are pseudo-octahedral. Note that unlike NH_3, the amide ion can act as a bridging ligand, since it has both sufficient basicity and a second lone pair of electrons. It follows that NH^{2-} and N^{3-} also form effective bridges, as, for example, in the $[Ru_2Cl_8(H_2O)_2N]^{3-}$ anion:[27]

$$[Ru(NO)Cl_5]^{2-} \xrightarrow[HCl]{SnCl_2} [(H_2O)(Cl)_4Ru-N-Ru(Cl)_4(H_2O)]^{3-} \quad (17.68)$$

It is also possible for two different donor atoms of a covalently bonded unit to bridge metal centers. An example is the Rh^I complex

where COD is the bidentate ligand cyclooctadiene and the bridging atoms are the nitrogens of two imidazole rings.[28]

Mixed-valence complexes, in which the oxidation states of the metals are different, are particularly interesting from the standpoint of oxidation–reduction.[29] Consider the system

$$[(H_3N)_5Ru-N\equiv C-C\equiv N-Ru(NH_3)_5]^{n+}$$

At least three different cations, $n = 4, 5$, and 6, are known. For $n = 4$, spectroscopic data are consistent with two Ru^{II} centers. The $n = 6$ ion

26. Flood, M. T.; Ziolo, R. F.; Earley, J. E.; Gray, H. B. *Inorg. Chem.* **1973**, *12*, 2153.

27. Cleare, M. J.; Griffith, W. P. *J. Chem. Soc. A* **1970**, 117.

28. Kaiser, S. W.; Saillant, R. B.; Butler, W. M.; Rasmussen, P. G. *Inorg. Chem.* **1976**, *15*, 2681.

29. Cruetz, C. *Prog. Inorg. Chem.* **1982**, *30*, 1; Wong, K. Y.; Schatz, P. N. *Prog. Inorg. Chem.* **1981**, *28*, 369; Meyer, T. J. *Acc. Chem. Res.* **1978**, *11*, 94.

contains two Ru^{III} moieties. Both localized (Ru^{II} and Ru^{III}) and delocalized (two $Ru^{II^{1/2}}$) possibilities exist for $n = 5$. Infrared data favor the latter.

Bibliography

Basolo, F.; Johnson, R. *Coordination Chemistry*; Science Reviews: Northwood, 1987.

Katakis, D.; Gordon, G. *Mechanisms of Inorganic Reactions*; Wiley: New York, 1987.

Atwood, J. D. *Inorganic and Organometallic Reaction Mechanisms*; Brooks/Cole: Pacific Grove, CA, 1985.

Wilkens, R. G. *The Study of Kinetics and Mechanism of Reactions of Transition Metal Complexes*; Allyn & Bacon: Boston, 1974.

Tobe, M. L. *Inorganic Reaction Mechanisms*; Nelson: London, 1972.

Taube, H. *Electron Transfer Reactions of Complex Ions in Solution*; Academic: New York, 1970.

Basolo, F.; Pearson, R. G. *Mechanisms of Inorganic Reactions*, 2nd ed.; Wiley: New York, 1967.

Sykes, A. G. *Kinetics of Inorganic Reactions*; Pergamon; London, 1966.

Langford, C. H.; Gray, H. B. *Ligand Substitution Processes*; Benjamin: New York, 1965.

Questions and Problems

1. Explain the significance of the Irving–Williams series in your own words. Discuss the reasons behind the ordering, and give examples.

2. Consider the hydrolysis reaction

$$[Cr(H_2O)_5X]^{2+} + H_2O \xrightarrow{\ -X^-\ } [Cr(H_2O)_6]^{3+}$$

   (a) Do you expect the equilibrium constant to be greater for X = F or X = I? Defend your answer.
   (b) Should the addition of acid have a greater effect on the position of equilibrium for X = F or for X = I? Why?

3. When the reaction described by equation (17.15) is performed so that the equilibrium concentration of Cl^- is 6 M, the concentrations of $CoCl_4^{2-}$ and $[Co(H_2O)_6]^{3+}$ are approximately equal to one another.
   (a) Estimate K_{eq} for the reaction.
   (b) Rationalize the fact that $K_{eq} > 1.0$.

4. The preparation of $[Fe(bipy)_3]^{2+}$ (bipy = bipyridine) according to the equation

$$FeCl_2 + 3bipy \longrightarrow [Fe(bipy)_3]Cl_2$$

fails when a solvation type of reaction is attempted. When the reaction is carried out in ethanol solvent, however, the indicated product is obtained. Explain.

5. Explain the chelate effect in your own words, and rationalize it on the basis of thermodynamics.

6. Give the probable product(s) for the following reactions. Indicate the number of possible isomers in each case.

   (a) $[Rh(NH_3)_4]^+ + CH_3I \longrightarrow$

   (b) $Ir(P\phi_3)_2(CO)Cl + H_2 \longrightarrow$

   (c) $[Pt(H)(CH_3)Cl_4]^{2-} \xrightarrow{\Delta}$

7. When $CH_3Rh(P\phi_3)_3$ is heated in an atmosphere of H_2, the formation of methane is observed.
   (a) Identify the ultimate rhodium-containing product.
   (b) Sketch the probable intermediate for this reaction.

8. Explain why the rate of H_2O exchange in aqueous solution is greater for:
   (a) Cs^+ than for K^+ (b) Cr^{2+} than for Cr^{3+}
   (c) Zn^{2+} than for Ga^{3+} (d) V^{3+} than for V^{2+}
   (e) Hg^{2+} than for Zn^{2+} (f) $[Fe(H_2O)_6]^{2+}$ than for $[Fe(CN)_5H_2O]^{3-}$

9. Use Figure 17.3 and Taube's definition to classify the following cations as labile or inert toward ligand exchange: Al^{3+}, Cr^{2+}, Cr^{3+}, Mn^{2+}, Rh^{3+}, Ni^{2+}, and Cu^{2+}.

10. Compare the predictions of Taube to those of Basolo and Pearson concerning lability toward ligand exchange.
    (a) The two methods give different predictions for which d electron configuration(s)?
    (b) Which of the methods appears to be more accurate? Defend your answer.

11. The rate of H_2O exchange by $[M(H_2O)_6]^{2+}$ is greater for $M = Fe$ than for $M = Ru$ by a huge amount (a factor of roughly 10^8). Why?

12. Relatively little is known about the mechanism of ligand substitution for Cu^{2+} complexes. The reason is related to experimental difficulty. Speculate.

13. Explain why the rate of hydrolysis for $[Co(NH_3)_5X]^{2+}$ ions increases in the order $X = F^- < Cl^- < Br^- < I^-$. Your answer should be consistent with the postulated mechanism.

14. Predict how the rate constant for the anation reaction

    $$[Co(NH_3)_5(H_2O)]^{3+} + Cl^- \longrightarrow [Co(NH_3)_5Cl]^{2+} + H_2O$$

    changes as the pH of the solution is increased. Explain your reasoning.

15. Argue that the data of Table 17.5 are consistent with the notion that the mechanism of substitution of SO_4^{2-} at $[Cr(H_2O)_6]^{3+}$ is of the S_N1CB type. Be specific and give details.

16. The rate constants for the hydrolysis of $[Cr(H_2O)_5OAc]^{2+}$ are $k_1 = 4.1 \times 10^{-7} s^{-1}$ and $k_2 = 7.70 \times 10^{-4}$ L/mol·s. Calculate the pH at which the rates of the two competing mechanisms are equal.

17. Explain in your own words why the addition of base increases the rate of hydrolysis of $[Co(NH_3)_5Cl]^{2+}$.

18. For each of the following hypothetical situations concerning ligand substitution reactions, decide whether a or d activation is indicated:
    (a) H_2O displacement from the complex $[ML_5(H_2O)]^{2+}$ is faster by S^{2-} than by Cl^-.
    (b) Substitution by PR_3 for R = Me, Et, and i-Pr all occur at the same rate with a certain substrate.
    (c) Hydrolysis of $[ML_5X]^{2-}$ is more rapid for L = Et_3N than for L = Me_3N.

19. Write the platinum-containing product for each of the following reactions, and indicate its stereochemistry:

    (a) $[Pt(NH_3)Cl_3]^- + NO_2^- \longrightarrow$
    (b) $[Pt(CN)Cl_3]^{2-} + NO_2^- \longrightarrow$
    (c) $PtCl_4^{2-} + 2P\phi_3 \longrightarrow$

20. Consider the species $Pt(py)(NH_3)(CN)Cl$.
    (a) How many stereoisomers are possible for this formula?
    (b) One isomer can be prepared from $PtCl_4^{2-}$ via the sequence

    $$PtCl_4^{2-} \xrightarrow{\text{py}} \xrightarrow{NH_3} \xrightarrow{CN^-} \begin{array}{c} py \\ \diagdown \\ Cl \end{array} Pt \begin{array}{c} NH_3 \\ \diagup \\ CN \end{array}$$

    Explain why this isomer results.
    (c) Five other three-reaction sequences are possible. Give the expected product for each.

21. For Fe^{II}/Fe^{III} electron exchange in aqueous solution, $k_1 = 4.0$ and $k_2 = 3 \times 10^3$ at 25°C. The hydrolysis constant K_h for $[Fe(H_2O)_6]^{3+}$ is 6.5×10^{-3} at that temperature. Which exchange process—inner- or outer-sphere—dominates in 0.1 M Fe^{3+} solutions at pH = 7.0?

22. Use the Marcus equation (17.64) and data presented in the chapter to estimate the rate constant for electron transfer from $[Fe(CN)_6]^{4-}$ to MnO_4^-. The K_{eq} for this reaction is 2.5×10^3.

23. Consider electron transfer from $[Cr(H_2O)_6]^{2+}$ to $[Ru(NH_3)_6]^{3+}$.
    (a) On the basis of crystal field theory, do you expect K_{eq} to be greater or less than 1.0 for this reaction? Explain.
    (b) The observed rate constant is 2×10^2 L/mol·s. Estimate K_{eq} using the simplified Marcus equation. Is the result consistent with your prediction?

24. One product of the reduction of $[Co(NH_3)_5(NCS)]^{2+}$ by $[Cr(H_2O)_6]^{2+}$ is $[Cr(H_2O)_5SCN]^{2+}$. What does this indicate about the reaction mechanism? Sketch the probable intermediate.

25. Predict which is faster—the oxidation of $[Ti(H_2O)_6]^{3+}$ by $[Co(NH_3)_5NCS]^{2+}$ or by $[Co(NH_3)_5SCN]^{2+}$. Defend your answer.

26. A compound gives an elemental analysis consistent with the formulation $Pt(NH_3)_2Cl_2$; however, it contains "free" (ionically bonded) chloride ions and has twice the expected formula weight. Write the most probable structure.

*27. M. L. Tobe and co-workers have described some square planar substitution reactions that occur by a dissociative process (*Inorg. Chem.* **1984**, 23, 4428). One such reaction is

$$Pt\phi_2(DMSO)_2 + bipy \longrightarrow Pt\phi_2(bipy) + 2\,DMSO$$

(a) How is this reactant prepared?

(b) Sketch the structure of the reactant complex. Discuss this structure in terms of the geometry and the coordinated atoms.

(c) This reaction was studied in two different solvents. Why?

*28. N. J. Curtis, G. A. Lawrance, and R. van Eldik used a sophisticated experimental approach to study the aquation of $[Cr(NH_3)_5Cl]^{2+}$ and related complexes (*Inorg. Chem.* **1989**, *28*, 329).

(a) Briefly describe their method.

(b) What mechanism do they propose?

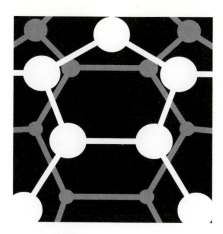

18

Organotransition Metal Chemistry and Catalysis

This chapter deals with compounds that have one or more bonds between a transition metal and carbon. Once considered novelties, such species have come to be of great importance, and an enormous amount of both pure and applied research is presently being conducted on the chemistry of organometallics. A major focus of this work involves the catalysis of organic reactions; a brief summary of the primary areas of catalytic study is given in Sections 18.6–18.10.

Organometallic compounds can be loosely divided into two types, depending on whether the M–C bonds are of the σ or π variety. Of course, the bonds to ligands such as CO and CN^- have both components—the ligands donate σ electron pairs via carbon, and are also π acceptors (Chapter 16). Some organic groups donate electrons exclusively from localized π bonds. In still other cases, the interactions involve delocalized (often cyclic) π systems.

18.1 Compounds Containing Metal–Carbon σ Bonds

Considerable variation occurs in the structures and bonding of species having M–C σ linkages. The most commonly encountered frameworks are summa-

Table 18.1 Classification of the common types of metal–carbon σ bonded species

Linkage	Name	Example
M—C— (alkyl)	Alkyl, aryl	EtHgBr, Scϕ_3, WMe$_6$
M—C— (acyl, C=O)	Acyl	MeCCo(CO)$_3$ (with C=O)
M—C—C metallacycle	Metallacycle	$(\phi_3 P)_4$Ru ∠ SiMe$_2$
M=C (alkylidene)	Alkylidene (Schrock carbene)	$(C_5H_5)_2$Ta—Me / CH$_2$
M=C—X: (lone pair on X)	Carbene	(CO)$_5$Cr=C—OMe / ϕ
M≡C—	Carbyne (alkylidyne)	I(CO)$_4$W≡Cϕ
M—C—M (μ)	μ-Alkylidene	H$_5$C$_5$, CH$_2$, NO / Fe——Fe / ON, C$_5$H$_5$
M—C—M (M$_3$)	μ_3-Alkylidyne	[(CO)$_3$Co]$_3$CMe

rized in Table 18.1. One important subgroup is the *homoleptic* alkyls and aryls—complexes in which all substituents are identical (ie, MR$_n$). A sampling of the best-known homoleptics is given in Table 18.2.

Modes of Metal–Carbon σ Bond Formation

Transition metal alkyls and their derivatives can be synthesized by many of the same types of reactions as the main group organometallics (Chapters 12 and 13). We describe six of the most general methods below.

Table 18.2 Some homoleptic metal alkyls of the first transition series

$R = CH_3$	$R = C_6H_5$	$R = C_6F_5$	$R = CH_2\phi$	$R = CH_2SiMe_3$	Others
	ScR_3				$Sc(C{\equiv}C\phi)_3$
	TiR_2				
			TiR_2		
			TiR_3	TiR_3	
TiR_4	TiR_4		TiR_4	TiR_4	$Ti(CH_2CMe_3)_4$
	VR_3			VR_3	$V(mes)_3$
	VR_4			VR_4	$V(mes)_4$
	CrR_3				
CrR_4			CrR_3^*		
				CrR_4	$Cr(i\text{-}Pr)_4$
			MnR_2	MnR_2	$Mn(C{\equiv}CR)_2$
					$Fe(mes)_2$
			CoR_2		
		NiR_2			$Ni(t\text{-}Bu)_2$
	CuR	$(CuR)_4$			
ZnR_2					$ZnEt_2$
					$Zn(CHMe_2)_2$

Note: Many anionic homoleptics are also known. * = THF complex; mes = mesityl, $2,4,6\text{-}Me_3C_6H_2$.
Source: Taken primarily from a compilation by Haiduc, I.; Zuckerman, J. J. *Basic Organometallic Chemistry*; Walter de Gruyter: New York, 1985; pp. 360–361.

Insertion of Metal Atoms into Covalent M–X Bonds (Direct Synthesis)[1]
Many, although not all, transition metals react with alkyl halides at or above room temperature. An organometallic compound was produced in this manner over 140 years ago.[2]

$$2Zn + 2EtI \xrightarrow{\Delta} [2EtZnI] \longrightarrow ZnEt_2 + ZnI_2 \qquad (18.1)$$

Certain acyl derivatives can be prepared by the insertion of a metal into the C–X bond of an acid halide.

$$Pd + CF_3COCl \longrightarrow CF_3C(O)PdCl \qquad (18.2)$$

Metal–Metal Exchange Recall the tendency for bonds to form between elements having large differences in electronegativity. This can be useful for converting one type of metal–carbon bond into another, especially if the incoming metal is the more electronegative of the two. Thus, since $\chi(Ti) > \chi(Mg)$, the formation of $MgCl_2$ provides a driving force for the reaction

$$4MeMgCl + TiCl_4 \longrightarrow TiMe_4 + 4MgCl_2 \qquad (18.3)$$

1. Allison, J. *Prog. Inorg. Chem.* **1986**, *34*, 627; Timms, P. L.; Turney, T. W. *Adv. Organomet. Chem.* **1977**, *15*, 53; Klabunde, K. J. *Acc. Chem. Res.* **1975**, *8*, 393; Rochow, E. G. *J. Chem. Educ.* **1966**, *43*, 58.
2. Frankland, E. *J. Chem. Soc.* **1848–1849**, *2*, 263.

As a result, Grignard and alkyllithium reagents are useful for the synthesis of many organotransition metal compounds. Examples include equations (18.3)–(18.5), each of which can be considered a nucleophilic displacement reaction.

$$WCl_6 + 6\,MeLi \longrightarrow WMe_6 + 6\,LiCl \tag{18.4}$$

$$Et_3PAuBr + C_6H_5Li \longrightarrow Et_3PAuC_6H_5 + LiBr \tag{18.5}$$

Metal–metal exchange also can be viewed as alkyl or aryl transfer, of course, depending on whether the focus of interest is on the metallic or the organic portion of the system.

Addition Across Carbon–Carbon Multiple Bonds Metal hydrides often can be induced to add to alkenes. A well-known example is

$$\begin{array}{c} Et_3P \quad\quad H \\ \diagdown Pt \diagup \\ Cl \diagup \quad \diagdown PEt_3 \end{array} + C_2H_4 \xrightarrow[\text{40 atm}]{95°C} \begin{array}{c} Et_3P \quad\quad CH_2CH_3 \\ \diagdown Pt \diagup \\ Cl \diagup \quad \diagdown PEt_3 \end{array} \tag{18.6}$$

Reactions of this type are often reversible, and, in fact, the product of equation (18.6) eliminates ethylene when heated to 180°C. (This is a β-hydride elimination reaction; see below.)

Addition across carbon–carbon triple bonds is also known.

$$AgF + F_3CC{\equiv}CCF_3 \longrightarrow CF_3C(Ag){=}C(F)CF_3 \tag{18.7}$$

$$HRh(P\phi_3)_2Cl_2 + C_2H_2 \longrightarrow H_2C{=}CHRh(P\phi_3)_2Cl_2 \tag{18.8}$$

Reactions of Anionic Complexes with Alkyl and Aryl Halides Metals in low oxidation states sometimes behave as nucleophiles. For example, many metal carbonyl anions (carbonylates) react with methyl iodide:

$$Na^+[M(CO)_n]^- + CH_3I \longrightarrow CH_3M(CO)_n + NaI$$
$$M = Mn,\ Co,\ W,\ etc. \tag{18.9}$$

Similarly, a series of symmetric bis(iron) complexes can be obtained from reactions between $[Fe(C_5H_5)(CO)_2]^-$ and dibromoalkanes.

$$2\,Na^+[Fe(C_5H_5)(CO)_2]^- + Br(CH_2)_nBr \xrightarrow{-2\,NaBr}$$
$$C_5H_5(CO)_2Fe{-}(CH_2)_n{-}Fe(CO)_2C_5H_5 \qquad n = 3\text{–}6 \tag{18.10}$$

Oxidative Addition As we saw in the previous chapter, alkyl halides undergo oxidative addition reactions with certain square planar complexes to produce octahedral products (see Figure 17.1). Related reactions are possible for other geometries as well. For example, CH_3I often reacts with trigonal bipyramidal complexes, with the elimination of one ligand:

$$Ru(P\phi_3)_2(CO)_3 \; + \; CH_3I \quad \xrightarrow{-CO} \quad \text{[structure]} \qquad \textbf{(18.11)}$$

Less commonly, it is possible to oxidatively add to molecules containing $C{=}C$ or $C{\equiv}C$ bonds, producing *metallacycles*:

$$Fe(CO)_5 + 2F_2C{=}CF_2 \quad \xrightarrow{-CO} \quad (CO)_4Fe\underset{CF_2-CF_2}{\overset{CF_2-CF_2}{<}} \qquad \textbf{(18.12)}$$

Intramolecular examples of such *cyclometallation* reactions are also known.[3] This typically requires C–H bond rupture within a ligand such as triphenylphosphine.

$$Ir(P\phi_3)_3Cl \quad \longrightarrow \quad \text{[structure]} \qquad \textbf{(18.13)}$$

A variation is oxidative insertion into a carbon–carbon single bond of a strained ring.[4]

$$PtCl_2 + \text{[cyclopropane]} \quad \longrightarrow \quad Cl_2Pt\text{[ring]} \quad \xrightarrow{+2py} \quad Cl_2(py)_2Pt\text{[ring]} \qquad \textbf{(18.14)}$$

Elimination The heating of acyls and related species may result in the elimination of a small molecule and the formation of a new M–C bond.

$$\overset{O}{\overset{\|}{H_3CC}}{-}Mn(CO)_5 \quad \xrightarrow[\Delta]{-CO} \quad H_3CMn(CO)_5 \qquad \textbf{(18.15)}$$

$$\text{[structure]} \quad \xrightarrow[\Delta]{-CO_2} \quad \text{[structure]} \qquad \textbf{(18.16)}$$

3. For a review, see Bruice, M. I. *Angew. Chem. Int. Ed. Engl.* **1977**, *16*, 73; *Angew. Chem.* **1977**, *89*, 75.

4. The interaction of $PtCl_2$ with cyclopropanes has been thoroughly studied. See Crabtree, R. H. *Chem. Rev.* **1985**, *85*, 245; and references cited therein.

Stabilities and Modes of Decomposition of Metal Alkyls

It was once believed that compounds containing metal–carbon bonds are inherently unstable. Later, it was thought that both diamagnetism and coordinate saturation were necessary for such species to be isolable. All this is now recognized as grossly incorrect.

A compound can be considered thermodynamically stable if it does not undergo self-decomposition under a given set of conditions. Thus, species having negative free energies of formation are stable with respect to decomposition to their elements. Disproportionation and redistribution reactions also must be considered; nevertheless, many organotransition metal compounds are stable in the thermodynamic sense. Others, such as $Cr(C_6H_6)_2$, have considerable kinetic stability.

However, stable compounds also can be highly reactive—in fact, high reactivity is a common problem in organometallic chemistry. Many species containing metal–carbon bonds are reactive toward components of the atmosphere [ie, are readily oxidized by O_2 and/or $H_2O(g)$]. For example, dimethylzinc ignites spontaneously in air.

$$(CH_3)_2Zn + 4O_2 \longrightarrow ZnO + 2CO_2 + 3H_2O \tag{18.17}$$

Three common modes of decomposition for compounds containing metal–carbon σ bonds are reductive elimination, α rearrangement, and β-hydride elimination.[5]

Reductive elimination[6] may occur by either an intermolecular or intramolecular pathway. An example of the latter is the reaction

$$\begin{array}{c} \phi_3P \diagdown \diagup H \\ Pt \\ \phi_3P \diagup \diagdown CH_3 \end{array} \xrightarrow{\Delta} Pt(P\phi_3)_2 + CH_4 \tag{18.18}$$

Isotopic labeling studies show that the elimination of CH_4 from $Os(CO)_4(H)Me$ is bimolecular:

$$2\,Os(CO)_4(H)Me \xrightarrow{50°C} (CO)_4(H)Os\text{–}Os(Me)(CO)_4 + CH_4 \tag{18.19}$$

For compounds in which the metal-bonded carbon is a carbonyl, α *rearrangement* to produce a dative $M \leftarrow CO$ interaction may occur:

$$\overset{R}{\underset{|}{M\text{–}C}}\!\!=\!\!\ddot{O}: \longrightarrow R\text{–}M\text{–}C\!\!\equiv\!\!O: \tag{18.20}$$

---

5. A fourth mode, α elimination, will be discussed in Section 18.2.

6. Millstein, D. *Acc. Chem. Res.* **1984**, 17, 221; Braterman, P. S.; Cross, R. J. *Chem. Soc. Rev.* **1973**, 2, 271.

An interesting example is the cobalt system described by equation (18.21), in which the acyl and methyl isomers are in equilibrium at room temperature:

$$\underset{\text{H}_3\text{C}}{\overset{\overset{\displaystyle O}{\overset{\displaystyle \|}{}}}{\text{C}}}\text{–Co(CO)}_3 \; \rightleftharpoons \; \text{H}_3\text{CCo(CO)}_4 \tag{18.21}$$

When this reaction occurs from left to right, it is elimination; going in the opposite direction, it is insertion. Many other small molecules are known to insert into M–C bonds. A few examples are given below.

$$\text{M–CR}_3 + \text{CO}_2 \longrightarrow \text{M–O–C(O)–CR}_3 \tag{18.22}$$

$$\text{M–CR}_3 + \text{SO}_2 \longrightarrow \text{M–O–S(O)–CR}_3 \tag{18.23}$$

$$\text{M–CR}_3 + \text{NO} \longrightarrow \text{M–O–N–CR}_3 \tag{18.24}$$

A general scheme for β-hydride elimination is

$$\text{M–CR}_2\text{–CR}'_2\overset{\displaystyle \text{H}}{\overset{\displaystyle |}{}} \longrightarrow \text{M–H} + \text{R}_2\text{C=CR}'_2 \tag{18.25}$$

Note that this is simply the reverse of M–H addition to carbon–carbon double bonds [eg, equation (18.6)]. Being an intramolecular process, it requires an approximately coplanar arrangement to bring the hydrogen close to the metal:

$$\text{M}\overset{\displaystyle \text{C}}{\underset{\displaystyle \text{H}}{\diagup}}\diagdown\text{C}$$

Species for which such an orientation is impossible are often stabilized; for example, the metallacycle

$$\text{L}_2\text{Pt}\diagdown\diagup$$

undergoes elimination of alkene very slowly in comparison to its acyclic analogue L_2PtEt_2.

Molecules for which α rearrangement and β-hydride elimination are discouraged show enhanced kinetic stability. Thus, it is not surprising that metal acyls tend to be less stable than structurally similar alkyls, or that species having methyl substituents (with no β hydrogens) are often more stable than the corresponding ethyl or propyl compounds. This helps explain why Table 18.2 contains a preponderance of alkyl substituents having tertiary Group 14 atoms in the β position (eg, $-CH_2CMe_3$, $-CH_2SiMe_3$, and $-CH_2\phi$). These substituents, of course, preclude β-hydride elimination. (The stabilization imparted by such groups probably has a steric component as well.)

For nonhomoleptic alkyls, the presence of certain other ligands promotes kinetic stability. Prominent among these are π acceptors such as CO, CN^-, and PR_3, metallocenes such as C_5H_5, and perfluorinated alkyl groups. The stabilization is probably due to the fact that none of these ligands are prone to dissociation. They are therefore effective at blocking potential sites for attack (ie, they inhibit intermolecular decomposition).

Metal–carbon σ bonds are also stabilized when part of a chelate linkage; an example is the product shown in equation (18.14). This stabilization results from the difficulty of intramolecular β-hydride elimination in the rigid ring system. Even three-membered rings are sometimes isolable,

$$
\begin{array}{c}
\text{M} \\
/ \ \backslash \\
\text{X}\!-\!\text{CH}_2
\end{array}
\qquad
\begin{array}{l}
\text{M} = \text{Mn(CO)}_4, \ \text{X} = \text{NMe}_2 \\
\text{M} = \text{Fe(H)(PMe}_3)_3, \ \text{X} = \text{PMe}_2 \\
\text{M} = \text{Pd(Cl)P}\phi_3, \ \text{X} = \text{SMe}
\end{array}
$$

as is the eight-membered metallacycle

$$
\begin{array}{ccc}
& \text{—Au—} & \\
\text{Me}_2\text{P} & & \text{PMe}_2 \\
& \text{—Au—} &
\end{array}
$$

Multiple Bonding Between Transition Metals and Carbon

In recent years, an astonishing number of species have been prepared and studied that contain double or triple bonds between a transition metal and carbon. Several categories are listed in Table 18.1.

The synthesis of such compounds is usually accomplished through elimination reactions. For example, consider the reaction

$$
\text{Ta(Cl)}_2\text{R}_3 + 2\,\text{RLi} \longrightarrow \text{TaR}_5 + 2\,\text{LiCl} \tag{18.26}
$$

When R = Me, the homoleptic product is obtained as indicated. If R is a large group such as neopentyl, however, the considerable steric hindrance causes decomposition via what is believed to be intramolecular elimination from one of the α carbons:[7]

$$R_3Ta \overset{CH_2CMe_3}{\underset{CH-CMe_3}{\langle H}} \longrightarrow R_3Ta{=}CHCMe_3 + CMe_4 \qquad (18.27)$$

The product is an *alkylidene* (or *Schrock carbene*).[8]

The evidence for multiple bonding includes shortened M–C bond distances, planarity at carbon, and ^{13}C NMR chemical shifts characteristic of sp^2 hybridization. Many alkylidenes have been reported in recent years. It is also possible to prepare bis(alkylidenes) by reactions such as

$$Nb(CH_2CMe_3)_4Cl \xrightarrow[-2CMe_4]{+2PMe_3} \begin{matrix} PMe_3 \\ | \\ Cl-Nb{\langle}^{CHCMe_3}_{CHCMe_3} \\ | \\ PMe_3 \end{matrix} \qquad (18.28)$$

If one of the substituents on carbon contains a lone pair (ie, $L_xM{=}C(R)X\colon$), then the species is referred to as a metal *carbene* (or *Fischer carbene*).[9] The first carbene to be reported resulted from the reaction of phenyllithium with $W(CO)_6$.[10] The initial step of the reaction involves nucleophilic attack at a carbonyl carbon:

$$W(CO)_6 + Li^+\phi^- \longrightarrow Li^+ \left[(CO)_5W{-}C{\langle}^O_\phi \right]^- \qquad (18.29)$$

Addition of a methylating agent then yields the carbene:

$$\left[(CO)_5W{-}C{\langle}^O_\phi \right]^- \xrightarrow{Me_3O^+BF_4^-} (CO)_5W{=}C{\langle}^{OMe}_\phi \qquad (18.30)$$

---

7. It is possible that hydrogen abstraction occurs after a single substitution (ie, for R_4TaCl). See Schrock, R. R. *J. Am. Chem. Soc.* **1974**, *96*, 6796.

8. For reviews, see Gallop, M. A.; Roper, W. R. *J. Organomet. Chem.* **1986**, *25*, 121; Brown, F. J. *Prog. Inorg. Chem.* **1980**, *27*, 1; Schrock, R. R. *Acc. Chem. Res.* **1979**, *12*, 98.

9. Dötz, K. H.; Fischer, H.; Hofmann, P.; Kreissl, F. R.; Schubert, U.; Weiss, K. *Transition Metal Carbene Complexes*; Verlag: Weinheim, 1983; Fischer, E. O. *Adv. Organomet. Chem.* **1976**, *14*, 1.

10. Fischer, E. O.; Maasböl, A. *Angew. Chem. Int. Ed. Engl.* **1964**, *3*, 580; *Angew. Chem.* **1964**, *76*, 645.

The reason for differentiating between alkylidenes and carbenes is that the latter exhibit resonance:

This resonance weakens the metal–carbon bond, as exhibited by an increased bond distance. In fact, in $(CO)_5W=C(\phi)OMe$, the $W=C$ bond distance is 205 pm, about 8% longer than the average of the M–CO bonds. (Dative metal–carbonyl interactions also have considerable double bond character; see below.) Conversely, there is significant shortening of the $M=C$ linkage(s) in alkylidenes. Bond distances provide a useful measure of the extent of M–C multiple bonding. This is apparent from Table 18.3, which contains data for a variety of tungsten–carbon multiple bonds.

Another significant difference between carbenes and alkylidenes is that the latter tend to be reactive toward nucleophiles (including Brønsted

Table 18.3 Tungsten–carbon bond distances as a function of bond order for selected compounds

Compound	Reference	$d(W-CO)$	$d(W-C)$	$d(W=C)$	$d(W\equiv C)$
$W(CO)_6$	a	206			
$\left[(CO)_5W-CH\begin{smallmatrix}OMe\\\phi\end{smallmatrix}\right]^-$	b		234		
$(CO)_5W=C\begin{smallmatrix}OMe\\\phi\end{smallmatrix}$	c	189		205	
$(CO)_5W=C\phi_2$	b	202		214	
$I(CO)_4W\equiv C\phi$	d	214			188
$\begin{smallmatrix}Me_3C-CH_2\\ \\Me_3C-CH\end{smallmatrix}W\equiv C-CMe_3$	e		226	194	176
$Cl(X)(CO)(py)_2W\equiv C\phi$	f	203			180

Note: Distances are in picometers. X = maleic anhydride.
References: (a) Arnesen, S. P.; Seip, H. M. *Acta Chem. Scand.* **1966**, *20*, 274; (b) Casey, C. P.; Burkhardt, T. J.; Bunnell, C. A.; Calabrese, J. C. *J. Am. Chem. Soc.* **1977**, *99*, 2127; (c) Mills, O. S.; Redhouse, A. D. *Angew. Chem. Int. Ed. Engl.* **1965**, *4*, 1082; *Angew. Chem.* **1965**, *77*, 1142; (d) Fischer, E. O.; Kreis, G.; Kreiter, C. G.; Müller, J.; Huttner, G.; Lorenz, H. *Angew. Chem. Int. Ed. Engl.* **1973**, *12*, 564; *Angew. Chem.* **1973**, *85*, 618; (e) Churchill, M. R.; Youngs, W. J. *Inorg. Chem.* **1979**, *18*, 2454; (f) Mayr, A.; Dorries, A. M.; McDermott, G. A.; Geib, S. J.; Rheingold, A. L. *J. Am. Chem. Soc.* **1985**, *107*, 7775.

bases), while carbenes are reactive toward electrophiles. Both kinds of reactions have been used to produce *carbynes* (*alkylidynes*), in which a metal–carbon triple bond is present.[11]

$$L_nM \overset{CH_2CMe_3}{\underset{CHR}{<}} \quad \xrightarrow{\text{Base}} \quad L_nM\equiv CR \;+\; CMe_4 \tag{18.31}$$

$$L_nM{=}C\overset{OMe}{\underset{R}{<}} \quad \xrightarrow[-L]{+BF_3} \quad FL_{n-1}M\equiv CR \;+\; MeOBF_2 \tag{18.32}$$

The π bonds of a carbyne are of the *p–d* type (see Figure 18.1).

Another important group of σ bonded organometallics has carbon atoms in bridging positions between two or more metals. The simplest of these cases $\left(M{-}\overset{|}{\underset{|}{C}}{-}M\right)$ are the *μ-alkylidenes.* (The Greek letter *μ* identifies a bridging atom, and distinguishes this type of compound from terminal $\left(M{=}C\overset{/}{\underset{\backslash}{}}\right)$ alkylidenes.) Thus, M–CH$_2$–M species are *μ-methylidenes*,

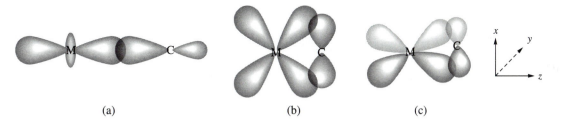

(a)   (b)   (c)

Figure 18.1 Metal–carbon bonding in metal carbynes, L$_n$M≡CR, assuming *sp*-hybridized carbon: (a) σ bond formed from overlap of C (*sp*) with M (*d$_{z^2}$*); (b) π bond from overlap of C (*p$_x$*) with M (*d$_{xz}$*); (c) π bond from overlap of C (*p$_y$*) with M (*d$_{yz}$*). The *z* axis is taken to be the bond axis.

11. Kim, H. P.; Angelici, R. J. *Adv. Organomet. Chem.* **1987**, *27*, 51; Fischer, E. O.; Kreis, G.; Kreiter, C. G.; Müller, J.; Huttner, G.; Lorenz, H. *Angew. Chem. Int. Ed. Engl.* **1973**, *12*, 564; *Angew. Chem.* **1973**, *85*, 618.

M–CH=CH–M compounds are μ-vinylidenes, etc. Triply bridged carbon

compounds $\left(M\text{–}CH\diagup^{M}_{\diagdown M} \right)$ are μ-alkylidynes.[12] Examples include

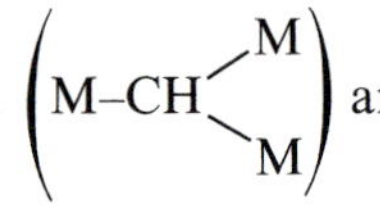

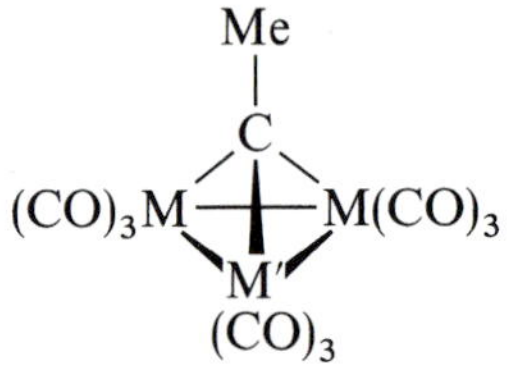

$$(CO)_3M \underset{\underset{(CO)_3}{M'}}{\overset{\overset{Me}{\overset{|}{C}}}{\diagup\!\!\!\!\diagdown}} M(CO)_3 \qquad \begin{array}{l} M = M' = Co,\ FeH \\ M = Co,\ M' = FeH \end{array}$$

The sequence

$$M\text{–}CH_3 \longrightarrow M\text{–}CH_2\text{–}M \longrightarrow M\text{–}CH\diagup^{M}_{\diagdown M}$$

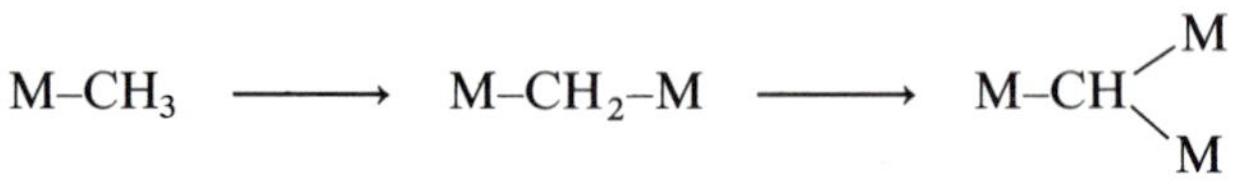

suggests that M_4C species might also exist. Such compounds are *carbides.* The bonding in μ-alkylidynes and carbides is best explained using cluster theory, and will be discussed in Chapter 19.

18.2   Metal Carbonyls and Their Derivatives

The most common of all carbon donor ligands is CO, and thousands of complexes are known that contain carbonyl groups. Recall that M–CO interactions include both dative σ ($M \leftarrow C$) and π ($M \rightarrow C$ back-bonding) contributions. This is shown in Lewis formulation by two resonance structures:

$$^\ominus\ddot{M}\text{–}C{\equiv}\overset{\oplus}{O}{:} \quad \longleftrightarrow \quad M{=}C{=}\ddot{O}{:}$$

The structure on the right is especially important for electron-rich metals (those in low oxidation states).

A list of homoleptic carbonyls is given in periodic table format in Table 18.4. It is apparent that metals near the center of each transition series (those of Groups 5–10) produce the most stable species. (The remaining metals form carbonyl-containing complexes, but only as anions or in the presence of

---

12.   For reviews of carbon-bridged systems, see Hahn, J. E. *Prog. Inorg. Chem.* **1984**, *31*, 205; and/or Herrmann, W. A. *Adv. Organomet. Chem.* **1982**, *20*, 159.

Table 18.4 Formulas of the best-known, neutral homoleptic metal carbonyls having 1–6 metal atoms

$V(CO)_6$	$Cr(CO)_6$	$Mn_2(CO)_{10}$	$Fe(CO)_5$	$Co_2(CO)_8$	$Ni(CO)_4$
			$Fe_2(CO)_9$	$Co_4(CO)_{12}$	
			$Fe_3(CO)_{12}$	$Co_6(CO)_{16}$	
$Nb(CO)_6$	$Mo(CO)_6$	$Tc_2(CO)_{10}$	$Ru(CO)_5$	$Rh_2(CO)_8$	
			$Ru_3(CO)_{12}$	$Rh_4(CO)_{12}$	
				$Rh_6(CO)_{16}$	
	$W(CO)_6$	$Re_2(CO)_{10}$	$Os(CO)_5$	$Ir_2(CO)_8$	
			$Os_2(CO)_9$	$Ir_4(CO)_{12}$	
			$Os_3(CO)_{12}$	$Ir_6(CO)_{16}$	
			$Os_5(CO)_{16}$		
			$Os_6(CO)_{18}$		

Note: Species that are not well-characterized and/or are stable only below room temperature are not listed.

other ligands.) This can be understood through MO theory. The stabilization created by metal–ligand back-bonding requires that the σ nonbonding orbitals of the metal are filled, which is not the case for the Group 3 or 4 metals. For the latter families (Groups 11 and 12), the *d* orbitals apparently are too stable (the $3d$–π^* energy difference is too great) for effective π overlap.

Trends in M–CO bond dissociation energies (Table 18.5) are of interest. There is a general increase in average bond energy going across the first transition series, with a maximum reached at nickel. (The homoleptics of Sc, Ti, V, Cu, and Zn are too unstable to permit measurement.) Bond strength appears to increase going down a family. This is thought to be a reflection of enhanced σ interaction, since back-bonding is greater for the $3d$ elements than for their heavier congeners.[13]

Table 18.5 Average M–CO bond dissociation energies for selected homoleptic carbonyls

Compound	$\bar{D}$, kJ/mol	Compound	$\bar{D}$, kJ/mol
$Cr(CO)_6$	108	$Re(CO)_5$	182
$Mo(CO)_6$	152	$Fe(CO)_5$	117
$W(CO)_6$	178	$Co(CO)_4$	136
$Mn(CO)_5$	99	$Ni(CO)_4$	147

Source: Taken from data compiled by Conner, J. A. *Top. Curr. Chem.* **1977**, *71*, 71.

13. Ziegler, T.; Tschinke, V.; Ursenbach, C. *J. Am. Chem. Soc.* **1987**, *109*, 4825.

The 18-Electron Rule

The formulas given in Table 18.4 provide clear evidence for the *18-electron rule* (sometimes called the *effective atomic number*, or *EAN*, *rule*).[14] This is a parallel to the 8-electron (octet) rule of main group chemistry, and can be understood in the same manner. Since a main group element has four valence orbitals, the complete occupancy of its valence shell requires 8 electrons. Similarly, transition metals have nine valence orbitals [the ns, np, and $(n - 1)d$ subshells], requiring a total of 18 electrons.

Recall that octahedral complexes have three σ nonbonding orbitals of t_{2g} symmetry. The total number of bonding electrons is, of course, independent of whether those orbitals are occupied. As a result, violations of the 18-electron rule are very common for octahedral complexes having σ-only (or π donating) ligands.[15]

For complexes containing π acceptor ligands such as CO, however, the t_{2g} level is π bonding and must be occupied for the bond energy to be maximized. As a result, the rule is generally adhered to. Consider the mononuclear carbonyls of Cr, Fe, and Ni:

For $Cr(CO)_6$: $6\ (Cr) + (6 \cdot 2)\ (CO) = 18e^-$

For $Fe(CO)_5$: $8\ (Fe) + (5 \cdot 2)\ (CO) = 18e^-$

For $Ni(CO)_4$: $10\ (Ni) + (4 \cdot 2)\ (CO) = 18e^-$

Metals having odd atomic numbers cannot reach a total of 18 electrons as neutral monomers. For example, $Mn(CO)_5$ has 17 valence electrons, while $Mn(CO)_6$ has 19. Each of these species is extremely reactive (the former toward reduction and the latter to oxidation), and both $[Mn(CO)_5]^-$ and $[Mn(CO)_6]^+$ are stable ions. The neutral free radical undergoes dimerization to $Mn_2(CO)_{10}$, in which the electron count reaches 18 through metal–metal bonding:

Per Mn of $Mn_2(CO)_{10}$: $7\ (Mn) + (5 \cdot 2)\ (CO) + 1\ (Mn–Mn) = 18e^-$

Similarly, "$Co(CO)_4$" is dimeric.

With a count of 17 valence electrons, $V(CO)_6$ is unusual in this regard. Its failure to dimerize is probably due to steric crowding (a metal–metal

---

14. Sidgwick, N. V. *J. Chem. Soc.* **1923**, 725; *Nature* **1923**, *111*, 8080; *The Electronic Theory of Valency*; Oxford University: London, 1927; Chapter 10.

15. The rule also fails for square planar complexes. The majority of stable square planar organometallics have 16 valence electrons, with one d orbital left vacant.

bonded dimer would have 7-coordinate vanadiums). As we would expect, $V(CO)_6$ is prone to 1-electron reduction.[16]

$$Na + V(CO)_6 \longrightarrow Na^+[V(CO)_6]^- \tag{18.33}$$

Synthetic Routes to Carbonyls[17]

A few metal carbonyls, the best-known being $Ni(CO)_4$ and $Fe(CO)_5$, can be prepared by simply heating the metal under a high pressure of CO. More commonly, chemical reduction is carried out in the presence of excess CO:

$$CrCl_3 + Al + 6CO \xrightarrow[140^\circ C]{150\ atm} Cr(CO)_6 + AlCl_3 \tag{18.34}$$

$$2CoCO_3 + 8CO + 2H_2 \xrightarrow{\Delta} Co_2(CO)_8 + 2H_2O + 2CO_2 \tag{18.35}$$

It is sometimes possible to use CO as the reducing agent:

$$Re_2O_7 + 17CO \xrightarrow{\Delta} Re_2(CO)_{10} + 7CO_2 \tag{18.36}$$

Higher carbonyls are typically prepared by temperature-induced or photochemically induced elimination:

$$2Fe(CO)_5 \xrightarrow{h\nu} Fe_2(CO)_9 + CO \tag{18.37}$$

Certain mixed species can be made in the same manner:

$$3Fe(CO)_5 + Ru_3(CO)_{12} \xrightarrow[\Delta]{-3CO} FeRu_2(CO)_{12} + Fe_2Ru(CO)_{12} \tag{18.38}$$

Bridging Carbonyl Groups

The geometry of $Co_2(CO)_8$ has been a subject of considerable study. The 18-electron rule is satisfied by a framework in which each cobalt bonds to four CO groups, with a metal–metal bond:

Per Co: $9(Co) + (4 \cdot 2)(CO) + 1(Co-Co) = 18e^-$

---

16. In addition to $V(CO)_6$, several cluster carbonyls [eg, $M_6(CO)_{16}$, where M = Co, Rh, and Ir] also violate the 18-electron rule. The bonding in such species is best understood through MO theory.

17. Abel, E. W.; Stone, F. G. A. *Quart. Rev.* **1970**, *24*, 498; King, R. B. *Organometallic Synthesis*; Academic: New York, 1965; Volume 1, pp. 82*ff*; Hileman, J. C. *Prep. Inorg. React.* **1964**, *1*, 77.

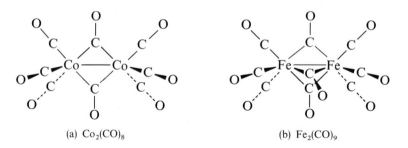

Figure 18.2
The structure of two metal carbonyls with bridging CO groups.

(a) $Co_2(CO)_8$ (b) $Fe_2(CO)_9$

This molecule is fluxional, however, and there is evidence for a structure in which two CO groups form bridges between the metal centers (Figure 18.2). The 18-electron rule is again obeyed:

Per Co: $9 \text{ (Co)} + (3 \cdot 2) \text{ (CO}_{terminal}) + (2 \cdot 1) \text{ (CO}_{bridge}) + 1 \text{ (Co–Co)} = 18e^-$

An important resonance structure for the bridged fragment is

$$
\begin{array}{c}
:\!O\!: \\
\| \\
C \\
M^{\diagup}\;\;{}^{\diagdown}M
\end{array}
$$

Note that bridging reduces the strength of the carbon–oxygen bond. As a result, terminal and bridging CO groups are readily differentiated by their different C–O stretching frequencies in infrared spectra (see Section 21.1).

Higher carbonyls (especially those of the $3d$ elements) often have bridged structures, even though the bonding requirements could be satisfied by a combination of only metal–metal bonds and terminal CO's. For example, we might expect $Fe_3(CO)_{12}$ to contain a triangle of $Fe(CO)_4$ groups:

$$
\begin{array}{c}
(CO)_4 \\
M \\
\diagup\;\;\diagdown \\
(CO)_4M\!-\!M(CO)_4
\end{array}
$$

However, in the solid state, three of the carbonyls occupy bridging positions.[18] The electron count remains 18.

Per Fe: $8 \text{ (Fe)} + (3 \cdot 2) \text{ (CO}_{terminal}) + (2 \cdot 1) \text{ (CO}_{bridge}) + (2 \cdot 1) \text{ (Fe–Fe)} = 18e^-$

Carbonyls of the $4d$ and $5d$ elements prefer nonbridged structures. For example, $Ru_3(CO)_{12}$ and $Os_3(CO)_{12}$ both have the triangular (D_{3h})

18. See Cotton, F. A. *Prog. Inorg. Chem.* **1975**, *21*, 1; and references cited therein.

geometry shown above. Similarly, $Fe_2(CO)_9$ has three bridging carbonyls, but $Os_2(CO)_9$ has only one. This is because the greater size of the $5d$ metal destabilizes M–C–M bridge bonding. [The average M–M bond distances are 288 pm in $Os_3(CO)_{12}$, but only 256 pm between the bridged irons in $Fe_3(CO)_{12}$.]

The geometries of selected metal carbonyls are given in Table 18.6 (p. 594).

It is possible to prepare very large compounds and anions that contain only metals and CO. Compounds having four through six metal atoms, such as $M_4(CO)_{12}$ (M = Co, Rh, and Ir), $M_5(CO)_{16}$ (M = Os), and $M_6(CO)_{16}$ (M = Co, Rh, and Ir), are well-established. These metal clusters will be discussed in Chapter 19.

More dramatic examples include the anions $[Rh_{14}(CO)_{25}]^{4-}$ and $[Rh_{22}(CO)_{37}]^{4-}$.[19] The covalent and metallic models begin to merge in these very large clusters. Thus, the metals of $[Rh_{22}(CO)_{37}]^{4-}$ occupy four layers (having 3, 6, 7, and 6 atoms, respectively) of a close-packed lattice (see Figure 18.3, p. 595).

Carbonyl Anions, Hydrides, and Other Derivatives[20]

Many carbonylate anions have been prepared and studied (see Table 18.7). These species are usually made by reacting metal carbonyls with bases, alkali metals, or other reducing agents. Representative examples are given by equations (18.33) and (18.39)–(18.41).

$$Fe(CO)_5 \xrightarrow{+3OH^-} [Fe(CO)_4]^{2-} + H_2O + HCO_3^- \qquad (18.39)$$

$$Co_2(CO)_8 + 2LiBHEt_3 \longrightarrow 2Li^+[Co(CO)_4]^- + H_2 + 2BEt_3 \qquad (18.40)$$

$$2Ni(CO)_4 \xrightarrow{Na/NH_3} [Ni_2(CO)_6]^{2-} + 2CO \qquad (18.41)$$

The first titanium and zirconium carbonylates were prepared by reduction, using potassium naphthalide in the presence of a crown ether or cryptand:[21]

$$TiCl_4(DME) + 6CO \xrightarrow{+4K^+C_{10}H_8^-} [Ti(CO)_6]^{2-} \qquad (18.42)$$

$$ZrCl_4(THF)_2 + 6CO \xrightarrow{+4K^+C_{10}H_8^-} [Zr(CO)_6]^{2-} \qquad (18.43)$$

19. Martinengo, S.; Ciani, G.; Sironi, A. *J. Am. Chem. Soc.* **1980**, *102*, 7564; Martinengo, S.; Ciani, G.; Sironi, A.; Chini, P. *J. Am. Chem. Soc.* **1978**, *100*, 7096.

20. Ellis, J. E. *Adv. Organomet. Chem.* **1990**, *31*, 1; Darensbourg, M. Y.; Ash, C. E. *J. Organomet. Chem.* **1987**, *27*, 1; Darensbourg, M. Y. *Prog. Inorg. Chem.* **1985**, *33*, 221.

21. Chi, K. M.; Frerichs, S. R.; Philson, S. B.; Ellis, J. E. *J. Am. Chem. Soc.* **1988**, *110*, 303; *Angew. Chem. Int. Ed. Engl.* **1987**, *26*, 1190; *Angew. Chem.* **1987**, *99*, 1203.

Table 18.6 Classification of the common geometries of carbonyls having 1–4 metal atoms

Formula	Point Group	Description	
$M(CO)_4$ M = Ni	T_d	Tetrahedral	
$M(CO)_5$ M = Fe, Ru, Os	D_{3h}	Trigonal bipyramidal	
$M(CO)_6$ M = V, Cr, Mo, W	O_h	Octahedral	
$M_2(CO)_8$ M = Rh, Ir	D_{3d}	Pseudo-trigonal bipyramidal	
$M_2(CO)_8$ M = Co	C_{2v}	Doubly bridged	
$M_2(CO)_9$ M = Ru, Os	C_{2v}	Singly bridged	
$M_2(CO)_9$ M = Fe	D_{3h}	Triply bridged	
$M_2(CO)_{10}$ M = Mn, Tc, Re	D_{4d}	Pseudo-octahedral	
$M_3(CO)_{12}$ M = Ru, Os	D_{3h}	Triangular (pseudo-octahedral)	
$M_3(CO)_{12}$ M = Fe	C_{2v}	Doubly bridged	
$M_4(CO)_{12}$ M = Ir	T_d	Tetrahedral (pseudo-octahedral)	
$M_4(CO)_{12}$ M = Co, Rh	C_{3v}	Triply bridged	

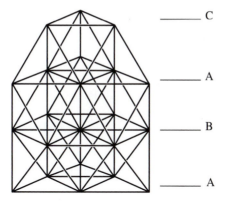

Figure 18.3 The structure of the $[Rh_{22}(CO)_{37}]^{4-}$ anion. [Reproduced with permission from Martinengo, S.; Ciani, G.; Sironi, A. *J. Am. Chem. Soc.* **1980**, *102*, 7564.]

Table 18.7 Some well-characterized carbonyl hydrides and carbonylate ions of the 3*d* metals

| | Number of Metal Atoms | | |
1	2	3	4
$[Ti(CO)_6]^{2-}$			
$[Zr(CO)_6]^{2-}$			
$HV(CO)_6$			
$[V(CO)_6]^-$			
$[V(CO)_5]^{3-}$			
$H_2Cr(CO)_5$			
$[HCr(CO)_5]^-$	$[HCr_2(CO)_{10}]^-$		
$[Cr(CO)_5]^{2-}$	$[Cr_2(CO)_{10}]^{2-}$		
$HMn(CO)_5$			
$[Mn(CO)_5]^-$			
$[Mn(CO)_4]^{3-}$			
$H_2Fe(CO)_4$		$H_2Fe_3(CO)_{11}$	
$[HFe(CO)_4]^-$	$[HFe_2(CO)_8]^-$	$[HFe_3(CO)_{11}]^-$	$HFe_4(CO)_{13}^-$
$[Fe(CO)_4]^{2-}$	$[Fe_2(CO)_8]^{2-}$	$[Fe_3(CO)_{11}]^{2-}$	
$HCo(CO)_4$			
$[Co(CO)_4]^-$			$[HCo_6(CO)_{15}]^-$
$[Ni(CO)_4]^{2-}$	$[Ni_2(CO)_6]^{2-}$		$[Ni_5(CO)_{16}]^{2-}$
			$[Ni_6(CO)_{12}]^{2-}$

Some carbonylate ions are strong Brønsted bases. As a result, the treatment of a neutral carbonyl with base sometimes yields the conjugate acid, with a metal–hydrogen bond [compare equations (18.44) and (18.45) to (18.39)]:

$$Cr(CO)_6 + 2OH^- \longrightarrow [HCr(CO)_5]^- + HCO_3^- \qquad (18.44)$$

$$Ru_3(CO)_{12} \xrightarrow[\text{(2) H}_3\text{O}^+]{\text{(1) NaBH}_4/\text{THF}} [HRu_3(CO)_{11}]^- \qquad (18.45)$$

The protonation or reduction of homoleptic carbonyls can be used to prepare certain carbonyl hydrides:

$$[V(CO)_6]^- + H^+ \longrightarrow HV(CO)_6 \qquad (18.46)$$

$$Co_2(CO)_8 + H_2 \longrightarrow 2HCo(CO)_4 \qquad (18.47)$$

Alkylation and acylation reactions are often facile as well:

$$[Mn(CO)_5]^- + CH_3I \xrightarrow{-I^-} CH_3Mn(CO)_5 \qquad (18.48)$$

$$[Mn(CO)_5]^- + CH_3C(O)Cl \xrightarrow{-Cl^-} CH_3C(O)Mn(CO)_5 \qquad (18.49)$$

It is also possible to incorporate halogens into metal carbonyl frameworks. Such reactions may or may not involve the elimination of carbon monoxide:

$$Mn_2(CO)_{10} + Br_2 \longrightarrow 2Mn(CO)_5Br \qquad (18.50)$$

$$Fe(CO)_5 + Br_2 \longrightarrow Fe(CO)_4Br_2 + CO \qquad (18.51)$$

You should be able to verify that all the metal-containing products of equations (18.39)–(18.51) conform to the 18-electron rule.

The mechanisms of metal carbonyl substitution reactions have been studied in detail. A commonly used nucleophile is triphenylphosphine:

$$M(CO)_n + P\phi_3 \longrightarrow M(CO)_{n-1}P\phi_3 + CO \qquad (18.52)$$

Such reactions typically follow first-order kinetics. They are believed to be initiated by dissociation of CO in a slow first step, followed by the rapid addition of $P\phi_3$ to the 16-electron intermediate. The general order of reactivity is[22]

$$Co(CO)_4, Mn(CO)_5 > V(CO)_6 > Ni(CO)_4 > Cr(CO)_6 > Fe(CO)_5$$

---

22. High temperatures ($>130°C$) are required to cause some of these reactions to occur at measurable rates. At such temperatures, there is considerable dissociation of $Co_2(CO)_8$ and $Mn_2(CO)_{10}$ to the monomers.

In addition to the rate law, at least two other types of experimental data are supportive of a dissociative mechanism. First, competition studies show little or no discrimination by the metal substrate among different nucleophiles. Also, the stereochemistries of the products observed in certain reactions are explained by positional scrambling (fluxionality) of the 5-coordinate intermediate.[23]

18.3 Other Carbonyl-like Ligands

Although the three diatomic compounds NO, N_2, and O_2 are not organic, they are sufficiently similar to CO to make it convenient to discuss their properties as ligands at this point.

Nitrosyl Complexes[24]

Compared to carbon monoxide, nitric oxide has an "extra" electron residing in a π^* molecular orbital of predominantly nitrogen character. It is therefore not surprising that NO is both similar to and different from CO as a ligand. Like CO, it generally binds to metals through its less electronegative element, and has an affinity for metals in low oxidation states. However, NO usually behaves as either a one- or three-electron donor. One way to rationalize this is to first assume the transfer of the antibonding electron of NO to the metal; the NO^+ group that results is isoelectronic with CO, and may donate two additional electrons.

The stable homoleptic nitrosyls and mixed nitrosyl–carbonyls have molecular formulas consistent with the 18-electron rule; examples include $Cr(NO)_4$, $Mn(CO)(NO)_3$, $Fe(CO)_2(NO)_2$, and $Co(CO)_3NO$. Nitric oxide acts as a three-electron donor in these cases; for example,

For $Cr(NO)_4$: $6 \, (Cr) + (4 \cdot 3) \, (NO) = 18e^-$

For $Mn(CO)(NO)_3$: $7 \, (Mn) + (1 \cdot 2) \, (CO) + (3 \cdot 3) \, (NO) = 18e$

Unlike carbonyls, metal nitrosyls usually cannot be prepared by direct union, because the high temperatures that are required cause decomposition of NO to $N_2 + O_2$. One general synthetic route is the displacement of CO groups by NO:

$$Mn_2(CO)_{10} + 2NO \longrightarrow 2Mn(CO)_4NO + 2CO \qquad \textbf{(18.53)}$$

$$Fe(CO)_5 + 2NO \longrightarrow Fe(CO)_2(NO)_2 + 3CO \qquad \textbf{(18.54)}$$

23. For a more detailed discussion and pertinent references, see Atwood, J. D. *Inorganic and Organometallic Reaction Mechanisms*; Brooks/Cole: Pacific Grove, CA, 1985; pp. 112–118.

24. Mingos, D. M. P.; Sherman, D. J. *J. Organomet. Chem.* **1989**, *34*, 293; Gladfelter, W. L. *J. Organomet. Chem.* **1985**, *24*, 41; Bottomley, F. D. *Acc. Chem. Res.* **1978**, *11*, 158.

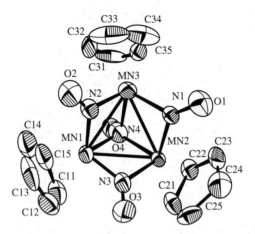

Figure 18.4 The structure of the trimanganese cyclopentadienyl nitrosyl $(C_5H_5)_3Mn_3(NO)_4$. There are three doubly bridging and one triply bridging NO groups. [Reproduced with permission from Elder, R. C. *Inorg. Chem.* **1974**, *13*, 1037.]

Another common approach is oxidative addition:

$$PtCl_4^{2-} + NOCl \longrightarrow [Pt(NO)Cl_5]^{2-} \tag{18.55}$$

Like the carbonyl ligand, NO can bridge two (or more) metals. A remarkable example is $(C_5H_5)_3Mn_3(NO)_4$, which has C_{3v} symmetry; it contains one triply bridging nitrosyl and three doubly bridging nitrosyls (Figure 18.4).[25]

N_2 and O_2: Homonuclear Diatomic Ligands

The ligating properties of N_2 and O_2 are of considerable interest, in large measure because they relate to important biological systems. Coordination to metals by N_2 is believed to be an early step in the process known as nitrogen fixation, and O_2 is bonded to Fe^{II} in oxyhemoglobin and oxymyoglobin (see Chapter 22).

As a general rule, these ligands bind less strongly to metals than CO, NO, or CN^-. This is probably because their HOMO's ($\pi_{x,y}$ for N_2 and $\pi^*_{x,y}$ for O_2) have symmetrical electron distributions. (Unlike polar diatomics, neither atom has excess negative charge.)

25. Elder, R. C. *Inorg. Chem.* **1974**, *13*, 1037.

The first dinitrogen complex was reported by Allen and Senoff:[26]

$$RuCl_3 + H_2N-NH_2 \xrightarrow{N_2} [Ru(NH_3)_5N_2]^{2+} \tag{18.56}$$

The same product is formed in the reaction

$$[Ru(NH_3)_5(H_2O)]^{2+} + N_2 \xrightarrow[-H_2O]{hv} [Ru(NH_3)_5N_2]^{2+} \tag{18.57}$$

The N_2 moiety can act as a two-atom bridge in binuclear complexes (ie, $M-N\equiv N-M'$), and in several other ways as well.[27]

Dioxygen also binds to metals in a variety of ways; the most common are shown in Figure 18.5. The oxidation states in such complexes can be ambiguous; it is sometimes unclear whether the ligand is best considered to be O_2, O_2^-, or O_2^{2-}.[28]

A common way to prepare metal–O_2 complexes is by oxidative addition. For example, Vaska's compound reversibly binds to molecular oxygen:

$$Ir(P\phi_3)_2(CO)Cl + O_2 \longrightarrow Ir(O_2)(P\phi_3)_2(CO)Cl \tag{18.58}$$

The oxygens occupy cis positions in the octahedral product (see Figure 17.1). The internuclear O–O distance is 130 pm, which is between the bond lengths in O_2^- (126 pm, bond order 1.5) and O_2^{2-} (149 pm, bond order 1.0).

Irreversible binding to produce bridged binuclear species is common. This occurs with aqueous Co^{II} in the presence of ammonia:

$$2Co^{II} + 10NH_3 + O_2 \longrightarrow [(H_3N)_5Co-O-O-Co(NH_3)_5]^{4+} \tag{18.59}$$

Figure 18.5
Framework geometries in dioxygen complexes; structures (a) and (f) are quite rare.

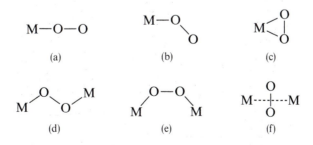

26.   Allen, A. D.; Senoff, C. V. *J. Chem. Soc. Chem. Commun.* **1965**, 621.

27.   For reviews, see Dilworth, J. R.; Richards, R. L. *Chem. Rev.* **1978**, *78*, 589; Zumft, W. G. *Struct. Bonding* **1976**, *29*, 1; Sellmann, D. *Angew. Chem. Int. Ed. Engl.* **1974**, *13*, 639; *Angew. Chem.* **1974**, *86*, 692.

28.   Karlin, K. D.; Gultneh, Y. *Prog. Inorg. Chem.* **1987**, *35*, 219; Taube, H. *Prog. Inorg. Chem.* **1986**, *34*, 607; Jones, R. D.; Hoffman, B. M.; Basolo, F. *J. Chem. Educ.* **1979**, *56*, 157.

The product might be viewed as two Co^{II} centers bridged by neutral O_2, as one Co^{II} and one Co^{III} bridged by superoxide, or as two Co^{III} centers bridged by peroxide (O_2^{2-}).

18.4   Donors Containing Localized π Bonds

C_2H_4 and Derivatives as Ligands[29]

There are a large number of complexes in which ethylene or some other olefin acts as a donor group. One example is *Zeise's salt*, which is produced by the reaction of $PtCl_4^{2-}$ with ethanol:

$$K_2[PtCl_4] + C_2H_5OH \longrightarrow K[Pt(C_2H_4)Cl_3] + H_2O + KCl \quad (18.60)$$

The structure of Zeise's salt shows the two carbon atoms to be equidistant from the metal and out of the $PtCl_3$ plane (Figure 18.6).

The mode of bonding in such species has been a topic of considerable study. It might be argued that Zeise's salt contains a metallacycle—that is, a three-membered C_2Pt ring held together by σ bonds:

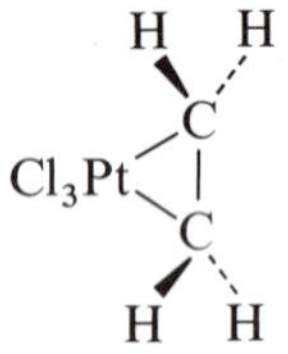

However, this view is not compatible with the carbon–carbon bond distance. If only σ bonds were present, then $d(C–C)$ should be close to the single-bond value of 154 pm; however, it is just 138 pm—only slightly longer than the double bond in ethylene (134 pm).

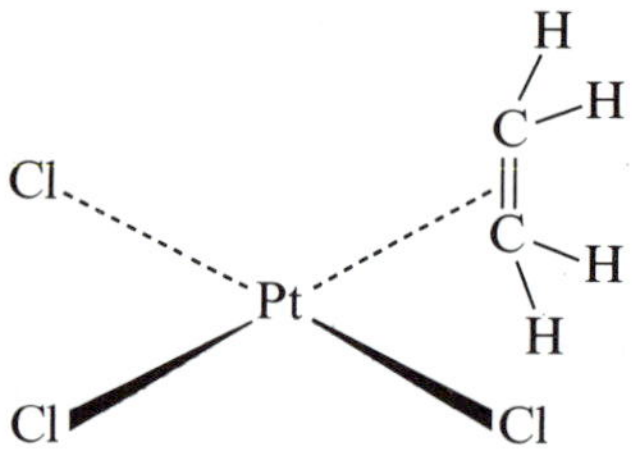

Figure 18.6
The structure of the anion of Zeise's salt, $[Pt(C_2H_4)Cl_3]^-$. The C–C bond axis is perpendicular to the $PtCl_3$ plane.

29.   Ittel, S. D.; Ibers, J. A. *Adv. Organomet. Chem.* **1976**, *14*, 33; Hall, D. I.; Ling, J. H.; Nyholm, R. S. *Struc. Bonding* **1973**, *15*, 3; Tolman, C. A. *Chem. Soc. Rev.* **1972**, *1*, 337.

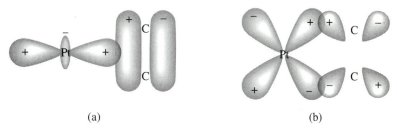

(a)   (b)

Figure 18.7 Orbital overlap in the $[Pt(C_2H_4)Cl_3]^-$ anion according to the Dewar–Chatt–Duncanson model: (a) dative overlap from the π orbital of C_2H_4 to a σ-type d_{z^2} metal orbital; (b) π back-bonding from the platinum d_{xz} to the π* orbital of C_2H_4.

A scheme more consistent with the observed bond distance and other experimental data (see below) is the Dewar–Chatt–Duncanson model.[30] The donated electrons are thought to originate from the π-type HOMO of C_2H_4, which overlaps with σ-type metal d orbitals. This $\pi \to \sigma$ interaction is augmented, in the same manner as metal–carbonyl bonds, by back-bonding (see Figure 18.7).

Alternatively, the interactions can be viewed via two resonance structures, as shown below:

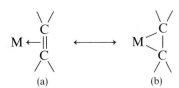

(a)   (b)

The C–C bond strength is between single and double, and the carbon atoms have hybridizations between sp^2 and sp^3. The relative contributions of structures (a) and (b) appear to vary with the nature of both the metal and the substituents on the carbons.

This bonding scheme is supported by experimental evidence other than bond lengths, including the following:

1. There is a decrease in the vibrational frequency of the C=C bond upon coordination, from about $1620 \, \text{cm}^{-1}$ in ethylene to $1525 \, \text{cm}^{-1}$ in $[Pt(C_2H_4)Cl_3]^-$.

2. The H–C–H bond angles in Zeise's salt are about 115°, intermediate between the ideal values for sp^2 and sp^3 hybridization.

30. Dewar, M. J. S. *Bull. Soc. Chim. Fr.* **1951**, *18*, C79; Chatt, J.; Duncanson, L. A. *J. Chem. Soc.* **1953**, 2939.

3. There is a significant trans influence (the cis and trans Pt–Cl bond distances are 230 and 234 pm, respectively), as would be expected if C_2H_4 behaves as a π acceptor ligand.

4. The carbon atoms exhibit electrophilic character. This is a reflection of resonance structure (b). (Alkenes are typically nucleophilic.) Thus, nucleophiles such as R_2N^- and CN^- attack at carbon rather than at the metal:

$$\text{M} \overset{C}{\underset{C}{\big\langle}} \; :X \longrightarrow \text{M–C–C–X} \tag{18.61}$$

This characteristic makes Zeise's salt and its palladium analogue useful catalysts for the conversion of ethylene to acetaldehyde (the *Wacker process*), a topic that will be considered in Section 18.9.

Tetracyanoethylene, $(CN)_2C{=}C(CN)_2$ (TCNE), binds even more strongly to metals than does C_2H_4. It is a potent π acceptor, because the electron-withdrawing cyano substituents make the alkene carbons relatively electron-poor. This is evidenced by the large change in the carbon–carbon bond distance upon coordination to a metal. (Since back-bonding populates antibonding orbitals, the bond lengthens.) Thus, while the C–C bond in Zeise's salt is only about 4 pm longer than in C_2H_4, it is about 16 pm longer in $Ir(P\phi_3)_2(CO)(Br)(TCNE)$ than in TCNE itself.

Metal–olefin bonds can be formed by direct reaction with a free metal (usually using a metal–vapor technique), by addition to a square planar complex, by ligand replacement, or by abstraction of a β hydrogen from an alkyl:

$$\text{Ni} + 3C_2H_4 \xrightarrow{\Delta} \text{Ni}(C_2H_4)_3 \tag{18.62}$$

$$\text{Ir}(P\phi_3)_2(\text{Cl})\text{CO} + C_2H_4 \longrightarrow \begin{array}{c} \text{CO} \\ \phi_3P{\cdots}\!\overset{|}{\underset{|}{\text{Ir}}}{\leftarrow}\!\overset{CH_2}{\underset{CH_2}{\parallel}} \\ \phi_3P{\diagup}\quad\text{Cl} \end{array} \tag{18.63}$$

$$\text{W(CO)}_6 + \text{TCNE} \xrightarrow{\Delta} (\text{CO})_5\text{W(TCNE)} + \text{CO} \tag{18.64}$$

$$C_5H_5(CO)_2FeCH_2CH_3 + \phi_3C^+ \xrightarrow{-\phi_3CH} [C_5H_5(CO)_2FeC_2H_4]^+ \tag{18.65}$$

Equation (18.63) represents an important type of reaction. The majority of isolable, diamagnetic organometallics either obey the 18-electron rule or are 16-electron, nonoctahedral systems. The interconversion of these two types of complexes is often facile. The most prominent examples, of course, are the classic oxidative addition–reductive elimination reactions involving square planar and octahedral complexes. However, many other 16-electron ↔ 18-electron transformations are known as well.

Metal–Alkyne Complexes[31]

The behavior of acetylene and other alkynes as ligands is similar to that of olefins, although alkyne complexes tend to be more reactive. The bonding is generally described in the same manner, with the resonance structure

$$M \diagdown \overset{\diagup}{\underset{\diagdown}{\overset{C}{\underset{C}{\|}}}}$$

making a major contribution to the hybrid. The carbon–carbon bond varies in length, but is often close to that in alkenes. For example, $d(C–C)$ in the tris(diphenylacetylene) complex $W(CO)(C_2\phi_2)_3$ is 130 pm. (Compare this to typical C–C distances of 134 pm in alkenes and 120 pm in alkynes.) Another structural feature of this complex (Figure 18.8, p. 604) is that the linearity of the C–C–C–C linkage is destroyed by ligation; the C–C–C bond angles are about 140°, intermediate between those expected for sp and sp^2 hybridization.

The reaction of Vaska's compound with bis(trifluoromethyl)acetylene is reversible at room temperature:

$$Ir(P\phi_3)_2(Cl)CO + F_3CC\equiv CCF_3 \rightleftharpoons \begin{matrix} & Cl & CF_3 \\ \phi_3P & | & C \diagup \\ & Ir \leftarrow \||| \\ \phi_3P & | & C \\ & CO & \diagdown CF_3 \end{matrix} \qquad (18.66)$$

Infrared spectroscopic data are of particular interest in this system. The $C\equiv C$ stretching frequency decreases by over 500 cm^{-1} upon coordination to the metal. Also, the frequency of the carbonyl stretch is about 75 cm^{-1} greater for the product than for the reactant. This suggests that Ir → CO

31. Raithby, P. R.; Rosales, M. J. *Adv. Inorg. Chem. Radiochem.* **1986**, *30*, 123; Sappo, E.; Tiripicchio, A.; Braunstein, P. *Chem. Rev.* **1983**, *83*, 203.

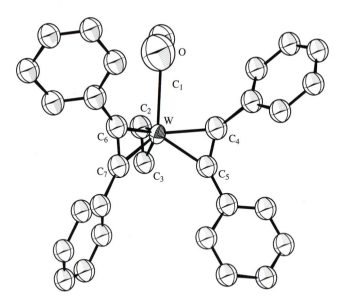

Figure 18.8
The structure of
$W(CO)(C_2\phi_2)_3$; the
phenyl groups
pendant to the C_2
and C_3 carbons are
omitted for clarity.
[Reproduced with
permission from
Laine, R. M.;
Moriarty, R. E.; Bau,
R. *J. Am. Chem. Soc.*
1972, *94,* 1402.]

back-bonding is reduced by the presence of the powerfully π accepting $F_3CC{\equiv}CCF_3$ ligand.

Alkyne complexes are stabilized by electron-withdrawing and/or sterically bulky groups on carbon. For example, the reaction

$$2PtCl_4^{2-} + 2(ac) \xrightarrow{-4Cl^-} \begin{array}{c} \text{(18.67)} \end{array}$$

where (ac) is an acetylenic ligand, gives stable products only when at least one substituent is a large group such as *t*-butyl. Other cases give intractable products, probably because of intermolecular reactions between coordinated ligands.

Allyl Complexes

The allyl radical ($\cdot C_3H_5$) can bind to metals in at least two ways—as a one- or three-electron donor:

$$M + \cdot CH_2{-}CH{=}CH_2 \longrightarrow M{-}CH_2{-}CH{=}CH_2 \qquad \textbf{(18.68)}$$

$$M + \cdot CH_2{-}CH{=}CH_2 \longrightarrow M{\longleftarrow}\begin{array}{c} CH_2 \\ \\\\ CH \\ \\\\ CH_2 \end{array} \qquad \textbf{(18.69)}$$

The operative mode is readily apparent if the structure of the complex is known. In the first case, only one carbon atom is adjacent to the metal, while in the second, the three carbons are equidistant from it. These modes are customarily differentiated in chemical formulas through the *hapto* system, where the Greek letter eta (η) is superscripted by an Arabic numeral equal to the number of carbons interacting with the metal. Thus, the complex $(CO)_5Mn-CH_2-CH=CH_2$ contains a monohapto (η^1) allyl group, while

$$(CO)_4Mn \longleftarrow \begin{matrix} CH_2 \\ \diagdown \\ CH \\ \diagup \\ CH_2 \end{matrix}$$

has a trihapto (η^3) ligand. The 18-electron rule is obeyed in each case.

Allyl groups have been incorporated into complexes by a variety of synthetic methods. A few examples are described by the following equations:

$$[Mn(CO)_5]^- + Cl-CH_2-CH=CH_2 \xrightarrow{-Cl^-} (CO)_5Mn-CH_2-CH=CH_2 \qquad (18.70)$$

$$HCo(CO)_4 + H_2C=CH-CH=CH_2 \longrightarrow (CO)_4Co-CH_2-CH=CH-CH_3 \qquad (18.71)$$

$$NiCl_2 + 2C_3H_5MgBr \longrightarrow Ni(C_3H_5)_2 + 2MgBrCl \qquad (18.72)$$

$$C_5H_5Co(CO)_2 + BrCH_2-CH=CH_2 \xrightarrow{-CO} [C_5H_5Co(CO)(C_3H_5)]^+Br^- \qquad (18.73)$$

Monohapto allyl groups often become trihapto upon elimination of another ligand. (The number of metal valence electrons is not changed by this process, since the C_3H_5 group changes from a one- to a three-electron donor.) This process may be reversible:

$$(CO)_5Mn(\eta^1\text{-}C_3H_5) \underset{hv}{\overset{hv}{\rightleftharpoons}} (CO)_4Mn(\eta^3\text{-}C_3H_5) + CO \qquad (18.74)$$

18.5 Complexes with Cyclic π Donors

Ferrocene

The cyclopentadienide anion, $C_5H_5^-$, has been used as a reagent for the synthesis of a huge number of complexes. The most famous of these is bis(cyclopentadienyl)iron, better known as ferrocene. This highly stable compound can be prepared by the reaction

$$FeCl_2 + 2Na^+C_5H_5^- \longrightarrow Fe(C_5H_5)_2 + 2NaCl \qquad (18.75)$$

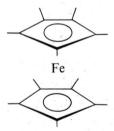

Figure 18.9 The "sandwich" structure of ferrocene, $Fe(C_5H_5)_2$. All ten Fe–C bond distances are equal; the two rings are rotated by 9° with respect to one another.

although it was initially synthesized (unexpectedly!) by less direct methods.[32]

Ferrocene has a "sandwich" structure. The ligands are pentahapto, with ten equal Fe–C distances (Figure 18.9). The rings are parallel and lie in a nearly eclipsed orientation (9° twist) with respect to one another.

The oxidation states in ferrocene are somewhat ambiguous. Based on equation (18.75), it seems reasonable to consider it as Fe^{II}-coordinated to two anionic ligands. However, up to this point all organic donors have been treated as neutral whenever possible. If the carbon-containing group is taken to be the free radical $\cdot C_5H_5$, then iron is in the 0 oxidation state. Similar ambiguity arises with allyl ($M^0/\cdot C_3H_5$ versus $M^+/C_3H_5^-$) and nitrosyl (M^0/NO versus M^+/NO^- and M^-/NO^+) ligands. The 18-electron rule is obeyed regardless of the choice.

$$Fe^0/\cdot C_5H_5: \quad 8\,(Fe) + (2 \cdot 5)\,(C_5H_5) = 18e^-$$

$$Fe^{2+}/C_5H_5^-: \quad 6\,(Fe) + (2 \cdot 6)\,(C_5H_5^-) = 18e^-$$

Numerous molecular orbital studies of ferrocene have been reported. The results differ somewhat, but generally resemble the diagram shown in Figure 18.10. Each metal valence orbital contributes to a bonding MO, and the nine bonding orbitals are just filled by the 18 electrons. Based on Figure 18.10, we would expect both $Mn(C_5H_5)_2$ and $Co(C_5H_5)_2$ (with 17 and 19 electrons, respectively) to be more reactive than ferrocene, and, in fact, they are (see below).

32.   Miller, S. A.; Tebboth, J. A.; Tremaine, J. F. *J. Chem. Soc.* **1952**, 632; Kealy, T. J.; Pauson, P. L. *Nature* **1951**, *168*, 1039.

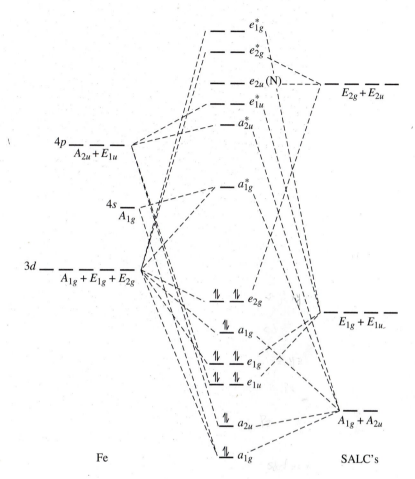

Figure 18.10
Qualitative molecular orbital diagram for ferrocene, $Fe(C_5H_5)_2$ (D_{5d} symmetry). [See Cotton, F. A. *Chemical Applications of Group Theory*, 3rd ed.; Wiley: New York, 1990; pp. 240*ff*].

Ferrocene undergoes a remarkable number of chemical reactions. Many can be classed as electrophilic aromatic substitutions and parallel known reactions of benzene and its derivatives; for example,

$$Fe(C_5H_5)_2 \xrightarrow[\text{AlCl}_3]{\text{MeCOCl}} C_5H_5FeC_5H_4C(O)Me \qquad (18.76)$$

$$Fe(C_5H_5)_2 \xrightarrow[\text{AlCl}_3]{\text{H}_2\text{C=CHR}} C_5H_5FeC_5H_4CH_2CH_2R \qquad (18.77)$$

$$Fe(C_5H_5)_2 \xrightarrow[\text{HF}]{\text{CH}_2\text{O}} C_5H_5FeC_5H_4CH_2OH \qquad (18.78)$$

The C–H protons are moderately acidic, and the monoanion is produced by reacting $Fe(C_5H_5)_2$ with alkyllithium reagents:

$$Fe(C_5H_5)_2 + n\text{-BuLi} \xrightarrow{-C_4H_{10}} Li^+[C_5H_5FeC_5H_4]^- \qquad \textbf{(18.79)}$$

This anion has, in turn, been used to produce many derivatives.

Other Metallocenes

Nearly all the d-block elements form complexes in which one or more C_5H_5 ligands are present. Although ferrocene-like bonding is the norm, the interactions are sometimes ionic. An example is $Sc(C_5H_5)_3$, which has a three-dimensional, ionic-type lattice (Sc^{3+}, $3C_5H_5^-$).

Chromocene, with only 16 valence electrons, is extremely reactive and difficult to handle in the laboratory. Molybdenum and tungsten conform to the 18-electron rule in the dihydrides $H_2M(C_5H_5)_2$. In these molecules, the cyclopentadiene rings are tilted, with both hydrogens on the opened side (Figure 18.11).

Figure 18.11
The structures of the cyclopentadienyl hydrides $H_2M(C_5H_5)_2$ (M = Mo and W).

Cobaltocene, with 19 electrons, undergoes several types of reactions that produce 18-electron products; for example,

$$2Co(C_5H_5)_2 + Br_2 \longrightarrow 2[Co(C_5H_5)_2]^+Br \qquad \textbf{(18.80)}$$

$$2Co(C_5H_5)_2 + CCl_4 \longrightarrow \text{Co} \qquad + [Co(C_5H_5)_2]^+Cl^- \qquad \textbf{(18.81)}$$

Like $Co(C_5H_5)_2$, nickelocene readily undergoes one-electron oxidation. The removal of a second electron can be accomplished electrochemically, but the dication undergoes rapid degradation. An interesting variation is the "triple-decker (club??) sandwich" cation $[Ni_2(C_5H_5)_3]^+$ (Figure 18.12), which is obtained from reactions of nickelocene with Lewis acids.

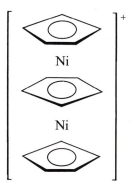

Figure 18.12
The structure of the "triple decker sandwich" cation, $[Ni_2(C_5H_5)_3]^+$. The three planar rings are parallel to one another.

All the carbon atoms of cyclopentadienyl-like units do not necessarily coordinate to the metal. For example, in $Hg(C_5H_5)_2$, the ligands are monohapto and there is a linear C–Hg–C linkage. Also, $C_5H_5Ni(C_5H_3C_2F_4)$ contains an allyl-like trihapto group:

See also the product of equation (18.81).

A number of mixed cyclopentadienyl–carbonyl and cyclopentadienyl–nitrosyl complexes are known. They are produced in various ways, as illustrated by the following equations (cp = C_5H_5):

$$cp_2V + 4CO \xrightarrow[120°C]{60\ atm} cpV(CO)_4 + C_5H_5 \qquad \textbf{(18.82)}$$

$$Co_2(CO)_8 + 2C_5H_6 \xrightarrow{hv} 2cpCo(CO)_2 + 4CO + H_2 \qquad \textbf{(18.83)}$$

$$cp_2Ni + NO \xrightarrow{25°C} cpNi(NO) + C_5H_5 \qquad \textbf{(18.84)}$$

$$Ni(C_5H_5)_2 + Ni(CO)_4 \xrightarrow{80°C} [cpNi(CO)]_2 + 2CO \qquad \textbf{(18.85)}$$

Conformity to the 18-electron rule is the norm for these mixed systems. For example, the 17-electron species $C_5H_5Ni(CO)$ dimerizes by metal–metal bond formation.

Per Ni: 10 (Ni) + 5 (C_5H_5) + 2 (CO) + 1 (Ni–Ni) = $18e^-$

Other Cyclic π Systems

Cyclobutadiene cannot be isolated at room temperature because of its thermal instability (although certain derivatives, particularly those with sterically bulky groups, can be obtained). However, isolable complexes such as $C_4H_4Fe(CO)_3$ and $C_4H_4CoC_5H_5$ can be prepared. The electron count in these compounds indicates that the C_4H_4 group acts as a four-electron donor, and structural studies mark it as a tetrahapto ligand.

The low chemical reactivity of benzene makes it a relatively poor donor. However, under the proper conditions it can be introduced into complexes as a six-electron, hexahapto ligand. For example, the preparation of dibenzenechromium from $CrCl_3$ and phenylmagnesium bromide has been described in detail.[33] Other examples of the synthesis of η^6-arene complexes are given by the following equations:

$$Cr(CO)_6 + C_6H_6 \xrightarrow{\Delta} Cr(C_6H_6)(CO)_3 + 3CO \tag{18.86}$$

$$Fe(C_5H_5)_2 + C_6H_6 + 2AlCl_3 \longrightarrow [C_5H_5FeC_6H_6]^+AlCl_4^- + C_5H_5AlCl_2 \tag{18.87}$$

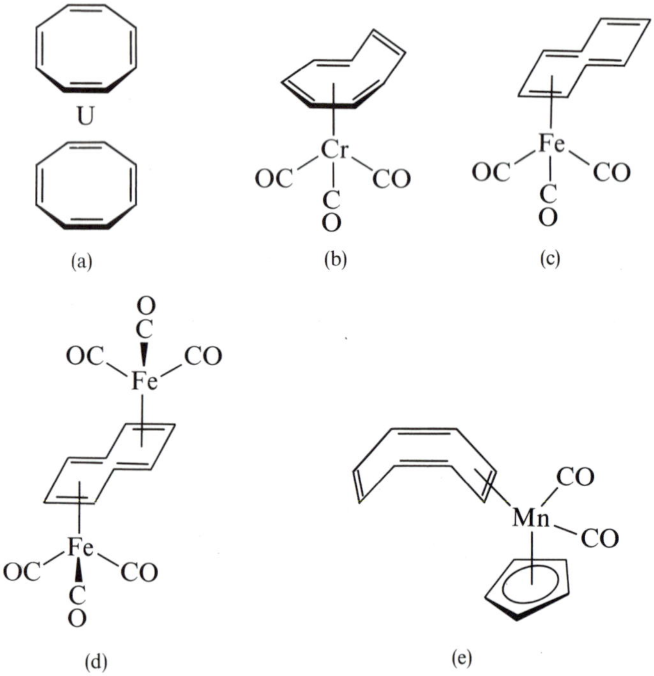

Figure 18.13
Examples of different modes of coordination by the cyclooctatetraene ligand:
(a) η^8 coordination in $U(C_8H_8)_2$;
(b) η^6 coordination in $C_8H_8Cr(CO)_3$;
(c) η^4 coordination in $C_8H_8Fe(CO)_3$;
(d) bridging η^4–η^4 coordination in $C_8H_8[Fe(CO)_3]_2$;
(e) η^2 coordination in $C_8H_8Mn(C_5H_5)(CO)_2$.

33. Fischer, E. O. *Inorg. Synth.* **1960**, *6*, 132.

Another important π donor is cyclooctatetraene, C_8H_8. This homocycle complexes strongly to the $4d$, $5d$, and f-block elements, usually as an η^8 group; its large size requires a large metal for effective orbital overlap. A well-known example is bis(cyclooctatetraene)uranium, $U(C_8H_8)_2$ (uranocene). When complexed to metals of the first transition series, a variety of bonding modes (but not usually octahapto) are observed for C_8H_8, including η^2, η^4, and η^6 (see Figure 18.13).

18.6   Organometallics and Catalysis—An Overview

As we mentioned in the chapter introduction, there is tremendous interest in the use of transition metals as catalysts for organic transformations. Here, we give a brief, general summary of catalysis; later, in Sections 18.7–18.10, the activation of four types of bonds (H–H, C–H, C=C, and C≡O) is considered.

Recall that a catalyst affects the forward and reverse rates of a chemical reaction, but does not alter the position of equilibrium; nor is it consumed by the process. Catalysts that increase reaction rates do so by providing an alternate reaction pathway—one that has a lower activation enthalpy than any available to the uncatalyzed system (Figure 18.14).[34]

Catalysts are commonly divided into two types, homogeneous and heterogeneous, depending on whether they operate in the same or in a

Figure 18.14
The influence of a catalyst on reaction rate. Catalysis provides an alternate reaction pathway that has a lower activation energy than the uncatalyzed route. [Reproduced with permission from Haim, A. *J. Chem. Educ.* **1989**, *66*, 935.]

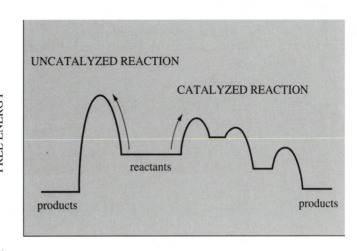

UNCATALYZED REACTION

CATALYZED REACTION

FREE ENERGY

reactants

products

products

REACTION COORDINATE

---

34.   For a relevant discussion, see Haim, A. *J. Chem. Educ.* **1989**, *66*, 935.

Table 18.8   Comparisons between homogeneous and heterogeneous catalysis

	Homogeneous	Heterogeneous
Phase of reaction	Solution	Gas–solid or solution–solid
Catalyst solubility	Soluble	Insoluble
Ease of separation of product	May be difficult	Easy
Ease of recovery of catalyst	May be difficult	Relatively easy
Catalytic site	Discrete complex	Active site on solid support

Note: The above are generalizations, and do not apply to all cases.

different phase from the reactants. This is significant because it leads to several other differences, some of which are summarized in Table 18.8. Because homogeneous catalysis results from reactions of discrete molecules, our discussion will emphasize that type.

The activity of an organometallic catalyst usually arises from any one of three types of transformations (or, more commonly, some combination of these):

1. *Ligand Elimination and Replacement*:   The "active form" of a catalyst is often coordinately unsaturated. For example, in the acceleration of certain reactions by *Wilkinson's catalyst*, $Rh(P\phi_3)_3Cl$, the initial step of the mechanism is believed to be the dissociation of a $P\phi_3$ ligand to give the highly reactive 14-electron intermediate $Rh(P\phi_3)_2Cl$ (see Section 18.9).

2. *Oxidative Addition and Reductive Elimination*:   Most oxidative addition reactions involve the cleavage of a bond of the incoming group. For example, the addition of H_2 to Vaska's complex causes rupture of the H–H bond (see Figure 17.1). This is precisely the type of behavior that catalyzes organic reactions, since the great strengths of C–C, C=C, C–H, and H–H bonds often cause the uncatalyzed reactions to be kinetically slow.

3. *Insertion*:   The insertion of a small molecule into a metal–ligand bond often weakens a bond of the inserting species. For example, in equation (18.6) the insertion of C_2H_4 into the Pt–H bond converts the C=C linkage to C–C. Similarly, insertion of $F_3CC{\equiv}CCF_3$ into AgF [equation (18.7)] reduces the carbon–carbon bond from triple to double, and the intramolecular insertion of CO into $MeCo(CO)_4$ [the reverse of equation (18.21)] reduces the formal C≡O linkage from a triple to a double bond.

18.7 Transition Metal Hydrides and H–H Bond Activation

Before we get into specific examples of the activation of molecular hydrogen, a general discussion of hydrido complexes of transition metals is in order.[35]

Several interesting binary metal–hydrogen complexes are known, including ReH_9^{2-}, TcH_9^{2-}, FeH_6^{4-}, and PtH_4^{2-}.[36] The structures of these species are consistent with expectations based on Chapter 15. For example, ReH_9^{2-} is a tricapped trigonal prism, while FeH_6^{4-} is octahedral. They are prepared under strongly reducing conditions by reactions such as

$$Na[ReO_4] \xrightarrow{\text{Na}} Na_2[ReH_9] \tag{18.88}$$

$$2\,Mg + Fe + 3\,H_2 \xrightarrow{\text{Pressure}} Mg_2[FeH_6] \tag{18.89}$$

Many other complexes are known that contain one or more hydrogens as ligands (formally as H^-). Typical M–H bond energies are on the order of 250 kJ/mol, about the same as the Sn–H bonds of SnH_4. Bond lengths are normally consistent with the sum of the covalent radii, and terminal hydrogens are stereochemically active. For example, $HMn(CO)_5$ has an approximately octahedral geometry about the manganese, with $d(\text{Mn–H}) = 160$ pm.

Like carbonyls, hydrogens often occupy bridging positions in organometallic complexes. There are two common bridging modes, dihapto (bridging two bonded metal atoms—that is, "protonating" a metal–metal bond) and trihapto (generally, centering a trimetallic face of a cluster). These are illustrated in Figure 18.15 (p. 614).

The protonation of bimetallic complexes to form $M\overset{H}{\diagup\diagdown}M$ species weakens the metal–metal bond. For example, in going from $[(CO)_5Cr–Cr(CO)_5]^{2-}$ to $[HCr_2(CO)_{10}]^-$ (Figure 18.15a), the chromium–chromium bond distance increases from 297 to 341 pm. This is indicative of reduced electron density between the metals, and is consistent with the notion of a three-center, two-electron bond (Chapter 3).

A variety of methods are used for the formation of M–H bonds. Five of the most common are listed on the following pages.

35. For reviews of various aspects of this topic, see Brookhart, M.; Green, M. H. L.; Wong, L.-L. *Prog. Inorg. Chem.* **1988**, *36*, 1; Crabtree, R. H.; Hamilton, D. G. *J. Organomet. Chem.* **1988**, *28*, 299; Latky, G. G.; Crabtree, R. H. *Coord. Chem. Rev.* **1985**, *65*, 1; Pearson, R. G. *Chem. Rev.* **1985**, *85*, 41.

36. Bronger, W.; Muller, P.; Schmitz, D.; Spittank, H. *Z. Anorg. Allg. Chem.* **1984**, *516*, 35; Didisheim, J.-J.; Zolliker, P.; Yvon, K.; Fischer, P.; Schefer, J.; Gubelmann, M.; Williams, A. F. *Inorg. Chem.* **1984**, *23*, 1953; Ginsberg, A. P.; Sprinkle, C. R. *Inorg. Synth.* **1972**, *13*, 219.

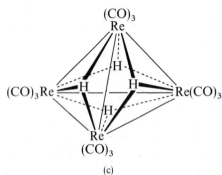

Figure 18.15
Bi-, tri-, and tetrametallic complexes containing bridging hydrogens:
(a) $[\mu_2\text{-}HCr_2(CO)_{10}]^-$;
(b) $(\mu_2\text{-}H)_3Re_3(CO)_{12}$;
(c) $(\mu_3\text{-}H)_4Re_4(CO)_{12}$.

1. **Protonation of Basic Complexes by Brønsted Acids:** Anionic complexes often can be protonated, particularly if the metal is coordinately unsaturated.

$$[Mn(CO)_5]^- \xrightarrow{\quad CF_3SO_3H \quad} HMn(CO)_5 \qquad (18.90)$$

$$C_5H_5Co(PR_3)_2 + HX \longrightarrow [HCo(C_5H_5)(PR_3)_2]^+ + X^-$$
$$X = OH, OMe, \text{etc.} \quad (18.91)$$

2. **Reactions of Hydridic Reagents with Metal Complexes:** Anions such as H^-, OH^-, and BH_4^- are often reactive toward metal centers because of their nucleophilicity. Example reactions include those described by equations (18.44), (18.45), and (18.92):

$$Ru_3(CO)_{12} \xrightarrow{\quad NaBH_4 \quad} Na[HRu_3(CO)_{11}] + BH_3CO \qquad (18.92)$$

3. **Oxidative Addition by a Heteronuclear H–X Bond:** Numerous examples of this type of reaction have already been cited. Often, but not always, X is a halogen. Silyl-, germyl-, and stannyl hydrides are also reactive.

$$Rh(P\phi_3)_3Cl + HSiCl_3 \xrightarrow{\quad -P\phi_3 \quad} HRh(P\phi_3)_2(Cl)(SiCl_3) \qquad (18.93)$$

4. ***Hydrogen Transfer from a Coordinated Ligand***: This category includes β-hydride elimination reactions and intramolecular oxidative additions, and as such has already been discussed.

5. ***"Direct Reaction" with*** H_2: Reactions of this type are particularly relevant to the present discussion, since they involve H–H bond activation and/or cleavage.

$$[Pt(SnCl_3)_5]^{3-} + H_2 \xrightarrow{-HSnCl_3} [HPt(SnCl_3)_4]^{3-} \qquad (18.94)$$

$$W(CO)_3[P(i\text{-}Pr_3)]_2 + H_2 \longrightarrow H_2W(CO)_3[P(i\text{-}Pr_3)]_2 \quad (18.95)$$

The product of equation (18.95) is remarkable in that the hydrogen–hydrogen bond is retained; that is, the H_2 molecule itself acts as a ligand.[37] Several dihydrogen complexes are now known. In $H_2W(CO)_3[P(i\text{-}Pr_3)]_2$ (Figure 18.16), the internuclear H–H distance is 84 pm, compared to 74 pm in $H_2(g)$. The infrared stretching frequency (ν_{H-H}) occurs at only 2655 cm^{-1} in the complex versus 4395 cm^{-1} in H_2. Hence, the H–H interaction is intact, but clearly weakened.

Figure 18.16
The complex $H_2W(CO)_3[P(i\text{-}Pr_3)]_2$, in which dihydrogen acts as a ligand. [Reproduced with permission from Kubas, G. J.; Ryan, R. R.; Swanson, B. I.; Vergamini, P. J.; Wasserman, H. J. *J. Am. Chem. Soc.* **1984**, *106*, 451.]

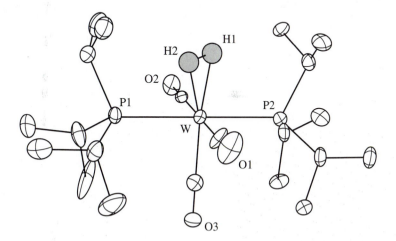

37. Kubas, G. J.; Ryan, R. R.; Swanson, B. I.; Vergamini, P. J.; Wasserman, H. J. *J. Am. Chem. Soc.* **1984**, *106*, 451. For a review, see Crabtree, R. H.; Hamilton, D. G. *Adv. Organomet. Chem.* **1988**, *28*, 299.

18.8 The Activation of Carbon–Hydrogen Bonds

Similar to the case of H_2, certain C–H bonds are weakened, and therefore activated, by the presence of a metal. A specific example involves the Ru^{II} complex *trans*-$Ru(dmpe)_2Cl_2$ (dmpe = bis-1,2-dimethylphosphino-ethane, $Me_2P–CH_2–CH_2–PMe_2$), which reacts with the naphthalide anion to form a Ru–H bond:

(18.96)

The product is believed to arise from the initial formation of a π complex,

followed by the rupture of a C–H bond and coordination of "H^-" to the metal.

Benzene undergoes hydrogen–deuterium exchange in the presence of certain complexes, most notably $PtCl_4^{2-}$. (As uncatalyzed exchange does not occur, the metal clearly promotes C–H bond activation.) The rate law for exchange is first-order in both C_6H_6 and $PtCl_4^{2-}$, and the process is inhibited by chloride ion. Several mechanisms have been suggested for this process, one of which (for aqueous solution) is given in Figure 18.17. In the scheme shown, benzene is incorporated as a π ligand, and deuterium becomes ligated by an oxidative addition–substitution sequence. Intramolecular insertion into the M–D bond, followed by the reductive elimination of C_6H_5D, completes the process.

A particularly interesting system is the interaction of alkanes with $PtCl_4^{2-}$. Such reactions can be made to follow exchange, dehydrogenation, or halogenation pathways, depending on the conditions.[38] The different carbon-containing products are probably formed from a common intermediate—the product of the oxidative addition of alkane to $PtCl_4^{2-}$:

$$PtCl_4^{2-} + RCH_2CH_3 \rightleftharpoons \left[\begin{array}{c} Cl \\ Cl \cdots \overset{|}{\underset{|}{Pt}} \cdots H \\ Cl \overset{}{\diagup} \overset{|}{\underset{Cl}{}} \diagdown CH_2CH_2R \end{array} \right]^{2-}$$

(18.97)

38. Shilov, A. E.; Shteinman, A. A. *Coord. Chem. Rev.* **1977**, *24*, 97.

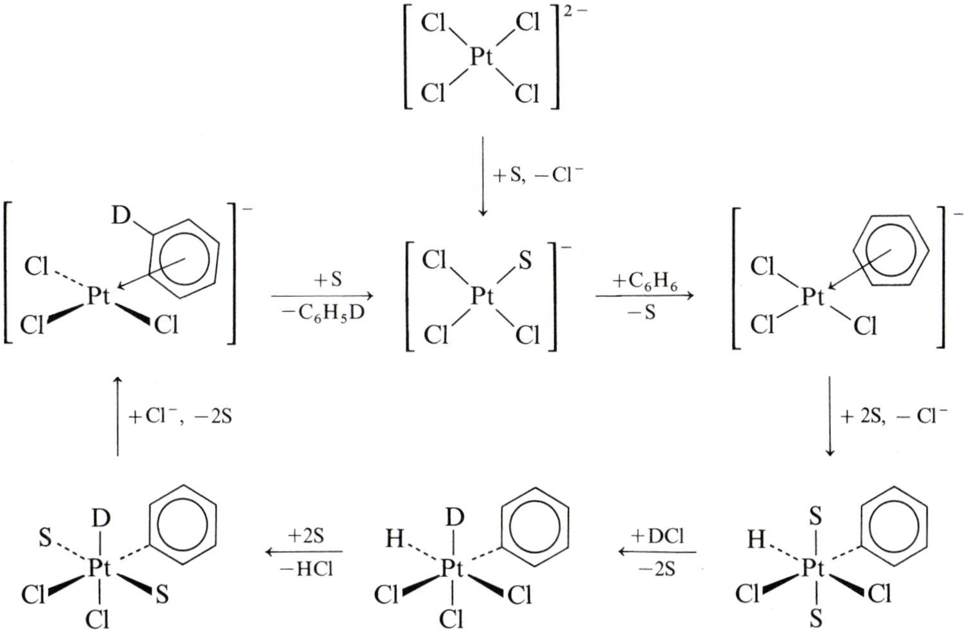

Figure 18.17   Proposed mechanism for $PtCl_4^{2-}$-catalyzed H–D exchange by benzene; S = solvent. [After Blackett, L.; Gold, V.; Rueben, D. M. E. *J. Chem. Soc. Perkin II* **1974**, 1869.]

Since this process is reversible, it can result in hydrogen–deuterium exchange if there is a source of deuterium. In that case, reductive elimination produces RCH_2CH_2D:

$$[Pt(Cl)_4(D)CH_2CH_2R]^{2-} \longrightarrow PtCl_4^{2-} + RCH_2CH_2D \quad \textbf{(18.98)}$$

Alternatively, the octahedral intermediate might undergo reductive elimination of HCl. Subsequent β-hydride elimination then gives dehydrogenation to an alkene:

$$[Pt(Cl)_4(H)CH_2CH_2R]^{2-} \xrightarrow{-HCl} [Pt(Cl)_3CH_2CH_2R]^{2-} \quad \textbf{(18.99)}$$

$$[Pt(Cl)_3CH_2CH_2R]^{2-} \longrightarrow [Pt(H)Cl_3]^{2-} + RCH{=}CH_2 \quad \textbf{(18.100)}$$

A third possibility is reductive elimination of the alkyl chloride:

$$[Pt(Cl)_4(H)CH_2CH_2R]^{2-} \longrightarrow$$
$$[Pt(H)Cl_3]^{2-} + RCH_2CH_2Cl \quad \textbf{(18.101)}$$

18.9 The Activation of C=C Bonds

The Hydrogenation of Olefins

The hydrogenation of alkenes using heterogeneous catalysts such as Raney nickel and palladium (on charcoal at elevated temperature and pressure) is a classic reaction of organic chemistry. A remarkable homogeneous complex for this purpose is Wilkinson's catalyst, $Rh(P\phi_3)_3Cl$.[39] This square planar, 16-electron species undergoes reversible oxidative addition with H_2:

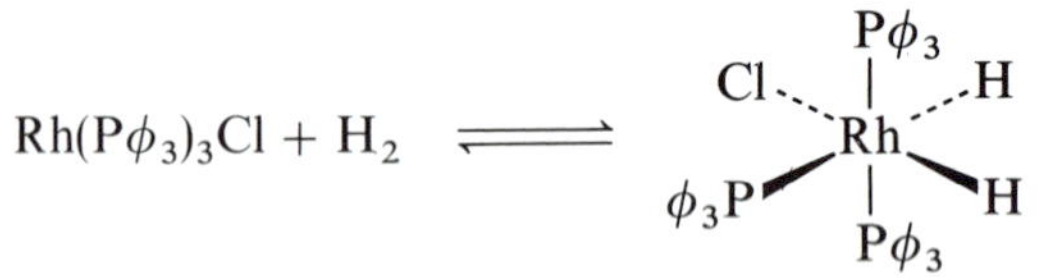

$$Rh(P\phi_3)_3Cl + H_2 \rightleftharpoons \begin{matrix} & P\phi_3 & \\ Cl\cdots & | & \cdots H \\ & Rh & \\ \phi_3P & | & H \\ & P\phi_3 & \end{matrix} \qquad (18.102)$$

The hydrogenation of alkenes occurs at room temperature and pressure in the presence of this complex. The reaction mechanism is complicated, and not entirely understood. A reasonable possibility is described by Figure 18.18.

The proposed sequence is: (1) ligand substitution of $P\phi_3$ by a solvent molecule; (2) oxidative addition of H_2 with concurrent loss of the solvent

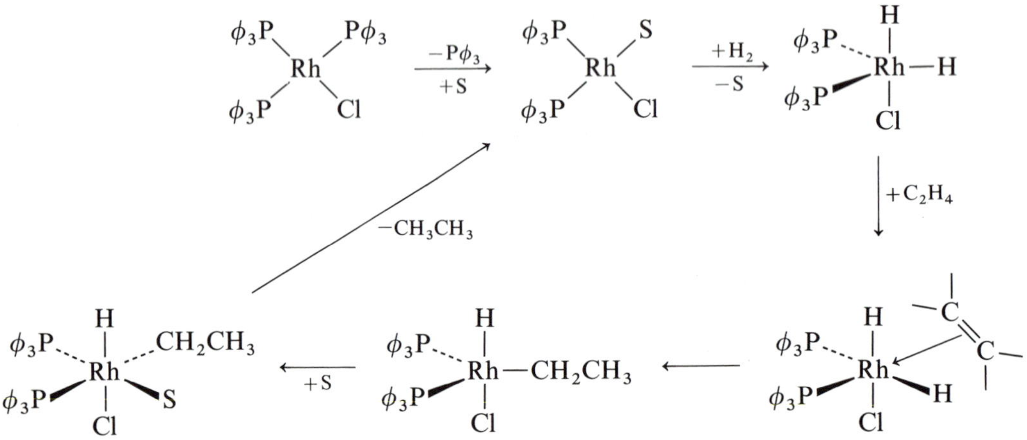

Figure 18.18 A plausible mechanism for the hydrogenation of C_2H_4, as catalyzed by $Rh(P\phi_3)_3Cl$ (Wilkinson's catalyst); S = solvent. Note the oxidative addition, insertion, and reductive elimination steps.

39. Osborn, J. A.; Jardine, F. H.; Young, J. F.; Wilkinson, G. *J. Chem. Soc. A* **1966**, 1711. See also, Jardine, F. H. *Prog. Inorg. Chem.* **1981**, *28*, 63.

ligand, giving a 5-coordinate Rh^{III} intermediate; (*3*) π coordination of the olefin to form an 18-electron, octahedral species; (*4*) C=C insertion into a rhodium–hydrogen bond, resulting in a rhodium–carbon σ bond and weakening the carbon–carbon linkage; (*5*) addition of solvent to the 16-electron complex from step 4; and (*6*) reductive elimination of C_2H_6.

The reverse reaction (dehydrogenation) also can be effected with the aid of metal catalysts under the proper conditions. For example, the conversion of cyclohexene to benzene is catalyzed by palladium acetate. This reaction is thought to involve the initial formation of a π complex, which undergoes deprotonation to yield acetic acid and the diene. The process then repeats, ultimately giving C_6H_6 (Figure 18.19).

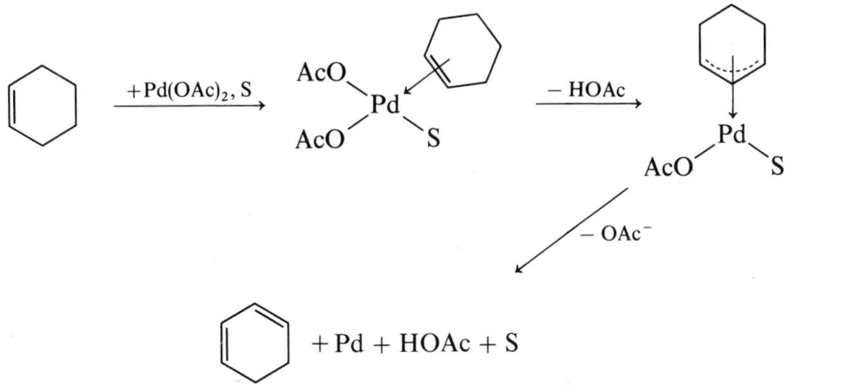

Figure 18.19   Proposed mechanism for the dehydrogenation of cyclohexene to benzene in the presence of palladium acetate; S = solvent. Only the elimination of 1 mol of H_2 is shown; the process repeats for the diene → benzene conversion. [See Brown, R. G.; Davidson, M. *J. Chem. Soc. A* **1971**, 1321.]

The Polymerization of Olefins[40]

Organometallic compounds have been used as catalysts for the polymerization of alkenes for many years. The simplest examples are *anionic polymerizations*, usually initiated by alkyllithium reagents:

$$R_2C{=}CR_2 + Li^+R'^- \longrightarrow Li^+[CR_2CR_2R']^- \qquad \textbf{(18.103)}$$

$$Li^+[CR_2CR_2R']^- + R_2C{=}CR_2 \longrightarrow Li^+[CR_2CR_2CR_2R']^- \quad \textbf{(18.104)}$$

$$Li^+[CR_2CR_2CR_2CR_2R']^- + R_2C{=}CR_2 \longrightarrow \text{etc.} \qquad \textbf{(18.105)}$$

40.   Jordan, R. F. *J. Chem. Educ.* **1988**, *65*, 285.

Superior results are obtained by the use of *Ziegler–Natta catalysts*,[41] which traditionally have been mixtures of Ti^{IV} and alkylaluminum reagents (eg, $TiCl_4$ and $AlEt_3$). These catalysts permit polymerization under comparatively mild conditions (typically, 20–50°C and 8–10 atm), and give remarkably stereospecific products. For example, by the appropriate choice of reaction conditions, the polymerization of 1,3-butadiene can be made to give any one of the 1,4-*cis*, 1,4-*trans*, or 1,2-butadiene polymers in greater than 98% purity.[42]

The polymerization mechanism is not known with certainty, but probably involves initial π coordination by the olefin followed by alkyl group migration. Such a scheme is shown in Figure 18.20.

Numerous variations of the Ziegler–Natta type of catalysts have been proposed. The $TiCl_4 + AlEt_3$ mixture is a heterogeneous catalyst, since it is insoluble in the solvents normally employed. The use of a titanium reagent such as $(C_5H_5)_2TiCl_2$ creates a homogeneous system. Also useful are "second-generation" catalysts such as $Mg^{II} + TiX_n$ mixtures, nonsupported Ti^{III} solutions, and certain chromium complexes.[43]

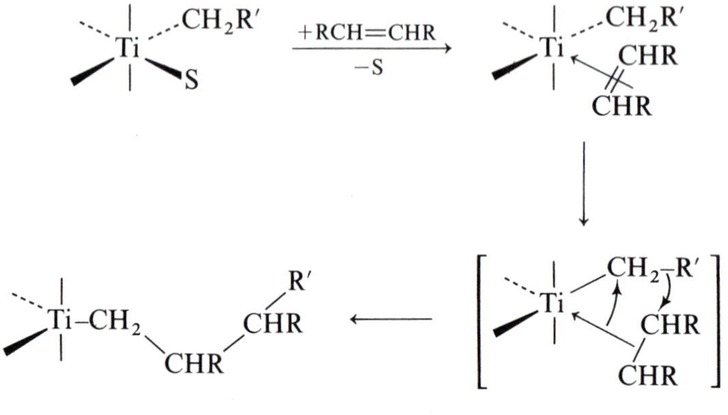

Figure 18.20
A generalized mechanism for Ziegler–Natta polymerization of the olefin RCH=CHR. Only the organic and solvent ligands are shown.

The Conversion of Alkenes to Aldehydes—The Wacker Process

Recall from Section 18.4 that the anion of Zeise's salt is a platinum complex containing ethylene as a ligand. The palladium analogue, $[Pd(C_2H_4)Cl_3]^-$, is difficult to isolate. It reacts with water in the following manner:

$$[Pd(C_2H_4)Cl_3]^- + H_2O \longrightarrow Pd + 3Cl^- + 2H^+ + CH_3CHO \qquad (18.106)$$

41. Ziegler, K.; Holzkamp, E.; Martin, H.; Breil, H. *Angew. Chem.* **1955**, *67*, 541; Natta, G. *J. Polym. Sci.* **1955**, *16*, 143.

42. Sinn, H.; Kaminski, W. *Adv. Organomet. Chem.* **1980**, *18*, 99.

43. See footnote 42; also Boor, J. *Ziegler–Natta Catalysis and Polymerizations*; Academic: New York, 1979.

This reaction has long been used for the synthesis of acetaldehyde on an industrial scale, and is known as the *Wacker process* after the German company that first utilized it. The mechanism bears similarity to those discussed earlier, and is summarized below.[44]

$$[Pd(C_2H_4)Cl_3]^- + H_2O \xrightarrow{-Cl^-} Cl_2(H_2O)PdC_2H_4 \qquad (18.107)$$

$$Cl_2(H_2O)PdC_2H_4 + H_2O \xrightarrow{-H^+} [Cl_2(H_2O)Pd–CH_2CH_2OH]^- \qquad (18.108)$$

$$Cl_2(H_2O)Pd–CH_2CH_2OH]^- \xrightarrow[-Cl^-]{+H_2O} [Cl(H_2O)_2Pd–CH_2CH_2OH \qquad (18.109)$$

$$Cl(H_2O)_2Pd–CH_2CH_2OH \xrightarrow{-H_2O} \underset{H_2O}{\overset{Cl}{\diagdown}} Pd \underset{H}{\overset{OH}{\diagup}} \qquad (18.110)$$

$$\underset{H_2O}{\overset{Cl}{\diagdown}} Pd \underset{H}{\overset{OH}{\diagup}} + H_2O \longrightarrow \underset{H_2O}{\overset{Cl}{\diagdown}} Pd \underset{OH_2}{\overset{CH(OH)CH_3}{\diagup}} \qquad (18.111)$$

$$\underset{H_2O}{\overset{Cl}{\diagdown}} Pd \underset{OH_2}{\overset{CH(OH)CH_3}{\diagup}} \longrightarrow \underset{H_2O}{\overset{Cl}{\diagdown}} Pd \underset{OH_2}{\overset{H}{\diagup}} + CH_3CHO \qquad (18.112)$$

$$\underset{H_2O}{\overset{Cl}{\diagdown}} Pd \underset{OH_2}{\overset{H}{\diagup}} \longrightarrow Pd + H^+ + Cl^- + 2H_2O \qquad (18.113)$$

$$[Pd(C_2H_4)Cl_3]^- + H_2O \longrightarrow Pd + 2H^+ + 3Cl^- + CH_3CHO$$

18.10   Metal-Catalyzed Reactions of Carbon Monoxide

The Water Gas Shift Reaction[45]

Recall from Chapter 12 that the passage of steam over red-hot coke can be utilized to generate hydrogen by either of the following two reactions:

$$C + H_2O \xrightarrow[\Delta]{} H_2 + CO \qquad (18.114)$$

$$C + 2H_2O \xrightarrow[\Delta]{} 2H_2 + CO_2 \qquad (18.115)$$

44. Backvall, J. E.; Åkermark, B.; Ljunggren, S. O. *J. Am. Chem. Soc.* **1979**, *101*, 2411.

45. Ford, P. C.; Rokicki, A. *J. Organomet. Chem.* **1988**, *28*, 139; Palmer, D. A.; van Eldik, R. *Chem. Rev.* **1983**, *83*, 651; Ibers, J. A. *Chem. Rev.* **1982**, *11*, 517.

The $H_2 + CO$ mixture produced by equation (18.114) is *synthesis gas*, or *water gas*, and is often used as a fuel. In other circumstances, it is desirable to maximize the amount of hydrogen obtained (eg, when it is to be used for the synthesis of ammonia); then the reaction described by equation (18.115) is preferred. However, this is difficult to accomplish because of the thermo-dynamics of the system. The reactions described by equations (18.114) and (18.115) both become spontaneous at about 980 K. The formation of CO is favored at still higher temperatures, so there is no way to obtain only CO_2 from $C + H_2O$ mixtures.

However, consider the coupling of equation (18.114) with the *water gas shift reaction*:

$$CO + H_2O \longrightarrow H_2 + CO_2 \tag{18.116}$$

Note that equations (18.114) and (18.116) sum to equation (18.115). Un-fortunately, the water gas shift reaction is a kinetically slow process. Con-siderable effort has therefore been expended toward finding a suitable catalyst. Metal oxides have been used as heterogeneous catalysts, and carbonyls such as $Fe(CO)_5$ act as homogeneous catalysts in basic solution. A plausible mechanism for the latter is

$$Fe(CO)_5 + OH^- \longrightarrow [HFe(CO)_4]^- + CO_2 \tag{18.117}$$

$$[HFe(CO)_4]^- + CO \longrightarrow Fe(CO)_5 + H^- \tag{18.118}$$

$$\underline{H^- + H_2O \longrightarrow OH^- + H_2} \tag{18.119}$$

$$H_2O + CO \longrightarrow H_2 + CO_2$$

Hydroformylation[46]

Another important industrial reaction of carbon monoxide is *hydroformyla-tion*. This reaction is competitive with the Wacker process in that it involves the conversion of an alkene to an aldehyde. However, in hydroformylation the product has one more carbon atom than the reactant alkene. Thus, the hydroformylation of ethylene produces propionaldehyde:

$$H_2C{=}CH_2 + CO + H_2 \longrightarrow H_3C\text{-}CH_2\text{-}CHO \tag{18.120}$$

Although such reactions are promoted by rhodium complexes such as $HRh(P\phi_3)_3(CO)$, the most commonly used industrial catalyst is still $Co_2(CO)_8$. The reaction mechanism is not known with certainty, but recent data are consistent with a free radical process, as summarized in Figure 18.21.

---

46. Tyler, D. R. *Prog. Inorg. Chem.* **1988**, *36*, 125; Pruett, R. L. *J. Chem. Educ.* **1986**, *63*, 196.

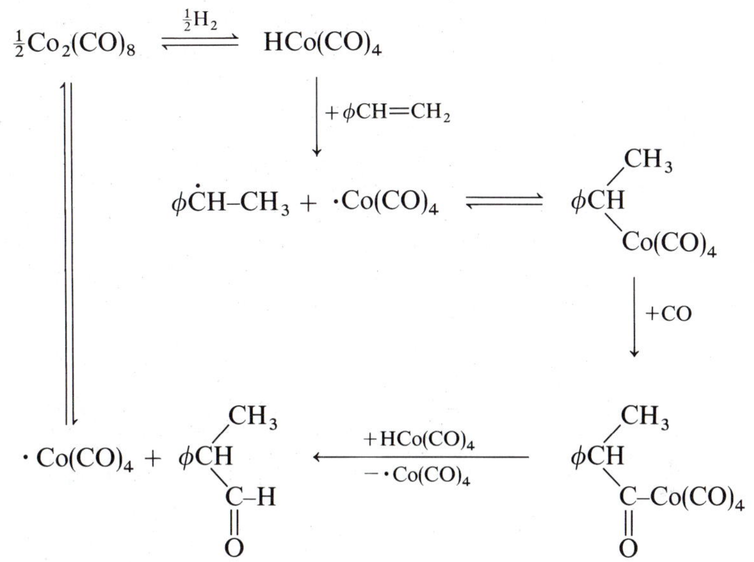

$$\tfrac{1}{2}Co_2(CO)_8 \;\; \underset{}{\overset{\tfrac{1}{2}H_2}{\rightleftharpoons}} \;\; HCo(CO)_4$$

Figure 18.21
A possible mechanism for hydroformylation catalysis by $Co_2(CO)_8$: The conversion of styrene to $C_6H_5CH(CH_3)CHO$ in the presence of H_2. [See Halpern, J. *Pure Appl. Chem.* **1986**, *58*, 575.]

$+\phi CH{=}CH_2$

$$\phi \overset{\cdot}{C}H{-}CH_3 \;+\; \cdot Co(CO)_4 \;\; \rightleftharpoons \;\; \phi CH \overset{CH_3}{\underset{Co(CO)_4}{}}$$

$+CO$

$$\cdot Co(CO)_4 \;+\; \phi CH \overset{CH_3}{\underset{\underset{O}{\overset{\|}{C}{-}H}}{}} \quad \underset{-\,\cdot Co(CO)_4}{\overset{+HCo(CO)_4}{\longleftarrow}} \quad \phi CH \overset{CH_3}{\underset{\underset{O}{\overset{\|}{C}{-}Co(CO)_4}}{}}$$

The Fischer–Tropsch Reaction

The *Fischer–Tropsch process* involves the metal- (or metal complex-) catalyzed reaction of carbon monoxide with hydrogen. Any of a number of carbon-containing compounds can be obtained, depending on the conditions. When nickel metal is used as a heterogeneous catalyst, methane is the primary product:

$$CO + 3H_2 \;\; \overset{Ni}{\longrightarrow} \;\; CH_4 + H_2O \tag{18.121}$$

Methanol is produced in the presence of a $Cu + ZnO$ mixed catalyst,

$$CO + 2H_2 \;\; \overset{Cu/ZnO}{\longrightarrow} \;\; CH_3OH \tag{18.122}$$

while cobalt metal tends to give higher alkanes:

$$nCO + (2n + 1)H_2 \;\; \overset{Co}{\longrightarrow} \;\; C_nH_{2n+2} + nH_2O \tag{18.123}$$

The reaction mechanism is a much debated topic. A detailed examination of the methanation process was carried out by isotopic labeling (^2H and ^{13}C).[47] The results suggest that the process probably involves the

---

47.   Kaminsky, M. P.; Winograd, N.; Geoffroy, G. L.; Vannice, M. A. *J. Am. Chem. Soc.* **1986**, *108*, 1315.

sequential hydrogenation of surface-held carbon. This is consistent with the original mechanism proposed by Fischer and Tropsch.[48]

Bibliography

Crabtree, R. H. *The Organometallic Chemistry of the Transition Metals*; Wiley: New York, 1988.

Powell, P. *Principles of Organometallic Chemistry*; Chapman and Hall: London, 1988.

Thayer, J. S. *Organometallic Chemistry: An Overview*; VCH: New York, 1988.

Bond, G. C. *Heterogeneous Catalysis*, 2nd ed.; Oxford University: New York, 1987.

Collman, J. P.; Hegedus, L. S.; Norton, J. R.; Finke, R. G. *Principles and Applications of Organotransition Metal Chemistry*; University Science: Mill Valley, CA, 1987.

Parkins, A. W.; Poller, R. C. *An Introduction to Organometallic Chemistry*; Oxford University: New York, 1986.

Yamamoto, A. *Organotransition Metal Chemistry*; Wiley: New York, 1986.

Atwood, J. D. *Inorganic and Organometallic Reaction Mechanisms*; Brooks/Cole: Pacific Grove, CA, 1985.

Haiduc, I.; Zuckerman, J. J. *Basic Organometallic Chemistry*; Walter de Gruyter: New York, 1985.

Lukehart, C. M. *Fundamental Transition Metal Organometallic Chemistry*; Brooks/Cole; Pacific Grove, CA, 1984.

Pearson, A. J. *Metallo-organic Chemistry*; Wiley: New York, 1985.

Masters, C. *Homogeneous Transition-Metal Catalysis*; Chapman and Hall: London, 1981.

Deganello, G. *Transition Metal Complexes of Cyclic Polyolefins*; Academic: New York, 1980.

Parshall, G. *Homogeneous Catalysis*; Wiley: New York, 1980.

Heck, R. F. *Organotransition Metal Chemistry: A Mechanistic Approach*; Academic: New York, 1974.

Questions and Problems

1. Suggest products for the following reactions:

   (a) $[Mn(CO)_5]^- + CH_3I \longrightarrow$ _____ + _____

   (b) $Pt(PMe_3)_2(Cl)CH_2CH_3 \xrightarrow{\Delta}$ _____ + _____

---

48. Fischer, F.; Tropsch, H. *Chem. Ber.* **1926**; *59*, 830.

(c) $Ir(PMe_3)_3Cl + \phi Li \longrightarrow$ _____ + _____

(d) $Pt(P\phi_3)_2\phi_2 + I_2 \longrightarrow$ _____

(e) $Rh(P\phi_3)_3CH_3 + HCl \longrightarrow$ _____ + _____

(f) $Cr(CO)_6 + en \xrightarrow{\Delta}$ _____ + _____

2. Write the metal-containing products for the following reactions:

   (a) $CH_3Pt(PEt_3)_2Cl + HCl \longrightarrow$

   (b) $[Co(CO)_4]^- + CH_3C(O)Cl \longrightarrow$

   (c) $Ir(P\phi_3)_2(CO)Cl + Me_3GeBr \longrightarrow$

   (d) $HRh(P\phi_3)_3 + C_2H_4 \rightleftharpoons$

3. For each of the following pairs, identify the species that has the greater thermal stability, and explain:

   (a) $CH_3Mn(CO)_5$ versus $CF_3Mn(CO)_5$

   (b) $CH_3Mn(CO)_5$ versus $CH_3CH_2Mn(CO)_5$

   (c) $[Co(CO)_4]^-$ versus $[Cu(CO)_3]^-$

4. The product indicated for equation (18.23) is the "kinetic isomer"—that is, the one that is first formed. However, rearrangement to form a metal–sulfur bond often occurs. Suggest a metal in a specific oxidation state for which such a rearrangement is likely, and defend your answer.

5. Write the equation representing the most probable mode of thermal rearrangement or decomposition for the following generalized complexes (L is a neutral ligand):

   (a) $CF_3C(O)M(CO)_x \xrightarrow{\Delta}$

   (b) $L_5MCH_2CH_3 \xrightarrow{\Delta}$

   (c) $CH_3Ir(H)L_4 \xrightarrow{\Delta}$

   (d) $Ta(CH_2CMe_2R)_5 \xrightarrow{\Delta}$

6. Consider the molecule: $(CO)_4Mn\begin{smallmatrix}\diagup CH_2 \\ | \\ \diagdown NMe_2\end{smallmatrix}$

   (a) Does this complex obey the 18-electron rule? Defend your answer.

   (b) Suggest a plausible synthesis.

7. Explain the difference between:

   (a) An alkylidene and a metal carbene

   (b) An alkylidene and an alkylidyne

   (c) A terminal alkylidene and a μ-alkylidene

8. The rhenium-containing product of the reaction below is a monovalent cation. Sketch its structure and briefly discuss the reaction mechanism.

$$C_5H_5(CO)_2Re{=}C\begin{smallmatrix}\diagup OMe \\ \diagdown \phi\end{smallmatrix} + BCl_3 \longrightarrow \ ??$$

9. Alkylidynes are often prepared by the reaction of base with an alkylidene [equation (18.31)]. Write a mechanism for that process.

10. The product of the reaction below is an alkylidene. Sketch its structure and briefly explain.

$$
\begin{array}{c}
\text{P}\phi_3 \\
\text{OC} \cdots \underset{\displaystyle \overset{|}{\underset{\text{P}\phi_3}{}}}{\overset{|}{\text{Os}}}\text{-CH}_3 + \text{HCl} \longrightarrow \text{??} \\
\text{Cl}
\end{array}
$$

11. Consider the reaction

$$
(C_5H_5)_2Ta\underset{CH_2}{\overset{CH_3}{\diagdown}} \xrightarrow{\text{Me}_3\text{SiBr}} \left[(C_5H_5)_2Ta\underset{CH_2SiMe_3}{\overset{CH_3}{\diagdown}} \right]^+ + Br^-
$$

Discuss the mechanism in terms of nucleophilic attack.

12. The bridged alkylidene $[(C_5H_5)(CO)_2Mn]_2CH_2$ is believed to contain a Mn–Mn bond. How does the 18-electron rule support this notion?

13. Suggest a structure that conforms to the 18-electron rule for:
    (a) $[HCr(CO)_5]^-$ (b) $Fe_2(CO)_8CH_2$ (c) $[Fe_3(CO)_{11}]^{2-}$
    (d) $Ir_2(CO)_8$ (e) $Ir_4(CO)_{12}$ (f) $Os_3(CO)_{12}$
    (g) $[MnFe(CO)_9]^-$ (h) $[MnFe_2(CO)_{12}]^-$

14. Explain why the bond dissociation energies for the reaction below follow the order $Co(CO)_4 < Ni(CO)_4 \gg Cu(CO)_3$.

$$
M(CO)_n \longrightarrow M(CO)_{n-1} + CO
$$

15. Write the metal-containing products for the following reactions. (The combining stoichiometries are not necessarily 1:1.)

    (a) $Mn_2(CO)_{10} + NO \longrightarrow$

    (b) $Co_2(CO)_8 + NO \longrightarrow$

    (c) $Ni(C_5H_5)_2 + NO \longrightarrow$

    (d) $[C_5H_5Ni(CO)]_2 + NO \longrightarrow$

16. The reaction of $W(CO)_6$ with methyllithium produces the anionic complex A. Treatment of A with methyl iodide gives the alkylidene B. Give the structures of A and B, and write balanced equations for their formation.

17. The compound $H_2Mo_2(CO)_8$ has D_{2h} symmetry and obeys the 18-electron rule. Does this species contain a metal–metal bond? If so, is it a single, double, triple, or quadruple bond? Defend your answer.

18. The carbonyl hydride $HCo(CO)_4$ is a strong Brønsted acid (100% dissociated) in water. Its pK_a in CH_3CN is 8.4 at 25°C. Rationalize the difference.

19. $[\mu\text{-}HCr_2(CO)_{10}]^-$ can be prepared by the treatment of $[HCr(CO)_5]^-$ with H_3O^+. Suggest a mechanism for this reaction, and identify any side products. [*Hint*: The 16-electron species $Cr(CO)_5$ is an intermediate.]

20. Suggest structures for the compounds $Fe_2(C_5H_5)_2(CO)_4$ and $Fe_2(C_5H_5)_2(NO)_2$. [*Hint*: The Fe–Fe distance in the nitrosyl is about 16 pm shorter than in the carbonyl.]

21. Develop a bonding scheme by which each metal in $(C_5H_5)_3Mn_3(NO)_4$ (Figure 18.4) obeys the 18-electron rule.

22. Isocyanides are similar to CO in their reactivity toward metals. Suggest a synthesis for the complex $Ni(CNMe)_4$ starting from:
    (a) Nickel metal   (b) $Ni(OH)_2$

23. Explain in your own words why N_2 and O_2 are poorer ligands than is NO.

24. Both $[\mu\text{-}(O_2)Co_2(CN)_{10}]^{5-}$ and $[\mu\text{-}(O_2)Co_2(CN)_{10}]^{6-}$ are known. Sketch the probable structures, and determine the best description for the bridging species (O_2, O_2^-, or O_2^{2-}) for each anion.

25. The reaction of Zeise's salt with concentrated acid gives an orange solid having the empirical formula $PtC_2H_4Cl_2$. Sketch its most likely structure.

26. The TCNE group is electron-withdrawing (at least in part) because of resonance. Write one or more relevant resonance structures and explain.

27. Suggest syntheses for the following, beginning with ferrocene as a reactant:
    (a) $C_5H_5FeC_5H_4CH_3$ (b) $C_5H_5FeC_5H_4C(O)CH_3$
    (c) $Fe(C_5H_4CO_2H)_2$ (d) $C_5H_5FeC_5H_4NO_2$
    (e) $C_5H_5FeC_5H_4SiMe_3$

28. The Ni–C bond distance in nickelocene is about 15 pm longer than the Fe–C distance in ferrocene. Is this as expected based on the relative sizes of nickel and iron? Rationalize.

29. Give a viable formula for both a neutral and an anionic complex that contain the following combinations:
    (a) Co, CO, and NO   (b) Fe, C_5H_5, and CO   (c) Ni, C_5H_5, and CO

30. A well-known organometallic compound has the molecular formula $(C_5H_5)_2Fe(CO)_2$. At first glance, this appears to be a case of the 18-electron rule run amok; however, it is actually obeyed. What is the structure?

31. Give the most likely values for x, y, and/or z (ie, predict the molecular formula) for the following; also, sketch a structure for each:
    (a) $(C_5H_5)_xCo(CO)_y$ (b) $(C_5H_5)_xNi(NO)_z$
    (c) $(C_5H_5)_xCr(CO)_y(NO)_z$ (d) The dimer of $(C_5H_5)_xFe(CO)_y$

32. Identify the $3d$ element M that forms the indicated compound for each of the following:

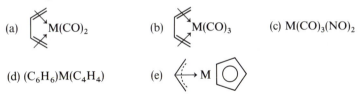

(a) $M(CO)_2$ (b) $M(CO)_3$ (c) $M(CO)_3(NO)_2$

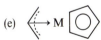

(d) $(C_6H_6)M(C_4H_4)$ (e)   M

33. Cyclooctatetraene (cot) reacts with $Fe(CO)_5$ in either of two ways, depending on the stoichiometry:

    (a) $Fe(CO)_5 + C_8H_8 \longrightarrow (CO)_3Fe(C_8H_8) + 2CO$

    (b) $2Fe(CO)_5 + C_8H_8 \longrightarrow [(CO)_3Fe]_2(C_8H_8) + 4CO$

    These iron–cot products obey the 18-electron rule. Give their structures.

34. Use group theory to construct a molecular orbital diagram for the FeH_6^{4-} anion. (Review the appropriate sections of Chapters 3 and 16 as needed.) How many net bonds are predicted?

35. The protonation of nickelocene produces a cation having the molecular formula $C_{10}H_{11}Ni^+$. Taking the 18-electron rule into account, suggest its structure.

36. Show that the hydrogenation mechanism given by Figure 18.18 sums to the overall equation

    $$H_2C=CH_2 + H_2 \longrightarrow CH_3CH_3$$

37. The reaction of $[Fe_2(CO)_8]^{2-}$ with CH_2I_2 gives a product having the molecular formula $Fe_2C_9H_2O_8$. Give structures for both the reactant anion and the product.

*38. A series of dimeric Cr^{III} alkyls has been reported by D. S. Richeson, J. F. Mitchell, and K. H. Theopold (*Organometallics* **1989**, *8*, 2570). Answer these questions after reading their paper.
    (a) It is stated in the introduction that very few Cr^{III} organometallics are known. Why is this the case?
    (b) What is the significance of the chromium–chromium internuclear distance (328.7 pm)?
    (c) How does the electron count of these species explain their behavior toward nucleophilic reagents?

*39. A novel complex containing O_2 as a ligand has been reported by J. W. Egan, B. S. Haggerty, A. L. Rheingold, S. C. Sendlinger, and K. H. Theopold (*J. Am. Chem. Soc.* **1990**, *112*, 2445).
    (a) The internuclear O–O distance in the complex is 126.2 pm. What conclusion can be drawn from that value?
    (b) What is unique about the geometry of this complex?

*40. M. S. Chinn, D. M. Heinekey, N. G. Payne, and C. D. Sofield have reported a complex containing H_2 as a ligand (*Organometallics* **1989**, *8*, 1824).
    (a) How was this complex synthesized?
    (b) Discuss the observation that this species is a potent Brønsted acid.
    (c) The C_5Me_5 group was used instead of C_5H_5. How might this influence the stability of the molecule?

Special Topics

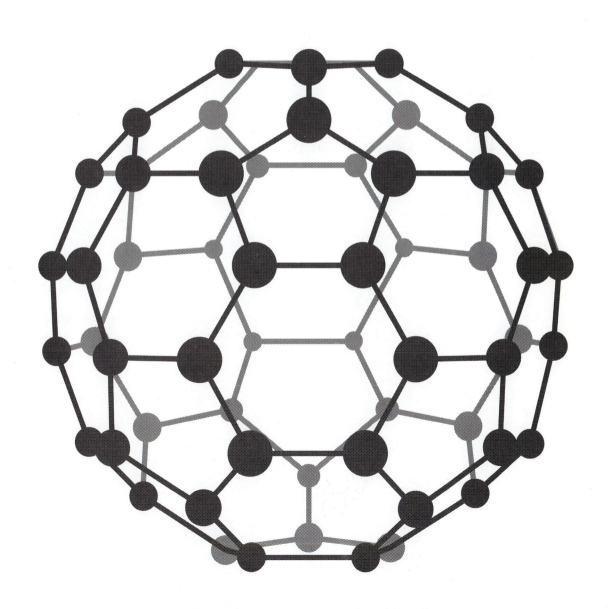

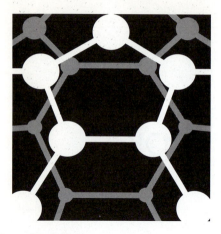

19

Inorganic Cages and Clusters

In this chapter we discuss multicentered, three-dimensional aggregates having 3–12 framework atoms. Such species provide a link between discrete molecules and macroscopic matter, so they are of great theoretical and practical importance.

The terms *cage* and *cluster* are both used to describe these species, depending on whether localized or delocalized covalent bonding is evidenced. To illustrate the difference, consider two dianions with geometries based on a trigonal bipyramid, Pb_5^{2-} and $B_5H_5^{2-}$:

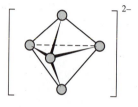

$\bigcirc$ = Pb or BH

A reasonable bonding model for Pb_5^{2-} can be developed by assuming that there is a covalent Pb–Pb bond along each of the nine edges of the polyhedron; this requires 18 of the 22 valence electrons. The remaining electrons are nonbonded lone pairs, one on each apical atom. A Lewis structure therefore can be drawn in which all five lead atoms have valence octets. Because of the electron-precise bonding (each line in the diagram represents a 2-electron bond), Pb_5^{2-} may be considered a *cage* anion.

For $B_5H_5^{2-}$, there are again 22 valence electrons; however, 10 of them are needed to form the five localized B–H bonds. Only 12 electrons remain—too few for the nine B–B nearest-neighbor interactions to be normal 2-electron bonds. These electrons are delocalized into multicentered molecular orbitals; thus, $B_5H_5^{2-}$ is a *cluster* anion.[1]

The polyhedra of greatest importance in this area of chemistry are pictured in Figure 19.1. They can be divided into two types—those that are *deltahedral* ("closed" structures in which all the faces are triangular) and

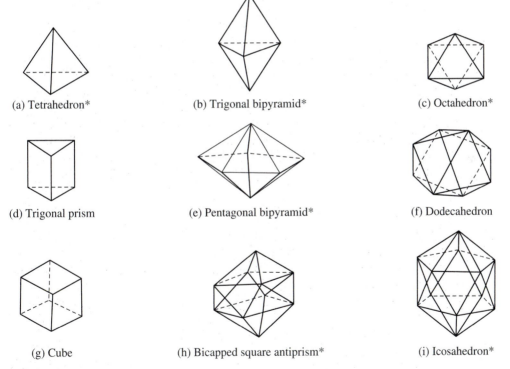

(a) Tetrahedron*

(b) Trigonal bipyramid*

(c) Octahedron*

(d) Trigonal prism

(e) Pentagonal bipyramid*

(f) Dodecahedron

(g) Cube

(h) Bicapped square antiprism*

(i) Icosahedron*

Figure 19.1
Common polyhedral frameworks for cage and cluster compounds
(* = deltahedron).

1. This distinction between cages and clusters is not always made. Thus, in conversation, as well as in the chemical literature, the two terms are often used interchangeably.

those that are not. The tetrahedron, trigonal bipyramid, octahedron, pentagonal bipyramid, bicapped square antiprism, and icosahedron, with 4, 5, 6, 7, 10, and 12 vertices, respectively, are all deltahedral.

Relationships between structure and bonding are explored in the next section.[2]

19.1 The $6n - 12$ (Deltahedral Cage) Rule

The distinction between deltahedra and other polyhedra is important from the standpoint of chemical bonding. Since the nearest neighbors define the edges of any polyhedron, the maximum possible number of localized two-electron bonds is equal to the number of edges. In addition, there is a specific relationship between the number of vertices, n, and the number of edges of deltahedra:

$$\text{Number of edges} = 3n - 6 \qquad (19.1)$$

As a result, a total of $6n - 12$ valence electrons maximizes the stability of a deltahedral framework.

As an illustration of the dependence of cage geometry on the electron count, we will consider the tetrahedral P_4 molecule. Twelve of the 20 valence electrons are bonding (the others being lone pairs), which is exactly the number required for the framework linkages ($6 \cdot 4 - 12 = 12$). Changing this number results in a change in geometry. For example, compare P_4 to a derivative, $P_4(NR_2)_2$ ($R = SiMe_3$). The conversion of P_4 to $P_4(NR_2)_2$ has the effect of adding 2 valence electrons to the cage system, since one framework P–P bond is converted to two exo-cage P–N bonds:

$$(19.2)$$

The rupture of the P–P bond causes the involved phosphorus atoms to separate, opening the framework.

Electron count predictions based on the number of polyhedral edges for cages are summarized in Table 19.1.

---

2. For a summary of the historical development of bonding models for cage and cluster compounds, see Mingos, D. M. P.; Johnston, R. L. *Struct. Bonding* **1987**, *68*, 29.

Table 19.1 Number of vertices and edges, and ideal number of electrons for localized bonding in some common three-dimensional frameworks

Structure	Vertices	Edges	Ideal Number of Electrons
Triangle	3	3	6
Tetrahedron	4	6	12
Trigonal bipyramid	5	9	18
Trigonal prism	6	9	18
Octahedron	6	12	24
Pentagonal bipyramid	7	15	30
Cube	8	12	24
Square antiprism	8	16	32
Pentagonal dodecahedron	8	30	60
Tricapped trigonal prism	9	18	36
Monocapped square antiprism	9	20	40
Bicapped square antiprism	10	24	48
Icosahedron	12	30	60

19.2 Triangular Arrays

Although not deltahedral, many three-vertex, triangular "cages" are known that conform to the $6n - 12$ rule. A facile example is cyclopropane, C_3H_6. Twelve of the 18 valence electrons are used to form the C–H bonds, leaving 6 for the three 2-electron carbon–carbon bonds along the edges of the triangle. As in P_4, the addition of electrons (eg, by hydrogenation to C_3H_8) causes the structure to open; in this case, a chain results.

A more interesting (to inorganic chemists, at least!) three-vertex aggregate is $Ru_3(CO)_{12}$. Recall from Chapter 18 that this molecule contains a triangle of ruthenium atoms, with pseudo-octahedral geometry about each metal:

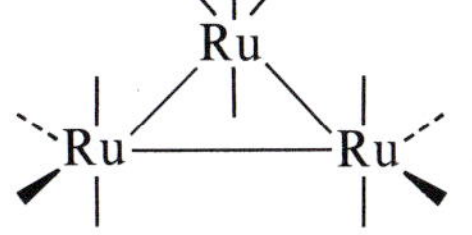

The Ru–CO linkages are dative, and therefore do not require any of the 24 total valence electrons of the rutheniums. If it is assumed that each metal

has three filled, σ nonbonding, t_{2g}-type orbitals, then there are $24 - 3 \cdot 6 = 6$ electrons available for the three Ru–Ru bonds. Thus, like the 18-electron rule, this model predicts $Ru_3(CO)_{12}$, its iron and osmium analogues, and valence isoelectronic species such as $H_2Os_3(CO)_{11}$ to be stable.

An interesting contrast is provided by Re_3Cl_9, which can be described as a triangle of $ReCl_3$ fragments, with D_{3h} symmetry (Figure 19.2). Each Re^{III} has 4 valence electrons, for a total of 12. (The Re–Cl interactions are assumed to be dative.) This is insufficient to enable occupancy of the t_{2g} subsets. Recall, however, that while these orbitals are σ nonbonding, they do engage in π interactions under the proper circumstances. In Re_3Cl_9, then, the 12 metal electrons allow for three Re=Re double bonds. This conclusion is supported by various kinds of experimental evidence. For example, the Re–Re bond distances of 249 pm are unusually short. (Compare to 275 pm in the free element.) Moreover, various addition reactions are known:

$$Re_3Cl_9 + 3HCl \longrightarrow [Re_3Cl_{12}]^{3-} + 3H^+ \qquad (19.3)$$

$$Re_3Cl_9 + 3P\phi_3 \longrightarrow Re_3(P\phi_3)_3Cl_9 \qquad (19.4)$$

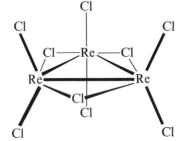

Figure 19.2
The structure of the trimer of rhenium(III) chloride, Re_3Cl_9.

Triosmium Clusters[3]

An important group of three-atom clusters have a core of three osmium atoms, and an impressive body of information has been accumulated concerning such species. Their chemistry is characteristic of that observed for cluster systems, and can therefore be used as a model.

The structures of the parent compound, $Os_3(CO)_{12}$, and three of its hydrido derivatives, $H_2Os_3(CO)_n$ ($n = 10$–12), are of special interest. Like $Ru_3(CO)_{12}$, triosmium dodecacarbonyl consists of three $Os(CO)_4$ units connected by metal–metal bonds. This structure conforms to the 18-electron

3. Deeming, A. J. *Adv. Organomet. Chem.* **1986**, *26*, 1; Johnson, B. F. G.; Lewis, J. *Adv. Inorg. Chem. Radiochem.* **1981**, *24*, 225.

rule. One way to show this is to determine the total number of valence electrons, and then add the number of electrons shared via metal–metal bonding; the result will be a multiple of 18 if the rule is obeyed. For example, $Os_3(CO)_{12}$ has 48 valence electrons (24 from the metals plus 2 from each ligand). Adding 6 shared electrons (three Os–Os bonds) gives 54—exactly 18 per metal atom.

The addition of two hydrogens to give $H_2Os_3(CO)_{12}$ increases the number of valence electrons from 48 to 50; hence, the number of Os–Os bonds decreases from three to two (Table 19.2). This severs the ring framework and results in a linear Os–Os–Os linkage; note the similarity to $C_3H_6 \rightarrow C_3H_8$.

Next, consider the removal of a carbonyl ligand from $H_2Os_3(CO)_{12}$ to give $H_2Os_3(CO)_{11}$ (48 valence electrons). Like $Os_3(CO)_{12}$, $H_2Os_3(CO)_{11}$ contains a triangle of osmium atoms. There are two $Os(CO)_4$ groups and one $Os(CO)_3$ group; the latter is associated with both a terminal and a

Table 19.2 Electron count and structural data for triosmium dodecacarbonyl and its hydrido derivatives

Compound	Valence e^-	Os–Os Bonds	Geometry	d(Os–Os), pm
$H_2Os_3(CO)_{12}$	50	2		—
$Os_3(CO)_{12}$	48	3		294*
$H_2Os_3(CO)_{11}$	48	3		299 291 286
$H_2Os_3(CO)_{10}$	46	4		282 281 268

Note: ● = $Os(CO)_3$; * = average value.

bridging hydrogen. These hydrogens undergo rapid exchange, probably via a doubly bridged intermediate.[4]

Removal of a second carbonyl group produces $H_2Os_3(CO)_{10}$, a 46-electron species. This molecule also has a triangle of osmiums; however, there is one short Os–Os linkage (268 pm) and two longer bonds (about 281 pm)—that is, one double and two single bonds.

The reaction chemistries of $Os_3(CO)_{12}$ and $H_2Os_3(CO)_{10}$ have been studied in detail. Six kinds of chemical transformations are described below.

Pyrolysis   The heating of $Os_3(CO)_{12}$ to 210°C for 12 hours produces a variety of higher carbonyls. These include $Os_5(CO)_{16}$ (7% of the product mixture), $Os_6(CO)_{18}$ (80%), $Os_7(CO)_{21}$ (10%), and $Os_8(CO)_{23}$ (2%). The detailed structures of the first three have been reported: $Os_5(CO)_{16}$ has a trigonal bipyramid of metal atoms, $Os_6(CO)_{18}$ is a capped trigonal bipyramid (or bicapped tetrahedron), and $Os_7(CO)_{21}$ is a capped octahedron (Figure 19.3).

Hydrogenation   Triosmium dodecacarbonyl reacts with H_2 under remarkably low pressure (1 atm) to initially form $H_2Os_3(CO)_{10}$. Longer reaction times produce the tetranuclear cluster $H_4Os_4(CO)_{12}$, which arises from the interaction of additional dihydrogen with $H_2Os_3(CO)_{10}$. Cluster degradation occurs when higher pressures of H_2 are used, giving the hydrido carbonyls $H_2Os(CO)_4$ and $H_2Os_2(CO)_8$.

Reactions with Base   The treatment of $Os_3(CO)_{12}$ with either OH^- or BH_4^- gives the $[HOs_3(CO)_{11}]^-$ anion. This species reacts with protonic

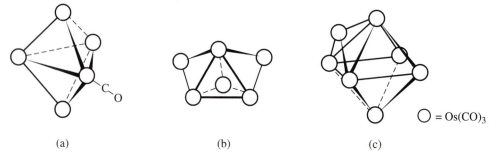

(a)   (b)   (c)

$\bigcirc = Os(CO)_3$

Figure 19.3   The framework structures of three osmium carbonyls: (a) $Os_5(CO)_{16}$, with a trigonal bipyramidal arrangement of osmiums; (b) $Os_6(CO)_{18}$, a capped trigonal bipyramid (or bicapped tetrahedron); (c) $Os_7(CO)_{21}$, a capped octahedron.

4.   Shapley, J. R.; Keister, J. B.; Churchill, M. R.; DeBoer, B. G. *J. Am. Chem. Soc.* **1975**, *97*, 4145.

hydrogen in either of two ways. Anhydrous Brønsted acids simply yield the conjugate $H_2Os_3(CO)_{11}$. Reaction with H_3O^+ produces the hydroxo species $HOs_3(CO)_{11}OH$.

$$[HOs_3(CO)_{11}]^- + H_3O^+ \longrightarrow HOs_3(CO)_{11}OH + H_2 \qquad (19.5)$$

Reactions with Halogens (Oxidative Addition) The interaction of $Os_3(CO)_{12}$ with I_2 gives the addition product $Os_3(CO)_{12}I_2$, which formally has two Os^I centers. Like $H_2Os_3(CO)_{12}$ (Table 19.2), $Os_3(CO)_{12}I_2$ contains a linear chain of osmiums. You should be able to verify that this is a 50-electron species, and that it obeys the 18-electron rule.

Substitutions by Other Nucleophiles Many reactions are known that amount to nucleophilic displacement of CO by other ligands on the $Os_3(CO)_{12}$ substrate. Some examples are given below.

$$Os_3(CO)_{12} + H_2O \xrightarrow{-CO} [Os_3(CO)_{11}OH_2]$$
$$\longrightarrow HOs_3(CO)_{11}OH \qquad (19.6)$$

$$Os_3(CO)_{12} + \phi C{\equiv}CH \xrightarrow[\Delta]{-2CO} Os_3(CO)_{10}(\phi C_2H) \qquad (19.7)$$

$$Os_3(CO)_{12} + H_2S \xrightarrow{-3CO} H_2Os_3(CO)_9S \qquad (19.8)$$

$$Os_3(CO)_{12} + Mn(CO)_5^- \xrightarrow{-CO} [Os_3Mn(CO)_{16}]^- \qquad (19.9)$$

Equations (19.8) and (19.9) represent *cluster expansion* reactions— the number of framework atoms increases from three to four in each case. Furthermore, the product of equation (19.9) is a *mixed-metal cluster*. Once very rare, such species are now commonplace.[5] Several other examples will be encountered later in the chapter.

Nucleophilic Addition Because $H_2Os_3(CO)_{10}$ is coordinately unsaturated, it typically reacts with nucleophiles via addition rather than substitution. For example, heating $H_2Os_3(CO)_{10}$ in a carbon monoxide atmosphere produces $H_2Os_3(CO)_{11}$. Additions also occur with reagents such as iso-cyanides and phosphines:

$$H_2Os_3(CO)_{10} + R\overset{\oplus}{N}{\equiv}\overset{\ominus}{C}{:} \longrightarrow H_2Os_3(CO)_{10}CNR \qquad (19.10)$$

$$H_2Os_3(CO)_{10} + P\phi_3 \longrightarrow H_2Os_3(CO)_{10}P\phi_3 \qquad (19.11)$$

---

5. For a review, see Gladfelter, W. L.; Geoffroy, G. L. *Adv. Organomet. Chem.* **1980**, *18*, 207.

19.3 Tetrahedral Arrays

The bonding in compounds and ions having tetrahedral frameworks can be readily understood through molecular orbital theory. The MO diagram for P_4 (Figure 19.4) is an appropriate starting point. There are six bonding orbitals, which belong to the a_1, t_2, and e representations in T_d symmetry. (The e level is only weakly bonding.) Complete occupancy of these orbitals requires 12 electrons, in agreement with the $6n - 12$ rule. Consistent with this, most stable tetrahedral clusters have exactly 12 skeletal electrons. A few tetrahedral 8-electron species have been reported, with one example being B_4Cl_4; the e level is vacant in such cases. It is likely that the σ bonding in B_4Cl_4 is supplemented by dative π interactions between chlorine and boron.

A large number of transition metal complexes are known that have tetrahedral framework geometries. An example is the carbonyl $Ir_4(CO)_{12}$. Using the same method of electron counting as we used earlier for $Ru_3(CO)_{12}$, the four metals each contribute 3 of their 9 valence electrons to the framework orbitals; the other 6 reside in σ nonbonding, t_{2g}-type atomic orbitals. Hence, the framework count is 12. (It also can be shown that the 18-electron rule is obeyed.)

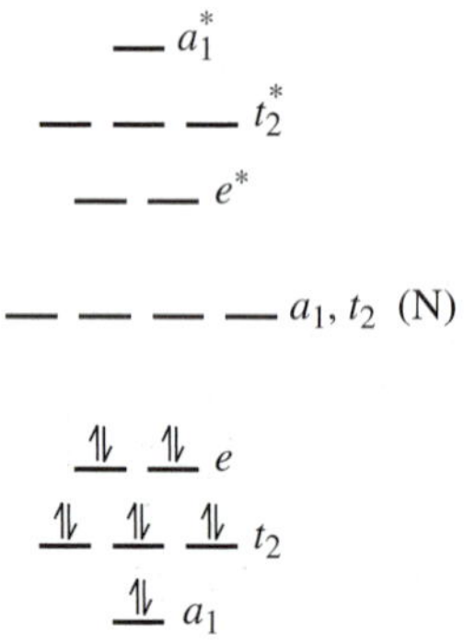

Figure 19.4
Molecular orbital energy levels for P_4. The six bonding MO's require 12 electrons for complete occupancy. N = nonbonding level. [See Kettle, S. F. A. *Theor. Chim. Acta* **1966**, *4*, 150.]

Isolobal Fragments[6]

The P_4, C_4H_4 (tetrahedrane), and $Ir_4(CO)_{12}$ molecules all have tetrahedral arrangements of atoms. Given the relationship between cage and cluster geometry and electron count, it is reasonable to consider the molecular fragments P, CH, and $Ir(CO)_3$ to be functionally equivalent for purposes of electron counting. In fact, these fragments are not only isoelectronic, but also

---

6. Hoffmann, R. *Angew. Chem. Int. Ed. Engl.* **1982**, *21*, 711; *Angew. Chem.* **1982**, *94*, 725; *Science* **1981**, *211*, 995.

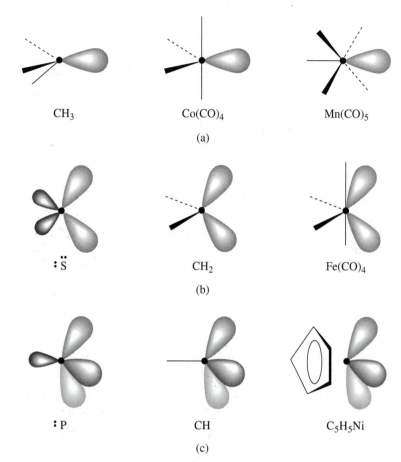

Figure 19.5
Some examples of isolobal molecular fragments: groups having (a) one, (b) two, and (c) three "active" orbitals, respectively.

similar in the shapes and directionalities of their frontier orbitals. This is shown schematically in Figure 19.5. The term *isolobal* has been applied to such fragments.[7] Other examples of isolobal groups are given in Table 19.3, at the top of the next page.

There is a great deal of experimental evidence to support the isolobal concept. For example, all five compounds conforming to the formula $Co_n(CO)_{3n}(CH)_{4-n}$ ($n = 0$–4) have been prepared, and each has a tetrahedral framework. The $Co_n(CO)_{3n}E_{4-n}$ (E = P and As) families are also known, and are isostructural with the $Co(CO)_3/CH$ series. The FeH unit is valence isoelectronic with Co, and species such as $H_3Fe_3(CO)_9CH$ and

7. Elian, M.; Chen, M. M. L.; Mingos, D. M. P.; Hoffmann, R. *Inorg. Chem.* **1976**, *15*, 1148; Halpern, J. *Disc. Farad. Soc.* **1968**, *46*, 7.

Table 19.3 Isolobal groups commonly encountered in molecular clusters

0	1	2	3
H_2	H		
CH_4	CH_3	CH_2	CH
Cl_2	Cl		
		S	
			P
$Cr(CO)_6$			
	$Mn(CO)_5$		
$Fe(CO)_5$		$Fe(CO)_4$	
	$C_5H_5Fe(CO)_2$		$C_5H_5Fe(CO)$
	$Co(CO)_4$		$Co(CO)_3$
$C_5H_5Co(CO)_2$		$C_5H_5Co(CO)$	
$Ni(CO)_4$		$Ni(CO)_3$	
	$C_5H_5Ni(CO)$		C_5H_5Ni

Note: The 0 column lists species that are coordinately saturated; members of the 1 column have one active lobe; etc.

$H_3Fe_3(CO)_9BH_2$ have been reported. A variety of other compounds can be placed into this category as well (Figure 19.6).

These tetrahedral arrays were synthesized by a variety of methods. A few examples are given here.

$$Co_2(CO)_8 + CHCl_3 \xrightarrow{\Delta} Co_3(CO)_9CH \quad \textbf{(19.12)}$$

$$Co_2(CO)_8 + AsCl_3 \xrightarrow{\Delta} Co_3(CO)_9As \quad \textbf{(19.13)}$$

$$Fe(CO)_5 + Co_2(CO)_8 + C_6H_5SH \xrightarrow{\Delta} FeCo_2(CO)_9S \quad \textbf{(19.14)}$$

$$C_5H_5(CO)_3M-M(CO)_3C_5H_5 + As_4 \longrightarrow [C_5H_5(CO)_2M]_2As_2$$
$$M = Mo, W \quad \textbf{(19.15)}$$

The molecules $Co_3(CO)_9S$ and $FeCo_2(CO)_9S$ differ by one framework electron, with the latter (another mixed-metal cluster) having the requisite 12. If the framework bonding were localized, then the transition from the Co_2Fe to the Co_3 triangle would add an electron to one of the metal–metal bonds; this should be reflected in an increase in one of the three M–M bond distances. In actuality, all three metal–metal bond lengths are altered. This is evidence for delocalization (atomic orbitals of all three metal atoms contribute to the HOMO and LUMO), and indicates that MO theory is the preferred method for interpreting the bonding in clusters.

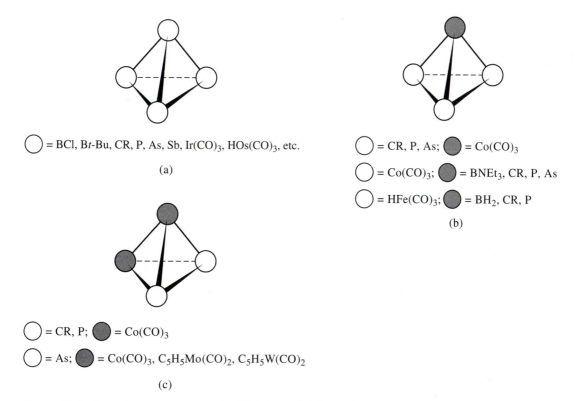

○ = BCl, Bt-Bu, CR, P, As, Sb, Ir(CO)$_3$, HOs(CO)$_3$, etc.

(a)

○ = CR, P, As; ● = Co(CO)$_3$

○ = Co(CO)$_3$; ● = BNEt$_3$, CR, P, As

○ = HFe(CO)$_3$; ● = BH$_2$, CR, P

(b)

○ = CR, P; ● = Co(CO)$_3$

○ = As; ● = Co(CO)$_3$, C$_5$H$_5$Mo(CO)$_2$, C$_5$H$_5$W(CO)$_2$

(c)

Figure 19.6 *Some known compounds with tetrahedral frameworks:* (a) A$_4$; (b) A$_3$B; and (c) A$_2$B$_2$ species. *(The listings are not intended to be comprehensive.)*

As we suggested earlier [see the example of P$_4$(NR$_2$)$_2$], four-atom aggregates with 14 framework electrons generally exhibit an open (often described as a *butterfly*) structure.[8] An example is the cation shown in Figure 19.7 (p. 642), [(μ-H)$_3$Os$_4$(CO)$_{12}$(CNMe)$_2$]$^+$. This species has five metal–metal bonds, with the internuclear Os–Os distances ranging from 283 to 315 pm; the sixth (nonbonded) distance is 434 pm. That there are five Os–Os bonds can be rationalized from the fact the total valence electron count is 62; hence, for the 18-electron rule to be obeyed there must be 4(18) − 62 = 10e$^-$ shared by the metals.

The anion [HFe$_4$(CO)$_{13}$]$^-$ is surprising in that it also exhibits a butterfly geometry.[9] (With an apparent total of 60 valence electrons, a tetrahedral framework is expected.) This species consists of four Fe(CO)$_3$

8. For a review, see Sappa, E.; Tiripicchio, A.; Carty, A. J. *Prog. Inorg. Chem.* **1987**, *35*, 437.

9. Manassero, M.; Sansoni, M.; Longoni, G. *J. Chem. Soc. Chem. Commun.* **1980**, 961.

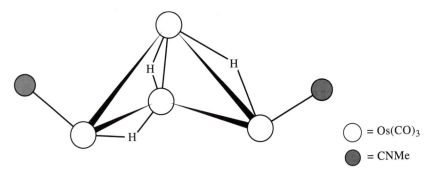

$= Os(CO)_3$

$= CNMe$

Figure 19.7 Simplified representation of the structure of the butterfly cluster $[(\mu\text{-}H)_3Os_4(CO)_{12}(CNMe)_2]^+$.

Figure 19.8
The structure of the cluster anion $[HFe_4(CO)_{13}]^-$. Note the bidentate CO group.

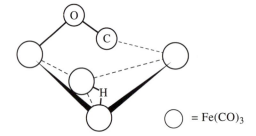

$= Fe(CO)_3$

groups plus an "odd" carbonyl that sits above the open face, with both its carbon and oxygen atoms in proximity to an iron (Figure 19.8). Thus, that CO group acts as a bidentate ligand (ie, a four-electron donor), and thereby provides the extra electron pair required for the butterfly structure.

19.4 Larger Cage Systems

Five- and Six-Atom Cages

A topic that has not yet been discussed, but is very important in cage bonding, is the constraint of orbital overlap on geometry. Electron-precise trigonal bipyramidal and octahedral cage compounds are rather rare, and the examples that can be cited generally have metals in the framework locations. This is probably because these geometries require vertex atoms with four bonding orbitals on the same side of the atom, as illustrated.

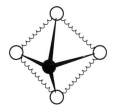 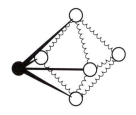

This appears to be viable only for atoms with valence s, p, and d orbitals (ie, transition or posttransition elements). Triangular and tetrahedral aggregates, which are more common, require only two or three bonding orbitals on any side.

Several penta- and hexametallic species can be rationalized as electron-precise cages, however. Among the trigonal bipyramidal cases, the example of Pb_5^{2-} was cited in the introduction to this chapter, and the valence isoelectronic Sn_5^{2-} and Bi_5^{3+} are known as well. The synthesis of these ions follows methods described in Chapter 11: the anions are prepared in some strongly basic, reducing solution (often liquid ammonia), and the cation in a strongly acidic medium (a superacid solution of AsF_5 in SO_2).[10]

$$Na_2M_5 \xrightarrow{\text{2,2,2-crypt}} 2Na(\text{2,2,2-crypt})^+ + [M_5]^{2-}$$
$$M = Sn \text{ or } Pb \quad \textbf{(19.16)}$$

$$10Bi + 9AsF_5 + 4SO_2 \xrightarrow{-3AsF_3} 2[(Bi_5)^{3+}(AsF_6)_3^- \cdot 2SO_2] \quad \textbf{(19.17)}$$

An array of six metals is found in the cation $Mo_6Cl_8^{4+}$. (The tungsten analogue and several derivatives have been reported as well.) The structure of this species (Figure 19.9) shows an octahedron of molybdenums surrounded by a cube of chlorines. Because of the spatial relationship between

Figure 19.9
The structure of the cation $Mo_6Cl_8^{4+}$, which has an octahedral metal framework.

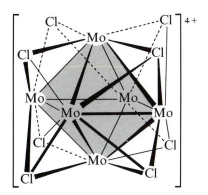

10. Burns, R. C.; Gillespie, R. J.; Luk, W.-C. *Inorg. Chem.* **1978**, *17*, 3596; Edwards, P. A.; Corbett, J. D. *Inorg. Chem.* **1977**, *16*, 903.

a cube and a concentric octahedron (Figure 2.7), this means that each chloride is centered above one of the triangular Mo_3 faces. Using the approach employed earlier for electron-precise metal aggregates such as Re_3Cl_9, the six Mo^{II} ions have 24 valence electrons, precisely the number required for Mo–Mo edge bonding ($6 \cdot 6 - 12 = 24$).

Seven-Atom Cages

A large number of compounds containing phosphorus, arsenic, or antimony exhibit seven-atom cages. The dominant framework can be described as a distorted, end-capped trigonal prism held together by nine localized bonds:

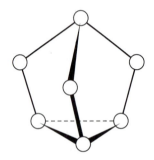

Four of the framework atoms—those in the basal and capping positions—are trivalent, and so are well-suited to occupancy by Group 15 elements. The bridge atoms may be divalent or, if substituents are present, tri- or tetravalent. This allows for considerable variation.

The "parent" species are the M_7^{3-} anions, which are produced by the action of bases on either the free element or the hydrides:

$$3P_4 + 6LiPH_2 \longrightarrow 2Li_3P_7 + 4PH_3 \qquad (19.18)$$

$$9P_2H_4 + 3n\text{-BuLi} \longrightarrow Li_3P_7 + 11PH_3 + 3C_4H_{10} \qquad (19.19)$$

The Lewis structure of M_7^{3-} (where M is a Group 15 element) conforms to the octet rule, with the three bridge atoms having negative formal charges:

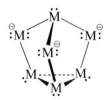

Derivatives can be prepared by electrophilic attack at the bridges.

$$Li_3P_7 + 3\,MeBr \longrightarrow Me_3P_7 + 3\,LiBr \qquad (19.20)$$

$$Li_3P_7 + 3\,Me_3SiCl \longrightarrow (Me_3Si)_3P_7 + 3\,LiCl \qquad (19.21)$$

$$(Me_3Si)_3P_7 + 3\,MeOH \longrightarrow H_3P_7 + 3\,Me_3SiOMe \qquad (19.22)$$

Many binary and ternary compounds are known in which divalent atoms (notably sulfur) occupy the bridging positions. For example, the direct reaction of P_4 or As_4 with elemental sulfur produces M_4S_3:

$$8\,M_4 + 3\,S_8 \xrightarrow{\ \Delta\ } 8\,M_4S_3 \qquad (19.23)$$

Combinations of Group 15 and 16 elements yield many variations on the above. Two additional examples are As_4S_4 and As_4S_5, with one and two "extra" sulfurs wedged between arsenic atoms in the basal plane.

Cubic Frameworks

Eight-atom systems are often cubic, particularly if they have enough valence electrons to permit completely localized bonding. Cubic structures are favored in part because the bond angles are reasonable for sp^3, sp^3d, and sp^3d^2 hybridization.

Oligomerization arising from adduct formation often yields cubic tetramers. For example, thallium(I) methoxide, "TlOMe," consists of distorted cubes in which each oxygen forms dative bonds to three thalliums (Figure 19.10).

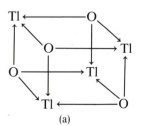
(a)

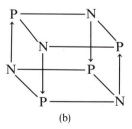
(b)

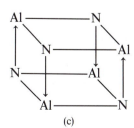
(c)

Figure 19.10 Three tetramers having cubic frameworks: (a) $(TlOMe)_4$; (b) $(F_3PNMe)_4$; (c) $(MeAlN\mathit{i}\text{-}Pr)_4$. Only the framework atoms are shown.

A second example is $(F_3PNMe)_4$, which is obtained by heating the corresponding four-membered ring:

$$2(F_3PNMe)_2 \longrightarrow (F_3PNMe)_4 \qquad \textbf{(19.24)}$$

The product can be considered an adduct in which one P_2N_2 ring eclipses another, forming four dative $N \rightarrow P$ bonds.

A remarkable variety of oligomers results from reactions of aluminum alkyls with amines:

$$\text{AlR}_3 + \text{R}'\text{NH}_2 \xrightarrow{\;-2RH\;} \frac{1}{x}(\text{RAlNR}')_x \qquad \textbf{(19.25)}$$

Depending on the substituents and reaction conditions, x may be 2, 3, 4, 6, 7, or 8. Cubic tetramers result if, for example, $R = H$ or Me and $R' = i\text{-Pr}$.

Among the transition metals, cubic tetrairon–tetrasulfur combinations are especially common. Some specific examples include $Fe_4(NO)_4S_4$ and $(i\text{-PrC}_5H_4)_4Fe_4S_4$. The Fe_4S_4 framework is of special interest because it is found in a variety of biological oxidation–reduction catalysts (see Chapter 22). It is therefore not surprising that the two synthetic clusters just mentioned survive reduction without undergoing decomposition.

$$Fe_4(NO)_4S_4 \xrightarrow{\;+ne^-\;} [Fe_4(NO)_4S_4]^{n-} \qquad n = 1 \text{ or } 2 \quad \textbf{(19.26)}$$

$$(i\text{-PrC}_5H_4)_4Fe_4S_4 \xrightarrow{\;+ne^-\;} [(i\text{-PrC}_5H_4)_4Fe_4S_4]^{n-} \qquad n = 1\text{–}4 \quad \textbf{(19.27)}$$

Under the proper circumstances, ionic salts can be formed by the simple mixing of two different cubic compounds.[11]

$$R_4Mo_4S_4 + R'_4Fe_4S_4 \longrightarrow [R_4Mo_4S_4]^+[R'_4Fe_4S_4]^-$$
$$R = i\text{-PrC}_5H_4; \; R' = NO, \text{ etc.} \quad \textbf{(19.28)}$$

Ten-Atom Cages: Adamantane-Based Structures

A common geometry for ten-atom electron-precise systems is exemplified by adamantane, $C_{10}H_{16}$ (Figure 19.11). Each framework atom has a tetrahedral environment.

---

11. Green, M. L. H.; Hamnett, A.; Qin, J.; Baird, P.; Bandy, J. A.; Prout, K.; Marseglia, E.; Obertelli, S. D. *J. Chem. Soc. Dalton* **1987**, 1811.

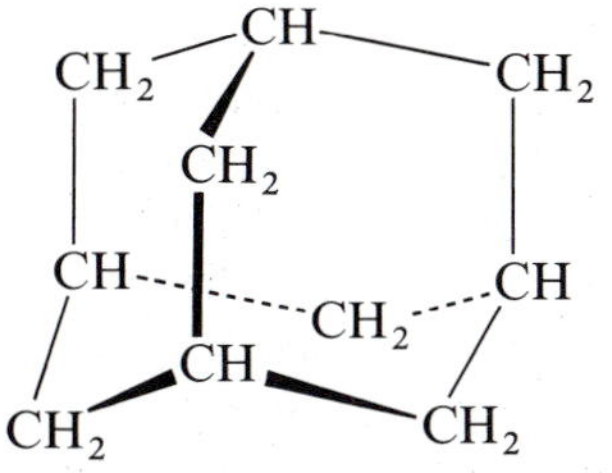

Figure 19.11
The structure of adamantane, $C_{10}H_{16}$. Each of the ten carbons has a tetrahedral arrangement of localized bonds.

An organometalloid analogue is $Si_{10}Me_{16}$, which is produced (along with several other ring compounds) by the reduction of $MeSiCl_3 + Me_2SiCl_2$ mixtures with sodium/potassium alloy. A related class of compounds has the formula $P_{10}R_6$; such species can be prepared by metallation reactions between PCl_3 and lithiophosphines.

$$6RPLi_2 + 4PCl_3 \longrightarrow P_{10}R_6 + 12LiCl \tag{19.29}$$

Many heteronuclear adamantane-like cages have been reported as well. Recall that P_4O_6 has such a structure (see Figure 14.8), and that up to four additional oxygens can be added as exo-cage substituents. (This does not alter the framework electron population.) The P_4S_n ($n = 6–10$) series is also known, as are As_4O_6 and As_4S_6. The $B_4S_{10}^{8-}$ anion of $Pb_4B_4S_{10}$ is both valence isoelectronic and isostructural with P_4O_{10}. Still another variation involves the formal replacement of the oxygen atoms by NR groups to give compounds such as $P_4(NR)_6$.

Isopoly and Heteropoly Systems[12]

Many metal-containing compounds and ions have three-dimensional units joined together by vertex, edge, or face sharing. We have already encountered several simple examples, such as the pyroacids and pyroanions (Chapters 10 and 14), in which the basic structural unit is the tetrahedron. For example, the conversion of the chromate ion to dichromate is accomplished by the removal of one oxygen for every two chromiums:

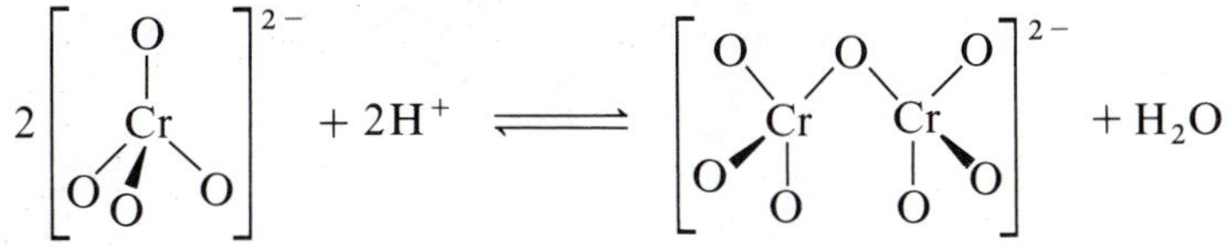

$$\tag{19.30}$$

The bridging oxygen represents a vertex shared by two tetrahedra.

12. Tytko, K.-H.; Glemser, O. *Adv. Inorg. Chem. Radiochem.* **1976**, *19*, 239; Kepert, D. L. In *Comprehensive Inorganic Chemistry*; Trotman-Dickenson, A. F., Ed.; Pergamon: Oxford, 1973; Volume 4, Chapter 51.

The sharing of vertices or edges by octahedral frameworks leads to a variety of extended structures. For example, $Nb_2F_{11}^-$ contains two NbF_6 octahedral units with a common vertex; U_2Cl_{10} has edge-sharing octahedra; and Nb_4F_{20} (Figure 15.6b) contains a square of vertex-shared octahedra.

Infinite structures can be obtained in the same manner. The ReO_3 lattice consists of ReO_6 octahedra in which each oxygen is shared by two rheniums; that is, each vertex is shared by adjacent octahedral units. (This lattice was described in a different way in Chapter 6; see Figure 6.13.)

Large but discrete binary oxides and oxyanions of vanadium, molybdenum, and tungsten are especially important in this regard. Such species are *isopoly* atoms or ions. A few examples of isopoly molybdates are $Mo_4O_{13}^{2-}$, $Mo_6O_{19}^{2-}$, $Mo_7O_{24}^{6-}$, $Mo_8O_{26}^{4-}$, and $Mo_8O_{27}^{6-}$. The structures of two of these anions are shown in Figure 19.12.

The term *heteropoly* applies if a third type of atom is present. A particularly well-known example is $Mo_{12}PO_{40}^{3-}$, which consists of a globular cluster of vertex- and edge-sharing MoO_6 units, linked in such a way as to create a tetrahedral cavity. The phosphorus resides in the cavity, stabilized by four P–O interactions. Other appropriately sized atoms may substitute for phosphorus, including Si, Ge, B, and Fe.[13]

Another important class of heteropoly ions is called the *tungsten bronzes*.[14] These are ternary systems having the formula M_xWO_3, where M is usually Na or K and $0 < x < 1$. The intense colors of these species have led to their use as paint pigments. The color can be controlled by the stoichiometry; thus, $Na_{0.9}WO_3$ is yellow, while $Na_{0.3}WO_3$ is blue-black. The structures of the tungsten bronzes involve vertex-shared WO_6 octahedra, and can be understood by analogy to the perovskite lattice (Figure 6.14); the body-centered position is populated by the Group 1 atom to the extent

Figure 19.12
The structures of two isopoly molybdate anions: (a) $Mo_7O_{24}^{6-}$; (b) $Mo_8O_{26}^{4-}$. [Reproduced with permission from Wells, A. F. *Structural Inorganic Chemistry*, 5th ed.; Clarendon: Oxford, 1984; p. 518.]

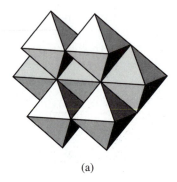

(a)

(b)

13. Canny, J.; Teze, A.; Thouvenot, R.; Herve, G. *Inorg. Chem.* **1986**, *25*, 2114.

14. Whittingham, M. S.; Dines, M. B. *Surv. Prog. Chem.* **1980**, *9*, 66.

of the stoichiometry. Because of the lattice vacancies, we might expect the bulk materials to have interesting electrical conductivities, and this is, in fact, the case.

19.5 The Bonding in Delocalized Clusters

The Electron Counting (Wade–Mingos) Rules

As in electron-precise cages, there is a definite relationship between the framework electron count and the geometry of delocalized clusters. However, the approach used for clusters is somewhat different. The method in most common use follows the *Wade–Mingos rules*, named after the individuals primarily responsible for their origin and development.[15]

These rules are based on molecular orbital theory. It can be shown that a closed polyhedron having n vertices ($n > 4$) has $n + 1$ bonding MO's, which are just filled by $2n + 2$ framework electrons. This $2n + 2$ *rule* is therefore the counterpart of the $6n - 12$ rule for deltahedral cages:

$$T = 2n + 2 \tag{19.31}$$

where T is the total number of delocalized electrons and n is the number of vertices. Specific correlations between the number of framework electrons and the predicted geometries are given in Table 19.4.

The contribution by each vertex unit to the delocalized molecular orbitals is determined as follows. For a main group element such as boron,

Table 19.4 Polyhedral electron count versus predicted structure according to the Wade–Mingos rules

Electron Count	Number of Vertices	Predicted Structure
12	5	Trigonal bipyramid
14	6	Octahedron
16	7	Pentagonal bipyramid
18	8	Dodecahedron
20	9	Tricapped trigonal prism
22	10	Bicapped square antiprism
24	11	Hexadecahedron
26	12	Icosahedron

15. Wade, K. *Adv. Inorg. Chem. Radiochem.* **1976**, *18*, 1; *J. Chem. Soc. Chem. Commun.* **1971**, 792; Mingos, D. M. P. *Chem. Soc. Rev.* **1986**, *15*, 31; *Nature Phys. Sci.* **1972**, *236*, 99. See also footnote 2.

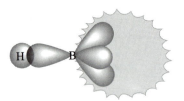

Figure 19.13
Orientation of the valence orbitals of a BH unit in a polyhedral borane cluster; the boron is assumed to have sp^3 hybridization.

To cluster
3 orbitals
2 electrons

three of the four valence orbitals (which are sometimes assumed to be sp^3 hybrids) are used to form the cluster MO's. The fourth is involved in exo-cluster bonding (in the case of boron, usually to a hydrogen atom). A BH unit has four valence electrons, two of which form the B–H bond. That leaves two electrons to be donated for the framework bonding (Figure 19.13). Thus, it is common to describe a BH fragment as a "two-electron donor." By the same reasoning, a CH (or CR) moiety is a three-electron donor. A sulfur having no external substituents (but with a lone pair) is a four-electron donor.

As an example of the application of the Wade–Mingos rules, consider the octahedral carborane $1,2\text{-}C_2B_4H_6$, whose structure is given in Figure 19.14a. The framework electron count is

$$(2 \cdot 3)\,(CH) + (4 \cdot 2)\,(BH) = 14e^-$$

Based on the $2n + 2$ rule, 14 electrons are the ideal number for an octahedral framework, and the experimental structure coincides with this prediction.

Another well-known carborane is $C_2B_4H_8$. Here, the two "extra" hydrogens are in bridging positions between adjacent borons, and their electrons must be included in the framework count:

$$(2 \cdot 3)\,(CH) + (4 \cdot 2)\,(BH) + 2 = 16e^-$$

Solving the $T = 2n + 2$ equation, $n = 7$, so a pentagonal bipyramidal framework is predicted.

Figure 19.14
The structures of two carboranes:
(a) $1,2\text{-}C_2B_4H_6$ (a closo cluster);
(b) $2,3\text{-}C_2B_4H_8$ (nido, with two bridging hydrogens).

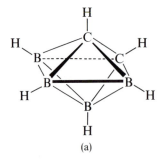

(a)

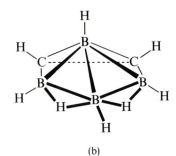

(b)

Since only six heavy atoms are present, however, one vertex must remain vacant. The experimental geometry is a pentagonal pyramid, which results from "decapping" one of the apical vertices (Figure 19.14b). Clusters that are one vertex removed from a *closo* (all vertices occupied) structure are described as *nido* (from the Greek word for "nestlike"). Species lacking two vertices are *arachno* (Greek for "weblike"). You should take the time to verify that B_4H_{10} is an arachno cluster, and sketch its structure.[16] The relationships among closo, nido, and arachno clusters are presented in Figure 19.15 (p. 652).

For transition metals, the nine valence orbitals are divided into three subgroups. Three orbitals are assigned to exo-cluster bonding, three are assumed to be σ nonbonding orbitals (requiring six valence electrons), and three participate in the delocalized framework bonding. Thus, the $Co(CO)_3$ fragment is a three-electron donor:

$$9 - 6(\text{nonbonding}) = 3e^-$$

(The six electrons comprising the cobalt–carbonyl bonds are all supplied by the ligand.) This analysis supports the isolobal concept—like the CH and P moieties, the $Co(CO)_3$ group contributes three electrons and three atomic orbitals to the cluster MO's.

The $Co(C_5H_5)$ fragment provides one less delocalized electron than does $Co(CO)_3$; that is, it is a two-electron donor. This is because the cyclopentadienyl ligand contributes only five electrons, and so does not completely fill the three exo-cluster orbitals; one of cobalt's electrons must be used for that purpose. Hence, $Co(C_5H_5)$ is framework isoelectronic with BH.

Experimental support for this notion has been provided by a comparative study of B_5H_9 and $C_5H_5CoB_4H_8$. In addition to being isostructural, these compounds exhibit similar reaction chemistries. Compare equations (19.32) and (19.33) to (19.34) and (19.35).[17]

$$B_5H_9 + NaH \longrightarrow Na^+B_5H_8^- + H_2 \tag{19.32}$$

$$2B_5H_9 + 2C_2H_2 \xrightarrow{\Delta} 2C_2B_4H_8 + B_2H_6 \tag{19.33}$$

$$C_5H_5CoB_4H_8 + NaH \longrightarrow Na^+C_5H_5CoB_4H_7^- + H_2 \tag{19.34}$$

$$2C_5H_5CoB_4H_8 + 2C_2H_2 \xrightarrow{\Delta} 2C_5H_5CoC_2B_3H_7 + B_2H_6 \tag{19.35}$$

Table 19.5 provides a chart of electron count contributions for groups common to cluster chemistry.

A few clusters are known that can be described as *hyperdeficient*; that is, they have fewer than $2n + 2$ delocalized electrons. An example is

---

16. Or see Figure 19.16b.

17. Weiss, R.; Bowser, J. R.; Grimes, R. N. *Inorg. Chem.* **1978**, *17*, 1522.

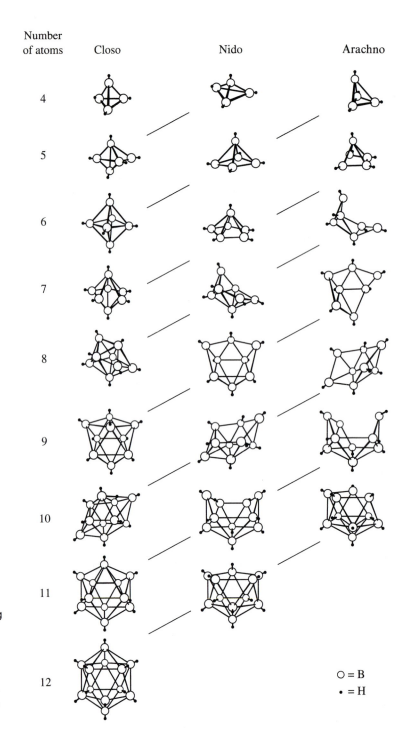

Number of atoms — Closo — Nido — Arachno

4
5
6
7
8
9
10
11
12

O = B
• = H

Figure 19.15
Structural relationships among closo, nido, and arachno boranes. [Reproduced with permission from Rudolph, R. W. *Acc. Chem. Res.* **1976**, *9*, 446.]

Table 19.5 Delocalized electron contributions by common polyhedral fragments according to the Wade–Mingos rules

Number of Electrons Donated			
1	2	3	4
H			
	BH	BH_2	
		CH	
SiR_3		SiR	
		P	
			S
$Mn(CO)_3$			
FeC_5H_5	$Fe(CO)_3$		
$Co(CO)_2$	CoC_5H_5	$Co(CO)_3$	
	$Ni(CO)_2$	NiC_5H_5	

$Os_6(CO)_{18}$, with an electron count of $6(2) = 12$. Here, the Wade–Mingos rules predict a five-atom geometry for a compound having six metal atoms.

Such hyperdeficient clusters often have capped structures. Thus, $Os_6(CO)_{18}$ is a capped trigonal bipyramid (see Figure 19.3b).[18] Consistent with the Wade–Mingos rules, that framework rearranges to an octahedron upon two-electron reduction to $[Os_6(CO)_{18}]^{2-}$.

The structures of ferrocene and other metallocenes also can be rationalized. According to the Wade–Mingos rules, $Fe(C_5H_5)_2$ is a nido cluster with a pentagonal bipyramidal framework:

The required electron count of 16 is readily obtained:

$$(5 \cdot 3)\,(CH) + (1 \cdot 1)\,(C_5H_5Fe) = 16e^-$$

Although the Wade–Mingos rules give remarkably accurate predictions for so simple a method, they are not inviolate. Certain polyhedra appear to be favored even when the electron count varies from the ideal. For

---

18. As indicated earlier, this structure also can be described as a bicapped tetrahedron; you should convince yourself that both characterizations are appropriate.

example, the eight-vertex clusters $(C_5H_5)_4Co_4B_4H_4$ and $(C_5H_5)_4Ni_4B_4H_4$ both adopt the closo dodecahedral geometry, in spite of the fact that the former has too few electrons (16 compared to the required 18), while the latter has too many (20).

Another limitation is that the method predicts the gross geometry, but not the finer points, of a given structure. The overall geometry of B_4H_{10} (octahedral-based arachno) is correctly predicted, but the disposition of the six "extra" hydrogens is not. (Experimentally, four hydrogens occupy bridging positions, while the others are terminal; see Figure 19.16b, below.) Nor does the method distinguish between bridging and terminal carbonyls [eg, the observation that $Co_4(CO)_{12}$ has bridging carbonyls in solution but $Rh_4(CO)_{12}$ does not, in spite of their identical framework electron counts]. More rigorous molecular orbital methods are useful for such cases.[19]

Next, we will turn our attention away from theory and toward the synthetic and reaction chemistry of boron-containing clusters.

19.6 The Boron Hydrides

Polyhedral boron–hydrogen compounds are of historical importance, since they were the first group of compounds to be recognized as clusters. The earliest boron hydrides were prepared by Stock and co-workers, who developed vacuum-line techniques to permit their isolation and study.[20] The *Stock series* contains the six compounds B_2H_6, B_4H_{10}, B_5H_9, B_5H_{11}, B_6H_{10}, and $B_{10}H_{14}$.

Diborane(6)

Diborane, B_2H_6, can be considered the parent member of the boron hydrides, since most of the others are prepared—either directly or indirectly—from its thermal decomposition. It can be synthesized by various acid–base and redox reactions; $NaBH_4$ is a convenient starting material.

$$2\,NaBH_4 + 2\,H_3PO_4 \longrightarrow B_2H_6 + 2\,NaH_2PO_4 + 2\,H_2 \qquad (19.36)$$

$$3\,NaBH_4 + BCl_3 \longrightarrow 2\,B_2H_6 + 3\,NaCl \qquad (19.37)$$

$$2\,NaBH_4 + I_2 \longrightarrow B_2H_6 + 2\,NaI + H_2 \qquad (19.38)$$

19. See, for example, Minot, C.; Criado-Sancho, M. *Nouv. J. Chim.* **1984**, *8*, 537; Evans, D. G. *J. Chem. Soc. Chem. Commun.* **1983**, 675; Mingos, D. M. P. *Pure Appl. Chem.* **1980**, *52*, 705; Lauher, J. W. *J. Am. Chem. Soc.* **1978**, *100*, 5305.

20. Stock, A. *Hydrides of Boron and Silicon*; Cornell University: Ithaca, NY, 1933.

The reaction chemistry of B_2H_6 has been thoroughly studied. A few of the major types of reactions are described below.

Adduct Formation (Nucleophilic Cleavage) The diborane molecule is readily cleaved by nucleophiles. The fragmentation may be either symmetric (to give two BH_3 groups) or asymmetric (forming BH_2^+ and BH_4^-):

$$B_2H_6 + 2CN^- \longrightarrow 2BH_3CN^- \tag{19.39}$$

$$B_2H_6 + 2NH_3 \longrightarrow [BH_2(NH_3)_2]^+BH_4^- \tag{19.40}$$

The product of equation (19.40) is called the *diammoniate of diborane*, and is noteworthy as a synthetic precursor to borazine (Chapter 13).

Oxidation–Reduction A host of reactions are known in which the borons of B_2H_6 are oxidized to B^{III}. Some examples are

$$B_2H_6 + 6H_2O \longrightarrow 2B(OH)_3 + 6H_2 \tag{19.41}$$

$$B_2H_6 + 3O_2 \longrightarrow B_2O_3 + 3H_2O \tag{19.42}$$

$$B_2H_6 + 6F_2 \longrightarrow 2BF_3 + 6HF \tag{19.43}$$

The great exothermicity of these reactions [eg, $\Delta H° = -2160 \text{ kJ/mol}$ for equation (19.42)] led military and civilian researchers to consider B_2H_6 for use as a rocket fuel additive. This stimulated the experimental study of the boranes in the 1950's and 1960's.

Addition Across Multiple Bonds

$$B_2H_6 + 2H_2C{=}CHR \longrightarrow 2H_2B{-}CH_2{-}CH_2R \tag{19.44}$$

When this reaction involves alkenes, it is often described as *hydroboration*. Such reactions are of great utility in organic synthesis; however, more easily handled boron reagents are normally used.

Thermal Decomposition As mentioned above, the thermal decomposition of B_2H_6 produces higher boranes. Temperatures in excess of 100°C are required. The first step of the reaction mechanism is thought to involve dissociation to the BH_3 monomer:

$$B_2H_6 \underset{\Delta}{\rightleftharpoons} 2BH_3 \tag{19.45}$$

This is followed by the formation of the unstable intermediate B_3H_9, which undergoes elimination of H_2:

$$BH_3 + B_2H_6 \rightleftharpoons B_3H_9 \qquad (19.46)$$

$$B_3H_9 \longrightarrow B_3H_7 + H_2 \qquad (19.47)$$

Higher Boranes

Continuing the decomposition process, higher boranes are formed sequentially via reactions such as the following:

$$B_3H_7 + BH_3 \longrightarrow B_4H_{10} \qquad (19.48)$$

$$B_4H_{10} + BH_3 \longrightarrow B_5H_{11} + H_2 \qquad (19.49)$$

$$B_5H_{11} \longrightarrow B_5H_9 + H_2 \qquad (19.50)$$

$$2B_5H_9 \longrightarrow B_{10}H_{14} + 2H_2 \qquad (19.51)$$

Several polyhedral boranes are pictured in Figure 19.16. The higher boranes generally mimic diborane in their physical and chemical properties, but become less volatile and less chemically reactive with increasing size. For example, B_2H_6 is gaseous well below room temperature (bp $= -93°C$), and it explodes on contact with air or water. Pentaborane is a colorless liquid at room temperature (boiling at about 60°C). Like B_2H_6, it explodes on contact with air, but it is easily handled using standard vacuum-line techniques. Decaborane is a white solid that decomposes only slowly in air.

Typical reactions of the higher boranes include adduct formation, displacement of terminal hydrogen by nucleophiles or electrophiles, chemical reduction, and coupling.

$$B_5H_9 + Et_3N \longrightarrow Et_3N \cdot B_5H_9 \qquad (19.52)$$

$$B_{10}H_{14} + 2Me_2S \longrightarrow (Me_2S)_2B_{10}H_{12} + H_2 \qquad (19.53)$$

$$B_5H_9 + I_2 \xrightarrow{AlCl_3} IB_5H_8 + HI \qquad (19.54)$$

$$B_{10}H_{14} + 2Na \longrightarrow 2Na^+[B_{10}H_{14}]^{2-} \qquad (19.55)$$

$$Na^+B_5H_8^- + B_5H_8Br \longrightarrow H_8B_5-B_5H_8 + NaBr \qquad (19.56)$$

The product of equation (19.56) is a *conjuncto* borane, with two clusters connected by a B–B σ bond.

Many other reactions derive from the Brønsted acidity of bridging hydrogens. Reactions with bases often result in deprotonation and the

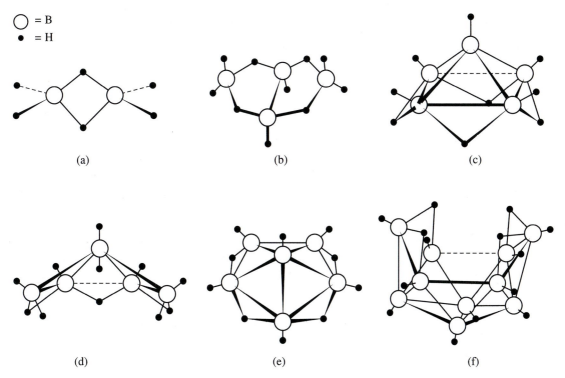

○ = B

● = H

(a)

(b)

(c)

(d)

(e)

(f)

Figure 19.16 The structures of Stock's polyhedral boranes: (a) B_2H_6; (b) B_4H_{10}; (c) B_5H_9; (d) B_5H_{11}; (e) B_6H_{10}; (f) $B_{10}H_{14}$.

formation of anions:

$$B_5H_9 + NaH \longrightarrow Na^+B_5H_8^- + H_2 \qquad (19.57)$$

$$B_{10}H_{14} + OH^- \longrightarrow B_{10}H_{13}^- + H_2O \qquad (19.58)$$

Such anions are often prepared in situ for subsequent reactions with electrophiles or metals [see equations (19.64), (19.67), and (19.68)].

It is sometimes possible to either increase or decrease the number of framework atoms. The former process is *cluster expansion*, while the latter is *cluster degradation*. Expansion often involves the use of B_2H_6 or an adduct of BH_3, while degradation is generally accomplished with bases; for example,

$$Li^+B_5H_8^- + B_2H_6 \longrightarrow B_6H_{10} + LiBH_4 \qquad (19.59)$$

$$B_{10}H_{14} + 2Et_3N \cdot BH_3 \longrightarrow (Et_3NH)_2[B_{12}H_{12}] + 3H_2 \qquad (19.60)$$

$$B_6H_{12} + Me_3P \longrightarrow B_5H_9 + Me_3P \cdot BH_3 \qquad (19.61)$$

19.7 Carboranes and Other Nonmetal Heteroboranes

In a variation of the hydroboration reaction, alkynes can be used to incorporate carbon atoms into polyhedral boranes or their anions or adducts. The reaction products are called *carboranes*.

$$B_5H_9 + MeC\equiv CMe \xrightarrow{\Delta} Me_2C_2B_4H_6 + \tfrac{1}{2}B_2H_6 \qquad (19.62)$$

$$(Me_2S)_2B_{10}H_{12} + C_2H_2 \xrightarrow{\Delta} C_2B_{10}H_{12} + 2Me_2S + H_2 \qquad (19.63)$$

Many carboranes have been prepared in this manner. For example, all members of the series $C_2B_nH_{n+2}$ ($n = 3$–10) have long been known. In accord with the Wade–Mingos rules, each has a closo geometry. Some specific examples are pictured in Figure 19.17.

More than one isomer is possible for all members of the $C_2B_nH_{n+2}$ series. As would be expected from the method of synthesis, the kinetic isomer (that which is initially formed) usually has adjacent carbons. However, in the more stable isomer(s), the carbons are separated. Thus, the product of equation (19.63) is the *ortho* (1,2) isomer. Heating to 450°C causes rearrangement to the *meta* (1,7) compound, and at about 625°C the *para* (1,12) isomer is produced.

A variety of other main group atoms have been incorporated into borane clusters, either in bridging positions or at one or more vertices. For

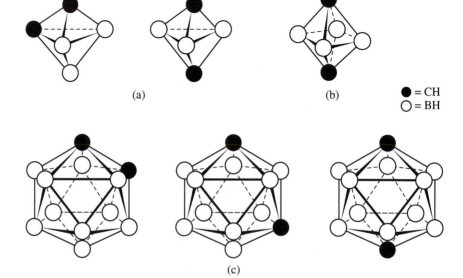

(a)

(b)

● = CH
○ = BH

Figure 19.17
Selected closo
carboranes:
(a) the 1,2- and
1,5-isomers
of $C_2B_3H_5$;
(b) 1,6-$C_2B_4H_6$;
(c) the three
isomers of $C_2B_{10}H_{12}$.

(c)

Figure 19.18
Some known
reactions of
thiaboranes.
[Reproduced with
permission from
Roesky, H. W., Ed.
*Rings, Clusters and
Polymers of the
Main Group and
Transition Elements;*
Elsevier: Amsterdam,
1989, p. 35.]

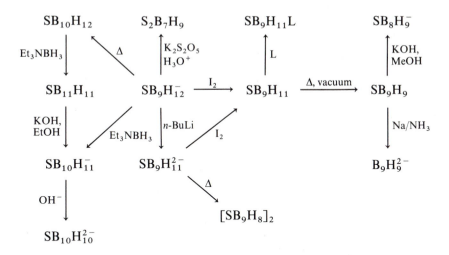

example, the $-SiR_3$ group can replace one of the bridging hydrogens of B_5H_9.

$$B_5H_9 \xrightarrow[\text{(2) Me}_3\text{SiCl}]{\text{(1) NaH}} \mu\text{-Me}_3\text{SiB}_5\text{H}_8 + \text{H}_2 + \text{NaCl} \qquad \textbf{(19.64)}$$

Among the nonmetals, a particularly large number of thiaboranes have been reported; this is primarily due to the efforts of Rudolph and co-workers.[21] Sulfur can be incorporated as a polyhedral atom via reagents such as $(NH_4)_2S_x$, $K_2S_2O_5$, and Na_2SO_3. The reactions of thiaboranes have been thoroughly studied, and some illustrative examples are summarized in Figure 19.18.

19.8 Metalloboranes and Metallocarboranes[22]

The incorporation of metals into borane and carborane clusters produces *metalloboranes* and *metallocarboranes*, respectively. The synthetic pathways to such species fall primarily into two broad types, depending on whether the initial substrate is a borane/carborane or a metal aggregate.

21. Canter, N.; Overberger, G. G.; Rudolph, R. W. *Organometallics* **1983**, *2*, 569; Pretzer, W. R.; Meneghelli, B. S.; Rudolph, R. W. *J. Am. Chem. Soc.* **1978**, *100*, 4626; Rudolph, R. W. *Acc. Chem. Res.* **1976**, *9*, 446.

22. Grimes, R. N. *Pure Appl. Chem.* **1987**, *59*, 847; Kennedy, J. D. *Prog. Inorg. Chem.* **1986**, *34*, 211; *Prog. Inorg. Chem.* **1984**, *32*, 519; Grimes, R. N. *Adv. Inorg. Chem. Radiochem.* **1983**, *26*, 55.

It is often possible to replace one or more ligands from a metal complex by a borane/carborane "chelate." For example, the high-temperature reaction between $C_2B_3H_5$ and $C_5H_5Co(CO)_2$ results in the displacement of two carbonyl ligands:

$$C_2B_3H_5 + C_5H_5Co(CO)_2 \xrightarrow{230°C} C_5H_5CoC_2B_3H_5 + 2CO \qquad \textbf{(19.65)}$$

Here, the cluster increases in size from five to six framework atoms. A similar reaction between B_5H_9 and $Fe(CO)_5$ leads to the replacement of a BH group by an isolobal $Fe(CO)_3$ fragment; this is therefore an insertion (or substitution), rather than an addition (cluster expansion), reaction:

$$B_5H_9 + Fe(CO)_5 \xrightarrow{220°C} (CO)_3FeB_4H_8 \qquad \textbf{(19.66)}$$

The products of equations (19.65) and (19.66) both have octahedral-based geometries (see Figure 19.19), as predicted by the Wade–Mingos rules:

For $C_5H_5CoC_2B_3H_5$: $\quad (3 \cdot 2)\,(BH) + (2 \cdot 3)\,(CH) + (1 \cdot 2)\,(C_5H_5Co) = 14e^-$

For $(CO)_3FeB_4H_8$: $\quad (4 \cdot 2)\,(BH) + (1 \cdot 2)\,[Fe(CO)_3] + (4)\,(H) = 14e^-$

Reactions between uncomplexed metal atoms and borane substrates are also known. For example, the cocondensation of Co(s), B_5H_9, and C_5H_6 at $-196°C$, followed by warming to room temperature, gives a mixture of metalloboranes including $(C_5H_5Co)_3B_3H_5$.[23]

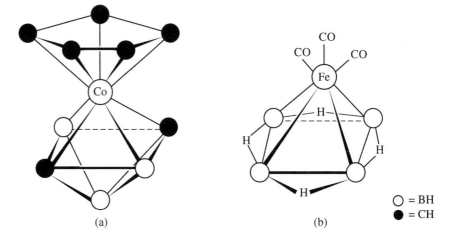

Figure 19.19
The structures of a metallocarborane and a metalloborane:
(a) $C_5H_5CoC_2B_3H_5$;
(b) $(CO)_3FeB_4H_8$.

○ = BH
● = CH

(a)   (b)

23. Zimmerman, G. J.; Hall, L. W.; Sneddon, L. G. *Inorg. Chem.* **1980,** *19,* 3642.

Another approach involves the preparation of a borane or carborane anion by the removal of a bridging hydrogen, followed by reaction with a metal halide. This may lead to metal bridging, substitution, and/or cluster expansion, depending on the situation. For example, the interaction of $B_5H_8^-$ with $Cu(P\phi_3)_3Cl$ gives $\mu\text{-}Cu(P\phi_3)_2B_5H_8$. The structure of this species is essentially that of B_5H_9 with one bridging hydrogen replaced by the metal fragment. A large number of metal and metalloid halides undergo similar reactions [see equation (19.64)].

Substitution is the primary reaction pathway when $CoCl_2$ reacts with $B_5H_8^- + C_5H_5^-$ mixtures. Here again, the metal atom is incorporated into the cluster, with a C_5H_5Co fragment replacing an isolobal BH group.

$$CoCl_2 + B_5H_8^- + C_5H_5^- \longrightarrow C_5H_5CoB_4H_8 \qquad (19.67)$$

If an excess of borane/carborane ligand is used, bis complexes may be obtained. The first metallocarborane was prepared in this manner (see Figure 19.20):[24]

$$Fe^{2+} + 2C_2B_9H_{11}^{2-} \longrightarrow [Fe(C_2B_9H_{11})_2]^{2-} \qquad (19.68)$$

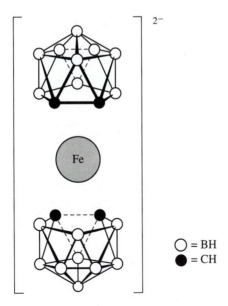

○ = BH
● = CH

Figure 19.20 The bis(dicarbollide)iron(II) complex, $[Fe(C_2B_9H_{11})_2]^{2-}$.

24. Hawthorne, M. F.; Young, D. C.; Wegener, P. A. *J. Am. Chem. Soc.* **1965**, *87*, 1818.

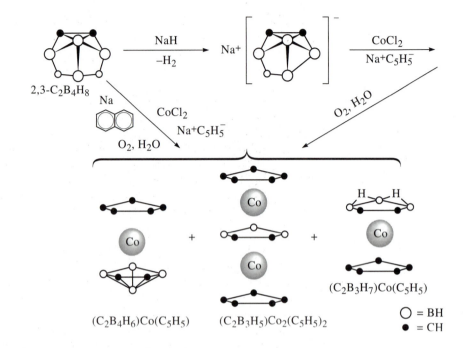

Figure 19.21
The CoCl$_2$ +
C$_2$B$_4$H$_7^-$ + C$_5$H$_5^-$
reaction system.
[Reproduced with
permission from
Grimes, R. N.; Beer,
D. C.; Sneddon, L.
G.; Miller, V.; Weiss,
R. *Inorg. Chem.*
1974, *13*, 1138.]

The *dicarbollide* ion, C$_2$B$_9$H$_{11}^{2-}$, and its interactions with various metal ions have been extensively studied by Hawthorne and co-workers.[25]

The CoCl$_2$ + C$_2$B$_4$H$_7^-$ + C$_5$H$_5^-$ system is especially interesting. A variety of compounds can be isolated from this reaction, including C$_5$H$_5$CoC$_2$B$_4$H$_6$ (an addition product), C$_5$H$_5$CoC$_2$B$_3$H$_7$ (an insertion product), and (C$_5$H$_5$)$_2$Co$_2$C$_2$B$_3$H$_5$ (combination addition and insertion), as shown in Figure 19.21. The C$_5$H$_5$CoC$_2$B$_3$H$_7$ molecule can be regarded as a metallocene complex in which the central cobalt has chelating C$_5$H$_5$ and C$_2$B$_3$H$_7$ ligands. In that sense it is analogous to ferrocene.

Metal-Rich Metalloboranes[26]

An alternative synthetic approach to metal–boron clusters is to allow a metal cluster to interact with a one- or two-boron reagent to produce a metal-rich metalloborane. For example,

$$Ru_3(CO)_{12} \xrightarrow{NaBH_4} Ru_3(CO)_9B_2H_6 \qquad (19.69)$$

$$H_2Os_3(CO)_{10} + Et_3N \cdot BH_3 \xrightarrow{\Delta} H_3Os_3(CO)_9BCO \qquad (19.70)$$

25. Callahan, K. P.; Hawthorne, M. F. *Adv. Organomet. Chem.* **1976**, *14*, 145.

26. Housecroft, C. E. *Polyhedron* **1987**, *6*, 1935.

The reaction described by equation (19.70) amounts to the insertion of boron into an Os–CO bond.

"Self-assembly" is also known. For example, the reaction of $[Co(CO)]_4^-$ with BBr_3 produces a tetrahedral array:

$$BBr_3 + 3[Co(CO)_4]^- + Et_3N \xrightarrow[-3Br^-]{-3CO} Co_3(CO)_9BNEt_3 \qquad (19.71)$$

Similarly, B_1 and Fe_1 reagents can be used to produce the tetrahedral compound $Fe_3(CO)_9BH_5$.[27]

19.9 Interstitial Clusters—Metal Carbides and Related Structures[28]

An increasing number of species are known in which a cluster framework surrounds a single atom or ion. For example, a polynuclear metal carbonyl may encapsulate a carbon atom, giving a *carbido cluster*. The first example of this phenomenon was reported by Dahl and co-workers, who obtained $Fe_5C(CO)_{15}$ from the thermolysis of $Fe_3(CO)_{12}$ in the presence of 1-hexyne.[29] The geometry of $Fe_5C(CO)_{15}$ (Figure 19.22) is that of a square pyramid, with the carbon centered in the basal plane. There are four Fe–C distances of 189 pm and one of 196 pm. The framework electron count consistent with this geometry (octahedral-based nido) is 14, which is obtained if carbon is assumed to be a four-electron donor.

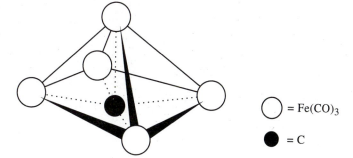

$\bigcirc$ = Fe(CO)$_3$

$\bullet$ = C

Figure 19.22
The carbido cluster $Fe_5C(CO)_{15}$.

27. Vites, J. C.; Housecroft, C. E.; Eigenbrot, C.; Buhl, M. L.; Long, G. J.; Fehlner, T. P. *J. Am. Chem. Soc.* **1986**, *108*, 3304.

28. Bradley, J. S. In *Metal Clusters*; Moscovits, M., Ed.; Wiley: New York, 1986; Chapter 5; Bradley, J. S. *Adv. Organomet. Chem.* **1983**, *22*, 1.

29. Braye, E. H.; Dahl, L. F.; Hubel, W.; Wampler, D. *J. Am. Chem. Soc.* **1962**, *84*, 4633.

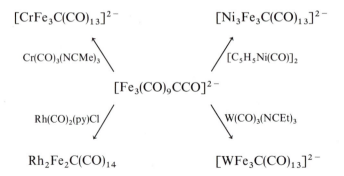

Figure 19.23
The synthesis of four mixed-metal carbido clusters from the $[Fe_3(CO)_9CCO]^{2-}$ anion.

In many carbides, the carbon is surrounded by an octahedron of metals (eg, in $[Fe_6C(CO)_{16}]^{2-}$ and $[Ru_6C(CO)_{16}]^{2-}$). This mimics the most common environment of carbon in binary transition metal carbide lattices. For this and other reasons, it has been suggested that carbido clusters might be useful models for bulk metals, refractory carbides, and other macroscopic materials.[30]

A systematic route to mixed-metal carbides, using the tetrahedral $[Fe_3(CO)_9C–CO]^{2-}$ anion as the starting material, has been developed by Shriver and co-workers. Interactions between this anion and the appropriate reagents yield a variety of four-, five-, and six-metal carbido clusters (Figure 19.23).[31]

Other main group atoms, especially phosphorus, have been incorporated into interstitial clusters. The preferred coordination geometry about phosphorus appears to be trigonal prismatic, although octahedral systems are also known.

Bibliography

Roesky, H. W., Ed. *Rings, Clusters and Polymers of the Main Group and Transition Elements*; Elsevier: Amsterdam, 1989.

Nugent, W. A.; Mayer, J. M. *Metal–Ligand Multiple Bonds*; Wiley: New York, 1988.

Woollins, J. D. *Non-Metal Rings, Cages and Clusters*; Wiley: New York, 1988.

Haiduc, I.; Sowerby, D. B., Eds. *The Chemistry of Inorganic Homo- and Heterocycles*; Academic: New York, 1987.

Moscovits, M., Ed. *Metal Clusters*; Wiley: New York, 1986.

Grimes, R. N., Ed. *Metal Interactions with Boron Clusters*; Plenum: New York, 1982.

---

30. See Hoffmann, R. *Pure Appl. Chem.* **1986**, *58*, 481; Mingos, D. M. P. *Chem. Soc. Rev.* **1986**, *15*, 31.

31. Hriljac, J. A.; Holt, E. M.; Shriver, D. F. *Inorg. Chem.* **1987**, *26*, 2943.

Wilkinson, G.; Stone, F. G. A.; Abel, E. W., Eds. *Comprehensive Organometallic Chemistry*; Pergamon: Oxford, 1982.

Johnson, B. F. G., Ed. *Transition Metal Clusters*; Wiley: New York, 1980.

Bevan, D. J. M.; Hagenmuller, P. *Nonstoichiometric Compounds: Tungsten Bronzes, Vanadium Bronzes, and Related Compounds*; Pergamon: Oxford, 1975.

Muetterties, E. L., Ed. *Boron Hydride Chemistry*; Academic: New York, 1975.

Grimes, R. N. *Carboranes*; Academic: New York, 1970.

Questions and Problems

1. Verify that the $6n - 12$ rule applies to the bicapped square antiprism (Figure 19.1h).

2. The reaction of $Os_3(CO)_{12}$ with excess I_2 destroys the Os_3 unit:

   $$Os_3(CO)_{12} + 3I_2 \longrightarrow 3Os(CO)_4I_2$$

   Suggest a mechanism for this reaction.

3. (a) The room-temperature reaction of $H_2Os_3(CO)_{10}$ with CH_2N_2 gives a μ-alkylidene having the molecular formula $Os_3C_{11}H_4O_{10}$. Predict its structure.
   (b) The thermal decomposition of the product from part (a) yields a μ-alkylidyne. Predict its structure. [Reference: Calvert, R. B.; Shapley, J. R. *J. Am. Chem. Soc.* **1977**, *99*, 5225.]

4. Consider the species $[(\mu\text{-}H)_3Os_4(CO)_{12}]^-$, in which the osmiums form a tetrahedral array. Sketch the structure, verify that the 18-electron rule is obeyed, and use the Wade–Mingos rules to demonstrate that this anion contains 12 delocalized framework electrons.

5. The compound $(CO)_9FeCo_2PMe$ has a tetrahedral framework.
   (a) Sketch this molecule.
   (b) Demonstrate that this species has 12 delocalized electrons.
   (c) The reaction of this compound with 2 moles of PMe_3 gives a product having a triangular framework. Predict its structure. [Reference: Planalp, R. P.; Vahrenkamp, H. *Organometallics* **1987**, *6*, 492.]

6. (a) Is $Co_3(CO)_9S$ diamagnetic or paramagnetic?
   (b) Do you expect this species to be more easily oxidized or reduced? Defend your answer.

7. (a) The reaction of $Fe(CO)_5$ with $Co_2(CO)_8$ gives an anion formulated as $[FeCo_3(CO)_{12}]^{n-}$. Give the most likely value of n, and explain.
   (b) Do you expect $[FeCo_2(CO)_9S]^?$ to be a cation, anion, or a neutral molecule? Defend your answer.

8. Explain why there was no mention of the 18-electron rule when the bonding in $Mo_6Cl_8^{4+}$ was discussed.

9. Explain in your own words why P_7^{3-} exhibits a capped trigonal pyramidal geometry, while $B_7H_7^{2-}$ forms a pentagonal bipyramid.

10. Draw a reasonable structure for:
    (a) $As_3Se_4^+$ (b) P_4S_9 (c) P_4S_4

11. The compound $Sb_4Co_4(CO)_{12}$ has a distorted cubic geometry, with antimony and cobalt atoms alternating around each face.
    (a) Demonstrate that the ideal number of electrons (24) are present for edge bonding in a cubic framework.
    (b) Does cobalt obey the 18-electron rule in this compound? Defend your answer.
    (c) The distortion is such that the Sb–Sb distance across each face diagonal is shortened:

    Suggest an explanation.

12. Give formulas consistent with the following descriptions; that is, assign values for a and b in the formula M_aX_b:
    (a) Two MX_6 octahedra sharing a common vertex
    (b) Two MX_6 octahedra sharing a common edge
    (c) Three MX_6 octahedra forming a triangle through vertex sharing
    (d) Four MX_6 octahedra forming a square through edge sharing

13. Give the Wade–Mingos electron count and describe the structure of:
    (a) $B_6H_6^{2-}$ (b) $C_2B_8H_{10}$ (c) CB_5H_7
    (d) $C_4B_2H_6$ (e) $(C_5H_5Ni)_2B_{10}H_{10}$ (f) $C_5H_5CoB_4H_7^-$
    (g) SB_9H_9

14. The triangular shape of $Ru_3(CO)_{12}$ was rationalized in the text via localized edge bonding. This structure also can be explained using the Wade–Mingos rules. Do so.

15. Classify as closo, nido, or arachno:
    (a) $B_6H_6^{2-}$ (b) B_6H_{10} (c) B_6H_{12} (d) $C_4B_8H_{12}$

16. The compound $H_nRu_6(CO)_{18}$ contains an octahedron of ruthenium atoms. Predict the value of n, and explain.

17. How many isomers are possible for the metalloborane $(C_5H_5)_2Ni_2B_4H_4$? Sketch the one that is the most stable.

18. Most of the boron hydrides have no known aluminum or gallium analogues. Rationalize.

19. Write the formulas of the boron-containing products for the following reactions of B_2H_6:
    (a) Symmetrical cleavage by Me_2S (b) Asymmetrical cleavage by NH_3
    (c) Oxidation by Cl_2 (d) Reaction with $H_3C-CH{=}CH-CH_3$

20. Predict the boron-containing products for the following reactions involving $B_{10}H_{14}$:

   (a) $B_{10}H_{14} + 2P\phi_3 \longrightarrow$

   (b) $B_{10}H_{14} + H_3CC\equiv CCH_3 \xrightarrow{\Delta}$

   (c) $B_{10}H_{14} \xrightarrow[\text{(2) } Et_4N^+Cl^-]{\text{(1) NaH}}$

21. Propose a plausible synthesis for:
   (a) $MeC_2B_4H_7$ from B_5H_9 (b) $[Fe(C_2B_9H_{11})_2]^{2-}$ from $C_2B_{10}H_{12}$
   (c) $Hg(B_5H_8)_2$ from B_5H_9

22. According to the Wade–Mingos rules, an icosahedral cluster should contain 26 framework electrons. Demonstrate that this is the case for $[\phi_3PCuC_2B_9H_{11}]^-$.

23. The ^1H NMR spectrum of one isomer of $C_5H_5CoB_4H_8$ shows two different kinds of bridging hydrogens. Which isomer?

24. Rationalize the bonding in $C_5H_5Co(C_4BH_5)$, in which the C_4B subunit is cyclic and planar, from the standpoint of:
   (a) The 18-electron rule   (b) The Wade–Mingos rules
   (c) Its relationship to ferrocene

25. Reexamine equation (19.70). Notice that the number of hydrogen atoms increases by one in going from the cluster reactant to the product. Rationalize.

26. Speculate on how the reagent cluster of Figure 19.23 might be prepared.

*27. The carborane $Et_2C_2B_5H_5$ was discussed in an article by J. S. Beck and L. G. Sneddon (*Inorg. Chem.* **1990**, *29*, 295). Answer these questions after reading their article.
   (a) How was this compound prepared?
   (b) Does its structure obey the Wade–Mingos rules?
   (c) Classify it as closo, nido, or arachno.
   (d) Rationalize its thermal behavior.

*28. The novel, six-atom cluster $Fe_3(CO)_9(BH)(CH)(CMe)$ was reported by X. Meng, T. P. Fehlner, and A. L. Rheingold (*Organometallics* **1990**, *9*, 534).
   (a) What is the framework electron count? Is the experimental geometry consistent with the Wade–Mingos rules?
   (b) What is a "borirene" ligand? What is it similar to?
   (c) What is unusual about the stereochemistry of this cluster?

*29. Can the complex $Os(CO_4)PMe_3$ act as a ligand? Read the report by H. B. Davis, F. W. B. Einstein, P. G. Glavina, T. Jones, R. K. Pomeroy, and P. Rushman (*Organometallics* **1988**, *8*, 1030).
   (a) Do $Os(CO_4)PMe_3$ and $Cr(CO)_5$ obey the 18-electron rule?
   (b) What is the justification for considering the $Os(CO_4)PMe_3$ group to be a ligand in the bimetallic complex $Os(CO_4)PMe_3Cr(CO)_5$?
   (c) How are the PMe_3 and $Cr(CO)_5$ groups oriented in this species? What is the most likely explanation?

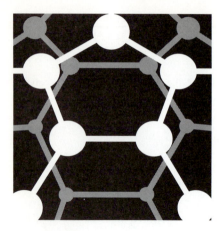

20

The Lanthanide, Actinide, and Transactinide Elements

Most of the elements of the periodic table have been discussed to one extent or another in Chapters 12–15. This chapter deals with "all the rest"—specifically, the *lanthanides* and *actinides* (the *f*-block elements) and the *transactinides* (members of the 6*d* series). An expanded periodic table locating these elements is shown in Figure 20.1.

Recall from Chapter 15 that there is not universal agreement on which elements should be considered transitional. Similarly, there is some question about where the lanthanide and actinide series begin and end. Part of the problem lies with the ground-state electron configurations of these elements. Since the *nd* and $(n-1)f$ orbitals typically lie close to one another in energy, mixed valence shell configurations often occur. This is evident from Table 20.1, in which the elements under consideration are grouped according to their configurations. Lanthanum, which might be expected to begin the series that bears its name, has no *f* electrons in its ground state. For this reason, the lanthanides are sometimes identified as the elements between cerium and lutetium (atomic numbers 58–71). If lanthanum is also included (as justified by its chemical similarity to the others), then there are 15 elements in the series.

Cs	Ba	La	Hf	Ta	W	Re	Os	Ir	Pt	Au	Hg	Tl	Pb	Bi	Po	At	Rn
55	56	57	72	73	74	75	76	77	78	79	80	81	82	83	84	85	86

Fr	Ra	Ac	Unq	Unp	Unh	Uns	Uno	Une	Uun	Uuu	Uub	Uut	Uuq	Uup	Uuh	Uus	Uuo
87	88	89	104	105	106	107	108	109	110	111	112	113	114	115	116	117	118

Uue	Ubn	Ubu
119	120	121

Figure 20.1
Expanded portion of the periodic table:
☐ = lanthanides (filling of the $4f$ orbitals);
☐ = actinides ($5f$);
■ = transactinides ($6d$). Syntheses of the elements beyond atomic number 109 are not yet generally recognized.

Ce	Pr	Nd	Pm	Sm	Eu	Gd	Tb	Dy	Ho	Er	Tm	Yb	Lu
58	59	60	61	62	63	64	65	66	67	68	69	70	71

Th	Pa	U	Np	Pu	Am	Cm	Bk	Cf	Es	Fm	Md	No	Lr
90	91	92	93	94	95	96	97	98	99	100	101	102	103

Ubb	Ubt
122	123

A different controversy involves the elements beyond lawrencium—that is, those having atomic numbers of 104 or greater. Disagreement over who should be recognized as the "discoverers" (ie, synthesizers, since these elements do not occur naturally) led to disagreement over what they should be named. IUPAC has suggested systematic names beginning with element 104 (Unq, for unnilquadium) and continuing through 105 (Unp, unnilpentium), 106 (Unh, unnilhexium), etc. (see Figure 20.1). This system provides a unique name and symbol not only for elements 104–106, but also for those yet to be prepared. For example, element 114 will be ununquadium (Uuq), and element 168 unhexoctium (Uho).

Table 20.1   Ground-state electron configurations of the lanthanide, actinide, and transactinide elements

Element(s)	At. Nos.	Configuration
La	57	$[Xe]6s^2(4f^0)5d^1$
Ce–Lu	58–71	$[Xe]6s^24f^n$ or $[Xe]6s^24f^{n-1}5d^1$
Ac	89	$[Rn]7s^2(5f^0)6d^1$
Th–Lr	90–103	$[Rn]7s^25f^n$ or $[Rn]7s^25f^{n-1}6d^1$
Unq	104	$[Rn]7s^25f^{14}6d^2$
Unp–	105–	$[Rn]7s^25f^{14}6d^n$

Note: The correct ground-state configurations of many of these elements are not known with certainty.

20.1 The Free Elements: Occurrence/Synthesis, Isolation, and Properties

The lanthanides have traditionally been referred to as the "rare earths," which is a misnomer. The abundance of cerium in the Earth's crust is over 60 ppm—more than three times greater than lithium. The other $4f$ elements occur in concentrations between 0.5 and 10 ppm, except for Pm, which has no stable isotopes. These abundances are roughly comparable to those of Cs, Be, Mo, and Ge.

The similar chemical properties of the lanthanides cause them to be found together in nature. They occur primarily as M^{3+} ions in oxide and phosphate ores, the most important of which is *monazite*. The isolation of individual elements relies primarily on ion-exchange chromatography (see below). Following their separation, the free metals can be obtained by electrolysis.

Preparation and Purification of the Actinides[1]

All isotopes of the actinides are radioactive. However, three of these elements occur in nature: thorium (about 8 ppm in the Earth's crust), uranium (2 ppm), and the less abundant protactinium. Traces of several of the others are found with these elements, formed mainly by the radioactive decay of Th and U.

Nuclear laboratory synthesis is necessary to obtain samples of any of the actinides (except thorium and uranium) in sufficient quantity for chemical study. This is usually accomplished by the bombardment of appropriate substrates with neutrons, deuterium nuclei, or alpha particles; a few specific examples are:

$$^{242}_{94}\text{Pu} + ^{1}_{0}n \longrightarrow ^{243}_{95}\text{Am} + ^{0}_{-1}e \tag{20.1}$$

$$^{238}_{92}\text{U} + ^{2}_{1}\text{H} \longrightarrow ^{240}_{93}\text{Np} \longrightarrow ^{240}_{94}\text{Pu} + ^{0}_{-1}e \tag{20.2}$$

$$^{253}_{99}\text{Es} + ^{4}_{2}\text{He} \longrightarrow ^{256}_{101}\text{Md} + ^{1}_{0}n \tag{20.3}$$

The free elements are usually prepared by chemical reduction of their oxides or fluorides.

$$\text{EsF}_4 + 4\text{Li} \longrightarrow \text{Es} + 4\text{LiF} \tag{20.4}$$

$$\text{UF}_4 + 2\text{Ca} \longrightarrow \text{U} + 2\text{CaF}_2 \tag{20.5}$$

$$\text{Am}_2\text{O}_3 + 2\text{La} \longrightarrow 2\text{Am} + \text{La}_2\text{O}_3 \tag{20.6}$$

An important method for the purification of these metals is the *van Arkel–DeBoer process*, in which the iodide salt is formed at relatively low

---

1. Spirlet, J. C.; Peterson, J. R.; Asprey, L. B. *Adv. Inorg. Chem.* **1987**, *31*, 1.

temperature, volatilized, and then thermally decomposed:

$$\text{M(impure)} + 2I_2 \xrightarrow{600°C} MI_4 \xrightarrow{1500°C} M + 2I_2 \tag{20.7}$$

As an example, through this process the purity of promethium metal can be increased from 93% to 99.7%.

Properties and Trends Among the Lanthanides

Relevant physical and chemical properties of the lanthanides are given in Table 20.2. These data can be used to rationalize much of their descriptive chemistry, and can be divided into two broad areas: ionization energy/oxidation state tendencies and size effects.

Ionization Energies and Oxidation States   As we would expect, there is a general increase in ionization energies going across the lanthanide series. The trend is less clear for IE_1 than for the sum of the first three ionizations (an important parameter because of the dominance of the M^{III} state). As can be

Table 20.2   Selected properties of the lanthanide elements

| Element | mp, °C | IE, eV | | $\mathscr{E}^0_{red}$ M^{3+}/M, V | r (M^{3+}), pm | MCl$_3$ (s) | |
		IE_1	IE_{1-3}			Coordination number	mp, °C
$_{57}$La	920	5.6	35.8	−2.37	106	9	860
$_{58}$Ce	798	5.5	36.5	−2.34	103	9	817
$_{59}$Pr	931	5.4	37.6	−2.35	101	9	786
$_{60}$Nd	1010	5.5	38.3	−2.32	100	9	758
$_{61}$Pm	1080	5.6	38.7	−2.29	98	9	682
$_{62}$Sm	1072	5.6	40.1	−2.30	96	9	678
$_{63}$Eu	822	5.7	41.8	−1.99	95	9	850
$_{64}$Gd	1311	6.1	38.9	−2.29	94	9	609
$_{65}$Tb	1360	5.8	39.2	−2.30	92	8	588
$_{66}$Dy	1409	5.9	40.4	−2.29	91	8	718
$_{67}$Ho	1470	6.0	40.6	−2.33	89	6	718
$_{68}$Er	1522	6.1	40.7	−2.31	88	6	776
$_{69}$Tm	1545	6.2	41.9	−2.31	87	6	824
$_{70}$Yb	824	6.3	43.5	−2.22	86	6	865
$_{71}$Lu	1656	5.4	40.3	−2.30	85	6	905

Source: Compiled primarily from data of Wells, A. F. *Structural Inorganic Chemistry*, 5th ed.; Clarendon: Oxford, 1984; p. 421; Sinha, S. P. *Struct. Bonding* **1976**, *25*, 69; Nugent, L. J. *J. Inorg. Nucl. Chem.* **1975**, *37*, 1767; and Johnson, M. D. A. *J. Chem. Soc. Dalton* **1974**, 1671.

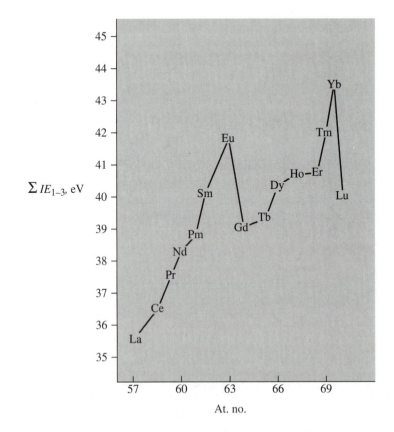

Figure 20.2
Plot of the sum of the first three ionization energies of the lanthanide elements as a function of atomic number. Data are from Table 20.2.

seen in Figure 20.2, deviations from the general trend are observed only for Eu/Gd and Yb/Lu, which involve half-filled and completely filled f subshells.

The M^{3+} ions have the configuration $[Xe](6s^o)(5d^o)4f^n$. This is consistent with the expectation that the $4f$ orbitals should be stabilized more rapidly than the $5d$ or $6s$ levels by the removal of electrons (see Chapter 1). The M^{III} state is dominant for all the lanthanides, as evidenced by the strongly negative potentials for the M^{III}/M and M^{III}/M^{II} couples. As a consequence, the free metals are strongly reducing, and are similar to the Group 1 and 2 elements in that regard. For example, they evolve H_2 when dropped into water.

$$2M + 6H_2O \longrightarrow 2M(OH)_3 + 3H_2 \qquad (20.8)$$

Other stable oxidation states include Ce^{IV}, Tb^{IV}, Eu^{II}, and Yb^{II}, all of which can be explained through electron configurations: They correspond to vacant ($4f^o$), half-filled ($4f^7$), or completely filled ($4f^{14}$) subshells. Thus, EuO, CeO_2, $EuCl_2$, and CeF_4 are relatively stable compared to the analogous compounds of the other lanthanides.[2] Even these exceptions are relative, however. For example, Ce^{III} is easily the thermodynamically stable state in aqueous solution:

$$Ce^{4+} \xrightarrow{\;+1.61\;} Ce^{3+} \xrightarrow{\;-2.48\;} Ce \quad (acid) \tag{20.9}$$

The disproportionation of Ce^{3+} is disfavored by over 4 V, and Ce^{4+} is reduced by water:[3]

$$4Ce^{4+} + 2H_2O \longrightarrow 4Ce^{3+} + 4H^+ + O_2 \quad \mathscr{E}^0 = +0.38 \text{ V} \tag{20.10}$$

Size Effects In accord with the usual periodic trends, metallic and ionic radii decrease going across the lanthanide series. The variation in M^{3+} radii is particularly regular (Table 20.2). This gives rise to several secondary effects. One is the variance in the hydration energies of these elements (see Figure 20.3). Also, the coordination numbers for the halide salts tend to decrease going across the series. Among the chlorides, $LaCl_3(s)$ forms a complex structure in which the La^{3+} ions are 9-coordinate. This lattice type is observed for the first eight (the largest) members of the series. In contrast, $TbCl_3$ and $DyCl_3$ have 8-coordinate metal ions, and the remaining (smallest) members of the series have octahedrally coordinated metals in their MCl_3 salts.

Given the observed trend in ionic radii, we might expect lattice energies to increase going across the series. However, this does not appear to be the case (at least as evidenced by the melting points of the trichlorides). Several factors act in the opposite direction, including lattice type (decreasing coordination numbers give decreased Madelung constants) and increasing ionization energies. (See the Born–Haber analysis of lattice energies given in Chapter 6.)

These differences in redox and complexation properties are utilized in the separation of the lanthanide cations. Monazite is a complex phosphate (MPO_4, where M is a mixture of trivalent metals). It dissolves in concentrated H_2SO_4, and thorium can be precipitated as ThO_2 by adjusting the pH of the resulting solution. The filtrate can be treated with an oxidizing agent (often OCl^-) to selectively oxidize cerium, precipitating it as CeO_2; the other lanthanides remain in the M^{III} state and in solution. Next, reduction (typically using zinc amalgam) is used to precipitate $EuSO_4$.

---

2. For more details concerning the lower oxidation states of these elements, see Meyer, G. *Chem. Rev.* **1988**, *88*, 93; Johnson, D. A. *Adv. Inorg. Chem. Radiochem.* **1977**, *20*, 1.

3. However, like many aqueous redox reactions having $\mathscr{E}^0 < +0.5$ V, this reaction is kinetically slow. Thus, Ce^{IV} solutions are useful oxidizing agents in quantitative analysis. See Manahan, S. E. *Quantitative Chemical Analysis*; Brooks/Cole: Pacific Grove, CA, 1986; pp. 346–348.

The lanthanides remaining in solution are most efficiently separated by ion-exchange chromatography, often using EDTA as a chelating agent. The equilibrium constant for complexation increases going across the series, with the range of values being a factor of about 10^4:

$$M^{3+} + EDTA^{2-} \longrightarrow [M(EDTA)]^+ \tag{20.11}$$
$$K_f = 2.0 \times 10^{15} \text{ for } La^{3+}$$
$$= 1.5 \times 10^{19} \text{ for } Lu^{3+}$$

Thus, lutetium, being the most strongly complexed, is eluted first, and the others follow in reverse order of atomic number.

20.2   Some Descriptive Chemistry of the Lanthanides[4]

Binary and Ternary Salts

We mentioned above that the free metals are quite reactive toward oxidation. As a result, direct reactions with halogens, O_2, or even N_2 (at elevated temperatures) give binary M^{III} salts:

$$2M + 3X_2 \longrightarrow 2MX_3 \tag{20.12}$$

$$4M + 3O_2 \longrightarrow 2M_2O_3 \tag{20.13}$$

$$2M + N_2 \xrightarrow{>800°C} 2MN \tag{20.14}$$

The dihalides can sometimes be prepared, especially under reducing conditions.

$$SmCl_3 \xrightarrow{Na/Hg} SmCl_2 + NaCl + Hg \tag{20.15}$$

$$2EuCl_3 + H_2 \xrightarrow{\Delta} 2EuCl_2 + 2HCl \tag{20.16}$$

$$2CeI_3 + Ce \xrightarrow{\Delta} 3CeI_2 \tag{20.17}$$

Binary hydrides and carbides are obtained by direct combination of the elements at elevated temperatures. The hydrides can be divided into two types. Those conforming to the formula MH_3 are saline (ie, ionic), and are similar to Group 1 and 2 hydrides such as NaH and CaH_2. Others have the

---

4.   Johnson, D. A. *J. Chem. Educ.* **1980**, *57*, 475; Karraker, D. *J. Chem. Educ.* **1970**, *47*, 424.

formula MH_2; they tend to be interstitial, and resemble the hydrides of the *d*-block elements. The carbides also appear to be interstitial.

The binary oxides and halides are similar in many ways to those of magnesium and calcium. This is not surprising, since the charge/size ratios of the M^{3+} ions are comparable to those of Mg^{2+} and Ca^{2+}. As an example of the chemical similarity, ternary compounds can be formed through reactions analogous to those undergone by CaX_2 and CaO.

$$M_2O_3 + 3H_2O \longrightarrow 2M(OH)_3 \tag{20.18}$$

$$M_2O_3 + 3CO_2 \longrightarrow M_2(CO_3)_3 \tag{20.19}$$

$$2MX_3 + 3SO_4^{2-} \longrightarrow M_2(SO_4)_3 + 6X^- \tag{20.20}$$

Hard–Soft Acid–Base Effects

We would expect the M^{3+} ions to prefer hard bases, and this is, in fact, the case. Thus, oxygen donors are favored over halides (except for F^-). As a result, the halides tend to be hygroscopic and/or quite soluble in water; solutions of $CeCl_3$ as concentrated as 4 M can be prepared. Also, certain types of reactions that are useful synthetic routes to other anhydrous metal halides fail for the lanthanides. For example, the treatment of cerium(III) oxide with HBr produces the oxybromide, rather than $CeBr_3$.

$$Ce_2O_3 \xrightarrow{HBr} \begin{matrix} \nearrow\!\!\!\!/ \;\; 2CeBr_3 + 3H_2O \\ \\ \searrow \;\; 2CeOBr + H_2O \end{matrix} \tag{20.21}$$

Comparisons between the sulfides and oxides are illustrative of hard–soft preferences. The M^{III} oxides and sulfides of all the lanthanides are known, with the oxides having greater thermal stabilities. However, sulfur is better able to stabilize the softer M^{II} state. Thus, europium is the only lanthanide that forms a stable monoxide, but nearly all form monosulfides (and in most case, monoselenides and monotellurides as well). Parallel behavior is observed for halides. All possible MX_3 binary compounds are known, but only fluorine is able to stabilize the tetravalent state (eg, in CeF_4, PrF_4, and TbF_4). In contrast, the bromides and iodides are preferred by M^{2+} ions; CeI_2, PrI_2, NdI_2, GdI_2, and TmI_2 are all well-characterized, but the corresponding fluorides are not.

Several of the diiodides are remarkable for their metal-like luster and high electrical conductivity. This has been explained by formulations such as $Ce^{3+}(I^-)_2e^-$. That is, these lattices are believed to contain M^{3+} ions (in spite of the stoichiometry), with two-thirds of the anionic sites occupied by iodide ions and the remainder by mobile, delocalized electrons.

The conductivity properties of a different type of lanthanide-containing salt have recently come under close scrutiny. The mixed oxide La_2CuO_4 forms an orthorhombic lattice (three unequal cell dimensions) at room temperature. This compound can be doped with Ca^{2+}, Sr^{2+}, and Ba^{2+} to give materials formulated as $La_{2-x}M_xCuO_{4-y}$, where $x \approx 0.2$ and $y \approx 0$. This substitution of $2+$ for $3+$ cations alters the lattice structure (to tetragonal) and causes a marked increase in the electrical conductivity; in fact, the resulting species behave as superconductors at low temperatures.[5]

Lanthanide Complexes

In contrast to the *d*-block elements, the lanthanides form complexes that are primarily electrostatic in nature; that is, the ion–dipole model appears to be generally superior to the Lewis acid–base (orbital overlap) theory for such species. This is due to the high charge densities of the lanthanide cations, as well as the reluctance of the $4f$ orbitals to participate in covalent bonding. Because the dominant forces are electrostatic, geometries are less well-defined (the directional requirements associated with orbital overlap are reduced), and ligand field effects are less important than for other transition metal complexes.

This is clearly evident from a plot of hydration energy versus atomic number (Figure 20.3). The variation in ΔH_{hydr} is remarkably steady for the $3+$ ions throughout the series. This is as expected for ion–dipole interactions (because of the steadily decreasing ionic radii), and is in contrast to the analogous plot for the $3d$ metals (see Figure 8.3).

The $[M(H_2O)_n]^{3+}$ complexes can be formed by simply dissolving halide or other salts in water. The coordination numbers in solution vary between about 6 and 9, depending on the identity of the metal and the method of measurement. The structures of several precipitated complexes have been reported. Both Nd and Sm form the hydrated bromate salts $M(BrO_3)_3 \cdot 9H_2O$, in which the metal ions are 9-coordinate; the water molecules form a tricapped trigonal prism about the metal. As discussed earlier, there is a general tendency for coordination number to decrease with decreasing cationic radius. Hence, the smaller Gd^{3+} ion gives $GdCl_3 \cdot 6H_2O$.

Among nonaqueous complexes, $K_3[La(NH_2)_6]$ has an octahedral environment about La^{III}, with six equidistant La–N interactions. However, the $[Ce(NO_3)_6]^{2-}$ anion in $(NH_4)_2[Ce(NO_3)_6]$ has 12-coordinate ceric ions. (Figure 20.4). The six NO_2^- groups form an approximate octahedron, but

5.   For further information, see Whangbo, M.-H.; Evain, M.; Beno, M. A.; Williams, J. M. *Inorg. Chem.* **1987**, *26*, 1829; Bednorz, J. G.; Muller, K. A. *Z. Phys.* **1986**, *64B*, 189.

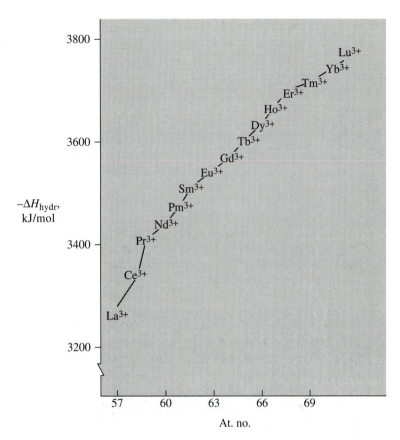

Figure 20.3 Variation of the experimental hydration enthalpies (referenced to $\Delta H_{hydr} = -1091$ kJ/mol for H^+) for the trivalent lanthanide ions. The steady increase is indicative of ion–dipole, rather than covalent, M^{3+}/H_2O interactions. [Data as compiled by Burgess, J. *Metal Ions in Solution*; Ellis Horwood: Chichester, 1978; pp. 182–183.]

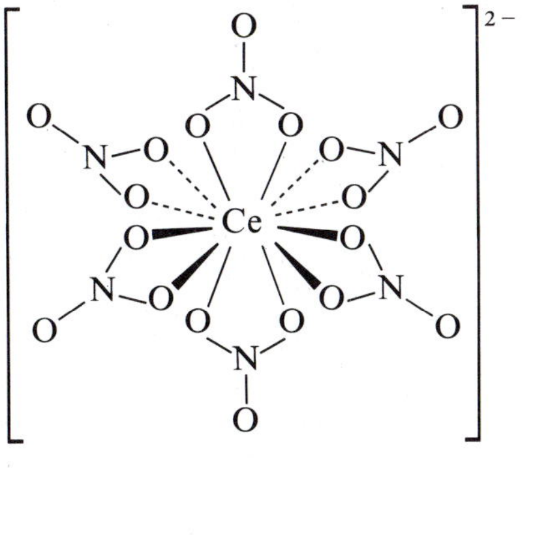

Figure 20.4
The structure of the $[Ce(NO_3)_6]^{2-}$ anion.

the Ce–N and Ce–O distances show that ligation occurs through the oxygens rather than the nitrogens. Similarly, $Ce(IO_3)_4$ has

$$M \overset{\displaystyle O}{\underset{\displaystyle O}{\diagdown}} X\text{–}O$$

subunits, which lead to a cubic environment about the metal.

The stabilities of lanthanide complexes are enhanced by the incorporation of chelates.[6] Given the hardness of the M^{3+} ions, it is not surprising that oxygen-donor chelates are especially effective. For example, the bidentate acac (acetylacetonate; see Appendix III) ligand is often used, as in $Ce(acac)_4$ and $La(acac)_3(H_2O)_2$. Common multidentate ligands include crown ethers, cryptands, and EDTA. Unlike the EDTA complexes of the $3d$ elements, those formed with lanthanide ions in aqueous solution often retain coordinated water molecules. Formulas such as $[M(EDTA)(H_2O)_n]^+$ ($n = 1$–3) are typical.

20.3 Compounds and Complexes of the Actinides

The amount of chemistry known for each actinide element is a function of its availability and cost. Thus, much more is known about thorium and uranium than the others, and, in general, the amount of published information decreases going across the series.

The early actinides (through uranium) bear greater resemblance to the d-block elements than to the lanthanides. That is because their $5f$ electrons are "active" (prone to involvement in chemical bonding), a result of the fact that they lie at higher energies than the lanthanide $4f$ electrons. However, since the effective nuclear charge experienced by these $5f$ electrons increases going across the series, their activity is significantly reduced with increasing atomic number. Hence, the heavier actinides are more like the lanthanides than they are like the $5d$ elements. For example, there is considerable similarity between $_{90}$Th ($[Rn]7s^26d^2$) and $_{72}$Hf ($[Xe]6s^25d^2$), $_{91}$Pa and $_{73}$Ta, and $_{92}$U and $_{74}$W. However, $_{98}$Cf is more similar to $_{66}$Dy than to $_{80}$Hg. Experimental evidence for this is found in oxidation states. The maximum state is the most important one for actinium through uranium (Ac^{III}, Th^{IV}, Pr^V, and U^{VI}, paralleling the $5d$ elements), after which the M^{IV} and M^{III} states dominate. This can be seen both in the binary oxides (Table 20.3) and in volt-equivalent diagrams (Figure 20.5).

---

6. Bünzli, J.-C. G.; Wessner, P. *Coord. Chem. Rev.* **1984**, *60*, 191.

Table 20.3 The best-known oxides of the early actinide elements

Oxidation State	$_{89}$Ac	$_{90}$Th	$_{91}$Pa	$_{92}$U	$_{93}$Np	$_{94}$Pu	$_{95}$Am	$_{96}$Cm
III	**Ac$_2$O$_3$**					Pu$_2$O$_3$	Am$_2$O$_3$	**Cm$_2$O$_3$**
IV		**ThO$_2$**	PaO$_2$	UO$_2$	**NpO$_2$**	**PuO$_2$**	**AmO$_2$**	CmO$_2$
				U$_4$O$_9$				
V			**Pa$_2$O$_5$**	U$_2$O$_5$	Np$_2$O$_5$			
				U$_3$O$_8$				
VI				UO$_3$				

Note: The most stable oxide for each element is shown in boldface type (U$_3$O$_8$ and UO$_3$ have approximately equal stabilities).

Uranium

Since uranium is the most important of the actinides, we will consider its chemistry in some depth. The primary natural source of uranium is *pitch-blende*, which has the empirical formula U$_3$O$_8$. It is a mixed oxide, containing both U^{IV} and U^{VI} (UO$_2 \cdot$ 2UO$_3$). This black ore can be dissolved in hot nitric

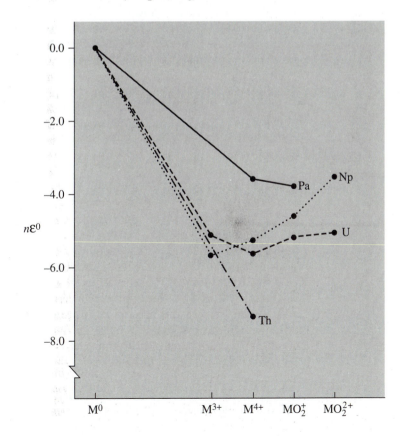

Figure 20.5
Volt-equivalent diagrams (acidic aqueous solution) for Th, Pa, U, and Np. Values are referenced to $n\mathcal{E}^0 = 0$ V for each free element.

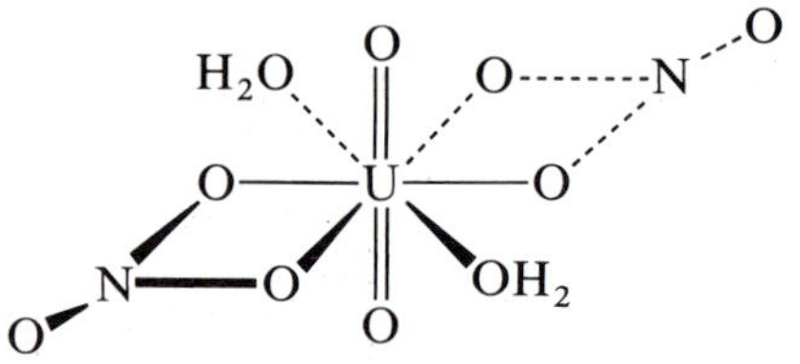

Figure 20.6
The structure of $UO_2(NO_3)_2 \cdot 2H_2O$, with hexagonal bipyramidal geometry about uranium.

or sulfuric acid, a process facilitated by forcing air into the mix to oxidize all uranium to U^{VI}. This leads to the precipitation of yellow salts of the uranyl (UO_2^{2+}) ion.

The linear UO_2^{2+} cation is quite common. A typical example is $UO_2(NO_3)_2 \cdot 2H_2O$ (Figure 20.6), in which each uranium is surrounded by eight oxygens—four from the NO_3^- ions, two from the H_2O's, and two from the UO_2^{2+} unit. The U–O distances are d(U–ONO$_2$), 252 pm; d(U–OH$_2$), 240 pm, and $d(UO_2^{2+})$, 176 pm. The significantly shorter distances in the UO_2^{2+} unit are suggestive of multiple bonding. The linearity results from the involvement of a $6p$ orbital (along with $5f$) in overlap with SALC's formed from a pair of oxygen $2p$ orbitals.[7]

Similar MO_2^{2+} cations are formed by Np, Pu, and Am. The relative stabilities decrease with increasing numbers of valence electrons, since the extra electrons populate antibonding orbitals:

$$UO_2^{2+} > NpO_2^{2+} > PuO_2^{2+} > AmO_2^{2+}$$

Uranium occurs in several different oxidation states in aqueous solution, with U^{IV} and U^{VI} being stable (Figure 20.5). However, U^V disproportionates in aqueous acid.

$$2UO_2^+ + 4H^+ \longrightarrow UO_2^{2+} + U^{4+} + 2H_2O \qquad \mathscr{E}^0 = +0.52 \text{ V} \qquad \textbf{(20.22)}$$

A number of binary uranium oxides and fluorides are known. The most stable oxide after U_3O_8 is UO_3. At least six polymorphs having this formula have been identified. The metal coordination number is either 6 or 7 in each case. One form has the ReO_3 structure (Figure 6.13), and another is suggestive of an ionic $UO_2^{2+}O^{2-}$ lattice. Also known is the monohydrate "$UO_3 \cdot H_2O$," which is more accurately formulated as $UO_2(OH)_2$. There are several polymorphs of this species, all of which have both short and long U–O bonds. For example, α-$UO_2(OH)_2$ has 8-coordinate uranium, with six U–O distances in one plane at 248 pm and two perpendicular U–O bonds (ie, a UO_2^{2+} subunit) at 171 pm.

---

7. Tatsumi, K.; Hoffmann, R. *Inorg. Chem.* **1980**, *19*, 2656.

Among the lower oxides, both UO_2 (fluorite structure) and UO (rock salt lattice) are known. Nonstoichiometric oxides also have been studied. For example, UO_2 will take up oxygen upon heating in air to give phases having the formula UO_{2+x}, where x is between 0 and 0.25. (That is, the limiting empirical formulas are UO_2 and U_4O_9.)

The uranium–fluorine system has been studied extensively, partly because UF_6 is readily volatilized (bp = 64°C at 1 atm). This makes it useful for the separation of uranium isotopes by gas diffusion, a fact of commercial and military significance; ^{235}U is a fissionable nucleus, but the more abundant ^{238}U is not.

The most important of the lower fluorides is UF_4, which is isostructural with CeF_4. This highly reactive compound can be prepared either by the thermal decomposition of $NH_4^+ UF_5^-$ or by the high-temperature fluorination of UO_2:

$$NH_4UF_5 \xrightarrow[\Delta]{} NH_4F + UF_4 \qquad \textbf{(20.23)}$$

$$UO_2 \xrightarrow[\Delta]{C_2Cl_4F_2} UF_4 \qquad \textbf{(20.24)}$$

Uranium(IV) fluoride can in turn be used to prepare a variety of derivatives (see Figure 20.7).

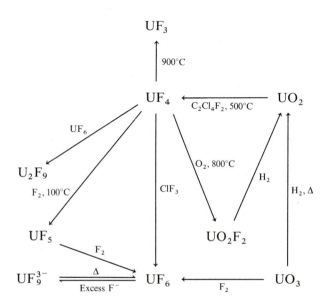

Figure 20.7
Some interconversions among uranium fluorides and oxides.

20.4 Organometallic Chemistry of the Lanthanides and Actinides[8,9]

Organometallic compounds of all the lanthanides, and of the actinides through at least berkelium, have been reported. Two generalizations can be made concerning these species:

1. The metal–carbon σ bonds are rather weak. As a result, few thermally stable homoleptic alkyls have been characterized. The carbonyls are also unstable, particularly in comparison to those of the d-block elements.

2. Complexes of cyclic π ligands are often quite stable. In such species, the bonding appears to be primarily ionic for the lanthanide metals, but covalent for the actinides.

Metal Alkyls, Aryls, and Carbonyls

The first compounds containing lanthanide–carbon σ bonds were reported in 1970.[10] They were prepared according to the equation

$$MCl_3 + 4Li^+C_6H_5^- \longrightarrow Li^+[M(C_6H_5)_4]^- + 3LiCl \qquad M = La, Pr \quad (20.25)$$

As for the d-block elements, the presence of certain substituents at carbon increases the thermal stability. For example, the C_6F_5 group was incorporated into an ytterbium complex by a metal–metal exchange reaction:[11]

$$Yb + Hg(C_6F_5)_2 \xrightarrow{\text{THF}} (C_6F_5)_2Yb(THF)_4 + Hg \qquad (20.26)$$

The first homoleptic thorium(IV) compound was prepared by the reaction of excess benzyllithium with $ThCl_4$,[12]

$$ThCl_4 + 4LiCH_2\phi \longrightarrow Th(CH_2\phi)_4 + 4LiCl \qquad (20.27)$$

8. For the lanthanides, see Schumann, H. In *Fundamental and Technological Aspects of Organo-f-Element Chemistry*; Marks, T. J.; Fragala, I. L., Eds.; D. Reidel: Dordrecht, 1985; p. 1; Schumann, H. *Angew. Chem. Int. Ed. Engl.* **1984**, *23*, 474; *Angew. Chem.* **1984**, *96*, 475; Evans, W. J. *J. Organomet. Chem.* **1983**, *250*, 217; Marks, T. J. *Prog. Inorg. Chem.* **1978**, *24*, 51.

9. For the actinides, see Fagen, P. J.; Maatta, E. A.; Manruquez, J. M.; Moloy, K. G.; Seyam, A. M.; Marks, T. J. In *Actinides in Perspective*; Edelstein, N. M., Ed.; Pergamon: Oxford, 1982; p. 433; Marks, T. J. *Science* **1982**, *217*, 989; *Prog. Inorg. Chem.* **1979**, *25*, 223.

10. Hart, F. A.; Massey, A. G.; Saran, M. S. *J. Organomet. Chem.* **1970**, *21*, 147.

11. Deacon, G. B.; Vince, D. G. *J. Organomet. Chem.* **1976**, *112*, C1.

12. Kohler, E.; Bruser, W.; Thiele, K.-H. *J. Organomet. Chem.* **1974**, *76*, 235.

and the homoleptic anions UR_6^{2-} and UR_8^{2-} (R = Me, CH_2SiMe_3, etc.) were produced from reactions of UCl_4 with excess alkyllithium reagent.[13]

All attempts to isolate lanthanide and actinide carbonyls have failed, and it is unlikely that $M(CO)_n$-type compounds are stable at room temperature. However, matrix isolation infrared spectroscopy has been used to study such species ($n = 1$–6) at 4 K. The results are suggestive of the involvement of f orbitals in the M–CO bonding, since Eu $(\ldots 4f^7)$ and Yb $(\ldots 4f^{14})$ differ in their spectral behavior from the other lanthanides. Andrews and Wayda have studied interactions between vapor-deposited Eu(s) and excess ethylene at 12 K. They found evidence for a 1:1 complex in which europium acts as a Lewis base via donation of f electron density into a π^* orbital of C_2H_4.[14]

Cyclopentadienyl Complexes

A host of $M(C_5H_5)_3$ compounds have been prepared, typically from the halides through reactions such as

$$MCl_3 + 3C_5H_5^- \longrightarrow M(C_5H_5)_3 + 3Cl^- \qquad (20.28)$$

As mentioned earlier, the bonding has a strong ionic component if M is a lanthanide element. For example, the $Sm(C_5H_5)_3$ lattice contains Sm^{3+} ions, each bound to two $C_5H_5^-$ groups; a third cyclopentadienide anion bridges adjacent $[Sm(C_5H_5)_2]^+$ units, forming an infinite chain. In contrast, the cyclopentadienyl complexes of the actinides resemble those of the d-block elements. Uranium forms both $U(C_5H_5)_3$ and $U(C_5H_5)_4$, which can be interconverted via the reaction

$$U + 3U(C_5H_5)_4 \xrightarrow{\Delta} 4U(C_5H_5)_3 \qquad (20.29)$$

The geometry about uranium in $U(C_5H_5)_4$ is approximately tetrahedral, and all twenty U–C bond distances are essentially equal (281 ± 2 pm; see Figure 20.8, p. 684).

Certain reactions of lanthanide–cyclopentadienyl complexes are indicative of their ionic character. The complexed $C_5H_5^-$ anion retains some Brønsted basicity, as evidenced by reactions such as the following:

$$M(C_5H_5)_3 + 3H_2O \longrightarrow M(OH)_3 + 3C_5H_6 \qquad (20.30)$$

$$M(C_5H_5)_3 + HCN \longrightarrow M(C_5H_5)_2CN + C_5H_6 \qquad (20.31)$$

13. Sigurdson, E. R.; Wilkinson, G. *J. Chem. Soc. Dalton* **1977**, 812.

14. Andrews, M. P.; Wayda, A. L. *Organometallics* **1988**, 7, 743.

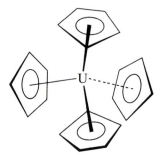

Figure 20.8
The solid-state structure of $U(C_5H_5)_4$, which has a tetrahedral arrangement of C_5H_5 rings.

Another characteristic of these complexes is their behavior toward $FeCl_2$. Covalent metal cyclopentadienyls are relatively unreactive toward metal–metal exchange; however, the (ionic) lanthanides readily form ferrocene.

$$2M(C_5H_5)_3 + 3FeCl_2 \longrightarrow 3Fe(C_5H_5)_2 + 2MCl_3 \qquad (20.32)$$

Mixed complexes containing both C_5H_5 and halide ligands are useful reagents. They can be prepared in several ways.

$$MCl_3 + 2Na^+C_5H_5^- \longrightarrow M(C_5H_5)_2Cl + 2NaCl \qquad (20.33)$$

$$2M(C_5H_5)_3 + MCl_3 \longrightarrow 3M(C_5H_5)_2Cl \qquad (20.34)$$

$$M(C_5H_5)_3 + HCl \longrightarrow M(C_5H_5)_2Cl + C_5H_6 \qquad (20.35)$$

A typical reaction of these mixed complexes is nucleophilic displacement, as described by the general equation

$$M(C_5H_5)_2Cl + M'X \longrightarrow M(C_5H_5)_2X + M'^+Cl^- \qquad (20.36)$$

$$M' = Li, Na$$

$$X = R, OR, RCO_2, NH_2, PR_2, \text{etc.}$$

Cyclooctatetraenyl and Related Complexes

The large size of cyclooctatetraene makes it an excellent ligand for the lanthanide and actinide cations. Uranocene (Figure 18.13a) and several anionic lanthanide complexes have been reported by Streitwieser and co-workers.[15]

$$UCl_4 + 2C_8H_8^{2-} \longrightarrow U(C_8H_8)_2 + 4Cl^- \qquad (20.37)$$

$$MCl_3 + 2C_8H_8^{2-} \longrightarrow [M(C_8H_8)_2]^- + 3Cl^- \qquad (20.38)$$

15. Hodgson, K. O.; Mares, F.; Starks, D. F.; Streitwieser, A. *J. Am. Chem. Soc.* **1973**, *95*, 8650; Streitwieser, A.; Müller-Westerhoff, U. *J. Am. Chem. Soc.* **1968**, *90*, 7364.

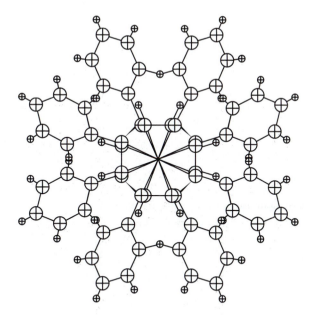

Figure 20.9
The structure of $U(\phi_4C_8H_4)_2$ (top view).
[Reproduced with permission from Templeton, D. H.; Walker, R. *Inorg. Chem.* **1976**, *15*, 3000.]

Both $U(C_8H_8)_2$ and its thorium analogue are air-sensitive; however, the incorporation of large substituents on carbon greatly reduces the reactivity. For example, 1,3,5,7-tetraphenylcyclooctatetraene can be used to prepare $U(\phi_4C_8H_4)_2$, a completely air-stable compound that also has exceptional thermal stability (to over 400°C). Its solid-state structure contains eclipsed, planar C_8 rings. The phenyl groups are rotated out of the C_8 planes, presumably to minimize the steric repulsions. There are two different uranium–carbon bond lengths; the U–Cϕ bonds are about 2% longer than the U–CH interactions (Figure 20.9).

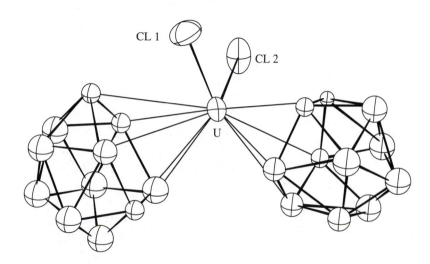

Figure 20.10
The structure of the bis(dicarbollide) complex $U(C_2B_9H_{11})_2Cl_2$.
[Reproduced with permission from Fronczek, F. R.; Halstead, G. W.; Raymond, K. N. *J. Am. Chem. Soc.* **1977**, *99*, 1769.]

Another dianion with a large open face is dicarbollide, $C_2B_9H_{11}^{2-}$ (Section 19.8). It also forms stable complexes with U^{4+}; one example is shown in Figure 20.10.

20.5 The Transactinide Elements

The ground-state electron configuration of $_{103}Lr$ is not known with certainty, but the most reasonable possibility is $[Rn]7s^25f^{14}6d^1$. The nine elements that follow Lr should have the configurations $[Rn]7s^25f^{14}6d^n$ ($n = 2$–10), and the experimental observation of nuclei having up to at least 109 protons has been claimed. The half-lives of all known isotopes of the transactinides are very short (typically 10^{-4}–10^1 s), with the stability decreasing with increasing atomic number (Figure 20.11).[16] Based on this trend, the elements

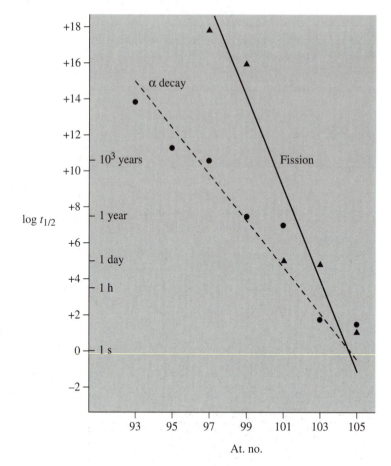

Figure 20.11

Plot of the logarithm of the half-life (in seconds) for fission and α decay for the transuranium elements. [See Fricke, B. *Struct. Bonding* **1975**, *21*, 89.]

---

16. Fricke, B. *Struct. Bonding* **1975**, *21*, 89.

beyond $_{105}$Unp should have half-lives of less than 1 s toward both α decay and spontaneous fission.

However, there is a theoretical basis for predicting a dramatic increase in nuclear stability in the vicinity of $_{114}$Uuq. The $^{298}_{114}$Uuq isotope has magic numbers of both neutrons and protons, and has therefore been a source of considerable interest. Seaborg has suggested two possible modes of nuclear synthesis:[17]

$$^{238}_{92}U + {}^{238}_{92}U \xrightarrow{?} {}^{298}_{114}Uuq + {}^{170}_{70}Yb + 8\,^{1}_{0}n \qquad (20.39)$$

$$^{238}_{92}U + {}^{136}_{54}Xe \xrightarrow{?} {}^{298}_{114}Uuq + {}^{72}_{32}Ge + 4\,^{1}_{0}n \qquad (20.40)$$

Element 114 would lie below lead in the periodic table ($\ldots 7p^2$), and various estimates have been made of its physical and chemical properties on that basis (see Table 20.4). From the table, we can see that $_{114}$Uuq is expected to have a higher melting temperature and to be larger and more easily oxidized than lead. It has also been suggested that, since the energy match between the $7s$ and $7p$ orbitals is not close, the importance of the tetravalent (sp^3 hybridized) state will be reduced. Hence, divalent (and perhaps hexavalent) behavior is likely.[18]

Table 20.4 Some predicted physical and chemical properties of the yet-to-be-synthesized element $_{114}$Uuq compared to those of lead

Property	$_{114}$Uuq		$_{82}$Pb
	Seaborg	Keller et al.	
At. wt, g/mol	298	298	207.2
d, g/cm^3	14	14	11.35
mp, °C	700	670	327
r_{met}, pm	185	185	175
r (M^{2+}), pm	131	120	133
IE_1, eV	8.5	8.5	7.42
IE_2, eV	16.8	16.8	15.03
$\mathscr{E}^0$ (M^{2+}/M), V	-0.8	-0.9	-0.126

Source: Estimates are from Seaborg, G. T. *J. Chem. Educ.* **1969**, *46*, 626; and Keller, O. L.; Burnett, J. L.; Carlson, T. A.; Nestor, C. W. *J. Phys. Chem.* **1970**, *74*, 1127.

---

17. Seaborg, G. *J. Chem. Educ.* **1969**, *46*, 626.

18. Fricke, B.; Waber, J. T. *J. Chem. Phys.* **1972**, *56*, 3726; Penneman, R. A.; Mann, J. B.; Jørgensen, C. K. *Chem. Phys. Lett.* **1971**, *8*, 321.

Looking beyond $_{114}$Uuq, the next magic number of protons is 184. The synthesis of $_{184}$Uoq is at best many years away, but speculation concerning its preparation and properties has already begun.[19]

Bibliography

Katz, J. J.; Seaborg, G. T.; Morss, L. R. *The Chemistry of the Actinide Elements*, 2nd ed.; Chapman and Hall: New York, 1986.

Freeman, A. J.; Keller, C., Eds. *Handbook on the Physics and Chemistry of the Actinides*; North-Holland: Amsterdam, 1985.

Marks, T. J.; Fragala, I. L., Eds. *Fundamental and Technological Aspects of Organo-f-Element Chemistry*; D. Reidel: Dordrecht, 1985.

Gschneidner, K. A.; Eyring, L., Eds. *Handbook on the Physics and Chemistry of the Rare Earths*; North-Holland: Amsterdam, 1984.

Erdos, P.; Robinson, J. M. *The Physics of Actinide Compounds*; Plenum: New York, 1983.

Edelstein, N. M., Ed. *Actinides in Perspective*; Pergamon: Oxford, 1982.

McCarthy, G. J.; Silber, H. B.; Rhyne, J. J., Eds. *The Rare Earths in Modern Science and Technology*; Plenum: New York, 1982.

Seaborg, G. T. *Transuranium Elements: Products of Modern Alchemy*; Academic: New York, 1978.

Bagnall, K. W., Ed. *Lanthanides and Actinides*; Butterworths: London, 1972.

Questions and Problems

1. Predict the ground-state electron configuration for:
   (a) Pr^{3+} (b) Eu^{2+} (c) Bk   (d) Bk^{4+}
   (e) Lr   (f) $_{107}$Uns   (g) Uns^{3+}

2. The relative abundances of the three naturally occurring actinides are Th > Pa < U. Rationalize. (A review of Section 7.1 may be useful.)

3. Suggest plausible nuclear syntheses for:
   (a) $_{100}^{252}$Fm from $_{98}^{249}$Cf   (b) $_{97}^{248}$Bk from $_{96}^{247}$Cm

4. From among the three elements Th, U, and Np, predict which one has:
   (a) The most stable $6p$ orbital   (b) The smallest IE_1
   (c) The largest metallic radius   (d) No significant M^V chemistry

---

19. See footnotes 16–18.

5. Explain in your own words why the elements having atomic numbers between 57 and 71 show marked similarities in their physical and chemical properties.

6. Identify lanthanide elements that form especially stable species having the following formulas, and briefly explain:
   (a) MF_4 (b) MI_2 (c) $[M(H_2O)_9]^{3+}$

7. A partial potential diagram for europium in acid is

$$Eu^{3+} \xrightarrow{\ -0.35\ } Eu^{2+} \xrightarrow{\ ?\ } Eu$$

with -1.98 connecting Eu^{3+} to Eu.

   (a) Calculate the standard reduction potential for the Eu^{2+}/Eu couple.
   (b) Is Eu^{2+} prone to disproportionation?
   (c) Examine Table 9.1. Find a reducing agent comparable in strength to Eu^{2+}.
   (d) Again using Table 9.1, find an oxidizing agent comparable in strength to Eu^{2+}.

8. The hydrolysis constant, pK_h, for Lu^{3+} is 6.6.
   (a) To what $3d$ element is this the most comparable? To what Group 2 element?
   (b) Calculate the pH of a 0.1 M Lu^{3+} solution.
   (c) Is the K_h of La^{3+} greater or smaller than that of Lu^{3+}? Rationalize.

9. Predict whether the position of equilibrium favors the reactants or the products for the following reactions. Briefly defend your answers.

   (a) $CeF_4 + LaF_3 \rightleftharpoons CeF_3 + LaF_4$
   (b) $LaBr_3 + AcF_3 \rightleftharpoons LaF_3 + AcBr_3$
   (c) $Ce^{4+}(aq) + Sm^{2+}(aq) \rightleftharpoons Ce^{3+}(aq) + Sm^{3+}(aq)$
   (d) $Nd_2O_3 + Yb_2S_3 \rightleftharpoons Nd_2S_3 + Yb_2O_3$

10. When Eu or Yb is added to anhydrous liquid ammonia, blue solutions characteristic of solvated electrons result, but that is not the case for the other lanthanides. Rationalize.

11. Explain in your own words how and why plots of ΔH_{hydr} versus atomic number differ for the $4f$ and the $3d$ elements.

12. Uranium has a hexagonal bipyramidal environment in the $[UO_2(C_2O_4)_3]^{4-}$ anion. Sketch the structure of this ion. Will all U–O bonds be of equal length? Why or why not?

13. The term "uranate" is loosely used, but often refers to species containing the UO_4^{2-} moiety (suggesting similarity to sulfate). An important difference between SO_4^{2-} and UO_4^{2-}, however, is that in the latter there are usually two different U–O bond distances. Discuss.

14. Predict products for the following reactions:

   (a) $U(C_5H_5)_3Cl + LiPEt_2 \longrightarrow$
   (b) $La(C_5H_5)_3 + Excess\ H_2O \longrightarrow$

(c) $(C_5H_5)_2LaCl + NaOC_2H_5 \longrightarrow$

(d) $Sm(C_5H_5)_3 + FeCl_2 \longrightarrow$

15. Consider the following enthalpies of formation: UF_4, -1509 kJ/mol; UF_6, -2197 kJ/mol; UCl_4, -1019 kJ/mol; UCl_6, -1092 kJ/mol. Calculate ΔH for the reaction $UX_6 \to UX_4 + X_2$ (X = F and Cl). Compare your two answers (ie, rationalize their relative values).

16. Several different modes of complexation by cyclooctatetraene were shown in Figure 18.13. However, in this chapter only one mode is evidenced. Which one is it, and why is this the case?

17. Give the atomic number of the lightest element expected to contain a ground-state:
    (a) $8s$ electron   (b) $5g$ electron   (c) $6f$ electron   (d) $7d$ electron

18. Calculate the n/p ratio of the doubly magic isotope of Uuq. Relate your answer to the discussion of optimal n/p values given in Chapter 7. Is that ratio reasonable?

19. (a) Give the atomic number of the next Group 18 element.
    (b) Seaborg suggested that that element might be a noble *liquid* at room temperature. Find appropriate experimental data to support or refute that notion.

20. Give the atomic number of the next alkali metal. After considering data for the other Group 1 metals, prepare a table similar to Table 20.4 in which you predict properties of that element. (Choose whatever properties you think are most relevant.)

*21. The different affinities of crown ethers for the lanthanide elements have been studied by J.-C. G. Bünzli and F. Pilloud (*Inorg. Chem.* **1989**, *28*, 2638).
    (a) How do the formation constants for complexation by 15-crown-5 vary throughout the lanthanides?
    (b) Based on Figure 7 of the article, which crown ether has the greatest formation constant with La^{III}? With Lu^{III}?
    (c) What is the explanation for these observations?

*22. The complex $UI_3(THF)_4$ has been prepared and studied (Clark, D. L.; Sattelberger, A. P.; Bott, S. G.; Vrtis, R. N. *Inorg. Chem.* **1989**, *28*, 1771).
    (a) How was the complex made? What is its geometry?
    (b) Give the product of its reaction with $Me_2NCH_2CH_2NH_2$.
    (c) The authors describe $UI_3(THF)_4$ as a useful precursor for other interesting species, and demonstrate by converting it to several derivatives. What species (other than those described) would you choose for reaction with it, and why?

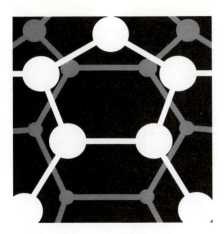

21

Instrumental Methods in Inorganic Chemistry

The purpose of this chapter is to provide a sense of which instrumental techniques are of greatest importance in contemporary inorganic chemistry, and to summarize the kinds of information obtained from each. Inorganic chemistry is so diverse that it is impossible to be comprehensive in this regard. To keep the chapter to a reasonable length, some techniques are given only a cursory treatment; others of considerable importance to chemists working in certain areas, such as mass spectrometry and X-ray diffraction, are omitted entirely.

It is assumed that you have a basic knowledge of the theory and instrumentation behind each method. Any of several general texts may be consulted for review as necessary.[1] For certain areas not normally covered in undergraduate instrumental analysis courses (eg, the electronic spectra of transition metal complexes), the relevant theory is discussed in more detail.

1. Willard, H. H.; Merritt, L. L.; Dean, J. A.; Settle, F. A. *Instrumental Methods of Analysis*, 7th ed.; Wadsworth: Belmont, CA, 1988; Rubinson, K. A. *Chemical Analysis*; Little, Brown: Boston, 1987; Christian, G. D.; O'Reilly, J. E. *Instrumental Analysis*, 2nd ed.; Allyn & Bacon: Boston, 1986; Skoog, D. A.; Leary, J. J. *Principles of Instrumental Analysis*, 4th ed.; Saunders: Philadelphia, 1992.

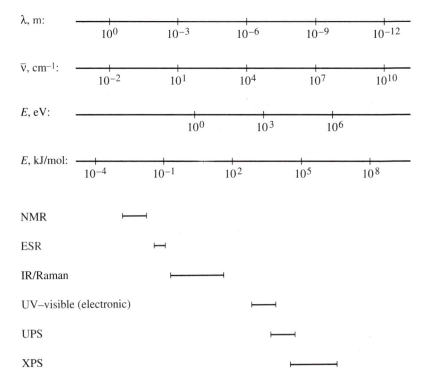

Figure 21.1
Wavelengths and approximate energy ranges associated with selected instrumental techniques.

In Sections 21.1–21.3, we will discuss techniques based on the absorption or emission of electromagnetic radiation. Beginning in Section 21.4, instrumental techniques based on magnetic properties are considered.

The regions of the electromagnetic spectrum include, in order of increasing energy, the radiowave, microwave, infrared, visible/ultraviolet, and X-ray domains. Each is associated with one or more spectroscopic methods (see Figure 21.1).

21.1   Vibrational (Infrared and Raman) Spectroscopy

As can be seen from Figure 21.1, infrared wavelengths fall between about 1 mm and 2.5 μm. This corresponds to an energy range of about 0.1–50 kJ/mol, 0.001–0.5 eV, or 10–4000 cm^{-1}.[2] These energies are generally insufficient to cause electronic transitions or ionizations or to rupture chemical bonds. Instead, their absorption typically results in bond deformations. Such vibrational motion is divided into two modes: *stretching* (oscillations in internuclear distances) and *bending* (vibrations involving changes in bond angles).

---

2. Factors for conversions among various energy units are given in Appendix I.

Bond Stretching in Diatomics[3]

An equation fundamental to vibrational spectroscopy is

$$\bar{v} = \frac{1}{2\pi c}\left(\frac{k}{\mu}\right)^{1/2} \tag{21.1}$$

where $\bar{v}$ is the vibrational frequency in wavenumbers, k is the *force constant* (indirectly related to the bond strength; see below), and μ is the reduced mass [$\mu = m_a \cdot m_b/(m_a + m_b)$, where m_a and m_b are the masses of the involved atoms]. Thus, the experimental vibrational frequency of $F_2(g)$ ($892\ cm^{-1}$) corresponds via equation (21.1) to a force constant of 4.46×10^5 dyn/cm (or 4.46 mdyn/Å).

An isotope effect is inherent in the reduced mass term. Consider the three species H_2, HD, and D_2, with reduced masses of 0.500, 0.667, and 1.00 g/mol, respectively. The three force constants are approximately equal, but the vibrational frequencies are very different:

Species	*k, mdyn/Å*	$\bar{v},\ cm^{-1}$
H_2	5.70	4395
HD	5.73	3817
D_2	5.74	3118

Such isotope effects are useful in structural analysis (see below).

Experimental vibrational frequencies and force constants for a variety of diatomics are given in Table 21.1 (p. 694). Although not a perfect indicator of bond strength, the variation in force constants is usually consistent with expectations based on bond energies. For example, it can be seen from Figure 21.2 (p. 695) that the relationship between $D(H–X)$ and k is approximately linear for the hydrogen halides. Series such as O_2^+, O_2, O_2^- also can be cited.

However, the dissociation energy of Cl_2 is about 60% greater than that of F_2, but fluorine's force constant is the larger of the two. This is because the force constant is actually a measure of the *rate of energy change* with changing internuclear distance, rather than of the bond energy itself. (Said another way, it is the degree of curvature of the potential energy well, rather than its depth.) The force constant for F_2 is large because the strength of the F–F interaction is unusually sensitive to the internuclear distance—a consequence of the severe adjacent atom lone-pair repulsions in that molecule.

---

3. Goodfriend, P. L. *J. Chem. Educ.* **1987**, *64*, 753.

Table 21.1 Vibrational frequencies and force constants for selected diatomic molecules and ions

Homonuclear Diatomics	$\bar{v}$, cm^{-1}	k, mdyn/Å	Heteronuclear Diatomics	$\bar{v}$, cm^{-1}	k, mdyn/Å
H_2^+	2297	1.56	LiH	1406	1.02
H_2	4395	5.70	NaH	1172	0.78
Li_2	351	0.25	KH	985	0.56
Na_2	159	0.17	HF	4138	9.66
K_2	93	0.10	HCl	2991	5.16
Rb_2	57	0.08	HBr	2650	4.12
Cs_2	42	0.07	HI	2310	3.12
N_2	2360	22.98	OH$^-$	3637	7.35
P_2	780	5.56	LiF	867	2.27
O_2^+	1865	16.39	LiCl	569	1.11
O_2	1580	11.77	LiBr	512	0.99
O_2^-	1097	5.68	LiI	433	0.72
S_2	726	4.98	NO$^+$	2273	22.76
F_2	892	4.45	NO	1880	15.55
Cl_2	546	3.19	NO$^-$	1350	7.03
Br_2	319	2.46	CO	2138	18.47
I_2	215	1.76	CN$^-$	2080	16.49

Note: Values are for diatomics containing the most abundant isotope of each element.
Source: Data taken primarily from that compiled by Nakamoto, K. *Infrared and Raman Spectra of Inorganic and Coordination Compounds*, 4th ed.; Wiley: New York, 1986; pp. 101–106.

Small Polyatomics: Normal Modes of Vibration[4]

Consider the possible vibrational deformations that a water molecule might undergo. Any movement can be described by assigning a set of *x*, *y*, and *z* coordinates to each atom. (This amounts to resolving vectors in three-dimensional space.) Thus, there are (3 atoms) × (3 dimensions) = 9 total *degrees of freedom*. However, three of these are *translational* (they change the location of the molecule without altering the internuclear distances or bond angle), while three others are *rotational*. The number of *normal vibrational modes* is therefore (3 × 3) − 6 = 3. In general, the normal modes equal $3n - 6$ (where *n* is the number of atoms) for nonlinear and $3n - 5$ for linear species.

One restriction on these vibrational modes is that they may not change the center of gravity of the molecule or ion. Given that limitation, the three modes for H_2O can be uniquely determined, and are diagrammed in Figure 21.3a. The first and third modes (v_1 and v_3) involve changes in the bond

---

4. Lacey, A. R. *J. Chem. Educ.* **1987**, *64*, 756.

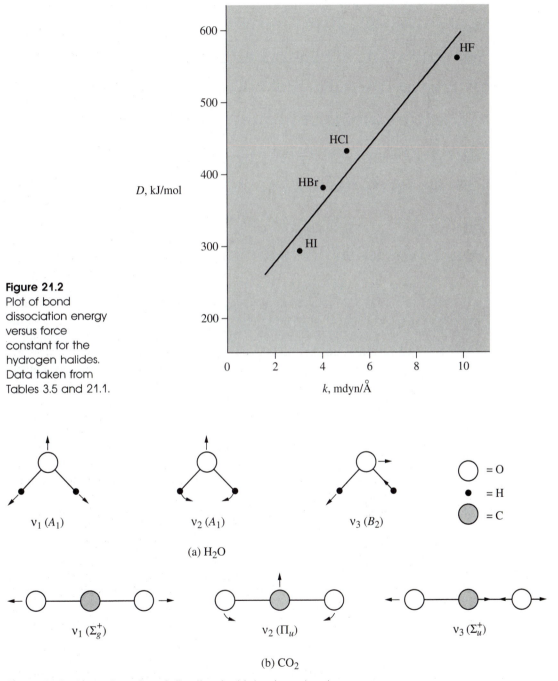

Figure 21.2
Plot of bond dissociation energy versus force constant for the hydrogen halides. Data taken from Tables 3.5 and 21.1.

Figure 21.3 Normal modes of vibration for triatomic molecules. ν_1 = symmetric stretch; ν_2 = bend; ν_3 = asymmetric stretch. The bending mode is doubly degenerate in CO_2 and other linear triatomics.

Table 21.2 Energies of the vibrational modes of H_2O as a function of sample state and isotope of hydrogen

Species	$v_1(A_1)$, cm^{-1}	$v_2(A_1)$, cm^{-1}	$v_3(B_2)$, cm^{-1}
H_2O(g)	3657	1595	3756
H_2O(l)	3450	1640	3615
H_2O(s)	3400	1620	3220
D_2O(g)	2671	1178	2788
D_2O(s)	2495, 2336	1210	2432
T_2O(g)	—	996	2370

Source: Data as compiled by Nakamoto, K. *Infrared and Raman Spectra of Inorganic and Coordination Compounds*, 4th ed.; Wiley: New York, 1986; p. 115.

distances, while v_2 is a pure bending mode. Thus, v_2 requires the lowest energy of the three. The locations of the infrared bands corresponding to these vibrations depend on the isotopes and on the physical state (since intermolecular interactions influence bond energies); see Table 21.2.

The symmetries of these vibrations can be assigned with the aid of the C_{2v} character table (Appendix IV) in the manner described in Chapter 2. This is important for the following reason: It can be shown via quantum mechanics and group theory that all molecular vibrations are not equally likely to occur; *selection rules* determine the probability that photons of a given energy will be absorbed. As a result, vibrational modes are often described as "active" or "allowed" (relatively probable) or as "inactive" or "forbidden" (relatively improbable). Vibrations are allowed in infrared spectroscopy if they involve a change in the dipole moment. In Raman spectroscopy, they are allowed if a change in the molecular polarizability occurs.[5] Since many vibrations affect the dipole but not the polarizability (or vice versa), these two techniques are, to a large extent, complementary.

Character tables provide a facile way to identify allowed versus forbidden vibrations. The x, y, and z vectors on the right side of the appropriate character table identify the infrared-active symmetries. For example, in the C_{2v} point group, the x vector transforms into the B_1 irreducible representation, y into B_2, and z into A_1. Vibrations having A_1, B_1, or B_2 symmetry are therefore infrared-active. For Raman spectroscopy, the binary functions (xy, xz, yz, x^2, y^2, z^2, and/or their combinations or differences) identify the active representations. The binary functions span all the irreducible representations in C_{2v} symmetry. As it happens, then, all the normal

---

5. These selection rules are discussed in Cotton, F. A. *Chemical Applications of Group Theory*, 3rd ed.; Wiley: New York, 1990; Chapter 10.

vibrations of H_2O are both infrared- and Raman-active, which is an unusual situation.

Linear triatomics such as CO_2 have $(3 \times 3) - 5 = 4$ fundamental modes (Figure 21.3b). One of these (the "symmetric stretch," v_1) is inactive in the infrared, while v_2 and v_3 are active. (Use the $D_{\infty h}$ character table to verify this.) Two of the fundamentals—the bending modes, having Π_u symmetry—are degenerate.

The N_2O_5 molecule has been studied by vibrational spectroscopy.[6] The solid-state structure of this compound is consistent with the ionic formulation $NO_2^+NO_3^-$ (see Figure 14.7). This is supported by the spectra, which contain bands near 1400 and 1050 cm^{-1}; these absorptions correspond to the symmetric stretching modes of the NO_2^+ and NO_3^- ions, respectively. (The higher vibrational frequency for NO_2^+ is consistent with its greater average bond strength.) Among the other absorptions, the bending mode of NO_2^+ appears at 534 cm^{-1}. The symmetry of that vibration (Π_u) should cause it to be infrared-active but Raman-inactive. However, it has been found to have a high intensity in both spectra. This is taken as evidence that the O–N–O linkage is slightly nonlinear. (For any bond angle except 180°, the point group is C_{2v} rather than $D_{\infty h}$.)

Infrared data for several triatomic systems are given in Table 21.3. As for diatomics, there is a general correlation between force constants and bond strengths. For example, the fundamental vibrations of CO_2 and SF_2 lie at higher energies and yield larger force constants than do the comparable bands of CS_2 and SCl_2.

Table 21.3 Vibrational frequencies for selected triatomics

Linear Triatomics	v_1, cm^{-1}	v_2, cm^{-1}	v_3, cm^{-1}	Angular Triatomics	v_1, cm^{-1}	v_2, cm^{-1}	v_3, cm^{-1}
BeF_2	680	345	1555	SnF_2	593	197	571
$BeCl_2$	390	250	1135	PbF_2	531	165	507
XeF_2	497	213	555	$PbCl_2$	297	—	321
CO_2	1337	667	2349	SF_2	825	358	799
CS_2	658	397	1533	SCl_2	518	208	526
NNO	2224	589	1285	NO_2	1318	749	1610
I_3^-	114	52	145	NH_2^-	3270	1556	3323
				H_2S	2615	1183	2627
				O_3	1135	716	1089

Source: Data taken primarily from that compiled by Nakamoto, K. *Infrared and Raman Spectra of Inorganic and Coordination Compounds*, 4th ed.; Wiley: New York, 1986; pp. 110–115.

---

6. Wilson, W. W.; Christie, K. O. *Inorg. Chem.* **1987**, *26*, 1631.

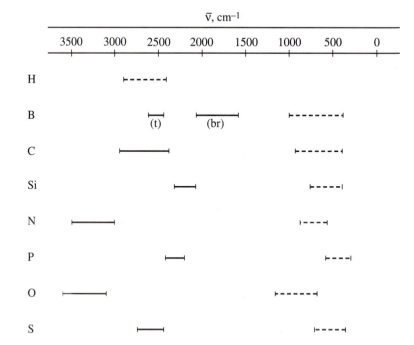

Figure 21.4
Group stretching frequencies for selected bonds to hydrogen and chlorine: ——— = H; – – – = Cl; (t) = terminal; (br) = bridging.

Group Frequencies

The complexity of normal coordinate analysis increases rapidly with increasing numbers of atoms. As a result, normal coordinate analysis is not feasible for large molecules of low symmetry. In such cases, *functional group frequencies* provide useful structural information. (This is the normal use of infrared spectroscopy in organic chemistry.) Group frequencies for inorganic hydrides and chlorides are shown in Figure 21.4. A range is given for each bond type, because the location of a given vibrational band varies with the nature of the bonding. For example, in borane and heteroborane clusters, bridging hydrogens (B–H–B units) typically have stretching frequencies between 1600 and 2100 cm^{-1}; hence, they are readily distinguished from terminal B–H stretching vibrations (2450–2650 cm^{-1}).

For difficult problems, isotope effects can be used in conjunction with group frequencies. As an example, silicon substituted "amides" have two different isomeric forms—the amide and imidate structures:

$$\text{Amide} \quad \underset{?}{\rightleftharpoons} \quad \text{Imidate} \tag{21.2}$$

We might expect vibrational spectroscopy to readily differentiate between these possibilities. Unfortunately, the C=O and C=N group frequencies fall into approximately the same region (1600–1750 cm^{-1}), as do Si–N and Si–O stretching vibrations (850–1000 cm^{-1}). The problem can be solved by isotopic labeling. The spectra of the ^{14}N- and ^{15}N-labeled compounds (Me$_3$Si)$_2$NC(O)R (R = H and Me) differ for certain bands.[7] The formamide (R = H) gives a band at 1659 cm^{-1} regardless of the nitrogen isotope; hence, it has been assigned to the C=O stretching vibration. However, a lower-energy absorption moves from 983 cm^{-1} (^{14}N) to 961 cm^{-1} (^{15}N), identifying it as a silicon–nitrogen vibration. Thus, the data are indicative of the amide isomer for R = H. The same approach gives a different result for the acetamide, whose structure is more correctly written Me$_3$SiN=C(Me)OSiMe$_3$.

Metal Carbonyls and Related Species[8]

The vibrational spectra of metal carbonyls have been studied in detail, since the C–O stretching region is a useful indicator of the character of the

$$\overset{\displaystyle O}{\overset{\displaystyle \|}{}}$$

metal–ligand bonding. Bridging carbonyls (M–C–M in Lewis formulation) absorb in approximately the same area as C=O double bonds in organic compounds (typically 1700–1900 cm^{-1}), while the frequencies for terminal carbonyls lie at higher energies (1800–2100 cm^{-1}; compare to 2138 cm^{-1} for free carbon monoxide).

The C–O stretching frequency decreases as the M–C bond strength increases through metal-to-ligand π back-bonding. This can be understood through two resonance structures:

$$^{\ominus}\ddot{M}\!\!-\!\!C\!\!\equiv\!\!\overset{\oplus}{O}\!: \quad\longleftrightarrow\quad M\!\!=\!\!C\!\!=\!\!\ddot{O}\!:$$

The relationship between back-bonding and band location is evident in Table 21.4. Metal-to-ligand π donation increases with increasing negative charge on the metal, as in the isoelectronic series Ni(CO)$_4$, [Co(CO)$_4$]$^-$, [Fe(CO)$_4$]$^{2-}$. This is reflected by both a decrease in v_{CO} and an increase in v_{MC}. Octahedral carbonyls and cyanides show similar trends.

Normal coordinate analysis is useful for interpreting the vibrational spectra of symmetrical carbonyls.[9] This method often permits distinction

7. Yoder, C. H.; Copenhafer, W. C.; DuBeshter, B. *J. Am. Chem. Soc.* **1974**, *96*, 4283; Yoder, C. H.; Bonelli, D. *Inorg. Nucl. Chem. Lett.* **1972**, *8*, 1027.

8. Braterman, P. S. *Struct. Bonding* **1976**, *26*, 1; *Metal Carbonyl Spectra*; Academic: New York, 1975.

9. Darensbourg, M. Y.; Darensbourg, D. J. *J. Chem. Educ.* **1974**, *51*, 787; *J. Chem. Educ.* **1970**, *47*, 33.

Table 21.4 Stretching frequencies for some metal carbonyl and metal cyanide complexes

Compound/Ion	ν_{CO}, cm^{-1}	ν_{MC}, cm^{-1}	Ion	ν_{CN}, cm^{-1}
$[V(CO)_6]^-$	2020, 1894, 1858	460, 393, 374	$[Mn(CN)_6]^{3-}$	2129, 2112
$Cr(CO)_6$	2119, 2027, 2000	440, 391, 379	$[Mn(CN)_6]^{4-}$	2082, 2066, 2060
$Mo(CO)_6$	2120, 2025, 2000	391, 381, 367	$[Fe(CN)_6]^{3-}$	2135, 2130, 2118
$W(CO)_6$	2126, 2021, 1998	426, 410, 374	$[Fe(CN)_6]^{4-}$	2098, 2062, 2044
$[Mn(CO)_6]^+$	2192, 2125, 2095	412, 390, 384	$[Co(CN)_6]^{3-}$	2150, 2137, 2129
$[Fe(CO)_4]^{2-}$	1788	644, 464	$[Ir(CN)_6]^{3-}$	2167, 2143, 2130
$[Co(CO)_4]^-$	2002, 1890	556, 431		
$Ni(CO)_4$	2131, 2058	421, 368		

Source: Taken from data compiled by Nakamoto, K. *Infrared and Raman Spectra of Inorganic and Coordination Compounds*, 4th ed.; Wiley: New York, 1986; pp. 273 and 292–293.

between geometric isomers—in general, the more symmetric the species, the smaller the number of allowed vibrations. For example, in $M(CO)_4L_2$ complexes, the cis isomer (idealized C_{2v} point group) is predicted to give four symmetry-allowed infrared bands in the CO stretching region, while the trans (D_{4h}) isomer gives only one such band.

21.2 Electronic (Ultraviolet–Visible) Spectroscopy

The absorption of an appropriate quantum of energy may result in the movement of an electron from one orbital to another (or, more properly stated, may alter the electronic state of a system). The required energy may fall into the near-infrared, visible, or ultraviolet region. The $3s$ and $3p$ orbitals of atomic sodium are relatively close in energy, and the $3s \rightarrow 3p$ electron transition lies in the visible region. For molecular transitions of the $\sigma \rightarrow \sigma^*$ type, the bonding–antibonding energy gap is usually greater, so the corresponding absorption bands are generally found in the ultraviolet.

As in vibrational spectroscopy, quantum mechanics can be used to identify selection rules. Such rules provide information about the probability of a given electronic transition, which in turn determines the band's intensity. The selection rules for electronic spectroscopy are listed below:

1. A transition is allowed only if it involves a change in the orbital quantum number (specifically, $|\Delta\ell| = 1$). Thus, $s \rightarrow p$ and $d \rightarrow p$ are examples of allowed transitions, but $p \rightarrow p$ and $d \rightarrow d$ are forbidden. This explains some familiar chemical phenomena, such as the great intensity of the yellow-orange color that results when a sodium-containing material is placed in a flame; the allowed $3s \rightarrow 3p$ transition occurs at 590 nm, in

the visible region. Conversely, the less intense colors of many transition metal salts result from orbitally forbidden $d \to d$ absorptions.

2. Only one-electron transitions are allowed. Thus, for calcium, the $3s^2 \to 3s^1 3p^1$ transition is allowed, but $3s^2 \to 3p^2$ is forbidden.

3. Spin multiplicity is maintained for allowed transitions ($\Delta S = 0$). That is, an electron may not "flip its spin" while undergoing excitation:

Spin allowed   Spin forbidden

Because of this rule, the only accessible excited states from a doublet ground state are also doublet states, and while singlet $\to$ triplet transitions are sometimes observed, their band intensities are low.

4. For species having an inversion center, the *parity* changes during an allowed transition. That is, $g \to u$ and $u \to g$ transitions are allowed, while $g \to g$ and $u \to u$ transitions are forbidden.[10] This is often called *Laporte's rule*. A practical consequence is that highly symmetric species normally absorb less strongly than similar, less symmetric ones. For example, the $d \to d$ absorption bands of the octahedral complex $[Ni(NH_3)_6]^{2+}$ are doubly forbidden (by both the orbital and parity rules), and so are quite weak. Replacement of one NH_3 group by a different ligand to give $[Ni(NH_3)_5 L]^{n+}$ yields a complex with no inversion center, so the band strengths are usually enhanced.

As a very general guide, each violation of a selection rule reduces the band intensity by a factor of 10^1–10^2. A completely allowed transition typically has a molar absorptivity of 10^4 L/mol·cm or greater. A value of about 10^2 is more typical of a $d \to d$ transition unless it is also forbidden by spin and/or parity rules; in that case, the absorptivity may drop to 10 L/mol·cm or less.

The Spectra of Transition Metal Complexes

The primary uses of ultraviolet–visible spectroscopy in inorganic chemistry involve electron transitions in metal complexes. To understand the absorption bands that arise for such species, consider the MO diagram shown in

10. The symbols g and u stand for the German words *gerade* (even) and *ungerade* (odd); see Section 2.5 for a more complete explanation.

Figure 21.5
Simplified MO
diagram and types
of possible electron
transitions for the
hypothetical d^1
complex $[TiL_6]^{3+}$:
(a) $d \rightarrow d$ transition;
(b) metal-to-ligand
charge transfer;
(c) ligand-to-metal
charge transfer.

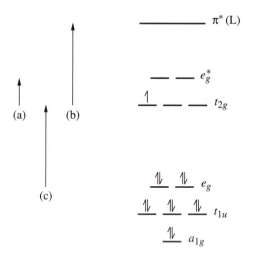

Figure 21.5 for the hypothetical octahedral complex TiL_6^{3+}.[11] Three kinds of transitions are illustrated. The lowest-energy possibility (a) involves excitation from the HOMO to the LUMO (the $t_{2g} \rightarrow e_g^*$ transition). Since both of these orbital levels are primarily of metal d orbital parentage, this transition is forbidden by the orbital selection rule, and the absorption band will be relatively weak.

If the ligand L is of the π acceptor type, then a second kind of transition [labeled (b), from the t_{2g} to the π^* level] might occur. This transition changes the metal–ligand charge distribution, since electron density is transferred from an orbital of high metal character to one of high ligand character. Hence, this is a *metal-to-ligand charge transfer* band. It is allowed by the orbital selection rule ($\Delta \ell = 1$, assuming that π^* has p-type parentage). Such bands tend to be strong ($\epsilon = 10^3 – 10^4$ L/mol·cm), and are usually found at higher energy than the $d \rightarrow d$ transition(s).

Excitation from an occupied orbital other than the HOMO is sometimes observed. An example is the $e_g \rightarrow t_{2g}$ transition (c) shown in Figure 21.5, which alters the charge density in the opposite direction from (b); that is, it is *ligand-to-metal charge transfer*. Such bands may or may not be observed for a given complex, since they are very sensitive to the ionization energy of the ligand. For example, $[Cr(NH_3)_6]^{3+}$ does not give any charge transfer (CT) bands within the normal range of spectrophotometric study. However, in $[Cr(NH_3)_5Cl]^{2+}$ there is a CT absorption at about 34,400 cm^{-1}. (The ionization energy of Cl$^-$ is considerably less than that of NH$_3$.) Similar bands are observed in the corresponding bromide and iodide complexes, with the energy maxima decreasing in the order Cl$^-$ > Br$^-$ > I$^-$.

---

11. Ti^{3+} (d^1) complexes exhibit Jahn–Teller distortion, and so are not truly octahedral.

Oxide and sulfide ions also have low ionization energies. As a result, metal oxides, metal sulfides, and complexes having O^{2-} or S^{2-} as ligands often exhibit charge transfer bands. Two well-known examples are the chromate and permanganate ions. The fact that their CT absorptions occur partly in the visible region explains the intense colors (yellow for CrO_4^{2-} and purple for MnO_4^-) of these anions.

d^1 and d^9 Complexes: Spectra Influenced by Tetragonal Distortion

Based on Figure 21.5 we might expect an "octahedral" d^1 complex to exhibit only one $d \rightarrow d$ transition; the conventional symbolism for that band would be $^2T_{2g} \rightarrow {}^2E_g$.[12] However, hexacoordinate d^1 complexes are subject to Jahn–Teller distortion (Chapter 16), which lowers the local symmetry to D_{4h}. Thus, the qualitative crystal field theory diagram is

$$\underline{\hspace{2cm}} \quad z^2 \; (A_{1g})$$

$$\underline{\hspace{2cm}} \quad x^2 - y^2 \; (B_{1g})$$

$$\underline{\hspace{1.5cm}} \; \underline{\hspace{1.5cm}} \quad xz, \, yz \; (E_g)$$

$$\underline{\overset{1}{\hspace{1.5cm}}} \quad xy \; (B_{2g})$$

As a result, $d \rightarrow d$ transitions of three different energies are possible. In $[Ti(H_2O)_6]^{3+}$, the lowest-energy band $(^2B_{2g} \rightarrow {}^2E_g)$ lies in the infrared region. The $^2B_{2g} \rightarrow {}^2B_{1g}$ and $^2B_{2g} \rightarrow {}^2A_{1g}$ transitions are poorly resolved, producing the spectrum shown in Figure 21.6. The strong absorption beginning at about 31,000 cm^{-1} arises from ligand-to-metal charge transfer.

The primary factors that determine the locations of the absorption maxima of metal complexes are, of course, the same as those that determine crystal field splittings. As such, they were discussed in Chapter 16.

Comparable octahedral d^n and d^{10-n} complexes normally give similar spectra. For example, the spectra of hexacoordinate d^9 species are similar to those of hexacoordinate d^1. However, the ground- and excited-state designations are reversed; that is, the (idealized) octahedral d^1 transition is $^2T_{2g} \rightarrow {}^2E_g$, while for d^9 it is $^2E_g \rightarrow {}^2T_{2g}$. The "hole" analogy is often used to explain this. The d^1 configuration has one electron and nine holes, while d^9 has nine electrons and one hole.

There is also a complementary relationship between octahedral and tetrahedral systems. Recall that the qualitative crystal field theory diagram

---

12. Spectroscopists often write the excited state in front with a backward arrow; for example, $^2E_g \leftarrow {}^2T_{2g}$.

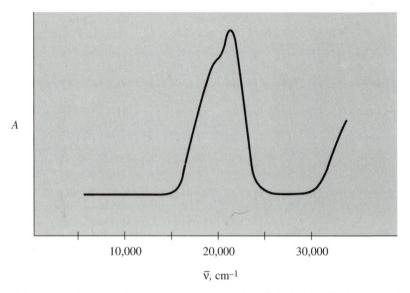

A

10,000   20,000   30,000

$\bar{\nu}$, cm^{-1}

Figure 21.6 The electronic absorption spectrum of $[Ti(H_2O)_6]^{3+}$. The band centered at 20,300 cm^{-1} is asymmetric because of Jahn–Teller distortion; the transition beginning at about 31,000 cm^{-1} is due to ligand-to-metal charge transfer.

for tetrahedral species is inverted compared to the octahedral case. This can be represented pictorially by an *Orgel diagram* (Figure 21.7).[13]

d^2–d^8 Complexes: Spectra Influenced by Russell–Saunders Coupling

For complexes having two through eight d electrons, the d orbital energy levels depend both on the ligand geometry and on electron–electron interactions (Russell–Saunders coupling). This can be qualitatively understood by considering a subconfiguration of the d^2 octahedral case, $t_{2g}^1 e_g^1$. This subconfiguration is split into two energy states, depending on which orbitals are occupied. For example, $(d_{xy})^1(d_{x^2-y^2})^1$ lies at a different (higher) energy than $(d_{xy})^1(d_{z^2})^1$, because the former concentrates electron density in the xy plane, maximizing the repulsion. It can be shown that $(d_{xy})^1(d_{x^2-y^2})^1$ contributes to a T_{1g} symmetry state, while $(d_{xy})^1(d_{z^2})^1$ has T_{2g} symmetry.[14]

13. Orgel, L. E. *J. Chem. Phys.* **1955**, *23*, 1004.

14. See Murrell, J. N.; Kettle, S. F. A.; Tedder, J. M. *The Chemical Bond*, 2nd ed.; Wiley: New York, 1985; Chapter 12.

Figure 21.7
Orgel diagram for certain octahedral and tetrahedral configurations. On the left side of the diagram are represented the energy states for d^9 octahedral and d^1 tetrahedral complexes; the right side represents the d^1 octahedral and d^9 tetrahedral configurations. Energy increases along the vertical axis.

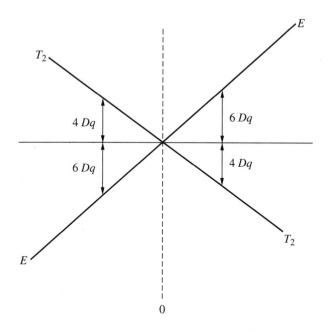

All possible combinations for the d^2 configuration in the O_h point group are given in Table 21.5.

A systematic approach for problems of this type is to first evaluate the free ion terms, and then consider the effect of ligand field theory on each. Using the method described in Chapter 1, we find that coupling in a free ion having a d^2 configuration gives rise to a ^3F ground state. The excited states are ^3P, ^1G, ^1D, and ^1S. The singlet states can be ignored for our purposes, since they are accessible from the ground state only through spin-forbidden (triplet → singlet) transitions. The ^3F–^3P energy difference results from differences in interelectronic repulsions, and can be expressed in terms of the *Racah parameter*, B. From the definitions in common use,

Table 21.5 The possible d orbital occupancies and resulting triplet energy states for the d^2 configuration in O_h symmetry

	xy	xz	yz	z^2	$x^2 - y^2$
xy	S				
xz	$T_{1g}(F)$	S			
yz	$T_{1g}(F)$	$T_{1g}(F)$	S		
z^2	$T_{1g}(P)$	T_{2g}	T_{2g}	S	
$x^2 - y^2$	T_{2g}	$T_{1g}(P)$	$T_{1g}(P)$	A_{1g}	S

Note: S = singlet state.

Table 21.6 Experimental values of the Racah parameter, B, for some first-row transition metal ions

Free Ion	$d^n, n =$	B, cm^{-1}	Free Ion	$d^n, n =$	B, cm^{-1}
Ti^{2+}	2	695	Fe^{3+}	5	1015
V^{3+}	2	861	Fe^{2+}	6	917
V^{2+}	3	755	Co^{3+}	6	1065
Cr^{3+}	3	918	Co^{2+}	7	971
Cr^{2+}	4	810	Ni^{2+}	8	1030
Mn^{2+}	5	860			

Source: Sutton, D. *Electronic Spectra of Transition Metal Complexes*; McGraw-Hill: New York, 1968.

$E(^3P) - E(^3F) = 15B$. Values of B for free ions are obtained spectroscopically, and are compiled for common cations of the first-row transition elements in Table 21.6.

For complexes, B is typically about 70–90% of the free ion value. The reduced repulsion is attributed to the expansion of the electron cloud in the presence of ligands, a consequence of the fact that the electrons reside in molecular rather than atomic orbitals. (They are delocalized over two or more nuclei.) This is sometimes referred to as the *nephelauxetic effect*.[15]

Free ion states having degeneracies greater than 1 may be split by symmetry constraints. For example, in the O_h point group, an F state (degeneracy = 7) is split into T_{1g} (degeneracy = 3), T_{2g} (3), and A_{2g} (1) states. Table 21.7 lists other free ion splittings for octahedral and tetrahedral geometries. Crystal field theory is used to assign relative energies to these

Table 21.7 Symmetry states arising from free ion splitting of the d orbitals of octahedral and tetrahedral complexes

Term	Degeneracy	O_h
S	1	A_{1g}
P	3	T_{1g}
D	5	$E_g + T_{2g}$
F	7	$A_{2g} + T_{1g} + T_{2g}$
G	9	$A_{1g} + E_g + T_{1g} + T_{2g}$

Note: The states for tetrahedral symmetry are the same as for octahedral, except the g and u subscripts are omitted.

15. Schaffer, C. E.; Jørgensen, C. K. *J. Inorg. Nucl. Chem.* **1958**, *8*, 143; see also, Jørgensen, C. K. *Prog. Inorg. Chem.* **1962**, *4*, 73.

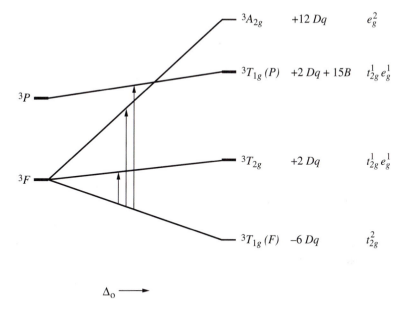

Figure 21.8

Correlation diagram for the triplet states of octahedral d^2 complexes. The free ion terms are at the extreme left; the strength of the ligand field increases going from left to right.

triplet states in terms of Δ_o and B. The $^3T_{1g}(F)$ ground state lies at $-6\ Dq$, $^3T_{2g}$ at $+2\ Dq$, $^3A_{2g}$ at $+12\ Dq$, and $^3T_{1g}(P)$ at $+2\ Dq + 15B$.[16]

A *correlation diagram* is useful for understanding the spectral implications. Figure 21.8 provides such a diagram for the octahedral d^2 case. The free ion terms are given on the left side of the figure. The far right side corresponds to a very strong ligand field; at that extreme the effect of coupling is small compared to that of symmetry. Connecting lines are then drawn between the related energy states. The result is a plot of the energies of these states as a function of the ligand field strength.

Based on Figure 21.8, three spin-allowed transitions are possible: from the $^3T_{1g}(F)$ ground state to the $^3T_{2g}$, $^3T_{1g}(P)$, and $^3A_{2g}$ excited states. Therefore, three $d \rightarrow d$ bands are expected in the ultraviolet–visible spectrum. The lowest-energy transition is $^3T_{1g}(F) \rightarrow {}^3T_{2g}$. The second band is $^3T_{1g}(F) \rightarrow {}^3A_{2g}$ for weak field ligands and $^3T_{1g}(F) \rightarrow {}^3T_{1g}(P)$ for strong field ligands. The $^3T_{1g}(F) \rightarrow {}^3A_{2g}$ band results from a two-electron transition ($t_{2g}^2 \rightarrow e_g^2$), so its intensity is relatively low.

Another way to present this information is through a Tanabe–Sugano diagram (Figure 21.9).[17] In such diagrams, the baseline always corresponds

16. For a complete explanation, see Douglas, B. E.; Hollingsworth, C. A. *Symmetry in Bonding and Spectra*; Academic: Orlando, FL, 1985; p. 264.

17. Tanabe, Y.; Sugano, S. *J. Phys. Sci. (Japan)* **1954**, *7*, 753, 766; see also, Konig. E.; Kremer, S. *Ligand Field Diagrams*; Plenum: New York, 1977.

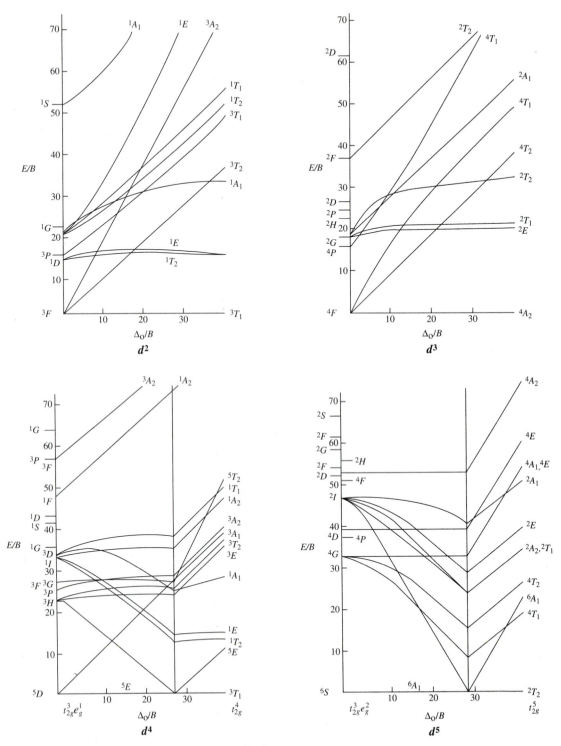

Figure 21.9 Tanabe–Sugano diagrams for d^2–d^8 metal ions in an octahedral environment.

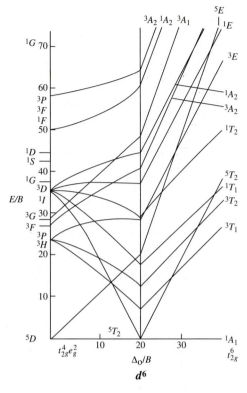

d^6

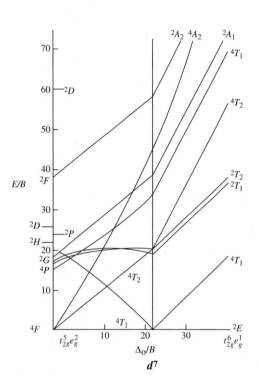

d^7

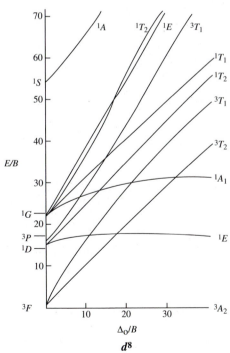

d^8

to the ground state, and a plot is constructed of E/B versus Δ_o/B. Tanabe–Sugano diagrams are especially useful for the d^4–d^7 cases, in which the ground state changes at some ligand field strength (the *high spin–low spin crossover*). The ground state remains the baseline; the crossover point is designated by a vertical line.

To illustrate the use of correlation and Tanabe–Sugano diagrams, we will consider the electronic spectrum of $[V(H_2O)_6]^{3+}$ with reference to Figures 21.8 and 21.9. When trapped in a crystalline alum matrix, $[V(H_2O)_6]^{3+}$ exhibits absorption bands at 17,400 cm^{-1} (v_1) and 25,700 cm^{-1} (v_2); both have intensities typical of $d \rightarrow d$ transitions ($\epsilon \approx 10^0$–10^1 L/mol·cm). A charge transfer band appears at much higher energy. Based on Figure 21.8, the first $d \rightarrow d$ band must arise from the $^3T_{1g}(F) \rightarrow {}^3T_{2g}$ transition. Thus, $8\,Dq$ is approximately equal to 17,400 cm^{-1}; this gives $Dq \approx 2175$ cm^{-1}, and $\Delta_o = 10\,Dq \approx 22,000$ cm^{-1}.[18] The second band might be due to either the $^3T_{1g}(F) \rightarrow {}^3T_{1g}(P)$ or the $^3T_{1g}(F) \rightarrow {}^3A_{2g}$ transition, depending on the relative magnitudes of Δ_o and B. The latter absorption should be at $18\,Dq$ (about 40,000 cm^{-1}), which is much higher than the observed value for v_2. Thus, v_2 is more reasonably assigned to $^3T_{1g}(F) \rightarrow {}^3T_{1g}(P)$.

The Tanabe–Sugano diagram (d^2) supports this assignment. Taking 90% of the free ion value (Table 21.6), $\Delta_o/B \approx 21,750/775 \approx 28$. A vertical line drawn upward from that point intersects the $^3T_{1g}(P)$ trace at $E/B \approx 37$, so that transition is predicted to occur at about $37(775) \approx 28,600$ cm^{-1}, in good agreement with v_2. The $^3T_{1g} \rightarrow {}^3A_{2g}$ band should be at about $E/B = 53$ (41,000 cm^{-1}). It is obscured by the charge transfer band.

A second example is provided by $[Ni(H_2O)_6]^{2+}$. The d^8 configuration is complementary to d^2, so similarities are evident in the V^{III} and Ni^{II} cases. The free ion ground state is again ^3F, and there is one triplet excited state (^3P). The F term is split by octahedral symmetry into $A_{2g} + T_{1g} + T_{2g}$ (Table 21.7). However, the energy levels are inverted compared to d^2, so $^3A_{2g}$ is the ground state. The triplet excited states are $^3T_{1g}(F)$, $^3T_{2g}$, and $^3T_{1g}(P)$; their relative energies are summarized in Figure 21.10.

The hexaaquanickel(II) ion gives electronic transitions at 8500, 15,400, and 26,000 cm^{-1}. The first band (v_1) is $^3A_{2g} \rightarrow {}^3T_{2g}$, and corresponds to $\Delta_o = 10\,Dq$. The Dq unit therefore equals about 850 cm^{-1} for this complex. The $^3A_{2g} \rightarrow {}^3T_{1g}(F)$ transition should then lie at about 15,300 cm^{-1} ($18\,Dq$), in excellent agreement with the location of v_2. The v_3–v_2 energy difference (10,600 cm^{-1}) equals $15B$, from which a B value of 707 cm^{-1} is obtained. This is about 69% of the free ion value.[19]

18. This is only an approximation, because the treatment has been simplified in several ways. For example, Jahn–Teller effects have been ignored.

19. Here again, these calculations are only approximate; certain secondary effects are ignored. For a more rigorous treatment, see Schlafer, H. L.; Gliemann, G. *Basic Principles of Ligand Field Theory*; Wiley: New York, 1969; and references cited therein.

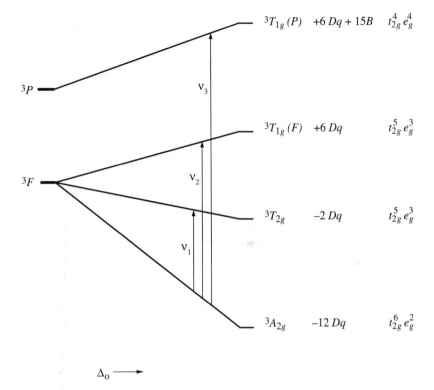

Figure 21.10
Correlation diagram for the octahedral d^8 configuration. The triplet free ion terms are at the extreme left; ligand field strength increases going from left to right.

21.3 Photoelectron Spectroscopy

In photoelectron spectroscopy a sample is bombarded with monochromatic radiation having enough energy to cause ionization. The kinetic energy (KE) of the ejected electron is then compared to the source energy, and the difference between the two is assumed to equal the negative of the orbital energy (*Koopmans' theorem*).[20]

$$-\epsilon = h\nu - \text{KE} \tag{21.3}$$

Either of two types of sources are used. In *ultraviolet photoelectron spectroscopy* (UV PES, or UPS), the ionizing radiation is provided by an ultraviolet source—normally a helium lamp, which emits photons at 21.21 and 40.81 eV. Such energy is sufficient to ionize only valence electrons. If an

20. Koopmans, T. *Physica (Utrecht)* **1934**, *1*, 104.

X-ray source is employed (often the K_α line of chromium at 5415 eV), the technique is *X-ray photoelectron spectroscopy* (XPS). (The older acronym ESCA, for *electron spectroscopy for chemical analysis*, is also used.) The ionization of core electrons is usually studied by XPS. Since the helium source is more monochromatic than are X-ray sources, resolution is usually superior in UPS.

UPS Studies[21]

The experimental data obtained from UPS experiments are usually correlated with molecular orbital diagrams. As an example, consider the spectrum of $H_2O(g)$. Four ionizations are observed, at 12.6, 13.1, 14.8, and 18.5 eV. The origins of these bands are apparent from the MO diagram (Figure 3.19b). The first two ionizations correspond to electron removal from the nonbonding $2p$ level of oxygen. The third and fourth bands are from the b_1 and $1a_1$ molecular orbitals; both consist of a series of lines.

Bonded electrons are, of course, associated with two or more nuclei. Their potential energies are therefore influenced by changes in internuclear distances, as in vibrational (stretching) oscillations. Thus, if the fine structure is resolved (if individual maxima can be discerned within the band envelope), then the spacing between adjacent peaks provides information about the vibrational frequencies of the cation produced upon ionization. In the H_2O spectrum, the 18.5 eV band consists of a series of lines spaced at 2990 cm^{-1} (about 0.37 eV). This corresponds to the symmetric stretching vibration of H_2O^+. The low vibrational energy compared to neutral H_2O (Table 21.2) is consistent with the loss of bond energy upon removal of a bonding electron.

A comparison between the MO diagram and the photoelectron spectrum of N_2 is also instructive. There are five ionization bands. (The $1s$ band at 412 eV is only observed when an X-ray source is used.) As shown in Figure 21.11, the 17 eV band differs in appearance from the others; it is extremely broad and has considerable fine structure. The distance between adjacent lines within its envelope is about 1800 cm^{-1}, which corresponds to the stretching vibration of N_2^+. (Compare to 2360 cm^{-1} for N_2.) Such band shapes are more characteristic of π than of σ orbitals; hence, it arises from the ionization of an electron from the $\pi_{x,y}$ level.[22]

The spectrum of $Ni(CO)_4$ contains bands at 8.9 and 9.8 eV, which are assigned to ionizations from the T_2 and E levels (primarily metal $3d$ in character). The intensity ratio of these two peaks is 3:2, which is consistent

21. Cowley, A. H. *Prog. Inorg. Chem.* **1979**, *26*, 45; DeKock, R. L.; Lloyd, D. R. *Adv. Inorg. Chem. Radiochem.* **1974**, 16, 66.

22. For a further explanation, see Eland, J. H. D. *Photoelectron Spectroscopy*, 2nd ed.; Butterworths: London, 1984; p. 10.

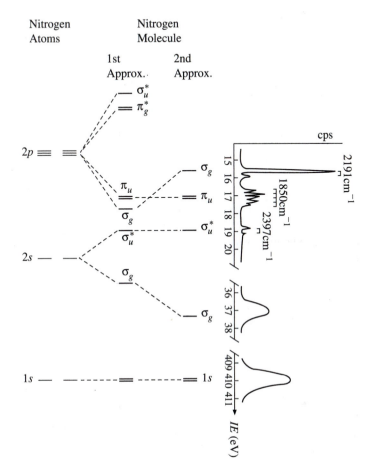

Figure 21.11
The MO diagram and combined UPS/XPS spectrum of N_2. Note the fine structure in the doubly degenerate π_u band. [Reproduced with permission from Bock, H.; Molliere, P. D. *J. Chem. Educ.* **1974**, *51*, 506.]

with the orbital degeneracies. The spectrum also contains higher-energy bands. Those beginning at about 14 eV are attributed to ionization from ligand orbitals (see Figure 21.12, p. 714).

For the octahedral carbonyls $M(CO)_6$ (M = Cr, Mo, and W) the HOMO is a triply degenerate t_{2g} level, which produces an ionization band at about 8.5 eV. The location of this band provides information about back-bonding, since the t_{2g} level is stabilized by $(M \rightarrow L)\pi$ interaction. Such studies have been extended by the substitution of CO for other ligands (see Table 21.8).

X-Ray Photoelectron Spectroscopy

One possible use of XPS by inorganic chemists is for qualitative analysis. The ionization of $1s$ electrons is appropriate for this purpose, since their

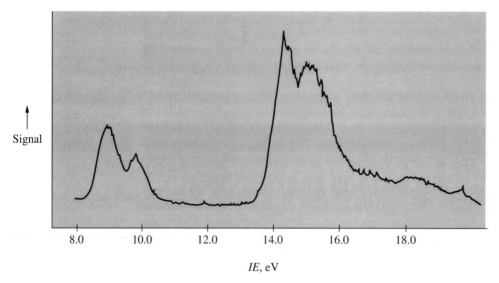

Figure 21.12 The UPS spectrum of $Ni(CO)_4$. The bands between 8 and 11 eV result from ionizations from the T_2 and E levels; those at higher energies are due to electron removal from ligand orbitals.

orbital energies are quite sensitive to the nuclear charge. For example, the 1s *binding energies* for elements having 3–9 protons range from 58 eV for lithium to 694 eV for fluorine (Figure 21.13). Unfortunately, because peak heights and integrated band intensities are affected by factors other than concentration, this method is limited to semiquantitative determinations.

Orbital energies vary with the chemical environment. It is therefore possible to gain information about electron densities, oxidation states, etc.,

Table 21.8 Lowest-energy UPS band(s) for compounds having the general formula $M(CO)_5L$ (M = Cr, Mo, and W)

M	L	IE, eV	M	L	IE, eV
Cr	CO	8.40	W	CO	8.56
Cr	PH_3	7.90, 8.03	W	NH_3	7.54, 7.55, 8.06
Cr	NH_3	7.56, 7.85	W	NMe_3	7.41, 7.62, 7.96
Cr	NMe_3	7.45, 7.76	W	PMe_3	7.46, 7.64, 7.90
Mo	CO	8.50			

Note: For nonhomoleptic complexes, the reduced symmetry (C_{4v} versus O_h) removes the degeneracy of the t_{2g} level; hence, more than one band is observed in these cases.

Source: Data taken from that compiled by Cowley, A. H. *Prog. Inorg. Chem.* **1979**, *26*, 45.

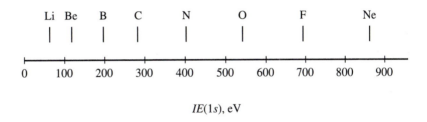

Figure 21.13
The 1s binding energies for the elements of the first short period.

through XPS. For example, the N (1s) and O (1s) binding energies presented in Table 21.9 demonstrate a clear correlation between oxidation state and ionization energy.

An especially interesting case is that of *trans*-$[Co(en)_2(NO_2)_2]NO_3$, a complex salt that has nitrogens in three different chemical environments. The N (1s) region of the X-ray photoelectron spectrum is shown in Figure 21.14 (p. 716). As the electron density about an atom increases, the increased electron–electron repulsions destabilize the core energy levels and thereby produce a shift to lower ionization energy. Therefore, the most electron-poor nitrogen in this salt (that in NO_3^-) gives rise to the highest-energy N (1s) peak. It is also apparent that the band intensities are approximately in the theoretical ratio of 4:2:1.

A detailed XPS study of over 100 complexes was carried out by Feltham and Brant, who used their data to establish a table of ligand group shifts.[23] An approximately linear relationship was observed between the ligand electronegativities and the metal binding energies.

Table 21.9 The 1s binding energies for selected compounds of nitrogen and oxygen

Compound	N (1s)	O (1s)	Compound	N (1s)	O (1s)
N_2	409.9		CH_3OH		538.9
NH_3	405.6		CH_3CHO		537.6
N_2H_4	406.1		NO	410.7	543.3
O_2		543.1	NO_2	412.9	541.3
H_2O		539.7	N_2O	408.6,	541.2
CO_2		542.1		412.5	
SO_2		539.6			

Note: Energies are given in electron volts.
Source: Taken from data compiled by Ghosh, P. K. *Introduction to Photoelectron Spectroscopy*; Wiley: New York, 1983; p. 55.

23. Feltham, R. D.; Brant, P. *J. Am. Chem. Soc.* **1982**, *104*, 641.

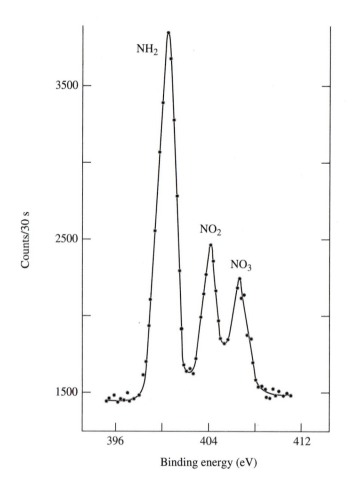

Figure 21.14
The nitrogen 1*s* region of the photoelectron spectrum of *trans-*[Co(en)$_2$(NO$_2$)$_2$]NO$_3$. [Reproduced with permission from Hendrickson, D. N.; Hollander, J. M.; Jolly, W. L. *Inorg. Chem.* **1969**, *8*, 2642.]

21.4 Magnetic Susceptibility[24]

Recall that the term *paramagnetic* describes a species having one or more unpaired electrons, while *diamagnetic* systems have all electrons paired. These possibilities can be differentiated experimentally by several instrumental tools, one of which is the *Faraday balance* (diagrammed in Figure 21.15). This apparatus makes use of the fact that paramagnetic samples are drawn into an external magnetic field, while diamagnetics are weakly repelled by such an environment. Thus, in a typical procedure, a sample is weighed first in the presence and then in the absence of a magnetic field. Paramagnetic

24. Spencer, B.; Zare, R. N. *J. Chem. Educ.* **1988**, *65*, 277; O'Connor, C. G. *Prog. Inorg. Chem.* **1982**, *29*, 203; Gerloch, M. *Prog. Inorg. Chem.* **1979**, *26*, 1.

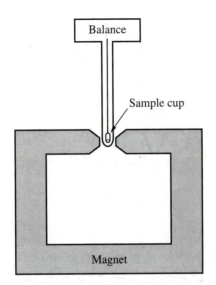

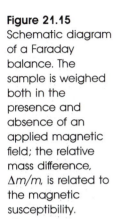

Figure 21.15
Schematic diagram of a Faraday balance. The sample is weighed both in the presence and absence of an applied magnetic field; the relative mass difference, $\Delta m/m$, is related to the magnetic susceptibility.

materials have a greater apparent mass in the former case, while the reverse is true for diamagnetics. The relative mass difference ($\Delta m/m$) is a measure of the sample's *magnetic susceptibility*—the tendency for magnetic moments to align with an external field.

The susceptibility of a bulk sample may be expressed in several ways. The susceptibility per unit volume (or just the volume susceptibility) is symbolized by κ; χ_g is the gram susceptibility; χ_m is the molar susceptibility; and χ'_m is the corrected (for diamagnetism) molar susceptibility. The units of these parameters are not straightforward. The cgs unit (centimeter-gram-second unit) is used for gram and molar susceptibilities.[25]

The diamagnetic correction is usually accomplished via a set of additive constants, called *Pascal's constants*, as given in Table 21.10 (p. 718). An example of the use of this table will be given later in this section.

The corrected molar susceptibility is related to the *magnetic moment*, μ, of a molecule or ion through the equation

$$\chi'_m = \frac{N\mu^2}{3kT} \tag{21.4}$$

where N is Avogadro's number, k is the Boltzmann constant, and T is the absolute temperature. The magnetic moment has units of Bohr magnetons (BM). By use of the appropriate conversion factors, equation (21.4) converts to

$$\mu = 2.83(\chi'_m T)^{1/2} \tag{21.5}$$

25. For a discussion of magnetic units, see Pass, G.; Sutcliffe, H. *J. Chem. Educ.* **1971**, *48*, 180.

Table 21.10 Diamagnetic correction factors (Pascal's constants) for selected ions and molecules

Cations	$\delta \times 10^{-6}$, cgs	Anions	$\delta \times 10^{-6}$, cgs	Other Common Ligands	$\delta \times 10^{-6}$, cgs
Li^+	−1.0	F^-	−9.1	H_2O	−13
Na^+	−6.8	Cl^-	−23.4	NH_3	−18
K^+	−14.9	Br^-	−34.6	C_2H_4	−15
Rb^+	−22.5	I^-	−50.6	en	−46
Cs^+	−35.0	O^{2-}	−12.0	acac	−52
Mg^{2+}	−5.0	OH^-	−12.0	py	−49
Ca^{2+}	−10.4	CN^-	−13.0	bipy	−105
Zn^{2+}	−15.0	NO_3^-	−18.9		
Hg^{2+}	−40.0	SO_4^{2-}	−40.1		
NH_4^+	−13.3	ClO_4^-	−32.0		

Note: The structures of the ligands denoted by en, acac, py, and bipy are given in Appendix III.
Source: Data taken primarily from Jolly, W. L. *The Synthesis and Characterization of Inorganic Compounds*; Prentice-Hall: Englewood Cliffs, NJ, 1970; p. 371.

The significance of μ is that it relates directly to the number of unpaired electrons through the so-called spin-only equation

$$\mu_{SO} = 2[S(S + 1)]^{1/2} \tag{21.6}$$

where S is the spin quantum number of the species in question. Since each unpaired electron contributes $\frac{1}{2}$ to the overall spin, it follows that

$$\mu_{SO} = [n(n + 2)]^{1/2} \tag{21.7}$$

where n is the number of unpaired electrons. Thus, there is a predicted spin-only magnetic moment for any given number of unpaired electrons (Table 21.11).

Table 21.11 Spin-only magnetic moments for systems having 0–7 unpaired electrons

n	μ_{SO}, BM	n	μ_{SO}, BM
0	0.00	4	4.90
1	1.73	5	5.92
2	2.83	6	6.93
3	3.87	7	7.94

Note: As calculated from equation (21.7).

The above discussion is somewhat oversimplified, since several other factors influence the magnetic moment of a molecule or ion.[26] However, the magnetic behaviors of a variety of species can be rationalized, at least to a first approximation, through equation (21.7).

The experimental moments of the hexaaqua complexes of some $3d$ cations are given in Table 21.12. The agreement between theory and experiment is imperfect; however, certain conclusions can be drawn. For example, the Cr^{II}, Mn^{II}, Fe^{II}, and Co^{II} complexes all must be high spin. (If this is not clear, calculate the theoretical μ_{SO} values for the low spin cases and compare them to the experimental data.)

We will use the $[Mn(CN)_6]^{4-}$ ion, as found in the complex salt $K_4[Mn(CN)_6] \cdot 3H_2O$, to illustrate the calculations involved in a magnetic susceptibility determination. Most octahedral Mn^{II} complexes are high spin. However, the strong field CN$^-$ ligands might be expected to induce spin pairing. The experimental gram susceptibility of this salt is about 4.4×10^{-6} cgs. The calculations then proceed as follows:

$$\chi_m = \chi_g(MW) = 4.4 \times 10^{-6}(421.5) = 1.85 \times 10^{-3} \text{ cgs}$$

Table 21.12 Experimental and calculated (spin-only) magnetic moments for salts containing hexaaqua complexes of the first transition series

Species	d^n, $n =$	μ_{exp}, BM	μ_{calc}, BM
$Cs[Ti(H_2O)_6](SO_4)_2 \cdot 6H_2O$	1	1.8	1.73
$NH_4[V(H_2O)_6](SO_4)_2 \cdot 6H_2O$	2	2.7	2.83
$K[Cr(H_2O)_6](SO_4)_2 \cdot 6H_2O$	3	3.8	3.88
$[Cr(H_2O)_6]Cl_3$	3	4.08	3.88
$[Cr(H_2O)_6]SO_4$	4	4.8	4.90
$K_2[Mn(H_2O)_6](SO_4)_2$	5	5.9	5.92
$(NH_4)_2[Fe(H_2O)_6](SO_4)_2$	6	5.25	4.90
$[Fe(H_2O)_6]SO_4 \cdot H_2O$	6	5.22	4.90
$[Co(H_2O)_6]Cl_2$	7	4.87	3.88
$(NH_4)_2[Co(H_2O)_6](SO_4)_2$	7	5.1	3.88
$[Ni(H_2O)_6](NO_3)_2$	8	3.24	2.83
$(NH_4)_2[Ni(H_2O)_6](SO_4)_2$	8	3.2	2.83
$(NH_4)_2[Cu(H_2O)_6](SO_4)_2$	9	1.9	1.73

Note: Calculated values are for high spin complexes.
Source: Data taken from that compiled by Jolly, W. L. *The Synthesis and Characterization of Inorganic Compounds*; Prentice-Hall: Englewood Cliffs, NJ, 1970; p. 375.

---

26. See Boudreaux, E. A.; Mulay, L. N., Eds. *Theory and Applications of Molecular Paramagnetism*; Wiley: New York, 1976.

Next, the diamagnetic correction must be made. Using Table 21.10, this amounts to

$$\sum \delta = [4\delta(K^+) + 6\delta(CN^-) + 3\delta(H_2O)] \times 10^{-6}$$
$$\approx [4(-15) + 6(-13) + 3(-13)] \times 10^{-6}$$
$$= -1.8 \times 10^{-4} \text{ cgs}$$

Then,

$$\chi'_m = \chi_m - \sum \delta = (1.85 \times 10^{-3}) - (-1.8 \times 10^{-4})$$
$$= 2.03 \times 10^{-3} \text{ cgs}$$

The room-temperature spin-only magnetic moment is therefore

$$\mu_{SO} = 2.83(\chi'_m T)^{1/2} = 2.83[(2.03 \times 10^{-3})298]^{1/2} = 2.2 \text{ BM}$$

which corresponds via equation (21.7) and the quadratic formula to about 1.4 unpaired electrons per formula unit. The high spin d^5 configuration predicts five unpaired electrons, and the low spin case only one, so the latter is clearly more consistent with the experimental data.

Temperature and Macroscopic Effects

Equation (21.4) indicates an inverse relationship between χ'_m and the absolute temperature. This was experimentally observed by Pierre Curie, and the equation

$$\chi'_m = \frac{C}{T} \tag{21.8}$$

is known as the *Curie law*. The constant C varies with the species. For substances that obey equation (21.8), a plot of χ'_m versus $1/T$ gives a straight line having slope C and passing through the origin. The decrease in susceptibility with increasing temperature is primarily due to thermal motion, which randomizes the directions of the magnetic moments.

Many paramagnetic species exhibit a variation of the Curie law. Specifically, the relationship

$$\chi'_m = \frac{C}{T - \theta} \tag{21.9}$$

where θ is another constant (having units of temperature), may apply. This is called *Curie–Weiss behavior*, and is believed to result from intermolecular magnetic interactions. In that sense it is a bulk, rather than a molecular, property.

21.5 Nuclear Magnetic Resonance Spectroscopy

In theory, the NMR experiment can be performed for any nucleus for which I, the spin quantum number, is nonzero. The ^1H nucleus is still the most commonly studied, and some commercial spectrometers are limited to ^1H experiments. However, instruments with multinuclear capability have become common research tools over the past two decades.

Data for the nuclei most frequently used by inorganic chemists are compiled in Table 21.13 (p. 722). The ^{19}F and ^{31}P nuclei are especially convenient for study because of their 100% isotopic abundances, and also because their spin quantum number is $\frac{1}{2}$. (Nuclei with $I > \frac{1}{2}$ have nonzero quadrupole moments. This promotes rapid relaxation, and often results in broad spectral lines.) A considerable body of information is available on these nuclei, so we will use them as examples throughout the remainder of this section.

Chemical Shifts

Chemical shift data provide information concerning the electron density about a given nucleus. The resonance frequency of hydrogen is relatively insensitive to its environment, and virtually all ^1H resonances fall within about 40 ppm of one another. Most other nuclei show larger variations. For example, the chemical shifts in ^{19}F NMR vary by over 800 ppm. Data for a variety of binary compounds and ions are presented in periodic table format in Table 21.14 (p. 723).

The data in the table illustrate certain trends common to all NMR-active nuclei:

1. The fluorine nucleus becomes more shielded as the electronegativity of the heteroatom decreases. Thus, the ^{19}F resonance moves to higher field going down a family (as in the series CF_4, SiF_4, GeF_4). Similarly, going across a period (eg, BF_3, CF_4, NF_3), the increased central atom electronegativity has a deshielding effect on the substituent fluorines.

2. Hybridization effects are often evident. For example, in going from PF_3 (sp^3) to PF_5 (sp^3d) the ^{19}F nuclei are shielded. This is consistent with the fact that sp^3d hybrid orbitals are less electronegative than sp^3 (see Chapter 1). Similar trends are seen for SF_4 versus SF_6 and SeF_4 versus SeF_6. (However, other factors such as charge may offset this; compare SiF_4 to SiF_6^{2-}.)

3. Because they contain geometrically nonequivalent fluorines, two different resonances can be observed for SF_4, ClF_3, and BrF_5. In SF_4, the axial fluorines resonate at $+70$ ppm and the equatorials at $+116$ ppm. The locations of these peaks are consistent with the relative electronegativities of the sp^2 and pd subhybridizations. Other molecules (eg,

Table 21.13 Relevant parameters for selected NMR-active nuclei

Nucleus	Natural Abundance, %	I	Relative Frequency, MHz	Relative Receptivity
^{1}H	99.985	$\frac{1}{2}$	100.0	100.0
^{2}H	0.015	1	15.4	0.00015
^{10}B	19.6	3	10.7	0.39
^{11}B	80.4	$\frac{3}{2}$	32.1	13
^{13}C	1.11	$\frac{1}{2}$	25.1	0.018
^{14}N	99.6	1	7.2	0.10
^{15}N	0.37	$\frac{1}{2}$	10.1	0.00039
^{17}O	0.037	$\frac{5}{2}$	13.6	0.0011
^{19}F	100.0	$\frac{1}{2}$	94.1	83
^{23}Na	100.0	$\frac{3}{2}$	26.5	9.3
^{27}Al	100.0	$\frac{5}{2}$	26.1	21
^{29}Si	4.7	$\frac{1}{2}$	19.9	0.037
^{31}P	100.0	$\frac{1}{2}$	40.5	6.6
^{51}V	99.8	$\frac{7}{2}$	26.3	38
^{55}Mn	100.0	$\frac{5}{2}$	24.7	18
^{59}Co	100.0	$\frac{7}{2}$	23.6	28
^{63}Cu	69.1	$\frac{3}{2}$	26.5	6.5
^{75}As	100.0	$\frac{3}{2}$	17.2	2.5
^{103}Rh	100.0	$\frac{1}{2}$	3.2	0.0032
^{107}Ag	51.8	$\frac{1}{2}$	4.0	0.0035
^{127}I	100.0	$\frac{5}{2}$	20.1	9.5
^{129}Xe	26.4	$\frac{1}{2}$	27.8	0.57
^{133}Cs	100.0	$\frac{7}{2}$	13.2	4.8
^{183}W	14.4	$\frac{1}{2}$	4.2	0.0011
^{195}Pt	33.8	$\frac{1}{2}$	21.4	0.34
^{197}Au	100.0	$\frac{3}{2}$	1.7	0.0026

Note: Resonance frequencies are relative to 100.0 MHz for ^{1}H. Receptivity is a measure of signal strength for a given concentration of element, relative to 100.0 for ^{1}H.

Source: Data taken from compilations by Ebsworth, E. A. V.; Rankin, D. W. H.; Cradock, S. *Structural Methods in Inorganic Chemistry*, 2nd ed.; Blackwell: Oxford, 1991; pp. 31–34.

Table 21.14 ^{19}F chemical shifts for some binary compounds and ions

BF_3 -133	CF_4 -67	NF_3 $+140$		
BF_4^- -150		NF_4^+ $+215$		
	SiF_4 -165	PF_3 -36	SF_4 $+116, +70$	ClF_3 $+114, +2$
	SiF_6^{2-} -128	PF_5 -78	SF_6 $+70$	
	GeF_4 -178		SeF_4 $+62$	BrF_3 -24
			SeF_6 $+50$	BrF_5 $+270, +140$
			TeF_4 -27	IF_5 $+60$
				IF_7 $+164$

Note: Values are in parts per million relative to $CFCl_3$. The shifts for PF_5, SeF_4, BrF_3, TeF_4, IF_5, and IF_7 are averages (fluxionality; see text).
Source: Compiled from data given by Mooney, E. K. *An Introduction to ^{19}F NMR Spectroscopy*; Heyden: London, 1970.

SeF_4) give only one resonance at room temperature because of rapid positional scrambling (fluxionality; see pp. 153–154).

The ^{19}F shifts of several xenon fluorides were reported by Schumacher and Schrobilgen.[27] All members of the series $Xe(OTeF_5)_nF_{4-n}$ and $OXe(OTeF_5)_nF_{4-n}$ were studied. The $-OTeF_5$ (teflate) group is strongly electron-withdrawing, as evidenced by the deshielding of the xenon-bonded fluorines with increasing n (Table 21.15, p. 724).

Spin–Spin Coupling

Spin–spin coupling between proximate nuclei is a valuable source of structural information. The number of lines a given resonance is split into is a function of the spin quantum number(s) of any nuclei coupled to it. A useful equation in this respect is

$$\text{Number of lines} = 2nI + 1 \tag{21.10}$$

27. Schumacher, G. A.; Schrobilgen, G. J. *Inorg. Chem.* **1984**, *23*, 2923.

Table 21.15 ^{19}F chemical shifts for some $OTeF_5$-substituted xenon fluorides

Compound	δ, ppm	Compound	δ, ppm
XeF_4	-15.66	$XeOF_4$	$+101.59$
$F_3Xe(OTeF_5)$	$+5.87$ (2)	$XeOF_3(OTeF_5)$	$+106.78$ (1)
	-11.98 (1)		$+103.00$ (2)
$cis\text{-}F_2Xe(OTeF_5)_2$	-8.58	$cis\text{-}XeOF_2(OTeF_5)_2$	$+112.59$
$trans\text{-}F_2Xe(OTeF_5)_2$	$+10.59$	$trans\text{-}XeOF_2(OTeF_5)_2$	$+108.24$
$FXe(OTeF_5)_3$	$+16.36$		

Note: Chemical shifts are for the xeron-bonded fluorines, referenced to $CFCl_3$; numbers in parentheses are the relative peak areas.
Source: Data taken from Schumacher, G. A.; Schrobilgen, G. J. *Inorg. Chem.* **1984**, *23*, 2923.

where n is the number of equivalent coupling nuclei. Thus, a singlet resonance is observed if I (or n) $= 0$. For the $I = \frac{1}{2}$ case, the right side of equation (21.10) simplifies to $n + 1$.

Consider the ^{19}F spectra of $^{12}CF_4$, ^1HF, $^{14}NF_3$, and $^{11}BF_3$. In this series, the spin quantum number of the heteroatom increases in units of $\frac{1}{2}$ from $I = 0$ (^{12}C) to $\frac{3}{2}$ (^{11}B). The idealized spectra that result are shown in Figure 21.16. The individual lines of a given resonance have equal intensities, as is always the case for $n = 1$.

Line intensities are unequal when $n > 1$. For coupling nuclei having $I = \frac{1}{2}$, the familiar patterns derived from Pascal's triangle (1:2:1 triplets for $n = 2$, 1:3:3:1 quartets for $n = 3$, etc.) result. If I is greater than $\frac{1}{2}$, then the patterns can be predicted by a treatment similar to that shown in Figure 21.17, where the relative intensities of the five-line pattern for a resonance coupled to two equivalent $I = 1$ nuclei are seen to be 1:2:3:2:1. [An example

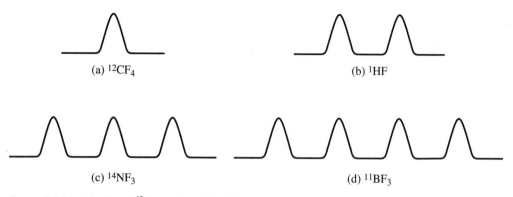

(a) $^{12}CF_4$ (b) ^1HF

(c) $^{14}NF_3$ (d) $^{11}BF_3$

Figure 21.16 Idealized ^{19}F spectra of four binary compounds. The spin quantum numbers of the coupling nuclei are $I = 0$, $\frac{1}{2}$, 1, and $\frac{3}{2}$ for ^{12}C, ^1H, ^{14}N, and ^{11}B, respectively.

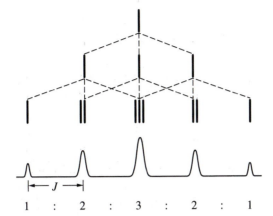

Figure 21.17
Derivation of the
NMR line pattern for
coupling to two
equivalent $I = 1$
nuclei.

$$1 \quad : \quad 2 \quad : \quad 3 \quad : \quad 2 \quad : \quad 1$$

of such a case is the ^{31}P spectrum of $RP(^{14}NR_2)_2$, where R represents a noncoupling nucleus.][28]

Resonances arising from nuclei that are simultaneously coupled to two or more different types of nuclei are more complex. In such situations it is necessary to know the relative magnitudes of the coupling constants in order to predict the appearance of the spectrum. Consider the ^{19}F spectrum of PHF_2, for which the coupling constants are 1140 Hz for $^1J_{PF}$ and 42 Hz for $^2J_{HF}$.[29] Interaction with the ^{31}P nucleus splits the fluorine resonance into a doublet. Each of these peaks is further split by coupling to ^1H, producing a four-line "doublet of doublets" pattern (Figure 21.18a).

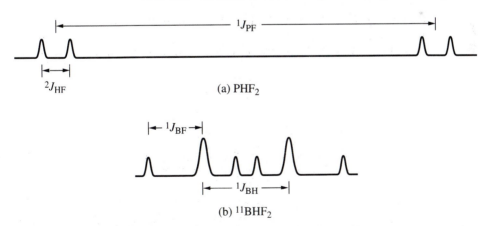

Figure 21.18   Idealized NMR spectra: (a) ^{19}F spectrum of PHF_2 and (b) ^{11}B spectrum of BHF_2.

28.   For a more extensive discussion of coupling, see Orcutt, R. H. *J. Chem. Educ.* **1987**, *64*, 763.

29.   Recall that J is the general symbol for a coupling constant, and that superscripts denote the number of bonds separating the coupled nuclei.

A more complex example is the ^{11}B spectrum of BHF_2, with approximate coupling constants of 85 Hz for $^1J_{BF}$ and 140 Hz for $^1J_{BH}$. The result is a six-line "doublet of triplets" spectrum, but it is complicated by the fact that the coupling constants are sufficiently similar to cause the triplets to overlap (Figure 21.18b).

NMR with Paramagnetics: Contact Shifts

The presence of a paramagnetic material in an NMR experiment generally has two important effects:

1. There are large changes in the chemical shifts, with the amount of change being related to the proximity to the paramagnetic center. These are referred to as *contact shifts*.

2. Line widths are broadened as a result of the rapid relaxation promoted by the interaction of the nuclear spin with the unpaired electron(s). Such broadening can make it difficult to obtain spectra.

As an example of how paramagnetism influences NMR behavior, consider a series of tris(acetylacetonate) complexes of trivalent $3d$ cations:

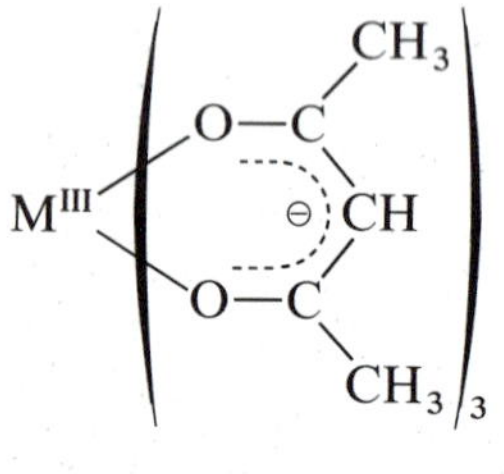

The chemical shifts and peak widths of such species have been studied by both ^1H and ^2H NMR, and relevant data are compiled in Table 21.16. Comparing the ^1H spectrum of the free ligand to that of the diamagnetic (low spin d^6) Co^{III} complex, the methyl resonance shifts to higher field by about 3.5 ppm upon coordination to the metal; the peak width (about 1 Hz) is normal for ^1H NMR. For paramagnetic complexes, however, large contact shifts (20–50 ppm) and severe line broadening are observed. The peak widths are less affected in the ^2H spectra of the corresponding deuterium-enriched ($-CD_3$) complexes. This is a general phenomenon, and makes ^2H NMR a useful option in systems for which high-quality ^1H spectra cannot be obtained.

Table 21.16 ^1H and ^2H NMR data for some trivalent $M(acac)_3$ complexes

M	d^n, $n =$	^1H (Me), ppm	$W_{1/2}$, Hz	
			^1H	^2H
V	2	−45.04	46	<3
Cr	3	−40.20	1400	35
Mn	4	−25.20	96	10
Fe	5	−21.80	36	26
Co	6	−2.22	1	2

Note: $W_{1/2}$ is the peak width at half height.
Source: Johnson, A.; Everett, G. W. *J. Am. Chem. Soc.* **1970**, *92*, 6705.

Not surprisingly, it is possible to measure magnetic moments by NMR. In the Evans method,[30] the ^1H chemical shift of the methyl resonance of *t*-BuOH (often 2% in H_2O) is measured as a function of the concentration of paramagnetic material (ie, the concentration of unpaired electrons).

Contact shifts are useful for structure elucidation. The addition of a small amount of a paramagnetic impurity tends to increase the range of chemical shifts, and thereby minimize accidental equivalence and/or overlap. The best-known examples are the lanthanide shift reagents used in ^1H studies of organic compounds. A different type of application is useful for ^{29}Si experiments. Because the relaxation of the ^{29}Si nucleus is unusually slow, the signal-to-noise ratio in Fourier transform experiments is often poor. The addition of paramagnetic material enhances the relaxation rate, and thereby improves the spectrum.

Dynamic NMR: Chemical Exchange and Fluxionality

The time scale of the NMR experiment is about 10^{-3} s; that is, for an individual structure to be observed by NMR, it must have a lifetime of 10^{-3} s or longer. (The value varies depending on the conditions, but 10^{-3} is appropriate for the discussion that follows.) This is quite slow compared to most other types of spectroscopy. For example, the time scales of vibrational and electronic spectroscopy are both shorter than 10^{-12} s. Because of this, NMR is the method of choice for the study of many dynamic systems.

In accord with the VSEPR prediction, ClF_3 is T-shaped (C_{2v} point group). Thus, there are two different fluorine environments, axial and equatorial. However, its high-temperature ^{19}F spectrum exhibits only a sharp

30. Evans, D. F. *J. Chem. Soc.* **1959**, 2003.

singlet. This suggests that the fluorines undergo positional interchange on the order of 10^3 times per second or faster.

The rate of interchange can be slowed by cooling the system. The signal begins to broaden as the temperature is reduced; the maximum broadness is reached at T_c, the *coalescence temperature*. On cooling to below T_c the broadened singlet gradually separates into two resonances, a 1:2:1 triplet at relatively low applied field ($+114$ ppm) and a 1:1 doublet at higher field ($+2$ ppm). The average chemical shift of the low-temperature peaks (taking into account their 2:1 intensity ratio) is approximately equal to that of the high-temperature singlet. Thus, the coupling pattern, relative intensities, and peak locations are all consistent with the C_{2v} structure.

The exchange process in ClF_3 is believed to be intramolecular (the Berry pseudorotation; see Chapter 5). Intermolecular exchange also can be followed by NMR. For example, a room-temperature mixture of $SnCl_4$ and $SnBr_4$ gives five resonances in the ^{119}Sn spectrum, corresponding to the presence of all five of the possible species $SnBr_nCl_{4-n}$ ($n = 0$–4). The integrated intensities indicate that the distribution of these compounds is purely statistical; that is, no thermodynamic preference is shown at room temperature.

Nuclear magnetic resonance also has been used to study ligand exchange by $[Ti(H_2O)_6]^{3+}$ in aqueous solution.[31] Two resonances are observed in the ^{17}O spectrum—one for free and the other for ligated H_2O. Changes in the chemical shifts of these resonances as a function of temperature can be related via the appropriate equations to the rate of exchange. Using both variable temperature and variable pressure conditions, it is possible to measure the rate constant (found to be 1.81×10^{-5} s^{-1} at 25°C), and to demonstrate that the mechanism of ligand exchange is associative in nature.

21.6 Electron Spin Resonance Spectroscopy

An unpaired electron has a spin quantum number of $\frac{1}{2}$. Therefore, like nuclei such as ^1H, ^{19}F, and ^{31}P, nonequivalent spin states can be induced by the application of a magnetic field. Thus, the electron equivalent to NMR is *electron spin resonance* (ESR), also known as *electron paramagnetic resonance* (EPR), spectroscopy.

A major difference between NMR and ESR is in the appearance of the spectra. It is conventional for ESR results to be presented as derivative, rather than absorption spectra (see Figure 21.19).

An equation fundamental to ESR is

$$hv = g\beta H_o \qquad (21.11)$$

31. Hugi, A. D.; Helm, L.; Merbach, A. E. *Inorg. Chem.* **1987**, *26*, 1763.

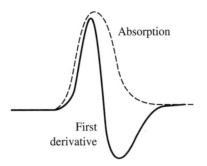

Figure 21.19
Signal plotted as an absorption and as a first derivative. The former is conventional in NMR, and the latter in ESR spectroscopy.

where v is the resonance frequency, β is the Bohr magneton (9.274×10^{-21} erg/G), and H_o is the magnetic field strength in gauss. The unitless g *value* of a free electron is 2.0023. The g value changes with the environment, and is analogous to the NMR chemical shift.

Hyperfine Splitting

As in NMR, interactions between unpaired electrons and nearby nuclei having $I > 0$ result in coupling. Thus, the ESR spectrum of the hydrogen atom (^{1}H·) is a doublet because of coupling to the proton, while that of ^{2}H· is a 1:1:1 triplet. This is as predicted from equation (21.10).

^{1}H: Number of lines $= 2(1)(\tfrac{1}{2}) + 1 = 2$

^{2}H: Number of lines $= 2(1)(1) + 1 = 3$

The peak separation (508 G for ^{1}H· and 78 G for ^{2}H·) is the coupling constant or, more properly, the *hyperfine splitting*.

Longer-range coupling is also possible. For example, the spectrum of the methyl radical consists of four lines (in 1:3:3:1 intensity ratios) due to coupling to the three equivalent hydrogens. The corresponding ^{13}CH$_3$· spectrum is a doublet of quartets (strong coupling to the $I = \tfrac{1}{2}$ carbon nucleus, with weaker coupling to the more distant hydrogens).

Coordination compounds provide more difficult examples. The VO^{2+} cation has an unpaired electron, and hence gives rise to an ESR signal. The spin quantum number of ^{51}V is $\tfrac{7}{2}$; the hyperfine splitting therefore produces a spectrum comprising eight lines of equal intensity. This basic pattern persists in many vanadium compounds, but is sometimes complicated by additional coupling. For example, the spectrum of the chelate complex

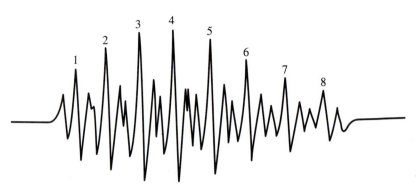

Figure 21.20 The ESR spectrum of VO[S$_2$P(Et)OMe]$_2$. Each of the eight main resonances is split into a 1:2:1 triplet by long-range coupling to phosphorus. [Reproduced with permission from Lorenz, D. R.; Johnson, D. K.; Stoklosa, H. J.; Wasson, J. R. *J. Inorg. Nucl. Chem.,* **1974**, *36*, 1184.]

is presented in Figure 21.20. There are twenty-four lines, since each of the eight main lines is split into a 1:2:1 triplet by coupling to the two equivalent ^{31}P nuclei. The hyperfine splitting values are 92.2 G (to ^{51}V) and 36.1 G (to ^{31}P).

Electron spin resonance is useful for the study of mixed-valence compounds. Consider the radical anion of the C$_5$H$_5$Co(CO) dimer:

$$\left[C_5H_5Co \underset{\underset{O}{\overset{\|}{C}}}{\overset{\overset{O}{\overset{\|}{C}}}{\diagdown\diagup}} CoC_5H_5 \right]^{\cdot -}$$

This is a 17/18 electron system. An eight-line ESR spectrum is expected if the unpaired electron is localized on either cobalt. (The spin quantum number of ^{59}Co is $\frac{7}{2}$.) However, what is actually observed is a fifteen-line spectrum, with the relative intensities being consistent with coupling to two equivalent $I = \frac{7}{2}$ nuclei.[32] Thus, the unpaired electron density either must be delocalized over both cobalts, or the electron must undergo exchange between the metal centers so rapidly that only an average ESR signal is observed.

Bibliography

Drago, R. S. *Physical Methods for Chemists*, 2nd ed.; Saunders: Philadelphia, 1991.

Ebsworth, E. A. V.; Rankin, D. W. H.; Cradock, S. *Structural Methods in Inorganic Chemistry*, 2nd ed.; Blackwell: Oxford, 1991.

32.   Schore, N. E.; Ilenda, C. S.; Bergman, R. G. *J. Am. Chem. Soc.* **1977**, *99*, 1781.

Mason, J., Ed. *Multinuclear NMR Spectroscopy*; Plenum: New York, 1987.

Yoder, C. H.; Schaeffer, C. D. *Introduction to Multinuclear NMR*; Benjamin/Cummings: Menlo Park, CA, 1987.

Carlin, R. *Magnetochemistry*, Springer-Verlag: New York, 1986.

Nakamoto, K. *Infrared and Raman Spectra of Inorganic and Coordination Compounds*, 4th ed.; Wiley: New York, 1986.

Douglas, B. E.; Hollingsworth, C. A. *Symmetry in Bonding and Spectra*; Academic: Orlando, FL, 1985.

Eland, J. H. D. *Photoelectron Spectroscopy*, 2nd ed.; Butterworths: London, 1984.

Gorenstein, D. G., Ed. *Phosphorus-31 NMR: Principles and Applications*; Academic: Orlando, FL, 1984.

Lever, A. B. P. *Inorganic Electronic Spectroscopy*, 2nd ed.; Elsevier: Amsterdam, 1984.

Strommen, D. P.; Nakamoto, K. *Laboratory Raman Spectroscopy*; Wiley: New York, 1984.

Ghosh, P. K. *Introduction to Photoelectron Spectroscopy*; Wiley: New York, 1983.

Poole, C. P. *Electron Spin Resonance*, 2nd ed.; Wiley: New York, 1983.

Kaplan, J. I.; Fraenkel, G. *NMR of Chemically Exchanging Systems*; Academic: New York, 1980.

Harris, D. C.; Bertolucci, M. *Symmetry and Spectroscopy*; Oxford University; New York, 1978.

Harris, R. K.; Mann, B. E., Eds. *NMR and the Periodic Table*; Academic: New York, 1978.

Rabalais, J. W. *Principles of Ultraviolet Photoelectron Spectroscopy*; Wiley: New York, 1977.

Braterman, P. S. *Metal Carbonyl Spectra*; Academic: New York, 1975.

Jørgensen, C. K. *Modern Aspects of Ligand Field Theory*; North-Holland: Amsterdam, 1971.

Jolly, W. L. *The Synthesis and Characterization of Inorganic Compounds*; Prentice-Hall: Englewood Cliffs, NJ, 1970.

Questions and Problems

1. Explain in your own words the molecular phenomena that are studied by infrared/Raman, electronic, and photoelectron spectroscopies. Why are different sources needed for these different instrumental techniques?

2. The stretching frequency for $^1H-^{35}Cl$ is 2991 cm^{-1}. Use equation (21.1) to qualitatively predict whether the corresponding frequency for $^2H-^{35}Cl$ is at higher or lower energy.

3. Consider the NH_3 molecule (C_{3v} point group).
   (a) How many vibrational degrees of freedom are there?
   (b) To what irreducible representation does the symmetric stretching mode belong?
   (c) Do you expect that mode to be active or inactive in the infrared? In the Raman spectrum? Defend your answers.

4. For each of the following pairs, predict which species has the lower average C–O stretching frequency and explain.
   (a) $Fe(CO)_5$ or $Fe_2(CO)_9$ (b) $Fe(CO)_5$ or $[Fe(CO)_4]^{2-}$
   (c) $Cr(CO)_5P\phi_3$ or $[Cr(CO)_5Cl]^-$

5. The stretching vibration of $CO(g)$ is found at 2138 cm^{-1}. This band is almost always displaced to lower wavenumber upon coordination of CO to a metal.
   (a) Explain why this is the case.
   (b) An exception is the complex $[Cu(CO)_4]^+$, for which ν_{CO} occurs at about 2200 cm^{-1}. Discuss.

6. The C–N stretching frequencies of the silver complexes $[Ag(CN)_x]^{1-x}$ ($x = 2$–4) exhibit a consistent decrease with increasing x. Why?

7. Use selection rules to decide which of each of the following pairs of transitions gives the more intense UV-visible band. Give a specific reason in each case:
   (a) The $4s \rightarrow 4p$ or $4s \rightarrow 4d$ transition of potassium
   (b) A $d \rightarrow d$ or a charge transfer band of NiI_4^{2-}
   (c) A $d \rightarrow d$ transition of NiI_4^{2-} or $[Ni(NH_3)_6]^{2+}$

8. The compound $MnCl_2$ is nearly colorless, while $MnCl_3$ is brown, with a color intensity typical of $d \rightarrow d$ transitions. Explain.

9. Hexaamminecobalt(III) is colored, while $[Ir(NH_3)_6]^{3+}$ is colorless. Both are low spin d^6 complexes. Rationalize.

10. Use the appropriate Tanabe–Sugano diagram (Figure 21.9) to predict the number of spin-allowed $d \rightarrow d$ transitions for:
    (a) CoF_6^{4-} (b) CrF_6^{3-} (c) CrF_6^{4-} (d) $[Cr(CN)_6]^{3-}$

11. The $^3A_{2g} \rightarrow {}^3T_{2g}$ transition in the electronic spectrum of $[Ni(NH_3)_6]^{2+}$ occurs at 10,700 cm^{-1}.
    (a) Calculate the value of Dq for this complex.
    (b) Predict the location of the $^3A_{2g} \rightarrow {}^3T_{1g}$ (F) band.
    (c) Use your answer to part (b) to predict the location of the $^3A_{2g} \rightarrow {}^3T_{1g}$ (P) band. Assume that B for $[Ni(NH_3)_6]^{2+}$ is 75% of the free ion value.

12. (a) In XPS, the N (1s) ionization band lies at lower energy than O (1s). Explain why this is the case.
    (b) In UPS, the lowest-energy ionization band of N_2 lies at higher energy than that in O_2. Explain.

13. (a) The UPS spectrum of H_2 shows one peak at 15.4 eV. The ionization energy of atomic hydrogen is 13.6 eV. Discuss.
    (b) The H_2 band exhibits fine structure, but that for H· does not. Why?

14. The vibrational fine structure for the π_u ionization band of N_2 indicates that the stretching frequency of N_2^+ is 1800 cm^{-1}.
    (a) Explain in your own words how this conclusion is reached.
    (b) Rationalize the fact that the stretching frequency of N_2^+ is found at lower energy than that of N_2.
    (c) The UPS spectrum of O_2 shows that the vibrational frequency of O_2^+ is greater than that of O_2. Discuss.

15. The UPS spectrum of N_2 (Figure 21.11) provides experimental support for the notion of s–p_z mixing. Explain.

16. The C (1s) ionization bands of CH_4, CF_4, CO_2, and CH_3OH are found at 301.8, 297.5, 292.3, and 290.7 eV (not necessarily in that order). Match each compound with its C (1s) binding energy, and defend your logic.

17. Sketch the N (1s) region of the X-ray photoelectron spectrum of the complex salt $[Pt(NH_3)_4][Pt(NO_2)(CN)_5]$. Estimate both the approximate ionization energies and relative band intensities.

18. Give the spin-only magnetic moment of:
    (a) Cu^{2+} (b) Cr^{3+} (c) Y^{3+} (d) Sm^{2+}
    (e) Ho^{3+} (f) U^{4+} (g) Tetrahedral Ni^{II} (h) Octahedral Ni^{II}

19. The experimental magnetic moment of ferrous ammonium sulfate hexahydrate, $(NH_4)_2[Fe(H_2O)_6](SO_4)_2$, is 5.25 BM at room temperature. Calculate χ_m', χ_m, and χ_g.

20. The corrected molar susceptibility of $K_3[Fe(CN)_6]$ is about 1.8×10^{-3} cgs at room temperature.
    (a) Calculate χ_g.   (b) Calculate μ.
    (c) Is this a high or low spin complex? Defend your answer.
    (d) Assume that the Curie law is obeyed. Predict the corrected molar susceptibility at liquid nitrogen temperature (77 K).

21. A compound has corrected molar susceptibilities of 3.00×10^{-3} cgs at 298 K and 6.00×10^{-3} cgs at 153 K. Assuming Curie–Weiss behavior, calculate the value of the Weiss constant, θ.

22. Consider the following ^{11}B chemical shifts (referenced to $BF_3 \cdot OEt_2$): BH_4^-, -40 ppm; BF_4^-, -2 ppm; BF_3, $+9$ ppm; BBr_3, $+39$ ppm; BMe_3, $+86$ ppm. Are these values internally consistent? (That is, do they follow the expected qualitative order?) Discuss.

23. Qualitatively sketch the following NMR spectra; consult Table 21.13 as necessary.
    (a) ^{14}N spectrum of NH_3 (b) ^1H spectrum of $^{14}NH_3$
    (c) ^{11}B spectrum of BH_4^- (d) ^{31}P spectrum of P_2H_4
    (e) ^{11}B spectrum of $H_3B \cdot PMe_3$ (assume that $^1J_{PB} \gg {}^2J_{BH}$)
    (f) Low-temperature ^{19}F spectrum of SF_4

24. The ^{31}P NMR spectrum of P_4S_3 consists of a 1:3:3:1 quartet and a 1:1 doublet. The intensity ratios are 1:3. Demonstrate that this is consistent with the molecular structure. (Review Section 19.4 if necessary.)

25. The ^{31}P NMR spectrum of H_3PO_2 consists of a 1:2:1 triplet. The peak separations are consistent with $^1J_{PH}$ coupling. Give the correct structure.

26. (a) Sketch the expected ^{19}F NMR spectrum of XeF_2. The spin quantum number of ^{131}Xe is $\frac{3}{2}$.
    (b) Sketch the expected xenon-decoupled ^{19}F spectrum of the XeF_5^+ cation. Use VSEPR theory as a structural basis.
    (c) [Not for the faint of heart!] Sketch the expected undecoupled ^{19}F spectrum of XeF_5^+. What assumption(s) must be made?

27. The low-temperature ^{19}F NMR spectrum of SeF_4 consists of two 1:2:1 triplets. Only a singlet resonance is observed at higher temperatures. Explain.

28. Sketch the predicted ESR spectrum of:
    (a) $\cdot OH$ (b) A ^{14}N atom   (c) $\cdot PH_2$
    (d) $^{52}Cr^{3+}$ (I for ^{52}Cr is 0)   (e) $^{55}Mn^{2+}$ (I for ^{55}Mn is $\frac{5}{2}$)

29. The ESR spectrum of the molybdenum complex $[Mo(CN)_x]^{3-}$ consists of a single line. The corresponding ^{13}C-enriched species gives a nine-line spectrum. Suggest the probable value of x, and explain your reasoning.

30. In Section 16.1, we noted that the ESR spectrum of $IrCl_6^{2-}$ gives evidence for the delocalization of unpaired electron density onto the ligands. Having now read this chapter, what specific aspect of the ESR spectrum do you believe leads to this conclusion?

*31. D. A. Saulys, J. Castillo, and J. A. Morrison used ^{19}F NMR to study exchange reactions of haloboranes (*Inorg. Chem.* **1989**, *28*, 1619). Answer the following after reading their paper.
    (a) Why was NMR a good tool for this study?
    (b) Sketch the spectra of

    Does NMR readily differentiate between these isomers?
    (c) Rationalize the following ^{19}F chemical shifts: F_2B-BF_2, $+26.6$ ppm; $F_2B-B(F)Cl$, $+21.7$ ppm; F_2B-BCl_2, $+17.8$ ppm.

*32. The geometry of the teflate ion ($OTeF_5^-$) has been examined by infrared and Raman spectroscopy and by X-ray diffraction (Miller, P. K.; Abney, K. D.; Rappe, A. K.; Anderson, O. P.; Strauss, S. H. *Inorg. Chem.* **1988**, *27*, 2255).
    (a) Before reading the article cited, predict the shape and point group of $OTeF_5^-$.
    (b) The actual structure is shown in Figure 1 of the paper. What evidence is there for partial Te–O π bonding?
    (c) The infrared spectra reported include the following bands: for $^{16}OTeF_5^-$, 867, 645, 635, and 576 cm^{-1}; for $^{18}OTeF_5^-$, 825, 644, 635, and 576 cm^{-1}. Discuss the significance.

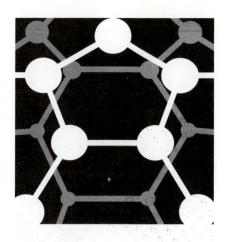

22

An Introduction to Bioinorganic Chemistry

Which of the chemical elements are necessary to sustain life? What are their specific functions? In what chemical forms are they found in vivo? The role of the inorganic elements in biological processes has received steadily increasing emphasis since about 1970.[1] The primary topics of what is often called *bioinorganic* chemistry are introduced in this chapter. After a general section on the biologically important elements, we will review some specific molecules of current research interest that relate to the functions of plants (chlorophylls and nitrogenases) and animals (zinc enzymes, various iron-containing species, and vitamin B_{12}).

22.1 Essential Elements, Toxins, Abundance, and Availability[2]

The first sentence of this chapter poses an intriguing question. A definitive answer has been difficult to come by, in part because the mere presence of

---

1. The first symposium on bioinorganic chemistry was held in 1970 at the Virginia Polytechnic Institute. For a summary of the proceedings, see Gould, R. F., Ed. *Bioinorganic Chemistry*; Advances in Chemistry 100; American Chemical Society: Washington, DC, 1971.

2. Frieden, E. *J. Chem. Educ.* **1985**, *62*, 917; Mertz, W. *Science* **1981**, *213*, 1332.

735

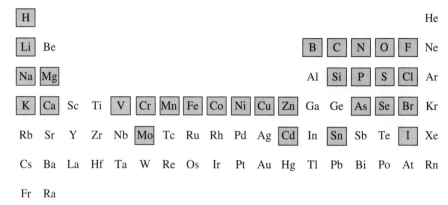

Figure 22.1
Thirty elements believed essential to life.

an element within an organism does not prove it to be necessary for the organism's survival. The best available evidence suggests that at least 30 elements are required by all life forms. It is an interesting exercise to try to guess their identities, and you are encouraged to do so before proceeding.

The 30 required elements are identified in Figure 22.1. For what purposes are these elements used? Four of the major categories relate to (*1*) structure (the proteinaceous elements; calcium for bones, teeth, and shells; etc.); (*2*) energy utilization and storage (eg, phosphorus in the adenosine phosphates); (*3*) material transport and storage (eg, of O_2 by certain iron and copper complexes); and (*4*) catalysis (primarily, complexes of magnesium and the transition metals). The relationships between the chemical properties of elements and the needs of organisms are fascinating, and we will discuss several such correlations in this chapter.

Most of the nonessential elements (and some of the others as well) are toxic above some concentration limit. Toxicity usually arises in one of three ways:[3]

1. ***Blockage of an Active Site of Some Biomolecule.***   Numerous π acceptor ligands, including CO, CN^-, H_2S, and PH_3, bind strongly to the iron centers of hemoglobin and myoglobin. This blocks that coordination site from the weaker ligand O_2, and destroys the ability of those molecules to serve as oxygen carriers.

2. ***Metal–Metal Displacement.***   A nonessential metal having chemical properties similar to an essential metal is often toxic. For example, barium (in soluble forms) is poisonous to humans, partly because it displaces calcium. Similarly, the toxicity of cadmium is due in part to its ability to displace zinc from certain enzymes and thereby deactivate

3. Ochiai, E.-I. *J. Chem. Educ.* **1974**, *51*, 235.

them. The relationship between toxic beryllium and essential magnesium is believed to be similar.

3. ***Modification of Molecular Structure.*** The nitrogen, oxygen, and sulfur atoms of proteins can serve as ligands for metal ions. However, the hydrogen bonding and other intermolecular interactions that determine the structural conformation of a protein are usually altered upon coordination to a metal. The result is a different molecular shape and, in many cases, deactivation.

A general order of elemental toxicity is Te > Se > Be > V > Cd > Ba > Hg > Tl > As > Pb > Sn > Ni > etc. This order changes somewhat, depending on the organism.

What causes some elements to be essential, while others are nonessential and/or toxic? For example, could a life form evolve that required barium, and for which calcium might then be a poison? The answer lies in relative abundances. There is a close correlation between the elements that are the most plentiful and those that are necessary for life. In the context of Darwin's natural selection, a species incorporating Ca^{2+} into its skeleton can proliferate much more rapidly than one utilizing Ba^{2+} for the same purpose, since the natural abundance of calcium is more than 100 times that of barium.

Environmental abundances (seawater and crustal) and human concentrations for a variety of elements are given in Figure 22.2. It is clear that the relationship described above holds. Moreover, several of the exceptions can readily be explained. For example, the crustal abundances of aluminum and silicon are very high, but neither is very important in biochemistry. This

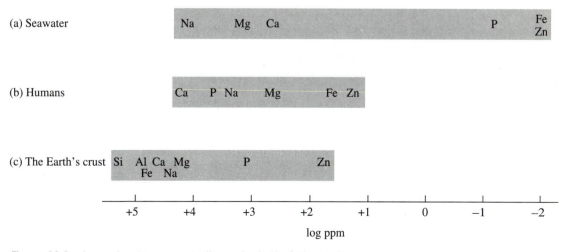

Figure 22.2 Approximate concentrations of selected elements.

illustrates the subtle but important difference between abundance and availability. Both of these elements exist mainly in rocks and minerals, either as unreactive oxides or as oxyanions; as such, they have high natural abundances, but are unavailable for use by living organisms.

22.2   Photosynthesis[4]

Photosynthesis is the use of solar energy by plant cells for the synthesis of cell components. The organic products of photosynthesis are carbohydrates, and the carbon source is CO_2. The generalized redox equation below can be written,

$$2H_2Z + CO_2 \longrightarrow \frac{1}{x}(CH_2O)_x + H_2O + 2Z \tag{22.1}$$

where H_2Z is a hydrogen source–reducing agent and Z is its oxidized form. This reagent is H_2S in the so-called *sulfur bacteria*, and an organic alcohol for certain other bacteria. In the most familiar cases (the green plants), water is the source of hydrogen and O_2 is the by-product. Hence, photosynthesis [equation (22.2)] is complementary to the process that enables muscle contraction in mammals.

$$H_2O + CO_2 \longrightarrow \frac{1}{x}(CH_2O)_x + O_2 \tag{22.2}$$

Studies have shown that the molecular oxygen is produced from H_2O rather than CO_2. This suggests that the oxidation half-reaction must be

$$2H_2O \longrightarrow 4H^+ + O_2 + 4e^- \tag{22.3}$$

The standard potential for this oxidation is -1.23 V. The higher pH of physiological conditions makes it somewhat less disfavored, but still negative ($\mathscr{E}_{phys} = -0.82$ V).

In order for the overall process to sum to equation (22.2), the reduction half-reaction must be

$$2CO_2 + 4H^+ + 4e^- \longrightarrow \frac{2}{x}(CH_2O)_x + O_2 \tag{22.4}$$

When glucose is the product, the physiological potential of the overall reaction is about -1.2 V. The need for an energy source is therefore obvious.

---

4.   Bishop, M. B.; Bishop, C. B. *J. Chem. Educ.* **1987**, *64*, 302.

The photosynthesis process can be divided into four sequential steps:

1. Solar energy is absorbed, primarily by biomolecules known as *chlorophylls* and *carotenoids*.

2. Electron transfer occurs from excited-state chlorophyll molecules to and from a series of redox agents, including a *ferredoxin* (see Section 22.5). This is accomplished within what is described as *photosystem I*.

3. The oxidation of water to O_2 occurs, being catalyzed by a structurally complex, manganese-containing enzyme. This is *photosystem II*.

4. Carbon dioxide is reduced to carbohydrate.

The most definitive "inorganic" studies to date have dealt mainly with the first and third steps; they are therefore emphasized in the following discussion.

The Chlorophylls[5]

Chlorophylls are green pigments that are highly efficient at absorbing radiant energy in the visible and near-ultraviolet regions. Many such molecules are known. One of the best-characterized, chlorophyll *a*, is pictured in Figure 22.3. Its formula is $C_{55}H_{72}MgN_4O_5$, which corresponds to a molecular mass of 894 daltons.

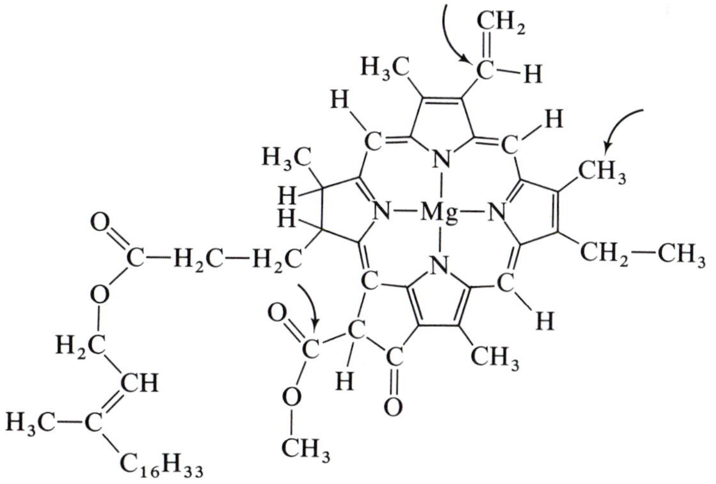

Figure 22.3
The structure of chlorophyll *a*. Other chlorophylls vary mainly in their substituents at the positions indicated by the arrows.

5. Strouse, C. E. *Prog. Inorg. Chem.* **1976**, *21*, 159.

The structure of chlorophyll *a* has three parts:

1. The Mg^{2+} ion is 4-coordinate. (A water molecule acts as a fifth coordination site in aqueous solution.) Four coplanar nitrogens serve as Lewis bases. The metal is raised above the ligand plane by about 40 pm, giving local C_{4v} (square pyramidal) geometry.

2. The porphyrin ring system (Figure 15.5b) forms a tetradentate cavity for the metal. There are four C_4N rings, which are interconnected by methine bridges. The various chlorophylls differ mainly in their substituents at three positions, shown by arrows in Figure 22.3.

3. A long-chain, organic *phytyl* group dangles from one ring. That group contains a total of 20 carbons, linked to the rest of the molecule via an ester functional group.

When chlorophyll *a* is dissolved in an organic solvent, its absorption spectrum exhibits intense $\pi \rightarrow \pi^*$ transitions having maxima near 420 and 660 nm. (The broadness of these bands results in at least some absorbance at all visible wavelengths.) However, a more complex spectrum, with maxima as high as 685 nm, is observed when chlorophyll is studied within living cells. The difference results from intermolecular associations. Chlorophyll *a* dimerizes in nonpolar solvents, probably by coordination of a carbonyl oxygen to the metal ion of the second molecule. Higher oligomers having the formula $(chl \cdot H_2O)_n$ are thought to exist in aqueous solution (see Figure 22.4). The individual molecules are linked by hydrogen bonding between the co-

Figure 22.4
Postulated structure of the dimer of chlorophyll *a*. [Reproduced with permission from Katz, J. J. In *Inorganic Biochemistry*; Eichhorn, G. L., Ed.; Elsevier: New York, 1973; Volume 2, p. 1022.]

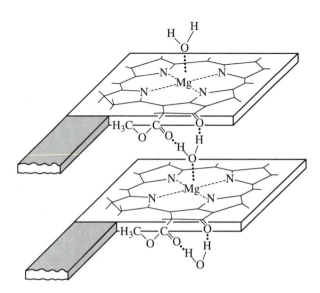

ordinated H_2O and carbonyl oxygens, producing a linear chain. This oligomerization appears to be crucial to the catalytic activity—the photo-electron "hops" from one molecule to the next in the chain. The light-trapping unit of photosystem I contains about 200 chlorophyll and about 50 carotenoid molecules.

Possible Structure–Function Correlations

It is interesting to speculate on the relationship between the structure and function of chlorophyll *a*. For example, given the organic portion of the molecule, how can the "choice" of magnesium, rather than some other metal, be rationalized? At least four reasons may be suggested:

1. Magnesium is among the most abundant of all metals.

2. The Mg^{2+} ion is a hard Lewis acid, and therefore has an affinity for the nitrogen bases.

3. The size of Mg^{II} is such that it nicely fits the porphyrin cavity. It is small enough to permit strong metal–ligand interactions, but large enough to cause it to lie above, rather than in, the plane of nitrogens. This facilitates interaction with H_2O and, hence, oligomerization.

4. Unlike many divalent metals, Mg^{II} has essentially no reduction chemistry. An easily reduced metal center might capture the excited electron via charge transfer. This would prevent the intermolecular electron donation required in photosystem I.

The porphyrin ring stabilizes the complex in several ways. The conjugated double bonds make the organic portion of the molecule stable to photodecomposition. The chelate effect adds to the stability of the inorganic region. Another benefit of the extended conjugation is that it shifts the $\pi \rightarrow \pi^*$ transitions into the visible region (see Table 22.1). This is important, because the highest-intensity wavelengths of the solar energy reaching Earth

Table 22.1 The effect of conjugation on the $\pi \rightarrow \pi^*$ transitions of some organic species

Compound	Number of Conjugated C=C Bonds	λ_{max}, nm
Ethene	1	165
1,3-Butadiene	2	217
1,3,5-Hexatriene	3	253
β-Carotene	11	465
Chlorophyll *a*	10 (including C=N linkages)	660

are in the visible domain. Furthermore, *p*-type conjugation gives structural rigidity. This limits the amount of absorbed energy lost to thermal vibration.

The phytyl group provides a hydrophobic region to complement the hydrophilic magnesium–porphyrin section, giving the chlorophylls both water and lipid solubility.

The Role of Manganese in Photosystem II[6]

It has been known for some time that the oxidation of H_2O to O_2 in photosystem II involves catalysis by a manganese-containing enzyme having a molecular weight of about 25,000 daltons. Although the exact structure of this enzyme is presently unknown, significant progress has been made toward understanding the catalytic mechanism.

Electron spin resonance has been an important tool for studying the metal centers of photosystem II. The spin quantum number of ^{55}Mn is $\frac{5}{2}$. This results in complex hyperfine splittings and causes the ESR spectra to be difficult to interpret. On the other hand, once the spectra are understood, the amount of structural information gleaned from the hyperfine interactions (along with relaxation effects) is considerable.

The spectrum obtained for the reduced form of the enzyme is consistent with a cluster containing four ligated Mn^{II} ions surrounded by a protein. It is now believed that the four metals plus four oxygens form a cubic subunit (Figure 22.5). During the course of the postulated catalysis mechanism, this subunit is oxidized in four sequential steps, ultimately producing a Mn^{III} cluster:

$$[Mn_4O_4] + 2H_2O \longrightarrow [Mn_4O_6] + 4H^+ + 4e^- \tag{22.5}$$

An adamantanelike structure is proposed for this oxidized subunit, with the manganese ions still forming a tetrahedral array (Figure 22.5). This is an attractive feature of the mechanism, since the movement of atoms is minimized.

The overall coordination geometry about each metal is thought to be tetrahedral, with the fourth ligand supplied by the surrounding protein. Subsequent O=O bond formation and the elimination of molecular oxygen returns the system to the reduced state:

$$[Mn_4O_6] \longrightarrow [Mn_4O_4] + O_2 \tag{22.6}$$

Note that equations (22.5) and (22.6) sum to the expected redox half-reaction.

6.   Brudvig, G. W.; Crabtree, R. H. *Prog. Inorg. Chem.* **1989**, *37*, 99; Sawyer, D. T.; Bodini, M. E.; Willis, L. A.; Riechel, T. L.; Magers, K. D. *Adv. Chem. Ser.* **1977**, *162*, 330.

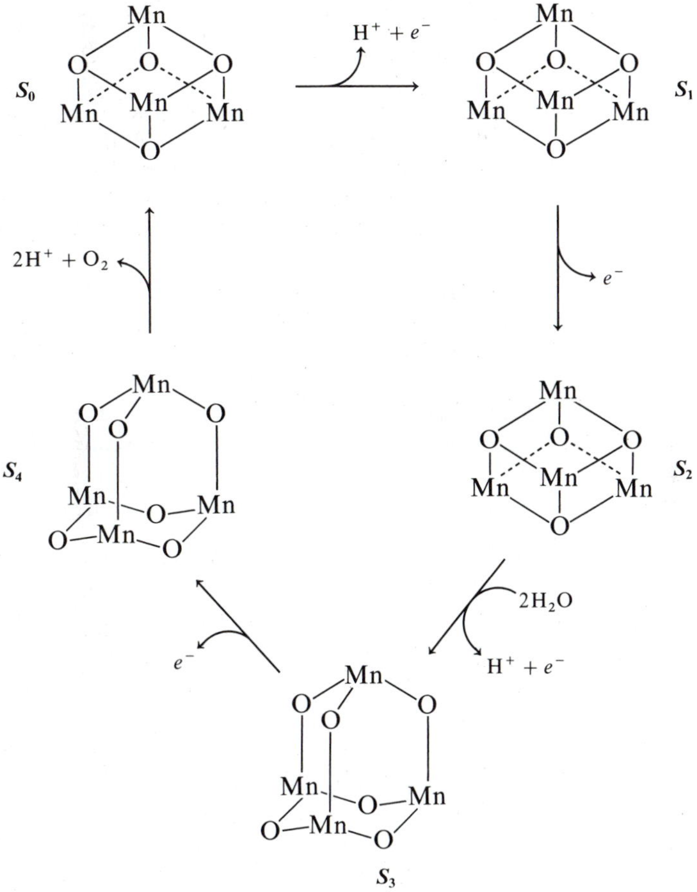

Figure 22.5
Proposed
mechanism of
oxidation–
reduction for the
manganese-
containing cluster of
photosystem II. The
cubic S_0 is the most
reduced, and the
adamantanelike S_4
the most oxidized,
of the five states.
[Reproduced with
permission from
Brudvig, G. W.;
Crabtree, R. H. *Proc.*
Nat. Acad. Sci.
1986, *83*, 4586.]

22.3 Nitrogenases[7]

Recall that the low chemical reactivity of molecular nitrogen is a consequence of the great strength of the N≡N triple bond (942 kJ/mol). Thus, although nitrogen is the most abundant element in the Earth's atmosphere, it is unavailable for use by most living things in its natural state.

There are exceptions, however. Certain blue-green algae, yeasts, and bacteria[8] are capable of reducing N_2 to NH_4^+, a form amenable to the

7. Henderson, R. A.; Leigh, G. J.; Pickett, C. J. *Adv. Inorg. Chem. Radiochem.* **1983**, *27*, 198; Nelson, M. J. *Adv. Inorg. Biochem.* **1982**, *4*, 1; Swedo, K. B.; Enemark, J. H. *J. Chem. Educ.* **1979**, *56*, 70.

8. Nitrogen fixation occurs in many legumes (peas, beans, clover, soybeans, etc.) through a cooperative relationship between the plant and the *Rhizobium* and other bacteria.

production of amino acids. Such organisms manufacture an enzyme known as *nitrogenase* for that purpose. The half-reaction below is representative of the process:

$$N_2 + 8H^+ + 6e^- \xrightarrow{\text{Nitrogenase}} 2NH_4^+ \tag{22.7}$$

The ability of this enzyme to hydrogenate N_2 under ambient conditions is remarkable, considering that the industrial conversion of N_2 to NH_3 involves both very high temperature (typically, 700°C) and pressure (30 atm).

Like most enzymes, the structure of nitrogenase varies somewhat, depending on the source. Common features include the following:

1. Nitrogenase actually consists of two separate proteins; neither is catalytically active without the other. The larger of the two has a molecular mass of about 230,000 daltons. It contains between 24 and 36 irons, an equal number of sulfide ions, and 2 molybdenums. The smaller protein (about 60,000 daltons) has an Fe_4S_4 cluster unit (probably a cube).

2. These enzymes appear to operate in an *anaerobic* atmosphere (ie, in the absence of O_2). This is probably necessary to avoid oxidation to nitrogen oxides and H_2O, and is consistent with the reducing environment required by equation (22.7).

3. The presence of certain small molecules inhibits the process; these include CO and, surprisingly, H_2.

Possible Mechanisms of Action

Neither the overall structure nor the structure of the active site of nitrogenase are known. Based on spectroscopic evidence, primarily ESR and EXAFS (*extended X-ray absorption fine structure*[9]), several different catalytic mechanisms have been proposed. One involves the initial coordination of N_2 to a molybdenum center:

$$(1) \quad [Mo] + :N{\equiv}N: \longrightarrow [Mo]-N{\equiv}N:$$

This is followed by the reductive addition of hydrogen (perhaps as H^+) in several steps. A possible sequence is

$$(2) \quad [Mo]-N{\equiv}N: \xrightarrow{+2H^+, 2e^-} [Mo]-NH{=}NH$$

$$(3) \quad [Mo]-NH{=}NH \xrightarrow{+2H^+, 2e^-} [Mo]-NH_2-NH_2$$

---

9. For discussions of the use of EXAFS in bioinorganic chemistry, see Cramer, S. P.; Hodgson, K. O. *Prog. Inorg. Chem.* **1979**, *25*, 1; Eisenberger, P.; Kincaid, B. M. *Science* **1978**, *200*, 1441.

The addition of two more hydrogens causes the release of ammonia, returning the molybdenum to its initial state:

(4) $[Mo]-NH_2-NH_2 \xrightarrow{+2H^+, 2e^-} [Mo] + 2NH_3$

The Fe_4S_4 clusters are thought to facilitate the electron transfer. (Synthetic clusters of this type undergo facile oxidation–reduction; see Chapter 19.)

Several features of this mechanism are consistent with previously discussed chemistry. For example, N_2 and CO behave similarly as ligands (σ donors and π acceptors). However, the HOMO (lone-pair) orbital lies at lower energy in N_2, and the overall charge distribution is more symmetric; hence, N_2 is a more reluctant donor. This explains the deactivation of the catalyst by carbon monoxide, which displaces N_2 from the active site. Also, in going from step (1) to step (4) above, the N–N bond is gradually weakened, explaining the activation of molecular nitrogen into a biologically useful form.

Inhibition by H_2 can be rationalized as resulting from the interruption of the process between steps (2) and (3). Supporting evidence for this notion comes from isotopic labeling studies.[10] When nitrogenase is placed under an atmosphere containing either N_2, H_2, and D_2O or N_2, D_2, and H_2O, HD is produced. The scrambling of the isotopically distinct hydrogens suggests a transition state such as

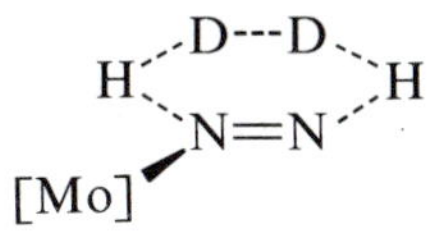

This is consistent with (but does not prove) the existence of an intermediate containing an Mo–NH=NH linkage.

Other mechanisms for N_2 activation have been suggested. For example, it is possible that molecular nitrogen acts as a bridging ligand between two metal centers:

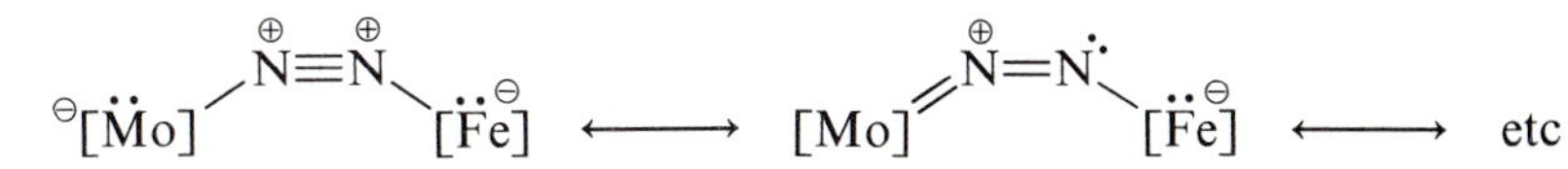

The addition of H_2 (or $2H^+ + 2e^-$) across the N≡N multiple bond would then give, in sequence, bridging diazine, bridging hydrazine, and ultimately the release of ammonia.

10. Stiefel, E. I. *Prog. Inorg. Chem.* **1977**, *22*, 1.

Dinitrogen Complexes: Model Systems[11]

A variety of complexes have been synthesized that mimic the behavior of nitrogenase. For example, several bis(dinitrogen) complexes produce NH_3 upon treatment with strong acid. An example is *trans*-$Mo(triphos)(N_2)_2(P\phi_3)$.[12] The reaction of this species with anhydrous HCl or HBr follows the equation

$$2[Mo^0]-(N_2)_2 \xrightarrow{\text{HX}} 2[Mo^{III}] + 3N_2 + 2NH_3 \qquad (22.8)$$

Model systems containing molybdenum and iron centers connected by one or more bridging ligands also have been reported. In particular, many mixed-metal Fe–S–Mo clusters are now known. A few of the known structural variations are pictured in Figure 22.6.

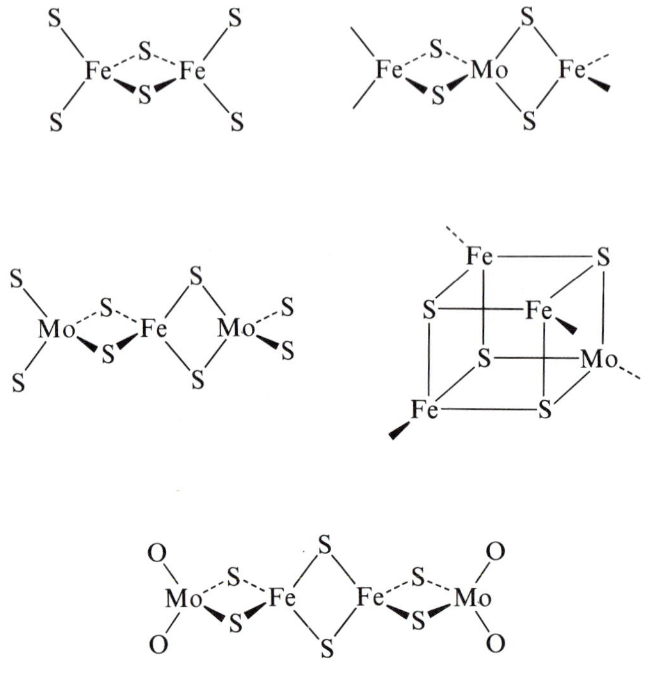

Figure 22.6

The frameworks of some known iron–sulfur–molybdenum cluster compounds and ions.

11. Holm, R. H. *Chem. Soc. Rev.* **1981**, *10*, 455.

12. Baumann, J. A.; Bossard, G. E.; George, T. A.; Howell, D. B.; Koczon, L. M.; Lester, R. K.; Noddings, C. M. *Inorg. Chem.* **1985**, *24*, 3568.

22.4   Zinc-Containing Enzymes[13]

A remarkable number of zinc-containing enzymes—well over 100 in humans alone—have been identified. The most important are summarized in Table 22.2.

Table 22.2   The functions of some well-characterized zinc-containing enzymes

Enzyme	Catalytic Function
Carbonic anhydrase	Conversion of CO_2 to HCO_3^-
Carboxypeptidase A and B	Digestion of proteins
Alkaline phosphatase	Conversion of phosphate esters to alcohols
Aldolase	Degradation of fructose
Alcohol dehydrogenase (coenzyme)	Dehydration of alcohols to aldehydes
Lactate dehydrogenase	Dehydration of lactic acid
DNA and RNA polymerase	Synthesis of polynucleotides

The broad utilization of zinc by living organisms is remarkable considering its moderate abundance and availability. Based on the functions of many of its enzymes, it can be speculated that the popularity of Zn^{II} is due to its ability to act as a Lewis acid without engaging in either oxidation or reduction.

The most studied of all zinc enzymes are carbonic anhydrase and the carboxypeptidases, which we will describe next.

Carbonic Anhydrase[14]

The reaction

$$CO_2 + H_2O \longrightarrow H^+ + HCO_3^- \tag{22.9}$$

is essential to many organisms, in part because of the role of bicarbonate ion in the buffering of mammalian blood. The uncatalyzed reaction is very slow ($k = 9.5 \times 10^{-2}$ L/mol·s at 25°C), which is not surprising given the strong C=O and H–O bonds of the reactants. The rate constant shows a phenomenal increase, to greater than 5×10^7 L/mol·s, in the presence of

13.   Bertini, I.; Luchinat, C.; Monnanni, R. *J. Chem. Educ.* **1985**, *62*, 924; Spiro, T. G., Ed. *Zinc Enzymes*; Wiley: New York, 1983; Prince, R. H. *Adv. Inorg. Chem. Radiochem.* **1979**, *22*, 349.

14.   Silverman, D. N.; Lindskog, S. *Acc. Chem. Res.* **1988**, *21*, 30; Hay, R. W. *Inorg. Chim. Acta* **1980**, *46*, L115.

carbonic anhydrase (CA). The catalyzed reaction is essentially diffusion-controlled, with a turnover rate of about 500,000 $CO_2 \rightarrow HCO_3^-$ conversions per enzyme molecule per second.

The enzyme contains one metal, surrounded by a nearly spherical (globular) protein. The molecular mass is about 29,000 daltons, varying somewhat with the source. As would be expected from the coordination chemistry of Zn^{II}, the metal has tetrahedral coordination. The ligands are three nitrogen atoms (of histidine residues) and a water molecule. An important feature of the structure is a "dimple," which exposes the active site (the zinc ion), so that the coordinated water lies at the end of what is often described as a *channel*; a schematic representation is given in Figure 22.7.

Various experiments have shown that a metal is required for enzymatic activity, but that it need not be zinc. The *apoenzyme* (the enzyme minus the metal ion) can be prepared by the reaction of CA with a chelating agent; it is catalytically inactive. The apoenzyme will incorporate other divalent metals, including cobalt, nickel, manganese, lead, and mercury. Cobalt^{II}–CA has approximately half the activity of the zinc enzyme; the Mn^{II} derivative is about 8% as active; and the others are either inactive or barely catalytic (see Table 22.3). Significantly, the metal ions that show at least some activity are those that most closely resemble Zn^{II} in size and geometric preference. (For example, both Mn^{II} and Co^{II} commonly form tetrahedral complexes.)

Figure 22.7
Schematic representation of the structure of carbonic anhydrase. The ball represents Zn^{II}; cylinders symbolize helices, and arrows represent the β structure. [Reproduced with permission from Kannan, K. K.; Notstrand, B.; Fridborg, K.; Lövgren, S.; Ohlsson, A.; Petef, M. *Proc. Nat. Acad. Sci.* **1975**, *72*, 51.]

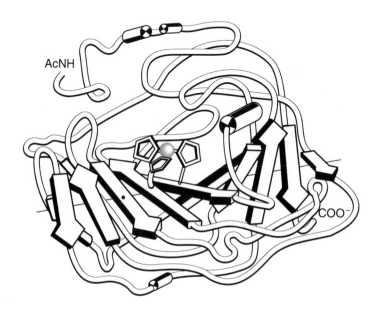

AcNH

COO⁻

Table 22.3 The relative catalytic activities of apocarbonic anhydrase, apocarboxypeptidase A, and some of their metal-containing derivatives

Carbonic Anhydrase–Metal	Activity	Carboxypeptidase A–Metal	Activity
Apoenzyme	0	Apoenzyme	0
Zn^{II}	100	Zn^{II}	100
Co^{II}	50	Co^{II}	160
Mn^{II}	8	Ni^{II}	107
Ni^{II}	0	Mn^{II}	8
Cd^{II}	0	Cu^{II}	0
Hg^{II}	0	Cd^{II}	0
Pb^{II}	0	Hg^{II}	0
		Pb^{II}	0

Note: Activities are referenced to 100 for the zinc-containing enzyme.
Source: Data taken from that compiled by Ochiai, E.-I. *Bioinorganic Chemistry: An Introduction*; Allyn & Bacon: Boston, 1977; p. 366.

The enzymatic action is believed to result from hydrolysis. Cleavage of an O–H bond is promoted by the charge redistribution that takes place upon coordination:[15]

$$[Zn]-\overset{\oplus}{\underset{\cdot\cdot}{O}}\overset{H}{\underset{H}{<}} + H_2O \rightleftharpoons H_3O^+ + [Zn]-\overset{\cdot\cdot}{\underset{\cdot\cdot}{O}}:\underset{H}{\diagdown} \qquad (22.10)$$

The effect of deprotonation, of course, is to greatly increase the basicity of the coordinated oxygen, facilitating nucleophilic addition to the Lewis acid CO_2:

$$[Zn]-\overset{\cdot\cdot}{\underset{\cdot\cdot}{O}}:\underset{H}{\diagdown} + :\overset{\cdot\cdot}{O}=C=\overset{\cdot\cdot}{O}: \longrightarrow [Zn] + \left[H-\overset{\cdot\cdot}{\underset{\cdot\cdot}{O}}-C\overset{\overset{\displaystyle :\overset{\cdot\cdot}{O}:}{\diagup}}{\underset{\underset{\displaystyle O:}{\diagdown}}{}} \right]^{-} \qquad (22.11)$$

(It is also possible that deprotonation occurs during or after the attack on CO_2.) The vacant coordination site is then occupied by another water molecule, completing the catalytic cycle.

The activity of carbonic anhydrase is inhibited by a variety of ligands, including CN^-, H_2S, and Cl^-, probably because of the blockage of the active site by such species.

15. Lindskog, S. *Adv. Inorg. Biochem.* **1982**, *4*, 115; see also, Haim, A. *J. Chem. Educ.* **1989**, *66*, 935.

The Carboxypeptidases[16]

There are two known carboxypeptidases (often abbreviated CPA and CPB). They serve as digestive enzymes, accomplishing the conversion of proteins to their constituent amino acids. This involves the "snipping" of terminal residues, and requires hydration to convert the residues to the free acids; the general equation below applies:

$$\sim HN-\underset{\underset{R}{|}}{CH}-\overset{\overset{O}{\|}}{C}-NH-\underset{\underset{R'}{|}}{CH}-\overset{\overset{O}{\|}}{C}-OH \xrightarrow{H_2O} \sim HN-\underset{\underset{R}{|}}{CH}-\overset{\overset{O}{\|}}{C}-OH + H_2N-\underset{\underset{R'}{|}}{CH}-\overset{\overset{O}{\|}}{C}-OH$$

(22.12)

Carboxypeptidase A and B are complementary catalysts. The former acts on residues containing aromatic and/or hydrophobic side chains such as ϕ and $CH_2\phi$, while CPB prefers positively charged side chains (eg, $CH_2CH_2CH_2CH_2NH_3^+$).

Carboxypeptidase A is an egg-shaped protein with 307 amino acid residues. The molecular mass is about 35,000 daltons. There are several similarities to carbonic anhydrase; for example, the zinc again has tetrahedral coordination. In CPA the ligands are one oxygen and two nitrogen donors from the protein chain plus a water molecule. There is again a depression, in this case called a *cleft*, that exposes the active site. The apoenzyme is catalytically inactive, but will coordinate to metals other than zinc. The Co[II] and Ni[II] enzymes are actually more active catalysts toward peptides than Zn–CPA (Table 22.3).

The carboxypeptidases also catalyze the hydrolysis of amino acid esters; this provides a useful model system for mechanistic study. It is believed that the initial step of the catalyzed process involves chelation by the substrate, forming a 6-coordinate intermediate:

$$H_2N-\underset{\underset{R}{|}}{CH}-\overset{\overset{O}{\|}}{C}-OR' + [Zn]-OH_2 \longrightarrow$$

(22.13)

16. Lipscomb, W. N. *Acc. Chem. Res.* **1982**, *15*, 232.

The carbonyl carbon of the intermediate is strongly electrophilic, and is susceptible to attack by the coordinated water molecule (or, if hydrolysis is assumed, by OH^-). This leads, after dissociation from the metal, to the observed products. An attractive feature of this mechanism is that it nicely explains the high activity of Co–CPA and Ni–CPA, since those two metals are more prone to octahedral coordination than zinc.

22.5   Iron in Biosystems

Iron is the most abundant metal in living organisms. Humans contain about 60 mg iron per kilogram of body mass, most of which is involved in oxygen transport. Other well-studied functions of ferraproteins include the catalysis of a variety of oxidation–reduction reactions and iron transport and storage. The primary iron-containing biomolecules are listed in Table 22.4.

Much of the biochemistry of iron relates in some manner or another to the half-reaction

$$Fe^{3+} + 1e^- \longrightarrow Fe^{2+} \qquad \mathscr{E}^0 = +0.771 \text{ V} \qquad \textbf{(22.14)}$$

since both the chemical properties and reactivity of iron are strongly dependent on its oxidation state. For example, the binding of O_2 to hemoglobin (Fe^{II}) is reversible, but that to methemoglobin (structurally similar, but with Fe^{III}) is not.

The solubilities of iron salts and complexes also vary with the oxidation state. Iron(III) salts have low aqueous solubilities, a fact of biochemical

Table 22.4   Functions and distributions of some important iron-containing biomolecules

Species	Function	Distribution
Transferrin	Transport of Fe	Animals
Siderochromes	Transport of Fe	Bacteria and fungi
Ferritin	Storage of Fe	Animals
Hemosiderin	Storage of Fe	Animals
Cytochromes	Redox enzymes	Plants and animals
Rubredoxins	Redox enzymes	Bacteria
Ferredoxins	Redox enzymes	Plants and bacteria
Nitrogenase	N_2 fixation	Plants and bacteria
Myoglobin	O_2 transport	Animals
Hemoglobin	O_2 transport	Animals
Hemerythrin	O_2 transport	Marine worms

significance. The recommended daily allowance of iron in the diet (in which much of the iron occurs as Fe^{III}) is about 18 mg, although the body actually needs only about 2 mg of new iron per day; the difference reflects the fact that iron is poorly ingested. It is also relevant that the body stores excess iron as Fe^{III}, but most of the active species contain Fe^{II}.

Iron Transport and Storage[17]

The transfer of iron from the digestive tract to the blood (either for subsequent use or for long-term storage) is accomplished in mammals by *transferrin*, a protein having a mass of about 80,000 daltons. Transferrin makes up about 2% of the blood serum solids in humans. It contains two irons in the high spin Fe^{III} state. The coordinations appear to be through the surrounding protein, plus one HCO_3^- ligand. The Fe^{III}–apotransferrin formation constant is quite large under physiological conditions (on the order of 10^{30}), but there is an extreme pH dependence; no complexation is observed at a pH of 4 or below. This is consistent with the observation that three hydrogen ions are released upon complexation. These hydrogens protonate a bicarbonate ion and, probably, two anionic sites of the surrounding protein.

In mammals, the long-term storage of excess iron involves *ferritin*. This large species (about 900,000 daltons) consists of a protein surrounding a *core* (up to 420,000 daltons) that contains a variable amount of iron; up to 4500 irons can be accommodated per molecule. Ferritin is nearly spherical, with a diameter of roughly 12,500 pm; the core cavity is thought to be about 7500 pm in diameter. An approximate formulation for the core is $Fe(H_2PO_4) \cdot 8Fe(OH)_3$. The structure is unknown, but it is probably similar to that of hydrated ferric oxide; EXAFS data suggest infinite sheets of iron atoms, with each metal octahedrally coordinated to six oxygens.[18]

Heme Redox Agents: The Cytochromes[19]

The term *cytochrome* means "*pigment of a cell.*" There are three major types of cytochromes, designated *a*, *b*, and *c*; each has known variations (eg, cytochrome a_3). The cytochromes act as electron transfer agents for respiratory metabolism; that is, they manage the transfer of electrons from substrates to O_2. This is accomplished through changes in the metal

17. Aisen, P.; Listowsky, I. *Ann. Rev. Biochem.* **1980**, *49*, 357.

18. Heald, S. M.; Stern, E. A.; Bunker, B.; Holt, E. M.; Holt, S. L. *J. Am. Chem. Soc.* **1979**, *101*, 67.

19. Sutin, N. *Adv. Chem. Ser.* **1977**, *162*, 156.

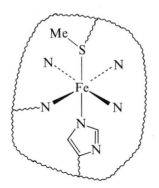

Figure 22.8
Schematic diagram of the structure of cytochrome *c*; ⁓= protein.

oxidation state (Fe^{II}/Fe^{III}). The suitability of the cytochromes for this task is demonstrated by both their ubiquitous nature (they are found in all living plants and animals) and their immunity to the evolutionary process. Evidence exists that cytochrome *c* has been manufactured by living species for well over a billion years!

Cytochrome *c* is the most studied of the three types. Its structure is shown schematically in Figure 22.8. Like hemoglobin and myoglobin (see Section 22.6), it is a *heme* protein—that is, an iron–porphyrin complex attached to a protein chain. In addition to the tetradentate porphyrin, the iron is coordinated to the surrounding protein at trans sites through nitrogen (histidine) and sulfur (methionine) donors. The protein chain is short compared to most other metalloenzymes (104 amino acid residues), giving a relatively low molecular mass of about 12,000 daltons.

The cytochromes act in sequence. Cytochrome *a* is the strongest oxidizing agent of the three ($\mathscr{E}^0_{red} = +0.24$ V), while cytochrome *b* is the weakest ($\mathscr{E}^0_{red} = +0.04$ V). Thus, electrons from the oxidized substrate (glucose) are initially transferred to cytochrome *b*, which relays them to cytochrome *c*, which in turn passes them on to cytochrome *a*. A complex species known as *cytochrome oxidase*, which contains two cytochrome *a* units and also a copper atom, provides the ultimate link to O_2.

Nonheme Redox Agents: The Doxins[20]

Two other groups of iron-containing redox enzymes are worthy of mention—the *rubredoxins* and the *ferredoxins*. They are especially important in the electron transfer aspects of photosynthesis and nitrogen fixation.

Rubredoxins and ferredoxins have iron coordinated to four sulfur donors, which are subdivided into two types, *labile* and *nonlabile*. Labile

20. Bezkorovainy, A. *Biochemistry of Non-Heme Iron*; Plenum: New York, 1980; Sweeney, W. V.; Rabinowitz, J. C. *Ann. Rev. Biochem.* **1980**, *49*, 139.

sulfur is ligated S^{2-} (typically in a bridging position between two or three irons), while nonlabile sulfur is part of an amino acid residue (cysteine or methionine). The distinction arises from their different behavior under laboratory conditions; labile sulfur is easily converted to $H_2S(g)$ by treatment with strong acid.

The rubredoxins are bacterial enzymes. They contain one iron, tetrahedrally coordinated to nonlabile sulfurs of a small protein (typically consisting of about 55 residues). They are weaker reducing agents than the cytochromes, with standard reduction potentials between -0.05 and -0.10 V.

The ferredoxins (found in plant chloroplasts and certain bacteria) are more complex. At least three categories are known, containing two, four, and eight irons, respectively. There is exactly one labile sulfur per iron in each case. The diiron ferredoxins have 4-coordinate irons linked by bridging sulfides (Figure 22.9a).

The tetrairon ferredoxins contain iron–sulfur cubes; this results in approximately tetrahedral coordination about the metal. Three sites are occupied by labile sulfurs from the adjacent vertices of the cube, with the fourth donor being from a cysteine residue of the protein. The eight-iron ferredoxins contain two Fe_4S_4 cubes, which are connected at two positions by branches of the protein chain (Figure 22.9b).

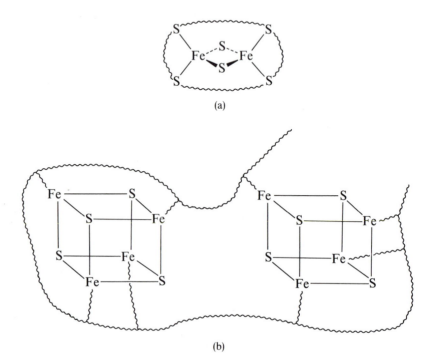

(a)

(b)

Figure 22.9
Schematic diagram of the structure of (a) a two-iron and (b) an eight-iron ferredoxin; ∿ = protein.

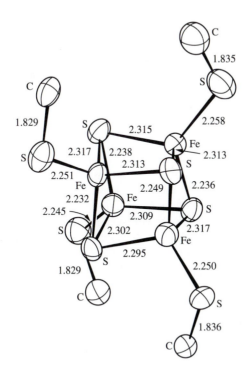

Figure 22.10
The structure of the model ferredoxin $[S_4Fe_4(SCH_2\phi)_4]^{2-}$; the bond distances are in angstrom units.
[Reproduced with permission from Herskovitz, T.; Averill, B. A.; Holm, R. H.; Ibers, J. A.; Phillips, W. D.; Weiher, J. F. *Proc. Nat. Acad. Sci.* **1972**, *69*, 2437.]

Many model systems have been synthesized in efforts to understand the chemistry of these iron–sulfur cubes.[21] One of the best-known is the complex anion $[S_4Fe_4(SCH_2\phi)_4]^{2-}$ (Figure 22.10). The spectroscopic properties of this species are quite similar to those of the four-iron ferredoxins. Interestingly, all the irons are both symmetrically and spectroscopically equivalent in spite of their nonintegral oxidation state of $+2.5$; this is indicative of magnetic interaction between the metals (antiferromagnetism).

22.6   Oxygen Carriers: Hemoglobin and Related Species[22]

Large animals require a mechanism for the transport of O_2 from the lungs to the muscles, and they manufacture an appropriate biomolecule for that purpose. The copper-containing protein *hemocyanin* is used by mollusks and arthropods. Two heme complexes, *hemoglobin* and *myoglobin*, are utilized in higher mammals and humans.

21.   Nakamura, A.; Ueyama, N. *Adv. Inorg. Chem.* **1989**, *33*, 39.

22.   Senozan, N. M.; Hunt, R. L. *J. Chem. Educ.* **1982**, *59*, 173.

These structurally related molecules operate in a complementary manner. Hemoglobin (Hb) gives blood its red color. It carries O_2 from the lungs to muscle cell boundaries. The oxygen is then transferred to myoglobin (Mb), which provides for short-term oxygen storage and facilitates the reaction between O_2 and glucose (metabolism).

Structures

The myoglobin molecule (shown in Figure 22.11) is nearly spherical, with a molecular mass of about 17,000 daltons. It contains a single, centrally located iron, coordinated to the four coplanar nitrogens of a porphyrin macrocycle known as *protoporphyrin IX* (PIX). The iron lies about 40 pm above the plane, giving local C_{4v} symmetry. An imidazole (histidine) nitrogen of the surrounding *globin* (globular protein) acts as a fifth ligand. The sixth coordination site is occupied by O_2 in the oxygenated form (MbO_2), and is vacant when deoxygenated.

The magnetic properties of Mb are intriguing. The deoxygenated, square pyramidal form is paramagnetic, with four unpaired electrons per molecule. However, MbO_2 is diamagnetic; this requires that the iron have a low spin

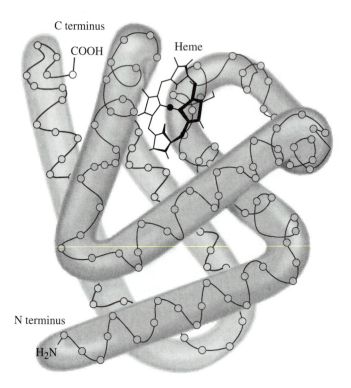

Figure 22.11
Schematic diagram of the structure of myoglobin.

configuration, and that the HOMO of O_2 be singly rather than doubly degenerate. (If this is not clear, see the MO diagram of O_2 given in Figure 3.9.) The Fe–O–O bond angle in MbO_2 is 115°, which is consistent with the valence bond formulation

$$
\underset{[Fe]}{\overset{\ominus}{\ddots}} \overset{\oplus}{\underset{}{\ddot{O}}} = \ddot{O}: \qquad \longleftrightarrow \qquad \underset{[Fe]}{} \overset{\oplus}{\underset{}{\ddot{O}}} - \ddot{\underset{}{O}} \overset{\ominus}{:}
$$

The metal–ligand bonding is therefore monohapto.

Hemoglobin is an approximate tetramer of Mb, with four heme groups and four globin proteins; hence, it can coordinate up to four oxygen molecules simultaneously. The subunits are not identical; instead, they are divided into two types (α and β) of two each, which differ in their protein chains. The subunits come together to form a roughly tetrahedral array. They are held in place by six electrostatic interactions ("salt links") between $-NH_3^+$ substituents of one protein and $-CO_2^-$ groups of another.

Each α and β globin chain contains 141 amino acid residues. Of particular significance is the sixth position of the β chain, which is a glutamic acid residue in normal Hb but a valine residue in victims of sickle-cell anemia. This subtle difference leads to a different folding of the protein chains, and results in a crescent instead of spherical shape. Unfortunately, the flattened shape impairs the O_2 carrying ability of the molecule.

Deactivation

The ability of hemoglobin and myoglobin to reversibly bind to oxygen can be destroyed either by chemical oxidation or by blockage of the sixth coordination site.

Deoxygenated Mb is readily oxidized to Fe^{III} to produce a species known as *metmyoglobin*, which is unreactive toward O_2. It is also possible to separate the heme portion of Hb or Mb from the surrounding protein. The isolated heme is rapidly oxidized in air to Fe^{III}. The initial product is a peroxo-bridged bimetallic complex, which slowly decomposes at ambient temperature:

$$
2[Fe^{II}] + O_2 \longrightarrow [Fe^{III}]-O-O-[Fe^{III}] \tag{22.15}
$$

Molecular oxygen is a relatively poor ligand. While this is a crucial aspect of the system (the reversibility would be lost if it were bound too strongly), it permits the facile and rapid replacement of O_2 by stronger ligands. In particular, π acceptors are dangerous in this regard. Thus, CO, CN^-, PR_3, and H_2S are able to deactivate Hb and Mb, and so are acutely toxic.

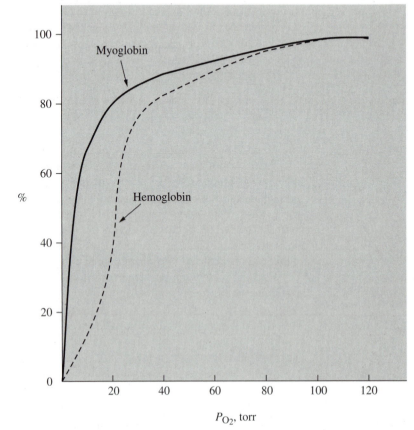

Figure 22.12
Plots of oxygen-binding ability (as percentage bound) versus partial pressure of O_2 for myoglobin and hemoglobin at pH = 7.6. The partial pressure of O_2 is roughly 40 torr in muscle tissue and 100 torr in the lungs.

Cooperativity in Hemoglobin

A plot showing the abilities of Mb and Hb to take up molecular oxygen as a function of the partial pressure of O_2 is given in Figure 22.12.

The most significant aspects of this graph are described below:

1. At the O_2 concentration of muscle tissue (about 40 torr) the affinity of Mb for oxygen is greater than that of Hb. Thus, the transfer reaction below is favored.

$$Hb(O_2)_n + nMb \longrightarrow nMbO_2 + Hb \qquad (22.16)$$

2. The reactivity of Hb toward O_2 is dependent on pH, being inversely related to the hydrogen ion concentration (the *Bohr effect*).[23] This is

23. Bohr, C.; Hasselbalch, K.; Krogh, A. *Skand. Arch. Physiol.* **1904**, *16*, 402; see also, Senozan, N. M.; Hunt, R. L. *J. Chem. Educ.* **1982**, *59*, 173.

consistent with the observation that about $\frac{1}{2}$ mol of H^+ is added for every mole of O_2 released during deoxygenation. These hydrogen ions probably protonate a histidine residue of a β chain.

3. As expected from equilibrium theory, the complexation of both Mb and Hb by O_2 increases with increasing oxygen concentration. The Mb line is consistent with a first-order dependence on $[O_2]$; that is, the expression below applies:

$$K_{eq}(Mb) = \frac{[MbO_2]}{[Mb] \cdot [O_2]} \qquad (22.17)$$

However, the relationship of Hb to O_2 is more complex, with a sigmoidal curve being observed. The "best-fit" mass action expression is

$$K_{eq}(Hb) \approx \frac{[HbO_2]}{[Hb] \cdot [O_2]^{2.8}} \qquad (22.18)$$

This nonintegral dependence on the oxygen concentration is indicative of different metal–ligand affinities, depending on the number of attached oxygens; that is, the ability of Hb to add O_2 is enhanced when one or more oxygen molecules are already coordinated:

$$Hb \xrightarrow[K_1]{O_2} HbO_2 \xrightarrow[K_2]{O_2} Hb(O_2)_2 \qquad K_1 < K_2 \qquad (22.19)$$

This is a *cooperativity effect*, and makes Hb a more efficient oxygen carrier at high O_2 concentrations (in vivo, in the lungs) than Mb. Conversely, Hb is less efficient in oxygen-poor environments such as the muscles, a fact that facilitates the $HbO_2 \rightarrow MbO_2$ transfer.

The cooperativity effect is thought to arise from a change in the protein conformation upon binding to O_2.[24] Recall that there is a change from the high to low spin state upon oxygenation of Hb. This is accompanied by a change in ionic radius of the iron (Table 6.7), which allows it to lie more nearly in the porphyrin plane. The movement of the iron alters the location of the attached histidine ligand relative to the rest of the protein, resulting in a different tertiary structure (a refolding of the globin chain) of the protein. The letter designations T (tense) and R (relaxed) are sometimes used to describe the deoxygenated and oxygenated states, respectively.

---

24. Perutz, M. F. *Adv. Chem. Ser.* **1980**, *191*, 201; *Sci. Am.* **1978**, *239*, 92.

An X-ray diffraction study of doubly oxygenated hemoglobin, $Hb(O_2)_2$ (oxygenated at the α chains), has relevance in this regard. As expected, the oxygenated irons were found to be more nearly in the porphyrin plane than the others. However, the difference was small (roughly 20 versus 30 pm above the plane in the two cases). The movement of the globin upon oxygenation was limited; that is, the structure was intermediate between the *T* and *R* states.[25]

Model Hemes and Related Species[26]

Many synthetic oxygen carriers have been reported. In general, researchers have attempted to produce complexes that mimic the reversible metal–O_2 binding in Mb and Hb. Certain early analogues were iron porphyrins with imidazole ligands. Several such species bind to molecular oxygen at low temperatures but are prone to oxidation, with the formation of peroxobridged species [equation (22.15)]. This problem has been overcome by the *picket fence hemes*, in which dimerization is prevented by the presence of large substituents on the porphyrin macrocycle. An example of such a complex is shown in Figure 22.13.

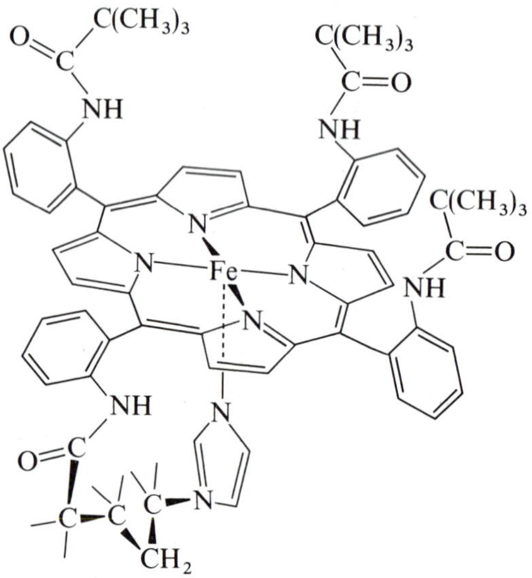

Figure 22.13
A synthetic, picket fence heme complex.
[Reproduced with permission from Collman, J. P. *Acc. Chem. Res.* **1977**, *10*, 265.]

25. Brozozowski, A.; Derewenda, Z.; Dodson, E.; Dodson, G.; Grabowski, M.; Liddington, R.; Skarzynski, T.; Vallely, D. *Nature* **1984**, *307*, 74.

26. Suslick, K. S.; Reinert, J. J. *J. Chem. Educ.* **1985**, *62*, 975; Collman, J. P.; Suslick, K. S. *Pure Appl. Chem.* **1978**, *50*, 951; McLendon, G. L.; Martell, A. E. *Coord. Chem. Rev.* **1976**, *19*, 1; Basolo, F.; Hoffmann, B. M.; Ibers, J. A. *Acc. Chem. Res.* **1975**, *8*, 384.

Certain synthetic cobalt(II) porphyrins (*coboglobins*) are excellent imitators of Hb and Mb. They bind reversibly to O$_2$, and are less prone to oxidation than the synthetic hemes.[27] (The Co^{III}/Co^{II} reduction potential is much more positive than that of Fe^{III}/Fe^{II}.)

22.7 Vitamin B$_{12}$ and Its Derivatives[28]

Vitamin B$_{12}$ is a cobalt-containing enzyme (actually, a coenzyme, since it acts in concert with others) with a molecular mass of about 1200 daltons. It is essential for the production of red blood cells, and a B$_{12}$ deficiency in humans leads to the condition known as *pernicious anemia*.

Cobalamins are vitamin B$_{12}$ derivatives that have the general structure shown in Figure 22.14 (p. 762). The Co^{III} has a pseudo-octahedral geometry. Four of the donor atoms are the nitrogens of a tetradentate *corrin* ring, a chelate somewhat reminiscent of the porphyrins. Corrins differ from porphyrins in that they lack one ring carbon, have only six conjugated π bonds, and contain different side groups.

The fifth coordination site in cobalamins is a histidine nitrogen. The sixth ligand is variable; it is usually H$_2$O in aqueous solution, CN$^-$ when isolated by conventional laboratory procedures, and 5'-deoxyadenosine (Figure 22.14) in vivo. The latter contains a cobalt–carbon bond, making it one of only two known naturally occurring organometallic compounds. (The other, methylcobalamin, is discussed at the end of this section.)

Cobalt(III) cobalamin (sometimes labeled vitamin B$_{12a}$) undergoes one- and two-electron reduction to produce vitamin B$_{12r}$ and vitamin B$_{12s}$, respectively. A few of its best-characterized derivatives are summarized in Table 22.5 (p. 762).

Vitamin B$_{12}$ coenzyme catalyzes three distinct types of reactions: redox, exchange, and methylation. Vitamin B$_{12s}$ contains Co^I, and is, not surprisingly, a potent nucleophile and/or reducing agent:

$$\text{Vitamin B}_{12s} \quad\begin{array}{l} \xrightarrow{\text{HC} \equiv \text{CH}} \text{[Co]–CH}=\text{CH}_2 \quad\quad \textbf{(22.20)} \\[2mm] \xrightarrow{\text{BrC} \equiv \text{CH}} \text{[Co]–C} \equiv \text{CH} \quad\quad \textbf{(22.21)} \\[2mm] \xrightarrow{\text{H}_2\text{C}=\text{CH–CO}_2\text{H}} \text{[Co]–CH}_2\text{CH}_2\text{CO}_2\text{H} \quad \textbf{(22.22)} \end{array}$$

27. Stevens, J. C.; Jackson, P. J.; Schammel, W. P.; Christoph, G. C.; Busch, D. H. *J. Am. Chem. Soc.* **1980**, *102*, 3283.

28. Toscano, P. J.; Marzilli, L. G. *Prog. Inorg. Chem.* **1984**, *31*, 105; Dolphin, D., Ed. *B*$_{12}$; Wiley: New York, 1982; Murakami, Y. *Adv. Chem. Ser.* **1980**, *191*, 178; Babior, B. M., Ed. *Cobalamin*; Wiley: New York, 1975; Brown, D. G. *Prog. Inorg. Chem.* **1973**, *18*, 177.

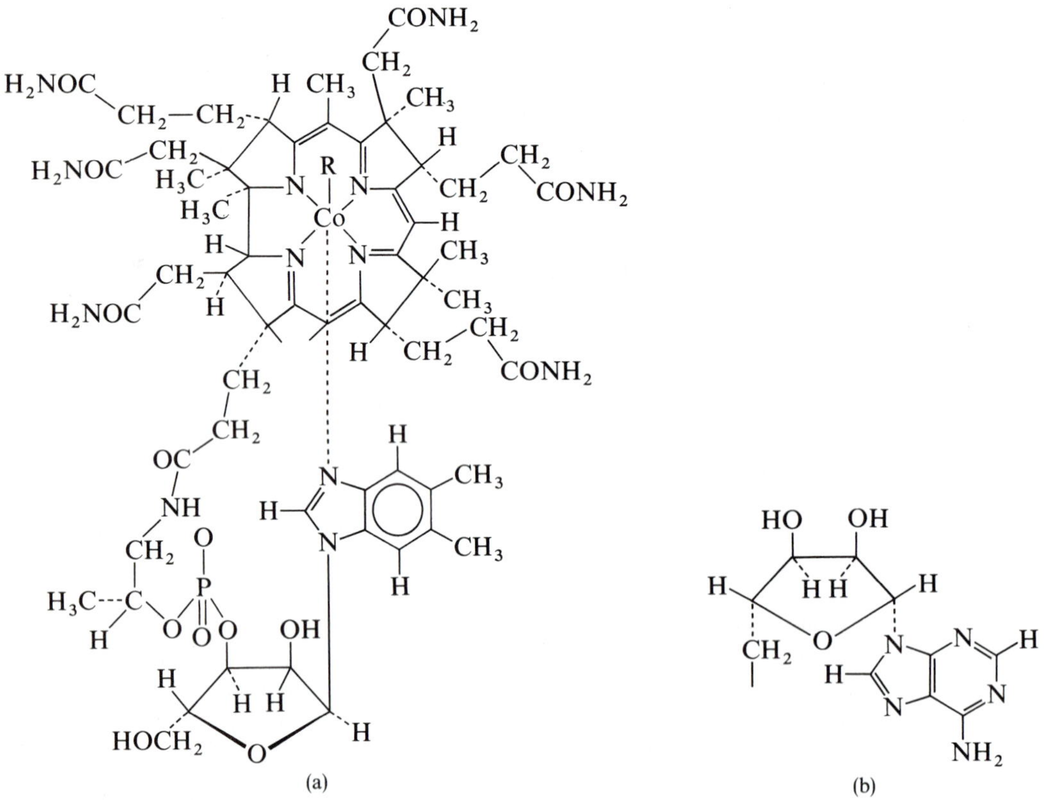

Figure 22.14
(a) The structure of the vitamin B_{12} coenzyme;
(b) in vivo, R = 5′- deoxyadenosine. (See Table 22.5 for a list of related species.)

Table 22.5 The common names and structures (after Figure 22.14) of some well-known cobalamins

R =	Oxidation State	Common Name
5′-Deoxyadenosine	Co^{III}	Vitamin B_{12a}
5′-Deoxyadenosine	Co^{II}	Vitamin B_{12r}
5′-Deoxyadenosine	Co^{I}	Vitamin B_{12s}
CN^-	Co^{III}	Cyanocobalamin
H_2O	Co^{III}	Aquacobalamin
OH^-	Co^{III}	Hydroxocobalamin
CH_3	Co^{III}	Methylcobalamin

Vitamin B_{12s} reduces a variety of metal ions, including Eu^{III}, Ti^{IV}, and U^{VI}, in one-electron steps; kinetic studies of such reactions suggest an outer-sphere mechanism.[29]

Especially intriguing is the catalysis of exchange reactions, in which substituents on adjacent carbons of an appropriate substrate are interchanged:

$$
\underset{\overset{|}{R}}{\overset{\overset{H}{|}}{\text{w--C--C--w}}} \xrightarrow{\text{Vitamin } B_{12}} \underset{\overset{|}{H}}{\overset{\overset{R}{|}}{\text{w--C--C--w}}} \tag{22.23}
$$

An example of such a reaction is the conversion of methylmalonyl coenzyme A to succinyl coenzyme A:

$$
\underset{\sim S}{\overset{O}{\diagdown}}\text{C--CH}\underset{\diagdown CH_3}{\overset{\diagup CO_2H}{}} \xrightarrow{\text{Vitamin } B_{12}} \underset{\sim S}{\overset{O}{\diagdown}}\text{C--CH}_2\text{CH}_2\text{CO}_2\text{H} \tag{22.24}
$$

Reactions of this type are generally thought to occur through a free radical mechanism in which the first step is homolytic cleavage of the Co–C bond. The alkyl radical then abstracts a hydrogen atom from the substrate. However, heterolytic Co–C cleavage to form a carbanion is also possible.[30]

Methylcobalamin is a potent methylating agent. It can be prepared in the laboratory by the treatment of vitamin B_{12s} with CH_2N_2 or CH_3I. An interesting aspect of its reactivity is its action on water-insoluble mercury compounds. Industrial pollution has created deposits of mercury-containing species in the sediments of rivers and oceans. Anaerobic bacteria living in these sediments generate methylcobalamin as a defense mechanism, in order to convert the mercury to CH_3HgX (X = CH_3, SMe, Cl, etc.)—compounds that are sufficiently volatile and/or soluble to permit their elimination from the local environment of the bacteria. Unfortunately, the result is to discharge these toxins into the aquatic food chain, and cases of humans and animals poisoned by eating mercury-laden fish are well-documented.

Bibliography

Que, L., Ed. *Metal Clusters in Proteins*; American Chemical Society: Washington, DC, 1988.

29. Chithambarathanu Pillai, G.; Ghosh, S. K.; Gould, E. S. *Inorg. Chem.* **1988**, *27*, 1868.

30. For a review of mechanistic studies of vitamin B_{12}-induced rearrangements, see Halpern, J. *Adv. Chem. Ser.* **1980**, *191*, 165.

Ochiai, E.-I. *General Principles of Biochemistry of the Elements*; Plenum: New York, 1987.

Frieden, E., Ed. *Biochemistry of the Essential Ultratrace Elements*; Plenum: New York, 1984.

Hay, R. W. *Bio-Inorganic Chemistry*; Ellis Horwood: Chichester, 1984.

Spiro, T. G., Ed. *Zinc Enzymes*; Wiley: New York, 1983.

Addison, A. W.; Cullen, W. R.; Dolphin, D.; James, B. R., Eds. *Biological Aspects of Inorganic Chemistry*; Wiley: New York, 1977.

Fersht, A. R. *Enzyme Structure and Mechanism*; Freeman: San Francisco, 1977.

Ochiai, E.-I. *Bioinorganic Chemistry*: An Introduction; Allyn & Bacon: Boston, 1977.

Underwood, E. J. *Trace Elements in Human and Animal Nutrition*, 4th ed.; Academic: New York, 1977.

Eichhorn, G. L., Ed. *Inorganic Biochemistry*; Elsevier: New York, 1973.

Questions and Problems

1. It was suggested by Isaac Asimov in his essay "Life's Bottleneck" that the element having the greatest concentration factor (the ratio of its concentration in life forms to its environmental concentration) is phosphorus.
   (a) Based on Figure 22.2, do you agree?
   (b) Suggest reasons why this might be the case.

2. Arsenic is notorious for its toxicity. The poisonous nature of its compounds results in large measure from the fact that they are soft Lewis bases. Discuss.

3. Although CN^- inhibits the O_2 carrying ability of hemoglobin and myoglobin, its rapid and acute toxicity is due primarily to its interaction with cytochrome oxidase. Speculate on the type of chemical interaction involved.

4. Reexamine the general order of elemental toxicity given in Section 22.1.
   (a) Account for the presence of tellurium and selenium at the top of this list.
   (b) Suggest a reason for the high toxicity of beryllium.

5. A common treatment for heavy metal poisoning is to administer an appropriate dosage of EDTA or some other chelating agent.
   (a) Explain the chemical basis for this treatment.
   (b) British anti-Lewisite [BAL, $HS-CH_2-CH(SH)-CH_2OH$] is a particularly useful antidote for mercury, lead, and arsenic poisoning. Why?

6. Explain why boron is relatively little-used by living organisms in spite of its high elemental abundance.

7. We stated in this chapter that H_2O is the source of molecular oxygen in photosynthesis. What specific type of experiment do you believe was performed to determine this?

8. Under physiological conditions, $\mathscr{E}$ for the oxidation of H_2O [equation (22.3)] was given as -0.82 V. Use the standard reduction potential of O_2 and the Nernst equation to show that this value is reasonable. Indicate any assumptions you make.

9. Chlorophyll *a* has an absorption maximum at about 660 nm.
   (a) To what energy (in kilojoules per mole) does this wavelength correspond?
   (b) What volt-equivalence ($n\mathscr{E}^0$) is provided if all of that light energy is converted to electric potential?
   (c) Is this sufficient for the oxidation of glucose? ($\mathscr{E}$ for this four-electron oxidation is about -1.2 V under physiological conditions.)
   (d) How might your answer to part (c) relate to the oligomerization of chlorophyll in vivo?

10. Assign an oxidation number to each manganese atom shown in Figure 22.5.

11. The larger protein of nitrogenase contains between 24 and 36 irons and an equal number of sulfurs. Both of those numbers are divisible by 4. What might this suggest?

12. Reexamine the mechanism suggested for nitrogen fixation. Note that the product of the second step is given as $[\text{Mo}]\text{–NH}\!=\!\text{NH}$.
   (a) It might also be $[\text{Mo}]\text{–N}\!=\!\text{NH}_2$. Write appropriate resonance structures for this linkage.
   (b) Based on general bonding principles, which possibility do you prefer? Why?
   (c) Make a similar evaluation of $[\text{Mo}]\text{–NH}_2\text{–NH}_2$ versus $[\text{Mo}]\text{–NH–NH}_3$ as the product of step (3).

13. The infrared spectra of $\text{Mo–N}\!\equiv\!\text{N–Fe}$ bridged complexes typically exhibit an absorption in the $1930\text{–}1945 \text{ cm}^{-1}$ region.
   (a) Speculate on the bond responsible for this absorption.
   (b) What does the band location suggest about the nature of the bonding?

14. Alkaline phosphatase is a zinc-containing enzyme that catalyzes reactions such as

$$O_2N\!-\!\!\bigcirc\!\!-\!OPO_3^{2-} \xrightarrow{\ H_2O\ } O_2N\!-\!\!\bigcirc\!\!-\!OH + HPO_4^{2-}$$

As its name suggests, it is most active in basic solution.
   (a) What is the probable role of zinc in this catalysis?
   (b) Speculate on the relationship between pH and reactivity.

15. Compared to most other transition metals, Zn^{II} complexes undergo very rapid ligand substitution reactions.
   (a) Why? (Review relevant portions of Chapter 17 if necessary.)
   (b) Discuss the significance of this with regard to enzymatic action.

16. Among metalloenzymes, those containing zinc are among the most difficult to characterize by common instrumental techniques such as electronic spectroscopy, magnetic susceptibility, and ESR spectroscopy. Explain.

17. Use the formulation and molecular mass given for the ferritin core to estimate the maximum number of irons that can be accommodated within that core. Is the value given in the text (4500) reasonable?

18. Compare the structure of cytochrome c to that of myoglobin. What is the most significant difference between the two in terms of coordination chemistry? How does this difference relate to the functions of these biomolecules?

19. (a) Based on its MO diagram, do you expect O_2 to be a weak or strong field ligand? Explain.
    (b) Recall that deoxygenated myoglobin is high spin, while MbO_2 is low spin. Is this consistent with your answer to part (a)? If not, explain the inconsistency. [*Hint*: A structural change takes place upon coordination by O_2.]

20. Do you expect CN^- to bind more strongly to hemoglobin or to methemoglobin? Why?

21. Consider the redistribution reaction

    $$2\,Hb(O_2) \longrightarrow Hb + Hb(O_2)_2$$

    Do you expect K_{eq} to be greater or less than 1.0? Why?

22. We stated in this chapter that the Bohr effect is consistent with the observation that H^+ is added to Hb as O_2 is released.
    (a) Explain the Bohr effect in your own words.
    (b) Explain the above statement on the basis of Le Châtelier's principle.

23. Consider "chromoglobin," a species identical to Mb but containing Cr^{II} instead of Fe^{II}. Predict how the chemical properties might be altered by the substitution of chromium.

24. There is some evidence that cobalt(II) might actually form more efficient complexes for the transport of oxygen than does iron(II). Did nature make a mistake? Suggest a reason why iron might be the preferred metal even if its complexes are slightly inferior as oxygen carriers.

25. It has been speculated that the coordination number of cobalt in vitamin B_{12s} is 4. Is there a sound chemical basis for such a notion? What is it?

*26. Cobalt–carbon bond disruption energies have been studied by P. J. Toscano, A. L. Seligson, M. T. Curran, A. T. Skrobutt, and D. C. Sonnenberger (*Inorg. Chem.* **1989**, *28*, 166).
    (a) Why is the thermochemical energy of this type of bond of interest to bioinorganic chemists?
    (b) An indirect method was used. Describe it.
    (c) Is a steric effect suggested by their results? Explain.

*27. A complex having an $MoFe_6S_6$ core has been suggested as a model for nitrogenase (Eldredge, P. A.; Bryan, R. F.; Sinn, E.; Averill, B. A. *J. Am. Chem. Soc.* **1988**, *110*, 5573).
    (a) Why do the authors feel their complex is a good model?
    (b) Which is the more likely site for oxidation, Fe or Mo? Explain.
    (c) What are the limitations of this complex as a nitrogenase model?

*28. A ruthenium–porphyrin complex has been found to be reactive toward H_2 (Collman, J. P.; Wagenknecht, P. S.; Hembre, R. T.; Lewis, N. S. *J. Am. Chem. Soc.* **1990**, *112*, 1294).

(a) What is the biochemical significance?

(b) Was direct evidence obtained for ligation by H_2?

(c) What is the evidence for such an interaction?

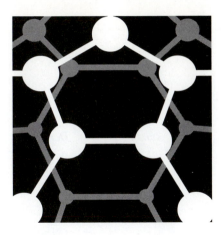

APPENDIX I

SI Units, Constants, and Conversion Factors

Système Internationale (SI) units are commonly used throughout most of the world and are slowly gaining popularity in the United States.[1] The units used in this book are a mixture of the SI units and other, more traditional ones. [For example, pressures are given in torr or atmospheres (atm), and the Celsius temperature scale is usually used.]

The seven basic SI units and their symbols are as follows:

Quantity	*Name*	*Symbol*
Length	Meter	m
Mass	Kilogram	kg
Amount of substance	Mole	mol
Time	Second	s
Temperature	Kelvin	K
Electric current	Ampere	A
Luminous intensity	Candela	cd

---

1. Adamson, A. W. *J. Chem. Educ.* **1978**, *55*, 634; Norris, A. C. *J. Chem. Educ.* **1971**, *48*, 797.

There are also several derived SI units. Those of greatest importance to practicing chemists are listed below:

Quantity	*Name*	*Symbol*	*Derivation*
Force	Newton	N	$kg \cdot m/s^2$
Energy	Joule	J	$N \cdot m$
Power	Watt	W	J/s
Pressure	Pascal	Pa	N/m^2
Frequency	Hertz	Hz	$1/s$
Electric charge	Coulomb	C	$A \cdot s$
Electric potential	Volt	V	W/A
Electric resistance	Ohm	Ω	V/A

All these units can be modified by prefixes, which multiply the values by specified powers of 10. These include the following:

Prefix	*Symbol*	*Multiplier*
Pico	p	10^{-12}
Nano	n	10^{-9}
Micro	μ	10^{-6}
Milli	m	10^{-3}
Centi	c	10^{-2}
Kilo	k	10^3
Mega	M	10^6
Giga	G	10^9
Tera	T	10^{12}

Some other frequently used constants are given below.

Constant	*Value*
Mass of a proton	1.6726×10^{-27} kg $= 1.007276$ amu
Mass of a neutron	1.6749×10^{-27} kg $= 1.008665$ amu

Constant	*Value*
Mass of an electron (at rest)	9.1096×10^{-31} kg $= 5.485803 \times 10^{-4}$ amu
Charge of an electron	1.6022×10^{-19} C $= 4.8030 \times 10^{-10}$ esu
Planck's constant	6.6262×10^{-34} J·s
Speed of light (vacuum)	2.9979×10^{8} m/s
Avogadro's number	6.0221×10^{23} mol^{-1}
Boltzmann's constant	1.3806×10^{-23} J/K
Permittivity (vacuum)	8.8542×10^{-12} C^2/m·J
Gas constant	8.3143 J/mol·K $= 8.2053 \times 10^{-2}$ L·atm/mol·K
Faraday's constant	9.6487×10^{4} C/mol
Bohr magneton	9.2741×10^{-21} erg/G

It is often necessary to convert from one unit of energy, distance, etc., to another. A number of such relationships are given below:

Energy

1 eV $= 96.487$ kJ/mol $= 1.6022 \times 10^{-22}$ kJ/atom $= 8.0657 \times 10^{3}$ cm^{-1}

1 kJ/mol $= 1.0364 \times 10^{-2}$ eV $= 83.594$ cm^{-1} $= 0.23901$ kcal/mol

1 cm^{-1} $= 1.2398 \times 10^{-4}$ eV $= 1.1963 \times 10^{-2}$ kJ/mol

1 kcal/mol $= 4.1840$ kJ/mol

Distance

1 angstrom (Å) $= 10^{-10}$ m $= 10^{-8}$ cm $= 10^{-4}$ μm $= 10^{-1}$ nm $= 10^{2}$ pm

(Other conversions are easily obtained using the table of prefixes given above.)

Atomic Mass Units

1 amu $= 1.6606 \times 10^{-27}$ kg

1 kg $= 6.0221 \times 10^{26}$ amu

Temperature

$$K = 273.16 + \,^{\circ}C$$

Pressure

$1\ \text{atm} = 760\ \text{torr} = 1.0132 \times 10^5\ \text{Pa}$

$1\ \text{torr} = 1.3158 \times 10^{-3}\ \text{atm} = 1.3332 \times 10^2\ \text{Pa}$

$1\ \text{Pa} = 9.8692 \times 10^{-6}\ \text{atm} = 7.5010 \times 10^{-3}\ \text{torr}$

Electric Charge

$1\ \text{coulomb (C)} = 2.9979 \times 10^9\ \text{esu}$

$1\ \text{electrostatic unit (esu)} = 3.3357 \times 10^{-10}\ \text{C}$

Dipole Moment

$1\ \text{debye (D)} = 3.336 \times 10^{-30}\ \text{C·m}$

$1\ \text{C·m} = 2.998 \times 10^{29}\ \text{D}$

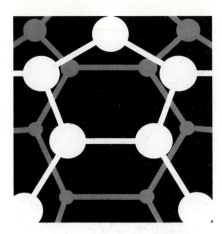

APPENDIX II

Nomenclature

The nomenclature used in this book is generally consistent with IUPAC convention; exceptions reflect common usage. A summary of the recommended nomenclature for inorganic compounds is provided below. The organization of this summary, and also some of the examples used, are taken from IUPAC's "Preamble and Rules of the 1971 Report of the Commission on the Nomenclature of Inorganic Chemistry."[1]

Binary Compounds

In writing formulas for binary compounds, metals are written before nonmetals. Metals are generally listed in alphabetical order of the symbols. For example,

$$Li_2O \qquad ZnCl_2 \qquad CsNa$$

1. International Union of Pure and Applied Chemistry. *How to Name an Inorganic Substance*; Pergamon: Oxford, 1977; see also, Block, B. P.; Powell, W. N.; Fernelius, W. C. *Inorganic Chemical Nomenclature*; American Chemical Society: Washington, DC, 1990.

For nonmetals, metalloids, and selected metals, the priority sequence $Rn > Xe > Kr > B > Si > C > Pb > As > P > N > H > Te > Se > S > At > I > Br > Cl > O > F$ is followed. Hence,

SiC, but CS_2

CH_4 and NH_3, but H_2Se and HAt

Cl_2O, but OF_2

The names of binary compounds end in the suffix -ide.[2] Prefixes are used only when necessary to avoid ambiguity.

SiC, silicon carbide

CS_2, carbon disulfide (The prefix di- is used to distinguish CS_2 from CS, carbon monosulfide.)

H_2Se, hydrogen selenide

Certain trivial names have been accepted in light of long-standing usage. These include the following:

H_2O, water	PH_3, phosphine
NH_3, ammonia	AsH_3, arsine
N_2H_4, hydrazine	SbH_3, stibene

Other volatile, binary hydrides are assigned the -ane suffix. If more than one heteroatom is present, a prefix is used as well. Examples include

GeH_4, germane

Si_2H_6, disilane

B_5H_{11}, pentaborane(11) [The number in parentheses distinguishes B_5H_{11} from B_5H_9, pentaborane(9).]

Polyatomic Ions and Radicals

Cations that are formed by protonation of neutral molecules end in -onium.

H_3O^+, hydronium	$N_2H_5^+$, hydrazonium
NH_4^+, ammonium	PH_4^+, phosphonium

---

2. Unfortunately, many ternary systems have the same suffix.

Simple oxyanions receive prefixes based on the relative number of oxygens present. The *-ite* suffix implies fewer oxygens than *-ate*:

$$NO_2^-, \text{ nitrite} \qquad NO_3^-, \text{ nitrate}$$

If necessary, the prefixes *hypo-* (still fewer oxygens) and *per-* (still more oxygens) are used in conjunction with the suffixes. For the oxychloride anions,

$$ClO^-, \text{ hypochlorite} \qquad ClO_2^-, \text{ chlorite}$$

$$ClO_3^-, \text{ chlorate} \qquad ClO_4^-, \text{ perchlorate}$$

More complex oxyanions contain prefixes to identify the number of heteroatoms present. Structural prefixes such as *cyclo-* are sometimes added in italics.

disulfate   *cyclo*-triphosphate

Certain covalently bonded moieties are considered to be "common radicals." As such, they are assigned special names having the suffix *-yl*:

HO, hydroxyl	SO, sulfinyl (or thionyl)
CO, carbonyl	SO_2, sulfonyl (or sulfuryl)
NO, nitrosyl	S_2O_5, disulfuryl
NO_2, nitryl	SeO, seleninyl
PO, phosphoryl	SeO_2, selenonyl
ClO, chlorosyl	CrO_2, chromyl
ClO_2, chloryl	UO_2, uranyl
ClO_3, perchloryl	NpO_2, neptunyl

This leads to names such as

$$NOF, \text{ nitrosyl fluoride} \qquad NO_2F, \text{ nitryl fluoride}$$

Acid and Double Salts

Acid salts contain three-word names, with the middle word "hydrogen." Prefixes are used as needed. For example,

$NaHSO_4$, sodium hydrogen sulfate

KH_2PO_4, potassium dihydrogen phosphate

(In contrast to American custom, IUPAC does not recommend a space after the word "hydrogen." That is, the name sodium hydrogensulfate is in more rigorous conformity to the IUPAC system.)

Double salts are both formulated and named by alphabetizing the ions of like charge. This occasionally reverses the order in which the elements are listed (note the example of $NaRbCO_3$). A space separates each word of the name.

$AlNa(SO_4)_2$, aluminum sodium sulfate

$NaRbCO_3$, rubidium sodium carbonate

$PbClF$, lead chloride fluoride

Coordination Compounds and Ions

In writing formulas, the central atom is written first, followed by the ligands. Square brackets properly enclose the complex, although in practice they are often omitted for straightforward cases.

$[Co(NH_3)_6]^{3+}$ $Na_2[PtCl_4]$ (or Na_2PtCl_4)

In naming coordination compounds, the ligands are alphabetized (ignoring prefixes) and given before the metal(s). The metal oxidation state is given in parentheses (the *Stock system*). Alternatively, the charge of the complex ion may be written using Arabic numbers (the *Ewens–Bassett system*). The former method is used in this book. Anionic complexes receive the suffix *-ate*.

$Mn(CO)_5P\phi_3$, pentacarbonyl(triphenylphosphine)manganese

$[Pt(py)_4]Cl_2$, tetrapyridineplatinum(II) chloride or tetrapyridineplatinum(2+) chloride

$K_4[Fe(CN)_6]$, potassium hexacyanoferrate(II) or potassium hexacyanoferrate(4−)

The prefixes *bis-*, *tris-*, *tetrakis-*, *pentakis-*, etc., are used when two normal prefixes (*di-*, *tri-*, *tetra-*, etc.) would otherwise run together. For example,

$[Pd(PMe_3)_4](NO_3)_2$, tetrakis(trimethylphosphine)palladium(II) nitrate

Anionic ligands are identified by the suffix *-o*. The names of some common anionic ligands are listed below:

H^-, hydrido	N_3^-, azido	O_2^{2-}, peroxo
F^-, fluoro	CN^-, cyano	OH^-, hydroxo
Cl^-, chloro	NO_3^-, nitrato	S^{2-}, thio
Br^-, bromo	O^{2-}, oxo	SH^-, mercapto
I^-, iodo	O_2^-, superoxo	SO_4^{2-}, sulfato

The following names are illustrative of these rules:

$[NiI_4]^{2-}$, tetraiodonickelate(II)

$(NH_4)_3[FeCl_6]$, ammonium hexachloroferrate(III)

$Mn(CO)_5I$, pentacarbonyl(iodo)manganese(I)

Certain neutral ligands have special names:

H_2O, aqua	NH_3, ammine
CO, carbonyl	NO, nitrosyl

Hence,

$[Co(NH_3)_6]Br_3$, hexaamminecobalt(III) bromide

$[Co(NH_3)_4Cl_2]Cl$, *cis*-tetraamminedichlorocobalt(III) chloride

$[Co(H_2O)_6]SO_4$, hexaaquacobalt(II) sulfate

$Cr(NO)_4$, tetranitrosylchromium

For ambidentate ligands, isomers are often differentiated by giving the symbol of the ligating atom in italics and parentheses.

$[Rh(NH_3)_5SCN]^{2+}$, pentaammine(thiocyanato-*S*)rhodium(III)

$[Rh(NH_3)_5NCS]^{2+}$, pentaammine(thiocyanato-*N*)rhodium(III)

However, special names are recognized for SCN^- and NO_2^-:

M–SCN, thiocyanato   M–NCS, isothiocyanato

M–NO$_2$, nitro   M–ONO, nitrito

Abbreviations have been established for certain other (mostly polydentate) ligands. These are summarized in Appendix III.

Greek letters are used in writing formulas and names for certain types of complexes. The letter η (eta) designates the number of coordinated atoms in, for example, a π ligand.[3]

Bridging ligands are identified by the Greek letter μ. The following examples are illustrative:

Co(CO)$_2$

Dicarbonyl (η^5-cyclopentadienyl)cobalt

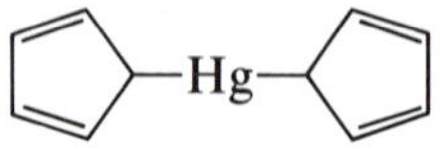

—Hg—

Bis(η^1-cyclopentadienyl)mercury

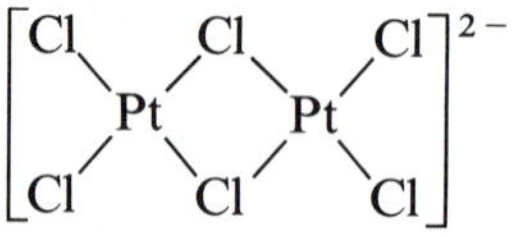

$$\begin{bmatrix} Cl & Cl & Cl \\ & Pt & Pt & \\ Cl & Cl & Cl \end{bmatrix}^{2-}$$

Di-μ-chlorobis(dichloroplatinum(II))

---

3. Cotton, F. A. *J. Am. Chem. Soc.* **1968**, *90*, 6230.

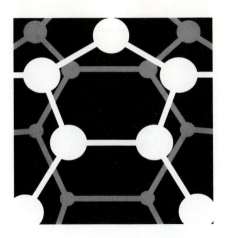

APPENDIX III

Structures and Abbreviations for Some Common Ligands

A number of ligands have well-established abbreviations. In some cases (the majority of which are given below), these abbreviations have been approved by IUPAC. The use of lowercase letters is recommended for ligand abbreviations.[1]

Name	Abbreviation	Structure
Acetylacetonate ion	acac	$\left[\text{Me–C–CH–C–Me} \right]^{-}$ with two $=O$ groups
Bipyridine	bipy	(structure of two pyridine rings joined, each with N)

---

1. However, capital letter abbreviations are commonly used for several ligands, including DMF, DMSO, and EDTA.

Name	*Abbreviation*	*Structure*
Bis(diphenylphosphino)-ethane	diphos	$\phi_2 P–CH_2–CH_2–P\phi_2$
Cyclooctatetraene	cot	
Cyclopentadienyl	cp	
Dimethylformamide	dmf	$Me_2N–COH$
Dimethylglyoxime	dmg	
Dimethylsulfoxide	dmso	Me_2SO
Ethylenediamine	en	$H_2N–CH_2–CH_2–NH_2$
Ethylenediaminetetra-acetic acid	edta	$(HO_2C–CH_2)_2N–CH_2–CH_2–N(CH_2–CO_2H)_2$
Phenanthroline	phen	
Pyridine	py	
Terpyridine	terpy	
Triaminotriethylamine	tren	$(H_2N–CH_2–CH_2)_3N$

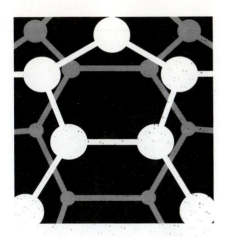

APPENDIX IV

Character Tables for Some Important Point Groups

C_1	E
A	1

C_s	E	σ_h		
A'	1	1	x, y, R_z	x^2, y^2, z^2, xy
A''	1	-1	z, R_x, R_y	yz, xz

C_i	E	i		
A_g	1	1	R_x, R_y, R_z	$x^2, y^2, z^2, xy, xz, yz$
A_u	1	-1	x, y, z	

C_2	E	C_2		
A	1	1	z, R_z	x^2, y^2, z^2, xy
B	1	-1	x, y, R_x, R_y	yz, xz

C_3	E	C_3	C_3^2		
A	1	1	1	z, R_z	$x^2 + y^2, z^2$
E	$\begin{cases} 1 \\ 1 \end{cases}$	$\begin{matrix} \epsilon \\ \epsilon^* \end{matrix}$	$\left.\begin{matrix} \epsilon^* \\ \epsilon \end{matrix}\right\}$	$(x, y), (R_x, R_y)$	$(x^2 - y^2, xy), (yz, xz)$

$\epsilon = e^{(2\pi i)/3}$; $\epsilon^* = \epsilon$ with $-i$ replacing i

C_4	E	C_4	C_2	C_4^3		
A	1	1	1	1	z, R_z	$x^2 + y^2, z^2$
B	1	-1	1	-1		$x^2 - y^2, xy$
E	$\begin{cases} 1 \\ 1 \end{cases}$	$\begin{matrix} i \\ -i \end{matrix}$	$\begin{matrix} -1 \\ -1 \end{matrix}$	$\left.\begin{matrix} -i \\ i \end{matrix}\right\}$	$(x, y), (R_x, R_y)$	(yz, xz)

C_5	E	C_5	C_5^2	C_5^3	C_5^4		
A	1	1	1	1	1	z, R_z	$x^2 + y^2, z^2$
E_1	$\begin{cases} 1 \\ 1 \end{cases}$	$\begin{matrix} \epsilon \\ \epsilon^* \end{matrix}$	$\begin{matrix} \epsilon^2 \\ \epsilon^{2*} \end{matrix}$	$\begin{matrix} \epsilon^{2*} \\ \epsilon^2 \end{matrix}$	$\left.\begin{matrix} \epsilon^* \\ \epsilon \end{matrix}\right\}$	$(x, y), (R_x, R_y)$	(yz, xz)
E_2	$\begin{cases} 1 \\ 1 \end{cases}$	$\begin{matrix} \epsilon^2 \\ \epsilon^{2*} \end{matrix}$	$\begin{matrix} \epsilon^* \\ \epsilon \end{matrix}$	$\begin{matrix} \epsilon \\ \epsilon^* \end{matrix}$	$\left.\begin{matrix} \epsilon^{2*} \\ \epsilon^2 \end{matrix}\right\}$		$(x^2 - y^2, xy)$

$\epsilon = e^{(2\pi i)/5}$

C_6	E	C_6	C_3	C_2	C_3^2	C_6^5		
A	1	1	1	1	1	1	z, R_z	$x^2 + y^2, z^2$
B	1	-1	1	-1	1	-1		
E_1	$\begin{cases} 1 \\ 1 \end{cases}$	$\begin{matrix} \epsilon \\ \epsilon^* \end{matrix}$	$\begin{matrix} -\epsilon^* \\ -\epsilon \end{matrix}$	$\begin{matrix} -1 \\ -1 \end{matrix}$	$\begin{matrix} -\epsilon \\ -\epsilon^* \end{matrix}$	$\left.\begin{matrix} \epsilon^* \\ \epsilon \end{matrix}\right\}$	$\begin{matrix} (x, y), \\ (R_x, R_y) \end{matrix}$	(xz, yz)
E_2	$\begin{cases} 1 \\ 1 \end{cases}$	$\begin{matrix} -\epsilon^* \\ -\epsilon \end{matrix}$	$\begin{matrix} -\epsilon \\ -\epsilon^* \end{matrix}$	$\begin{matrix} 1 \\ 1 \end{matrix}$	$\begin{matrix} -\epsilon^* \\ -\epsilon \end{matrix}$	$\left.\begin{matrix} -\epsilon \\ -\epsilon^* \end{matrix}\right\}$		$(x^2 - y^2, xy)$

$\epsilon = e^{(\pi i)/3}$

C_7	E	C_7	C_7^2	C_7^3	C_7^4	C_7^5	C_7^6		
A	1	1	1	1	1	1	1	z, R_z	$x^2 + y^2, z^2$
E_1	$\begin{cases} 1 \\ 1 \end{cases}$	$\begin{matrix} \epsilon \\ \epsilon^* \end{matrix}$	$\begin{matrix} \epsilon^2 \\ \epsilon^{2*} \end{matrix}$	$\begin{matrix} \epsilon^3 \\ \epsilon^{3*} \end{matrix}$	$\begin{matrix} \epsilon^{3*} \\ \epsilon^3 \end{matrix}$	$\begin{matrix} \epsilon^{2*} \\ \epsilon^2 \end{matrix}$	$\left.\begin{matrix} \epsilon^* \\ \epsilon \end{matrix}\right\}$	$\begin{matrix} (x, y), \\ (R_x, R_y) \end{matrix}$	(xz, yz)
E_2	$\begin{cases} 1 \\ 1 \end{cases}$	$\begin{matrix} \epsilon^2 \\ \epsilon^{2*} \end{matrix}$	$\begin{matrix} \epsilon^{3*} \\ \epsilon^3 \end{matrix}$	$\begin{matrix} \epsilon^* \\ \epsilon \end{matrix}$	$\begin{matrix} \epsilon \\ \epsilon^* \end{matrix}$	$\begin{matrix} \epsilon^3 \\ \epsilon^{3*} \end{matrix}$	$\left.\begin{matrix} \epsilon^{2*} \\ \epsilon^2 \end{matrix}\right\}$		$(x^2 - y^2, xy)$
E_3	$\begin{cases} 1 \\ 1 \end{cases}$	$\begin{matrix} \epsilon^3 \\ \epsilon^{3*} \end{matrix}$	$\begin{matrix} \epsilon^* \\ \epsilon \end{matrix}$	$\begin{matrix} \epsilon^2 \\ \epsilon^{2*} \end{matrix}$	$\begin{matrix} \epsilon^{2*} \\ \epsilon^2 \end{matrix}$	$\begin{matrix} \epsilon \\ \epsilon^* \end{matrix}$	$\left.\begin{matrix} \epsilon^{3*} \\ \epsilon^3 \end{matrix}\right\}$		

$\epsilon = e^{(2\pi i)/7}$

C_8	E	C_8	C_4	C_2	C_4^3	C_8^3	C_8^5	C_8^7		
A	1	1	1	1	1	1	1	1	z, R_z	$x^2 + y^2, z^2$
B	1	-1	1	1	1	-1	-1	-1		
E_1	$\begin{cases} 1 \\ 1 \end{cases}$	$\begin{matrix} \epsilon \\ \epsilon^* \end{matrix}$	$\begin{matrix} i \\ -i \end{matrix}$	$\begin{matrix} -1 \\ -1 \end{matrix}$	$\begin{matrix} -i \\ i \end{matrix}$	$\begin{matrix} -\epsilon^* \\ -\epsilon \end{matrix}$	$\begin{matrix} -\epsilon \\ -\epsilon^* \end{matrix}$	$\left.\begin{matrix} \epsilon^* \\ \epsilon \end{matrix}\right\}$	$\begin{matrix} (x, y), \\ (R_x, R_y) \end{matrix}$	(xz, yz)
E_2	$\begin{cases} 1 \\ 1 \end{cases}$	$\begin{matrix} i \\ -i \end{matrix}$	$\begin{matrix} -1 \\ -1 \end{matrix}$	$\begin{matrix} 1 \\ 1 \end{matrix}$	$\begin{matrix} -1 \\ -1 \end{matrix}$	$\begin{matrix} -i \\ i \end{matrix}$	$\begin{matrix} i \\ -i \end{matrix}$	$\left.\begin{matrix} -i \\ i \end{matrix}\right\}$		$(x^2 - y^2, xy)$
E_3	$\begin{cases} 1 \\ 1 \end{cases}$	$\begin{matrix} -\epsilon \\ -\epsilon^* \end{matrix}$	$\begin{matrix} i \\ -i \end{matrix}$	$\begin{matrix} -1 \\ -1 \end{matrix}$	$\begin{matrix} -i \\ i \end{matrix}$	$\begin{matrix} \epsilon^* \\ \epsilon \end{matrix}$	$\begin{matrix} \epsilon \\ \epsilon^* \end{matrix}$	$\left.\begin{matrix} -\epsilon^* \\ -\epsilon \end{matrix}\right\}$		

$\epsilon = e^{(\pi i)/4}$

C_{2v}	E	C_2	$\sigma_v(xz)$	$\sigma_v'(yz)$		
A_1	1	1	1	1	z	x^2, y^2, z^2
A_2	1	1	-1	-1	R_z	xy
B_1	1	-1	1	-1	x, R_y	xz
B_2	1	-1	-1	1	y, R_x	yz

C_{3v}	E	$2C_3$	$3\sigma_v$		
A_1	1	1	1	z	$x^2 + y^2, z^2$
A_2	1	1	-1	R_z	
E	2	-1	0	$(x, y), (R_x, R_y)$	$(x^2 - y^2, xy), (xz, yz)$

C_{4v}	E	$2C_4$	C_2	$2\sigma_v$	$2\sigma_d$		
A_1	1	1	1	1	1	z	$x^2 + y^2, z^2$
A_2	1	1	1	-1	-1	R_z	
B_1	1	-1	1	1	-1		$x^2 - y^2$
B_2	1	-1	1	-1	1		xy
E	2	0	-2	0	0	$(x, y), (R_x, R_y)$	(xz, yz)

C_{5v}	E	$2C_5$	$2C_5^2$	$5\sigma_v$		
A_1	1	1	1	1	z	$x^2 + y^2, z^2$
A_2	1	1	1	-1	R_z	
E_1	2	$2\cos 72°$	$2\cos 144°$	0	$(x, y), (R_x, R_y)$	(xz, yz)
E_2	2	$2\cos 144°$	$2\cos 72°$	0		$(x^2 - y^2, xy)$

C_{6v}	E	$2C_6$	$2C_3$	C_2	$3\sigma_v$	$3\sigma_d$		
A_1	1	1	1	1	1	1	z	$x^2+y^2,\ z^2$
A_2	1	1	1	1	-1	-1	R_z	
B_1	1	-1	1	-1	1	-1		
B_2	1	-1	1	-1	-1	1		
E_1	2	1	-1	-2	0	0	$(x, y), (R_x, R_y)$	(xz, yz)
E_2	2	-1	-1	2	0	0		(x^2-y^2, xy)

C_{2h}	E	C_2	i	σ_h		
A_g	1	1	1	1	R_z	$x^2,\ y^2,\ z^2,\ xy$
B_g	1	-1	1	-1	R_x, R_y	$xz,\ yz$
A_u	1	1	-1	-1	z	
B_u	1	-1	-1	1	x, y	

C_{3h}	E	C_3	C_3^2	σ_h	S_3	S_3^5		
A'	1	1	1	1	1	1	R_z	$x^2+y^2,\ z^2$
E'	$\begin{cases}1\\1\end{cases}$	$\begin{matrix}\epsilon\\\epsilon^*\end{matrix}$	$\begin{matrix}\epsilon^*\\\epsilon\end{matrix}$	$\begin{matrix}1\\1\end{matrix}$	$\begin{matrix}\epsilon\\\epsilon^*\end{matrix}$	$\begin{matrix}\epsilon^*\\\epsilon\end{matrix}\Big\}$	(x, y)	(x^2-y^2, xy)
A''	1	1	1	-1	-1	-1	z	
E''	$\begin{cases}1\\1\end{cases}$	$\begin{matrix}\epsilon\\\epsilon^*\end{matrix}$	$\begin{matrix}\epsilon^*\\\epsilon\end{matrix}$	$\begin{matrix}-1\\-1\end{matrix}$	$\begin{matrix}-\epsilon\\-\epsilon^*\end{matrix}$	$\begin{matrix}-\epsilon^*\\-\epsilon\end{matrix}\Big\}$	(R_x, R_y)	(xz, yz)

$\epsilon = e^{(2\pi i)/3}$

C_{4h}	E	C_4	C_2	C_4^3	i	S_4^3	σ_h	S_4		
A_g	1	1	1	1	1	1	1	1	R_z	$x^2+y^2,\ z^2$
B_g	1	-1	1	-1	1	-1	1	-1		$x^2-y^2,\ xy$
E_g	$\begin{cases}1\\1\end{cases}$	$\begin{matrix}i\\-i\end{matrix}$	$\begin{matrix}-1\\-1\end{matrix}$	$\begin{matrix}-i\\i\end{matrix}$	$\begin{matrix}1\\1\end{matrix}$	$\begin{matrix}i\\-i\end{matrix}$	$\begin{matrix}-1\\-1\end{matrix}$	$\begin{matrix}-i\\i\end{matrix}\Big\}$	(R_x, R_y)	(xz, yz)
A_u	1	1	1	1	-1	-1	-1	-1	z	
B_u	1	-1	1	-1	-1	1	-1	1		
E_u	$\begin{cases}1\\1\end{cases}$	$\begin{matrix}i\\-i\end{matrix}$	$\begin{matrix}-1\\-1\end{matrix}$	$\begin{matrix}-i\\i\end{matrix}$	$\begin{matrix}-1\\-1\end{matrix}$	$\begin{matrix}-i\\i\end{matrix}$	$\begin{matrix}1\\1\end{matrix}$	$\begin{matrix}i\\-i\end{matrix}\Big\}$	(x, y)	

C_{5h}	E	C_5	C_5^2	C_5^3	C_5^4	σ_h	S_5	S_5^7	S_5^3	S_5^9		
A'	1	1	1	1	1	1	1	1	1	1	R_z	$x^2+y^2,\ z^2$
E_1'	$\begin{cases}1\\1\end{cases}$	$\begin{matrix}\epsilon\\\epsilon^*\end{matrix}$	$\begin{matrix}\epsilon^2\\\epsilon^{2*}\end{matrix}$	$\begin{matrix}\epsilon^{2*}\\\epsilon^2\end{matrix}$	$\begin{matrix}\epsilon^*\\\epsilon\end{matrix}$	$\begin{matrix}1\\1\end{matrix}$	$\begin{matrix}\epsilon\\\epsilon^*\end{matrix}$	$\begin{matrix}\epsilon^2\\\epsilon^{2*}\end{matrix}$	$\begin{matrix}\epsilon^{2*}\\\epsilon^2\end{matrix}$	$\begin{matrix}\epsilon^*\\\epsilon\end{matrix}\Big\}$	(x,y)	
E_2'	$\begin{cases}1\\1\end{cases}$	$\begin{matrix}\epsilon^2\\\epsilon^{2*}\end{matrix}$	$\begin{matrix}\epsilon^*\\\epsilon\end{matrix}$	$\begin{matrix}\epsilon\\\epsilon^*\end{matrix}$	$\begin{matrix}\epsilon^{2*}\\\epsilon^2\end{matrix}$	$\begin{matrix}1\\1\end{matrix}$	$\begin{matrix}\epsilon^2\\\epsilon^{2*}\end{matrix}$	$\begin{matrix}\epsilon^*\\\epsilon\end{matrix}$	$\begin{matrix}\epsilon\\\epsilon^*\end{matrix}$	$\begin{matrix}\epsilon^{2*}\\\epsilon^2\end{matrix}\Big\}$		$(x^2-y^2,\ xy)$
A''	1	1	1	1	1	-1	-1	-1	-1	-1	z	
E_1''	$\begin{cases}1\\1\end{cases}$	$\begin{matrix}\epsilon\\\epsilon^*\end{matrix}$	$\begin{matrix}\epsilon^2\\\epsilon^{2*}\end{matrix}$	$\begin{matrix}\epsilon^{2*}\\\epsilon^2\end{matrix}$	$\begin{matrix}\epsilon^*\\\epsilon\end{matrix}$	$\begin{matrix}-1\\-1\end{matrix}$	$\begin{matrix}-\epsilon\\-\epsilon^*\end{matrix}$	$\begin{matrix}-\epsilon^2\\-\epsilon^{2*}\end{matrix}$	$\begin{matrix}-\epsilon^{2*}\\-\epsilon^2\end{matrix}$	$\begin{matrix}-\epsilon^*\\-\epsilon\end{matrix}\Big\}$	(R_x,R_y)	$(xz,\ yz)$
E_2''	$\begin{cases}1\\1\end{cases}$	$\begin{matrix}\epsilon^2\\\epsilon^{2*}\end{matrix}$	$\begin{matrix}\epsilon^*\\\epsilon\end{matrix}$	$\begin{matrix}\epsilon\\\epsilon^*\end{matrix}$	$\begin{matrix}\epsilon^{2*}\\\epsilon^2\end{matrix}$	$\begin{matrix}-1\\-1\end{matrix}$	$\begin{matrix}-\epsilon^2\\-\epsilon^{2*}\end{matrix}$	$\begin{matrix}-\epsilon^*\\-\epsilon\end{matrix}$	$\begin{matrix}-\epsilon\\-\epsilon^*\end{matrix}$	$\begin{matrix}-\epsilon^{2*}\\-\epsilon^2\end{matrix}\Big\}$		

$\epsilon = e^{(2\pi i)/5}$

C_{6h}	E	C_6	C_3	C_2	C_3^2	C_6^5	i	S_3^5	S_6^5	σ_h	S_6	S_3		
A_g	1	1	1	1	1	1	1	1	1	1	1	1	R_z	$x^2+y^2,\ z^2$
B_g	1	-1	1	-1	1	-1	1	-1	1	-1	1	-1		
E_{1g}	$\begin{cases}1\\1\end{cases}$	$\begin{matrix}\epsilon\\\epsilon^*\end{matrix}$	$\begin{matrix}-\epsilon^*\\-\epsilon\end{matrix}$	$\begin{matrix}-1\\-1\end{matrix}$	$\begin{matrix}-\epsilon\\-\epsilon^*\end{matrix}$	$\begin{matrix}\epsilon^*\\\epsilon\end{matrix}$	$\begin{matrix}1\\1\end{matrix}$	$\begin{matrix}\epsilon\\\epsilon^*\end{matrix}$	$\begin{matrix}-\epsilon^*\\-\epsilon\end{matrix}$	$\begin{matrix}-1\\-1\end{matrix}$	$\begin{matrix}-\epsilon\\-\epsilon^*\end{matrix}$	$\begin{matrix}\epsilon^*\\\epsilon\end{matrix}\Big\}$	(R_x,R_y)	$(xz,\ yz)$
E_{2g}	$\begin{cases}1\\1\end{cases}$	$\begin{matrix}-\epsilon^*\\-\epsilon\end{matrix}$	$\begin{matrix}-\epsilon\\-\epsilon^*\end{matrix}$	$\begin{matrix}1\\1\end{matrix}$	$\begin{matrix}-\epsilon^*\\-\epsilon\end{matrix}$	$\begin{matrix}-\epsilon\\-\epsilon^*\end{matrix}$	$\begin{matrix}1\\1\end{matrix}$	$\begin{matrix}-\epsilon^*\\-\epsilon\end{matrix}$	$\begin{matrix}-\epsilon\\-\epsilon^*\end{matrix}$	$\begin{matrix}1\\1\end{matrix}$	$\begin{matrix}-\epsilon^*\\-\epsilon\end{matrix}$	$\begin{matrix}-\epsilon\\-\epsilon^*\end{matrix}\Big\}$		$(x^2-y^2,\ xy)$
A_u	1	1	1	1	1	1	-1	-1	-1	-1	-1	-1	z	
B_u	1	-1	1	-1	1	-1	-1	1	-1	1	-1	1		
E_{1u}	$\begin{cases}1\\1\end{cases}$	$\begin{matrix}\epsilon\\\epsilon^*\end{matrix}$	$\begin{matrix}-\epsilon^*\\-\epsilon\end{matrix}$	$\begin{matrix}-1\\-1\end{matrix}$	$\begin{matrix}-\epsilon\\-\epsilon^*\end{matrix}$	$\begin{matrix}\epsilon^*\\\epsilon\end{matrix}$	$\begin{matrix}-1\\-1\end{matrix}$	$\begin{matrix}-\epsilon\\-\epsilon^*\end{matrix}$	$\begin{matrix}\epsilon^*\\\epsilon\end{matrix}$	$\begin{matrix}1\\1\end{matrix}$	$\begin{matrix}\epsilon\\\epsilon^*\end{matrix}$	$\begin{matrix}-\epsilon^*\\-\epsilon\end{matrix}\Big\}$	(x,y)	
E_{2u}	$\begin{cases}1\\1\end{cases}$	$\begin{matrix}-\epsilon^*\\-\epsilon\end{matrix}$	$\begin{matrix}-\epsilon\\-\epsilon^*\end{matrix}$	$\begin{matrix}1\\1\end{matrix}$	$\begin{matrix}-\epsilon^*\\-\epsilon\end{matrix}$	$\begin{matrix}-\epsilon\\-\epsilon^*\end{matrix}$	$\begin{matrix}-1\\-1\end{matrix}$	$\begin{matrix}\epsilon^*\\\epsilon\end{matrix}$	$\begin{matrix}\epsilon\\\epsilon^*\end{matrix}$	$\begin{matrix}-1\\-1\end{matrix}$	$\begin{matrix}\epsilon^*\\\epsilon\end{matrix}$	$\begin{matrix}\epsilon\\\epsilon^*\end{matrix}\Big\}$		

$\epsilon = e^{(\pi i)/3}$

D_2	E	$C_2(z)$	$C_2(y)$	$C_2(x)$		
A	1	1	1	1		$x^2,\ y^2,\ z^2$
B_1	1	1	-1	-1	$z,\ R_z$	xy
B_2	1	-1	1	-1	$y,\ R_y$	xz
B_3	1	-1	-1	1	$x,\ R_x$	yz

D_3	E	$2C_3$	$3C_2$		
A_1	1	1	1		$x^2+y^2,\ z^2$
A_2	1	1	-1	$z,\ R_z$	
E	2	-1	0	$(x,y),\ (R_x,R_y)$	$(x^2-y^2,\ xy),\ (xz,\ yz)$

D_4	E	$2C_4$	$C_2(=C_4^2)$	$2C_2'$	$2C_2''$		
A_1	1	1	1	1	1		x^2+y^2, z^2
A_2	1	1	1	−1	−1	z, R_z	
B_1	1	−1	1	1	−1		x^2-y^2
B_2	1	−1	1	−1	1		xy
E	2	0	−2	0	0	$(x, y), (R_x, R_y)$	(xz, yz)

D_5	E	$2C_5$	$2C_5^2$	$5C_2$		
A_1	1	1	1	1		x^2+y^2, z^2
A_2	1	1	1	−1	z, R_z	
E_1	2	$2\cos 72°$	$2\cos 144°$	0	$(x, y), (R_x, R_y)$	(xz, yz)
E_2	2	$2\cos 144°$	$2\cos 72°$	0		(x^2-y^2, xy)

D_6	E	$2C_6$	$2C_3$	C_2	$3C_2'$	$3C_2''$		
A_1	1	1	1	1	1	1		x^2+y^2, z^2
A_2	1	1	1	1	−1	−1	z, R_z	
B_1	1	−1	1	−1	1	−1		
B_2	1	−1	1	−1	−1	1		
E_1	2	1	−1	−2	0	0	$(x, y), (R_x, R_y)$	(xz, yz)
E_2	2	−1	−1	2	0	0		(x^2-y^2, xy)

D_{2h}	E	$C_2(z)$	$C_2(y)$	$C_2(x)$	i	$\sigma(xy)$	$\sigma(xz)$	$\sigma(yz)$		
A_g	1	1	1	1	1	1	1	1		x^2, y^2, z^2
B_{1g}	1	1	−1	−1	1	1	−1	−1	R_z	xy
B_{2g}	1	−1	1	−1	1	−1	1	−1	R_y	xz
B_{3g}	1	−1	−1	1	1	−1	−1	1	R_x	yz
A_u	1	1	1	1	−1	−1	−1	−1		
B_{1u}	1	1	−1	−1	−1	−1	1	1	z	
B_{2u}	1	−1	1	−1	−1	1	−1	1	y	
B_{3u}	1	−1	−1	1	−1	1	1	−1	x	

D_{3h}	E	$2C_3$	$3C_2$	σ_h	$2S_3$	$3\sigma_v$		
A_1'	1	1	1	1	1	1		x^2+y^2, z^2
A_2'	1	1	−1	1	1	−1	R_z	
E'	2	−1	0	2	−1	0	(x, y)	(x^2-y^2, xy)
A_1''	1	1	1	−1	−1	−1		
A_2''	1	1	−1	−1	−1	1	z	
E''	2	−1	0	−2	1	0	(R_x, R_y)	(xz, yz)

D_{4h}	E	$2C_4$	C_2	$2C_2'$	$2C_2''$	i	$2S_4$	σ_h	$2\sigma_v$	$2\sigma_d$		
A_{1g}	1	1	1	1	1	1	1	1	1	1		$x^2+y^2,\ z^2$
A_{2g}	1	1	1	-1	-1	1	1	1	-1	-1	R_z	
B_{1g}	1	-1	1	1	-1	1	-1	1	1	-1		x^2-y^2
B_{2g}	1	-1	1	-1	1	1	-1	1	-1	1		xy
E_g	2	0	-2	0	0	2	0	-2	0	0	(R_x,R_y)	(xz,yz)
A_{1u}	1	1	1	1	1	-1	-1	-1	-1	-1		
A_{2u}	1	1	1	-1	-1	-1	-1	-1	1	1	z	
B_{1u}	1	-1	1	1	-1	-1	1	-1	-1	1		
B_{2u}	1	-1	1	-1	1	-1	1	-1	1	-1		
E_u	2	0	-2	0	0	-2	0	2	0	0	(x,y)	

D_{5h}	E	$2C_5$	$2C_5^2$	$5C_2$	σ_h	$2S_5$	$2S_5^3$	$5\sigma_v$		
A_1'	1	1	1	1	1	1	1	1		$x^2+y^2,\ z^2$
A_2'	1	1	1	-1	1	1	1	-1	R_z	
E_1'	2	$2\cos 72°$	$2\cos 144°$	0	2	$2\cos 72°$	$2\cos 144°$	0	(x,y)	
E_2'	2	$2\cos 144°$	$2\cos 72°$	0	2	$2\cos 144°$	$2\cos 72°$	0		(x^2-y^2,xy)
A_1''	1	1	1	1	-1	-1	-1	-1		
A_2''	1	1	1	-1	-1	-1	-1	1	z	
E_1''	2	$2\cos 72°$	$2\cos 144°$	0	-2	$-2\cos 72°$	$-2\cos 144°$	0	(R_x,R_y)	(xz,yz)
E_2''	2	$2\cos 144°$	$2\cos 72°$	0	-2	$-2\cos 144°$	$-2\cos 72°$	0		

D_{6h}	E	$2C_6$	$2C_3$	C_2	$3C_2'$	$3C_2''$	i	$2S_3$	$2S_6$	σ_h	$3\sigma_d$	$3\sigma_v$		
A_{1g}	1	1	1	1	1	1	1	1	1	1	1	1		$x^2+y^2,\ z^2$
A_{2g}	1	1	1	1	-1	-1	1	1	1	1	-1	-1	R_z	
B_{1g}	1	-1	1	-1	1	-1	1	-1	1	-1	1	-1		
B_{2g}	1	-1	1	-1	-1	1	1	-1	1	-1	-1	1		
E_{1g}	2	1	-1	-2	0	0	2	1	-1	-2	0	0	(R_x,R_y)	(xz,yz)
E_{2g}	2	-1	-1	2	0	0	2	-1	-1	2	0	0		(x^2-y^2,xy)
A_{1u}	1	1	1	1	1	1	-1	-1	-1	-1	-1	-1		
A_{2u}	1	1	1	1	-1	-1	-1	-1	-1	-1	1	1	z	
B_{1u}	1	-1	1	-1	1	-1	-1	1	-1	1	-1	1		
B_{2u}	1	-1	1	-1	-1	1	-1	1	-1	1	1	-1		
E_{1u}	2	1	-1	-2	0	0	-2	-1	1	2	0	0	(x,y)	
E_{2u}	2	-1	-1	2	0	0	-2	1	1	-2	0	0		

D_{2d}	E	$2S_4$	C_2	$2C_2'$	$2\sigma_d$		
A_1	1	1	1	1	1		$x^2+y^2,\ z^2$
A_2	1	1	1	-1	-1	R_z	
B_1	1	-1	1	1	-1		x^2-y^2
B_2	1	-1	1	-1	1	z	xy
E	2	0	-2	0	0	$(x,y),(R_x,R_y)$	(xz,yz)

D_{3d}	E	$2C_3$	$3C_2$	i	$2S_6$	$3\sigma_d$		
A_{1g}	1	1	1	1	1	1		$x^2 + y^2,\ z^2$
A_{2g}	1	1	-1	1	1	-1	R_z	
E_g	2	-1	0	2	-1	0	(R_x, R_y)	$(x^2 - y^2,\ xy),\ (xz,\ yz)$
A_{1u}	1	1	1	-1	-1	-1		
A_{2u}	1	1	-1	-1	-1	1	z	
E_u	2	-1	0	-2	1	0	(x, y)	

D_{4d}	E	$2S_8$	$2C_4$	$2S_8^3$	C_2	$4C_2'$	$4\sigma_d$		
A_1	1	1	1	1	1	1	1		$x^2 + y^2,\ z^2$
A_2	1	1	1	1	1	-1	-1	R_z	
B_1	1	-1	1	-1	1	1	-1		
B_2	1	-1	1	-1	1	-1	1	z	
E_1	2	$\sqrt{2}$	0	$-\sqrt{2}$	-2	0	0	(x, y)	
E_2	2	0	-2	0	2	0	0		$(x^2 - y^2,\ xy)$
E_3	2	$-\sqrt{2}$	0	$\sqrt{2}$	-2	0	0	(R_x, R_y)	$(xz,\ yz)$

D_{5d}	E	$2C_5$	$2C_5^2$	$5C_2$	i	$2S_{10}^3$	$2S_{10}$	$5\sigma_d$		
A_{1g}	1	1	1	1	1	1	1	1		$x^2 + y^2,\ z^2$
A_{2g}	1	1	1	-1	1	1	1	-1	R_z	
E_{1g}	2	$2\cos 72°$	$2\cos 144°$	0	2	$2\cos 72°$	$2\cos 144°$	0	(R_x, R_y)	$(xz,\ yz)$
E_{2g}	2	$2\cos 144°$	$2\cos 72°$	0	2	$2\cos 144°$	$2\cos 72°$	0		$(x^2 - y^2,\ xy)$
A_{1u}	1	1	1	1	-1	-1	-1	-1		
A_{2u}	1	1	1	-1	-1	-1	-1	1	z	
E_{1u}	2	$2\cos 72°$	$2\cos 144°$	0	-2	$-2\cos 72°$	$-2\cos 144°$	0	(x, y)	
E_{2u}	2	$2\cos 144°$	$2\cos 72°$	0	-2	$-2\cos 144°$	$-2\cos 72°$	0		

D_{6d}	E	$2S_{12}$	$2C_6$	$2S_4$	$2C_3$	$2S_{12}^5$	C_2	$6C_2'$	$6\sigma_d$		
A_1	1	1	1	1	1	1	1	1	1		$x^2 + y^2,\ z^2$
A_2	1	1	1	1	1	1	1	-1	-1	R_z	
B_1	1	-1	1	-1	1	-1	1	1	-1		
B_2	1	-1	1	-1	1	-1	1	-1	1	z	
E_1	2	$\sqrt{3}$	1	0	-1	$-\sqrt{3}$	-2	0	0	(x, y)	
E_2	2	1	-1	-2	-1	1	2	0	0		$(x^2 - y^2,\ xy)$
E_3	2	0	-2	0	2	0	-2	0	0		
E_4	2	-1	-1	2	-1	-1	2	0	0		
E_5	2	$-\sqrt{3}$	1	0	-1	$\sqrt{3}$	-2	0	0	(R_x, R_y)	$(xz,\ yz)$

S_4	E	S_4	C_2	S_4^3		
A	1	1	1	1	R_z	$x^2 + y^2,\ z^2$
B	1	-1	1	-1	z	$x^2 - y^2,\ xy$
E	$\begin{cases}1\\1\end{cases}$	$\begin{matrix}i\\-i\end{matrix}$	$\begin{matrix}-1\\-1\end{matrix}$	$\begin{matrix}-i\\i\end{matrix}$	$(x, y), (R_x, R_y)$	$(xz,\ yz)$

S_6	E	C_3	C_3^2	i	S_6^5	S_6		
A_g	1	1	1	1	1	1	R_z	$x^2 + y^2,\ z^2$
E_g	$\begin{cases}1\\1\end{cases}$	$\begin{matrix}\epsilon\\\epsilon^*\end{matrix}$	$\begin{matrix}\epsilon^*\\\epsilon\end{matrix}$	$\begin{matrix}1\\1\end{matrix}$	$\begin{matrix}\epsilon\\\epsilon^*\end{matrix}$	$\begin{matrix}\epsilon^*\\\epsilon\end{matrix}$	(R_x, R_y)	$(x^2 - y^2,\ xy),$ $(xz,\ yz)$
A_u	1	1	1	-1	-1	-1	z	
E_u	$\begin{cases}1\\1\end{cases}$	$\begin{matrix}\epsilon\\\epsilon^*\end{matrix}$	$\begin{matrix}\epsilon^*\\\epsilon\end{matrix}$	$\begin{matrix}-1\\-1\end{matrix}$	$\begin{matrix}-\epsilon\\-\epsilon^*\end{matrix}$	$\begin{matrix}-\epsilon^*\\-\epsilon\end{matrix}$	(x, y)	

$\epsilon = e^{(2\pi i)/3}$

S_8	E	S_8	C_4	S_8^3	C_2	S_8^5	C_4^3	S_8^7		
A	1	1	1	1	1	1	1	1	R_z	$x^2 + y^2,\ z^2$
B	1	-1	1	-1	1	-1	1	-1	z	
E_1	$\begin{cases}1\\1\end{cases}$	$\begin{matrix}\epsilon\\\epsilon^*\end{matrix}$	$\begin{matrix}i\\-i\end{matrix}$	$\begin{matrix}-\epsilon^*\\-\epsilon\end{matrix}$	$\begin{matrix}-1\\-1\end{matrix}$	$\begin{matrix}-\epsilon\\-\epsilon^*\end{matrix}$	$\begin{matrix}-i\\i\end{matrix}$	$\begin{matrix}\epsilon^*\\\epsilon\end{matrix}$	$(x, y),$ (R_x, R_y)	
E_2	$\begin{cases}1\\1\end{cases}$	$\begin{matrix}i\\-i\end{matrix}$	$\begin{matrix}-1\\-1\end{matrix}$	$\begin{matrix}-i\\i\end{matrix}$	$\begin{matrix}1\\1\end{matrix}$	$\begin{matrix}i\\-i\end{matrix}$	$\begin{matrix}-1\\-1\end{matrix}$	$\begin{matrix}-i\\i\end{matrix}$		$(x^2 - y^2,\ xy)$
E_3	$\begin{cases}1\\1\end{cases}$	$\begin{matrix}-\epsilon^*\\-\epsilon\end{matrix}$	$\begin{matrix}-i\\i\end{matrix}$	$\begin{matrix}\epsilon\\\epsilon^*\end{matrix}$	$\begin{matrix}-1\\-1\end{matrix}$	$\begin{matrix}\epsilon^*\\\epsilon\end{matrix}$	$\begin{matrix}i\\-i\end{matrix}$	$\begin{matrix}-\epsilon\\-\epsilon^*\end{matrix}$		$(xz,\ yz)$

$\epsilon = e^{(\pi i)/4}$

$C_{\infty v}$	E	$2C_\infty^\phi$	$\cdots$	$\infty\sigma_v$		
$A_1 \equiv \Sigma^+$	1	1	$\cdots$	1	z	$x^2 + y^2,\ z^2$
$A_2 \equiv \Sigma^-$	1	1	$\cdots$	-1	R_z	
$E_1 \equiv \Pi$	2	$2\cos\Phi$	$\cdots$	0	$(x, y), (R_x, R_y)$	$(xz,\ yz)$
$E_2 \equiv \Delta$	2	$2\cos 2\Phi$	$\cdots$	0		$(x^2 - y^2,\ xy)$
$E_3 \equiv \Phi$	2	$2\cos 3\Phi$	$\cdots$	0		
$\cdots$	$\cdots$	$\cdots$	$\cdots$	$\cdots$		

$D_{\infty h}$	E	$2C_\infty^\phi$	$\cdots$	$\infty\sigma_v$	i	$2S_\infty^\phi$	$\cdots$	∞C_2		
Σ_g^+	1	1	$\cdots$	1	1	1	$\cdots$	1		x^2+y^2, z^2
Σ_g^-	1	1	$\cdots$	-1	1	1	$\cdots$	-1	R_z	
Π_g	2	$2\cos\Phi$	$\cdots$	0	2	$-2\cos\Phi$	$\cdots$	0	(R_x, R_y)	(xz, yz)
Δ_g	2	$2\cos 2\Phi$	$\cdots$	0	2	$2\cos 2\Phi$	$\cdots$	0		(x^2-y^2, xy)
$\cdots$	$\cdots$	$\cdots$	$\cdots$	$\cdots$	$\cdots$	$\cdots$	$\cdots$	$\cdots$		
Σ_u^+	1	1	$\cdots$	1	-1	-1	$\cdots$	-1	z	
Σ_u^-	1	1	$\cdots$	-1	-1	-1	$\cdots$	1		
Π_u	2	$2\cos\Phi$	$\cdots$	0	-2	$2\cos\Phi$	$\cdots$	0	(x, y)	
Δ_u	2	$2\cos 2\Phi$	$\cdots$	0	-2	$-2\cos 2\Phi$	$\cdots$	0		
$\cdots$	$\cdots$	$\cdots$	$\cdots$	$\cdots$	$\cdots$	$\cdots$	$\cdots$	$\cdots$		

T	E	$4C_3$	$4C_3^2$	$3C_2$		
A	1	1	1	1		$x^2+y^2+z^2$
$E\begin{cases}\\\\\end{cases}$	1	ϵ	ϵ^*	1		$(2z^2-x^2-y^2, x^2-y^2)$
	1	ϵ^*	ϵ	1		
T	3	0	0	-1	$(R_x, R_y, R_z), (x, y, z)$	(xy, xz, yz)

$\epsilon = e^{(2\pi i)/3}$

T_d	E	$8C_3$	$3C_2$	$6S_4$	$6\sigma_d$		
A_1	1	1	1	1	1		$x^2+y^2+z^2$
A_2	1	1	1	-1	-1		
E	2	-1	2	0	0		$(2z^2-x^2-y^2, x^2-y^2)$
T_1	3	0	-1	1	-1	(R_x, R_y, R_z)	
T_2	3	0	-1	-1	1	(x, y, z)	(xy, xz, yz)

T_h	E	$4C_3$	$4C_3^2$	$3C_2$	i	$4S_6$	$4S_6^5$	$3\sigma_h$		
A_g	1	1	1	1	1	1	1	1		$x^2+y^2+z^2$
A_u	1	1	1	1	-1	-1	-1	-1		
$E_g\begin{cases}\\\\\end{cases}$	1	ϵ	ϵ^*	1	1	ϵ	ϵ^*	1		$(2z^2-x^2-y^2, x^2-y^2)$
	1	ϵ^*	ϵ	1	1	ϵ^*	ϵ	1		
$E_u\begin{cases}\\\\\end{cases}$	1	ϵ	ϵ^*	1	-1	$-\epsilon$	$-\epsilon^*$	-1		
	1	ϵ^*	ϵ	1	-1	$-\epsilon^*$	$-\epsilon$	-1		
T_g	3	0	0	-1	3	0	0	-1	(R_x, R_y, R_z)	(xy, xz, yz)
T_u	3	0	0	-1	-3	0	0	1	(x, y, z)	

$\epsilon = e^{(2\pi i)/3}$

O	E	$6C_4$	$3C_2(=C_4^2)$	$8C_3$	$6C_2$		
A_1	1	1	1	1	1		$x^2+y^2+z^2$
A_2	1	-1	1	1	-1		
E	2	0	2	-1	0		$(2z^2-x^2-y^2, x^2-y^2)$
T_1	3	1	-1	0	-1	$(R_x, R_y, R_z), (x, y, z)$	
T_2	3	-1	-1	0	1		(xy, xz, yz)

O_h	E	$8C_3$	$6C_2$	$6C_4$	$3C_2(=C_4^2)$	i	$6S_4$	$8S_6$	$3\sigma_h$	$6\sigma_d$		
A_{1g}	1	1	1	1	1	1	1	1	1	1		$x^2+y^2+z^2$
A_{2g}	1	1	-1	-1	1	1	-1	1	1	-1		
E_g	2	-1	0	0	2	2	0	-1	2	0		$(2z^2-x^2-y^2, x^2-y^2)$
T_{1g}	3	0	-1	1	-1	3	1	0	-1	-1	(R_x, R_y, R_z)	
T_{2g}	3	0	1	-1	-1	3	-1	0	-1	1		(xy, xz, yz)
A_{1u}	1	1	1	1	1	-1	-1	-1	-1	-1		
A_{2u}	1	1	-1	-1	1	-1	1	-1	-1	1		
E_u	2	-1	0	0	2	-2	0	1	-2	0		
T_{1u}	3	0	-1	1	-1	-3	-1	0	1	1	(x, y, z)	
T_{2u}	3	0	1	-1	-1	-3	1	0	1	-1		

I	E	$12C_5$	$12C_5^2$	$20C_3$	$15C_2$		
A	1	1	1	1	1		$x^2+y^2+z^2$
T_1	3	$\frac{1}{2}(1+\sqrt{5})$	$\frac{1}{2}(1-\sqrt{5})$	0	-1	$(x, y, z), (R_x, R_y, R_z)$	
T_2	3	$\frac{1}{2}(1-\sqrt{5})$	$\frac{1}{2}(1+\sqrt{5})$	0	-1		
G	4	-1	-1	1	0		
H	5	0	0	-1	1		$(xy, xz, yz, x^2-y^2, 2z^2-x^2-y^2)$

I_h	E	$12C_5$	$12C_5^2$	$20C_3$	$15C_2$	i	$12S_{10}$	$12S_{10}^3$	$20S_6$	15σ		
A_g	1	1	1	1	1	1	1	1	1	1		$x^2+y^2+z^2$
T_{1g}	3	$\frac{1}{2}(1+\sqrt{5})$	$\frac{1}{2}(1-\sqrt{5})$	0	-1	3	$\frac{1}{2}(1-\sqrt{5})$	$\frac{1}{2}(1+\sqrt{5})$	0	-1	(R_x, R_y, R_z)	
T_{2g}	3	$\frac{1}{2}(1-\sqrt{5})$	$\frac{1}{2}(1+\sqrt{5})$	0	-1	3	$\frac{1}{2}(1+\sqrt{5})$	$\frac{1}{2}(1-\sqrt{5})$	0	-1		
G_g	4	-1	-1	1	0	4	-1	-1	1	0		
H_g	5	0	0	-1	1	5	0	0	-1	1		$(2z^2-x^2-y^2,$ $x^2-y^2, xy,$ $xz, yz)$
A_u	1	1	1	1	1	-1	-1	-1	-1	-1		
T_{1u}	3	$\frac{1}{2}(1+\sqrt{5})$	$\frac{1}{2}(1-\sqrt{5})$	0	-1	-3	$-\frac{1}{2}(1-\sqrt{5})$	$-\frac{1}{2}(1+\sqrt{5})$	0	1	(x, y, z)	
T_{2u}	3	$\frac{1}{2}(1-\sqrt{5})$	$\frac{1}{2}(1+\sqrt{5})$	0	-1	-3	$-\frac{1}{2}(1+\sqrt{5})$	$-\frac{1}{2}(1-\sqrt{5})$	0	1		
G_u	4	-1	-1	1	0	-4	1	1	-1	0		
H_u	5	0	0	-1	1	-5	0	0	1	-1		

Table 9-1

Ion	Ionic Radius	Ion	Name
H^+		Cu^+	copper(I), or cuprous
Li^+	60 pm	Cu^{2+}	copper(II), or cupric
Na^+	95 pm	Fe^{2+}	iron(II), or ferrous
K^+	133 pm	Fe^{3+}	iron(III), or ferric
Cs^+	169 pm	Cr^{2+}	chromium(II), or chromous
Ag^+	126 pm	Cr^{3+}	chromium(III), or chromic
Mg^{2+}	65 pm	Hg_2^{2+}	mercury(I), or mercurous
Ca^{2+}	99 pm	Hg^{2+}	mercury(II), or mercuric
Sr^{2+}	113 pm	NH_4^+	ammonium
Ba^{2+}	135 pm	OH^-	hydroxide
Zn^{2+}	74 pm	HCO_3^-	bicarbonate, or hydrogen carbonate
Cd^{2+}	97 pm	CO_3^{2-}	carbonate
Ni^{2+}	69 pm	NO_3^-	nitrate
Al^{3+}	50 pm	NO_2^-	nitrite
H^-	208 pm	PO_4^{3-}	(ortho)phosphate
F^-	136 pm	SO_4^{2-}	sulfate
Cl^-	181 pm	SO_3^{2-}	sulfite
Br^-	195 pm	ClO_4^-	perchlorate
I^-	216 pm	ClO_3^-	chlorate
O^{2-}	140 pm	ClO_2^-	chlorite
S^{2-}	184 pm	ClO^-	hypochlorite
		$Cr_2O_7^{2-}$	dichromate
		CrO_4^{2-}	chromate
		MnO_4^-	permanganate

ded to form neutral atoms:

$$1s < 2s < 2p < 3s < 3p < 4s < 3d < 4p < 5s < 4d < 5p < 6s < 5d = 4f < 6p < 7s < 6d = 5f$$

These principles are at the heart of the *periodic law*, according to which the properties of the elements

The chemical behavior of an atom is based primarily on

Index

$$\mathscr{F} = (1.602 \times 10^{-19} \text{ C/electron})(6.022 \times 10^{23} \text{ electrons/mol}) = 9.65 \times 10^4 \text{ C/mol}$$

$$\Delta G = -n\mathscr{F}E$$

The dependence of a cell potential on the concentrations of the reactan
from the known dependence of ΔG upon concentration, (16-6).

$$E = -\frac{\Delta G}{n\mathscr{F}} = -\frac{\Delta G° + RT \ln Q}{n\mathscr{F}}$$

rxn rate $k = Ae^{-E_a/RT}$
E=energy of activation

ACIDS AND BASES

Arrhenius concept

According to the classical definition as formulated by Arrhenius, an *acid* is a substance which can yield H^+ in solution. $HClO_4$ and HNO_3, which are completely ionized in water into H^+ and ClO_4^- and into H^+ and NO_3^-, respectively, are called *strong acids*. $HC_2H_3O_2$, acetic acid, and HNO_2, nitrous acid, are only partly ionized into H^+ and $C_2H_3O_2^-$ and into H^+ and NO_2^-, and these substances are called *weak acids*. The dissociation of a *weak acid* is reversible in aqueous solutions and may be described by an equilibrium constant, usually designated as K_a. Thus

$$HC_2H_3O_2 \rightleftharpoons H^+ + C_2H_3O_2^- \qquad K_a = \frac{[H^+][C_2H_3O_2^-]}{[HC_2H_3O_2]} \qquad (17\text{-}1)$$

Similarly, a *base* is a substance which can yield OH^-. NaOH, a *strong base*, is completely ionized in water into Na^+ and OH^-; even relatively insoluble hydroxides, such as $Ca(OH)_2$, give solutions within their limited solubility range which are completely ionized. A solution of ammonia, NH_3, in water also produces hydroxide ions, and so at one time was thought to consist of NH_4OH in solution. Since the OH^- concentration is only a few percent of the ammonia concentration, ammonia is considered a *weak base*.

Table 19-1. Standard Electrode Potentials at 25 °C

Reaction	$E°/V$
$F_2 + 2e^- \rightarrow 2F^-$	2.87
$S_2O_8^{2-} + 2e^- \rightarrow 2SO_4^{2-}$	1.96
$Co^{3+} + e^- \rightarrow Co^{2+}$	1.92
$H_2O_2 + 2H^+ + 2e^- \rightarrow 2H_2O$	1.763
$Ce^{4+} + e^- \rightarrow Ce^{3+}$ (in 1 M $HClO_4$)	1.70
$MnO_4^- + 8H^+ + 5e^- \rightarrow Mn^{2+} + 4H_2O$	1.51
$Cl_2 + 2e^- \rightarrow 2Cl^-$	1.358
$Tl^{3+} + 2e^- \rightarrow Tl^+$	1.25
$MnO_2 + 4H^+ + 2e^- \rightarrow Mn^{2+} + 2H_2O$	1.23
$O_2 + 4H^+ + 4e^- \rightarrow 2H_2O$	1.229
$Br_2 + 2e^- \rightarrow 2Br^-$	1.065
$AuCl_4^- + 3e^- \rightarrow Au + 4Cl^-$	1.002
$Pd^{2+} + 2e^- \rightarrow Pd$	0.915
$Ag^+ + e^- \rightarrow Ag$	0.799 1
$Fe^{3+} + e^- \rightarrow Fe^{2+}$	0.771
$O_2 + 2H^+ + 2e^- \rightarrow H_2O_2$	0.695
$I_2(s) + 2e^- \rightarrow 2I^-$	0.535
$Cu^+ + e^- \rightarrow Cu$	0.520
$Fe(CN)_6^{3-} + e^- \rightarrow Fe(CN)_6^{4-}$	0.361
$Co(dip)_3^{3+} + e^- \rightarrow Co(dip)_3^{2+}$	0.34
$Cu^{2+} + 2e^- \rightarrow Cu$	0.34
$Ge^{2+} + 2e^- \rightarrow Ge$	0.247
$PdI_4^{2-} + 2e^- \rightarrow Pd + 4I^-$	0.18
$Sn^{4+} + 2e^- \rightarrow Sn^{2+}$	0.15
$Ag(S_2O_3)_2^{3-} + e^- \rightarrow Ag + 2S_2O_3^{2-}$	0.017
$2H^+ + 2e^- \rightarrow H_2$	0.000 0
$Ge^{4+} + 2e^- \rightarrow Ge^{2+}$	0.00
$Pb^{2+} + 2e^- \rightarrow Pb$	-0.126
$Sn^{2+} + 2e^- \rightarrow Sn$	-0.14
$Ni^{2+} + 2e^- \rightarrow Ni$	-0.257
$Tl^+ + e^- \rightarrow Tl$	-0.336
$Cd^{2+} + 2e^- \rightarrow Cd$	-0.403
$Fe^{2+} + 2e^- \rightarrow Fe$	-0.44
$Zn^{2+} + 2e^- \rightarrow Zn$	$-0.762 6$
$Na^+ + e^- \rightarrow Na$	-2.713
$Li^+ + e^- \rightarrow Li$	-3.040